Handbooks of Sociology and Social Research

Series Editor

Richard Serpe, Department of Sociology, Kent University, Kent, USA

The handbook series includes the latest and up-to-date overviews on topics that are of key significance to contemporary sociological and related social science research, including recent topics and areas of scholarship. Several of the volumes discuss important topics from an interdisciplinary social science perspective, covering sociology, anthropology, psychology, and psychiatry. This prestigious series includes works by some of the top scholars in their fields. These foundational works seek to record where the field has been, to identify its current location, and to plot its course for the future.

Will Kalkhoff • Joseph Dippong
Rengin B. Firat
Editors

Handbook of Neurosociology

Second Edition

Springer

Editors
Will Kalkhoff
Dept of Sociology
Kent State University
Kent, OH, USA

Joseph Dippong
University of North Carolina at Charlotte
Charlotte, NC, USA

Rengin B. Firat
Graduate School of Leadership and Change
Antioch University
Yellow Springs, OH, USA

ISSN 1389-6903 ISSN 2542-839X (electronic)
Handbooks of Sociology and Social Research
ISBN 978-3-031-95614-0 ISBN 978-3-031-95615-7 (eBook)
https://doi.org/10.1007/978-3-031-95615-7

This Springer imprint is published by the registered company Springer Nature Switzerland AG
The registered company address is: Gewerbestrasse 11, 6330 Cham, Switzerland

For our friend and mentor David Franks, in recognition of his substantial contributions that have profoundly shaped the field of neurosociology.

Contents

Part III Mind, Self, Interaction, and Sociality

Part IV Culture, Society, and Social Problems

Introduction: The Past, Present, and Future of Neurosociology

1

Russell K. Schutt

Abstract

The 25 chapters in this second edition of the *Handbook of Neurosociology* represent the breadth and depth of advances in research and theory in the last decade. This impressive body of work makes it clear that sociological understanding of human emotion, social interaction, social problems, and cultural variation cannot be considered complete without a foundation that includes insights from neurosociology. Nonetheless, neurosociology remains a marginal topic in sociology, as reflected in it having the smallest of the American Sociological Association's 53 sections and a meager fraction of the articles published in sociology's peer reviewed journals. This introductory chapter first identifies the basis of this disinterest by reviewing the history of the disciplinary divorce between sociology and evolutionary biology; it then identifies the more recent developments that have removed the grounds for divorce and describs the new directions represented in this book.

R. K. Schutt (✉)
Beth Israel Deaconess Medical Center, Harvard Medical School, Boston, MA, USA
e-mail: rschutt@bidmc.harvard.edu

In his preface to the first edition of the *Handbook of Neurosociology*, Douglas S. Massey (2013) noted sociology's historical resistance to investigating biological roots of social behavior—despite his own presidential address elucidating the interdisciplinary connection (Massey, 2001). However, Massey (2013) also believed that the 2004 formation of the American Sociological Association's Section on Evolution, Biology, and Society represented a turning point in disciplinary recognition. The first edition's contents made it clear that the scholarly foundation for such a turning point had been created and "that the field of sociology is definitely moving in this direction" (Massey, 2013:*vi*).

The contents of this second edition of the *Handbook of Neurosociology*, including 15 brand new chapters authored almost entirely by sociologists, prove that this movement continued during the next decade. The chapters show that "unsubstantiated assumptions about the nature of human beings" (Massey, 2013:*vi*) can now be replaced with a substantial body of scientific research on the neurobiology of cognition, emotion, and behavior, on the bidirectional relationship between human behavior and the social environment, and on the role of the environment in shaping genetic influence. Those who read through the second edition's 25 chapters will be rewarded with an interdisciplinary understanding of human emotion, social interaction, social problems, and cultural variation that simply was

W. Kalkhoff et al. (eds.), *Handbook of Neurosociology*, Handbooks of Sociology and Social Research, https://doi.org/10.1007/978-3-031-95615-7_1

not possible in earlier decades. They will appreciate the evolutionary origins of human social proclivities and will be prepared to combat simplistic beliefs and common misunderstandings about the implications of human biology for human society.

Yet despite the evident progress represented in the following chapters, the Evolution, Biology, and Society section continues as the smallest of the ASA's 53 sections, with its 67 members in 2024 corresponding to only 7% of the membership of the largest section—Sociology of Culture (933 members)—and to just 15% of the mean 2024 section size of 444 (American Sociological Association, 2024). Moreover, at least since 2008, when EBS membership was 37% as large as the mean section size, the trend has been downward. Several tweaks to the section's name have not lessened the membership decline, although some may hope that the section's omission of the word "evolution" from its name as of 2024 will increase its appeal.

A review of the peer reviewed journal literature reinforces the impression of marginality. Between 1987 and 2024, only 35 English-language articles in peer-reviewed disciplinary journals included the words biology AND sociology AND evolution in their abstract (20 since 2000) (Sociological Abstracts). By comparison, between 1987 and 2024, 410 articles included the words race AND sex AND class in their abstracts. An inspection of six leading introductory sociology texts reveals no mention of biology (although another introductory text by Rosemary Hopcroft uses an evolutionary perspective).

It is therefore important not just to introduce the contents of this second edition, but to reach out to the many sociologists who remain indifferent to its many insights. If, as those of us contributing to this volume believe, neurosociology is an essential part of our discipline's foundation, such outreach can only further the discipline's progress. If, as Jonathan H. Turner (Franks & Turner, 2013:4) reminded us in his introduction to the first edition, interdisciplinary engagement fuels scientific advance, neurosociology can generate fresh insights about core sociological concerns. David Franks (Franks & Turner, 2013) struck a hopeful note in the introduction to the first edition by pointing to the large audience for a session on the social nature of the brain at the ASA's 2010 annual meeting; we hope that those who turn to this volume to learn about the latest advances will also use it to at least tweak the interest of even more sociologists about neurosociology's value.

But the challenge we face can only be understood if we first examine its origins in the controversies that accompanied sociology's emergence 150 years ago. That historical perspective will help clarify the disciplinary fault lines and illuminate the bases of scholarly progress. I will therefore begin this introduction by reviewing sociology's early years and identifying the bases for the disciplinary divorce. I will then highlight the changes that have made interdisciplinary progress possible and introduce the scholarly advances represented in each of the book's chapters that have nullified the grounds for divorce.

Disciplinary Divorce

Sociology "is altogether selfish," *The New York Times* announced in an editorial on March 9, 1883. Yale Professor William Graham Sumner had offended the editorial board's sensitivities by identifying "survival of the fittest" as "the great law of progress in human society" in a talk to the Nineteenth Century Club. Three years earlier, the *Times* had reported on Professor Sumner's success in assigning English philosopher Herbert Spencer's (1874) *The Study of Sociology* as a text over the initial objections of Yale President Noah Porter (Marsden, 2021); now it inveighed against the central tenet of Spencer's sociological theory. As the nation's first sociology professor, Sumner's choice of a book that had just been serialized in the new *Popular Science Monthly* and had been authored by the "colossus of Victorian intellectual life" made that decision a newsworthy event (Carneiro, 1974:544).

The *Times* editorial further explained that at the same event Columbia Professor Isaac L. Rice had ridiculed "this so-called law" as "based on a lame analogy from the world of organic life":

> The late Mr. Darwin once mumbled something about "natural selection," by which he seems to have meant that organized life moves from lower to higher by an automatic stirpiculture [selective breeding].

Instead, Professor Rice had argued, "the true object of sociology is to work for the brotherhood of the whole human race, for the benefit of all."

But it was Herbert Spencer, not Isaac Rice, who had "first awakened in England, America, France, Italy and Russia a wide interest" in the discipline of sociology (Carneiro, 1974:546; Giddings, 1900:543). And it was Herbert Spencer who coined the term "survival of the fittest" and convinced Darwin to substitute it for "natural selection." Many sociologists followed Spencer's lead. Writing in the *American Journal of Sociology* (AJS) in 1895 (p. 309), John Hobson argued that "every intelligent society" should impose "sterility upon the physically unfit of each generation." Britain's Sociological Society invited Darwin's cousin and leading eugenicist, Francis Galton, to speak at its May 1904 meeting at the School of Economics (London University). The *American Journal of Sociology* then published the talk, "Eugenics: Its Definition, Scope, and Aims," as its lead article in 1904, with invited commentary by eighteen sociologists and other opinion leaders, including H. G. Wells (a critic) and G. Bernard Shaw (an admirer).

After Darwin published *The Origin of Species* (1859), the primary implication of evolutionary biology for sociology was thus seen as competition between individuals, with the fittest surviving and the less fit ultimately eliminated. In the words of ethologist Richard Dawkins (1976), for whom this perspective remained an article of faith more than a century later, government institutions were needed to regulate and temper Homo sapiens' drive for individual advantage. Left to humans' own "natural" proclivities, societies would fracture into a perpetual war of all against all.

But the disciplinary tide of social Darwinism began to recede. Lester Frank Ward, the founder of American sociology and the first president of the American Sociological Society (1906), rejected the laissez faire philosophy of both Spencer and Sumner, criticizing Sumner's (1974/1883) *What Social Classes Owe to Each Other* as "a sort of final wail against the modern practices of states and people which run counter to these theories" (Russett, 1976:104). The A.S.S.'s fourth president, Charles Horton Cooley (1902:14–15) began *Human Nature and the Social Order* by acknowledging the "evolutionary point of view," but then emphasizing that social progress "does not require any change in heredity, but is a development of knowledge, arts, and institutions that takes place in the social process with little or no alteration of the germ-plasm."

Although William Graham Sumner was elected as the A.S.S. 1908–1909 president and Edward A. Ross, the 1914–1915 A.S.S. president endorsed Wisconsin's eugenic sterilization law (Degler, 1991:46), the *AJS* published in 1920 three articles on "heredity and opportunity" by Erville B. Woods of Dartmouth College that disputed the basic premises of social Darwinism. Reviewing the "very spirited controversy over the relative influence of 'nature and nurture'" by "biosociologists and sociologists of the ordinary sort," Woods (1920a:1–3) rejected sociologists' "extravagant" "professions of indebtedness to biology" and such "absurdities in social theory" as "the attempt to base an ethics of rapacity and greed" on social Darwinism (see also Woods, 1920b). What was needed instead of eugenics was "humanizing group sentiments" (Woods, 1920a:20). In *Primitive Behavior* (1937), 1927 A.S.S. President W. I. Thomas emphasized the unity of mankind, pointing to many examples of comparable abilities in native peoples and "the white man," and rejecting Herbert Spencer's contrary claims (p. 782) and the supposed evidence for racial differences in abilities (and see Degler, 1991:15–16, 89–90).

The discipline's founders in Europe also turned away from Spencer and social Darwinism. Durkheim criticized Spencer's sociology throughout his first doctoral thesis—*The Division of Labor in Society* (1933/1895)—and charted a disciplinary course that abjured Spencer's blending of utilitarianism and "biologism" (Nisbet, 1965). Durkheim (1974/1953:198) averred in his inaugural lecture on sociology at the University of Bordeaux that "there has to be something in the social domain which escapes biological inquiry in order to justify sociology" and rejected Spencer's assumption that competition between self-interested individuals is the basic force underlying societal progress (Durkheim, 1933/1895:196). Weber (1947:93–94) termed biological (and psychological) influences as "not understandable" influences from a sociological perspective that should be treated as "given data" by sociologists (Weber, 1978:8).

Thus, by the second decade of the twentieth century, these leading sociologists charted a course for the discipline that turned explicitly and decisively away from social Darwinism and Spencer's notion of the "survival of the fittest." Nor did biologists discern a reason to use biology to help explain the human social world, as the modern synthesis (Huxley, 1943) resulting from the integration of Mendelian genetics with Darwin's evolutionary theory posited a unidirectional and deterministic process of genetic variation allowing natural selection. The disjuncture between a biology that privileged individualist competition and a sociology that focused on human altruism and social connection did not allow for sustained cooperation. To repurpose a phrase used by paleontologist and popular science writer Stephen Jay Gould (1997) in relation to science and religion, the disciplines of biology and sociology were seen as "non-overlapping magesteria."

Interdisciplinary Progress

But while the disciplinary divorce solved some problems, it exacerbated others. Biologists recognized that nature is not always "red in tooth and claw," even though this suggested evolutionary pressures beyond the survival of the fittest paradigm. Some recognized the importance of Darwin's inclusion of two chapters on human altruism in *Descent of Man*, even though he had not subsequently drawn attention to this remarkable addition to the evolutionary story. The discovery that epigenetic processes allow the external environment to stimulate or suppress gene expression was transforming the nature-nurture debate. Growing recognition that neuroplasticity allows the brain to adapt to its environment throughout the life course suggested flexibility in biological influence on human development.

Some sociologists also found a basis for renewed interest in the sociology-biology nexus. The new subfield of social neuroscience recognized that much of the human brain is engaged in creating and maintaining social connections, in turn highlighting the need to explain how human social proclivities evolved (Schutt et al., 2015). Increased attention to the influence of emotion on human behavior and decision-making implied that explanations of social connection and social support had to take account of this fundamentally more biological element. The exceptional growth of the prefrontal cortex during the evolution of Homo sapiens laid the foundation for abstract reasoning and enhanced executive functioning, but it did not sever the biologically based bond between reason and emotion (Damasio, 1994).

It is now possible for sociologists to recognize the discipline's biophobia (Freese et al., 2003; Kalkhoff et al., 2011) for what it is: an impediment to understanding people as they are and human society as it is. Each section of this Handbook provides substantial evidence of how those who have been breaching the barriers to interdisciplinary engagement have been advancing the discipline. The first section reviews the developments in evolutionary theory and neuroscientific research methods that provide the foundation for the progress reflected in the rest of the book. Chapters in section two then establish the unique role of emotions in the brain's evolution and their integral role in human thought and behavior. The third section on mind, self, interaction, and sociality is the core of the Handbook, with chapters that deepen understanding of how evolution produced such a socially oriented brain and then demonstrate how this perspective applies to traditional sociological concepts. The last section examines variation across cultures and in relation to social problems from this evolutionarily informed vantage point.

New Directions

As the human brain's central role in social connection has been recognized across disciplines, different labels have reflected different perspectives: sociobiology, evolutionary psychology, neurosociology, and social neuroscience. For sociological theorizing to advance, the boundaries of and bridges between the empirical phenomena implied by these labels need to be identified. This is the first focus of the Theory and Methods section, as David Franks differentiates neurosociology from social neuroscience by carefully distinguishing the more sociological concept of role-taking from the psychological concept of theory of mind. Jonathan Turner then furthers construction of the appropriate interdisciplinary Venn diagram by differentiating neurosociology from evolutionary psychology and sociobiology, while also critiquing the General Social Science Model that treats biology as relatively unimportant for understanding human behavior. The next introductory chapters elaborate the theoretical foundations for neurosociology in evolutionary analysis and neuroscience. Norman, Restrepo, and Berntson stress the value of social neuroscience as an interdisciplinary, multi-level perspective required to explain social processes. Gabe Ignatow adds a history of science perspective on the entry of social neuroscience into sociology. The section concludes with Joshua Pollock and colleagues' overview of neurobiological research methods and their current limitations with respect to the study of social capacities and social connections.

The centrality the Handbook gives to the process of human evolution, as reflected in this first section, stems from a temporal leap in the sociological imagination. From the discipline's emergence in the last decades of the nineteenth century until the end of the 20th, most sociologists began the analysis of societal development with the transition from feudal to nascent industrial patterns of social organization. To begin sociological analysis with human evolution itself is to multiply the relevant timeframe for sociology by more than one hundred times and to bring within the scope of sociological concern phenomena previously considered solely within the purview of other disciplines.

And in the long process of hominin evolution, the role of emotion is at least *primus inter pares*. For too long, emotion has been considered a marginal topic in sociology. Unique human capacities were seen as reflecting a much larger prefrontal cortex that overrode the more emotionally charged limbic system. From this perspective, Western rationality had replaced emotionally charged decisions with rational calculation and a bureaucratic style of social organization. Durkheim (1995/1912) recognized the role of emotion in his concept of collective effervescence, but even this was seen as a bonding process in primitive societies. Popular culture assumed that more educated, higher status members of society are socialized to prevent their emotions from influencing their reasoning. Emotionality and rationality could best be understood as at opposite ends of a continuum.

The Handbook's second section demonstrates convincingly that the study of emotion instead needs to play a central role in sociology and that emotionality is misconceived as inapposite to rationality. Neuroscience research has identified emotional connections throughout the brain and demonstrated that they play an essential role both in human behavior and human reasoning. In fact, evolutionary research has revealed that the expansion of emotions preceded enlargement of the neocortex and enabled human ancestors to sustain the cooperative social groups that increased the survival odds of an otherwise defenseless primate. This research also explains how emotions solve the paradox of the human propensity for group living despite a lack of genes ("bioprogrammers" in Turner's language) like those that induce herding in other species, and despite the conflictual consequences of the panoply of negative emotions. Emotions are therefore introduced in this section as the foundation for understanding both the evolution of the human brain and of human functioning.

Jonathan Turner emphasizes in the first chapter (7) in this section that human emotions are biologically, not culturally based and that they emerged in the hominin line at least 4 million years before the evolution of language, the biological prerequisite for culture. But enhanced emotional capacity is only part of the human evolutionary story. In the next chapter (8), Turner details the key pre-adaptations that preceded the emergence of hominins from their great ape lineage, including grasping hands, a sophisticated sensory system, and neurological change facilitating language development and more flexible patterns of social organization. Turner also describes the emotions that evolved in humans to facilitate social control: shame, guilt, and alienation and explains the strengthening of the pre-adult mother-child bond and then the emergence of the nuclear family as steps in support of a more prolonged infancy and juvenility.

Warren TenHouten then argues in Chap. 9 that a "three-level science of mind, brain, and society" requires analysis of the full human emotional palette and its functions. He reviews different conceptions of this palette and the evidence that particular emotions are lateralized between the brain's right and left hemispheres.

Objections to including neurobiological processes in sociological explanations have rested at least in part on the argument that some essential forms of social behavior could not be connected to biology because they have no physical foundation. Symbolic interaction theory and Goffman's dramaturgical perspective on human interaction prioritize the meanings people give to social experience, rather than the experiences themselves. The challenge to a neurobiological connection is compounded by considerable cross-cultural variation in social patterns, as well as by extensive variability in forms of social relations within single societies.

David D. Franks begins the Mind, Self, Interaction, and Sociality section by arguing that this supposed disjuncture is unwarranted. Instead, split brain research supports symbolic interaction's expectation that actors' stated motives are post hoc rationalizations for their actions, rather than rational explanations. Neurological research has validated symbolic interaction theory's assumption that what is perceived and remembered are not direct perceptions, but rather the results of a neural process that filters and interprets those perceptions.

Alexandra Maryanski (Chap. 11) then leads readers through 65 million years of primate evolution to explain how anthropoid evolution laid the foundation for cortical expansion in the hominoid (ape and human) lineage and then the appearance of the Homo lineage 2.3 to 1.8 million years ago. Environmental change forms the bedrock of Maryanski's evolutionary analysis, with attention to how apes were challenged by the changing African climate in ways that furthered increasing capacities and ultimately led to the evolution of humans as both highly social and exceptionally adaptable. "The integration of culture with biology is what allows for this [adaptability]---a truly unique capacity that along with our one-of-a-kind sociality is the true 'secret of the human mind.'" Jonathan Turner next guides a whirlwind tour of the ways that the brain, "working through interpersonal behaviors, makes society and culture possible": role-taking and role-making and similar behaviors that sustain structures, rituals, different types of situations, identities, and motives. Alexander Shkukro (Chap. 13) continues the focus on roles and role expectations but adds a comparison of different theories of social roles and identifies the potential benefits for sociology of instead grounding role theory in neurosociology. In a related way, Richard Niemeyer explains how a neurosociological perspective can enrich identity theory and specify identity processes. Rosemary Hopcroft's review of the theory of mind and its relation to neural structure and process exemplifies the process and merits of conceptual clarification.

Rejecting the traditional view of the brain as a passive "bottom-up" processor of sensory information, Cheadle, Davidson-Turner, and Goosby (Chap. 16) see the brain as a predictive modeler that develops "'top-down' 'generative models' tested and refined against 'bottom-up' sensory errors." This "Active Inference Framework" (SIF) improves understanding of how sociocultural environments shape and are shaped by "predictive neurosociological actors." As the action-perception cycle continues, prediction errors about the external environment are minimized and actions taken become both more impactful and less likely to be harmful to the actor.

Matthew Brashears and Brea Perry extend Dunbar's (1992) classic analysis of the association between group size and intelligence by explaining how "the need to manage the social has defined our species and appears to have been responsible for the development of our impressive level of intelligence." They identify the specific brain regions involved in the bidirectional relationship between social connection and brain growth and they distinguish the underappreciated role of the mind in creating schemas that allow efficient management of social network information that would otherwise exceed the brain's capacity. Dippong and Overton then add to this analysis the concept of interpersonal synchrony and show that observed synchrony in social interaction is reflected in neural as well as vocal synchrony. The research they review also provides a compelling example of the use of brain imaging to study social connection that was introduced in the chapter on methods by Pollack and colleagues.

Since Franz Boas and Margaret Mead, cross-cultural variation in human behavior and social organization has been used as evidence of the innate flexibility of humans and the absence of a biological "blueprint" connoting an innate human nature (cf. Christakis, 2019). Deniz Buyukgok and Steve Hitlin begin the Handbook's last section, Culture, Society, and Social Problems, by showing that, contra Boas and Mead, neural plasticity allows cultural differences to affect the brain in ways that shape child development and intergroup relations. Abrutyn and Uscola then demonstrate that failing to recognize the pain of social loss as being linked to a biosocial understanding of "the affectual nature of the human … is to deny something fundamental." Yulia Shkurko's provocative analysis of human-AI interaction argues that even more fundamental is the social brain's misattribution of human characteristics to the computer-based side of this interchange. This "evolutionary mismatch" instead creates an illusion of social interaction that can erode prosocial behavior, although this erosion can be slowed if not completely forestalled by requiring that AI systems use "Human-Centered" principles that keep humans in control.

The last four chapters focus, respectively, on the problems of persistent inequality, health inequities, aggressive behavior, and conduct problems. Each uses a neurosociological perspective to elaborate a conceptual antinomy involving the problem. Jeff Davis contrasts predictable and uncertain environments to explain the effects of persistent inequality. In predictable environments, the brain's PFC and ACC regulate decision-making and control input from the brain structures associated with socioemotional experience, while in uncertain environments—more common in lower status groups—increased dopamine activates these structures and inhibits the functioning of the PFC, with behavioral results being increased aggression and sexual activity. Bridget Goosby and Jacob Cheadle contrast the effects of linear and complex systems on immune system dynamics. The former involve additive, predictable effects, but the latter create "feedback loops and cascading effects at both structural and individual levels" that lead to stress proliferation. Goosby and Cheadle use racism to illustrate a complex system, showing how it raises the risk of inflammatory conditions and so of chronic physical and mental illness. Jason Isbell, Stefan M. Goetz, Pranjal H. Mehta, and Justin M. Carré identify the effects of biological and environmental risk factors on human aggression, highlighting the roles of specific genes and hormones and of violent media exposure and interpersonal rejection and then suggesting relevant psychological interventions and productive research directions. Kalina J. Michalska and Rengin B. Firat review evidence for the contributions of both neurocognitive and neurodevelopmental processes and multi-level social factors to conduct problems, while also recognizing racial biases in both assessment of conduct problems and interpretation of related neuroscience data. Their concluding cautionary note bears repetition: "this focus on neurobiology often neglects subjective experiences and sociocultural factors, leading to biased interpretations based on race or ethnicity and posing risks to vulnerable populations".

This Handbook represents a new future for sociological engagement with evolutionary biology. The development of group selection theory, the recognition of neuroplasticity, and the discovery of epigenetics have created a turning point for sociology. We can confidently leave behind the dark past of social Darwinism and eugenic policies and instead look forward to enriching our analyses of human social connection with an understanding of how our social needs and capacities evolved and helped to make us the dominant species on the planet.

References

American Sociological Association. (2024). https://www.asanet.org/wp-content/uploads/2024/09/Section-Membership-Totals-Comparison-2024.pdf

Carneiro, R. L. (1974). Herbert Spencer's "The Study of Sociology" and the rise of social science in America. *Proceedings of the American Philosophical Society, 118*(6), 540–554.

Christakis, N. A. (2019). *Blueprint: The evolutionary origins of a good society*. Little, Brown Spark.

Cooley, C. H. (1902). *Human nature and the social order*. C. Scribner's Sons.

Damasio, A. R. (1994). *Descartes' error: Emotions, reason and the human brain*. HarperCollins.

Darwin, C. (1859). *On the origin of species by means of natural selection, or the preservation of favoured races in the struggle for life* (J. W. Burrow, Ed.). Penguin, 1968.

Degler, C. N. (1991). *In search of human nature: The decline and revival of Darwinism in American social thought*. Oxford University Press.

Dunbar, R. (1992). Neocortex size as a constraint on group size in primates. *Journal of Human Evolution, 22*(6), 469–493.

Durkheim, É. (1933/1895). *The division of labor in society* (G. Simpson, Trans.). Macmillan.

Durkheim, É. (1974/1953). *Sociology and philosophy* (D. F. Pocock, Trans.). Free Press.

Durkheim, É. (1995/1912). *The elementary forms of religious life* (Karen E. Fields, Trans. with an Introduction). The Free Press.

Dawkins, Richard. 2016 [1976]). The Selfish Gene, 40th anniversary edition. Oxford, UK: Oxford University Press.

Franks, D. D., & Turner, J. H. (2013). Introduction: Summaries and comments. In D. D. Franks & J. H. Turner (Eds.), *Handbook of Neurosociology* (pp. 1–5). Springer.

Freese, J., Li, J.-C. A., & Wade, L. D. (2003). The potential relevances of biology to social inquiry. *Annual Review of Sociology, 29*, 233–256.

Galton, F. (1904). Eugenics: Its definition, scope, and aims. *American Journal of Sociology, 10*(1), 1–25.

Giddings, F. H. (1900). Modern sociology. *International Monthly, 2*, 536–554.

Gould, S. J. (1997). Nonoverlapping magisteria. *Natural History, 106*, 16–22. and 60–62.

Huxley, J. (1943). *Evolution: The modern synthesis*. Harper and Brothers.

Kalkhoff, W., Dippong, J., & Gregory Jr, S. W. (2011). The biosociology of solidarity. *Sociology Compass, 5*(10), 936–948.

Marsden, G. M. (2021). Prologue III. Yale embattled: Noah Porter versus William Graham Sumner (1880). In *From the soul of the American University revisited*. Oxford University Press.

Massey, D. S. (2001). A brief history of human society: The origin and role of emotion in social life. *American Journal of Sociology, 67*, 1–29.

Massey, D. S. (2013). Preface. In D. D. Franks & J. H. Turner (Eds.), *Handbook of neurosociology* (pp. v–vii). Springer.

Nisbet, R. A. (1965). *Émile Durkheim*. Prentice-Hall.

Russett, Cynthia Eagle. (1976). Darwin in America: The Intellectual Response 1865-1912. San Francisco: W. H. Freeman and Company.

Schutt, R. K., Seidman, L. J., & Keshavan, M. S. (Eds.). (2015). *Social neuroscience: Brain, mind, and society*. Harvard University Press.

Spencer, H. (1874). *The study of sociology*. Henry S. King.

Sumner, W. (1974/1883). *What social classes owe to each other*. Caxton.

The New York Times (Editorial Board). (1883). The selfish sciences. *The New York Times*, March 9.

Thomas, W. I. (1937). *Primitive behavior, an introduction to the social sciences*. McGraw-Hill.

Weber, M. (1947). *The theory of social and economic organization*. (A. M. Henderson & T. Parsons, Trans.). The Free Press.

Weber, M. (1978). *Economy and society: An outline of interpretative sociology*. University of California Press.

Woods, E. B. (1920a). Heredity and opportunity. *American Journal of Sociology, 26*(1), 1–21.

Woods, E. B. (1920b). Heredity and opportunity. *American Journal of Sociology, 26*(2), 146–161.

Part I

Theory and Methods

Why We Need Neurosociology as Well as Social Neuroscience: Or—Why Role-Taking and Theory of Mind Are Different Concepts

2

David D. Franks

Abstract

In Chap. 3, David Franks demonstrates why neurosociology and social neuroscience can be seen as complimentary to each other. Once again, neurosociology diverges from time-honored academic traditions, in this case shedding what T. D. Kemper referred to as our fortress mentality. Ironically to some, this complimentarity can only be achieved efficiently by being very clear about the units of analysis that distinguish the disciplines. This would be an interactional unit of analysis in sociology and an individual one in psychology. G. H. Mead's role-taking and his four-staged theory of the act exemplifies the more voluntaristic sociological unit of analysis while learning theory, applicable to all mammals, distinguishes the psychological one. Suggestions are made to what kind of cross-disciplinary research could be conducted that would contribute to both fields.

In this chapter, I compare George Herbert Mead's concept of role-taking in neurosociology with psychology's social neuroscience notion of theory of mind (ToM). (See Chap. 12.) My goal is to illustrate how the unit of analysis in the two fields is different and yet show how each has great potential for an innovative sharing of their substance and methodologies. I will also suggest some specific ways that the differences can complement each other. Next I use a hypothesis about how Mead's concept of role-taking is connected with social structure and power to show how typical instruments of social neuroscience could be used in replicating and verifying this relationship. I argue that in order to come close to explaining a satisfactory portion of the variance in such a study, we need both perspectives. In this case, neurosociology is useful in formulating hypotheses to be tested by social neuroscience using Read Montague's (Montague et al. 2002) fMRI method of "Hyperscanning" which allows persons in separate scanners to interact with each other simultaneously.

History of the Terms Neurosociology and Social Neuroscience

John Cacioppo and Gary Berntson used the term "social neuroscience" in 1992 to describe psychology's inquiry into the social nature of our brains (Cacioppo and Bernston 1992). Twenty years earlier, neurosurgeon Joseph Bogen and Warren TenHouten, a sociologist, coined the term

D. D. Franks (✉)
Department of Sociology, Virginia Commonwealth University, Richmond, VA, USA

W. Kalkhoff et al. (eds.), *Handbook of Neurosociology*, Handbooks of Sociology and Social Research, https://doi.org/10.1007/978-3-031-95615-7_2

"neurosociology" (Bogan et al. 1972: 49). Bogen had worked with Gazzaniga on the first split-brain research. Here they referred to "a confluence of neurologic and sociologic observations."

The term social neuroscience was comparatively standardized and accepted in the field of psychology by the beginning of this century while at that time neurosociology was accepted only with ambivalence by the field of sociology. There had been a vigorous battle over E. O. Wilson's reductionist version of sociobiology, so sociology had significantly more issues to sort out in the area of social and biological connections than did psychology.

This raises the question of why neurosociology is needed at all when a more developed and vigorously growing field called social neuroscience already exists. One way to answer this is to look at how the two fields deal with similar but different concepts and yet keep within their distinct academic boundaries.

For example, although there are similarities in the sociological concept of "role-taking" and the concept of "theory of mind," a closer look will reveal differences that define the boundaries between the two fields and suggest why their differences should be maintained. The social neuroscience theory of mind, or "ToM," refers to the everyday person's informal "theories" of the minds of others as to what they are thinking, feeling and motivated to do.[1] ToM has nothing to do with the observing scientist's general theory of mentation though that may seem to be implied to the uninitiated.

Distinguishing Between the Two Fields Using Role-Taking and ToM

Role-taking, as developed by G. H. Mead, contains much more than "assumptions made about the mind of another" even if this "much more" is not always appreciated by sociologists. Exploring the full implications of role-taking can shed light on how the fields differ and why both fields can work together in mutually beneficial ways. Assuming a pre-established communicative context of shared symbols, role-taking refers to the capacity for one's verbal gestures to "call out" in one's self the same response that one's gesture *calls out in the other*. Put differently, the self assumes the perspective of the other and in effect apprehends what the other hears. Experiencing the self from the other's standpoint is called "taking the role of the other." This social and reciprocal shift creates intersubjectivity as well as fitting the actor into the social context and generating subsequent communication (See Mead 1934: 254–55). A major difference in role-taking and ToM is that in role-taking, actors experience not only what they *think* the other is thinking, but experience their own *selves* from their construction of the other's *different cognitive* perspective and emotional predispositions. The actor actually incorporates the imagined response of the other into his own emerging behavior. Frequently sociologists fail to emphasize the fact that actors have to *take on different social perspectives from their own* and then use these to assess how they will appear to others. This stresses interactional processes in a way that ToM does not necessarily do.

This process has important theoretical implications. Foremost is that it offers one of the few voluntaristic theories of the *self*-control of social conduct.[2] This alone would distinguish it from social neuroscience which is more often embedded in a different theory of social control,

[1]Valerie Stone (2006: 106) is more specific: ToMs require "metarepresentations." Some inferred mental states involve metarepresentational abilities and some do not. According to her, for example, inferring another's emotional state does not have to involve representing a person's mental state as a ToM. Insofar as mirror neurons simulate emotions unconsciously, this may well be correct.

[2]Read Montague, a social neuroscientist, is no stranger to voluntaristic frameworks as indicated by his work on social trust. In this context, he uses more of a rational choice psychology than learning theory. Talcott Parson's theory of social action is also voluntaristic but Parsons is working on a structural level rather than a Meadian social psychological one.

namely, a more individualistic learning theory. Granted, Mead's framework is compatible with learning theory since it applies to any mammal including human beings, but his intent is to go beyond that to what is distinctively human. Role-taking assumes two persons in interaction mutually influencing each other. We become social in a thoroughly sociological sense when *the other person's anticipated response is incorporated into one's own developing lines of action.* Another difference for sociologists using Mead's theory of the act is that we are *pulled* along in a teleological fashion by our anticipations of our own consummation of the act. Learning theory assumes the actor's motivation comes from external stimuli that happen *to* the person while Mead sees motivation as emanating from the actor him or herself—thus its voluntaristic quality.

Finally, role-taking meets the critical defining nature of a distinctively sociological concept because two actors who are role-taking can produce emergent behavioral outcomes that are not reducible to the persons taken singularly (Franks 2010).

It bears repeating that this does not make neurosociology and social neuroscience incompatible in any way because role-taking, which captures what is distinctively human, is built on top of learning theory.

Some Ways Role-Taking and Power Can be Explored Experimentally

The articulation of role-taking with the structural concept of power helped to move it outside of the exclusive confines of Mead's foundational symbolic interaction. For example, in 1972 Thomas et al. examined the ability of family members to predict accurately the responses of others in the family. They asked members to predict how each other family member would respond to a series of vignettes. An inverse relationship was found between the social status held by family members and each family member's accuracy in predicting the responses of other members. The most accurate role-takers were little girls, then little boys, with the least accurate role-takers being fathers and after them, mothers (See Franks 1989).

This research emphasizes an important distinction between *interpersonal* resources, which stem from one's social position and those stemming from one's *personal* capacities. Resources accruing from the self-system are available to all persons depending on their individual competencies in interpersonal relations. *Power* relates to the potential for control given by the impersonal structural position, in contrast to *influence* which is based on one's interpersonal skills and personality. The research by Thomas et al. suggests other investigations for neurosociology and social neuroscience, which separate personal dimensions from social structural ones in order to study their relationships. For example, would departmental chairpersons in a university be more accurate role-takers of deans than vice versa? Would deans be more accurate role-takers of vice presidents than of their departmental chairpersons? Exploring these questions would provide us with some clear demarcations of the social structural and the personal.

To date, however, these theoretically important findings have not been replicated although some unpublished attempts to test the hypothesis have been made. This leads us to ask how social neuroscience can help in this regard.

What Social Neuroscience Can Offer Sociological Research on Role-Taking and Power

On the social neuroscience side, the amount of research and replication is far greater than in neurosociology, further demarcating the differences in the two fields. The important places in the brain where a neurosociologist might look for correlates of role-taking start with the parts that are newest from an evolutionary point of view thus, one might expect uniquely human behavior to be related to uniquely human areas of the brain. That would lead us to the neocortex and particularly to the prefrontal lobes. There we would find dense, two-way paths to the thalamus, amygdala, and other subcortical structures.

In like fashion, social neuroscientists trace the significant evidence for relatively dedicated brain systems that are significantly active when subjects have to think about what others are thinking. These findings remain constant in numerous different research contexts and in ways that would be the envy of empirically oriented social psychologists from sociology. As Mitchell et al. (2006: 65) phrase it:

> Data from neuroimaging and patient studies…provide an efficient means for addressing the question of whether social cognition relies on its own set of mental processes or piggybacks on other more general processes of memory, inference, planning, and so forth.

They present a wealth of evidence that "social cognition" relies on a discrete set of brain regions and that this imaging data carries over specifically to "mind reading". Predominantly, this includes the medial prefrontal cortex, but as is usual with the brain, many other regions also contribute critically such as the temporoparietal junction, amygdala, superior temporal junction, and the orbitofrontal cortex. These results do not answer the important questions about how these areas work, or even if they should be conceived of as "areas," but from an evolutionary viewpoint, they do strongly suggest that mind-reading, however one looks at it, is a necessary brain process for human survival and that these processes are vital parts of current social behavior.[3] Despite their equivocal status, the results do allow some exploratory guidelines of what brain phenomena might be related to role-taking.

One methodological strategy in the context of communication involves isolating mind-reading tasks from non-mind-reading tasks and then showing how distinguishing between the two makes a difference in brain activity which seems relatively "dedicated" to this social process (See Mitchell et al. 2006: 66). For example, participants who were scanned with fMRIs were given three semantic categories to consider: names of a number of known people (Frank, John, Mary), names of fruits followed by names of articles of clothing. Each participant was presented with an adjective such as curious, bored or sad for people, pitted for fruit, and woolen for clothes. Then they were asked if the adjective could be appropriately applied to each item. The researchers found that judgments about the mental states of the named people were associated with modulations in a different brain area than were found in the responses about the inanimate objects. The areas used in contemplating persons were the MPFC, ventrolateral PFC and the right temporoparietal juncture, superior temporal sulcus, and the fusiform gyrus. As the authors (2006: 66) stated:

> (Subjects) making semantic differences about (mental) characteristics of other people appeared to engage a qualitative set of cognitive processes than did similar decisions about inanimate objects. (Parens. added).

Other studies, some using stories and cartoons that could only be understood by attributing mental states to their characters, found consistent correlations with greater activity in the medial PFC during tasks requiring the subjects to consider the mental states of others.

Empirically Testing the Role-Taking and Power Hypothesis

Given the differences between the units of analysis of sociology and psychology, it comes as little surprise that the variations in brain activity in intrapersonal rather than *inter*personal cogitations have been given most of the attention in social neuroscience. This is also to be expected when the major data collection techniques are scanners that isolate people from one another.

[3] In the first chapter of Cacioppo et al. (*Social Neuroscience* 2006), Berntson is more cautious than the quotes above and than other authors in the book. He warns that it is too early to tell whether social processes reflect special brain processes. "We do not know enough about either social psychological process or brain mechanisms to answer this question at the present time. Brain localization can inform neuropsychological theories, but meaningful neurological theories will not be about places nor will…they be couched in the language of space. Rather they will have to incorporate fundamental underlying processes that subserve social processes." Charles Kaplan in a personal communication suggests replacing "areas" with "circuitry."

However, as noted before, Montague (Montague et al. 2002) has developed imaging devices wherein several persons can communicate with each other across the space of the two "hyper-scanners." Unfortunately, the technology does not allow for normal verbal interaction. It must be limited to movements of the fingers on keyboards and the subject's head cannot move which takes on significance when we consider the important role that body posture plays in human communication.[4]

Even with these limitations, hyperscanning might well be used for the purpose of testing hypotheses derived from Mead's concept of the role-taking process and especially in the context of role-taking and power. Social neuroscience has given us the consistent finding that more brain energy as measured by blood flow is needed when a person is involved in constructing theories of mind than when ToM is not a part of the person's activity. It would be expected that a nurse communicating with a well-known doctor would be more pressured to role-take than would a participant answering the question of whether subjective adjectives applied to people are more appropriate than adjectives applied to inanimate objects. In such settings, nothing is at stake personally. Identifying the brain processes differentially involved and comparing them with the previous social neuroscience findings might well benefit both fields. On the other hand, scanned indicants of brain activity in measures of doctors who are role-taking with nurses may be expected not to involve the same intensity of brain processing. In no sense would this imply exhaustive evidence but it would be a start.

Conclusions

The hypotheses about role-taking and power demonstrate differences in the ostensibly similar concepts of role-taking and ToM. At the same time, they also illustrate how the fields can be useful to each other. A joint study of the conditions where situational factors cause more brain activity as opposed to conditions where personality factors predominate would be a case in point. Another old and important question concerns the part that projection enters into mind reading. Those who study mirror neurons contend that understanding others' behavior is fostered because we actually *do* in our motor cortexes the behaviors we observe in others. There seems to be a great deal of evidence that such is the case (Iacoboni 2008).[5] In contrast, questions as to whether the medial PFC contributes differently when one takes the perspective of a role different from one's own also deserve attention. There is much to explore here and discovering some of the secrets of the human brain would move neurosociology to the next level.

[4]Charles Kaplan tells me in a personal communication that he is experimenting with virtual reality solutions to this problem and that Read Montague is aware of this.

[5]But see Philip Gerrands in John Cacioppo and Berntson, *Handbook of Social Neuroscience* (2009). He is not so sure of this evidence but he does not go in depth into his reasons in this particular source.

References

Bogan, J. E., Dezure, R., TenHouten, W. D., & Marsh, J. F. (1972). The other side of the brain IV: The A/P ratio. *Bulletin of the Los Angeles Neurological Societies, 37*, 49–61.

Berntson, G. G., & Cacioppo, J. T. (2009). *Handbook of Neuroscience for the Behavioral Sciences*. John Wiley & Sons.

Cacioppo, J. T., & Bernston, C. C. (1992). Social psychological contributions to the decade of the brain: Doctrine of multilevel analysis. *American Psychologist, 47*, 1019–1028.

Cacioppo, J. T., Visser, P. S., & Pickett, C. L. (2006). *Social Neuroscoence: People thinking about other people*. The MIT Press.

Franks, D. D. (1989). Role-taking, social power and imperceptiveness: The analysis of rape. In D. Norman (Ed.), *Studies in symbolic interaction* (pp. 229–259). JAI Press.

Franks, D. D. (2010). *Neurosociology: The nexus between neuroscience and social psychology*. Springer.

Iacoboni, M. (2008). *Mirroring people: The new science of how we connect with others*. Farrar, Straus and Giroux.

Mead, G. H. (1934). In C. Morris (Ed.), *Mind, self and society*. Chicago University Press.

Mitchell, J. P., Mason, M. F., Macrea, C. M., & Banaji, M. R. (2006). Thinking about others: The neural substrates of social cognition. In J. Cacioppo, P. S. Visser, & C. L. Pickett (Eds.), *Social neuroscience: People thinking about people* (pp. 63–82). The MIT Press.

Montague, P. R., Burns, G. S., Cohen, J. D., et al. (2002). Hyperscanning: Simultaneous *f*MRI during linked social interactions. *NeuroImage, 16*(4), 1159–1164.

Stone, V. (2006). Theory of mind and the evolution of social intelligence. In J. T. Cacioppo, P. S. Visser, & C. L. Pickett (Eds.), *Social neuroscience: People thinking about people* (pp. 103–130). MIT Press.

Thomas, D. L., Franks, D. D., & Calonico, J. M. (1972). Role-taking and power in social psychology. *American Sociological Review, 37*, 605–614.

Neuro-Sociology as an Alternative to Evolutionary Psychology and Sociobiology

3

Jonathan H. Turner

Abstract

The incursion of biologically-based analyses of social phenomena from biologists and, more recently, from evolutionary psychologist has often made sociologists defensive. Hence, for the most part, sociologists have tended to reject biological analysis, *per se*, which is clearly an over-reaction to both sociobiology and evolutionary psychology. This chapter proposes that an alternative for evolutionary analysis focusing on the evolution of hominin and human brains. This alternative employs (1) *cladistic analysis* and (2) *comparative neuro-anatomy* that can overcome the reluctance of many sociologists to brining biology into sociology. Moreover, an evolutionary slant can complement evolutionary analysis in sociology more generally. This analysis can also provide a much better source of data (on the neurology of the evolution of human brain) that can correct for excesses in many analyses in sociobiology and evolutionary psychology. And, in so doing, it is possible to *develop a sociology on the biology of human behavior, and equally important, on patterns of social organization*. By understanding how and why the human brain evolved and became so large can inform sociological analysis of all facets and levels of the socio-cultural universe.

The emergence of *neuro-sociology* offers yet another opportunity for sociologists to study human biology comparatively by studying the brain of humans closest primate relatives, the great apes. For, *virtually everything about the emergence of human societies is driven by the human brain,* inherited from the distant ancestors of contemporary great apes. This paper advocates such comparative analysis, while offering a critique of approaches in biology and psychology that, in my view, take sociology and, indeed all of social science, astray.[1]

Furthermore, evolutionary analysis that is coupled with neuro-sociology can offer yet one more approach by which to explain humans and their creation, *patterns of social organization*. The approach that I advocate allows us to see how societies and, indeed, how the entire *socio-cultural universe* emerged, becoming the third universe, emerging out of the physical and biological universes. Indeed, the emergence of the socio-cultural universe of humans is only possible by the evolution of the mammalian brain by

J. H. Turner (✉)
Department of Sociology, University of California, Riverside, Riverside, CA, USA
e-mail: jonathan.turner@ucr.edu

[1]Nonetheless, it should be mentioned that some very accomplished sociologists have adopted many of the assumptions of sociobiology and evolutionary psychology and, indeed, have produced very useful findings that can be supplement by the approach that I am advocating.

W. Kalkhoff et al. (eds.), *Handbook of Neurosociology*, Handbooks of Sociology and Social Research, https://doi.org/10.1007/978-3-031-95615-7_3

natural selection. Selection is not goal oriented, of course, but it can be *directional* in that if one path randomly pursued by natural selection increases the fitness of an organism (to reproduce and survive in a given habitat), it will continue along this path as long as it promotes fitness. And so, understanding fully the dynamics of human societies (and the societies of other highly intelligent animals) must be evolutionary. Human evolution in both its simplest and most complex forms requires that sociologists understand how selection was working for at least six million years along the *hominin*[2] line to create humans and the first human societies. Most of the important characteristics of hominins leading to humans revolve around changes in the neurology of hominins, thus making neuro-sociology an essential tool for explaining the dynamics of the socio-cultural universe.

The Challenge Posed by Evolutionary Psychology and Sociobiology vs. the Sociological Alternative

The Critique of the General Social Science Model

In the early 1990s, a new field in psychology was created: evolutionary psychology (e.g., Barkow et al., 1992). This approach accepted most of the tenets of sociobiology, to be examined shortly, but it also posed a challenge to many of the comfortable assumptions of many social sciences, especially sociology. The critique of The General Social Science Model (GSSM) by early evolutionary psychologists is still relevant today, even with some penetration of biology into the social sciences. This GSSM model assumes that (a) humans are born as "blank states," (b) human brains are general purpose computers, (c) culture and socialization into culture program *all* human behaviors, (c) culture is free to vary any trait in any direction, and (d) biology is relatively unimportant in understanding behavior.

In contrast, the assumptions of the evolutionary psychology model include: (a) Humans are born with a bundle of emotional, motivational, and cognitive adaptations during their evolution, (b) the brain is a collection of modular, domain-specific processes, (c) behavior is the result of interactions between evolved psychological 1 mechanisms and cultural and environmental influences, (d) culture itself is based on a universal human nature and is constrained by it; and (d) an analysis of interactions between nature and nurture is important to understanding behavior.

Sociology's resistance to evolutionary analysis in general, and to analysis of the biology of behavior and social organization more specifically, is captured in this notion of a *General Social Science Model.* The evolutionary psychology model has elements of insight but many are obvious overstatements. The general critique that sociologists do not pay enough attention to the biology is certainly well taken, as is the notion that culture and socialization are the only ways to understand human behavior and social organization. And, for many, evolutionary psychology was right in critiquing the assumption in the GSSM that biology is no longer important because, today, it is culture and social structure that drive behavior. The chapters in this book and other recent books (see, for example, Turner & Machalek, 2018, Turner, 2021, and Turner & Maryanski, 2024) seek to overcome this critique; nevertheless, most sociologists still view biology suspiciously.

The over-reach by evolutionary psychologists is, however, highly problematic in assuming that the brain is highly modular and that critical elements of the human brain evolved mostly in the late *Pleistocene*. The other assumptions are overstated, but they are important because ignoring biology reduces the capacity of sociologists to

[2]The *hominin line* is composed of those evolving apes carrying the genes of the ancestors of current great apes that, over six millions years, led to evolution of several subspecies of humans around 500,000 years ago.

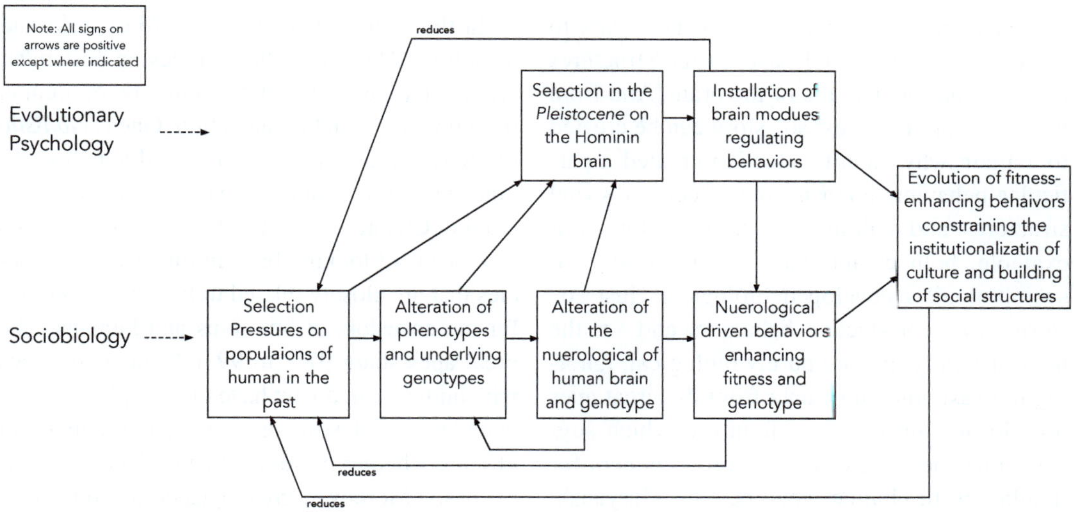

The upper portion of the model highlights the general argument of evolutionary psychology, while the lower part outlines the argument of sociobiology when it addresses neurological isssue. Evolutionary psychology appears to accept most of the assumptions of sociobiology, adding the emphasis of modification of the brain structures which in turn regulate genetically controlled behaviors that allow for the building out of the socio-cultural universe (societies). There are several problems with this whole set of approaches. First, research tends to do "reverse engineering," assuming that prevalent behaviors are genetically driven, thereby obviating the power of culture and social structure to constrain and even force behaviors, often against behavioral propensities that have a genetic basis. Second, evolutionary psychology initially emphasized that the late *Pleistocene* is when these changes in the brain occurred. Changes had been occuring for at least seven to eight million years of hominin evolution, at the least. Third, the principle changes to the brain have revolved around enlargement of all brain structures, with relativley few "new" structures or modules added to the brains of hominins and hence humans. Fourth, the "modules" or more comprensive structures of human neurology driving behaviors as not clearly specified in much evolutionary psychology and sociobiology. Hence, in the end, these approaches make many assumptions that not empirically proven, that employ a methodology that does not start with the evolution of the brain but simply assumes that any prevalent behvavior must have a genetic basis, and that does not employ a strategy where they can actually look at the brains, or reasonable proxies of the brains, of the great ape ancestors of today's great apes and their ancestors who are also the ancestors of hominin and, hence can provide a sense for what the brain of humans' ancestors was like. The result tends to be "just so" stories of why a brain structure (not fully specified) controlling certain behaviors evolved in response to often unspecified selection pressures.

Fig. 3.1 The questionable logic and methodology of evolutionary psychology and sociobiology when addressing human behavior, social structure, and culture

understand fully the socio-cultural universe. In Fig. 3.1, I have drawn a simple model that summarizes not only evolutionary psychology, but also sociobiology, from which evolutionary psychology drew its inspiration. The basic problem with both approaches is that even hard-wired behaviors installed by natural selection cannot alone explain the nature of human behaviors and their institutionalization in culture and social structures in human societies. For once social structure and culture exist, they do indeed constrain behaviors, even those that are hard-wired in the human genome. In fact, many social structures and cultural forces *go against* humans evolved biological nature—again something that should be clearly evident by the many pathologies evident in humans and their societies.

Another problem of both sociobiology[3] and evolutionary psychology[4] is the emphasis on behavior rather than social structural and cultural forces. When only behavior is analyzed, the explanatory logic of both fields took the following form: universal behaviors among humans are likely driven by the genome, and

[3]See Barkow et al.(1992); Barrett et al. (2002); Buss (1999); Cosmides and Tooby (1992); Tooby and Cosmides (2005); Dawkins (1976); Pinker (2002).

[4]See Hamilton (1964, 1971, 1984); Williams (1966, 1985); Wilson (1975, 1978); Wilson and Wilson (2007).

so, to explain these behaviors, it is necessary to discover the hard-wired basis in the structures of brain and the genome of the brain. And from this assumption, a "just so" story can be created to explain why natural selection created a particular behavior, ignoring the effects of social structures and cultures on all behaviors. For example, both evolutionary psychologists and sociobiologists developed biological "just so" stories for reproductive behaviors and for the nuclear family that are purely biological, ignoring the vast amount of data from the great apes that do not form nuclear families, which suggests that there were few bio-programmers for families in the human genome (see Maryanski & Turner, 1992, 2024). Instead, the nuclear family is mostly a cultural creation, although the evolution of hominin and then human neurology is relevant to understanding why such is the case (see Turner, 2021; Turner & Maryanksi, 2024 and reference therein).

As Fig. 3.1 emphasizes, evolutionary psychology argued early on that the brain was critical to understanding behavior, which they felt was regulated by *discrete modules* that evolved to solve adaptive problems. Thus, behaviors that were widespread among humans were assumed, without much (if any) data from neurology, to be evolved because they enhanced fitness and that the critical period was the late *Pleistocene*, the epoch during which hominins evolved. It was also assumed that each adaptive behavior was regulated by a discrete module, with natural selection creating a module in the brain for each behavior increasing fitness of hominins and eventually humans (the last of the hominins). A typical explanation was that a behavior that is common or even universal must have a biological basis in a module in the brain *without*, however, specifying the module. These kinds of arguments have receded, but they had the effect of giving sociologists some motivation to search for structures in the brain producing behaviors leading to the formation of social structures. And, the more socio-neurology looked at the brain, a very different picture of its structure and operation appeared, compared to the *ad hoc* explanations of evolutionary psychology (see Turner, 2000, 2021; Turner & Maryanski, 2024).

In the approach advanced by Maryanski and myself (2024), two methodologies, in particular, are used to understand the biology of human behavior and social organization. One is *cladistic analysis*, outlined in Chapters 7 and 8 in this volume; another is comparative neuro-anatomy, outlined in Chapters 7, 8, and 12. Cladistic analysis is a means of looking back in time by using species that are closely related to humans as proxies for the ancestors of hominins and humans. The great apes share 97% to 99 + % of their genes with humans but do not have the complications of complex social structures and culture that can obscure whether or not a behavior has a biological basis. Moreover, great apes *do not* form permanent groups beyond the mother-child bond, which is virtually universal among mammals. And so, while some behaviors have a biological basis, group structures beyond the mother-infant bond (which ends with puberty) are *human creations* arising from their higher level of cognitive capacities. Yet, great apes are highly skilled at reading each other in interpersonal relations, but groups are mostly temporary "hook-ups" that soon break apart. But these interpersonal skills do indeed have a biological basis because both great apes and humans, descendants of common ancestors, reveal these capacities during interaction rituals with conspecifics (Goffman, 1967; Collins, 1975; Turner, 2002). The genetic similarity of humans' closest relations—the great apes (chimpanzees, gorillas, and orangutans)—would suggest that the ancestors of these contemporary great apes and the ancestors of humans shared many of these genetically-based behavioral capacities that, during the evolution of hominins, *were enhanced and used to construct more permanent groups* and, eventually, large scale structures evident today.

The second methodology—*comparative neuro-anatomy*—compares great ape brains with human brains, operating under the assumption that features of the human brain also evident in the great ape brains were inherited from the ancestors of today's great apes. When such an exercise is performed (see Turner, 2000, 2002, 2021; and Turner & Maryanski, 2024), a very different picture emerges. Past and even more recent analyses in evolutionary psychology and sociobiology over-emphasize the unsupported assumption that widespread behaviors among humans *must have an evolutionary basis lodged in the human genome*. Comparative neuro-anatomical analysis of human and great ape brains offers a much better way to assess these assumptions of evolutionary psychology. In several chapters in this volume, I have provided details about what these two methodologies—*cladistics* and *comparative neuro-anatomy*—which, from a sociologists viewpoint, is considerably more useful in understanding the biology of human behavior and social organization than is possible with evolutionary psychology or sociobiology as they are employed today.[5]

Still, sociobiology and evolutionary psychology have provided some very useful insights and, perhaps equally important, have motivated some sociologists to find alternative methodologies that also challenge the General Social Science Model. Indeed, many of the chapters in this book provide various approaches to understanding the evolution and operation of the human brain that do not fall prey to the problems of many arguments in evolutionary psychology and sociobiology.

In this chapter, then, my goal is to outline in general how evolution, driven by *natural selection* and the other forces of evolution (*mutations*, *gene flow*, and *genetic drift*), worked on the brains of the ancestors of great apes and, then, on species of hominins leading to the emergence of first humans around 500,000 years ago. The story to be told is, in the end, a useful tool in socio-neurology because it provides *information on how the neurology of humans evolved over millions of years*. This information, as I hope to show, challenges many of the assumptions of evolutionary psychology and sociobiology.

Some Elementary Considerations in an Evolutionary Approach to Neuro-Sociology

First, at the biological level, evolution is, as Darwin argued, "descent with modification." Natural selection is the driving force of evolution, although mutations, genetic drift, and gene flow are also involved because they explain the sources of variation on which natural selection goes to work. Evolution is an inherently conservative process that modifies existing *phenotypical structures* and underlying *genotypes* of members of a species (often viewed as a *gene pool*). *Mutations* can be a more dramatic force in altering phenotypes and genotypes but most mutations are harmful, and large mutations are almost universally harmful because they create contradictions at the biological level on the interdependencies among biological structures (Fisher, 1930). Thus, arguments about emergent modules arising for specific functions are overdrawn and go against evolutionary theory as conceptualized in the Modern Synthesis.

Second, natural selection often operates on *spandrels* (Gould & Lewontin, 1979), *expatations*, or the original term (which I still prefer), *pre-adaptations*. These labels all denote the simple fact that many phenotypical characteristics of humans evolve as byproducts of natural selection and, often, have no effects on behavior or organization when they first evolved but are later critical features that are selected upon when the environments of a species change. Probably, the most important pre-adaptation among the ape ancestors of humans was when natural selection shifted the dominance of the sense modalities from olfaction[6] (smelling) to vision during the initial evolution of those mammals that would evolved into primates after the demise of the dinosaurs 60+ million years ago. In this shift, new brain structures were created in and around the *inferior parietal lobe* where the neurological

[5] For early comparative anatomy comparing great ape and human brains, see: Stephan (1983), Stephan and Andy (1969, 1977); Stephan et al. (1988); Stephen et al. (1970); Stephan et al. (1981).

[6] Most mammals are olfactory dominant. Primates are visually dominant, a shift that was more fitness enhancing in three dimensional habitats—the forests of Africa and later South American—where primates evolved.

centers for touch (*parietal lobe*), sound (*temporal lobe*), and *vision* (occipital lobe) meet in the neo-cortex [the *olfactory bulb* for smell is located in the sub-cortex where emotions are generated]. This rewiring of the brain, as a side effect of this shift in sensory system of primates, produced a pre-adaptation in great apes for *language comprehension*. This remained dormant for as much as 60 million years until natural selection randomly hit upon these dormant structures to rework late hominin neurology for speech production. In turn, this allowed for symbolic culture among the most intelligent primates, the hominin ancestors[7] of today's great apes. Without this pre-adaption already present, it is unlikely that natural selection could have worked to create the capacity for speech and, hence, culture. But with this pre-adaptation just "sitting there" in the neo-cortex for millions of years, it became possible with selection on other areas of the brain affecting speech production for spoken language to emerge some 1.6 million years ago (Turner & Maryanski, 2024). Indeed, the most complex part of this adaptation was, in essence, a byproduct of selection for a more efficient way to move around the arboreal habitat, since seeing ones way around the trees is more efficient than smelling (the dominant sense modality in most mammals) one ways around the trees, where a misstep can lead to death by gravity.

Third, evolution is often *punctuated* (i.e., moving rapidly) and *directional* (i.e., moving in a given direction). Moreover, selection works on the *tail ends of the distribution of phenotypical* features (and their underlying genotypes). If one side of a Bell curve on particular traits is fitness enhancing, selection will become directional over generations, with each generation being somewhat more biased to the favored side of the Bell curve. As a result, selection can move rapidly over several generations altering the favored side of the trait increasing fitness. For example, in human neurological evolution, the most prominent feature of the last two million years was increases in the size of both the neo-cortex and sub-cortical cortex. In fact, comparing great apes brains with those of humans look very much the same, *except for size*. There are some new structures, to be sure, but not nearly as many as early evolutionary psychology seemed to propose, essentially a new "module" for each new behavior.

Fourth, most mutations are harmful, as noted earlier, and the larger the mutation, the less likely is this mutation to be fitness enhancing. And, for the evolution of neurological structures, mutations are generally even more harmful because of the connectivity among all brain structures in a system of interdependences. Thus, a large change in one area would disrupt other areas of the brain. Far more likely is *directional selection* in the brain, as was the case as the sub-cortical areas hominins began to grow. Yet this happened very slowly, thereby sustaining the integration of the sub-cortex and its relations with the neo-cortex. Dropping new modules into the brain would almost always not be fitness enhancing, but increasing the size of either the neo-cortex or sub-cortex over time—in the case of hominins, six million years—proved to be adaptive in the long run. The growth of the hominin brain from around 450 cm^3 at 2.5 million years ago to almost 900 cm^3 at 1.5 years ago was, in evolutionary terms, still rather rapid for such a complex structure (Turner & Maryanski, 2024: 102).[8] Moreover, the growth was phased, with the sub-cortical areas growing first, thereby setting the stage for neo-cortical growth. This allowed for emotional tags to a larger and more complex inventory of cognitions in the neo-cortex. Since cognitions

[7]It took five to six million years for this pre-adaptation to emerge because of the complexity of the changes needed to rework the brain to download braining thinking (which is more gestalt-based) vial *Broca's Area* (Falk, 2007) and to re-aligned the many structures involved in speech production of the vocal tract, lips, tongue, and muscles controlling these. And it only occurred for great apes because there was clearly a threshold that none of the other 200+ species of primates possessed in the past and today. There was, then, another condition for unleashing the potential of this pre-adaptation: high levels of intelligence, which as is clear was dramatically enhanced during hominin evolution.

[8]See for details: Holloway (1974, 1979, 1996, 1998, 2015); Holloway and Post (1982); and Holloway et al. (2004).

can only be remembered by being tagged with emotions, first in the *hippocampus* and later in the *front lobe* (Damasio, 1994; LeDoux, 1991, 1993, 1996), growth of the sub-cortex had to occur before significant growth in the neo-cortex. In most of this growth, there were relatively few new "modules" but, instead, gradual and directional growth over a long period of time.

Fifth, in the case of the brain, which is the most significant aspect of hominin evolution, once evolution begins to exhibit directional selection on the size of key structures, this increases the likelihood of selection on other structures of the brain that enhance the effects of the directional inertia. For example, brain growth was rather slow during the first four million years of hominin evolution, increasing by perhaps only 100 cm^3 of the size of present day chimpanzees. Furthermore, most of this growth, I believe, was *sub-cortical on the emotion centers of the brain,* thereby laying the groundwork for enlarging the neo-cortex. This is because a large neo-cortex without the requisite complexity of emotions to tag cognitions would simply be an empty warehouse consuming proteins and calories but not cognitive capacities or intelligence.

Sixth, there emerged roadblocks to further evolution on the brain in a given direction. For example, the early growth of emotion centers encouraged early neo-cortical growth, but for the neo-cortex to reach human proportions, the preadaptation for language could only be unleashed when and if natural selection could increase hominin capacities for speech production.[9] If raised in an environment where human speech occurs, great apes can understand human language at about the level of a three year old child (Rumbaugh & Washburn, 2003).[10] Indeed, the association cortices in and around the inferior parietal lobe could not produce spoken language *until natural selection could create necessary structures for highly articulated sequences of speech production.* Selection began to work on what is often termed "Broca's hump," which is a slight enlargement on the left side of the brain around where the temporal and parietal lobe meet near the front lobe. Selection on this hump—an existing structure that was simply enlarged to what is now termed *Broca's Area*—was one necessary modification of an existing structure insofar as it allowed hominins to "download" brain thinking in gestalts into sequential speech (Falk, 2007). The converse of downloading gestalts into sequences of speech is "uploading" experiences via the sense modalities (often sequentially experienced) into the brain's way of thinking. This structure was, in large part, already extant among hominins because it is also evident among great apes and, hence, their common ancestors with hominins. This structure is now termed *Wernieke's area*, and it exists in the area of the brain around the inferior parietal lobe uploading into the brain visual, haptic, and auditory information. Hence, selection did not have to grow this area dramatically, but instead, fine-tune an existing structure to upload sensory information into the brain more efficiently.

The biggest obstacle for speech production, however, was the structures among great apes for auditory or vocal production. These structures are not set up for finely articulated sequences of talk in great apes (and, hence, hominins), although they can produce simple calls that can carry meanings. But the juxtaposition of the structures for speech-like vocalization—lips, tongue, larynx, pharynx, and the muscles regulating these—had to be significantly re-aligned, involving many genes. This realignment appears to have taken millions of years to occur—perhaps as many as five to six million years. And, if selection had not found a "solution" to the limitations of vocalization of hominins, humans would not have evolved. But once speech was possible, the brain could now grow significantly and rapidly, at a punctuated speed, because with speech, symbolic culture is possible. And, with symbolic culture, cognitions can become complex and more readily remembered, thereby making enlargement of the neo-cortex fitness enhancing. By 1.5 million years ago (Turner, 2021; Turner & Maryanski, 2024), selection had probably made the breakthrough because the brain growth of hominins accelerated rapidly to the human mea-

[9] See Geschwind (1965a, 1965b).

[10] See also Savage-Rumbaugh et al. (1978, 1988, 1993).

sure in less than a million years, which is fast in evolutionary terms for such a complex structure as the brain.

Seventh, as is evident in the above, the evolution of the brain did not involve so much as creating new structures in the brain as enlarging existing structures and selecting on one side of structures that are bilateral in the brain, which is most of the key brain structures. Thus, as was the case for *Broca's area* on the left side of most brains, selection enlarged it and perhaps also altered it in some way more than by size alone. However, since the structure was bilateral, a new capacity for speech production was added to an area that was probably very much like its counterpart on the right side of the brain, which is devoted rhythmic utterances. In fact, when some people lose the ability to talk because of damage to *Broca's area*, they can sometimes learn to communicate by putting their words into a song, such as "Jingle Bells," apparently using the opposite (i.e., right) side of the brain for speech. Thus, again rather than generating entirely new "modules," selection was working to alter existing structures. In fact, there are very few entirely new modules in the human brain compared to the great ape brain, and hence the hominin brain. In my own research on emotions, relatively recent data on the *amygdala*—which is the ancient area of the brain devoted to producing anger and fear as emotions—shows that the lateral nucleus part of the *amygdala* exists in both great apes and humans but has little to do with anger or fear among the latter. Rather, in humans, the *lateral amygdala* is specially dedicated to assessing the context and appropriateness of behaviors (Barger et al., 2007, 2012, 2014), Thus, here is a case where an area of the human brain was usurped for an entirely different function and, thus, might be seen as a new structure. Still, it is a structure that was already present in the legacy of great apes, and so it is not a new structure, *per se*. Again, natural selection did not install a mutation but rather altered the operation of an existing structure in probably what is the most ancient part of the mammalian brain: the center of emotions for sustaining life by activating the flight-or-fight responses.

The Evolution of Hominin Neurology

Humans ultimately evolved from the small, burrowing mammals that were able to survive the "age of the dinosaurs" by scurrying around in and out of holes in the ground where they could hide from danger. There is a reason that medical research can use mice and rats as proxies for other mammals in testing the effects of drugs. Natural selection, even though it drives the existence of all life forms, is a rather conservative and slow process. A good example is that humans share over 90% of their genes with the first mammals, rodents. Indeed, all mammals are closely related but have, of course, differentiated over the last 60 million years. Once free of the danger posed by dinosaurs, and after the destruction of the comet that kill off many life forms, mammals literally had a world of potential niches in which to survive, with natural selection working to alter structures to increase fitness in many different niches. Part of this process was altering brains, but the basic structure remained, with changes mostly revolving around brain size and selection on particular structures affecting locomotion and other traits for diverse niches. The mammals that ascended the trees to form the primate universe—differentiated into pre-monkeys (pro-simians), monkeys, and apes—have similar phenotypes because they were adapting to the same basic habitat: life in the trees. As emphasized earlier, the shift from olfaction (smell) as the dominant sense modality to vision created the most important pre-adaptation for hominin survival. This conversion to visual dominance and the neurology of *speech comprehension* remained dormant for perhaps 60 million years. Yet, this hidden secret would in the end be what allowed humans to evolve. Without this hard-wired, but hidden capacity for language, *humans would not have evolved.* But once unlocked by natural selection over millions of years, it would make late hominins and then humans the most emotional and intelligent animals on earth.

Early on, apes were more intelligent than monkeys and prominins and dominated them.

But by 20 million years ago, monkeys were dominant in that they far out numbered great apes in species and controlled the verdant areas of the forests in Africa. Great apes were forced to the marginal feeding areas high in the forests, probably because monkeys could eat unripe fruit and thus outcompete great apes who cannot easily digest unripe fruit. Thus, intelligence does not always determine dominance. Indeed, less than 5% of primate species are great apes; the other 95% of the primate order is composed of monkeys and prosimians. Earth is "the planet of the apes" because of the overpopulation on humans. The rest of the apes are endangered species, mostly the result of the overpopulation of humans and the final verdict is not in for humans in the present era. Yet, in the long run, the greater intelligence of great apes was another a pre-condition for unlocking the capacity for speech. Monkeys and prosimians have the same structures allowing for visual dominance as do great apes, but they do not have the same level of intelligence. Thus, it was the pre-adaptation in and around the parietal lobe, *plus* more intelligence, that allowed great apes but not monkeys or prosimians, to eventually develop the capacity for speech. Indeed, monkeys have much better vocal system for finely tuned speech, but they cannot unlock the key pre-adaptation for language comprehension and production for lack of the requisite intelligence.

The emergence of hominins some 6–8 million years ago, as they split off from the ancestors of the ancestors of current great apes, initiated the evolution a number of species of hominins that would lead to the emergence of various subspecies of humans some 500,000 years ago. There were, however, two periods when humans, regardless of their intelligence and capacity to speak and develop culture, were close to extinction (Amos & Hoffman, 2010). In fact, humans today are so closely related genetically because we are all descendants of very small populations that came quite close to the edge of extinction. Thus, intelligence does not assure survival, a lesson that remains true today as humans overpopulate the earth and dramatically alter the ecologies on which all life forms depend.

The evolution of the large neo-cortex of the human brain was not, I would argue, the reason that hominins survived as the forests in Africa were receding. A popular account is that a large neo-cortex allowed hominins to withstand the environmental challenges that forced an animal built for life in the trees onto the ground and, increasingly, out onto more open-country habitats—bush-lands, secondary forests, savanna, and even deserts. Alternatively, the key to hominin survival was, as I have emphasized endlessly, was selection on the *sub-cortical* areas of the hominin brain. This made our ancient ancestor *more emotional and able to forger stronger social bonds and thereby overcome the critical weakness of a great ape on the African savanna: the lack of permanent group structures to organize food gathering and to mount defenses to the many predators populating open-country habitats.*[11] Great apes live in loosely organized communities, with no permanent group structures beyond mother-offspring bonds. Even those bonds are broken forever for all mothers and daughters when daughters at puberty migrate to another community, never to return. The same is true of male orangutans and gorillas, but not chimpanzee males who remain in their natal community. Still, there are no permanent groups among chimpanzees, and certainly no nuclear families in great-ape communities because all sex between males and females is promiscuous. Hence, paternity of offspring is never known. Yet, the lack of group structures does not mean that great apes are not capable of highly coordinated interactions. Indeed, as other chapters in this book point out (see chapter 7, 8, and 12), the closest genetic cousins of humans, chimpanzees, can read the emotions of others very well, engage in greeting and parting rituals, pick up an interaction that occurred weeks earlier, and at times even engage in carnival like rituals looking very much like collective gathering of humans (see Maryanski, 1986, 1987, 1992, 1993, 1995, 1996, 1997, 2013).

Thus, long before the hominin brain grew much beyond that of a chimpanzee (for at least 3

[11] See Turner (2000, 2002, 2007, 2021).

to 4 million years), natural selection had begun to enlarge the *sub-cortical areas of the hominin brain*. This increased their emotionality, thereby increasing the likelihood of hominins moving out into open-country habitats—at first for only short periods of time but, eventually after millions of years, for long periods of time, or even permanently (Turner, 2000, 2007). If selection had not hit on this strategy for increasing bonds and solidarity via dramatically heightened emotional capacities, humans would never have evolved. However, once the emotional palette of hominins expanded with later hominins, this emotional base could support neo-cortical growth since, as mentioned earlier, memory or intelligence are dependent on the attachment of emotions to cognitions (Damasio, 1994). This interaction proved to be fitness enhancing, and so it continued. However, as mentioned, the interaction was episodically blocked by natural selection to rework the vocal tract for articulated speech and, to a lesser degree, to create a structure facilitating the downloading of brain thinking (in gestalts) into sequential speech (via *Broca's area*). Once unlocked, speech could produce symbolic culture, which in turn would make the final push on increasing sub-cortical and neo-cortical areas of the brain to the maximum size possible and still get babies through a woman's birth canal.

Thus, the evolution of human neurology, especially the structure and size of sub-cortical and neo-cortical areas, was ultimately started by slow enlargement of the sub-cortical areas of the brain in response to the biggest problems facing great apes moving out of the forests: *the lack of permanent group structures* and *the lack of bio-programmers for stable group organization* to coordinate resource collection and defense against predators in more open-country habitats.

The Strength of Generalized Selection

It is somewhat ironic that the *specialized niches* in which great apes evolved (terminal feeding areas of forests of Africa) created the potential for capacities to adapt to a wide variety of niches—indeed, virtually any place on earth. Great apes have some important characteristics that bias selection, the most important of which are: (a) large brains compared to most other mammals, (b) few bio-programmers for group-level formations and, hence, lack of nuclear families and other stable group formations, (c) generalized body plans revolving around dexterous hands and feet, capacities for brachiation, vision as the dominant sense modality, and capacities for knuckle walking and, when needed, bi-pedal postures. While body plans, neurology, organizational propensities, reproductive cycles, and interpersonal capacities evolved in a specialized set of niches, there was the potential for these features to make hominins capable of adapting to a wide variety of niches, *once they could begin to form group-level structures for foraging and hunting beyond the forest canopy*.

The neurology of great apes was critical because selection began to first work on emotion centers to make hominins more social and able to form groups and then on the neo-cortex to make hominins more intelligent. Beginning about 1.6 million years ago, this resulted in dramatic growth in the hominin brain that would eventually lead natural selection to "discover" the already partially-wired capacity for understanding speech that ultimately made the formation of symbolic culture possible. But, as emphasized, selecting on the brain—a tightly organized system—requires that neuro-structures and their organization remain fundamentally intact, while at the same time enlarging key brain structures to increase emotionality and intelligence. This is why the brains of great apes and humans are so similar in basic structure, except for their size and a few structures (e.g., *Broca's area*, the *lateral portion of the amygdala*). The lack of bio-programmers was, in the end, a key element in the evolution of humans' emotions, intelligence, and speech production because it did *not lock hominins into a particular form of organization at the genetic level*. Rather, by using a generalized capacity like increased emotionality (culminating in *guilt* and *shame*), increased cognitive capacities via natural selection produced a series of *generalized capacities that would allow humans to adapt to all virtually niches capable of supporting life on the planet.*

Thus, human neurology is ancient in its basic structure—many millions of years old—but selection leading to enlargement of *existing* neurological structures,[12] with some rewiring, is what led to increased emotionality and intelligence. These capacities, in turn, led to speech and culture, which eventually allowed hominins and then humans to construct diverse forms of social structures and cultures, thereby facilitating their adaptation to a wide variety of habitats. Still, even with these highly generalized capacities, humans have had at least two close to extinction episodes (Amos & Hoffman, 2010) in their history. As noted earlier, this is why humans are more closely related genetically than any of the 200+ species of primates. Eventually, humans began to create larger social structures and more complex symbolic cultures that increased fitness but, at the same time, increased the potential for self-made avenues of extinction, namely the capacity to kill millions of conspecifics and many other species via warfare and human-caused environmental changes. Indeed, as Roger Brown 1965), emphasized, "[T]he smart money is on the insects in the long run." High levels of genetically controlled societal organization of very large populations of life forms with very short life spans gives natural selection endless opportunities to make critical adjustments to their genomes. By contrast, over-population, warfare, and human-caused dramatic ecological change could potentially work against the evolutionary trends of a species with a relatively long life span and perhaps too much intelligence and/or emotions—capacities that ironically may ultimately work against the long-term survival of humans.

Why Knowing More of the Biology of Humans Is Important

Recognizing the biological bases of human behavior, interaction, and social organization increases the scope of theorizing and research in the social sciences. Conversely, knowing that a particular trait *does not have a biological* basis suggests that it evolved because of socio-cultural selection pressures and, hence, can be modified without encountering resistance from genetically controlled bio-programmers. Moreover, given that enhancing emotions has driven virtually all of neurological evolution of hominins means that much behavior and organization occurs as an outcome of enhanced emotions, which *all* have a biological basis. Indeed, the centrality of emotions indicates that the biology of emotional arousal affects virtually every aspect of human action and organization and is, therefore, a driving force of the entire socio-cultural universe.

Thus, much of what sociologists view as biologically based is, only indirectly biological, because emotions affect all aspects of human behavior, interaction, and social organization. Virtually nothing is free of emotions since thinking, acting, and organizing are all driven by emotions generated in the sub-cortex as it activates body systems,[13] outlined in other chapters 7, 8 and 12) of this volume. For example, it is often assumed that the nuclear family is universal and hence biologically based, when in fact cladistic analysis tells us that, since great apes and hence humans' ancestors do not have nuclear families, the nuclear family *is constructed by agency* and *emotional arousal* between mating pairs. One implication of this emotionally loaded structure is that instability of this structure should be expected because it is basically driven by emotions, rather than dedicated bio-programmers. The fact that great apes only organize permanent groupings at the community level suggests that this is the only social structure that has a genetically controlled effect on human social organization. Thus, human

[12] See Semendeferi et al. (2002); Sherwood (2007); Sherwood et al. (2005, 2008); Spocter et al. (2010).

[13] Neurotransmitter, endocrine/hormone, skeletal, and automatic nervous systems).

built the first societies from emotionally-charged attachments between mating partners leading to the nuclear family, held together by emotions and regulated by culture (e.g., kinship norms). Indeed, human societies could eventually evolve to their present proportions because humans *lacked bio-programmers for group formation*, with the result that groups were constructed and then linked together to form organizations and institutional domains. Only the bio-programmer for community organization was operative compared to most other mammals where group structures are almost always biological based. Moreover, human culture was only possible because of the enlargement of the neo-cortex, with the larger neo-cortex only possible with enhanced emotions driven by emotions. And so, the biology of humans is very much evident in enabling the building up of socio-cultural formations, and of course tearing them down through warfare and internal conflict among sub-populations. And so, understanding how and why humans can create mini-societies (e.g., nomadic hunting and gathering) and mega-societies (post-industrial societies of millions and in two cases billions of inhabitants) is the result of the growth of the brain (sub- and neo-cortex) enhancing emotions and cognitive capacities (and eventually language and culture), which provide a biological basis driving for the elaboration of culture and social structures. Understanding this biology, then, is critical for explaining that part of the socio-cultural universe is constructed by human thought and agency (both activated and driven by emotions.

Aside from the effects of emotions on the neurology and sociology of humans and patterns of social organization, there are more direct bio-programmers affecting human behavior. In Table 3.1, I have listed briefly behaviors that among humans have a biological basis, as revealed by cladistic analysis. Looking down the list, many of the behavior capacities that are hard-wired in great apes and hence humans pertain reading gestures of others visually, role taking, and empathizing. Other behavioral capacities pertain to rhythmic synchronization of gestures during interactions; another biologically based propensity in great apes and hence the common ancestors with hominins is the use of rituals to heighten the sense of solidarity. Still another capacity is to see self as an object. Thus, humans have genetically driven capacities for fine-tuned interaction by reading others' gestures signaling emotions and indicating lines of conduct with others. Since humans have sub-cortical emotion centers about twice the size of those in chimpanzees and neo-cortex that is three time the size of chimpanzees, these interactive propensities in humans are much more refined, especially since they can also communicate with spoken language along with reading as chimpanzees do "body language" reveal emotions. With humans' larger brains, they can have a more developed sense of self and identity than great apes. Add to these capacities are the biological ability to engage in exchange of resources and to experience positive or negative emotions with exchanges are viewed as fair or, alternatively, unfair. Much theorizing on human behavioral and interaction should, therefore, be biologically based, with the consequence that symbolic-interactionist theories, identity theories, ritual theories, and exchange theories need to recognize that these behaviors are all, to a great extent, genetically driven. And, as such, they will *always be in play*, when seeking to verify self or identity, receive resources in fair exchanges, or engage in collective, highly

Table 3.1 Hard-wired behavioral capacities and propensities great apes

1. Reading faces and eyes of conspecifics
2. Following eye movements to determine likely interests of conspecifics
3. Ability of infants to imitate Orofacial movements of caretakers
4. Empathy and role-taking
5. Reading of body language to coordinate action and interaction
6. Rhythmic synchronization of communication
7. Collective emotional effervescence
8. Learn and use language at the level of a three-year-old human child.
9. Capacity for self awareness, or self conception or identity
10. Reciprocity in the give and take (exchange) of resources
11. Calculations of fairness in exchanges of resources
12. Capacity to experience empathy with conspecifics

ritualized, and emotional interactions promoting solidarities. Indeed, given that this set of behavioral capacities is built into humans evolved neurology, sociologists can no longer see these behaviors as purely cultural; rather, they are part of the inherited biology from the common ancestors of great apes and, hence, must be studied, to some degree, from an evolutionary and biological perspective.

Thus, agency and construction of social relations in many types of social structures and their cultures often has a biological/neurological basis, which most sociology does not fully recognize, and this is why *neuro-sociology* can add a great deal to research and theories of human behavior, interaction, and organization. While social structures and their culture can become emergent properties, many of the actions and interactions creating these emergent properties are biological based and, as such, push behavior and interaction consistently in certain directions. And so, to fully understand human action, interaction, and organization, knowing what has a biological basis and what does not (e.g., group formations) can increase the sophistication of explanatory theories in sociology. Indeed, for too long, sociologists have under-emphasized the genetic basis of some of the fundamental processes involved in human interaction and the building up of social structures and their cultures. Thus, cladistics when combined with comparative neuro-anatomy can do what sociobiology and evolutionary psychology have failed to do—generate a clear poofs about what is biologically based in the human socio-cultural universe.

Conclusion

My goal in this chapter was to stress that neuro-sociology offers an alternative to sociobiology and evolutionary psychology that is more in tune with the interests of sociologists. Hominin evolution revolved around the problem of *getting better organized, or die*. Much of the neurology for interpersonal behavior was evident in the ancestors of great apes and hominins. And once natural selection found a way to unlock the neurological capacity for speech comprehension and production, the stage was set for the eventual evolution of human societies which, for many hundreds of thousands of years were small hunting and gathering bands. But the highly generalized neurology and body plans of hominins always had the potential to form larger socio-cultural systems beyond the hunting and gathering band. And so, by 10,000 years ago, societies became larger, primarily because of the evolution of human neurology. This neurology, while fundamentally the same as great ape neurology, reveals what simple enlarging of neuro-structures can do, often rather rapidly, as selection begins to push in a particular direction that proves fitness enhancing.

Understanding how this neurology evolved by directional selection revolving around growth and selective rewiring of the brain gives us new insights into how humans and human societies could evolve. And this kind of neurological analysis does not lead to a focus on narrow behaviors, *per se*, but on behavioral capacities that allow for varying patterns of social organization, eventually on the scale of billions of large-bodied and highly intelligent primates, humans. Thus, neuro-sociology has the potential to bring new insights about human social organization not possible by standard sociobiological analysis or by evolutionary psychology, or by sociological analysis that sees speech and culture as somehow obviating the impact of hard-wired behavioral propensities.

References

Amos, W., & Hoffman, J. I. (2010). Evidence that two main bottleneck events shaped modern human genetic diversity. *Proceedings of the Royal Society, 277*, 131–137.

Barger, N., Hanson, K. L., Teffer, K., Schenker-Ahmed, N. M., & Semendeferi, K. (2014). Evidence for evolutionary specialization in human limbic structures. *Frontiers in Human Neurosciences, 8*, 1–17. https://doi.org/10.3389/fnhum.2014.00277

Barger, N., Stefanacci, L., Schumann, C., Sherwood, C. C., Annese, J., Allman, J. M., Buckwalter, J. A., Hof, P. R., & Semendeferi, K. (2012). Neuronal populations in the basolateral nuclei of the amygdala are differentially in humans compared with apes: A stereological study. *Journal of Comparative Neurology, 520*, 3035–3054.

Barger, N., Stefanacci, L., & Semendeferi, K. (2007). A comparative volumetric analysis of the amygdaloid complex and division of the human and ape brain. *American Journal of Physical Anthropology, 134*, 392–403.

Barkow, H., Cosmides, L., & Tooby, J. (Eds.). (1992). *The adapted hind: Evolutionary psychology and the generation of culture*. Oxford University Press.

Barrett, L., Dunbar, R., & Lycett, J. (Eds.). (2002). *Human evolutionary psychology*. Princeton University Press.

Brown, R. (1965). *Social psychology*. The Free Press.

Buss, D. (1999). *Evolutionary psychology: The new science of the mind*. Allyn and Bacon.

Collins, R. (1975). *Conflict sociology: Toward an explanatory science*. Academic Press.

Cosmides, L., & Tooby, J. (1992). Cognitive adaptations for social exchange. In *The adapted mind: Evolutionary psychology and the generation of culture*.

Damasio, A. (1994). *Descartes' error: Emotion, reason, and the human brain*. Putnam.

Dawkins, R. (1976). *The selfish gene*. Oxford University Press.

Falk, D. (2007). The evolution of broca's area. IBRO history of neuroscience. http://www.ibro.info/Pub/Pub_Main_Display.asp?LC_Docs_ID=3145

Fisher, R. A. (1930). *The genetical theory of natural selection*. Clarendon.

Geschwind, N. (1965a). Disconnection syndromes in animals and man, Part I. *Brain, 88*, 237–294.

Geschwind, N. (1965b). Disconnection syndromes in animals and man, Part II. *Brain, 88*, 585–644.

Goffman, E. (1967). *Interaction ritual: Essays in race-to-race behavior*. Anchor Books.

Gould, S. J., & Lewontin, R. C. (1979). The spandrels of San Marco and the Panglossian paradigm: A critique of the adaptationist programme. https://doi.org/10.1098/rspb.1979.008.

Hamilton, W. D. (1964). The Genetical evolution of social behaviour. *Journal of Theoretical Biology, 7.*: Part I, 1–16; Part II, 17–52.

Hamilton, W. D. (1971). Geometry for the selfish herd. *Journal of Theoretical Biology, 31*, 295–311.

Hamilton, W. D. (1984). Innate social aptitudes of man: An approach from evolutionary genetics. In R. Fox (Ed.), *Biosocial anthropology* (pp. 135–155). Wiley.

Holloway, R. L. (1974). On the meaning of brain size. Review of J. J. Jerison's 1973 evolution of the brain and intelligence. *Science, 184*, 677–679.

Holloway, R. L. (1979). Brain size, allometry, and reorganization: Toward a synthesis. In M. E. Hahn, G. Jensen, & B. C. In Dudek (Eds.), *Development and evolution of brain size: Implications* (pp. 59–88). Academic Press.

Holloway, R. L. (1996). Evolution of the human brain. In A. Lock & C. Peters (Eds.), *Handbook of human symbolic evolution* (pp. 74–116). Clarendon.

Holloway, R. L. (1998). Relative size of the human Corpus callosum redux: Statistical smoke and mirrors? *Behavioral and Brain Sciences, 21*, 333–335.

Holloway, R. L. (2015). The evolution of the hominid. In B. W. Henke & I. Tattersall (Eds.), *Handbook of paleoanthropology* (pp. 1961–1987). Springer.

Holloway, R. L., Broadfield, D. C., & Yuan, M. S. (2004). *The human fossil record, volume three: Brain endocasts—The paleoneurological evidence*. Wiley-Liss.

Holloway, R. L., & Post, D. G. (1982). The relativity of relative brain measures and hominid mosaic evolution. In E. Armstrong & D. Falk (Eds.), *Primate brain evolution: Methods and concepts* (pp. 57–72). Plenum Press.

LeDoux, J. E. (1991). Neuroscience commentary: The emotional brain. *Journal of NIH Research, 3*, 49–51.

LeDoux, J. E. (1993). Emotional memory systems in the brain. *Behavioral Brain Research, 58*, 69–79.

LeDoux, J. E. (1996). *The emotional brain: The mysterious underpinnings of emotional life*. Simon & Schuster.

Maryanski, A. (1986). African ape social structure: A comparative analysis. PhD dissertation, University of California, Irvine.

Maryanski, A. (1987). African ape social structure: Is there strength in weak ties? *Social Networks, 9*, 191–215.

Maryanski, A. (1992). The last common ancestor: An ecological-network model on the origins of human sociality. *Advances in Human Ecology, 2*, 1–32.

Maryanski, A. (1993). The elementary forms of the first proto-human society: An ecological/social network approach. *Advances in Human Evolution Ecology, 2*, 215–241.

Maryanski, A. (1995). African ape social networks: A blueprint for reconstructing early hominid social structure. In J. Steele & S. Shennan (Eds.), *The archaeology of human ancestry: Power, sex and tradition*. Routledge.

Maryanski, A. (1996). Was speech an evolutionary afterthought? In B. M. Velichkovsky & D. M. Rumbaugh (Eds.), *Communicating meaning: The evolution and development of language* (pp. 121–131). Erlbaum.

Maryanski, A. (1997). The origin of speech and its implications for optimal size of human groups. *Critical Review, 12*, 233–249.

Maryanski, A. (2013). The secret of the hominin mind: An evolutionary story in D. In D. Franks & J. H. Turner (Eds.), *Handbook of neurosociology* (pp. 257–287). Springer.

Maryanski, A., & Turner, J. H. (1992). *The social cage: Human nature and the evolution of society*. Stanford University Press.

Pinker, S. (2002). *The blank slate: The modern denial of human nature*. Penguin.

Rumbaugh, D., & Washburn, D. G. (2003). *Intelligence of apes and other rational beings*. Cambridge University Press.

Savage-Rumbaugh, S. E., Murphy, J., Sevcik, R. A., Brakke, K. E., Williams, S. L., & Rumbaugh, D. M. (1993). *Language comprehension in the ape and child*. University of Chicago Press.

Savage-Rumbaugh, S. E., Rumbaugh, D. M., & Boysen, S. (1978). Symbolic communication between two chimpanzees (pan troglodytes). *Science, 201*(4356), 641–644.

Savage-Rumbaugh, S. E., Sevcik, R. A., & Hopkins, W. D. (1988). Symbolic cross-modal transfer in two species of chimpanzees. *Child Development, 59*, 617–625.

Semendeferi, K., Lu, A., Schenker, N., & Damasio, H. (2002). Humans and great apes share a large frontal cortex. *Nature Neuroscience, 5*, 272–276.

Sherwood, C. C. (2007). The evolution of neuron types and cortical histology in apes and humans. In T. M. Preuss & J. H. Kaas (Eds.), *Evolution of nervous systems, volume 4: Primates* (pp. 355–378). Academic Press.

Sherwood, C. C., Holloway, R. L., Semendeferi, K., & Hof, P. R. (2005). Is prefrontal white matter enlargement a human evolutionary specialization? *Nature Neuroscience, 8*, 537–538.

Sherwood, C. C., Subiaul, F., & Zawidzki, T. W. (2008). A natural history of the human: Tracing evolutionary changes in brain and cognition. *Journal of Anatomy, 212*, 426–454.

Spocter, M. A., Hopkins, W. D., Garrison, A. R., Bauernfeind, A. L., Stimpson, C. D., Hof, P. R., & Sherwood, C. C. (2010). Wernicke's area homologue in chimpanzees (*pan troglodytes*) and its relation to the appearance of modern human language. *Proceedings Biological Sciences, 277*(1691), 2165–2174. https://doi.org/10.1098/rspb.2010.0011

Stephan, H. (1983). Evolutionary trends in limbic structures. *Neuroscience and Biobehavioral Reviews, 7*, 367–374.

Stephan, H., & Andy, O. J. (1969). Quantitative comparative neuroanatomy of primates: An attempt at phylogenetic interpretation. *Annals of the New York Academy of science, 167*, 370–387.

Stephan, H., & Andy, O. J. (1977). Quantitative comparison of the amygdala in insectivores and primates. *Acta Anatomica, 98*, 130–153.

Stephan, H., Baron, G., & Frahm, H. (1988). Comparative size of brains and brain components. In H. Steklis & J. Erwin (Eds.), *Comparative primate biology* (Vol. 4, pp. 1–37). Alan R. Liss.

Stephan, H., Bauchot, R., & Andy, O. J. (1970). Data size of the brain and the various brain parts in insectivores and primates. In C. R. Noback & W. Monta (Eds.), *Advances in primatology, volume 1: The primate brain*. Appleton.

Stephan, H., Frahm, H., & Baron, G. (1981). New and revised data on volumes of brain structures in insectivores and primates. *Folia Primatological, 35*, 1–29.

Tooby, J., & Cosmides, L. (2005). Conceptual foundations of evolutionary psychology. In D. Buss (Ed.), *The handbook of evolutionary psychology* (pp. 5–67). Wiley.

Turner, J. H. (2000). *On the origins of human emotions: A sociological inquiry into the evolution of human affect*. Stanford University Press.

Turner, J. H. (2002). *Face to face: Toward a theory of interpersonal behavior*. Stanford University Press.

Turner, J. H. 2007. Human emotions: A sociological theory. Abingdon, U.K.: Routlege.

Turner, J. H. (2021). *On human nature: The biology and sociology of what made us human*. Routledge.

Turner, J. H. and Machalek, (2018). The new evolutionary sociology. New York: Routledge.

Turner, J. H. and Maryanski, A. (2024). Biosocial evolutionary analysis: Reconciling biology, psychology, and sociology. Cheltenham, UK: Edward Elgar.

Williams, G. C. (1966). *Adaptation and natural selection: A critique of some current evolutionary thought*. Princeton University Press.

Williams, G. C. (1985). A Defense of reductionism in evolutionary biology. In R. Dawkins & M. Ridley (Eds.), *Oxford surveys in evolutionary biology* (pp. 1–27). Oxford University Press.

Wilson, E. O. (1975). Sociobiology: The new synthesis. Harvard University Press.

Wilson, E. O. (1978). *On human nature*. Harvard University Press.

Wilson, D., & Wilson, E. O. (2007). Rethinking the theoretical foundations of sociobiology. *Quarterly Review of Biology, 82*(4), 327–348.

Social Neuroscience: Bridging Social and Biological Levels of Analysis

4

Greg J. Norman, Anita Restrepo, and Gary G. Berntson

Abstract

Maneuvering the complexities of social relations has required that brains of social mammals develop specialized circuitry to detect, process, and manage this information. This chapter discusses the field of social neuroscience, an interdisciplinary field focused on the behaviors and circuits underlying social life. Integrative multilevel analysis that draws from the variety of disciplines involved in the study of social neuroscience is essential for the development of comprehensive models of social behavior that take into account interactions between environmental and genetic factors in determining this behavior. This chapter provides a brief overview of the hierarchical and heterarchical organization of neural structures and how these structures influence and shape social perception and behavior. In particular, different levels of the neural hierarchy—the neuraxis—vary in the complexity, integration, and automaticity of their neural organization and information processing capacities. These structural and functional differences impact underlying approach and withdrawal tendencies, which influence interactions with the environment, perception and evaluation of social information, and subsequent behavior. A multilevel analysis in social neuroscience can uncover the complex interplay between the social environment and the biological structures—as well as their physical underpinnings—that make up and explain social life.

Introduction

The brains of social mammals are equipped with specialized neural circuits that detect, process, and manage socially relevant information. In some instances, these networks evolve by repurposing older mechanisms involved in phenomena like reward and pain processing (Parkinson & Wheatley, 2015; Landa et al., 2020). In other cases, they have developed to meet the complex demands of language and understanding others' thoughts, known as theory of mind (Spunt & Adolphs, 2017; Shultz & Dunbar, 2022). Managing intricate social interactions requires significant cognitive resources, which is thought to have driven the evolutionary increase in brain size seen in primates (Shultz & Dunbar, 2022). Supporting this idea, studies have found that in

G. J. Norman (✉) · A. Restrepo
Department of Psychology, The University of Chicago, Chicago, IL, USA
e-mail: gnorman@uchicago.edu; ar277@uchicago.edu

G. G. Berntson
Department of Psychology, The Ohio State University, Columbus, OH, USA
e-mail: berntson.2@osu.edu

W. Kalkhoff et al. (eds.), *Handbook of Neurosociology*, Handbooks of Sociology and Social Research, https://doi.org/10.1007/978-3-031-95615-7_4

primates, species with larger social groups tend to have a larger neocortex relative to their brain size (Dunbar & Shultz, 2007). This "social brain" enables us to pick up on subtle social cues—like body language, eye contact, and tone of voice—from the wealth of sensory information in social situations (Adolphs, 2009; Molapour et al., 2021). Importantly, much of this processing happens without our conscious awareness, and we often notice it only when something goes wrong. For instance, damage to certain brain areas can significantly alter how social information is processed. Conditions like frontotemporal dementia—which affects the frontal and temporal brain regions—and injuries to the ventromedial prefrontal cortex, a key area for regulating emotions and making social decisions, are linked to increased antisocial behavior (Mendez, 2022; Lockwood et al., 2024). Conversely, damage to the amygdala—a small brain region involved in processing emotions like fear and anxiety—can impair the ability to understand others' emotions and lessen responses to negative or threatening situations (Adolphs et al., 1994; Berntson et al., 2007). Thus, the brains of social mammals are highly specialized for processing information essential to social communication and behavior.

Social neuroscience is an interdisciplinary field that examines the complex social structures, behaviors, and their neurobiological mechanisms. It focuses on social networks, the individuals within them, and the neural, hormonal, and genetic mechanisms that support these networks. The field encompasses a wide range of disciplines, including neuroscience, psychology, immunology, and ethology, which can be challenging for researchers. However, this multilevel approach allows for the refinement of concepts in one area, like social psychology, by integrating insights from another, such as neuroscience. This method fosters novel ideas and enables experimental tests of theories that might otherwise be hard to distinguish, enhancing the comprehensiveness and relevance of the research. By bridging different levels of analysis, social neuroscience deepens our understanding of human social behavior, facilitating the development of more accurate models of social interactions and their biological mechanisms.

Conducting this type of intensive interdisciplinary research is a relatively new trend. Historically, biological and social analyses were often seen as separate, even conflicting, paths to understanding behavior. In the early twentieth century, psychology developed distinct schools of thought on how to explain human nature. The biopsychological perspective focused on the physiological basis of behavior, often ignoring environmental context (Berntson & Cacioppo, 2000). In contrast, social psychology emphasized how environmental factors influence behavior, frequently downplaying biological aspects (Cacioppo & Berntson, 1992). These distinctions are understandable because human biology is based on tangible anatomy and genetics, while the social environment involves abstract concepts like relationships, groups, societies, and cultures. Recognizing and bridging this divide is crucial for a comprehensive understanding of human social behavior, as it allows us to integrate biological and social perspectives. Some saw this divergence as an unbridgeable gap between social and biological approaches and even as a sign that psychology might decline as a unified discipline (Scott, 1991). There has long been debate over the roles of environmental versus genetic factors in psychology and behavior, but it's now clear that this either/or thinking is misleading. Both genetics and environment significantly influence behavior, and importantly, their effects are not always simply additive. Interactions between genes and environment can lead to emergent outcomes (Dick, 2011; Manuck & McCaffery, 2014). The same genetic makeup can result in different behaviors depending on these interactions, and different individuals may respond differently to the same environment. Recognizing the interplay between genetics and environment is vital for understanding human social behavior, as it highlights how complex interactions shape our actions and relationships.

Just as biological sciences were constrained for much of the twentieth century, social scientists largely ignored biological analyses. Historical events like two world wars, a great depression, and social injustices made social and cultural forces seem urgent enough that social scientists couldn't wait for biological explanations. Despite this, by the late twentieth century, enough interdisciplinary evidence had accumulated that both biological and social scientists began studying aspects of the "social brain" in nonhuman primates (Dunbar, 1998; Pawlowski, 1998) and humans (Cacioppo & Berntson, 1992). Individuals with amygdala damage, for example, show decreased emotional responses to negative or threatening social stimuli (Tranel et al., 2006; Berntson et al., 2007), have difficulty using gaze direction as a visual cue (Akiyama et al., 2007), and struggle to recognize and understand others' emotions (Adolphs et al., 2005; Cardinale et al., 2021). Damage to specific areas of the temporal lobe on both sides of the brain can leave individuals unable to recognize familiar faces (Damasio, 1985). These findings underscore the crucial role of specific brain mechanisms involved in facilitating complex social behavior and highlight how understanding them can provide insights into human social interactions.

It is now evident that integrating multiple levels of analysis can greatly improve our understanding of complex social behavior. The multilevel analyses in social neuroscience involve interdisciplinary approaches, connecting measures, concepts, and theories across different organizational levels. While integrating information from these levels is challenging due to the complexity of biological and social systems, it's only through this kind of research that we can fully understand the human mind and behavior. This integrative approach is essential for advancing our comprehension of how biological processes underpin social interactions, ultimately leading to a more complete picture of human social behavior.

Levels of Neural Organization

To fully grasp the complexities of social behavior, it is essential to understand the foundational structures of the nervous system and how they are organized. The nervous system operates across multiple levels, from simple reflex circuits to intricate networks responsible for higher-order functions. This multilevel organization allows for both rapid, automatic responses and complex, adaptable behaviors necessary for navigating social environments. In the section below, we will discuss aspects of these neural levels of organization and how they relate to understanding the mind and behavior.

Neural Hierarchies and Heterarchies

Multilevel perspectives of neuronal organization have been emphasized by both scientists and philosophers, with the nineteenth-century neurologist John Hughlings Jackson being particularly influential. In his essay "Evolution and Dissolution of the Nervous System", Jackson (1884) laid the groundwork for a multilevel characterization of neuronal organization. Jackson argued that the evolutionary emergence of higher levels of neuronal organization does not involve the replacement or displacement of lower levels. Instead, evolutionary development entails a re-representation and elaboration of functions at progressively higher levels of the nervous system. While higher levels are characterized by more complex networks capable of sophisticated functions, they remain highly dependent on lower levels. For instance, the critical spinal networks for stepping and related locomotor reflexes form essential lower processing circuits that support outputs from higher motor systems (e.g., motor and premotor cortices). According to Jackson, the proper interpretation of brain injuries should not be primarily defined by the functions that are lost but rather by the reversion (dissolution) of those functions to lower levels of neural organization.

The neuraxis features hierarchical organizations, with simple reflex-like circuits at the lowest levels, such as the brainstem and spinal cord, and more integrative neural networks at higher levels (Berntson et al., 1993; Norman et al., 2014; Smith et al., 2017). The relatively simple neural circuitry at the lower levels is crucial for survival, enabling rapid computations and motor outputs. This rapid response is adaptive because, in some situations, a quick but imperfect reaction can be more beneficial for survival than a slower, more elaborate response. As environmental challenges become more complex, more integrated neuronal processing at higher levels becomes adaptive, bringing higher-level analytical and response mechanisms into play. Additionally, learned anticipatory processes aid in strategically avoiding challenges before they arise.

However, the complexity of neural organization goes beyond a simple hierarchical model (Fig. 4.1). Additional neuroarchitectural complexities exist due to long descending pathways that bypass intermediate levels and directly synapse onto lower levels of the neuraxis (Wakana et al., 2004; Levinson et al., 2023). This type of organization is evidenced by direct, long descending projections from higher neuraxial systems to lower motor neurons, effectively bypassing intermediate levels. This pattern is documented not only in somatomotor systems but also in the autonomic nervous system (Berntson et al., 2017). The presence of long ascending and descending pathways, combined with lateral interconnections between levels, has been described as a neural heterarchy (Berntson et al., 1993; Norman et al., 2014; Berntson et al., 2019).

Heterarchical organization patterns incorporate elements of hierarchical systems, where higher levels communicate continuously with lower-level systems through intermediate levels, but they also feature the capacity for direct interactions across widely separated levels via direct connections. These direct neuronal projections from higher brain systems to lower-level systems enable higher computational networks to function without being constrained by intermediate-level organizations. This arrangement allows for cognitive and behavioral flexibility when needed, while still permitting intermediate-level processing when necessary.

The multiple levels of organization and their associated functional flexibility come with a drawback: heterarchical organization can create the potential for functional conflicts between distinct levels of processing. For instance, an organism might need to inhibit a pain withdrawal reflex to achieve a higher-order goal, such as rescuing someone from danger. Understanding these organizational principles is key to deciphering the neural mechanisms underlying complex behaviors, including social cognition and emotional regulation, which are crucial for navigating the intricacies of social life.

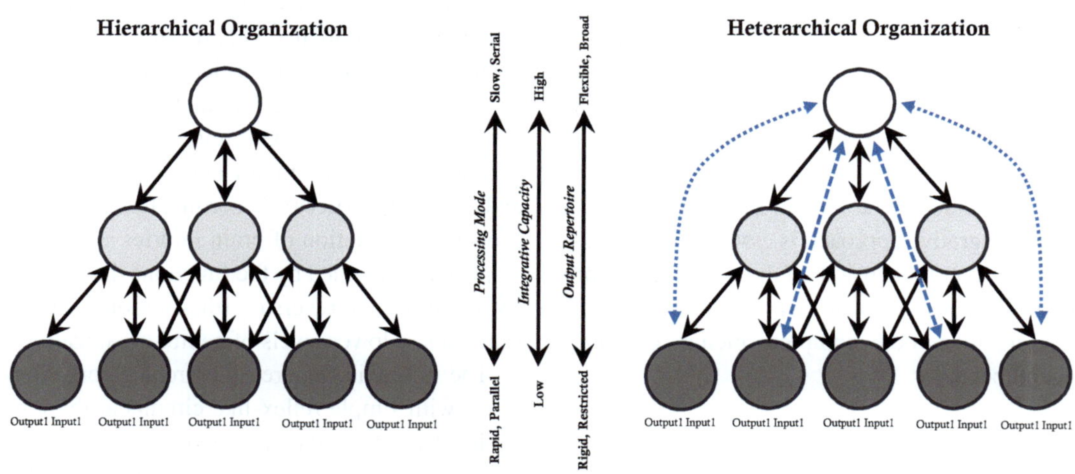

Fig. 4.1 Hierarchical and heterarchical organizations. A heterarchy differs from a hierarchy in the existence of long ascending and descending pathways that span intermediate levels and by the presence of lateral interactions. The ascending/descending pathways allow higher level systems to communicate directly with low levels, in addition to the indirect route through intermediate levels. Levels also vary with regards to their in their functional properties (e.g. processing mode, integrative capacity, and output repertoire)

Levels of Evaluative Function

As we delve deeper into the nervous system's organization, it becomes evident how different levels contribute to evaluative functions—processes that determine how organisms assess and respond to stimuli. These functions range from basic reflexes to complex decision-making processes, all of which are fundamental for social behavior.

Lessons from Lower-Level Neural Representations

At the most basic level of the central nervous system, spinal reflexes represent some of the simplest forms of neural organization. These reflexes are rapid, involuntary responses to stimuli that enable swift and efficient reactions to the environment. For example, the flexor withdrawal reflex allows an organism to quickly retract a limb in response to painful stimuli, providing immediate protection from harm. This reflex operates through nociceptive signals that activate flexor neuron pools via interneuron circuits within the spinal cord (Lundberg, 1979; Sandrini et al., 2005). While spinal reflexes can function independently of higher brain regions, they also lay the groundwork for more complex behaviors. Appetitive reflexes—those encouraging approach or engagement—are evident in spinal extensor reflexes. For instance, a thrust reflex in the extensor muscles of the arm occurs in response to contact on the palm, facilitating interaction with the environment (Sherrington, 1906). These approach reflexes complement avoidance reflexes, providing a basic framework for how organisms interact with their surroundings.

Importantly, flexor (withdrawal) and extensor (approach) reflexes are not entirely separate; they interact through mechanisms like reciprocal innervation, where activating one set of muscles inhibits the opposing set. This coordination ensures efficient and purposeful movement. Such interactions are fundamental for understanding how basic neural circuits contribute to more complex behaviors, including social interactions where approach and avoidance dynamics are crucial. One key observation is that avoidance reactions, such as the flexor withdrawal reflex, are generally stronger than approach reactions at all levels of the neuraxis, including spinal reflexes. This asymmetry is adaptive; failing to avoid harmful stimuli can have severe consequences, so natural selection favors robust avoidance mechanisms. This principle extends to higher levels of the nervous system, where negative stimuli often elicit stronger responses than positive ones—a phenomenon known as the negativity bias (Cacioppo & Berntson, 1999; Norris, 2021). Understanding this bias helps explain why negative social experiences can have a profound impact on behavior and relationships. Conversely, in low-activation contexts without strong negative stimuli, organisms tend to exhibit approach behaviors—a tendency referred to as the positivity offset (Cacioppo & Berntson, 1999; Norris et al., 2011; Gollan et al., 2016). This inclination toward exploration and engagement is fundamental for social bonding and cooperation. Recognizing these basic neural dispositions provides insight into how organisms balance exploration and avoidance, a dynamic essential for navigating social environments.

As we move to intermediate levels of neural organization, such as the brainstem, these basic reflexes become more elaborate. Studies on decerebrate animals—those with disconnected cerebral hemispheres—reveal that the brainstem alone can support a surprising range of behaviors. These animals can maintain posture, move around, consume food and water when presented, groom themselves, and display basic emotional responses like aggression or fear in response to harmful stimuli (Adams, 1979; Berntson & Micco, 1976). Similarly, in humans with conditions like anencephaly or hydranencephaly, where higher brain regions are absent or underdeveloped, infants still exhibit basic reflexes and responses to stimuli. They show flexor and extensor reflexes, stepping movements, and suckling behaviors. While these individuals lack higher cognitive functions, their preserved brainstem activities demonstrate the foundational role of lower neural representations in generating basic evaluative and behavioral responses.

Brainstem circuits also mediate more complex approach-avoidance behaviors related to taste and pleasure. For example, orofacial expressions in response to sweet or bitter tastes are well-conserved across mammals and can occur without cortical involvement. Positive tastes elicit intake responses like licking and swallowing, while negative tastes trigger rejection responses such as gaping or tongue protrusion (Berridge & Grill, 1984; Steiner et al., 2001). Importantly, these approach and avoidance systems in the brainstem are partially independent and can be co-activated. An organism might simultaneously exhibit both intake and rejection responses to a stimulus that is both sweet and bitter, reflecting a complex underlying state that cannot be fully captured by a simple positive-to-negative scale. This complexity at lower neural levels provides a foundation for understanding the nuanced emotional responses observed in social interactions where individuals routinely experience both positive and negative emotions simultaneously (Larsen & McGraw, 2014).

Understanding these lower-level neural representations offers valuable insights into more complex processes relevant to social neuroscience and behavior. The fundamental principles of approach and avoidance, reciprocal inhibition, and the asymmetry between positive and negative responses are echoed in higher brain functions that mediate complex social interactions. They inform how organisms evaluate social cues, make decisions, and regulate emotions. By studying these basic neural mechanisms, we gain a deeper appreciation of the building blocks that support the sophisticated social behaviors seen in humans.

Higher-Level Representations

Building upon the foundational principles observed in lower-level neural systems, we find that as we move to higher regions of the brain, the complexity and sophistication of evaluative processes increase dramatically. While the spinal cord and brainstem can generate basic approach and avoidance behaviors in response to stimuli, they lack the flexibility and adaptability seen in organisms with intact higher brain regions. For example, decerebrate animals may consume palatable foods when presented but do not actively seek them out, demonstrating dependence on immediate stimuli rather than engaging in goal-directed behaviors (Berntson & Micco, 1976; Berntson et al., 1993). The emergence of higher brain structures—such as the limbic system and neocortex—marks a significant evolution in evaluative processing (MacLean, 1985). These regions enable more complex functions like planning, strategizing, and anticipation, freeing the organism from the constraints of the immediate environment. This advancement allows for sophisticated social behaviors, as individuals can interpret social cues, predict others' actions, and adjust their behavior accordingly.

As the cognitive demands of evaluative processes expand at these higher levels, so does the complexity of the underlying neural networks. Unlike the straightforward approach and avoidance circuits at lower levels, higher-level processes involve intricate and distributed systems that cannot be easily divided into positive and negative components. They require the integration of memory, attention, and associative learning to navigate complex social environments. One important distinction at this level is between the "liking" and "wanting" aspects of motivation, as described by Berridge (2023). "Liking" refers to the hedonic pleasure derived from a stimulus, evident in immediate, reflexive responses. In contrast, "wanting" involves an attentional focus and goal-seeking behavior toward a desired outcome, relying on the enhanced computational capacity of higher brain regions. This distinction is crucial for understanding human social behavior, as it underpins our ability to pursue long-term goals, form relationships, and engage in cooperative activities.

The neuroarchitecture supporting these higher evaluative processes includes complex networks such as the nucleus accumbens (nACC) and the amygdala. Historically, the nACC has been associated with reward and positive emotional states. Studies have shown that stimulation of the nACC can elicit feelings of pleasure and desire, and dopamine release in this area is linked to rewarding stimuli (Robinson & Berridge, 2003; Berridge & Kringelbach, 2015). However, recent research reveals that the nACC is not a simple reward center; different regions within it mediate the "liking" and "wanting" aspects of motivation, and it can also be involved in processing negative stimuli (Berridge, 2019; Levita et al., 2020). This complexity highlights how higher-level neural representations contribute to the nuanced emotional experiences that influence social interactions. Similarly, the amygdala plays a critical role in processing fear and negative emotions. It enables rapid detection and response to threats through subcortical pathways that bypass extensive cortical processing (LeDoux, 1996; Bertini & Làdavas, 2021). This allows for quick reactions to danger, essential for survival. Understanding the function of the amygdala and related structures is important for social neuroscience because it influences how we perceive and respond to social threats, anxiety, and fear-related behaviors. Higher-level evaluative processes also exhibit hemispheric lateralization. Research indicates that the right hemisphere is slightly more involved in processing negative emotions and avoidance behaviors, while the left hemisphere is more engaged in positive emotions and approach behaviors (Davidson, 2004; Harmon-Jones & Gable, 2018). This specialization may enhance our ability to navigate complex social situations by efficiently managing both positive and negative social cues. For instance, positive stimuli induce greater activation in the left hemisphere, and damage to this area is often associated with depression and negative affect (Davidson, 1998).

By examining these higher-level neural representations, we gain valuable insights into the neural basis of sophisticated social behaviors. The complex interplay between various brain regions underlies our ability to experience nuanced emotions, form social bonds, and engage in cooperative and competitive interactions. Recognizing the intricate neuroarchitecture of these evaluative processes is essential for developing a comprehensive understanding of human social behavior and for addressing social and psychological disorders. Understanding how higher brain regions integrate and elaborate upon the basic evaluative functions observed at lower levels allows us to appreciate the full spectrum of human social cognition and emotion. It emphasizes that our social behaviors are rooted not just in advanced cognitive abilities but also in fundamental neural processes shared across species. This perspective is crucial for social neuroscience, as it bridges the gap between simple neural mechanisms and the complex social phenomena that characterize human life.

Multilevel Mechanisms of Social Behavior

Exploring the multilevel mechanisms of social behavior involves examining how biological systems interact with environmental and social factors to shape complex behaviors. Two examples that illustrate this interplay are the roles of oxytocin and vasopressin in social bonding and the phenomenon of loneliness and its impact on perception and health outcomes.

The Role of Oxytocin and Vasopressin

Social neuroscience reveals that social behavior is shaped by a complex interplay of neural, physiological, psychological, and environmental factors. Among these, the neuropeptides oxytocin and vasopressin have emerged as pivotal regulators of social bonds and interactions. They serve as crucial links between molecular signaling and behavioral outcomes, influencing how we form and maintain social relationships. Importantly, oxytocin and vasopressin not only mediate specific behaviors but also adjust an organism's responses to different social contexts, underscoring the need for an integrated approach in understanding social neuroscience.

Oxytocin and vasopressin are structurally similar peptides that play critical roles in mediating a variety of social behaviors, including social recognition, parental care, and reproductive activities (Carter et al., 2008; Hammock, 2015). These neuropeptides influence both basic and complex behaviors across species, highlighting their evolutionary importance in shaping adaptive social responses. For instance, social recognition—essential for forming long-term social bonds—requires complex neural processing involving these neuropeptides (Gabor et al., 2012). Notably, both oxytocin and vasopressin contribute to species-specific social behaviors in rodents. Prairie voles, for example, serve as a model for understanding pair-bonding processes. In these animals, oxytocin and vasopressin act on the mesolimbic dopamine reward system to reinforce social bonding between mates—a process absent in montane voles, which do not form long-term pair bonds (Lim & Young, 2004). This difference illustrates how variations in the distribution of oxytocin and vasopressin receptors in the brain can lead to distinct social behaviors (Johnson et al., 2016).

The role of oxytocin and vasopressin in social bonding and attachment extends to humans, demonstrating the generalizability of these neuropeptides across species. In humans, oxytocin has been implicated in regulating maternal behavior, emotional empathy, and trust-building, while vasopressin is linked to aggressive behaviors and territoriality in social contexts (Carter, 2014). These findings illustrate the multilevel nature of social behavior, where molecular processes influence complex behaviors through the modulation of neural circuits involved in emotion regulation, stress responses, and social cognition. Additionally, the oxytocinergic system shows remarkable sensitivity to early social experiences, emphasizing the integration of biological and environmental factors across development. For example, early childhood neglect or parental separation can impair the typical increase in oxytocin levels triggered by social interactions, potentially influencing an individual's capacity to form secure attachments later in life (Fries et al., 2005; Müller et al., 2019). This sensitivity underscores the feedback loops between social environments and biological processes, where social experiences shape neurochemical responses that, in turn, influence social behaviors.

Conversely, positive social interactions, such as receiving social support, lead to increased oxytocin levels (Colonnello et al., 2017). The release of oxytocin during supportive social interactions helps buffer against stress by dampening activity in stress-related brain regions, including the hypothalamus and amygdala, and promoting activation in reward-related regions like the ventral striatum (Zhao et al., 2019). This suggests that oxytocin not only modulates social bonding but also plays a protective role in stress regulation and emotional well-being (IsHak et al., 2011; Hostinar et al., 2014; Colonnello et al., 2017). This process highlights how the biological substrates underlying social behavior are dynamically responsive to social environments, reinforcing the bidirectional nature of these relationships.

The role of oxytocin and vasopressin in social behavior is also influenced by genetic factors. Variants of the oxytocin receptor gene (OXTR) have been associated with individual differences in social behaviors, including sensitivity to social rejection and experiences of loneliness (Lucht et al., 2009; McQuaid et al., 2015). Moreover, epigenetics work has shown that experience dependent methylation of oxytocin receptor genes is associated with social anxiety (Ziegler et al., 2015), problems with facial and emotional recognition, and social cognitive deficits in Autistic Spectrum Disorder (Maud et al., 2018). This connection between genomics, neuropeptide signaling, and social behavior exemplifies the multilevel nature of social neuroscience, where biological factors interact with environmental influences to shape individual differences in social cognition and behavior.

In summary, oxytocin and vasopressin play central roles in mediating social behavior through their influence on both lower-level neurochemical processes and higher-order social cognition. These neuropeptides demonstrate how biological mechanisms are finely tuned to social environments, with social experiences influencing neurochemical responses that, in turn, shape complex social behaviors. The study of oxytocin and vasopressin within the framework of social neuroscience underscores the importance of a multilevel approach, integrating biological, psychological, and social factors to provide a comprehensive understanding of social behavior. By examining these processes across levels of analysis, researchers can gain deeper insights into how the social brain functions and how evolutionary pressures have shaped the neurobiological substrates underlying human sociality.

The Role of Loneliness

Loneliness serves as a profound example of how social experiences and biological processes are intricately connected, operating across multiple levels of organization. Evolutionary theories suggest that loneliness functions as a biologically ingrained signal, alerting individuals to the dangers of social isolation—which historically threatened survival and reproduction (Cacioppo et al., 2011). In this context, loneliness acts as a powerful motivator for social reconnection, driving individuals to restore social bonds. However, the experience of loneliness is not merely a social or psychological phenomenon; it activates fundamental biological systems, underscoring the critical importance of social relationships for overall well-being. Indeed, at the neural level, loneliness engages brain regions associated with both pain and reward, emphasizing its evolutionary significance. The anterior cingulate cortex (ACC), a key region implicated in processing physical pain, also plays a role in the experience of social rejection and loneliness (Eisenberger, 2012). This overlap highlights the shared neural circuitry that processes aversive experiences, whether physical or social. Social rejection activates neural circuits that increase sensitivity to perceived social threats, leading to heightened vigilance in social interactions (Cacioppo et al., 2011). This increased sensitivity can create a feedback loop, where lonely individuals become more attuned to potential social threats, further amplifying their sense of isolation. This self-reinforcing cycle of loneliness is one of its most debilitating aspects. Prolonged loneliness triggers defensive behaviors, such as increased wariness of others' intentions or avoidance of social interactions to prevent further rejection (Cacioppo & Hawkley, 2009). While these defensive responses may have been adaptive in short-term survival contexts, they can exacerbate perceived social isolation in modern environments and have catastrophic impacts on health and well-being (Hawkley, 2022). This pattern mirrors the negativity bias discussed earlier, where avoidance reactions can dominate in contexts perceived as threatening.

Biologically, the effects of loneliness extend beyond neural circuits involved in pain and reward processing to broader physiological systems. The perception of social isolation activates the hypothalamic-pituitary-adrenal (HPA) axis and autonomic nervous system, initiating stress responses that elevates cortisol levels and cardiovascular activation (Cacioppo et al., 2011). Chronic exposure to unregulated loneliness is linked to adverse health outcomes, including increased inflammation, impaired immune function, and higher risk of cardiovascular and cerebrovascular diseases (Cacioppo et al., 2011; Hawkley et al., 2012; Hawkley, 2022). These physiological consequences underscore the deep interconnection between social experiences and physical health, illustrating how social disconnection can impact well-being on multiple levels. Moreover, loneliness can alter gene expression, particularly in genes involved in immune response regulation. Research shows that socially isolated individuals exhibit changes in their immune cell gene expression profiles, with an increase in pro-inflammatory gene expression and a decrease in genes responsible for antiviral responses (Slavich & Cole, 2013). This shift places lonely individuals at a heightened risk for inflammation-related diseases and infections, further linking social disconnection to long-term health outcomes. These findings reveal the biological underpinnings of loneliness as a multilevel phenomenon, where social experiences directly shape physiological functioning ranging from the expression of genes to the regulation of leukocytes.

At the neural level, oxytocin plays a role in modulating responses to social isolation.

Oxytocin is critical for fostering social attachment and promoting prosocial behaviors such as trust and cooperation (Carter et al., 2008). However, in the context of loneliness, the oxytocinergic system can become dysregulated, reducing an individual's ability to seek out and maintain social connections. Studies suggest that loneliness is associated with altered oxytocin receptor functioning, impairing the neurochemical pathways that typically support social bonding (McQuaid et al., 2015). This dysregulation further reinforces the cycle of social withdrawal, as individuals experiencing loneliness may struggle to engage in behaviors that could restore social bonds. The complexity of loneliness is further evidenced by its impact on social cognition. Socially isolated individuals tend to display a cognitive bias toward interpreting ambiguous social cues as threatening, exacerbating their feelings of social rejection and reinforcing avoidance of social interactions (Cacioppo & Hawkley, 2009). This cognitive bias may stem from heightened activity in brain regions like the amygdala, which, as discussed earlier, is responsible for processing emotional and social threats. Thus, loneliness alters both the perception of social reality and the underlying neural mechanisms that shape these perceptions.

From a multilevel perspective, loneliness exemplifies how social and psychological states are deeply influenced by underlying biological processes, which in turn shape broader social behaviors. The social and biological components of loneliness interact in a reciprocal feedback loop: social disconnection triggers neurobiological changes, which subsequently affect social behavior and further reinforce feelings of loneliness. This dynamic underscores the need for an integrative framework in social neuroscience that accounts for the mutual influence of biological and social factors.

In conclusion, loneliness highlights the multilevel nature of social neuroscience, where social experiences are intricately linked to biological processes at neural, physiological, and molecular levels. The interaction between social isolation, stress responses, immune function, and neurochemical pathways demonstrates the complexity of loneliness as both a psychological and biological phenomenon. By examining loneliness through a multilevel lens, researchers can gain a deeper understanding of how social environments shape the brain and body, ultimately influencing long-term health and well-being.

Conclusion

Advances in social neuroscience underscore the necessity of adopting a multilevel framework to truly understand the complexities of human behavior. As this chapter illustrates, psychological constructs cannot be fully explained by examining any single level of analysis. These emergent processes and associated behaviors arise from intricate interactions between biological systems and social environments. For example, while neural circuits in the prefrontal cortex and limbic system influence social behavior, these interactions alone are insufficient to explain the nuances of behaviors like trust-building and conflict resolution. Instead, integrating neural, psychological, and social levels of analysis reveals the dynamic interplay that gives rise to these complex social behaviors.

Furthermore, biological systems actively shape social behavior while being shaped by social experiences. The relationship between social and biological processes is profoundly bidirectional, as illustrated by phenomena like oxytocin regulation and the effects of loneliness on gene expression. For instance, while oxytocin and vasopressin regulate social bonding and various other social behaviors, loneliness can reshape these very biological processes, influencing everything from emotional regulation to physiological health. This bidirectional interplay pushes social neuroscience beyond reductionism, moving toward a more holistic and integrative model where levels of analysis mutually inform and refine each other. This framework is essential for comprehending the emergent properties of the social brain as a dynamic, interacting system in which social experiences and biological mechanisms are continuously engaged in a dynamic feedback loop.

This cross-disciplinary approach reveals that neither biological substrates nor social contexts alone can explain the complexity of human social behavior; both must be examined together to fully appreciate the emergent properties of the social brain. In this way, multilevel, cross-disciplinary research has begun to dismantle the artificial academic barriers that have historically separated biological and social sciences. Integrating fields as diverse as molecular biology, psychology, and sociology may seem daunting, but it is essential for achieving a comprehensive understanding of the human mind and behavior. Social neuroscience provides a theoretical and methodological framework for integrating these fields, promoting meaningful reductionism and reintegration. The goal is not to reduce complex phenomena to isolated components but to understand how constructs at multiple levels of analysis mutually inform and shape one another. Multilevel analysis allows researchers to uncover how the social environment influences biological systems and vice versa, highlighting the emergent, dynamic nature of social behaviors.

Ultimately, social neuroscience offers a unique lens through which to understand the full complexity of human behavior. By bridging biological, psychological, and social domains, this field provides a pathway toward uncovering the intricacies of human cognition, behavior, and neurobiology. Through this integrated multilevel approach, researchers can better understand the social brain, revealing how evolutionary pressures, neurobiological processes, and social environments converge to drive human behavior. Such insights offer a robust framework for future research, paving the way for deeper exploration into the emergent properties that define our social lives.

References

Adams, D. B. (1979). Brain mechanisms for offense, defense, and submission. *Behavioral and Brain Sciences, 2*(2), 201–213.

Adolphs, R. (2009). The social brain: Neural basis of social knowledge. *Annual Review of Psychology, 60*, 693–716.

Adolphs, R., Gosselin, F., Buchanan, T. W., Tranel, D., Schyns, P., & Damasio, A. R. (2005). A mechanism for impaired fear recognition after amygdala damage. *Nature, 433*(7021), 68–72.

Adolphs, R., Tranel, D., Damasio, H., & Damasio, A. (1994). Impaired recognition of emotion in facial expressions following bilateral damage to the human amygdala. *Nature, 372*(6507), 669–672.

Akiyama, T., Kato, M., Muramatsu, T., Umeda, S., Saito, F., & Kashima, H. (2007). Unilateral amygdala lesions hamper attentional orienting triggered by gaze direction. *Cerebral Cortex, 17*(11), 2593–2600.

Berntson, G. G., Bechara, A., Damasio, H., Tranel, D., & Cacioppo, J. T. (2007). Amygdala contribution to selective dimensions of emotion. *Social Cognitive and Affective Neuroscience, 2*(2), 123–129.

Berntson, G. G., Boysen, S. T., & Cacioppo, J. T. (1993). Neurobehavioral organization and the cardinal principle of evaluative bivalence. *Annals of the New York Academy of Sciences, 702*, 75–102.

Berntson, G. G., & Cacioppo, J. T. (2000). Psychobiology and social psychology: Past, present, and future. *Personality and Social Psychology Review, 4*(1), 3–15.

Berntson, G. G., Cacioppo, J. T., & Bosch, J. A. (2017). From homeostasis to allodynamic regulation. In J. T. Cacioppo, L. G. Tassinary, & G. G. Berntson (Eds.), *Handbook of psychophysiology* (4th ed., pp. 401–426). Cambridge University Press.

Berntson, G. G., Gianaros, P. J., & Tsakiris, M. (2019). Interoception and the autonomic nervous system: Bottom-up meets top-down. In M. Tsakiris & H. De Preester (Eds.), *The interoceptive mind: From homeostasis to awareness* (pp. 3–23). Oxford University Press.

Berntson, G. G., & Micco, D. J. (1976). Organization of brainstem behavioral systems. *Brain Research Bulletin, 1*(5), 471–483.

Berridge, K. C. (2019). Affective valence in the brain: Modules or modes? *Nature Reviews Neuroscience, 20*(4), 225–234.

Berridge, K. C. (2023). Separating desire from prediction of outcome value. *Trends in Cognitive Sciences, 27*(3), 228–240.

Berridge, K. C., & Grill, H. J. (1984). Isohedonic tastes support a two-dimensional hypothesis of palatability. *Appetite, 5*(3), 221–231.

Berridge, K. C., & Kringelbach, M. L. (2015). Pleasure systems in the brain. *Neuron, 86*(3), 646–664.

Bertini, C., & Làdavas, E. (2021). Fear-related signals are prioritised in visual, somatosensory and spatial systems. *Neuropsychologia, 150*, 107698.

Cacioppo, J. T., & Berntson, G. G. (1992). Social psychological contributions to the decade of the brain: Doctrine of multilevel analysis. *American Psychologist, 47*(8), 1019–1028.

Cacioppo, J. T., & Berntson, G. G. (1999). The affect system: Architecture and operating characteristics. *Current Directions in Psychological Science, 8*(5), 133–137.

Cacioppo, J. T., & Hawkley, L. C. (2009). Perceived social isolation and cognition. *Trends in Cognitive Sciences, 13*(10), 447–454.

Cacioppo, J. T., Hawkley, L. C., Norman, G. J., & Berntson, G. G. (2011). Social isolation. *Annals of the New York Academy of Sciences, 1231*(1), 17–22.

Cardinale, E. M., Reber, J., O'Connell, K., Turkeltaub, P. E., Tranel, D., Buchanan, T. W., & Marsh, A. A. (2021). Bilateral amygdala damage linked to impaired ability to predict others' fear but preserved moral judgements about causing others fear. *Proceedings of the Royal Society B: Biological Sciences, 288*(1943), 20202651.

Carter, C. S. (2014). Oxytocin pathways and the evolution of human behavior. *Annual Review of Psychology, 65*, 17–39.

Carter, C. S., Grippo, A. J., Pournajafi-Nazarloo, H., Ruscio, M. G., & Porges, S. W. (2008). Oxytocin, vasopressin and sociality. *Progress in Brain Research, 170*, 331–336.

Colonnello, V., Petrocchi, N., Farinelli, M., & Ottaviani, C. (2017). Positive social interactions in a lifespan perspective with a focus on opioidergic and oxytocinergic systems: Implications for neuroprotection. *Current Neuropharmacology, 15*(4), 543–561.

Damasio, A. R. (1985). Prosopagnosia. *Trends in Neurosciences, 8*(4), 132–135.

Davidson, R. J. (1998). Anterior electrophysiological asymmetries, emotion, and depression: Conceptual and methodological conundrums. *Psychophysiology, 35*(5), 607–614.

Davidson, R. J. (2004). Well-being and affective style: Neural substrates and biobehavioral correlates. *Philosophical Transactions of the Royal Society B: Biological Sciences, 359*(1449), 1395–1411.

Dick, D. M. (2011). Gene–environment interaction in psychological traits and disorders. *Annual Review of Clinical Psychology, 7*, 383–409.

Dunbar, R. I. M. (1998). The social brain hypothesis. *Evolutionary Anthropology, 6*(5), 178–190.

Dunbar, R. I. M., & Shultz, S. (2007). Evolution in the social brain. *Science, 317*(5843), 1344–1347.

Eisenberger, N. I. (2012). The neural bases of social pain: Evidence for shared representations with physical pain. *Psychosomatic Medicine, 74*(2), 126–135.

Fries, A. B., Ziegler, T. E., Kurian, J. R., Jacoris, S., & Pollak, S. D. (2005). Early experience in humans is associated with changes in neuropeptides critical for regulating social behavior. *Proceedings of the National Academy of Sciences, 102*(47), 17237–17240.

Gabor, C. S., Phan, A., Clipperton-Allen, A. E., Kavaliers, M., & Choleris, E. (2012). Interplay of oxytocin, vasopressin, and sex hormones in the regulation of social recognition: Hormones and cognition. *Behavioral Neuroscience, 126*(1), 97–109.

Gollan, J. K., Hoxha, D., Hunnicutt-Ferguson, K., Norris, C. J., Rosebrock, L., Sankin, L., & Cacioppo, J. T. (2016). Twice the negativity bias and half the positivity offset: Evaluative responses to emotional information in depression. *Journal of Behavior Therapy and Experimental Psychiatry, 52*, 166–170.

Hammock, E. A. D. (2015). Developmental perspectives on oxytocin and vasopressin. *Neuropsychopharmacology, 40*(1), 24–42.

Harmon-Jones, E., & Gable, P. A. (2018). On the role of asymmetric frontal cortical activity in approach and withdrawal motivation: An updated review of the evidence. *Psychophysiology, 55*(1), e12879.

Hawkley, L. C. (2022). Loneliness and health. *Nature Reviews Disease Primers, 8*(1), 22.

Hawkley, L. C., Cole, S. W., Capitanio, J. P., Norman, G. J., & Cacioppo, J. T. (2012). Effects of social isolation on glucocorticoid regulation in social mammals. *Hormones and Behavior, 62*(3), 314–323.

Hostinar, C. E., Sullivan, R. M., & Gunnar, M. R. (2014). Psychobiological mechanisms underlying the social buffering of the hypothalamic–pituitary–adrenocortical axis: A review of animal models and human studies across development. *Psychological Bulletin, 140*(1), 256–282.

IsHak, W. W., Kahloon, M., & Fakhry, H. (2011). Oxytocin role in enhancing Well-being: A literature review. *Journal of Affective Disorders, 130*(1–2), 1–9.

Jackson, J. H. (1884). The Croonian lectures on evolution and dissolution of the nervous system. *British Medical Journal, 1*(1214), 660–663.

Johnson, Z. V., Walum, H., Jamal, Y. A., Xiao, Y., Keebaugh, A. C., Inoue, K., & Young, L. J. (2016). Central oxytocin receptors mediate mating-induced partner preferences and enhance correlated activation across forebrain nuclei in male prairie voles. *Hormones and Behavior, 79*, 8–17.

Landa, A., Fallon, B. A., Wang, Z., Duan, Y., Liu, F., Wager, T. D., & Peterson, B. S. (2020). When it hurts even more: The neural dynamics of pain and interpersonal emotions. *Journal of Psychosomatic Research, 128*, 109881.

Larsen, J. T., & McGraw, A. P. (2014). The case for mixed emotions. *Social and Personality Psychology Compass, 8*(6), 263–274.

LeDoux, J. E. (1996). *The emotional brain: The mysterious underpinnings of emotional life*. Simon & Schuster.

Levinson, S., Miller, M., Iftekhar, A., Justo, M., Arriola, D., Wei, W., Hazany, S., Avecillas-Chasin, J. M., Kuhn, T. P., Horn, A., & Bari, A. A. (2023). A structural connectivity atlas of limbic brainstem nuclei. *Frontiers in Neuroimaging, 1*, 1009399.

Levita, L., Bois, C., Healey, A., & Smyllie, E. (2020). The role of the nucleus accumbens in regulating aversive and rewarding behaviors in humans. *The Journal of Neuroscience, 40*(4), 842–848.

Lim, M. M., & Young, L. J. (2004). Vasopressin-dependent neural circuits underlying pair bond formation in the monogamous prairie vole. *Neuroscience, 125*(1), 35–45.

Lockwood, P. L., Cutler, J., Drew, D., Abdurahman, A., Jeyaretna, D. S., Apps, M. A., ... & Manohar, S. G. (2024). Human ventromedial prefrontal cortex is necessary for prosocial motivation. *Nature human behaviour, 88*(7), 1403–1416.

Lucht, M. J., Barnow, S., Sonnenfeld, C., Rosenberger, A., Grabe, H. J., Schroeder, W., & Rosskopf, D. (2009). Associations between the oxytocin receptor gene (OXTR) and affect, loneliness and intelligence in normal subjects. *Progress in Neuro-Psychopharmacology and Biological Psychiatry, 33*(5), 860–866.

Lundberg, A. (1979). Multisensory control of spinal reflex pathways. In R. Granit & O. Pompeiano (Eds.), *Reflex control of posture and movement* (pp. 11–28). Elsevier.

MacLean, P. D. (1985). Brain evolution relating to family, play, and the separation call. *Archives of General Psychiatry, 42*(4), 405–417.

Manuck, S. B., & McCaffery, J. M. (2014). Gene–environment interaction. *Annual Review of Psychology, 65*, 41–70.

Maud, C., Ryan, J., McIntosh, J. E., & Olsson, C. A. (2018). The role of oxytocin receptor gene (OXTR) DNA methylation in human social and emotional functioning: A systematic narrative review. *BMC Psychiatry, 18*, 154.

McQuaid, R. J., McInnis, O. A., Matheson, K., & Anisman, H. (2015). Distress of ostracism: Oxytocin receptor gene polymorphism confers sensitivity to social exclusion. *Social Cognitive and Affective Neuroscience, 10*(8), 1153–1159.

Mendez, M. F. (2022). Behavioral variant frontotemporal dementia and social and criminal transgressions. *The Journal of Neuropsychiatry and Clinical Neurosciences, 34*(4), 328–340.

Molapour, T., Hagan, C. C., Silston, B., Wu, H., Ramstead, M., Friston, K., & Mobbs, D. (2021). Seven computations of the social brain. *Social Cognitive and Affective Neuroscience, 16*(8), 745–760.

Müller, L. E., Bertsch, K., Bülau, K., Herpertz, S. C., & Buchheim, A. (2019). Emotional neglect in childhood shapes social dysfunctioning in adults by influencing the oxytocin and the attachment system: Results from a population-based study. *International Journal of Psychophysiology, 136*, 73–80.

Norman, G. J., Berntson, G. G., & Cacioppo, J. T. (2014). Emotion, somatovisceral afference, and autonomic regulation. *Emotion Review, 6*(2), 113–123.

Norris, C. J. (2021). The negativity bias, revisited: Evidence from neuroscience measures and an individual differences approach. *Social Neuroscience, 16*(1), 68–82.

Norris, C. J., Larsen, J. T., Crawford, L. E., & Cacioppo, J. T. (2011). Better (or worse) for some than others: Individual differences in the positivity offset and negativity bias. *Journal of Research in Personality, 45*(1), 100–111.

Parkinson, C., & Wheatley, T. (2015). The repurposed social brain. *Trends in Cognitive Sciences, 19*(3), 133–141.

Pawłowski, B. (1998). Neocortex size, social skills and mating success in primates. *Behaviour, 135*(3), 357–368.

Robinson, T. E., & Berridge, K. C. (2003). Addiction. *Annual Review of Psychology, 54*, 25–53.

Sandrini, G., Serrao, M., Rossi, P., Romaniello, A., Cruccu, G., & Willer, J. C. (2005). The lower limb flexion reflex in humans. *Progress in Neurobiology, 77*(6), 353–395.

Scott, T. R. (1991). A personal view of the future of psychology departments. *American Psychologist, 46*(9), 975–976.

Sherrington, C. S. (1906). *The integrative action of the nervous system*. Yale University Press.

Shultz, S., & Dunbar, R. I. M. (2022). Socioecological complexity in primate groups and its cognitive correlates. *Philosophical Transactions of the Royal Society B: Biological Sciences, 377*(1860), 20210296.

Slavich, G. M., & Cole, S. W. (2013). The emerging field of human social genomics. *Clinical Psychological Science, 1*(3), 331–348.

Smith, R., Thayer, J. F., Khalsa, S. S., & Lane, R. D. (2017). The hierarchical basis of neurovisceral integration. *Neuroscience & Biobehavioral Reviews, 75*, 274–296.

Spunt, R. P., & Adolphs, R. (2017). A new look at domain specificity: Insights from social neuroscience. *Nature Reviews Neuroscience, 18*(9), 559–567.

Steiner, J. E., Glaser, D., Hawilo, M., & Berridge, K. C. (2001). Comparative expression of hedonic impact: Affective reactions to taste by human infants and other primates. *Neuroscience & Biobehavioral Reviews, 25*(1), 53–74.

Tranel, D., Gullickson, G., Koch, M., & Adolphs, R. (2006). Altered experience of emotion following bilateral amygdala damage. *Cognitive Neuropsychiatry, 11*(3), 219–232.

Wakana, S., Jiang, H., Nagae-Poetscher, L. M., Van Zijl, P. C., & Mori, S. (2004). Fiber tract–based atlas of human white matter anatomy. *Radiology, 230*(1), 77–87.

Zhao, Z., Ma, X., Geng, Y., Zhao, W., Zhou, F., Wang, J., & Becker, B. (2019). Oxytocin differentially modulates specific dorsal and ventral striatal functional connections with frontal and cerebellar regions. *NeuroImage, 184*, 781–789.

Ziegler, C., Dannlowski, U., Bräuer, D., Stevens, S., Laeger, I., Wittmann, H., et al. (2015). Oxytocin receptor gene methylation: Converging multilevel evidence for a role in social anxiety. *Neuropsychopharmacology, 40*(6), 1528–1538.

Theorists as *Arbitrageurs*: How Cognitive Neuroscience Entered Sociology

5

Gabe Ignatow

Abstract

Since the 1990s several influential publications have explored the implications of cognitive neuroscience concepts and methods for sociology. Rather than argue for the relevance of specific concepts such as schemas, implicit attitudes, dual process theory or embodied knowledge for sociology, this chapter reflects on the success of neurosociology (and with it cognitive sociology and research on culture and cognition) since the 1990s and contrasts that success with the relative stagnation of sociological theory over the same period. My argument is twofold. First, features of the Big Science research landscape explain both the decline of traditional subfields such as sociological theory and the emergence of hybrid subfields including neurosociology. Second, sociological theorists since the 1990s have performed theoretical arbitrage, importing concepts from cognitive neuroscience into sociology. Analysis of Web of Science citation data supports these arguments. Identifying which areas of sociology were most influenced by cognitive neuroscience, which areas were influenced first, and who were the *arbitrageurs* who brought cognitive neuroscience into sociology, I show that theorists operating within social psychology, cultural sociology, and sociological theory subfields imported cognitive neuroscience into mainstream sociology. In doing so they accentuated preexisting philosophical divisions within sociology between constructionist and naturalist/realist camps and redefined the role of sociological theorist from abstract theoretician to theoretical *arbitrageur*.

G. Ignatow (✉)
Department of Sociology, University of North Texas, Denton, TX, USA
e-mail: ignatow@unt.edu

Introduction: The Emergence of Neurosociology and the Crisis of Sociological Theory

Since the 1990s a series of influential publications have explored the implications of the emergence of cognitive neuroscience for general sociological theory (DiMaggio, 1997; Lizardo, 2004; Turner, 2002, 2018; Ignatow, 2007; see Dubash & Brett, 2023: 5) and for sociological methodology (Hunzaker & Valentino, 2019; Shepherd, 2011, 2019; Vaisey & Miles, 2014; Vaisey & Valentino, 2018). Over this same period sociological theory has been widely understood to be in a state of stagnation or perhaps even crisis. In his 1995 monograph *Sociological Theory: What Went Wrong?* Mouzelis attempted to rescue sociological theory from what he saw as its state of malaise. In 2006 Turner identified a generation gap in sociological theorizing a consequence of

W. Kalkhoff et al. (eds.), *Handbook of Neurosociology*, Handbooks of Sociology and Social Research, https://doi.org/10.1007/978-3-031-95615-7_5

which was that theoretical innovation in sociology had effectively ended in the 1990s. In a 2015 American Sociological Association newsletter article Szelenyi suggested that sociological theory had been on a "downward slope since the 1980s," while in European sociological circles the lack of conceptual renewal, innovation and excitement since the generation of Beck, Baumann, Giddens and Bourdieu has been a topic of debate for well over a decade (Outhwaite, 2009). Sociological theory certainly appears to be less theoretically innovative and relevant within the discipline and in the public sphere than it was a generation ago (Halford & Savage, 2017).

In this chapter I reflect on the success of neurosociology (and with it cognitive sociology and sociological research on culture and cognition) since the 1990s, contrast that success with the relative stagnation (and public irrelevance) of sociological theory over this same period, and ask what lessons can be learned from the history of the migration of concepts and methods from cognitive neuroscience into sociology (cf. Dubash & Brett, 2023). My argument is twofold. First, features of the contemporary Big Science research landscape explain both the decline of traditional subfields such as sociological theory and the emergence of domain-spanning hybrid subfields including neurosociology. Second, sociological theorists since the 1990s have performed theoretical arbitrage, importing concepts from cognitive neuroscience into sociology, and in so doing have to some degree redefined the institutional role of sociological theory within the discipline. I use Web of Science (WoS) citation data to explore how cognitive neuroscience influenced sociology, finding that theory, culture, and social psychology subfields were influenced first and that theorists operating within these subfields effectively imported cognitive neuroscience into mainstream sociology. These projects of theoretical arbitrage accentuated preexisting philosophical divisions within sociology between constructionist and naturalist/realist camps as evidenced by the contrast between sociological subfields such as sexuality, gender, and medical sociology that almost never engage with cognitive neuroscience, and subfields such as social psychology and cultural sociology that consistently engage with cognitive neuroscience. I further argue that the arbitrage pattern seen from the late 1990s to the present between cognitive neuroscience and sociology indicates a shift in the institutional role of the sociological theorist from abstract theoretician to theoretical *arbitrageur* starting in the late 1960s and continuing to the present day. This shift, while not universal, can be seen in the careers of theoretically oriented sociologists including Bourdieu, Turner, DiMaggio, Cerulo, Lizardo, Vaisey, Ignatow, Shepherd, Leschziner, Kaidesoja, and Vila-Henninger, among many others.

Sociological Theory in the Big Science Research Ecosystem

Several features of the contemporary research landscape help us to understand both the decline of traditional subfields such as sociological theory and the emergence of hybrid research subfields including neurosociology. Since the second world war the global system of science and higher education has expanded extraordinarily rapidly (Frank & Meyer, 2007; Meyer et al., 2007) and higher education systems have been reorganized along business lines, taking on features of contemporary capitalism (Bailey, 2011; Palfreyman & Tapper, 2014; Readings, 1996; Troschitz, 2017). One consequence of this reorganization has been a disincentivization of risky theoretical work in favor of derivative empirical research programs. Fuller's (2009) account of the rise of quantitative evaluation of academic 'outputs' shows how national university audit measures have incentivized empirical research programs that allow individual researchers to advance their careers by demonstrating consistent productivity and 'impact' to university administrators (Subramaniam et al., 2014). A predictable negative consequence of this national and global "audit culture" (Fuller, 2009) is a lack of daring theoretical innovation. In this climate in which consistent empirical research output with at least potential practical applications is rewarded with grants, tenure and promotions, the market value of abstract sociological theories has predictably declined, and theoretically oriented sociologists are less likely to professionally identify as sociological theorists per se (Lamont, 2004: 10, 14, 16). In the mid-2000s Lamont, reflecting on sociologists' turn away from Parsonian grand theory, suggested that the American Sociological Association's "Culture Section and the Comparative Historical Sociology Section…have become de facto 'satellite' theory sections, i.e.…a large number of sociologists interested in theory have become involved in these sections, both as theoretical producers and consumers" (2004: 16). She further suggested that

> a number of young sociologists interested in theory may have come to define themselves primarily as cultural sociologists, perhaps because this field allowed them to pursue theoretical interests within a context more favorable to empirical research, and without having to deal with old theoretical dichotomies that have come to appear increasingly obsolete (e.g., between micro/macro, symbols/structure, objective/subjective, etc.). It has also allowed them to escape the age-old polarization between doing theory for theory's sake versus doing theoretically informed empirical research (a polarization that was much more salient to the older generations of theorists than it is to the younger generation, which tends to combine theoretical and empirical training—often quantitative and qualitative). This hypothesis may be supported by the fact that graduate students represent only 26 percent of the Theory Section membership, but 33 percent of the members of the Comparative Historical Sociology Section, and 39 percent of the Culture Section…"[G]rand theory" entirely disembodied from empirical research seems increasingly to be a thing of the past… Thus, grand theory certainly remains a crucial part of a good sociological education, and of intellectual literacy more generally, but is, in practice, probably increasingly conceived as a complement to middle range theory and empirical research (p. 16).

The contemporary research landscape certainly continues to favor steady empirical research output over abstract theorizing. But the sheer scale of Big Science and wide availability of Web-based research database search tools also provide unprecedented opportunities for theoretically oriented domain-spanning research.

Theory *Arbitrage*

Led most prominently by Pierre Bourdieu (Lizardo, 2004), beginning in the 1970s and accelerating through the 1990s a significant number of theoretically oriented sociologists turned away from what Bourdieu had termed 'theoretical theory.' By the early 2000s, many theoretically oriented sociologists were interested in Bourdieu and came to

> define themselves as cultural sociologists. Many research departments wanting to hire theorists did in fact hire cultural sociologists…However, the Bourdieu they consumed was not channeled through the omnivoriousness hypothesis, but through *Outline of a Theory of Practice* (Bourdieu, 1977), and through a Bourdieu more heavily influenced by the Durkheim of the *Elementary Forms of Religious Life*, a Bourdieu concerned with agency, classification, representations, and the interaction between symbolic and other aspects of power. Although less consistently organized into a self-conscious and self-referential literature, those teaching and producing this kind of scholarship… remained a deep-seated and powerful presence in the field. (Lamont, 2012: 234)

Instead of 'writing theory' (Reed & Alexander, 2009: 22) independent of empirical research, these Durkheim- and Bourdieu-influenced sociologists increasingly engaged in intellectual arbitrage, importing theoretical concepts from cognitive science, neuroscience and related fields into social science. This style of theorizing, performed for the most part by theoretically oriented sociologists engaged in empirical research arbitraging theoretical concepts from extrasociological fields into sociology, harkens back to the work of Durkheim, Mauss and other early sociologists. It is well known that Durkheim's early influences included the philosophers Renouvier and Hamelin (Lukes, 1973: 54–58) as well as Fustel de Coulanges and Emile Boutrox (LaCapra, 1972: 30). Early in his career Durkheim studied in Germany directly under the pioneering experimental psychologist Wilhelm Wundt (LaCapra, 1972: 33), and was influenced by the psychologists Ribot and Espinas (Mauss, 1979/1935: 12–13). Mead, although claimed by sociology, was not a sociologist at all but a philosopher and psychologist by training (Cook,

1993). Bourdieu's influences included the medieval historian Panofsky and cognitive psychologist Piaget (Lizardo, 2004), and he was profoundly a product of French structuralism, a movement that "allowed psychologists, linguists, anthropologists, and sociologists to speak a common theoretical language for almost two decades" (Lizardo, 2009: 715; also Dosse, 1999). More recently Vaisey effectively arbitraged the psychologist Jonathan Haidt's dual process theory into sociology (Vaisey, 2021; Miles et al., 2019). Haidt's signature paper was published in 2001 and drew heavily from his advisor Rozin's work (Haidt, 2001; Rozin et al., 1999) as well as from Zajonc's earlier intuitionist psychology (Zajonc, 1980). Lizardo (2004, 2007, 2009) arbitraged contemporary research from psychology and cognitive neuroscience into sociology, arguing that research on the human Mirror Neuron System (MNS), a specialized yet flexible "neurocognitive structure primarily charged with forming fairly abstract representations of practical action upon objects in the world performed by other actors" (Lizardo, 2007: 330), supports Bourdieusian practice theory and especially Bourdieu's notions of "bodily generalization" (Bourdieu, 1977: 175; 1990: 89) and habitus. He suggests that mirror neurons may be the basis of both action imitation and action understanding. Ignatow (2007) arbitraged theories of embodied knowledge from cognitive neuroscience (e.g. DaMasio, 2006) into sociology and more recently arbitraged research on the Microbiome-Gut-Brain axis from contemporary biology into sociology (2021, 2022), drawing implications for Bourdieusian theory and for empirical sociological research on social bonding, violence and political extremism.

In the sociology of science the phenomenon of researchers importing concepts form fields outside their own has been variously characterized as brokerage (Fleming et al., 2007), domain-spanning (Leahey & Moody, 2014), or recombinant innovation (Weitzman, 1998), but I suggest that in interpretive and historical fields such as sociology this process is more aptly characterized as theoretical or intellectual arbitrage. Intellectual arbitrage refers to researchers 'purchasing' concepts and methods in one field in which they have come to be well accepted, and repackaging and 'reselling' them in their home field in which they are less well known and are potentially valued as significant innovations. What the arbitrage concept adds to closely related concepts of brokerage, domain-spanning and recombinant innovation is recognition of the differing value of information between the purchasing and selling fields. Arbitrage opportunities increase when concepts and methods have become well established within the purchasing field; the availability of secondary literatures containing published debates and literature reviews on a given theory or method lowers the time commitment required to search unfamiliar topics and to master and adequately represent literature from multiple specialized subfields (Leahey & Moody, 2014: 228). A close reading of Bourdieu (Lizardo, 2004), DiMaggio (1997), Vaisey (2009), Lizardo (2009), Ignatow (2007, 2022), and many other cognitive sociologists and neurosociologists shows them to be engaging in arbitrage more than brokerage or domain-spanning, as the flow of ideas and methods most often occurs following a significant time lag between the theoretical or methodological cognitive neuroscience innovation and its arbitrage into sociology and is more often than not unidirectional from cognitive neuroscience into sociology (Ignatow, 2014; Lizardo, 2014; Vaisey, 2021). There is, however, a 'social neuroscience' subfield of cognitive neuroscience that is at least potentially open to engaging with theories and methods from cognitive sociology (Ibáñez et al., 2017).

The notion that contemporary sociological theorists operate as crypto-theorists from within substantive research areas (2004, 2012) rather than as theory writers *per se* (Reed & Alexander, 2009) and perform theoretical arbitrage with research from other disciplines may appear problematic. However, I suggest that the theoretical arbitrage concept does not imply that sociology is a lower status discipline than cognitive neuroscience any more than, for example, geology is a lower status discipline than physics and chemistry, two fields from which it consistently arbitrages theories and methods (Frodeman, 1995). Sociology may not be the 'queen of the sciences' as per Comte, but it does operate at a higher level of analysis than does cognitive neuroscience (Kaidesoja, 2013). So while the social sciences may occasionally make basic discoveries about social relationships and organizations, social sciences are sciences of higher-order social and cultural phenomena with emergent properties that are irreducible to life science and cognitive neuroscience mechanisms that are analyzable at lower levels of analysis. As such it is to be expected that arbitrage would tend to occur in one direction.

Web of Science Citation Analysis

The most influential publications that have arbitraged cognitive neuroscience theories and research into sociology are reasonably well known (Dubash & Brett, 2023: 5); what is less well understood is (A) which areas of sociology have been most influenced by cognitive neuroscience, and (B) patterns of influence between subfields. To explore these patterns I analyze high-impact sociology journal articles published in Clarivate Web of Science (WoS), a database containing over 63 million article records from tens of thousands of academic journals. The WoS is widely considered to be among the most useful databases for understanding citation patterns over time and across research areas (Visser et al., 2021).

I searched the WoS core collection for sociology articles only, excluding all other research areas. The period covered for all searches is 1900 to April 12, 2023. Subjects selected for searches (Tables 5.1, 5.2, 5.3) were based on the American Sociological Association's 2023 list of 53 sections, condensed into major sociological subjects rather than section titles (e.g. the subject *race* rather than the section titles *Race, Gender and Class* or *Racial and Ethnic Minorities*). However, for very general subjects such as culture, education, and theory, I used the academic subfield to either replace or supplement the subject (e.g. *sociology of education* substituted for *education*). The final list, which is by no means comprehensive but rather representative of a range of sociological subjects that can be condensed into one or two words that are not overly general in their meaning, includes 15 subjects: *culture* and *cultural sociology*, *economic sociology*, *environmental sociology*, *organizations*, *medical sociology*, *public sociology*, *race*, *religion*, *sexuality*, *social psychology*, *sociology of education*, *sociology of gender*, *theory*, and *sociological theory*.

Web of Science searches were conducted for topic words and phrases in titles and abstracts of

Table 5.1 H-Index for subject and (cogni* or neurosci*)

Subject	H-Index	First year
Theory	55	1967
Culture	51	1989
Race	34	1991
Organizations	34	1991
Religion	21	1991
Social psychology	12	1996
Cultural sociology	10	2009
Sociological theory	10	1993
Sociology of education	4	2003
Economic sociology	4	2002
Sexuality	4	2004
Public sociology	3	2007
Environmental sociology	3	2009
Medical sociology	3	2016
Sociology of gender	0	N/A

Table 5.2 % Results (subject and neurosci*) / subject

Subject	% Results	First year
Cultural sociology	2.78	2013
Social psychology	1.99	2001
Sociology of education	0.98	2015
Public sociology	0.57	2007
Sociological theory	0.52	2010
Theory	0.27	2001
Economic sociology	0.21	2021
Culture	0.17	2008
Religion	0.12	2003
Race	0.10	2013
Organizations	0.09	2001
Sexuality	0.07	2021
Environmental sociology	0	N/A
Medical sociology	0	N/A
Sociology of gender	0	N/A

Table 5.3 % Results (subject and (cogni* or neurosci*)) / subject

Subject	% Results
Social psychology	10.60
Cultural sociology	8.33
Sociology of education	5.88
Sociological theory	4.56
Economic sociology	3.95
Theory	3.52
Culture	3.32
Medical sociology	3.09
Public sociology	2.86
Religion	2.57
Race	2.2
Organizations	1.9
Environmental sociology	1.83
Sexuality	0.95
Sociology of gender	0

sociology journal articles only. The topic searches included each subject (theory, culture, race etc.) and either *cognitive* (lemmatized as *cogni** to include cognitive, cognition etc.) or *neuroscience* (lemmatized as *neurosci** to include neuroscience, neuroscientist etc.) (Tables 5.1 and 5.3), or only subject and *neurosci** (Table 5.2). I chose to search by subject and *neurosci** (Table 5.2) because of sociologists' frequent use of cognitive science vocabulary (e.g. cognition, schemas) with minimal engagement with cognitive neuroscience research literatures. The subject and *neurosci** search (Table 5.2) is thus more targeted at articles that engage with cognitive neuroscience research than is the broader search (Tables 5.1 and 5.3).

Table 5.1 presents h-indexes for each subject and the lemmatizations of *cognition* or *neuroscience*. The widely used h-index represents the number of articles h with h citations; it thus represents both the size and impact of scholarly discussion about a subject or combination of subjects. So as Table 5.1 shows, from 1967–2023 55 sociology articles were published that contained both *theory* and either *cognition* or *neuroscience* that had at least 55 citations each. There were 52 such articles published between 1989 and 2023 for *culture* but only three articles for *medical sociology* and no articles for *sociology of gender*. Of course, there may have been articles that covered topics related to the sociology of gender, medical sociology, the sociology of education etc. that did not use these exact phrases in the title or abstract. Such articles would not be captured in this analysis, so the results presented here should be interpreted with caution.

Table 5.2 shows the impact of cognitive neuroscience on each subject by calculating the percent of topics from article abstracts and titles that contain the subject and *neurosci** over the total number of topics with the subject only. In other words, of all the sociology articles that mention *cultural sociology* in the title or abstract published between 1900 and April 2023, 2.78% (10/360) of these articles also mention *neuroscience*; 1.99% (6/302) of *social psychology* articles also mention *neuroscience*; and 0.98% (3/306) of *sociology of education* articles mention *neuroscience*. Percentages fall dramatically from there: the remaining subjects had either one *neuroscience* mention (*public sociology, sociological theory, theory, economic sociology, race, sexuality*) or zero mentions (*environmental sociology, medical sociology, sociology of gender*). Several articles from the search results were critical of neuroscience. For example the title of a sociology of education article published by Cuthbert in *Sociology Compass* in 2015 was "Neuroscience and Education—An Incompatible Relationship," and a 2021 article by Schmitz in *Frontiers of Sociology* mentioned sexuality and cautioned against "neurobiological determinism." So while Table 5.1 shows that there has been intense engagement with cognitive neuroscience in sociological research related to *theory* and *culture*, and a relatively superficial use of cognitive science language across many sociological subjects, Table 5.2 shows that within sociology there has been intense engagement with neuroscience only in *cultural sociology* and *social psychology*.

Table 5.3 presents the results from the more inclusive search for sociology articles that mention each subject and either *cogni** or *neurosci** as a percentage of all articles that contain each subject. So 10.6% of social psychology articles also mention either cognition or neuroscience in the title or abstract. The percentage is 8.33% for cultural sociology, 5.88% for the sociology of education, and 4.56% for sociological theory. For sexuality the percentage is less than one, and for sociology of gender it is zero.

Taken together, Tables 5.1, 5.2, 5.3 show that cognition and neuroscience concepts entered sociology through the research areas of theory, cultural sociology, and social psychology. There is intense engagement with cognitive science and neuroscience concepts within these areas, somewhat less intense, often superficial engagement in other subfields, and silence or resistance in several areas including sexuality and gender.

The last column in Tables 5.1 and 5.2, *First year*, represents the first year of publication of a WoS sociology article with a given combination of subject and *cogni** or *neurosci** in the abstract or title. *First year* allows us to examine the historical pattern of the migration of cognitive neuroscience concepts into sociology.

The 1960s–1980s: Cognitive Science Enters Sociological Theory

How did cognitive neuroscience concepts first enter sociology? *First year* in Table 5.1 shows that *theory* was the first subject associated with *cogni** (in 1967), and *culture* was second (in 1989). The theoretical articles include studies of cognitive dissonance by Geschwender (1967) and Capel (1971). A 1978 article by Warner, cited by Swidler (1986), Lizardo (2016), and Strand and Lizardo (2022), is particularly supportive of this chapter's main arguments. Warner's article is a theoretical critique of Parsons in which Warner argued for a theory of action that "acknowledges explicitly the analytically independent role of the cognitive element in action" (p. 1317). In this way, although he did not engage with cognitive neuroscience per se, he did make an explicit theoretical argument for cognitive sociology almost two decades before DiMaggio's seminal 1997 culture and cognition article. Two other articles from this period were late-1980s studies, the first of expectation states theory and social cognition by Knottnerus (1988), the second of cross-cultural cognitive development by Tenhouten (1989). Consistent with Lamont's observations about the Durkheim-Bourdieu influence on sociological theory and cultural sociology more than 20 years later (Lamont, 2012), both Knottnerus and Tenhouten can be fairly characterized as Durkheimians (e.g. Knottnerus, 1986; Tenhouten, 2016).

The 1990s: Cognitive Science Enters Mainstream Sociology

In the 1990s there were 47 WoS sociology articles published with both *theory* and either *cognition* or *neuroscience* as topics; ten of these were published in 1999. There were 11 articles in the 1990s with *sociological theory* as a topic and 37 with *culture* as a topic; 14 of these were published in 1999. There were smaller numbers of *race*, *organizations*, *religion* and *social psychology* articles, and many of these articles used cognitive science vocabulary (schemas and so on) without intense engagement with cognitive neuroscience research. Nevertheless, we can see a mainstreaming of cognitive neuroscience concepts and methods in this period, with the most intense engagement occurring in theory- and culture-oriented research.

2000–2023: The Emergence of Neurosociology

A 2001 *American Sociological Review* article by Owen-Smith is a red herring in Table 5.2. It is the first WoS sociology article with *neuroscience* as a topic as well as *theory*, *social psychology*, and *organizations*. But Owen-Smith's article was an ethnographic study of a neuroscience laboratory that had nothing to do with cognitive neuroscience research. There have been 35 WoS sociological theory articles with *neuroscience* as a topic published since 2000 and 31 *cultural sociology* articles. Since 2000 there have been six *neuroscience* articles that mention *race* and five that mention *public sociology*, although only one *public sociology* article engages meaningfully with cognitive neuroscience—a 2007 article by Pisati in *Sociologica: Italian Journal of Sociology*

Online, "Unity of Sociology, Unity of Science: Some Reflections on the Identity of Sociology."

In articles with *sexuality* as a topic there is superficial use of cognitive science language with the exception of LaFleur's Zerubavelian 2022 *Symbolic Interaction* article in which she proposed to "synthesize sexual social constructionism, radical feminism, and cognitive sociology in order to identify and analyze 'gender asymmetries' in heterosexual erotic attention" (p. 72).

Conclusions: *Arbitrageurs* as Theorists

Taken together, Tables 5.1, 5.2, 5.3 show that cognitive neuroscience concepts entered sociology through the subfields of theory, social psychology, and cultural sociology. There is intense engagement with cognition and neuroscience concepts within these subfields, somewhat less intense engagement in other subfields (economic sociology, sociology of education, organizations, religion), and silence or resistance in sexuality and gender subfields—with the exception of LaFleur's 2022 article. How can we explain the extreme differences in engagement with cognitive neuroscience across sociological subfields since the late 1960s? Fundamental epistemological and ontological commitments appear to play a role (Ignatow, 2014). Reed (2011) describes sociology as divided methodologically between hermeneutic qualitative methods and explanatory quantitative methods, but also ontologically and epistemologically between research performed in an interpretive epistemic mode and research performed in an explanatory realist mode.

Which of Reed's two modes is characteristic of neurosociology research? Neurosociology shares with cognitive anthropology (D'Andrade, 1986) and many other fields a set of assumptions about how best to evaluate truth and validity, what constitutes data, and how data should connect with theory. Neurosociology's implicit ontology is constructionist because it focuses on how social groups come to interpret the social world in particular ways. But its ontology is also implicitly realist, accepting that experiments and magnetic resonance imaging reveal empirical social cognitive phenomena. The findings of cognitive neuroscience are interpretations of reality but they are interpretations that assume that reality exists independent of interpretation. Neurosociology almost always presumes that a real world exists that is to some degree independent of our descriptions of it; in this way neuroscientists and neurosociologists are philosophical realists, but realists who are especially interested in social phenomena, such as culture and discourses, that are complex, intersubjectively constituted, and uniquely challenging to conceptualize and measure. This ontological and epistemological position is compatible with various versions of philosophical realism, including Elder-Vass's "realist constructionism" (2012) as well as Kaidesoja's "naturalized" realism (2013) which is explicitly grounded in empirical research findings from cognitive neuroscience. Sociological subfields including the sociology of gender and sexuality are not, to my knowledge, associated with realist constructionism, naturalized realism or any other pragmatic realist or naturalist ontology, but rather with forms of social constructionism that are less conducive to intense engagement with cognitive neuroscience.

In light of all of the above, the generation gap in sociological theory identified by Turner (2006) represents perhaps less of an historical anomaly than a response to shifting incentives and opportunities facing sociological theorists navigating the modern system of science and higher education. In becoming crypto-theorists and theoretical *arbitrageurs*, theorists operating within various sociological subfields have over the last 30 years inextricably intertwined the development of sociological theory with the emergence of cognitive sociology and neurosociology. Like Durkheim, Mauss, Mead and Bourdieu, contemporary theoretical arbitrageurs have written a new chapter in the history of sociological theory. As Lamont noted in 2004 and again in 2012, sociological theory is neither dead nor in a state of crisis; rather it has been reinvigorated by new developments in cognitive neuroscience and other fields by sociological theorists working within substantive research areas. With few exceptions the existence of crypto-theorists and theory arbitrage within sociology is not reflected in the institutional structures of contemporary sociology, and neither crypto-theorists nor theory arbitrage yet feature in sociological theory pedagogy. Perhaps future generations of sociologists would benefit from learning how theoretical innovation has occurred within, and from without, their chosen field (cf. Swedberg, 2012).

References

Bailey, M. (2011). *The assault on universities: A manifesto for resistance*. Pluto Press.

Bourdieu, P. (1977). *Outline of a theory of practice* (Vol. 16). Cambridge University Press.

Bourdieu, P. (1990). *The logic of practice*. Stanford university press.

Capel, W. C. (1971). Cognitive dissonance, reconstruction and negro civil rights: Applications of a theory. *Journal of Human Relations, 19*(2), 225–238.

Cook, G. A. (1993). *George Herbert Mead: The making of a social pragmatist*. University of Illinois Press.

Cuthbert, A. (2015). Neuroscience and education–An incompatible relationship. *Sociology Compass, 9*(1), 49–61.

D'Andrade, R. G. (1986). three scientific world views and the covering law model. In D. W. Fiske & R. A. Shweder (Eds.), *Metatheory in social science: Pluralisms and subjectivities* (pp. 19–39). University of Chicago Press.

Damasio, A. R. (2006). *Descartes' error*. Random House.

DiMaggio, P. (1997). Culture and cognition. *Annual Review of Sociology, 23*, 263–287.

Dosse, F. (1999). *Empire of meaning: the humanization of the social sciences*. U of Minnesota Press.

Dubash, S., & Brett, G. (2023). The diffusion of culture and cognition within and beyond sociology, 1997–2021. In *Sociological Forum* (Vol. 38, No. 2, pp. 575–584).

Elder-Vass, D. (2012). *The reality of social construction*. Cambridge University Press.

Fleming, L., Mingo, S., & Chen, D. (2007). Collaborative brokerage, generative creativity, and creative success. *Administrative Science Quarterly, 52*(3), 443–475.

Frank, D. J., & Meyer, J. W. (2007). University expansion and the knowledge society. *Theory and Society, 36*(4), 287–311.

Frodeman, R. (1995). Geological reasoning: Geology as an interpretive and historical science. *Geological Society of America Bulletin, 107*(8), 960–968.

Fuller, S. (2009). *The sociology of intellectual life: The career of the mind in and around academy*. Sage.

Geschwender, J. A. (1967). Continuities in theories of status consistency and cognitive dissonance. *Social Forces, 46*(2), 160–171.

Haidt, J. (2001). The emotional dog and its rational tail: A social intuitionist approach to moral judgment. *Psychological Review, 108*(4), 814.

Halford, S., & Savage, M. (2017). Speaking sociologically with big data: Symphonic social science and the future for big data research. *Sociology, 51*(6), 1132–1148.

Hunzaker, M. F., & Valentino, L. (2019). Mapping cultural schemas: From theory to method. *American Sociological Review, 84*(5), 950–981.

Ibáñez, A., Sedeño, L., & García, A. M. (Eds.). (2017). *Neuroscience and social science: The missing link*. Springer International Publishing.

Ignatow, G. (2007). Theories of embodied knowledge: New directions for cultural and cognitive sociology? *Journal for the Theory of Social Behaviour, 37*(2), 115–135.

Ignatow, G. (2014). Ontology and method in cognitive sociology. In *Sociological Forum* (Vol. 29, No. 4, pp. 990–994).

Ignatow, G. (2021). Cognitive sociology after relational biology. In *Sociological Forum* (Vol. 36, pp. 1253–1270).

Ignatow, G. (2022). The microbiome-gut-brain and social behavior. *Journal for the Theory of Social Behaviour, 52*(1), 164–182.

Kaidesoja, T. (2013). *Naturalizing critical realist social ontology*. Routledge.

Knottnerus, J. D. (1986). Emile Durkheim: His methodology and uses of history. *Journal of the History of the Behavioral Sciences, 22*(2), 128–139.

Knottnerus, J. D. (1988). A critique of expectation states theory: Theoretical assumptions and models of social cognition. *Sociological Perspectives, 31*(4), 420–445.

LaCapra, D. (1972). *Emile Durkheim: Sociologist and Philosopher*. Ithaca and London: Cornell University Press.

LaFleur, G. (2022). Desire/desirability: Gender asymmetries in heterosexual erotic attention. *Symbolic Interaction, 45*(1), 72–96.

Lamont, M. (2004). The theory section and theory satellites. *Perspectives: Newsletter of the ASA Theory Section, 27*, 1–17.

Lamont, M. (2012). How has Bourdieu been good to think with? The case of the United States. In *Sociological Forum* (Vol. 27, No. 1, pp. 228–237). Blackwell Publishing Ltd.

Leahey, E., & Moody, J. (2014). Sociological innovation through subfield integration. *Social Currents, 1*(3), 228–256.

Lizardo, O. (2004). The cognitive origins of Bourdieu's habitus. *Journal for the Theory of Social Behaviour, 34*(4), 375–401.

Lizardo, O. (2007). "Mirror neurons," collective objects and the problem of transmission: Reconsidering Stephen Turner's critique of practice theory. *Journal for the theory of social behaviour, 37*(3), 319–350.

Lizardo, O. (2009). Is a "special psychology" of practice possible? From values and attitudes to embodied dispositions. *Theory & Psychology, 19*(6), 713–727.

Lizardo, O. (2014). Beyond the Comtean schema: The sociology of culture and cognition versus cognitive social science. In *Sociological Forum* (Vol. 29, No. 4, pp. 983–989).

Lizardo, O. (2016). Cultural theory. In *Handbook of contemporary sociological theory* (pp. 99–120).

Lukes, S. (1973). *Emile Durkheim: His Life and Work*. London: Penguin.

Mauss, M. (1979). *Sociology and psychology: Essays* (Ben Brewster, Trans.). Routledge & Kegan Paul.

Meyer, J. W., Ramirez, F. O., Frank, D. J., & Schofer, E. (2007). Higher education as an institution. In *Sociology of higher education: Contributions and their contexts* (p. 187).

Miles, A., Charron-Chénier, R., & Schleifer, C. (2019). Measuring automatic cognition: Advancing dual-process research in sociology. *American Sociological Review, 84*(2), 308–333.

Outhwaite, W. (2009). Canon formation in late 20th-century British sociology. *Sociology, 43*(6), 1029–1045.

Owen-Smith, J. (2001). Managing laboratory work through skepticism: Processes of evaluation and control. *American Sociological Review, 66*(3), 427–452.

Palfreyman, D., & Tapper, T. (2014). *Reshaping the university: The rise of the regulated market in higher education*. Oxford University Press.

Pisati, M. (2007). Unità della sociologia, unità della scienza. Alcune riflessioni sull'identità disciplinare della sociologia. *Sociologica, 1*(1), 0–0.

Readings, B. (1996). *The university in ruins*. Harvard University Press.

Reed, I. A. (2011). *Interpretation and social knowledge: On the use of theory in the human sciences*. University of Chicago Press.

Reed, I., & Alexander, J. (2009). Social science as reading and performance: A cultural-sociological understanding of epistemology. *European Journal of Social Theory, 12*(1), 21–41.

Rozin, P., Lowery, L., Imada, S., & Haidt, J. (1999). The CAD triad hypothesis: a mapping between three moral emotions (contempt, anger, disgust) and three moral codes (community, autonomy, divinity). *Journal of Personality and Social Psychology, 76*(4), 574.

Schmitz, S. (2021). TechnoBrainBodies-in-Cultures: An intersectional case. *Frontiers in Sociology, 6*, 651486.

Shepherd, H. (2011). The cultural context of cognition: What the implicit association test tells us about how culture works 1. In *Sociological Forum* (Vol. 26, No. 1, pp. 121–143). Blackwell Publishing Ltd.

Shepherd, H. (2019). Methods for studying the contextual nature of implicit cognition. In *The Oxford handbook of cognitive sociology* (Vol. 367).

Strand, M., & Lizardo, O. (2022). Chance, orientation, and interpretation: Max Weber's neglected probabilism and the future of social theory. *Sociological Theory, 40*(2), 124–150.

Subramaniam, M., Perrucci, R., & Whitlock, D. (2014). Intellectual closure: A theoretical framework linking knowledge, power, and the corporate university. *Critical Sociology, 40*(3), 411–430.

Swedberg, R. (2012). Theorizing in sociology and social science: Turning to the context of discovery. *Theory and Society, 41*, 1–40.

Swidler, A. (1986). Culture in action: Symbols and strategies. *American Sociological Review, 51*(2), 273–286.

Szelenyi, I. (2015). The triple crisis of sociology. https://contexts.org/blog/the-triple-crisis-of-sociology/

TenHouten, W. D. (1989). Application of dual brain theory to cross-cultural studies of cognitive development and education. *Sociological Perspectives, 32*(2), 153–167.

TenHouten, W. D. (2016, June). Normlessness, anomie, and the emotions. In *Sociological Forum* (Vol. 31, No. 2, pp. 465–486).

Troschitz, R. (2017). *Higher education and the student: From welfare state to neoliberalism*. Routledge.

Turner, S. (2002). *Brains/practices/relativism: Social theory after cognitive science*. University of Chicago Press.

Turner, J. H. (2006). American sociology in chaos: Differentiation without integration. *The American Sociologist, 37*(2), 15–29.

Turner, S. P. (2018). *Cognitive science and the social: A primer*. Routledge.

Vaisey, S. (2009). Motivation and justification: A dual-process model of culture in action. *American journal of sociology, 114*(6), 1675–1715.

Vaisey, S. (2021). Welcome to the real world: Escaping the sociology of culture and cognition. In *Sociological Forum* (Vol. 36, pp. 1297–1315).

Vaisey, S., & Miles, A. (2014). Tools from moral psychology for measuring personal moral culture. *Theory and Society, 43*, 311–332.

Vaisey, S., & Valentino, L. (2018). Culture and choice: Toward integrating cultural sociology with the judgment and decision-making sciences. *Poetics, 68*, 131–143.

Visser, M., van Eck, N. J., & Waltman, L. (2021). Large-scale comparison of bibliographic data sources: Scopus, web of science, dimensions, Crossref, and Microsoft academic. *Quantitative Science Studies, 2*(1), 20–41.

Warner, R. S. (1978). Toward a redefinition of action theory: Paying the cognitive element its due. *American Journal of Sociology, 83*(6), 1317–1349.

Weitzman, M. L. (1998). Recombinant growth. *The Quarterly Journal of Economics, 113*(2), 331–360.

Zajonc, R. B. (1980). Feeling and thinking: Preferences need no inferences. *American Psychologist, 35*(2), 151.

Methods in Neurosociology

6

Joshua Pollock, Matthew Pfeiffer,
Graem Sigelmier, and Chloe Miller

Abstract

One primary challenge faced by aspiring neurosociologists is developing a solid understanding of the methodologies used in the field. This is even furthered by the highly technical nature of these approaches. This can result in a barrier to entry, especially for individuals with no background in these methods. With that in mind, this chapter provides an introduction and overview of the methods commonly used in neurosociology and is intended to be accessible to those who have little experience in the field. In the chapter, we highlight three prominent neuroscience methods that have played a significant role in expanding the field of neurosociology: electroencephalography (EEG), functional magnetic resonance imaging (fMRI), and functional near infrared spectroscopy (fNIRS). We offer explanations of their function, the analytical techniques associated with them, their strengths and limitations, and recommendations for implementation. We also discuss hyperscanning—the collection of neural data from multiple actors—and highlight its importance in the study of social interaction. Finally, we conclude this chapter by expanding upon the advantages and disadvantages of a neuroscientific approach to sociology and an emphasis on our goal of advancing the field of neurosociology.

Introduction

The growing field of neurosociology contributes to the bridge between sociological theories of social contexts and neuroscientific models of cognition. In combining the two disciplines, all neurosociology methodologies must be capable of investigating how the human brain influences the expansive and complicated set of forces that drive human relationships and organize the social world (i.e., a bottom-up approach). Thus, the primary role of neurosociology methodologies is to link neural functioning to human behavior through our understanding of social theory. In so doing, it is essential that neither discipline take priority over the other. This pairing has proven to be very effective as sociology and neuroscience methods complement one another (see Kalkhoff et al., 2016a). This is not a surprise, especially when one considers the goal of this partnership—to supplement our classical understanding of sociological theory or traditional research methods with neuroscientific methods to gain an even greater awareness of human behavior.

J. Pollock (✉) · M. Pfeiffer · G. Sigelmier · C. Miller
Department of Sociology and Criminology, Kent State University, Kent, OH, USA
e-mail: jpollo10@kent.edu; mpfeiff5@kent.edu; gsigelm2@kent.edu; cmill206@kent.edu

W. Kalkhoff et al. (eds.), *Handbook of Neurosociology*, Handbooks of Sociology and Social Research, https://doi.org/10.1007/978-3-031-95615-7_6

Conveniently, the technological developments in neuroscience in recent years have made it so that there is no better time to capitalize on the synthesis of these two disciplines. Accordingly, we as neurosociologists must take advantage of this opportunity.

With this in mind, we begin with an overview of neurosociological methodology. We then outline different methodological approaches in neurosociology, focusing largely on Electroencephalography (EEG), functional Magnetic Resonance Imagery (fMRI), and functional Near Infrared Spectroscopy (fNIRS). We start each subsection with an explanation of the method's design and measurement, followed by a discussion of its strengths and limitations. We end each subsection with a description of the different uses for each methodology, which span from the more traditional social-neuroscience individual-focused analyses to the more modern hyperscanning methodology that has become more prevalent in the past two decades. We conclude this chapter with a brief summary and an emphasis on the synergistic relationship between sociological theory and neuroscience methods.

Overview

When discussing methodology, it is important to posit the question "Why?" In this case, "Why bother learning new techniques and applying neuroscience to the field of sociology?" The answer to this question is found within the heart of sociology itself. Sociology studies social interaction and the way in which the social world is organized. Both have a strong mental component, yet this component is often overlooked by classical sociological theory (Cooley's (1902) looking-glass self and Mead's (1934) theories of the mind notwithstanding). Neuroscience, however, prioritizes this mental component. Accordingly, applying neuroscience methodologies to classical sociological methods helps to fill this methodological gap. It does so by providing a neural level of analysis to supplement more traditional sociological data, which offers sociologists the unique opportunity to test abstract theories (classic and novel) on multiple levels. These multiple levels are laid out by Tenhouten et al. (2023), who explains how neurosociologists employ three levels of analysis–(1) social relations, (2) mind and information processing, and (3) the nervous system and the brain. This three-level analysis creates the basis for any methodology in the field of neurosociology.

The field of neurosociology involves the use of neuroscience methodologies to understand how the social world is organized and how the organization of the social world influences neurological processes. The neuroscience methodologies described in this chapter demonstrate this synergistic approach. The unique strengths and ecological validity of these methods provide researchers with benefits that have been notably missing from classical sociological methodology (e.g., a measure devoid of social biases). Additionally, the different neuroscience techniques we discuss below have varied strengths and weaknesses, meaning some techniques are better suited to study certain social phenomena as opposed to others. One purpose of this chapter—in addition to emphasizing the versatility of neuroscience methods—is to highlight these areas.

One of the cornerstones of neurosociology is the analysis of group processes and intergroup relations. These concepts underly the construction of the social world and are often studied in the context of social interaction. Indeed, a considerable amount of research has been conducted on a wide variety of phenomena found within social interaction, including research on decision making, memory, cognition, and emotion, just to name a few. Notably, human brains are shaped and developed through social interaction (Hari et al., 2015), and this fundamental component of brain development shapes future interactions with the environment. These processes are often disregarded within sociology given traditional sociological methods are incapable of accounting for them. As a result, sociologists' assessments of how individuals interact with one another are usually at the mercy of self-report data and researcher observations. Unsurprisingly, this data is vulnerable to bias. Neuroscience methodologies, on the other hand, are not subject to bias in this manner. Thus, they can occupy a very important role in assessing the numerous social phenomena found within social interaction. Moreover, neuroscience methods can also assess implicit processes with extreme temporal precision. EEG, for example, can measure neural activity to the millisecond (i.e., outside of conscious awareness and employs time-course analysis that can be used to assess when different (social and neural) processes unfold. Given these methodological advantages, sociologists can delve deeper into their theories' assumptions and test previously untestable hypotheses. For example, the real-time dynamics of social interaction have been theorized about but never truly interrogated without the use of neuroscience methodology (Tenhouten et al., 2023). Put simply, neuroscience methods can help quantify what sociologists previously thought was unquantifiable.

Within sociology, testing theoretical assumptions and hypotheses through neuroscience methodology highlights the unique and synergistic relationship between them. Neuroscience methods offer valuable insight into neural correlates, or the neural activity that results from social stimuli. Neuroscience methods, though, mean little when not guided by theory. As noted in this chapter, many neuroscience analysis techniques rely heavily on a priori models. Accordingly, the blending of neuroscience methodology with sociology relies on sociological theory.

Sociological theory, meanwhile, reveals a considerable amount about neural functioning and how it relates to social interaction. So, sociological theory "drives" neuroscientific methodology in a complementary manner. Sociological theory identifies and explains the sociological phenomena of interest within social interaction, and neuroscience methodology examines the sociological phenomena of interest at multiple levels of social interaction (i.e., by way of neural processes). In this way, sociological theory and neuroscience methodology occupy opposite sides of the proverbial *synergistic coin.*

One important note here is the distinction between neurosociology and social neuroscience. Notably, both disciplines assess social mechanisms through brain correlates and use many of the same methodologies. Where they differ is in their focus: a neurosociologist focuses more on how brain function is affected by social interaction while a social neuroscientist focuses more on brain activity during social interaction (e.g., through patterns of blood-oxygen flow and electrical field activity) (TenHouten et al., 2023). Still, these two fields, while somewhat distinct in their focus, are becoming more and more intertwined, and their differences are increasingly fading away into a transdisciplinary grey area.

As evidenced by the increasing entanglement of neurosociology with social neuroscience, the field of neurosociology is relatively young and continuously evolving. Therefore, our aim with this chapter is to provide a simple overview of basic neuroscience methodologies, including their features, strengths, and limitations. Accordingly, we aspire to present this information in a way that non-specialists can comprehend. Ultimately, we hope this chapter will leave non-specialists feeling comfortable incorporating neuroscience methodologies into their own work. In accordance with these goals, we include streamlined explanations when discussing technical details, if applicable. We also prioritize the current, most frequently used approaches in neurosociology—specifically, Electroencephalography (EEG), functional Magnetic Resonance Imaging (fMRI), and functional Near Infrared Spectroscopy (fNIRS). We also outline other, less commonly used methods such as psychophysiological measurements and Transcranial Magnetic Stimulation (TMS). Finally, we provide an overview of hyperscanning (Montague et al., 2002)—the process of recording brain activity from multiple individuals at the same time—including its historical and modern-day significance in the study of social phenomena.

Electroencephalography

Methods

Electroencephalography (EEG) is a neuroscience method of recording and analyzing voltage fluctuations across the scalp (Kappenman & Luck, 2011). It is one of the primary methods used to directly record neural activity. This stands in direct contrast to methods that indirectly measure neural activity by using blood flow transitions as a proxy, such as fMRI or fNIRS, which are detailed later in this chapter. The primary strength of EEG paradigms lies in EEG's unrivaled temporal precision (Luck, 2014). Because EEG directly records neural activity, EEG can record the time course of neural activity with significant granularity; signature patterns in neural activity are measured in terms of mere milliseconds after a stimulus is shown (Ibáñez et al., 2012). In addition, EEG data can be collected in a laboratory or mobile environments at a relatively low cost while maintaining high temporal precision.

The EEG signal is a result of a combination of post-synaptic potentials in cortical pyramidal neurons in different regions of the brain (Kappenman & Luck, 2011). When firing together, these specialized neural cells create electrical signals that create a measurable outcome at the surface of the scalp. These signals can be influenced by deeper regions of the brain such as the amygdala, thalamus, or hippocampus (Kappenman & Luck, 2011), but the low spatial resolution of EEG does not allow for assessment of these areas. Most conclusions drawn from EEG analysis are non-structural in nature due to this limitation.

The EEG signal is recorded from sensors (electrodes) placed on the surface of the scalp. First utilized by Hans Berger (Jung & Berger, 1979) in 1924, EEG is an inexpensive, non-invasive, and passive method. Passive, in this case, refers to the continuous, real-time recording of neural activity from individuals without any direction from the researcher. In other words, once EEG data collection begins, no active intervention from the researcher is necessary. What is more, EEG caps are either wired or wireless, giving them flexibility to be applied in both laboratory and naturalistic settings.

EEG electrode arrays comprise anywhere from 10 to 256+ electrodes arranged in a standardized pattern across the scalp. Using a higher number of electrodes (high-density electrode arrays) offers more coverage, which can offer more spatial resolution; however, preparation time and costs are higher with larger scale setups. Accordingly, most neurosociology EEG setups balance these discordant factors by collecting data from between 16 and 64 electrodes per individual. Traditionally, electrodes are mounted in caps or grids for faster application, but not always (for a review, see Casson, 2019). EEG is recorded continuously with markers placed according to events of interest, permitting stimulus-response analysis. For this reason, most devices are paired with data collection software. Once data is collected, the process of analysis begins (Fig. 6.1).

Analysis

The EEG signal contains information on when neural processes occur (time domain) and the oscillations of the signal (frequency domain). These domains have traditionally been analyzed separately; however, time-frequency domain analyses are becoming more prevalent. Historically, most EEG analyses have utilized time-based analyses in the form of event related potentials (ERPs) (Cohen, 2014) or frequency-based analyses in the form of Fourier transformations (Delorme & Makeig, 2004; Gregory & Kalkhoff, 2007). As noted above, recent developments in the field have expanded upon this understanding and more advanced analysis techniques have been developed.[1] EEG has numerous dedicated software packages for analysis (e.g., EEGLab, Fieldtrip, Brainvision Analyzer). These programs traditionally have a standard analysis pipeline that is outlined for use.

[1]While beyond the scope of this chapter, a more detailed description of these advanced techniques is provided by Morales and Bowers (2022).

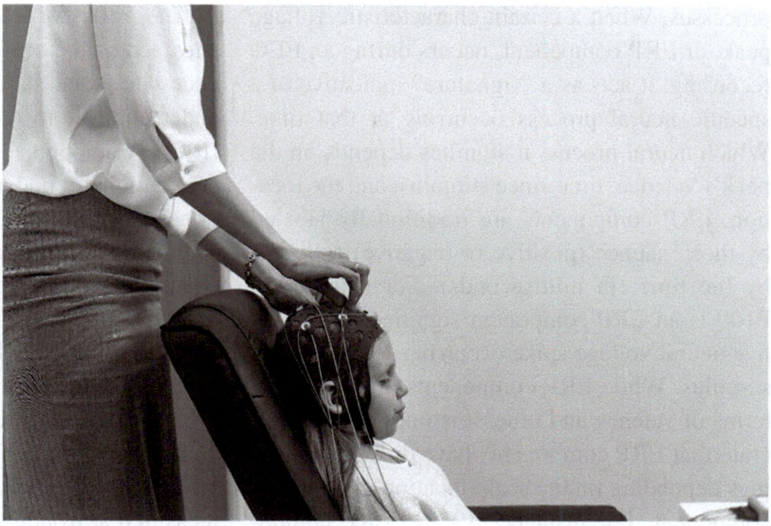

Fig. 6.1 A researcher applying a low-profile EEG cap to a young individual

Prior to analysis, several pre-processing steps are required, including filtering, re-referencing, and artifact rejection. This section of analysis is relatively standardized, with some variation seen in the literature (Kappenman & Luck, 2011). In short, researchers will usually import the continuous EEG data before filtering the data to identify important bandwidths and limit noise. Noise in an EEG analysis is akin to error in other more traditional analyses in the sense that it creates issues isolating true measurements. Therefore, neurosociologists strive to reduce it in all paradigms.

After filtering, re-referencing is performed. The purpose of a reference is to provide a relative measure to relate the neural signal of interest to. Typically, researchers re-reference to either (1) the electrodes located directly behind the ears over the mastoids or (2) to an average of all electrodes (Luck, 2014). Additionally, muscle activity and eye movement create extraneous noise in an EEG recording; these artifacts are then removed either through automated processes or visual inspection (Cohen, 2014). Finally, the data can be segmented around times of interest (i.e., into epochs). This is usually based on a priori theory and in response to a certain event or stimulus.

Time Domain Analysis

ERPs are the dominant approach in EEG analysis to analyze responses to specific events. ERPs are voltages generated in the brain as a direct response to a stimulus or specific event (Blackwood & Muir, 1990). They reflect the summed activity of the postsynaptic potentials when a large number of cortical pyramidal neurons fire in synchrony (i.e., while processing information) (Huang et al., 2015). In traditional ERP paradigms, stimuli are presented to participants many times; a participant may see a specific stimulus 30 times or more during an EEG recording session (Blackwood & Muir, 1990). After pre-processing, chunks of time around the stimuli, called epochs, are extracted from the recording. Artifacts are then removed from the data and the epochs are averaged together based on the experimental condition (e.g., control vs. experimental).

In the recorded and averaged epoch data, there may be negative or positive peaks in neural voltage observed at certain points of time after a stimulus at certain locations on the scalp. Past research using EEG in conjunction with fMRI has connected certain voltage peaks' characteristics (i.e., their polarity/valence, time since stimulus, and scalp location) with specific neural processes. When a certain characteristic voltage peak, or ERP component, occurs during an EEG recording, it acts as a "signature" indicative of a specific neural process occurring at that time. Which neural process it signifies depends on the peak's valence, time since stimulus, and the location. ERP components are traditionally labeled by their valence (positive or negative) followed by the time (in milliseconds). For example, a P100 is an ERP component signified by a positive neural voltage spike occurring 100ms after a stimulus. While ERP components are labeled in terms of valence and time, it is important to reiterate that ERP components have different meanings depending on the scalp location of that ERP component. For instance, a P300 ERP component recorded at the back of the scalp near the occipital lobe would have a different interpretation than a P300 ERP component recorded at the front of the scalp near the prefrontal cortex, even though they may have the same signature pattern. Decades of research have established ERPs at specific times, valences, and locations as reliable predictors of phenomenon in response to a stimulus. Below, we highlight a small subset of ERPs and their common applications which have been utilized in neurosociological methods.[2]

The P100 and N100 (100 millisecond post stimuli) are affected by emotional salience (Pourtois & Vuilleumier, 2006), relevance of the stimuli (Biehl et al., 2013), race when processing faces (Cunningham et al., 2012), and selective attention (Herrmann & Knight, 2001). These ERP components are traditionally recorded from the temporal and occipital areas on the sides and rear of the scalp, respectively.

Also recorded in temporal-occipital regions, the N170 is linked with the perception of human faces (Bentin et al., 1996). It can be measured to interrogate racial cues (Cunningham et al., 2012, Balas & Nelson, 2010; Herrmann et al., 2007), contextual effects (Ibáñez et al., 2011), and emotional variables (Schyns et al., 2007).

[2]For a more general and more specific discussion of ERPs, see Ibáñez et al. (2012) and Luck and Kappenman (2013), respectively.

The P300, which is recorded at a variety of sites across the scalp, has been linked to differences in social status (Gyurovski et al., 2018), and responses to faces varying in social status (Ito & Cacioppo, 2000). Notably, the P300 is conventionally understood as two different components–P3a and P3b. The P3a reflects initial evaluation and categorization of relevant material (Potts, 2004), and it can be modulated by whether an individual is watching another perform a complementary task (Sebanz et al., 2006). The P3b, meanwhile, is commonly associated with working memory representing or updating the task context (Polich, 2007).

The Feedback Related Negativity (FRN) is measured at frontal sites approximately 200–300 milliseconds post stimuli. It is linked to unexpected or undesired feedback (Miltner et al., 1997) and has shown modulation based on social expectancy violations (Sun & Yu, 2014), including identity nonverification (Kalkhoff et al., 2016b).

The Late Positive Potential (LPP) is sensitive to preceding emotional context (Schupp & Kirmse, 2021) and has shown some influence of in-group out-group biases (Hurtado et al., 2009). It is measured 300–700 milliseconds post stimuli from frontal sites.

Frequency Domain Analysis

Frequency domain analyses are commonly conducted using a mathematical equation known as a Fourier Transform (see Gregory & Kalkhoff, 2007) to break up the EEG signal into component frequencies and amplitudes. These transformations result in output data in the form of frequency, power, and phase. The frequency is the speed of an oscillation or rhythm, and it is measured in Hertz (Hz): the number of oscillations per second. The power is the amount of energy in a frequency band, which is typically expressed as squared amplitude. Phase is the amount of synchronization across several neurons (Morales & Bowers, 2022) and is commonly measured by intertrial coherence (ITC; also called the phase-locking factor or phase-locking value).

Neural oscillations are clustered in frequency bands, including delta (0.5–4 Hz), theta (4–7

Hz), alpha (8–12 Hz), beta (13–30 Hz), and gamma (30–90 Hz) (Teplan, 2002). EEG analysis in this area is typically carried out in one or more of these bands, based on past research. Delta oscillations are mostly absent in neurosociology research, but a summary of certain bands of focus in neurosociological research is provided below.

Theta oscillations have been associated with memory encoding and retrieval (Klimesch, 1999), emotions (Knyazev, 2007), and human emotional processing (Krause et al., 2000). Theta oscillations have been further linked with social exclusion (Cristofori et al., 2013) and speech coordination in hyperscanning (Kawasaki et al., 2013). More comprehensive reviews of the role of theta oscillations can be found elsewhere (e.g, Colgin, 2013).

Alpha oscillations are thought to represent an awakened but relaxed state and have large roles in decision making (Davis et al., 2011). They traditionally exhibit an inverse correlation with cognitive performance and are thought to reflect inhibition of non-task relevant cortical structures (Jensen & Mazaheri, 2010). A critical review of the function and role of alpha oscillations has been published by Başar (2012).

Additionally, when discussing alpha oscillations in the motor cortex, they are commonly referred to as mu. This is important, as the motor cortex is a key region in the mirror neuron system. This system is a collection of regions involved in viewing the actions of others and "mirroring" them (Rizzolatti, 2005). Accordingly, it is commonly assessed in social interaction. The motor neuron system, as measured by the alpha/mu band, has also shown in-group and out-group bias in ethnic comparisons (Gutsell & Inzlicht, 2010). Interestingly, higher levels of explicit prejudice have been shown to further exacerbate these differences (Gutsell & Inzlicht, 2010).

Beta oscillations are generated in both the posterior and frontal regions and are associated with active concentration and thinking (Lim et al., 2019). They have also been associated with the performance of motor tasks (Neuper & Pfurtscheller, 2001) and are commonly analyzed in dyadic interaction synchrony paradigms (Wikström et al., 2022; Balconi et al., 2019).

Gamma oscillations are linked with cortical activation (Merker, 2013) and have been linked to the higher processing of information (Fries et al., 2001). There is also evidence of gamma synchronization in hyperscanning paradigms, which are discussed later in this chapter (e.g., Kinreich et al., 2017).

Time-Frequency Analyses

Time and frequency analyses both have their weaknesses. Specifically, ERP analyses ignore data not tied to a specific stimulus and frequency analyses ignore temporal information (Morales & Bowers, 2022). Time-frequency measures combine both domains to address these issues. These analyses include time frequency power (signal strength) and inter-trial phase synchrony (signal consistency), among others (Cohen, 2014). While these analyses are beyond the scope of this chapter, Morales and Bowers (2022) provide a detailed overview.

Strengths

The greatest strengths of EEG paradigms are their temporal precision and direct measurement of electrical activity. Everyday social interactions require the rapid involvement of many neural processes, and no other commonly applied neurosociology methodology can rival the temporal precision of EEG. Indeed, scanning methods intended to track social interaction must be able to measure down to 10–20 milliseconds to assess real-time activity (e.g., speech, emotional expressions, gestures) (Hari & Kujala, 2009). Additionally, modern EEG setups are both mobile (with the advent of wireless EEG caps) and affordable (relative to other neurosociological paradigms). The low cost, mobile setups, and manageable setup time makes larger scale hyperscanning experiments (e.g., more than one dyad) a possibility as well.

Limitations

The primary limitation of EEG analysis is the lack of spatial resolution–that is the exact location of activity in the brain. In line with this, EEG paradigms suffer from the inverse problem. The sheer amount of information coming from the brain results in a nearly infinite number of neural generators creating a particular signal (Cohen, 2014). This limitation can be addressed by (1) higher density arrays (e.g., 64 channels or more) and source localization or (2) focusing on neural generators near the scalp. Having a priori hypotheses about signals from fMRI studies or combined EEG/fMRI studies can also address this (Grech et al., 2008). Nonetheless, these methods still do not allow EEG to rival hemodynamic methods (e.g., fMRI) in spatial resolution.

General Findings and Applications

EEG paradigms have become much more prevalent in the neuroscience field and in the blossoming field of neurosociology. Their low cost, low invasiveness, and mobility make them a very efficient paradigm. The temporal precision of this method also makes it well suited to studying dynamic social interactions, which unfold at very quick paces (Tenhouten et al., 2023).

The flexibility underlying EEG allows it to be applied with a variety of focuses that may not traditionally be thought of as possible with neuroscience methodology. Some examples of this include investigating cultural differences in facial perception (Sui et al., 2009) and education influencing trait violation information about individuals (Varnum et al., 2016). The flexible uses of EEG, such as modeling mirror neurons in relation to group biases (Gutsell & Inzlicht, 2013) or social interaction with the use of hyperscanning (e.g., Liu et al., 2021; Kinreich et al., 2017), cannot be understated. The prevalence of EEG hyperscanning in modeling social interaction scenarios involving guitar playing (Lindenberger et al., 2009), joint movements (Kawasaki et al., 2018), finger movement (Dumas et al., 2010), and real-time emotional experiences (Ding et al., 2018) demonstrate the efficacy of such an approach. In short, EEG is a flexible neurosociology approach that is very effective at interrogating temporally based hypotheses in either single or multi-actor paradigms.

Functional Magnetic Resonance Imagery (fMRI)

Methods

fMRI has long been the mainstay in neuroscience methodology (Logothetis, 2008). This methodology is intended to probe brain function and has excellent spatial resolution. The theory underlying fMRI research is as follows. First, neurons within certain regions of the brain are activated when presented with a stimulus or task. The increase in activity creates demands for more metabolic resources, so to address this demand oxygenated hemoglobin is directed to the area in need. This results in a greater ratio of oxygenated to deoxygenated hemoglobin in that area. Oxygenated and deoxygenated hemoglobin have different magnetic properties and therefore can be identified with magnetic resonance imagery. This is the basis for the blood oxygen level dependent (BOLD) signal used in fMRI methodology. The nature of this assessment is why fMRI requires the use of a very powerful superconductive magnet to provide a strong magnetic field. In the past, 1.5 and 3 Tesla (T) field strengths have predominantly been used, however modern approaches have used ultrahigh magnetic field strengths (7T) that have improved spatial resolution and specificity even more (Vu et al., 2017).

The concentrations of oxygenated and deoxygenated hemoglobin are assessed within small cubes, or voxels, of a predetermined size. The BOLD response is slow, starting 2–5 seconds after stimulus and peaking between 6–9 seconds (Kwong et al., 1992). However, the BOLD response can be measured from brief stimuli (i.e., as low as 0.5 seconds; see Bandettini et al., 1995) and is very consistent between repeated task performances.

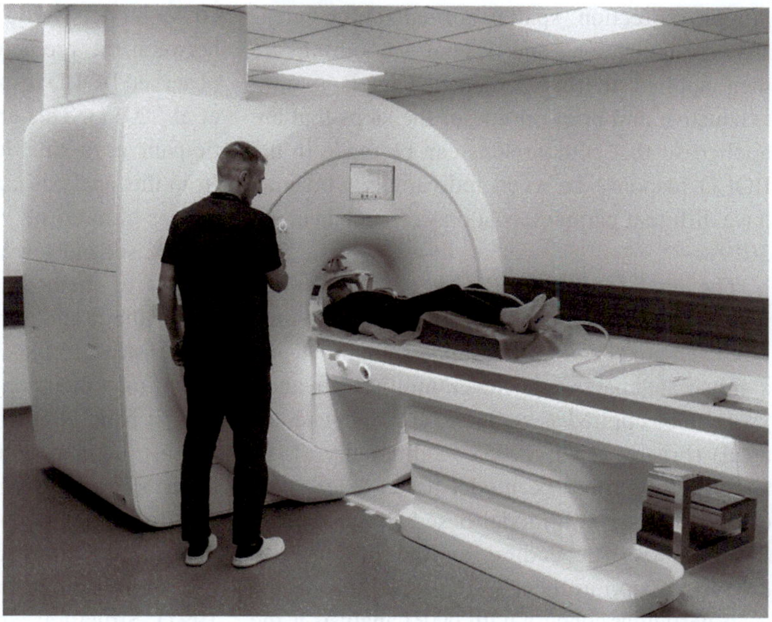

Fig. 6.2 An individual lying down inside an fMRI scanner while a technician looks on

Anatomical scans are typically done at the onset of an fMRI study with functional scans being conducted every 2 seconds following the onset of a task. These "images" or "scans," sometimes known as magnetic resonance (MR) volumes, are the primary target of analyses (Strother, 2006) (Fig. 6.2).

Task-Based fMRI

In task-based fMRI designs, participants are cued to perform a task while BOLD contrast images are collected for a fixed duration. Most task-based fMRI designs collect data from all conditions (e.g., control and experimental) in one sitting due to the variation in the BOLD response at different scans. Typically, experiments are done in a block trial or event-related (ER) trial (Dale, 1999; Miezin et al., 2000). In block trials, events of similar types are grouped together and presented in succession. In ER trials, different events can be presented in close proximity but at pseudorandom intervals. Currently, ER trials are the most used approach with fMRI in the social sciences (Berkman et al., 2014).

Resting State fMRI (rsfMRI)

Resting state fMRI approaches are used to assess networks hypothesized to fluctuate together. These areas are treated as nodes in a network (i.e., Resting State Networks [RSN]) (Greicius et al., 2003; Buckner et al., 2008). Participants in resting state fMRI studies are asked to remain still and avoid actively thinking. While not as commonly applied as task-based designs in neurosociology methodology, resting state fMRI has implications for the study of self-referential processing (SRP) and the ability to relate external stimuli, thoughts, emotions, or memories to oneself (Na et al., 2022). Critically, this gives neurosociologists a way to scientifically interrogate the "self".

Analysis

fMRI analysis has numerous associated software packages (see Behroozi & Daliri, 2012). Similar to EEG, there are several pre-processing steps where the functional scans are cleaned and normalized. Thoroughly describing each step is outside the scope of this chapter, and there have been several papers outlining the procedure (Strother, 2006; Wu et al., 2011; Power et al., 2014). Put briefly, pre-processing starts with a four-dimensional (4D) data set comprised of the different scans. Traditionally, slice timing correction, motion correction, spatial normalization, intensity normalization, and spatial smoothing are then applied to this data set. This is a non-exhaustive and brief outline of the steps that are applied so that comparisons can be made in the BOLD response between individuals who may have different parameters at onset (Esteban et al., 2019).

The goal of fMRI analysis is to determine which voxels are activated. It is common to analyze each voxel's time series independently using a univariate analysis; however, multivariate analyses can be applied as well (and are often encouraged) (Friston et al., 1995). Additionally, fMRI analyses can either be model based (McKeown et al., 1998) or non-model based (Beckmann & Smith, 2004), depending on the presence or absence of a priori theory.

A particular concern with fMRI analysis is the sheer number of statistical tests, as each voxel has one test associated with it (leading to possibly 100,000 tests per experiment). These issues of multiple comparison and the increased probability of Type II error are addressed with the use of voxel-based thresholding in the form of family-wise error rate, false-discovery rate, or cluster-based thresholding (Forman et al., 1995). These techniques are commonly included in analysis software for fMRI.

Empirical fMRI studies commonly analyze group-level outcomes by combining individual-level findings (Chen et al., 2012). This is common practice in other methodologies as well (e.g., EEG, fNIRS).

Task Based fMRI Analysis

Task-based analysis utilizes a procedure known as Statistical Parameter Mapping (SPM), which is based upon the general linear model (GLM) (Poline & Brett, 2012). Every voxel that is associated with the relationship between the stimulus and the function linking neural activity with fMRI activity (e.g., the hemodynamic response function [HRF]) is predicted to have a BOLD response. These predicted BOLD responses for each voxel are then input into a regression model as independent variables with the observed BOLD response as the dependent variable. This GLM model would then show the relationship between the predicted BOLD response and the observed BOLD response at each voxel (Berkman et al., 2014). This analysis is done for each participant as a first level analysis. Beta weights from this analysis can then be compared for conditions of interest.

The second level of fMRI analysis applies a variety of methods used to assess connections across the brain at the group level. These results are sensitive to the choice of analysis and distinguish functional connectivity (undirected associations) from effective connectivity (directed and causal relationships) (Friston, 2011). Psychophysiological Interaction (PPI) assesses how functional connectivity between a region of interest and the rest of the brain is affected by psychophysiological variables (Friston et al., 1997). Structural equation modeling (SEM) has been used to assess effective connectivity when researchers have an a priori model of causality (Büchel & Friston, 1997; Kline, 2011). Still another analysis technique—Granger causality mapping (GCM)—infers causality from time-lagged correlation (Friston et al., 2013).

Additionally, analysts use advanced machine learning approaches known as Multivoxel Pattern Analysis (MVPA), which use pattern-based algorithms to identify different mental states (Premi et al., 2016). Similar techniques (e.g., support vector machines, random forests, and deep learning) (Kuncheva & Rodríguez, 2010; Taspinar & Ozkurt, 2024) have utilized machine learning to extract meaningful patterns from fMRI data. Multivariate pattern analysis (MVPA), for example, utilizes machine learning to study patterns associated with certain stimuli (Yen et al., 2023). Unsurprisingly, then, many state-of-the-art approaches to fMRI analysis (most notably rsfMRI) are rooted in machine learning. What is more, recent approaches to large scale compilation of neuroimaging data and learning methods for machine learning models have continued to advance these techniques (Khosla, 2021). That said, even with these advances there are still concerns with this machine learning approach. Notably, there is no grounded truth for evaluation and subjectivity exists. Thus, care needs to be exhibited when utilizing these methods (Khosla, 2021).

Resting State fMRI Analysis

rsfMRI commonly uses correlational analyses based around a priori established regions of interest (Biswal et al., 1995). This region of interest may be a volume (or even a single voxel), which is compared to the rest of the voxels in the brain. This analysis is very standard, but it does require a priori established regions of interest. Alternative model-based techniques include Amplitude of Low Frequency Fluctuations (ALFF) or fractional ALFF. These techniques, along with regional homogeneity analysis (ReHo), reflect different properties of localized spontaneous activity and are usually used as complementary analyses.

Non-model-based analyses have employed exploratory techniques such as principal components analysis (PCA), independent components analysis (ICA), and clustering (Griffanti et al., 2014; Jutten & Herault, 1991; Mezer et al., 2009). Graph theory has also been applied frequently in rsfMRI (Wang et al., 2010) and in task-based fMRI (Cao et al., 2014). This models the brain as an interconnected series of nodes and looks for functional interaction of brain regions. Graph theory analyses are commonly utilized in functional connectivity analyses, where the understanding is that different regions of the brain are functionally connected and constantly share information with one another (Shahhosseini & Miranda, 2022). These functional connectivity analyses can be utilized to assess the brain as an overarching network and identify network-level changes.

Strengths

fMRI provides indirect measures of neural activation with unrivaled spatial resolution (Eisenberger et al., 2003). It is one of the most popular methods in neuroscience, and research questions involving where signals arise in the brain benefit greatly from fMRI paradigms. It is also a non-invasive method for studying deep brain structures—an advantage not shared by many other techniques.

Limitations

The major limitations of fMRI research are the immobility of the participant, high cost, and low temporal resolution. Participants are required to lay in a stable position during fMRI studies, so studying social factors in a naturalistic way is very difficult. Additionally, increases in the BOLD response are thought to be reflective of increased neural activity (as evidenced by multimodal approaches), but this cannot be confirmed by fMRI (Czeszumski et al., 2020).

Accordingly, there have been several critiques of fMRI approaches in recent years. Amongst others, there have been concerns with "voodoo correlations" that link brain regions to personality traits via inflated statistics (Vul et al., 2009; Lyon, 2017). There has also been literature showing issues with fMRI research where brain activation has been found in deceased animals (e.g., the "dead salmon") (Bennett et al., 2014). These issues highlight the most important part of neuroscience methodology: proper understanding and use of the technology. Many neuroscience critiques are largely due to misuse of the technology or analytical approach—not an inherent issue with the method, itself. In fact, when proper techniques are applied to the previous examples, these "voodoo correlations" and "dead salmon" findings were proven to simply be misapplications of the analysis technique.

General Findings and Applications

fMRI research is one of the most prevalent neuroscience methodologies. The unmatched spatial resolution of fMRI establishes it as a very powerful tool in any neurosocioiologists' toolbelt. Additionally, the ability to assess deep brain structures that are unable to be targeted by alternate methods (e.g., EEG, fNIRS) designates fMRI as an incredibly useful method in many different paradigms.

fMRI research has shown a variety of different uses, including investigating differences in perceptions of ingroup and outgroup members (e.g., Forbes et al., 2012; van Bavel et al., 2008) and predicting others' behaviors (Mitchell et al., 2005). fMRI research has also linked activity in the medial prefrontal cortex (mPFC) to taking the perspective of others (Mitchell et al., 2005; van Overwalle & Baetens, 2009). What is more, fMRI research has connected this research on ingroup-outgroup members and role-taking (Harris & Fiske, 2006). Indeed, when asked to take the perspective of ingroup members, areas of the mPFC used to infer their own attitudes are recruited. When asked to take the perspective of outgroup members, however, areas of the mPFC associated with stereotypes (social semantic information) are recruited. Research has also investigated class-based racial evaluations utilizing fMRI (Firat et al., 2017). This line of research identified two different neural structures underlying these evaluations: the ventromedial prefrontal cortex (vmPFC) for in-group members and the amygdala for social classes perceived as threats.

In sum, fMRI research has been at the forefront of analyzing the relationship between social perception and neural processing substrates. However, due to the limitations of fMRI research (e.g., correlational measures, the participant must be isolated in a large scanner), research into the basis of social interaction is difficult. Moreover, fMRI is a very cost-restricted approach. Nonetheless, as noted above, this research can be, and still is, being done. In short, fMRI is incredibly powerful when accessible, and its prevalence in the neuroscience literature emphasizes its importance. Whenever a neurosociologist hypothesizes about a specific brain area, especially deep neural substrates, fMRI is the optimal choice.

Functional Near-Infrared Spectroscopy (fNIRS)

Methods

fNIRS has seen rapid growth as a neurosociology methodology over the past two decades. Its predominant use has been in neurodevelopment (Pinti et al., 2020) and its strengths, such as higher temporal precision than fMRI and more spatial precision than EEG, have made it a very appealing addition to other domains (TenHouten et al., 2023). In fact, the number of fNIRS papers being published in journals doubled every 3.5 years between 1993 and 2012 (Boas et al., 2014) and research using fNIRS is still seeing rapid growth.

In fNIRS, light emitting diodes are placed on top of the head while affixed to a cap or headband in a pre-established array. These diodes emit light at near-infrared wavelengths. These light waves enter the very top layers of the cortex where they reflect off of (or are absorbed by) oxygenated or deoxygenated hemoglobin molecules. Oxygenated and deoxygenated molecules reflect light waves differently based on their molecular makeup. The reflected light waves then travel back to the scalp where sensors measure them and calculate levels of oxygenated and deoxygenated hemoglobin (Lloyd-Fox et al., 2010).

fNIRS designs offer a similar approach to fMRI with some key differences. As in fMRI, the output is in oxygenated and deoxygenated hemoglobin concentrations. However, subjects do not need to be immobile in a scanner and fNIRS is well suited for mobile data collection due to minimal interference in the presence of movement. Additionally, many modern fNIRS systems are wireless and battery powered. These are usually lightweight and allow for more naturalistic actions (Pinti et al., 2020).

Analysis

There is specialty software associated with fNIRS analysis (e.g., NIRSTORM, Fieldtrip, Homer2 and Homer3) and it is usually done in a standardized workflow. Preprocessing occurs first in which bad channels, or channels with insufficient data, are removed (Sappia et al., 2020). Second, artifacts are identified by either visual inspection or automated processes. These artifacts are then either removed or interpolated (Scholkmann et al., 2010). The data is then filtered such that the frequency of the hemodynamic response of interest is kept while other frequencies are removed. Different procedures can be applied here such as bandpass filtering (Pinti et al., 2018) or principal components analysis (Virtanen et al., 2007). Data may then be segmented into times of interest and baseline corrected, if appropriate. Finally, the optical data is converted into relative changes in oxygenated and deoxygenated hemoglobin using the modified Lambert-Beer Law (Baker et al., 2014). Simply put, this analysis relates changes in light transmission to tissue absorption.

The data is then averaged as determined by the research question, traditionally across trials per condition and then over subjects. Data is usually presented in change of oxygenated and deoxygenated hemoglobin over time. However, modern fNIRS toolboxes can also estimate changes in certain brain regions given their good array coverage of optic diodes.

Many of the same statistical analysis techniques applied in fMRI analyses have been used in fNIRS as well. The statistical parametric mapping approach (Friston et al., 1994) based on the GLM fits the fNIRS data with a combination of regressors based on a priori theory. Other data driven approaches such as independent components analysis (Patel et al., 2011) and principal components analysis (Tanaka et al., 2014) have also been used to analyze fNIRS data. Multivoxel Pattern Analysis (MVPA), commonly applied in fMRI, has also been applied to fNIRS data (Emberson et al., 2017). Of course, much like the other paradigms discussed in this chapter, much of the fNIRS analyses are model based, although there are options for naturalistic approaches when no a priori model is needed.

Still, there are some analyses specific to fNIRS. A modern algorithm known as Automatic Identification of Functional Events (AIDE; see Pinti et al., 2017) analyzes pre-processed fNIRS data and uses patterns in the data to identify events of interest. These events can then be linked back to behavior performed by the participant. While still relatively new, this analysis technique offers more flexibility in fNIRS analysis in naturalistic environments.

Strengths

Two technical reviews of fNIRS literature (Scholkmann et al., 2014; Torricelli et al., 2014) provide detailed strengths and limitations of the method, but we will summarize the main points here. fNIRS offers a mobile, non-invasive, cost-effective neurosociology method that still offers significant spatial resolution.

One of the most cited strengths of fNIRS is the tolerance of motion artifacts (Herold et al., 2017). Due to this strength and the non-invasiveness of the procedure, it can be applied in naturalistic environments. This allows assessment during tasks that require movement, such as walking or even talking. It is also relatively light (until many probe holders are used) and comfortable, which is why it is often used with children (Lloyd-Fox et al., 2010). Finally, fNIRS can assess the hemodynamic response in cortical neural regions of interest at a fraction of the cost of fMRI while still offering more ecological validity.

Limitations

One of the largest concerns in the fNIRS community is the lack of standardization of the preprocessing and analysis procedures (Hocke et al., 2018). The other two primary methods discussed in this chapter (EEG and fMRI) both have longstanding standardized procedures, and the lack of this standardization for fNIRS can cause interpretation confusion and issues of reliability (e.g., lack of reproducibility in results). Another limitation of fNIRS is the shallow penetration of the sensors, which, due to the nature of light penetration, cannot interrogate deep structures of the brain. Finally, as with fMRI, the fNIRS hemodynamic response acts as a proxy for neural activation at the cortical surface and does not directly measure neural activation (Pinti et al., 2020).

General Findings and Application

Extensive reviews have been done on fNIRS findings (e.g., Lloyd-Fox et al., 2010; Boas et al., 2014; Pinti et al., 2020). Overall, fNIRS has been applied to a variety of research domains, including the social sciences. The naturalistic setting allows for very flexible research topics such as emotional perception (Ding et al., 2014), facial perception (Balconi et al., 2015), affective touch (Bennett et al., 2014), and imitation behavior (Oliver et al., 2018). fNIRS has also been used extensively to contribute to our understanding of the social brain in studies with infants and children. These studies have found noteworthy connections between infants and mothers in social communication (Minagawa-Kawai et al., 2009), and differences in both social situations across cultures (Lloyd-Fox et al., 2014) and human action processing (Lloyd-Fox et al., 2015). Cheng et al. (2015) also used fNIRS hyperscanning to show task-related coherence in cooperation, especially in opposite sex partners. Burns and Lieberman (2020) highlight other important contributions of fNIRS into affective and social neuroscience, including research into social evaluation, social exclusion, and competition (Balconi & Vanutelli, 2016; Rosenbaum et al., 2018, 2018; Ruocco et al., 2010). They also highlight the importance of fNIRS in social interaction research, emphasizing the flexibility of the approach in assessing social interaction (e.g., Feldman et al., 2011; Schmidt et al., 2012).

fNIRS is a very powerful approach that doesn't have the temporal precision of EEG or the spatial precision of fMRI. However, this "middle of the road" method offers much better spatial precision than EEG and better temporal precision than fMRI. It is also very flexible and offers a great amount of ecological validity due to the ease of use and minimal invasiveness. In short, fNIRS should be used when spatial resolution is needed in a naturalistic setting (such as dynamic conversation). This is especially true if the hypothesized neural substrates are cortical, or near the surface of the brain, as fNIRS excels in that area.

Other Methods

Transcranial Magnetic Stimulation (TMS)

TMS is a non-invasive method using magnetic stimulation to create interference in brain regions near the scalp (Walsh & Cowey, 2000). In social science research, TMS is commonly used to create interference at the medial and lateral frontal cortex, which are associated with higher-order social processing (Kwan et al., 2007). Knoch et al. (2009), for example, found that disruption to the lateral frontal cortex decreased respective behavior in situations that call for it.

TMS paradigms are applied in cause-and-effect relationships where certain brain regions are associated with behavioral responses. Generally, lower spatial and temporal resolution hinder this technique, so it is frequently used in conjunction with EEG and fMRI (Ruff et al., 2009; Thut & Miniussi, 2009). Creating interference in brain regions allows us to see what would happen when a brain region is not working, and therefore establish a causal hypothesis. Correlational methods such as EEG and fMRI are unable to establish similar causal effects so using TMS in conjunction with EEG and fMRI is very effective.

Psychophysiological Measurements

While not acting as precise indices of neural functioning, psychophysiological measures still have significant contributions in neurosociology research. Physiological arousal is an effective measure of stress (Zellars et al., 2009) and fluctuates throughout the day. These fluctuations are generally associated with parasympatehetic and sympathetic activity in the autonomic nervous system. Such fluctuations can be measured by utilizing an electrocardiogram for measuring the heart (ECG) or an electromyogram for measuring skeletal muscles (EMG).

Common measurements include galvanic skin response (GSR), heart rate (HR), heart rate variability (HRV), blood pressure, breathing rate, and temperature (Dalmeida & Masala, 2021). Galvanic skin response (GSR) (Boucsein, 2012) is a measure of sympathetic activity while heart rate (HR) is predominantly used as a measure of the parasympathetic nervous system (Mendelowitz, 1999). Heart rate variability (HRV) is the variation in time interval between consecutive heart beats. These physiological measures are not direct measures of neural functioning, but they can and do act as proxy measurements of physiological arousal and are used in conjunction with traditional neurosociology measures (Park et al., 2022). Additionally, these methods are commonly non-invasive and portable which significantly contributes to improving the ecological validity of experimental designs utilizing them.

Hyperscanning

For some time, contributions from neuroscience to the field of sociology have been met with hesitancy. This is because it was not initially clear to sociologists how measuring neural activity could give any insight into social interaction and the organization of the social world. However, with time, researchers began to utilize traditional neuroscience paradigms that focused on the individual and how they perceived/reacted to the (social) world (Adolphs, 2010). Hypotheses investigating social perception, social cognition, and social regulation were therefore interrogated. Such contributions began to demonstrate how researchers' incorporation of neuroscience methodologies permitted a deeper understanding of sociological phenomenon. This was further illustrated in 2002 with the first successful "hyperscan" by Montague et al. (2002). By connecting two functional magnetic resonance imagery (fMRI) scanners and allowing communication over the Internet, Montague et al. (2002) successfully demonstrated the ability to collect neural data synchronously from two individuals during a social interaction task (i.e., hyperscanning). Now, over two decades later, hyperscanning has opened new avenues and outlets to bridge sociology and neuroscience. While the more traditional single participant neuroscience methods still exist, hyperscanning is more prevalent than ever given its capacity to simultaneously record the neural activity of multiple actors—something both sociologists and neuroscientists never thought possible (Babiloni & Astolfi, 2014). Hyperscanning, therefore, gives researchers a much more naturalistic way to investigate interpersonal relationships in a neurosociological framework. Accordingly, the importance of this method cannot be understated, which is why we address it in detail. We do so for each approach described in this chapter, as not all hyperscanning methods are created equal (i.e., they each have their own strengths/weaknesses).

Of note, while hyperscanning has revolutionized neurosociologists' approach to the study of social interaction (i.e., given its unique capacity to provide insight into the neural basis for dynamic social interaction), its use does not come without disadvantages. Hyperscanning is a complex method that frequently requires significant coordination and the synchronization of task measures. Moreover, consideration must also be paid to the fact that simultaneous activation may arise from other factors, such as a response to a common input (e.g., hearing the same words or seeing the same images) or the co-occurrence of physiological responses (e.g., breathing) (Hasson et al., 2012). Put simply, other factors can produce synchronization—not just social interaction. Finally, hyperscanning is not always necessary, as neurosociological questions can be answered by employing other, non-hyperscanning approaches, such as those described above.

EEG Hyperscanning

Hyperscanning is applied in many neurosociology settings (see fMRI, fNIRS sections), but it is most commonly applied in EEG paradigms (Nam et al., 2020). fMRI and fNIRS-based hyperscanning have the limitations of low temporal resolution due to the basis of hemodynamic analysis, whereas EEG has excellent temporal resolution (Koike et al., 2015). Social interactions that unfold at a fast pace require a method that is sensitive to it, and as such, hyperscanning EEG paradigms fit well. Different hyperscanning techniques are still used, however, and their applications are outlined and described below. A complete description of these techniques and their pitfalls is extensive (see Bastos & Schoffelen, 2016 for a review), so this chapter will highlight only the essentials.

In a traditional EEG hyperscanning paradigm two or more participants undergo EEG measurement as the same time. Their EEG recordings are time-locked, such that researchers are able to know precisely when something happens to each participant. There are several different analytical techniques within EEG hyperscanning approaches. One such approach utilizes an assessment of phase synchrony. Phase synchrony is the amount of overlap in the oscillation of two different signals. Measures of phase synchrony are computed from an estimate of the amplitude and oscillations for a specific frequency and a set of observations. One widely used method to assess this is with a coherence coefficient, which provides the amount of variance in one signal that is explained by the other (Bastos & Schoffelen, 2016). The assessment of phase synchrony provides researchers with a quantifiable measure of how synchronized two different brains are.

Hyperscanning designs have also adopted connectivity measures from single brain studies to assess inter-brain coupling. Adapted from single brain studies, two such connectivity measures are the phase-locking value (PLV) and phase-lag index (PLI). PLV is a measure of how synchronized two signals are (from 0 to 1) from different sources (in the case of hyperscanning—different brains). It is calculated by applying the computation of coherence (discussed above) to amplitude normalized Fourier transformed signals (Lachaux et al., 1999). PLV has been used to assess a variety of different tasks, including implicit coordination (Yun et al., 2012). PLI is a similar measure to PLV, although it evaluates the distribution of phase differences across observations (Stam et al., 2007). It can also be applied when assessing connectivity between source brains (Lindenberger et al., 2009; Sänger et al., 2012).

Connectivity measures all assess the synchrony between two neural signals originating from different brains when applied in a hyperscanning paradigm. Notably, the results of the different methods outlined above (PLV, PLI, etc.) are all referred to as inter-brain synchrony. Accordingly, it is critically important that researchers operationalize any measure of inter-brain synchrony in order to avoid confusion. Additionally, neurosociologitsts should exhibit care when interpreting synchrony results.

Correlations, or similarities, between different signals, can also be quantified in hyperscanning approaches. Correlation analyses performed between different aspects of brain waves can be used to assess a variety of sociological questions. For example, Liu et al. (2021) showed higher correlation between delta and theta bands during cooperation and lower correlation between alpha and beta during competition. Kinreich et al. (2017), meanwhile, showed higher correlation patterns in gamma waves between individuals when they interacted with partners as opposed to strangers.

EEG researchers have also applied data driven approaches to interrogate the complexity of inter-brain networks (Ciaramidaro et al., 2018). Modeling network connectivity using graph theory, for example, has been applied to leader/follower tasks (Sänger et al., 2013) and modulation of processes in dyadic interaction tasks (De Vico Fallani et al., 2010a).

Finally, information flow analysis focuses on the causal link between brains. Accordingly, it is best suited for when information flow and causal links are the intended quantitative assessment. This type of analysis often employs either partially directed coherence (Baccalá & Sameshima,

2001) or Granger causality (Granger, 1969; Schippers et al., 2010), which have been used to identify increases in causal links between brains in decision making tasks on defection (De Vico Fallani et al., 2010b) and punishment (Ciaramidaro et al., 2018).

fMRI Hyperscanning

The limitations of fMRI research (immobile subjects and high costs) cause difficulties with hyperscanning. This includes how interaction between individuals is hindered by their placement in separate scanners. While research has attempted to address this through the use of a dual-head coil (Renvall et al., 2020), issues still remain. That being said, the first successful hyperscanning setup was an fMRI study (Montague et al., 2002) using two scanners; and feasibility studies using scanners in distal locations have supported the efficacy of this methodology (King-Casas et al., 2005). What is more, new methodological approaches using online interactions have significantly improved this methodological approach.

Another limitation of fMRI research, low temporal resolution, also causes issues with hyperscanning. As previously discussed, the BOLD signal recorded by fMRI has a very low temporal frequency, whereas social interaction usually occurs at higher temporal frequencies (Misaki et al., 2021). Therefore, the correlation of BOLD signals likely does not reflect the same measures (phase association of oscillations) as EEG hyperscanning measures. Nonetheless, even with these considerations, fMRI hyperscanning paradigms are still effective methodologies in neurosociology, as evidenced by their application in numerous studies. Indeed, several studies have investigated social interaction and organization with fMRI (Spiegelhalder et al., 2014; Koike et al., 2019). For example, researchers have investigated social cooperation during a maze task (Krill & Platek, 2012) and social interactions during a game emphasizing cooperation in pattern building (Špiláková et al., 2019). Shaw et al. (2018) have also applied this paradigm to the ultimatum game, a cooperative game with a monetary payout.

Hyperscanning analysis of the BOLD response can be done with linear dependance analysis (Czeszumski et al., 2020), where regression coefficients associated with the BOLD response are used in correlational analyses. Intersubject correlation analysis (ISC) (Nastase et al., 2019), which correlates the time series of identified voxels in both participants, can be done as well. So, too, can higher level causal modeling in the form of Granger causality mapping—a technique also used in EEG analysis (Schippers et al., 2010).

In sum, hyperscanning in fMRI research has many advantages. Its full brain coverage and high spatial resolution is unrivaled. Moreover, the capacity for fMRI to assess brain areas associated with emotion—a component linked intrinsically with conversation and primarily found in the medial frontal lobes and limbic systems—is something that other methodologies simply cannot accomplish (i.e., due to the depth of these brain structures) (Robinson et al., 2009). These advantages, however, can be overshadowed by fMRI's low temporal resolution, which limits our capacity to conduct research on coupling and connectivity (Czeszumski et al., 2020).

fNIRS Hyperscanning

fNIRS offers a comparable hyperscanning design to EEG with better spatial precision and worse temporal precision. Wireless fNIRS recordings allow for more naturalistic experimental paradigms and the robust nature in the presence of movement artifacts significantly reduces the signal-to-noise ratio. Put another way, the primary strength of fNIRS is its ability to map neural activity at a structural level while still permitting natural interactions (Pinti et al., 2020)—something that is not possible in isolated EEG or fMRI hyperscanning techniques. What is more, its relative ease of use allows for even larger sets of synchronized participants such as triads (Jiang et al., 2015) or tetrads (Nozawa et al., 2016).

Early fNIRS hyperscanning studies (Funane et al., 2011) used simple tasks to prove the efficacy of the paradigm. Since then, fNIRS have been employed in a series of social interaction tasks (e.g., Scholkmann et al., 2013). Cui et al. (2012), for example, used fNIRS hyperscanning to assess cooperation in dyads. Interestingly, they found increased coherence in right superior frontal cortices in cooperation trials only (i.e., this was not evident in competition trials).

Wavelet transform coherence (WTC) is the most common hyperscanning analysis technique in fNIRS. It assesses the coherence of two signals and has been used to measure inter-brain synchrony in cooperative and competitive behaviors (Cui et al., 2012; Osaka et al., 2014; Wang et al., 2019), verbal communication (Jiang et al., 2015), non-verbal communication (Osaka et al., 2015), and imitation (Holper et al., 2012), amongst others. Much like EEG and fMRI, correlation analyses with fNIRS signals can be utilized in fNIRS hyperscanning (Funane et al., 2011; Liu et al., 2016).

In conclusion, fNIRS hyperscanning offers comparable ease of use to EEG while offering better spatial resolution. Moreover, if the social processes intended to be studied in a hyperscanning design are linked to a specific cortical area, then fNIRS is the optimal choice (Koike et al., 2015)—especially since it permits data collection during moment-to-moment interactions and in a natural context.

Multimodal Approaches

Each methodology reviewed in this chapter has its strengths and limitations. Therefore, caution should always be used when analyzing or interpreting different methods. There is a recognized benefit of using a multimodal approach, which means using multiple methods are the same time. The weaknesses associated with different methods (e.g. spatial precision with EEG or temporal precision with fMRI) can create difficulties for researchers. Multimodal approaches help to address these weaknesses by utilizing alternative techniques which do not share them. Studies using an EEG-fMRI approach have been utilized to investigate the brain at multiple spatial and temporal scales simultaneously (Chang & Chen, 2021). EEG-fNIRS measurements have also been collected (Chen et al., 2016). In fact, the ease of implementing fNIRS with EEG has led to many systems with both sensors in place in one array, which allows for concurrent data collection from both. Coincidentally, this approach has seen significant interest in recent years.

Conclusions

The field of neurosociology, while old by name (Tenhouten & Kaplan, 1973), is still developing, and sociologists are in a great position to benefit. Sociology has only just begun to employ neuroscience methods, and the results have already been positive. To date, sociologists have gained incredible insight and strengthened interdisciplinary bridges by employing neuroscience methods such as EEG and fMRI to investigate a multitude of social phenomena, including culture, discrimination, emotion, empathy, identity, morality, race, relationships, power, social class, and status processes (just to name a few) (Firat, 2021; Firat et al., 2017; Franks, 2014; Kalkhoff et al., 2016a, b, 2020).

Sociological explanations of behavior and social organization are tied directly to the methods used to collect information on these phenomena. Neuroscientific methods allow measurements of both conscious and unconscious thought and emotional processes in real-time without being subject to possible response bias (Kalkhoff et al., 2016a; Laureiro-Martinez et al., 2015). This allows for the advancement of traditional theory and offers new angles on old problems—two of the three clear benefits identified by Kalkhoff et al. (2016a) in using neuroscience methodology to supplement sociological theories. M any sociological theories have implicit assumptions about the brain and the mind that can now be tested via the use of neuroscientific methods. Examples of this approach have been seen in recent years with regards to how the brain represents and supports identity processes (Kalkhoff et al., 2016b) and how relationships develop via relational cohesion (Kalkhoff et al., 2016a). The third clear benefit identified by Kalkhoff et al. (2016a) is practice and translational science. The goal of translational science is to take basic research and apply it to promote human well-being (Kalkhoff et al., 2016a). This goal is strengthened through the development and testing of more precise and comprehensive theories (Turner, 2001). Neuroscience methodologies, specifically, aid in the achievement of this goal by promoting the development of more comprehensive sociological theories (that account for both conscious and unconscious phenomena), which, in turn, allow for better translational science. Considering the need to facilitate translational science is more paramount than ever, we believe the advancement of this goal is especially noteworthy.

Given that neuroscience methods supplement traditional sociological methods by filling important methodological gaps (e.g., non-reliance on self-report data) and can be applied to a variety of social phenomena due to their versatility, we believe this is just the beginning. We hope this chapter contributes to furthering such research by (a) emphasizing the symbiotic relationship between sociology and neuroscience, (b) detailing the strengths and limitations of several neuroscience methods, and (c) offering some guidance into method choice and application.

For this hope to come to fruition, it is important to dispel any myth that neuroscience methods will replace traditional sociological methods. The simple truth is that the introduction of neuroscience methods into sociology was never intended to supplant traditional sociological methods. The goal has always been to supplement them so that neurosociologists can gain valuable insights into human behavior. With this in mind, we do offer a bit of advice—to use the best and most powerful neuroscience methods alongside sociological theories that have persisted in the face of rigorous empirical testing. This is because newer, less empirically tested sociological theories may not have the foundation necessary for the introduction of neurological methods. So, while neuroscience methods may allow for novel approaches to assessing implicit biases, circumventing issues with self-report, or even interrogating sensitive topics such as prejudice, these advantages mean little when used in conjunction with minimally supported sociological theories. This, in combination with how the research question drives methodological selection, emphasizes that the incorporation of neuroscience methods into one's research is not always appropriate. Put simply, just because the methods are there does not always mean that they are applicable. That is, simply because one can, does not mean that they should (Willingham & Dunn, 2003).

There is also the discussion of neuroscience methodology limitations. Whilst this was discussed with regards to specific methods earlier in this chapter, there are specific concerns with general neuroscience approaches that have been addressed in recent years. The first of these concerns is the concept of a non-equitable neuroscience. Many neuroscience methods are not designed to handle phenotype variability (Webb et al., 2022). For example, EEG systems are not designed with thick, coarse hair in mind. This data has been identified as unusable due to high signal-to-noise ratios and therefore has resulted in the exclusion of black participants at much higher rates (Etienne et al., 2020). Choy et al. (2022) reviewed 81 EEG studies and found only 5 of them had black participants. Other neuroscience methodologies are not immune to this either. fMRI scanners are also not designed to handle big, afro-textured hair or sew-in hair extensions with metal tracks (La Scala et al., 2023). fNIRS systems have a similar problem, as their reliance on the known properties of scattering and the absorption of light in skin tissue introduces racial bias when collecting data from darker skin toned individuals (Webb et al., 2022). What is more, biased assumptions in neuroscience research have led to a much larger focus on Western, educated, industrialized, rich, and democratic populations (WEIRD; Henrich et al., 2010). Unsurprisingly, this has been connected to a lack of reproducibility in other samples (Webb et al., 2022). Accordingly, recent approaches, including novel EEG solutions, are attempting to address these biases and establish a more equitable neuroscience (Webb et al., 2022; La Scala et al., 2023). Still, issues persist.

Another note of limitation is that neuroscientific approaches are largely correlational in nature. The noted exceptions are manipulation (e.g., TMS or pharmacological agents) or lesion studies. In both cases, functioning of specific brain areas is halted or altered (in manipulation studies this is temporary and reversible). These approaches stand alone in establishing the necessity of a brain area for a specific function. Doing so allows for true causation to be established—something that is not possible with the other methods described above (e.g., EEG, fNIRS, fMRI). Lesion studies are usually done on someone who has incurred brain damage due to an external cause, damage, or specific disease (Beadle & Tranel, 2011). Neurosociology research has utilized lesion approaches to investigate the relationship between racial group members and neural responses (Firat et al., 2017) and has identified two distinct neural mechanisms that underlie class-based racial evaluations. Whilst these approaches can establish causation (i.e., by linking brain areas to specific functions), others cannot, so care must be exhibited in the interpretation of correlational neuroscience approaches.

Finally, researchers share concerns about the traditionally underpowered nature of neuroscience methods. In direct contrast to traditional sociological approaches, neuroscience approaches are limited in their sample sizes due to costs and the time needed for data collection. While these concerns persist, neuroscience techniques have accurate measurements of power and effect size. They also employ good practice on both analysis and recording, which reduces error (Nichols et al., 2017). Moreover, most communities using neuroscience techniques have adopted an "open science" approach to sharing of both data and analysis code, which greatly enhances the validity and reliability of these techniques (Nichols et al., 2017).

With that in mind, the biggest and final piece of advice that we can offer in pursuit of our goal of encouraging sociologists to employ neuroscience methods (*if applicable*) is to think of how these methods can supplement your existing research questions. Research questions drive methodology, and neuroscience methods are no exception. The versatility of neuroscience methods, however, makes this addition to traditional sociological methods much easier. Recall, each of the aforementioned methods are suited for different means, whether that be temporal (EEG), spatial (i.e., knowing what part of the brain activity occurs in following social stimuli) (fMRI), or a combination of both (fNIRS). This, in our opinion, is where the nature of neurosociology really shines—specifically, in the form of a collaborative effort of sociological knowledge and neuroscience methodology.

In conclusion, this chapter provided an overview of some of the most frequently used neuroscience methods and how they work. We have highlighted their strengths and their limitations as well as offered advice on best-use scenarios. Again, we always recommend caution in both the application of neuroscience methodologies and their interpretation. The benefits of tying sociology and neuroscience—two incredibly strong disciplines—together cannot be overlooked. This is why sociologists must capitalize on the strengths of both to gain a greater understanding of human behavior and promote growth in both fields. We are hopeful that this summary of neuroscience methods provides an outline to aspiring neurosociologists and offers guidance to accomplish just that.

References

Adolphs, R. (2010). Conceptual challenges and directions for social neuroscience. *Neuron, 65*(6), 752–767.

Babiloni, F., & Astolfi, L. (2014). Social neuroscience and hyperscanning techniques: Past, present and future. *Neuroscience & Biobehavioral Reviews, 44*, 76–93.

Baccalá, L. A., & Sameshima, K. (2001). Partial directed coherence: A new concept in neural structure determination. *Biological Cybernetics, 84*(6), 463–474.

Baker, W. B., Parthasarathy, A. B., Busch, D. R., Mesquita, R. C., Greenberg, J. H., & Yodh, A. G. (2014). Modified Beer-Lambert law for blood flow. *Biomedical Optics Express, 5*(11), 4053–4075.

Balas, B., & Nelson, C. A. (2010). The role of face shape and pigmentation in other-race face perception: An electrophysiological study. *Neuropsychologia, 48*(2), 498–506.

Balconi, M., Cassioli, F., Fronda, G., & Vanutelli, M. E. (2019). Cooperative leadership in hyperscanning. Brain and body synchrony during manager-employee interactions. *Neuropsychological Trends, 2019*(26), 23–44.

Balconi, M., Grippa, E., & Vanutelli, M. E. (2015). What hemodynamic (fNIRS), electrophysiological (EEG) and autonomic integrated measures can tell us about emotional processing. *Brain and Cognition, 95*, 67–76.

Balconi, M., & Vanutelli, M. E. (2016). Competition in the brain. The contribution of EEG and fNIRS modulation and personality effects in social ranking. *Frontiers in Psychology, 7*, 212981.

Bandettini, P. A., Binder, J. R., DeYoe, E. A., Rao, S. M., Jesmanowicz, A., Hammeke, T. A., et al. (1995). Functional MR imaging using the BOLD approach: Applications. *Diffusion and Perfusion Magnetic Resonance Imaging*, 335–349.

Başar, E. (2012). A review of alpha activity in integrative brain function: Fundamental physiology, sensory coding, cognition and pathology. *International Journal of Psychophysiology, 86*(1), 1–24.

Bastos, A. M., & Schoffelen, J. M. (2016). A tutorial review of functional connectivity analysis methods and their interpretational pitfalls. *Frontiers in Systems Neuroscience, 9*, 175.

Beadle, J., & Tranel, D. (2011). Social neuroscience: A neuropsychological perspective. In *The Oxford handbook of social neuroscience* (pp. 49–68).

Beckmann, C. F., & Smith, S. M. (2004). Probabilistic independent component analysis for functional magnetic resonance imaging. *IEEE Transactions on Medical Imaging, 23*(2), 137–152.

Behroozi, M., & Daliri, M. R. (2012). Software tools for the analysis of functional magnetic resonance imaging. *Basic and Clinical Neuroscience, 3*(5), 71–83.

Bennett, R. H., Bolling, D. Z., Anderson, L. C., Pelphrey, K. A., & Kaiser, M. D. (2014). fNIRS detects temporal lobe response to affective touch. *Social Cognitive and Affective Neuroscience, 9*(4), 470–476.

Bentin, S., Allison, T., Puce, A., Perez, E., & McCarthy, G. (1996). Electrophysiological studies of face perception in humans. *Journal of Cognitive Neuroscience, 8*(6), 551–565.

Berkman, E. T., Cunningham, W. A., & Lieberman, M. D. (2014). Research methods in social and affective neuroscience. In *Handbook of research methods in social and personality psychology* (pp. 123–158).

Biehl, S. C., Ehlis, A. C., Müller, L. D., Niklaus, A., Pauli, P., & Herrmann, M. J. (2013). The impact of task relevance and degree of distraction on stimulus processing. *BMC Neuroscience, 14*, 1–11.

Biswal, B., Zerrin Yetkin, F., Haughton, V. M., & Hyde, J. S. (1995). Functional connectivity in the motor cortex of resting human brain using echo-planar MRI. *Magnetic Resonance in Medicine, 34*(4), 537–541.

Blackwood, D. H. R., & Muir, W. J. (1990). Cognitive brain potentials and their application. *The British Journal of Psychiatry, 157*(S9), 96–101.

Boas, D. A., Elwell, C. E., Ferrari, M., & Taga, G. (2014). Twenty years of functional near-infrared spectroscopy: Introduction for the special issue. *Neuroimage, 85*, 1–5.

Boucsein, W. (2012). *Electrodermal activity*. Springer Science & Business Media.

Büchel, C., & Friston, K. J. (1997). Modulation of connectivity in visual pathways by attention: Cortical interactions evaluated with structural equation modelling and fMRI. *Cerebral Cortex (New York, NY: 1991), 7*(8), 768–778.

Buckner, R. L., Andrews-Hanna, J. R., & Schacter, D. L. (2008). The brain's default network: Anatomy, function, and relevance to disease. *Annals of the New York Academy of Sciences, 1124*(1), 1–38.

Burns, S. M., & Lieberman, M. D. (2020). The use of functional near infrared spectroscopy (fNIRS) for unique contributions to social and affective neuroscience. *Preprint*.

Cao, H., Plichta, M. M., Schäfer, A., Haddad, L., Grimm, O., Schneider, M., et al. (2014). Test–retest reliability of fMRI-based graph theoretical properties during working memory, emotion processing, and resting state. *Neuroimage, 84*, 888–900.

Casson, A. J. (2019). Wearable EEG and beyond. *Biomedical Engineering Letters, 9*(1), 53–71.

Chang, C., & Chen, J. E. (2021). Multimodal EEG-fMRI: advancing insight into large-scale human brain dynamics. *Current Opinion in Biomedical Engineering, 18*, 100279.

Chen, G., Saad, Z. S., Nath, A. R., Beauchamp, M. S., & Cox, R. W. (2012). FMRI group analysis combining effect estimates and their variances. *Neuroimage, 60*(1), 747–765.

Chen, L. C., Sandmann, P., Thorne, J. D., Bleichner, M. G., & Debener, S. (2016). Cross-modal functional reorganization of visual and auditory cortex in adult cochlear implant users identified with fNIRS. *Neural Plasticity, 2016*, 4382656.

Cheng, X., Li, X., & Hu, Y. (2015). Synchronous brain activity during cooperative exchange depends on gender of partner: A fNIRS-based hyperscanning study. *Human Brain Mapping, 36*(6), 2039–2048.

Choy, T., Baker, E., & Stavropoulos, K. (2022). Systemic racism in EEG research: Considerations and potential solutions. *Affective Science, 3*(1), 14–20.

Ciaramidaro, A., Toppi, J., Casper, C., Freitag, C. M., Siniatchkin, M., & Astolfi, L. (2018). Multiple-brain connectivity during third party punishment: An EEG hyperscanning study. *Scientific Reports, 8*(1), 6822.

Cohen, M. X. (2014). *Analyzing neural time series data: Theory and practice*. MIT Press.

Colgin, L. L. (2013). Mechanisms and functions of theta rhythms. *Annual Review of Neuroscience, 36*, 295–312.

Cooley, C. H. (1902). *Human nature and the social order*. Charles Scribner's Sons.

Cristofori, I., Moretti, L., Harquel, S., Posada, A., Deiana, G., Isnard, J., et al. (2013). Theta signal as the neural signature of social exclusion. *Cerebral Cortex, 23*(10), 2437–2447.

Cui, X., Bryant, D. M., & Reiss, A. L. (2012). NIRS-based hyperscanning reveals increased interpersonal coherence in superior frontal cortex during cooperation. *Neuroimage, 59*(3), 2430–2437.

Cunningham, W. A., Van Bavel, J. J., Arbuckle, N. L., Packer, D. J., & Waggoner, A. S. (2012). Rapid social perception is flexible: Approach and avoidance motivational states shape P100 responses to other-race faces. *Frontiers in Human Neuroscience, 6*, 140.

Czeszumski, A., Eustergerling, S., Lang, A., Menrath, D., Gerstenberger, M., Schuberth, S., et al. (2020). Hyperscanning: A valid method to study neural inter-brain underpinnings of social interaction. *Frontiers in Human Neuroscience, 14*, 39.

Dale, A. M. (1999). Optimal experimental design for event-related fMRI. *Human Brain Mapping, 8*(2–3), 109–114.

Dalmeida, K. M., & Masala, G. L. (2021). HRV features as viable physiological markers for stress detection using wearable devices. *Sensors, 21*(8), 2873.

Davis, C. E., Hauf, J. D., Wu, D. Q., & Everhart, D. E. (2011). Brain function with complex decision making using electroencephalography. *International Journal of Psychophysiology, 79*(2), 175–183.

De Vico Fallani, F., Costa, L. D. F., Rodriguez, F. A., Astolfi, L., Vecchiato, G., Toppi, J., et al. (2010a). A graph-theoretical approach in brain functional networks. Possible implications in EEG studies. In *Nonlinear biomedical physics* (Vol. 4, pp. 1–13). BioMed Central.

De Vico Fallani, F., Nicosia, V., Sinatra, R., Astolfi, L., Cincotti, F., Mattia, D., et al. (2010b). Defecting or not defecting: How to "read" human behavior during cooperative games by EEG measurements. *PLoS One, 5*(12), e14187.

Delorme, A., & Makeig, S. (2004). EEGLAB: An open source toolbox for analysis of single-trial EEG dynamics including independent component analysis. *Journal of Neuroscience Methods, 134*(1), 9–21.

Ding, X. P., Fu, G., & Lee, K. (2014). Neural correlates of own-and other-race face recognition in children: A functional near-infrared spectroscopy study. *Neuroimage, 85*, 335–344.

Ding, Y., Hu, X., Xia, Z., Liu, Y. J., & Zhang, D. (2018). Inter-brain EEG feature extraction and analysis for continuous implicit emotion tagging during video watching. *IEEE Transactions on Affective Computing, 12*(1), 92–102.

Dumas, G., Nadel, J., Soussignan, R., Martinerie, J., & Garnero, L. (2010). Inter-brain synchronization during social interaction. *PLoS One, 5*(8), e12166.

Eisenberger, N. I., Lieberman, M. D., & Williams, K. D. (2003). Does rejection hurt? An fMRI study of social exclusion. *Science, 302*(5643), 290–292.

Emberson, L. L., Zinszer, B. D., Raizada, R. D., & Aslin, R. N. (2017). Decoding the infant mind: Multivariate pattern analysis (MVPA) using fNIRS. *PLoS One, 12*(4), e0172500.

Esteban, O., Markiewicz, C. J., Blair, R. W., Moodie, C. A., Isik, A. I., Erramuzpe, A., et al. (2019). fMRIPrep: A robust preprocessing pipeline for functional MRI. *Nature Methods, 16*(1), 111–116.

Etienne, A., Laroia, T., Weigle, H., Afelin, A., Kelly, S. K., Krishnan, A., & Grover, P. (2020). Novel electrodes for reliable EEG recordings on coarse and curly hair. In *2020 42nd annual international conference of the IEEE engineering in medicine & biology society (EMBC)* (pp. 6151–6154). IEEE.

Feldman, R., Magori-Cohen, R., Galili, G., Singer, M., & Louzoun, Y. (2011). Mother and infant coordinate heart rhythms through episodes of interaction synchrony. *Infant Behavior and Development, 34*(4), 569–577.

Firat, R. B. (2021). A neurosociological theory of culturally and structurally situated cognition and ethnoracial stress. *Frontiers in Sociology, 6*, 695042.

Firat, R. B., Hitlin, S., Magnotta, V., & Tranel, D. (2017). Putting race in context: Social class modulates processing of race in the ventromedial prefrontal cortex and amygdala. *Social Cognitive and Affective Neuroscience, 12*(8), 1314–1324.

Forbes, C. E., Cox, C. L., Schmader, T., & Ryan, L. (2012). Negative stereotype activation alters interaction between neural correlates of arousal, inhibition and cognitive control. *Social Cognitive and Affective Neuroscience, 7*(7), 771–781.

Forman, S. D., Cohen, J. D., Fitzgerald, M., Eddy, W. F., Mintun, M. A., & Noll, D. C. (1995). Improved assessment of significant activation in functional magnetic resonance imaging (fMRI): Use of a cluster-size threshold. *Magnetic Resonance in Medicine, 33*(5), 636–647.

Franks, D. D. (2014). Emotions and neurosociology. In *Handbook of the sociology of emotions: Volume II* (pp. 267–281).

Fries, P., Reynolds, J. H., Rorie, A. E., & Desimone, R. (2001). Modulation of oscillatory neuronal synchronization by selective visual attention. *Science, 291*(5508), 1560–1563.

Friston, K. J. (2011). Functional and effective connectivity: A review. *Brain Connectivity, 1*(1), 13–36.

Friston, K. J., Buechel, C., Fink, G. R., Morris, J., Rolls, E., & Dolan, R. J. (1997). Psychophysiological and modulatory interactions in neuroimaging. *Neuroimage, 6*(3), 218–229.

Friston, K. J., Frith, C. D., Frackowiak, R. S., & Turner, R. (1995). Characterizing dynamic brain responses with fMRI: A multivariate approach. *Neuroimage, 2*(2), 166–172.

Friston, K. J., Holmes, A. P., Worsley, K. J., Poline, J. P., Frith, C. D., & Frackowiak, R. S. (1994). Statistical parametric maps in functional imaging: A general linear approach. *Human Brain Mapping, 2*(4), 189–210.

Friston, K., Moran, R., & Seth, A. K. (2013). Analysing connectivity with Granger causality and dynamic causal modelling. *Current Opinion in Neurobiology, 23*(2), 172–178.

Funane, T., Kiguchi, M., Atsumori, H., Sato, H., Kubota, K., & Koizumi, H. (2011). Synchronous activity of two people's prefrontal cortices during a cooperative task measured by simultaneous near-infrared spectroscopy. *Journal of Biomedical Optics, 16*(7), 077011–077011.

Granger, C. W. (1969). Investigating causal relations by econometric models and cross-spectral methods. *Econometrica: Journal of the Econometric Society, 37*(3), 424–438.

Grech, R., Cassar, T., Muscat, J., Camilleri, K. P., Fabri, S. G., Zervakis, M., et al. (2008). Review on solving the inverse problem in EEG source analysis. *Journal of Neuroengineering and Rehabilitation, 5*, 1–33.

Gregory, S. W., Jr., & Kalkhoff, W. (2007). Analyzing sequences of interactive voice data using fast Fourier transform analysis. *The Mathematical Sociologist, 11*(1), 4–8.

Greicius, M. D., Krasnow, B., Reiss, A. L., & Menon, V. (2003). Functional connectivity in the resting brain: A network analysis of the default mode hypothesis. *Proceedings of the National Academy of Sciences, 100*(1), 253–258.

Griffanti, L., Salimi-Khorshidi, G., Beckmann, C. F., Auerbach, E. J., Douaud, G., Sexton, C. E., et al. (2014). ICA-based artefact removal and accelerated fMRI acquisition for improved resting state network imaging. *Neuroimage, 95*, 232–247.

Gutsell, J. N., & Inzlicht, M. (2010). Empathy constrained: Prejudice predicts reduced mental simulation of actions during observation of outgroups. *Journal of Experimental Social Psychology, 46*(5), 841–845.

Gutsell, J. N., & Inzlicht, M. (2013). Using EEG mu-suppression to explore group biases in motor resonance. In *Neuroscience of prejudice and intergroup relations* (pp. 279–297).

Gyurovski, I., Kubota, J., Cardenas-Iniguez, C., & Cloutier, J. (2018). Social status level and dimension interactively influence person evaluations indexed by P300s. *Social Neuroscience, 13*(3), 333–345.

Hari, R., Henriksson, L., Malinen, S., & Parkkonen, L. (2015). Centrality of social interaction in human brain function. *Neuron, 88*(1), 181–193.

Hari, R., & Kujala, M. V. (2009). Brain basis of human social interaction: From concepts to brain imaging. *Physiological Reviews, 89*(2), 453–479.

Harris, L. T., & Fiske, S. T. (2006). Dehumanizing the lowest of the low: Neuroimaging responses to extreme out-groups. *Psychological Science, 17*(10), 847–853.

Hasson, U., Ghazanfar, A. A., Galantucci, B., Garrod, S., & Keysers, C. (2012). Brain-to-brain coupling: A mechanism for creating and sharing a social world. *Trends in Cognitive Sciences, 16*(2), 114–121.

Henrich, J., Heine, S. J., & Norenzayan, A. (2010). Most people are not WEIRD. *Nature, 466*(7302), 29–29.

Herold, F., Wiegel, P., Scholkmann, F., Thiers, A., Hamacher, D., & Schega, L. (2017). Functional near-infrared spectroscopy in movement science: A systematic review on cortical activity in postural and walking tasks. *Neurophotonics, 4*(4), 041403.

Herrmann, C. S., & Knight, R. T. (2001). Mechanisms of human attention: Event-related potentials and oscillations. *Neuroscience & Biobehavioral Reviews, 25*(6), 465–476.

Herrmann, M. J., Schreppel, T., Jäger, D., Koehler, S., Ehlis, A. C., & Fallgatter, A. J. (2007). The other-race effect for face perception: An event-related potential study. *Journal of Neural Transmission, 114*, 951–957.

Hocke, L. M., Oni, I. K., Duszynski, C. C., Corrigan, A. V., Frederick, B. D., & Dunn, J. F. (2018). Automated processing of fNIRS data—A visual guide to the pitfalls and consequences. *Algorithms, 11*(5), 67.

Holper, L., Scholkmann, F., & Wolf, M. (2012). Between-brain connectivity during imitation measured by fNIRS. *Neuroimage, 63*(1), 212–222.

Huang, W. J., Chen, W. W., & Zhang, X. (2015). The neurophysiology of P 300–An integrated review. *European Review for Medical & Pharmacological Sciences, 19*(8), 1480–1488.

Hurtado, E., Haye, A., González, R., Manes, F., & Ibáñez, A. (2009). Contextual blending of ingroup/outgroup face stimuli and word valence: LPP modulation and convergence of measures. *BMC Neuroscience, 10*, 1–21.

Ibáñez, A., Hurtado, E., Lobos, A., Escobar, J., Trujillo, N., Baez, S., et al. (2011). Subliminal presentation of other faces (but not own face) primes behavioral and evoked cortical processing of empathy for pain. *Brain Research, 1398*, 72–85.

Ibanez, A., Melloni, M., Huepe, D., Helgiu, E., Rivera-Rei, A., Canales-Johnson, A., et al. (2012). What event-related potentials (ERPs) bring to social neuroscience? *Social Neuroscience, 7*(6), 632–649.

Ito, T. A., & Cacioppo, J. T. (2000). Electrophysiological evidence of implicit and explicit categorization processes. *Journal of Experimental Social Psychology, 36*(6), 660–676.

Jensen, O., & Mazaheri, A. (2010). Shaping functional architecture by oscillatory alpha activity: Gating by inhibition. *Frontiers in Human Neuroscience, 4*, 186.

Jiang, J., Chen, C., Dai, B., Shi, G., Ding, G., Liu, L., & Lu, C. (2015). Leader emergence through interpersonal neural synchronization. *Proceedings of the National Academy of Sciences, 112*(14), 4274–4279.

Jung, R., & Berger, W. (1979). Fiftieth anniversary of Hans Berger's publication of the electroencephalogram His first records in 1924–1931. *Archiv für Psychiatrie und Nervenkrankheiten, 227*, 279–300.

Jutten, C., & Herault, J. (1991). Blind separation of sources, part I: An adaptive algorithm based on neuromimetic architecture. *Signal Processing, 24*(1), 1–10.

Kalkhoff, W., Melamed, D., Pollock, J., Miller, B., Overton, J., & Pfeiffer, M. (2020). Cracking the black box: Capturing the role of expectation states in status processes. *Social Psychology Quarterly, 83*(1), 26–48.

Kalkhoff, W., Thye, S. R., & Pollock, J. (2016a). Developments in neurosociology. *Sociology Compass, 10*(3), 242–258.

Kalkhoff, W., Serpe, R. T., Pollock, J., Miller, B., & Pfeiffer, M. (2016b). Neural processing of identity-relevant feedback. In *New directions in identity theory and research* (pp. 195–238).

Kappenman, E. S., & Luck, S. J. (2011). ERP components: The ups and downs of brainwave recordings. In S. J. Luck & E. S. Kappenman (Eds.), *The Oxford handbook of event-related potential components* (pp. 3–30). Oxford University Press.

Kawasaki, M., Kitajo, K., & Yamaguchi, Y. (2018). Sensory-motor synchronization in the brain corresponds to behavioral synchronization between individuals. *Neuropsychologia, 119*, 59–67.

Kawasaki, M., Yamada, Y., Ushiku, Y., Miyauchi, E., & Yamaguchi, Y. (2013). Inter-brain synchronization during coordination of speech rhythm in human-to-human social interaction. *Scientific Reports, 3*(1), 1692.

Khosla, M. (2021). *Machine learning in resting-state and naturalistic fMRI analysis*. Cornell University.

King-Casas, B., Tomlin, D., Anen, C., Camerer, C. F., Quartz, S. R., & Montague, P. R. (2005). Getting to know you: reputation and trust in a two-person economic exchange. *Science, 308*(5718), 78–83.

Kinreich, S., Djalovski, A., Kraus, L., Louzoun, Y., & Feldman, R. (2017). Brain-to-brain synchrony during naturalistic social interactions. *Scientific Reports, 7*(1), 17060.

Klimesch, W. (1999). EEG alpha and theta oscillations reflect cognitive and memory performance: A review and analysis. *Brain Research Reviews, 29*(2–3), 169–195.

Kline, R. B. (2011). Convergence of structural equation modeling and multilevel modeling. In *The SAGE handbook of innovation in social research methods* (pp. 562–589).

Knoch, D., Schneider, F., Schunk, D., Hohmann, M., & Fehr, E. (2009). Disrupting the prefrontal cortex diminishes the human ability to build a good reputation. *Proceedings of the National Academy of Sciences, 106*(49), 20895–20899.

Knyazev, G. G. (2007). Motivation, emotion, and their inhibitory control mirrored in brain oscillations. *Neuroscience & Biobehavioral Reviews, 31*(3), 377–395.

Koike, T., Sumiya, M., Nakagawa, E., Okazaki, S., & Sadato, N. (2019). What makes eye contact special? Neural substrates of on-line mutual eye-gaze: A hyperscanning fMRI study. *Eneuro, 6*(1).

Koike, T., Tanabe, H. C., & Sadato, N. (2015). Hyperscanning neuroimaging technique to reveal the "two-in-one" system in social interactions. *Neuroscience Research, 90*, 25–32.

Krause, C. M., Sillanmäki, L., Koivisto, M., Saarela, C., Häggqvist, A., Laine, M., & Hämäläinen, H. (2000). The effects of memory load on event-related EEG desynchronization and synchronization. *Clinical Neurophysiology, 111*(11), 2071–2078.

Krill, A. L., & Platek, S. M. (2012). Working together may be better: Activation of reward centers during a cooperative maze task. *PLoS One, 7*(2), e30613.

Kuncheva, L. I., & Rodríguez, J. J. (2010). Classifier ensembles for fMRI data analysis: An experiment. *Magnetic Resonance Imaging, 28*(4), 583–593.

Kwan, V. S., Barrios, V., Ganis, G., Gorman, J., Lange, C., Kumar, M., et al. (2007). Assessing the neural correlates of self-enhancement bias: A transcranial magnetic stimulation study. *Experimental Brain Research, 182*, 379–385.

Kwong, K. K., Belliveau, J. W., Chesler, D. A., Goldberg, I. E., Weisskoff, R. M., Poncelet, B. P., et al. (1992). Dynamic magnetic resonance imaging of human brain activity during primary sensory stimulation. *Proceedings of the National Academy of Sciences, 89*(12), 5675–5679.

La Scala, S., Mullins, J. L., Firat, R. B., & Michalska, K. J. (2023). Equity, diversity, and inclusion in developmental neuroscience: Practical lessons from community-based participatory research. *Frontiers in Integrative Neuroscience, 16*, 141.

Lachaux, J. P., Rodriguez, E., Martinerie, J., & Varela, F. J. (1999). Measuring phase synchrony in brain signals. *Human Brain Mapping, 8*(4), 194–208.

Laureiro-Martínez, D., Brusoni, S., Canessa, N., & Zollo, M. (2015). Understanding the exploration–exploitation dilemma: An fMRI study of attention control and decision-making performance. *Strategic management journal, 36*(3), 319–338.

Lim, S., Yeo, M., & Yoon, G. (2019). Comparison between concentration and immersion based on EEG analysis. *Sensors, 19*(7), 1669.

Lindenberger, U., Li, S. C., Gruber, W., & Müller, V. (2009). Brains swinging in concert: cortical phase synchronization while playing guitar. *BMC Neuroscience, 10*, 1–12.

Liu, N., Mok, C., Witt, E. E., Pradhan, A. H., Chen, J. E., & Reiss, A. L. (2016). NIRS-based hyperscanning reveals inter-brain neural synchronization during cooperative Jenga game with face-to-face communication. *Frontiers in Human Neuroscience, 10*, 82.

Liu, H., Zhao, C., Wang, F., & Zhang, D. (2021). Inter-brain amplitude correlation differentiates cooperation from competition in a motion-sensing sports game. *Social Cognitive and Affective Neuroscience, 16*(6), 552–564.

Lloyd-Fox, S., Blasi, A., & Elwell, C. E. (2010). Illuminating the developing brain: The past, present and future of functional near infrared spectroscopy. *Neuroscience & Biobehavioral Reviews, 34*(3), 269–284.

Lloyd-Fox, S., Papademetriou, M., Darboe, M. K., Everdell, N. L., Wegmuller, R., Prentice, A. M., et al. (2014). Functional near infrared spectroscopy (fNIRS) to assess cognitive function in infants in rural Africa. *Scientific Reports, 4*(1), 4740.

Lloyd-Fox, S., Wu, R., Richards, J. E., Elwell, C. E., & Johnson, M. H. (2015). Cortical activation to action

perception is associated with action production abilities in young infants. *Cerebral Cortex, 25*(2), 289–297.
Logothetis, N. K. (2008). What we can do and what we cannot do with fMRI. *Nature, 453*(7197), 869–878.
Luck, S. J. (2014). *An introduction to the event-related potential technique*. MIT Press.
Luck, S. J., & Kappenman, E. S. (Eds.). (2013). *The Oxford handbook of event-related potential components*. Oxford University Press.
Lyon, L. (2017). Dead salmon and voodoo correlations: should we be sceptical about functional MRI?. *Brain, 140*(8), e53–e53.
McKeown, M. J., Makeig, S., Brown, G. G., Jung, T. P., Kindermann, S. S., Bell, A. J., & Sejnowski, T. J. (1998). Analysis of fMRI data by blind separation into independent spatial components. *Human Brain Mapping, 6*(3), 160–188.
Mead, G. H. (1934). Mind, self, and society from the standpoint of a social behaviorist.
Mendelowitz, D. (1999). Advances in parasympathetic control of heart rate and cardiac function. *Physiology, 14*(4), 155–161.
Merker, B. (2013). Cortical gamma oscillations: the functional key is activation, not cognition. *Neuroscience & Biobehavioral Reviews, 37*(3), 401–417.
Mezer, A., Yovel, Y., Pasternak, O., Gorfine, T., & Assaf, Y. (2009). Cluster analysis of resting-state fMRI time series. *Neuroimage, 45*(4), 1117–1125.
Miezin, F. M., Maccotta, L., Ollinger, J. M., Petersen, S. E., & Buckner, R. L. (2000). Characterizing the hemodynamic response: Effects of presentation rate, sampling procedure, and the possibility of ordering brain activity based on relative timing. *Neuroimage, 11*(6), 735–759.
Miltner, W. H., Braun, C. H., & Coles, M. G. (1997). Event-related brain potentials following incorrect feedback in a time-estimation task: Evidence for a "generic" neural system for error detection. *Journal of Cognitive Neuroscience, 9*(6), 788–798.
Minagawa-Kawai, Y., Matsuoka, S., Dan, I., Naoi, N., Nakamura, K., & Kojima, S. (2009). Prefrontal activation associated with social attachment: Facial-emotion recognition in mothers and infants. *Cerebral Cortex, 19*(2), 284–292.
Misaki, M., Kerr, K. L., Ratliff, E. L., Cosgrove, K. T., Simmons, W. K., Morris, A. S., & Bodurka, J. (2021). Beyond synchrony: The capacity of fMRI hyperscanning for the study of human social interaction. *Social Cognitive and Affective Neuroscience, 16*(1–2), 84–92.
Mitchell, J. P., Banaji, M. R., & Macrae, C. N. (2005). The link between social cognition and self-referential thought in the medial prefrontal cortex. *Journal of Cognitive Neuroscience, 17*(8), 1306–1315.
Montague, P. R., Berns, G. S., Cohen, J. D., McClure, S. M., Pagnoni, G., Dhamala, M., et al. (2002). Hyperscanning: Simultaneous fMRI during linked social interactions. *Neuroimage, 16*(4), 1159–1164.
Morales, S., & Bowers, M. E. (2022). Time-frequency analysis methods and their application in developmental EEG data. *Developmental Cognitive Neuroscience, 54*, 101067.
Na, C. H., Jütten, K., Forster, S. D., Clusmann, H., & Mainz, V. (2022). Self-referential processing and resting-state functional MRI connectivity of cortical midline structures in glioma patients. *Brain Sciences, 12*(11), 1463.
Nam, C. S., Choo, S., Huang, J., & Park, J. (2020). Brain-to-brain neural synchrony during social interactions: A systematic review on hyperscanning studies. *Applied Sciences, 10*(19), 6669.
Nastase, S. A., Gazzola, V., Hasson, U., & Keysers, C. (2019). Measuring shared responses across subjects using intersubject correlation. *Social Cognitive and Affective Neuroscience, 14*(6), 667–685.
Neuper, C., & Pfurtscheller, G. (2001). Event-related dynamics of cortical rhythms: Frequency-specific features and functional correlates. *International Journal of Psychophysiology, 43*(1), 41–58.
Nichols, T. E., Das, S., Eickhoff, S. B., Evans, A. C., Glatard, T., Hanke, M., et al. (2017). Best practices in data analysis and sharing in neuroimaging using MRI. *Nature Neuroscience, 20*(3), 299–303.
Nozawa, T., Sasaki, Y., Sakaki, K., Yokoyama, R., & Kawashima, R. (2016). Interpersonal frontopolar neural synchronization in group communication: An exploration toward fNIRS hyperscanning of natural interactions. *Neuroimage, 133*, 484–497.
Oliver, D., Tachtsidis, I., & Hamilton, A. F. D. C. (2018). The role of parietal cortex in overimitation: A study with fNIRS. *Social Neuroscience, 13*(2), 214–225.
Osaka, N., Minamoto, T., Yaoi, K., Azuma, M., & Osaka, M. (2014). Neural synchronization during cooperated humming: A hyperscanning study using fNIRS. *Procedia-Social and Behavioral Sciences, 126*, 241–243.
Osaka, N., Minamoto, T., Yaoi, K., Azuma, M., & Osaka, M. (2015). How two brains make one synchronized mind in the inferior frontal cortex: fNIRS-based hyperscanning during cooperative singing. *Frontiers in Psychology, 6*, 162605.
Park, J., Shin, J., & Jeong, J. (2022). Inter-brain synchrony levels according to task execution modes and difficulty levels: An fNIRS/GSR study. *IEEE Transactions on Neural Systems and Rehabilitation Engineering, 30*, 194–204.
Patel, S., Katura, T., Maki, A., & Tachtsidis, I. (2011). Quantification of systemic interference in optical topography data during frontal lobe and motor cortex activation: An independent component analysis. In *Oxygen transport to tissue XXXII* (pp. 45–51). Springer.
Pinti, P., Aichelburg, C., Gilbert, S., Hamilton, A., Hirsch, J., Burgess, P., & Tachtsidis, I. (2018). A review on the use of wearable functional near-infrared spectroscopy in naturalistic environments. *Japanese Psychological Research, 60*(4), 347–373.
Pinti, P., Merla, A., Aichelburg, C., Lind, F., Power, S., Swingler, E., et al. (2017). A novel GLM-based method for the Automatic IDentification of functional

Events (AIDE) in fNIRS data recorded in naturalistic environments. *Neuroimage, 155*, 291–304.

Pinti, P., Tachtsidis, I., Hamilton, A., Hirsch, J., Aichelburg, C., Gilbert, S., & Burgess, P. W. (2020). The present and future use of functional near-infrared spectroscopy (fNIRS) for cognitive neuroscience. *Annals of the New York Academy of Sciences, 1464*(1), 5–29.

Polich, J. (2007). Updating P300: An integrative theory of P3a and P3b. *Clinical Neurophysiology, 118*(10), 2128–2148.

Poline, J. B., & Brett, M. (2012). The general linear model and fMRI: does love last forever? *Neuroimage, 62*(2), 871–880.

Potts, G. F. (2004). An ERP index of task relevance evaluation of visual stimuli. *Brain and Cognition, 56*(1), 5–13.

Pourtois, G., & Vuilleumier, P. (2006). Dynamics of emotional effects on spatial attention in the human visual cortex. *Progress in Brain Research, 156*, 67–91.

Power, J. D., Mitra, A., Laumann, T. O., Snyder, A. Z., Schlaggar, B. L., & Petersen, S. E. (2014). Methods to detect, characterize, and remove motion artifact in resting state fMRI. *Neuroimage, 84*, 320–341.

Premi, E., Cauda, F., Costa, T., Diano, M., Gazzina, S., Gualeni, V., et al. (2016). Looking for neuroimaging markers in frontotemporal lobar degeneration clinical trials: A multi-voxel pattern analysis study in granulin disease. *Journal of Alzheimer's Disease, 51*(1), 249–262.

Renvall, V., Kauramäki, J., Malinen, S., Hari, R., & Nummenmaa, L. (2020). Imaging real-time tactile interaction with two-person dual-coil fMRI. *Frontiers in Psychiatry, 11*, 516604.

Rizzolatti, G. (2005). The mirror neuron system and its function in humans. *Anatomy and Embryology, 210*(5), 419–421.

Robinson, S., Moser, E., & Peper, M. (2009). fMRI of emotion. In *fMRI techniques and protocols* (pp. 411–456).

Rosenbaum, D., Hilsendegen, P., Thomas, M., Haeussinger, F. B., Metzger, F. G., Nuerk, H. C., et al. (2018). Cortical hemodynamic changes during the trier social stress test: An fNIRS study. *NeuroImage, 171*, 107–115.

Rosenbaum, D., Thomas, M., Hilsendegen, P., Metzger, F. G., Haeussinger, F. B., Nuerk, H. C., et al. (2018). Stress-related dysfunction of the right inferior frontal cortex in high ruminators: An fNIRS study. *NeuroImage: Clinical, 18*, 510–517.

Ruff, C. C., Driver, J., & Bestmann, S. (2009). Combining TMS and fMRI: From 'virtual lesions' to functional-network accounts of cognition. *Cortex, 45*(9), 1043–1049.

Ruocco, A. C., Medaglia, J. D., Tinker, J. R., Ayaz, H., Forman, E. M., Newman, C. F., et al. (2010). Medial prefrontal cortex hyperactivation during social exclusion in borderline personality disorder. *Psychiatry Research: Neuroimaging, 181*(3), 233–236.

Sänger, J., Müller, V., & Lindenberger, U. (2012). Intra- and interbrain synchronization and network properties when playing guitar in duets. *Frontiers in Human Neuroscience, 6*, 312.

Sänger, J., Müller, V., & Lindenberger, U. (2013). Directionality in hyperbrain networks discriminates between leaders and followers in guitar duets. *Frontiers in Human Neuroscience, 7*, 234.

Sappia, M. S., Hakimi, N., Colier, W. N., & Horschig, J. M. (2020). Signal quality index: An algorithm for quantitative assessment of functional near infrared spectroscopy signal quality. *Biomedical Optics Express, 11*(11), 6732–6754.

Schippers, M. B., Roebroeck, A., Renken, R., Nanetti, L., & Keysers, C. (2010). Mapping the information flow from one brain to another during gestural communication. *Proceedings of the National Academy of Sciences, 107*(20), 9388–9393.

Schmidt, R. C., Morr, S., Fitzpatrick, P., & Richardson, M. J. (2012). Measuring the dynamics of interactional synchrony. *Journal of Nonverbal Behavior, 36*(4), 263–279.

Scholkmann, F., Holper, L., Wolf, U., & Wolf, M. (2013). A new methodical approach in neuroscience: assessing inter-personal brain coupling using functional near-infrared imaging (fNIRI) hyperscanning. *Frontiers in Human Neuroscience, 7*, 813.

Scholkmann, F., Kleiser, S., Metz, A. J., Zimmermann, R., Pavia, J. M., Wolf, U., & Wolf, M. (2014). A review on continuous wave functional near-infrared spectroscopy and imaging instrumentation and methodology. *Neuroimage, 85*, 6–27.

Scholkmann, F., Spichtig, S., Muehlemann, T., & Wolf, M. (2010). How to detect and reduce movement artifacts in near-infrared imaging using moving standard deviation and spline interpolation. *Physiological Measurement, 31*(5), 649.

Schupp, H. T., & Kirmse, U. M. (2021). Case-by-case: Emotional stimulus significance and the modulation of the EPN and LPP. *Psychophysiology, 58*(4), e13766.

Schyns, P. G., Petro, L. S., & Smith, M. L. (2007). Dynamics of visual information integration in the brain for categorizing facial expressions. *Current Biology, 17*(18), 1580–1585.

Sebanz, N., Knoblich, G., Prinz, W., & Wascher, E. (2006). Twin peaks: An ERP study of action planning and control in coacting individuals. *Journal of Cognitive Neuroscience, 18*(5), 859–870.

Shahhosseini, Y., & Miranda, M. F. (2022). Functional connectivity methods and their applications in fMRI data. *Entropy, 24*(3), 390.

Shaw, D. J., Czekóová, K., Staněk, R., Mareček, R., Urbánek, T., Špalek, J., et al. (2018). A dual-fMRI investigation of the iterated Ultimatum Game reveals that reciprocal behaviour is associated with neural alignment. *Scientific Reports, 8*(1), 10896.

Spiegelhalder, K., Ohlendorf, S., Regen, W., Feige, B., van Elst, L. T., Weiller, C., et al. (2014). Interindividual synchronization of brain activity during live verbal communication. *Behavioural Brain Research, 258*, 75–79.

Špiláková, B., Shaw, D. J., Czekóová, K., & Brázdil, M. (2019). Dissecting social interaction: Dual-fMRI reveals patterns of interpersonal brain-behavior relationships that dissociate among dimensions of social exchange. *Social Cognitive and Affective Neuroscience, 14*(2), 225–235.

Stam, C. J., Nolte, G., & Daffertshofer, A. (2007). Phase lag index: Assessment of functional connectivity from multi channel EEG and MEG with diminished bias from common sources. *Human Brain Mapping, 28*(11), 1178–1193.

Strother, S. C. (2006). Evaluating fMRI preprocessing pipelines. *IEEE Engineering in Medicine and Biology Magazine, 25*(2), 27–41.

Sui, J., Liu, C. H., & Han, S. (2009). Cultural difference in neural mechanisms of self-recognition. *Social Neuroscience, 4*(5), 402–411.

Sun, S., & Yu, R. (2014). The feedback related negativity encodes both social rejection and explicit social expectancy violation. *Frontiers in Human Neuroscience, 8*, 556.

Tanaka, H., Katura, T., & Sato, H. (2014). Task-related oxygenation and cerebral blood volume changes estimated from NIRS signals in motor and cognitive tasks. *NeuroImage, 94*, 107–119.

Taspinar, G., & Ozkurt, N. (2024). A review of ADHD detection studies with machine learning methods using rsfMRI data. *NMR in Biomedicine, 37*(8), e5138.

TenHouten, W. D., & Kaplan, C. D. (1973). *Science and its mirror image: A theory of inquiry*. Harper and Row.

TenHouten, W., Schussel, L., Gritsch, M. F., & Kaplan, C. D. (2023). Hyperscanning and the future of neurosociology. *Sociological Methodology, 53*(1), 139–157.

Teplan, M. (2002). Fundamentals of EEG measurement. *Measurement Science Review, 2*(2), 1–11.

Thut, G., & Miniussi, C. (2009). New insights into rhythmic brain activity from TMS–EEG studies. *Trends in Cognitive Sciences, 13*(4), 182–189.

Torricelli, A., Contini, D., Pifferi, A., Caffini, M., Re, R., Zucchelli, L., & Spinelli, L. (2014). Time domain functional NIRS imaging for human brain mapping. *Neuroimage, 85*, 28–50.

Turner, M. (2001). *Cognitive dimensions of social science*. Oxford University Press.

Van Bavel, J. J., Packer, D. J., & Cunningham, W. A. (2008). The neural substrates of in-group bias: A functional magnetic resonance imaging investigation. *Psychological Science, 19*(11), 1131–1139.

Van Overwalle, F., & Baetens, K. (2009). Understanding others' actions and goals by mirror and mentalizing systems: A meta-analysis. *Neuroimage, 48*(3), 564–584.

Varnum, M. E., Blais, C., & Brewer, G. A. (2016). Social class affects Mu-suppression during action observation. *Social Neuroscience, 11*(4), 449–454.

Virtanen, S., Antikainen, O., & Yliruusi, J. (2007). Uniformity of poorly miscible powders determined by near infrared spectroscopy. *International Journal of Pharmaceutics, 345*(1–2), 108–115.

Vu, A. T., Jamison, K., Glasser, M. F., Smith, S. M., Coalson, T., Moeller, S., et al. (2017). Tradeoffs in pushing the spatial resolution of fMRI for the 7T Human Connectome Project. *Neuroimage, 154*, 23–32.

Vul, E., Harris, C., Winkielman, P., & Pashler, H. (2009). Voodoo correlations in social neuroscience. *Perspectives on Psychological Science, 4*(3), 274–290.

Walsh, V., & Cowey, A. (2000). Transcranial magnetic stimulation and cognitive neuroscience. *Nature Reviews Neuroscience, 1*(1), 73–80.

Wang, C., Zhang, T., Shan, Z., Liu, J., Yuan, D., & Li, X. (2019). Dynamic interpersonal neural synchronization underlying pain-induced cooperation in females. *Human Brain Mapping, 40*(11), 3222–3232.

Wang, J., Zuo, X., & He, Y. (2010). Graph-based network analysis of resting-state functional MRI. *Frontiers in Systems Neuroscience, 4*, 1419.

Webb, E. K., Etter, J. A., & Kwasa, J. A. (2022). Addressing racial and phenotypic bias in human neuroscience methods. *Nature Neuroscience, 25*(4), 410–414.

Wikström, V., Saarikivi, K., Falcon, M., Makkonen, T., Martikainen, S., Putkinen, V., Cowley, B. U., & Tervaniemi, M. (2022). Inter-brain synchronization occurs without physical co-presence during cooperative online gaming. *Neuropsychologia, 174*, 108316.

Willingham, D. T., & Dunn, E. W. (2003). What neuroimaging and brain localization can do, cannot do and should not do for social psychology. *Journal of Personality and Social Psychology, 85*(4), 662.

Wu, C. W., Chen, C. L., Liu, P. Y., Chao, Y. P., Biswal, B. B., & Lin, C. P. (2011). Empirical evaluations of slice-timing, smoothing, and normalization effects in seed-based, resting-state functional magnetic resonance imaging analyses. *Brain Connectivity, 1*(5), 401–410.

Yen, C., Lin, C. L., & Chiang, M. C. (2023). Exploring the frontiers of neuroimaging: a review of recent advances in understanding brain functioning and disorders. *Life, 13*(7), 1472.

Yun, K., Watanabe, K., & Shimojo, S. (2012). Interpersonal body and neural synchronization as a marker of implicit social interaction. *Scientific Reports, 2*(1), 959.

Zellars, K. L., Meurs, J. A., Perrewé, P. L., Kacmar, C. J., & Rossi, A. M. (2009). Reacting to and recovering from a stressful situation: The negative affectivity-physiological arousal relationship. *Journal of Occupational Health Psychology, 14*(1), 11.

Part II

Emotions

The Evolved Neurology Generating Human Emotions

7

Jonathan H. Turner

Abstract

The evolution of the brain is examined, with an emphasis on the structures generating increasing variations in human emotions. Tables on types of human emotions are outlined, including primary and secondary emotions, as well as emotions activating defense systems. The body systems—e. g. autonomic nervous system, neurotransmitter and peptide systems, endocrine and vascular systems, musculoskeletal—system are described. And the feed backs involved in the activation of varying emotional are described. The result of these evolved systems that allowed the neo-cortex to grow, and eventually allow for the evolution of emotionally charged speech and culture that allow humans to create ever larger and more complex societies.

Humans are probably the most emotional animals on earth, an often obscured by the fact that humans are likely the most intelligent animals on earth. As I outline in detail in other chapters, intelligence is related to emotions because memories, thinking, and decision-making in the neo-cortex of mammals are only possible when emotions are attached to cognitions. Indeed, the greater the range of emotions that an animal has, the greater is the potential for a larger neo-cortex and, hence, greater levels of intelligence. In fact, the evolution of the *sub-cortex* where emotions are ultimately generated had to grow first before the neo-cortex of the ancestors of humans could grow. The now obsolete philosophical notion that emotionality and rationality (and intelligence more generally) are on opposites of a continuum could not possible be true, given the way the brain remembers and makes decisions. Thus, the evolution of the human neo-cortex was driven by the prior evolution and enlargement of key structures of the sub-cortex of great apes that were on the human line of evolution. These ancestors are termed *hominins*, and humans are the last hominin to evolve.

In this chapter, I will first examine what is known about the relationship between emotions and cognitions, and then outline how I think the range and intensity of humans emotions emerged, thereby making possible for the neo-cortex to grow. And, in other Chapters, I emphasize that emotions have been the driving force of growth in the neo-cortex of hominins, the late emergence of speech in the immediate species of hominins before humans, and then the emergence of symbolic culture. Indeed, if it were possible to have more data on the ancestors of all intelligent life forms, I would argue that *emotions are the driv-*

J. H. Turner (✉)
Department of Sociology, University of California, Riverside, Riverside, CA, USA
e-mail: jonathan.turner@ucr.edu

W. Kalkhoff et al. (eds.), *Handbook of Neurosociology*, Handbooks of Sociology and Social Research, https://doi.org/10.1007/978-3-031-95615-7_7

ing force behind the intelligence of all living forms in the universe—granted, a rather grandiose claim. Let me begin with an outline of the neurology of emotions, and then in later sections, outline my thoughts (an a number of others) as to how the sub-cortical portions of the human brain evolved in way that increase the range of hominin emotions over the last seven million years of hominin evolution.

The Neurology of Emotions[1]

There was early work by sociologists on brain mechanisms (e.g., Kemper, 1987; Smith & Stevens, 1977), but these approaches were not evolutionary. To understand emotions and brain neurology, I think it is always best to pursue an evolutionary approach and have a sense for how the traits of a species neuro-anatomy evolved. Indeed, I have argued several decades that we cannot fully understand the nature of human behavior and interaction within socio-cultural contexts without understanding the evolution of the neurology behind human emotions.

Comparative Neuro-anatomy

For over 30 years, I have followed Alexandra Maryanksi's strategy of comparing humans with their closest biological relatives: the great apes (orangutans, gorillas, and chimpanzees). Since humans share a very high proportion of their genes with great apes (99 + % with Chimpanzees, 98% + with Gorillas, and 96–97% with orangutans), humans and great apes share common ancestors. Given this fact, it becomes possible to compare great apes with human brains to determine how natural selection worked to alter the hominin and then human brain after the break off of hominins from the ancestors of the great apes. What is immediately evident is that the principle change in the human brain was its increased size at both the sub-cortical and neo-cortical level. Thus, in contrast to many arguments made within evolutionary psychology, there are surprisingly *few new structures* or "modules" in the human brain compared to the great ape brain. The hominin brain was enlarged over a long period of at least seven million years, with a few new modules, but more dramatically much enhanced connectivity among structures of the brain.[2]

Thus, in comparing the human brain to humans closest relatives—the great apes—there are several differences between great ape brains and human brains—even though both humans and great apes have *the same basic neurological architecture*. One difference is that there are more asymmetries in significant portions of the neo-cortex because bilateral structures on the left and right side of the brain have somewhat differentiated functions (Bradshaw & Nettleton, 1983; Needham, 1982; Sperry, 1982). This differentiation greatly increased connectivity between the two size of the brain through the *corpus collosum* (see middle portion of Fig. 7.1). Much of this asymmetry revolved around specialization for language comprehension (*Wernicke's area*) and for speech production (*Broca's area*) on the left side of the neo-cortex, while on the right side of Broca's area is devoted to pattern and temporal recognition of signals.

Another difference is the size of the *frontal lobe* where long-term memory, and other cognitive functions are carried out (Semendeferi et al., 2002; Ruff et al., 1997). Still, the difference in frontal lobe is not as great as once thought, although the pre-frontal lobe is more developed in humans than in other primates, and this area along with the prefrontal cortex is devoted to thinking, planning, and decision-making, using

[1] See also Turner (1999a, 1999b, 2000, 2007) for earlier analyses, as well as more recent analyses (e.g., Turner, 2021; and Turner & Maryanski, 2024).

[2] A moment's reflection reveals why such is the case. The brain is a complex system regulating just about everything in a life form. Dramatic changes via mutations would inevitably be harmful for such a complex system of interdependencies, and so, natural selection followed an easier path: enlarging all structures by selecting on the tails of the Bell curve distributing size of structure and, in so doing, the *integration among all of these structures* is retained. Still, the few structures that differentiate humans from great apes are revealing because they are what allowed humans to survive in new habitats and, in the end, build up complex social structures and their cultures.

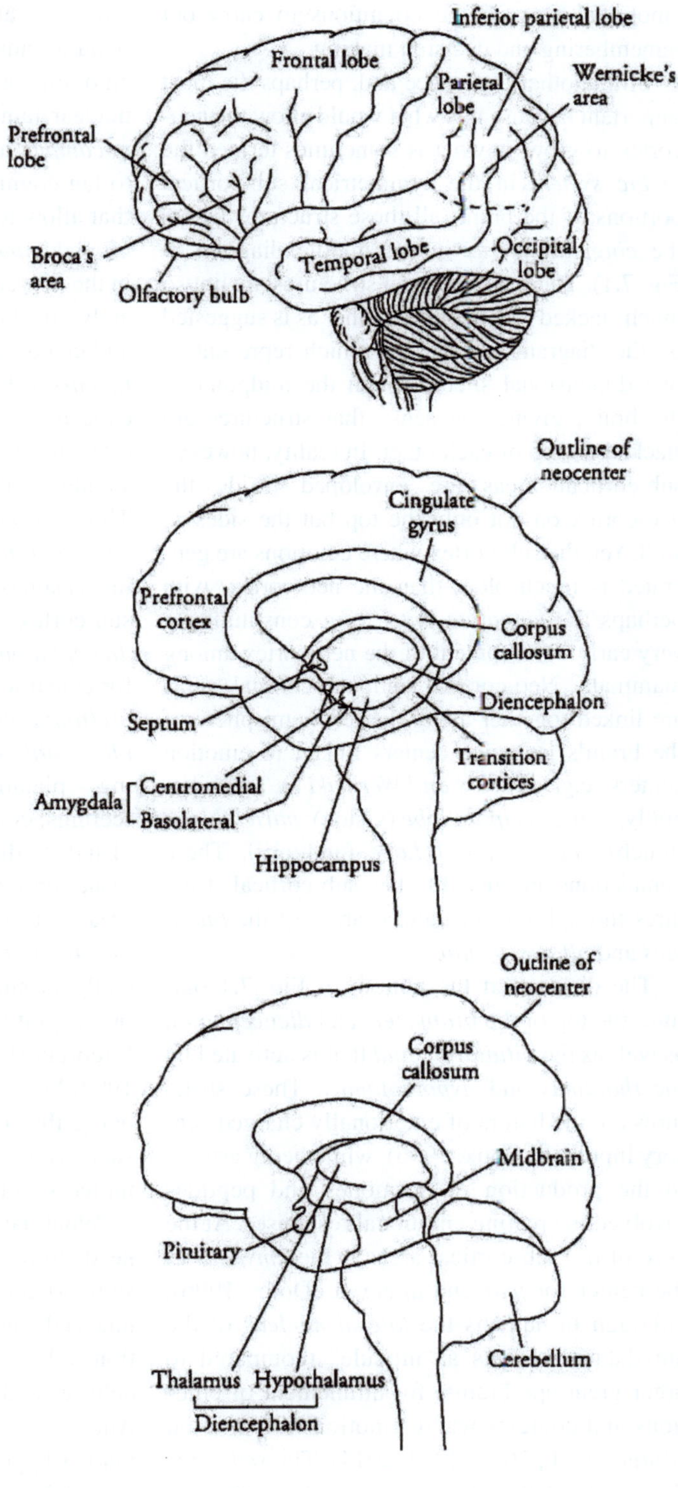

Fig. 7.1 The human brain and emotions

emotional tags on the cognitions to carry out remembering and decision making.

Still another difference and, perhaps the most important because it is what would allow the neo-cortex to grow, is what is sometimes termed the *limbic system* in the symmetrical sub-cortical portions of the brain (all those structures below the *cingulate gyrus* in the middle diagram in Fig. 7.1). These sub-cortical structures are not so much stacked on top of each other as is suggested by the diagram in Fig. 7.1, which represents a two dimensional slice through the midpoint of the brain, giving the sense that structures are stacked on top of each other. In reality, however, sub-cortical areas are enveloped inside the neo-cortex on not only the top but the sides as well. Yet, the sub-cortex where emotions are generated is much older than the neo-cortex, with perhaps the *cingulate outer ring* constituting a very early development of the neo-cortex among mammals. Neo-cortical and sub-cortical areas are linked together along the left hemisphere of the brain's language centers linked to emotion centers (e.g., *Broca's and Werneiki's*), and bilaterally, to the *occipital lobe* (vision), *parietal lobe* (touch*), and temporal Lobe* (auditory*).* These connections are linked to key sub-cortical structures through the specialized areas of the *thalamus* and *olfactory bulb.*

The diagram at the bottom of Fig. 7.1 outlines the top of the *brain stem* and *diencephalon* as well as the *pituitary gland* that is activated by the *thalamus* and *hypothalamus*. These structures are mediators of emotionally charged sensory inputs (LeDoux, 1996), while being critical to the production of hormones and peptides involved in creating emotional responses. At the core of the sub-cortical areas is the *amydala* as the center for *fear* and *anger* (LeDoux, 1996), although in humans the *lateral nucleus* of the amydala represents a "module" (compared to other great ape brains) for attunement of emotions and contexts where emotions are aroused (Barger et al., 2007, 2012, 2014). The *septum* is the center of the pleasure associated with sex, but in humans, there appear to be additional nuclei that generate nuclei adding pleasure beyond the sex act itself, such as emotions like "love" or at least a sense of commitment and bonding among sexual partners—a complex of emotions allowing for the evolution of the nuclear family among later hominins. The *hippocampus* and related *transition cortices* work to tag cognitions of experiences with emotions that allow them to be remembered at first in the *hippocampus* and, then, later in the *frontal lobe* in the neo-cortex, thereby making them immediately available in decision-making, planning, and calculation by the *frontal lobe* and *prefrontal cortex*. Indeed, "rationality" is built upon the capacity of emotions to *charge cognitions with affect* used to determine which options a person should decide to take when making a decision. Thus, *rationality is completely dependent on the emotional tags on cognitions* involved in thinking (Damasio, 1994). These structures of the sub-cortex—*hippocampal, transition cortices, diencphalon, amydala*, and *septum*—are critical for constructing lines of conduct. Layer 4 of the *prefrontal cortex*—sometimes termed the *granulofrontal cortex*—is an important in anticipating, planning, empathizing, and altruistic feelings, concern for welfare of self and others, visual reading of gestures (via connection to visual cortex), touch (via connections to motor areas), crying and laughing, (via connections to the *cingulate gyrus*). The *cingulate gyrus*, especially the anterior portion, is apparently involved in integrating other emotion systems and the forebrain (LeDoux, 1996; Devinski & Luciano, 1993; Vogt, 1993; MacLean, 1993), and in turn being the locus of key mammalian behaviors, such as the "separation cry" of infants from mothers and "playfulness" (MacLean, 1990).

What becomes clear, even as much more needs to be done to uncover the effects of these sub-cortical structures, is that complex and nuanced emotions are what drive human interaction and, in so doing so, affect the building up of culture and social structures. I draw from Antonio Damaso's (1994) notion of the "somatic marker hypothesis" trying to outline the neurology of human series if critical "body systems" driven by emotions that are also critical in determining the arousal of emotions used in cognitive processes.

Neuroanatomy and More Inclusive Body Systems Generating Emotions

1. ***The Autonomic Nervous Body System.*** The autonomic body system is composed of smooth muscles that control visceral responses accompanying emotional arousal. For example, as *fear* increases, the rate of heartbeats increases dramatically. Indeed, escalated respiration, muscle tension, dryness of mouth, sweating, and tenderness in stomach are generated though the autonomic system (ANS), roughly outlined in Fig. 7.2 where the brain systems involved in autonomic responses are listed. As the bold arrow suggests, the *hypothalamus* receiving inputs from other areas in the brain converts neural information into hormonal information by targeting areas of the pituitary gland that secretes hormones into the bloodstream. Once activated, the ANS operates as a critical feedback systems, which works through the *thalamus* to project hormones into other sub-cortical emotions systems and the neo-cortex (LeDoux, 1996). As Damasio (1994) and LeDoux (1996) outlined, these feedback processes have large effects on emotional arousal by activating other body systems that, in turn, generate their own outputs of emotional arousal. Thus, the ANS interacting with other body systems has very large effects on the initial arousal of emotions and then, through interactions with other body systems, the intensity and duration of emotional arousal.

Human emotions in all situations are mediated by the degree *ANS body system* is aroused, as well by the effects of ANS outputs on the other body systems. The same is true of the other body systems that feedback and feed-forward via the flow of hormones and other media as individuals assess and respond to situations. Of great importance is that the ANS system can be activated even when a person is not initially consciously aware of the emotional arousal; and so, in many ways the ANS system is an emotional "alarm"

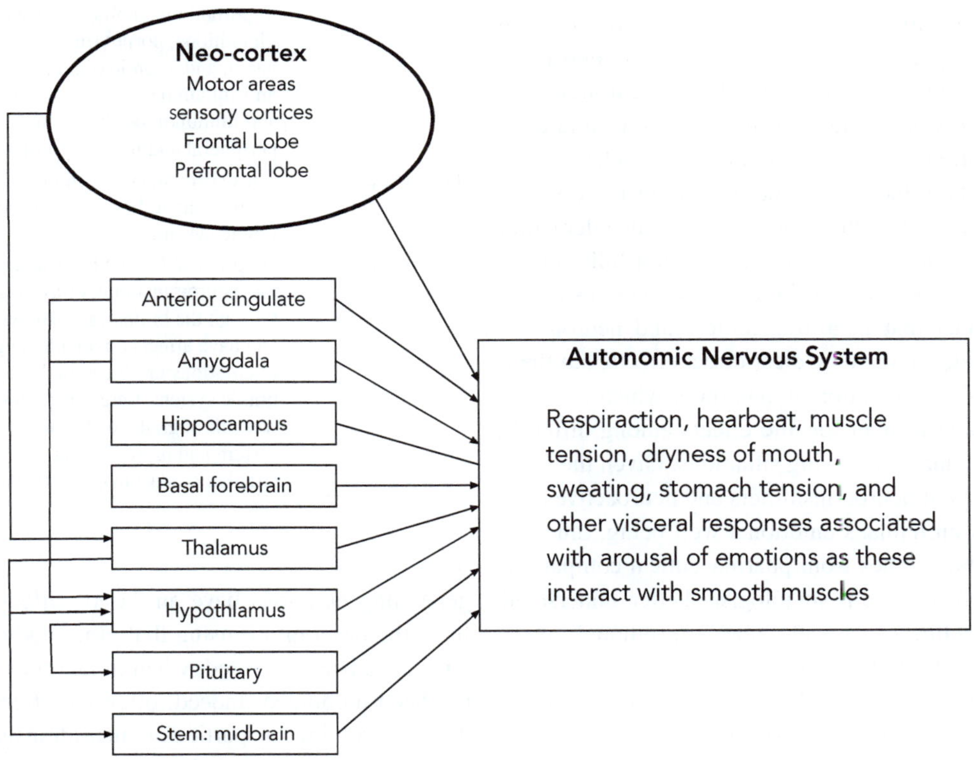

Fig. 7.2 Body-system 1—the autonomic nervous system (ANS)

that will eventually penetrate a person conscious sense of emotions being activated.

2. **The Neurotrasmitter and Nueroactive Peptide Body System (s).** The neotransmitters involved in emotional arousal are released by the midbrain portions of the brain stem, typically from activation of other emotional states and from influence the subgenual prefrontal cortex (Drevets et al., 1997). Moreover, the basal forebrain appears to release one neurotransmitter, acetylcholine (Ach, under stimulation from the thalamus). Over the last 20 years, research indicates that neurotransmitters are also activated outside of the brain stem proper. The key conclusion is that neurotransmitters and neo-active peptides have very large effects on emotional arousal. Table 7.1 and Fig. 7.3 list these effects.

Efforts to establish the specific effects of each neurotransmitter or neuroactive peptide on humans have proven somewhat illusive, although there are clear patterns. For example, the neurotransmitter, *serotonin*, is clearly involved in creating a sense of well-being and relaxation, while encouraging sleep. Or, the neurotransmitter, *dopamine*, clearly is involved as a stimulus to attention arousal and feeding. Still, such common effects of these and other neurotransmitters vary for each individual; and moreover, the interaction effects on neurotransmitters are not fully understood. The problem in sorting effects resides in the fact that neurotransmitters and neuroactive peptides have some clear direct effects, but there are also interaction among these which makes it difficult to map specific effects among different individuals in varying situations. Given the large effects of neurotransmitters and neuroactive peptides on humans emotional well being, much of the data comes from pharmaceutical companies, which may help sociologists better understand their effects in social situations, although much of this research is on addiction effects of some peptides that also block some of the emotional effects of other body systems.

The effects of nuerotransmitters and neuroactive peptides in relation to other body systems generating emotions have an "as if" effect that fools the brain into sensing that other body systems have been activated, when in fact they may not been mobilized. Indeed, often the effects of drugs involving a particular neurotransmitter and/or peptide can be shallow and often frustrating because individuals do not experience the full

Table 7.1 Effects of neurotransmitters and neuroactive peptides

	Effects on brain
Neurotransmitters	
Acertylcholine (*ACh*)	Cortical arousal, learning, and memory as these stimulate body systems; growing evidence that ACh is related to satisfaction
Monoamines dopamine	regulates motor and hypothalamic functions and stimulates most limbic systems
Noradrenaline (*norepinephrine*)	enhances ability of neurons to respond to inuts and stimulates arousal
Adrenaline (*epinephrine*)	Same as noradrenaline
Serotonin	regulates sleep-wake cycles and generations relaxation and pleasure
Histamine	acts as a modulator of several neurotransmitters in the brain and promotes wakefulness and perhaps is involved in addictive problems
Amino acids	
y-Aminoburyric acid (GABA)	works as inhibitory action that controls outputs of neurons
Glycine	is precursor for a variety of important metabolites such as glutathione, porphyrins, purines, haem, and as antioxidant, anti-inflammatory, cryoprotective, and immunomodulatory in peripheral and nervous tissues
Glutamate	Cause excitatory action in neurons, including those of the limbic system
Neuroactive peptides	are produced in the brain and act like neurotransmitters traveling through the brains vascular system and have affects on a wide range of emotions produced in the limbic system. Larger peptides work through the endocrine system and more general circulatory system of the body.

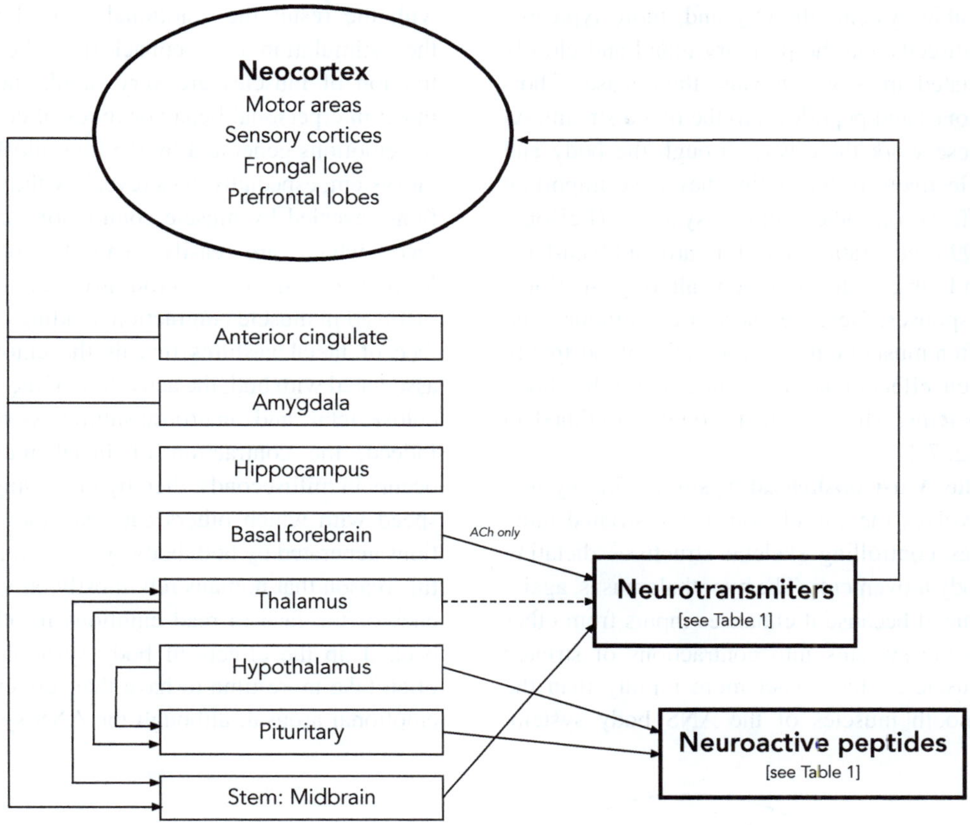

Fig. 7.3 Body system 2—nuerotransmitters and neuroactive peptides

interaction effects that come with the arousal of other body systems.

Having effects like neurotransmitters and neuro-active peptides also moderates connections in various patterns. They are produced in cell bodies, packaged in secretory granules, transported from cell-body terminals, and then released by the *hypothalamus*, *pituitary*, and associated areas of the brain (Heimer, 1995: 388–393). And, as they travel target areas, they also modulate the connections among neurons. Neuro-active peptides are included in analysis of neurotransmitters because they are released in the brain, although their interaction effects are not completely understood. Furthermore, other hormones and peptides operate through the more *general endocrine system*, at times coming back to the brain and have specific effects. Just how many peptides move around in this manner is not fully known. Also, peptides are slower to generate effects when compared to neurotransmitters, which are activated very rapidly. Thus, the effects of neuro-active peptides on emotions will typically take longer to be evident to an individual and others observing this individual. While slow to activate a given mood, the effects of peptides are longer lasting in their effects on emotions. Peptides are, of course, involved in addiction programs because they give individuals the somewhat illusionary emotions of pleasure, while suppressing other body systems generating emotions, which is part of their power to be addictive once individuals turn to addictive drugs.

3. **The Endocrine Systems and Inclusive Vascular System.** This body system involves the endocrine system, the flow of neuro-active peptides, and the vascular system of the body. The hypothalamus receives inputs of other

limbic systems directly and, more typically, indirectly via the pituitary gland and closely related areas, which cause the release of hormones and peptides into the blood stream. As these work there way through the body and selectively in the brain, they have important effects on other limbic systems (LeDoux, 1996) generating feed-forward and feed-back on limbic systems, thereby altering emotional responses. Because these are hormones and often must circulate through the bloodstream, their effects take more time than other body systems. This system is roughly outlined in Fig. 7.4.

4. **The Musculoskeletal System.** This system involves the stimulation of the striated muscles controlling skeletal structures dictating body movements. The hypothalamus is, again, crucial because it enhances inputs from other limbic systems into contractions of striated muscles, which react more rapidly than the smooth muscles of the ANS body system, with the result the emotional arousal from their stimulation and feedback from the contraction of muscles are very rapid. Indeed, much interpersonal behavior relies of cues of the emotions generated by the musculoskeletal system, especially cues reveal by the emotions revealed by muscle contractions in the face, which are easily read by others. Moreover, since neurotransmitters are involved in muscle contraction, reading of the face of facial gestures reveals the emotions associated with both the activation of the musculoskeletal and neurotransmitter systems. Indeed, the contraction of facial muscles occurs in milliseconds, thereby increasing the speed with which others can read the emotions generated by both body systems. It is for this reason that humans rely mostly on visual cues of the face to read emotions in others, backed up the effects of body systems that often take more time to have their effects on emotional arousal, although the ANS system

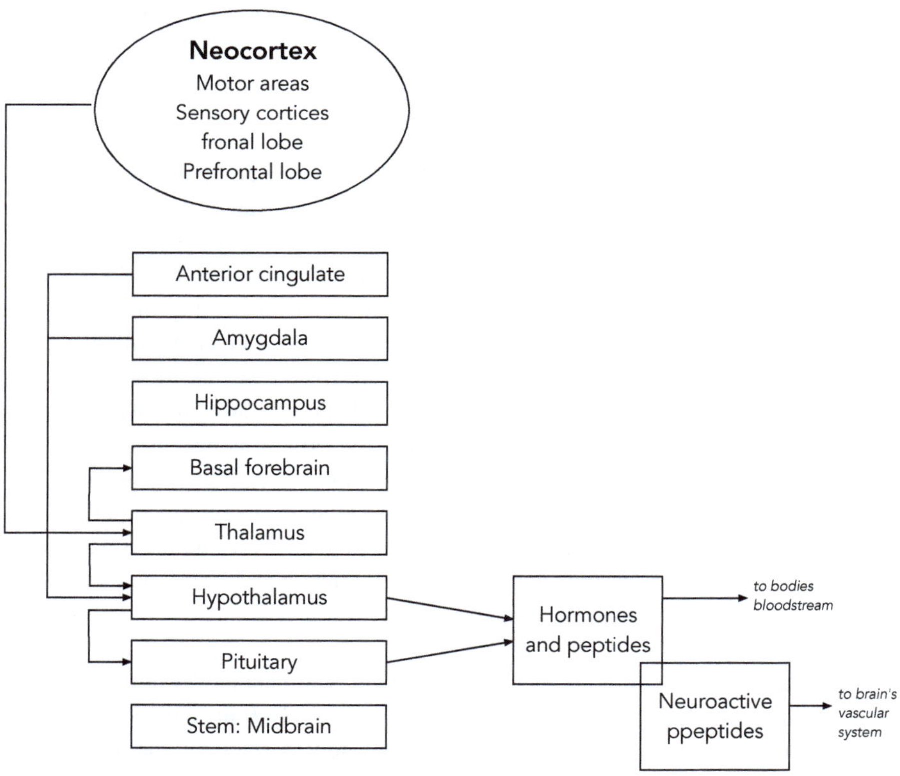

Fig. 7.4 Body-system 3—endocrine and general vascular system

also works rapidly but the emotions generated cannot always be immediately read by others (the person experiencing ANS effects is, of course, feeling the effects rapidly). This system is outline in Fig. 7.5.

The Body Feedback System

The systems involved in mobilizing the body emotionally play off each other in complex ways that are not fully understood. The four body systems briefly outlined above are discrete, in a way, but these systems are interconnected. For instance, the musculoskeletal and ANS are influenced by neurotransmitters; and neurotransmitters are augmented by neuro-active peptides; and endocrine processes are activated by the other body systems, and *vice versa*. This interdependence operates as a series of feed-forward and feed-back waves of activation and de-activations and thereby regulate to some degree emotional arousal. Indeed, these four systems feed back to the rest of the limbic system generating emotions, primarily through the *thalamus* that, in turn, can stimulate the neo-cortex and make emotional consciously experienced. Yet, these feedback processes can remain below consciousness, and as a result, an individual can remain unaware of the emotions being aroused by the four body systems, although others in a situation may be well aware of this individual's emotional arousal. Moreover, as LeDoux (1993a, 1993b, 1996) emphasized for the *amygdala*, animals have an "emotional memory system" that often is unconscious; and he argues that such is probably the case for other emotion-generating structures, although where these unconscious emotions are stored is not entirely clear. But, the interesting dynamic is that some emotional memories remain outside consciousness, as is often the case with the activation of defense mechanisms. Figure 7.6

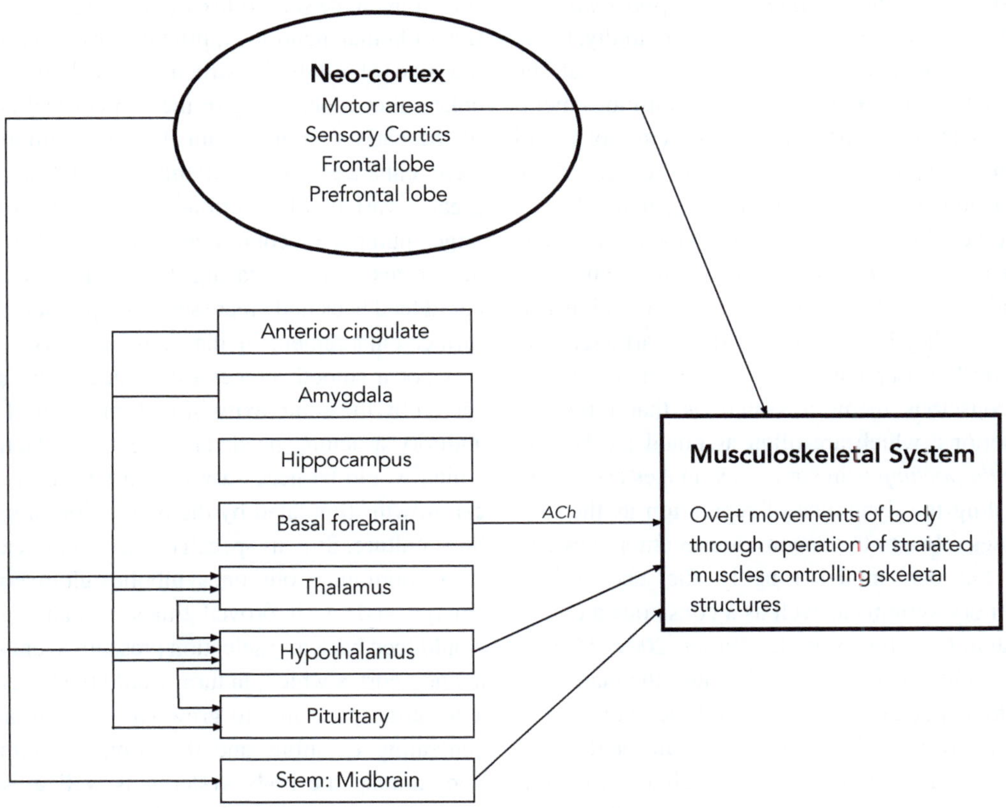

Fig. 7.5 Body system 4—the musculoskeletal activation

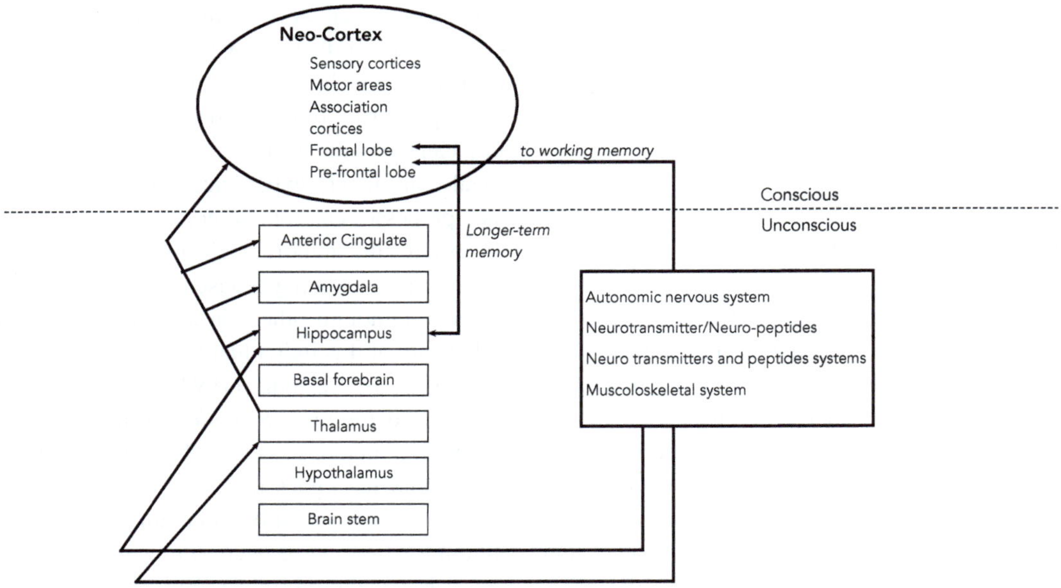

Fig. 7.6 The body feedback system, sub-cortex, and neo-cortex

offers a crude approximation of the feed-back dynamics from the four emotion-generating body systems via the *thalamus* and *hippocampus* to other sub-cortical systems and, potentially, to the neo-cortex and become conscious, but at the same time the existence of an unconscious memory system assures that some emotions will remain unconscious, at least to the person not experiencing the emotions being generated, even as others in this person's environment pick interpersonal cues on the nature of these emotions. And indeed, others will often inform an individual that they look mad, scared, embarrassed, etc. A further implication of the existence of an unconscious memory system is that repressed emotions, which are often assumed are kept in the *hippocampus* in some way, makes even more baffling the dynamics of repression as the processes relate to the unconscious memory system.

One can model roughly the unconscious memory systems as well as a conscious memory system (see, for example, Turner, 2007: 55–65], but I will not pursue these because the main goal of this chapter is to understand the evolution of the neurology of the limbic system, or the production of emotions, *per se*, which are critical to interaction processes and dynamics of the sociocultural formations in a society. Indeed, most analyses of the evolution of the brain focuses on the neo-cortex, but as I outline other chapters, selection on the sub-cortex was driving the evolution of human neurology, primarily by increasing the size of all brain structures, with the first enlargements occurring in the sub-cortical areas of the early hominin brain. For once emotions were enhanced during hominin evolution, the greater variety of emotions could allow for a larger number of cognitions to be stored in the neo-cortex, thus increasing its size that, in turn, could feed back and create selection pressures for further enlargement of the neo-cortex. But this was not a smooth processes; indeed, it became stalled by the complexity of realigning all of the auditory structures needed for speech production, which would be necessary for further neo-cortical growth stimulated by the production of symbolic culture. Still, as speech capacities increased in late hominin evolution, symbolic culture could emerge and, as it proved fitness enhancing, it would feed back as a selection pressure to enlarge the neo-cortex which, in turn, would feed back as a selection pressures to grow sub-cortical areas generating emotions and the complex connections among the body systems as well as sub-cortical and neo-cortical structures. The key to these changes, however, revolved around the sub-cortex getting larger and more inter-connected to produce a larger array or palette of emotions that could be used to tag cognitions and, thereby, set into motion selection to realign structures around the hominin vocal tract which in turn to allow for symbolic culture among late hominins. Thus, the evolution of the neurology of human emotions revolves around the body systems as outlined above, coupled with the growth of key sub-cortical structures where emotions are generated and the other sub-cortical structures, such as *thalamus* and *hippocampus*, that link specific limbic structures generating the primary emotions of *anger*, *fear*, *sadness*, and *happiness* in all high mammals, including hominins and their ancestors.

Growth of the Hominin Subcortex

Most higher-level mammals are highly emotional and all have the capacity to experience at the very least the four basic *primary emotions* of *anger*, *fear*, *sadness*, and *happiness*, with variants revolving around different levels of intensity of these and perhaps a few other emotions considered to be hard-wired in the neurology of mammals. Hominins were a higher mammal because they evolved from the great ape evolutionary line that is evident today in orangutans, gorillas, and chimpanzees. Over a 12 to 18 million year period, the ancestors of hominins were breaking off the line of the ancestors of today's non-human great apes, with the last break off occurring around 6–8 million years ago, creating another great ape line, hominins, that would at about 500,000 years ago evolve into humans. This last break occurred when the first hominins split off from the line ancestors of present-day chimpanzees, thereby initiating the hereditary line leading to the evolution of various subspecies of humans.

The split with the ancestors of today's orangutans, gorillas, and chimpanzees and the ancestors of humans occurred a long time ago (e.g., 12–14 million years ago with the ancestors of orangutans, 10–9 million years ago with the ancestors of gorillas, and 8–6 million years ago with ancestors of today's chimpanzees). But in evolutionary times, this is not such a long period. And the proof of their closeness to humans is the high proportion of the genes that they have in common, albeit across one additional chromosome pair for great apes. This level of genetic closeness allows for what is termed *cladisic analysis*: Using orangutans, gorillas, and chimpanzees as proxies to assess what the hominin ancestors of humans were like millions of years ago. In a sense, cladistic analysis allows us to look back in time and see reasonable approximations of the last common ancestor of humans' hominin ancestors because *humans today still have a very high proportion of the genes* of the ancestors of today's great apes from, as a rough guess, twelve, nine, and six million years ago. Thus, the neuro-anatomy and more general anatomy of great apes, on one side, and humans hominin ancestors, on the other side, can be seen as the phenotypical stock from which humans eventually evolved. Thus, the differences in the brain of great apes and humans today in their neuro-anatomy and general anatomy give us a good view of what natural selection did during the evolution of earlier hominins to the last hominin, humans. Moreover, the behavioral propensities and other key traits of today's great apes are a good proxy for hominin behavioral propensities and other key traits.[3] Again, it is important to emphasized that differences in neuro-anatomy and other traits between today's great apes and humans signal what natural selection did to the hominin line leading to humans. Moreover, those behavior traits and capacities that great apes have in common can tell us *what behavioral traits have a genetic basis* since they exist in all four great apes (labeling humans as a great ape, which is perhaps difficult to accept but a biological certainty).

[3]See for detailed analysis using cladisitic analysis: Maryanski and Turner (1992); Turner and Maryanski (2024); and Turner (2000, 2007, 2021).

What emerges in such comparison, when looking at the neuro-anatomy of the four great apes, is that they are very similar, except for the comparative size of brain structures and, less pronounced but still significant, the density of connections among structures of the brain of humans and the three other great apes. There are a few new structures, and a bit more asymmetry in the brain structures of humans across the right and left hemispheres (which accounts for some of the density of connections among brain structures). Still, otherwise, the brains are very similar, which is predicable given the genetic closeness among all four great apes (counting, as we should, humans as a great ape).

And so, when we compare the sub-cortical neo-cortical areas of the four great apes, as a general rule, sub-cortical areas of the human brain are twice as large as those in great apes, whereas neo-cortexes of the human brain are three times as large compared to the neo-cortexes of the other three great apes. Thus, as might be expected on such a complex and highly integrated structure as a mammalian brain, especially a large one (and despite the size differences between humans and the other great apes in brain structures, all four great apes have large brains for a mammal). Thus, humans' hominin ancestors probably had brains about the same size as great apes scaled to body size, which is correlated with brain size. Thus, the average brain volume of a chimpanzees is about 350–400 cm^3, whereas human brains today have mean average volume of around 1350cm^3.

And yet, the structure of the brains are not that different, which means that natural selection was simply enlarging brain structures rather than doing extensive rewiring or adding a host of new structures, which again make sense since most mutations creating new structure are generally harmful, and doubly harmful in a structure as integrated and complex a brain. And where there are somewhat new structures that exist in the brains of humans but not the other three great apes, they have pushed for additional connectivity in reworking structures for speech production and for enhanced emotionality.

Another source of structural change and connectivity revolves around selection usurping structures that are more symmetrical (on both sides of the brain) in non-human great apes and using the structure on one side for a enhancement or different function, thereby increasing the connectivity of the new structure on one side of the brain and its older counterpart on the other side of the brain with additional neurons through the *corpus callosum*.

Selection thus blindly pursued a strategy during the evolution that would be most fitness-enhancing: Enlarging virtually all key brain structures, but leaving them in place, with selection occurring on existing structures already in place and interconnected with other structures as was the case, for example, of converting *Broca's hump* into of *Broca's area* on the left side of the brain, while, leaving the same area on the right side of the brain intact (for interpreting and emitting in rhythmic patterns auditory responses). Interestingly, in some cases where people have lost the ability to speak because of damage to *Broca's area* on the left, they can speak if they *sing* their "talk" to a well-known song, indicating that they are using the older counterpart of *Broca's area* on the right side of the brain as a conduit now rhythmically embellished by a song. In contrast, *Wernieke's area* in the human the brain is bilateral and responsible for uploading all sensory inputs upload from sequential feeds into the brain for processing (into more gestalt formats compatible with how the human brain processes information, and then using *Broca's area* to decode and download the gestalt-based mode of the "brains way of thinking" into sequential speech. Thus, in the evolution of the hominin brain to the human brain over the last 6 million years and, especially over the last 1.5 million years, natural selection did not add entirely new structures to the brain but found a way to *alter areas of the brain* so as to facilitate changes in the brains of hominins.

This began early in hominin evolution when selection converted primates from olfactory dominant (with the *olfactory bulb* being the only sub-cortical sense receptor) to visual dominance that subordinates all sensory information. The *olfactory bulb* and the *occipital lobe* remain in the same approximate position and reveal pretty much the connections but with additional association cortices added in and around the *inferior parietal lobe* so that tactile sensory inputs (via the *inferior parietal lobe*) and auditory inputs (from the *temporal lobe*) can be subordinated to occipital lobe. So there is some new wiring but everything is in the same approximate location and the internal structure of the sensory lobes is the same, except they are enlarged with some new rewiring in and around the *inferior parietal lobe*. And, this new wiring lobes is probably responsible for subordinating olfaction to vision and, as an byproduct, creating the neurological capacity of great apes for language. Thus, again selection is not doing a wholesale remodel of the brain but working at the edges by reconnecting and re-routing existing neuo-nets among key structures of the brain.

Yet, even in working in this conservative ways, new neurological capacities were generated. For example, the rewiring of—the *association cortices* around the *inferior parietal lobe* for visual dominance created a dramatically new capacity for language, if further selection during hominin evolution could overcome the problems of aligning the structures needed for speech production. Great apes can learn English or any human language if raised in a language using environment, understanding human languages at about the level of a three-to-four year old human child. This capacity for language occurred as a consequence of two ways that natural selection worked to create the hominin to human brain. First was the rewiring in a around the *inferior parietal lobe*, and second was the enlargement of the great apes brain compared to the size of the brain of monkeys and pro-simians. The existing rewiring of monkey and prosimian brains was not enough for understanding languages or "speaking" via hand signals. For, monkeys and great apes have the same rewiring around the inferior parietal lobe for visual dominance, which obviously, would be more fitness-enhancing for animals living in three-dimensional environments (in the forest canopy). Smelling, the dominant sense modality in almost all mammals, but smelling oneself around a three dimensional environment is obviously not as efficient as vision in an ecological niche where one misstep can result in death by gravity. But, monkeys and prosimians cannot understand human language in the way that great apes can, nor can they "speak" via sign language communicated via the visual sense modality. The extra size of the great ape brain allowed new wiring around the *inferior parietal lobe* for visual dominance generated a new emergent property in the brain: the neurological capacity for language. Without this existing capacity installed at least 30 million years ago, humans probably would not have evolved.

This capacity for language among great apes thus remained dormant for millions of years right up to present day with great apes, until they were put in a language speaking environment of primatologists. The discovery that great apes were "understanding" occurred by chance but, once recognized, led primatologists to teach, initially, American sign language, which allows "speaking" to others via the visual rather than auditory neuro-nets in the temporal lobe. But, during hominin evolution, natural selection stumbled onto what primatologists did and began to select on those neuro-nets and structures critical to language use, especially speech production. Thus, perhaps as far back as 50 and certainly by 30 million years ago, this latent capacity for language was being installed with the conversion of olfactory to visually dominance among primates. By natural selection and only after a great deal of rewiring of the vocal track of hominins, which like that of all great apes was not conducive to speech production, did this pre-adaptation become subject to further selection, and eventually over the last 2.0 to 1.0 million years find the right re-alignment among vocal tract and surrounding structures to allow later hominins, perhaps as early as 1.6 million years ago to speak. And with speech would come symbolic culture, both of which would feedback and allow, and indeed push for, more complex emotions and enlargement of the late hominin neo-cortex.

The Significance of Emotions to the Rewiring of the Hominin and Then Human Brain

The Vulnerability of Great Apes in New Habitats

Hominins emerged when the forest of Africa were receding, opening up new open-country habitats and forcing the less fit inhabitants of the forests to adapt to these new habitats where a primate would be exposed to the dangers of predation. Indeed, descendants of earlier great apes like orangutans, gorillas, and chimpanzees cannot survive in savanna-like habitats that do not have trees to retreat into for sleeping at night and, thereby, escaping the reach of many predators that always inhabit open-country habitats.[4] The reason that great apes are vulnerable is that they do not have tight-knit social structures, such as permanent groups, that can coordinate movement through open-country habitats and, moreover, that can mount collective defense against predators as well collective food hunting away from the protection offered by trees. The individualism, weak ties, and lack of permanent group structures, especially the nuclear family, make today's great ape vulnerable when leaving the forest of today, and such as was the case 6–7 million years ago when hominins split for the ancestors of contemporary chimpanzees. Thus, the paramount vulnerability of great apes in the open country in Africa as the forests were receding is *the lack of high solidarity, stable group structures*. It would be up to blind natural selection, working on the neuro- and body-anatomy of hominins, who were initially very much like today's chimpanzees, to somehow by chance discover a way to increase the strength of bonds among hominins that would allow them to create groups of higher levels of group solidarity and stability to survive in a more predatory-ridden habitat like the African secondary forests, bushlands and savannas.

Monkeys, in contrast to great apes, have little trouble in open country because they have *genetically controlled bio-programmers* for group formations. And given that monkey groupings can be quite large, but highly stable because of female matrilines among related female kind and because of male dominance hierarchies, a group of monkeys can generally protect itself even in open-country habitats filled with predators. Chimpanzees cannot, even though they are bigger and smarter than monkeys, because they do not have any genetically controlled bio-programmers for group formation, except some genetically controlled propensities to live in larger communities of conspecifics but these communities are too loosely structured compared to permanent groups (although all great apes have mother-offspring relations until puberty but *do not* reveal a nuclear family because sexual relations between males and females are highly promiscuous). And so, how was natural selection to *randomly hit up a solution to this problem and somehow make great apes more social and more prone to form groups*?

The Enhancement of Emotions as an Adaptive Strategy

Natural selection hit upon an adaptive strategy by enlarging the sub-cortex. In turn, a larger palette of emotions would increase the effects of the sub-cortex on the body systems that are implicated in emotional arousal in both feed forward and feedback relations among the structures outlined in the middle and bottom panels of Fig. 7.1. I first began to believe several decades ago that the emotions were the driving force of the neurology of hominins is when I discovered, almost by

[4]In west Senegal, there is a small population of chimpanzees that do survive in a sanvana-desert like habitat and, unlike most other chimpanzees they found a way to organize into more cohesive groups for hunting and defense. They survive, it appears, by retreating at night not to a forest but stick-like trees with few leaves that dot the landscapes. Without these "trees" they probably would not have been able to survive and, on their own, figure out how to become better organized. All other chimpanzee populations in the world cannot leave the forest canopy into more open-country habitats. See: Baldwin (1979), Baldwin et al. (1982), McGrew (1981, 1992, 2010) McGrew et al. (1981), Tutin et al. (1981), Pruetz and Bertolani (2007).

Table 7.2 Relative size of brain components of apes and humans, compared to *Tenrecinae*

Brain component	Apes (Pongids)	Humans (Homo)
Neocortex	61.88	196.41
Diencephalon thalamus hypothalamus	8.57	14.76
Amygdala	1.85	4.48
Centromedial	1.06	2.52
Basolateral	2.45	6.02
Septum	2.16	5.48
Hippocampus	2.99	4.87
Transition cortices	2.38	4.43

Note: Numbers represent how many times larger than *Tenrecinae* each area of the brain is, with *Tenrecinae* representing a base of 1

Sources: Data from Stephan (1983), Stephan and Andy (1969, 1977); Stephan et al. (1988); and Eccles (1989)

chance, the data in Table 7.2. These were early measurements comparing great ape and human brains; and while the neo-cortex at three times the volume of great apes gets most the press, the doubling and sometimes more than doubling of humans sub-cortical areas also needed explanations. One explanation was that since cognitions can only be remembered when tagged with emotions, the sub-cortical areas generating emotions, *per se*, and in their interactions with the four body systems outlined in Figs. 7.2, 7.3, 7.4, 7.5, and 7.6 had to have evolved *before the neo-cortex could grow*. For, without emotional tags on cognitions, they cannot be retained in memory and decision-making.

In the 1980s, Alexandra Maryanski had done research on the networks of monkeys and great apes, outlining their differences in capacities and propensities to form stable groups.[5] And, in our first book together on the evolution of humans (Maryanski & Turner, 1992), it became clear that for great apes to survive in more open country savannas in Africa, emotions would be an important force generating more stable groups structure since a large sociological literatures documents that emotions and ritualized exchanges of emotions generate social solidarity among humans (e.g., Goffman, 1959, 1967; Collins, 2004). Also intriguing was the very large *septum*, which is the area generating emotions attached to sexual pleasure in mammals, was well over twice as large as it is in Chimpanzees. Chimpanzee societies are built around promiscuous sex, and this is why paternity can never be known and why we know that great apes, especially chimpanzee and orangutans, do not generate *anything like a nuclear family beyond attachments*. Females raise their children alone, with all female great apes leaving their mothers at puberty, migrating to a new community never to return. Among two species of great apes (orangutans and gorillas), males also leave their mother and natal community, never to return. Only male chimpanzees remain in their natal community and visit their mothers but to do form a permanent group with them. In return, the offspring lost to migration are replaced by female offspring migrating away from their natal communities into what will be their new adult community for their lifetime. Thus, there is *no enduring group built around kinship solidarity of fathers, mothers, and offspring in all great apes societies*. And, by the logic of cladistic analysis, the same had to be true for early hominins. How then did these early hominins survive, especially when pushed out of the forest canopy by ecological changes?

Part of the answer is that, earlier on, the first hominins probably stayed within the forests and did not wonder out onto the savanna, or even bushlands, because of the dangers of predation. Yet, the forest were shrinking in Africa over the last 10,000 years and, eventually, there was no choice but to move out from the forests, with a transition period moving out onto the savanna in daytime forays but always retreating the trees at night. But, at some point, some hominins would have to live *full-time away from the forests* and, then, we can ask: How did they survive without group formations and nuclear families? The answer is that selection probably had begun to select on sub-cortical areas, increasing the emotionality of hominins, with this extra dose of emotionality—as revealed by the doubling of sub-cortical emotions systems—providing for

[5]See Alexandra Maryanski (1986, 1987, 1992, 1993, 1995, 1996, 1997).

an alternative force for increasing social solidarity, commitments, and eventually more stable groups built from exchanges of positive emotions. This is not difficult conclusion to make because this is how humans generate solidarity today, and it is also the way that chimpanzees generate emotional arousal in their various festival-like gatherings where they hoot and holler, bang wooden logs in drum-like rhythms, and other wise act like humans at a rock festival or a collective celebration like Mardi Gras. And when it is over, they wander away and revert to their more solidarity ways in temporary gatherings or wandering in a community alone. So, these festivals probably have a genetic basis and, if emotions are ramped up by doubling the size of emotion centers, dramatically enhanced emotional states could generate emotional attachments and more cohesive groups foraging in more open-habitats filled with predators. And, perhaps these emotions could begin to allow them to form stable groups, and moreover, given that the *septum* of humans (where emotions associated with sex are generated) is so large that they could have the beginnings of *a nuclear family*, perhaps only temporary, of males and females with children forming a more stable conjugal group to protect young offspring. Indeed, given that chimpanzees get great pleasure in promiscuous sex, enlargement of the septum, *per se*, would seem like overkill, but it is likely that new emotions generated in this larger septum were of a different nature were evolving, such as emotions for commitment, caring, attachment, or even "love" beyond and above the pleasures of sex, *per se*. If such was the case, it is likely that enlarging the septum would be unnecessary and perhaps even dysfunctional to load up powerful and already emotional animals like chimpanzees with even more sex drive! But, as outlined earlier, selection took the route of selecting on more generalized attachments and emotions in creating and sustaining the structure organizing male, female, and offspring in a nuclear family, then enlarging septum would add a new dimension to great ape communities: stability of a viable nuclear family within a larger community formation.

The data in Table 7.2 also reveal that the *basolateral nuclei* in the *amygdala* account for almost all of the doubling of the volume of the *amygdala*. How could this be fitness enhancing if solidarity and group structures is what is needed? Why doubling the size of the structure generating negative emotions—*anger* and *fear*—work to produce solidarity? replace them with positive emotions promotion solidarity? Indeed, why would selection pursue this strategy of enlarging emotion centers when three of the basic primary emotions—anger, fear, and sadness—are negative? Much later than the data in like that in Table 7.2, used a different methodology—allometric measurements of nuclei in brain structures—revealed that doubling of the size of the *amygdala* was achieved by not the *basolateral nuclei* but rather by only the *lateral nucleolus* which *does not generate anger or fear* but, instead, activates attention to the more general context of interaction among conspecifics. Most revealing are the syndromes that are generated when the *lateral nucleolus* within the *amygdala* is damaged. One is Williams Syndrome, which is a situation of over-empathizing as is the case when a young boy with this syndrome sees a stranger who is sad and goes over to hugged distressed stranger. Thus, in the middle of the subcortical structures for anger and fear sits a structure promoting *empathy which leads to bonding*, with damage to this nucleolus making some people super-empathic. The other pathology is just opposite of Williams Syndrome, autism, where individuals have great difficulty experiencing empathy with others. Thus, natural apparently created a new module in the middle of the most ancient emotional system in mammals: the centers for *fear* and *anger*, which are critical for survival but not conducive to forming social bonds. But, as is the case in other parts of the brain, an existing structure is altered somewhat by adding new capacities while at the same time sustaining the original programming of the overall structure—in this case, the *amygdala*. And so, installing a structure in the center for negative that (1) increases capacities for empathy or for seeing context of behaviors and the emotions generated and perhaps (2) mitigates the power of anger or fear was one route for enhancing the solidarities, even in the face of negative emotions. Of course, this is not the silver bullet signaling the origins of solidarity but it does suggest how natural selection was working: altering existing structures through enlargement allowing for the enhancement of capacities to experience emotions that generate solidarity or, at the very least, mitigate the arousal of negative emotions that could disrupt solidarities. In either case, by reducing the power of negative emotion, which probably constitute three quarters of all primary emotions among hominins including humans, works to promote social solidarity, especially if the primary emotion of happiness can be enhanced and new variants generated by enlarging various sub-cortical systems that generate pleasure and happiness, activating those body-system that can do the same.

Thus, for the last 25 years that I have studied emotions, I have had this story in mind: Natural selection working of the neurology of hominins increased emotionality that, in turn, could *create just enough* solidarity and group structure for hominins to survive at first in more open-country secondary forests and bush lands, and then open savanna and desert ecologies. Selection was taking a conservative and less disruptive route—enlargement of structures, increased connectivity, and modification of *existing* portions of existing structures rather than creating entirely new structures. Comparing great ape brains with human brains reveals that the two brains are basically the same except for size, connectivity, and partial modification of structures generating new emotional capacities that, in turn, opened the door for enlarging the neo-cortex. If selecting on the brain is the only way to increase fitness, then this conservative approach would be the only viable strategy for natural selection working randomly without foresight or teleology.

Complex structures with many different interconnected systems cannot be radically changed without decreasing fitness. Yet, slower and modest alterations of structures over millions of years proved to be a fitness enhancing strategy for the hominin line because humans, for better or worse, now dominate the planet.

The fact that the hominin brain did not grow beyond the size of common chimpanzees for a long time suggests that the protection of the forests was enough for survival in early hominin evolution, but after four million years, the hominin brain did begin to grow, first about $50cm^3$ and soon $100cm^3$ in latter part of time that *Australopithecines* lived on the earth. This modest increase was mostly on the sub-cortex sitting in the center of the brain, pushing out on the neo-cortex that remained about the same size as the last common ancestor with chimpanzees.

For the neo-cortex to grow dramatically, selection had to first grow the sub-cortex because the *hippocampus* must tag cognitions with emotions if they are to be remembered and used in decisions making. Again, increasing the size of the neo-cortex without the capacity to expand the emotions used to tag a larger number of diverse cognitions would be impossible without a more diverse and complex set of emotions with which to tag these cognitions. Indeed, without a large range of emotions to tag a new kinds of cognitions in the neo-cortex, enlarging the neo-cortex, *per se*, would like building an empty warehouse that, in the brain, would consume valuable resources, such as calories and protein, which would not be fitness enhancing.

How, Then, Did the Range and Diversity of Human Emotions Evolve?

The Evolution of Variants of "Primary Emotions"

All emotions are generated by the neurological structures of the human brain. In a sense, the notion of "primary emotions" vs. "derived emotions" rests on a false assumption that some emotions are biological, while others are socio-cultural. The fact is *all* emotions among humans are both in this sense: all emotions have a biological basis, and all emotions have cultural labels. Still, the notion of primary or, in reality, the first emotions evident among humans is useful in order to understand what the evolution of the neurology generating emotions accomplished: an increasing range and diversity of emotions that humans can experience. Thus, in my view, the notion of "primary" is the emotions that the ancestors of hominins first possessed. All researchers on the issues of primary emotions agree on at least four primary emotions, and these are clearly evident in great apes and, hence by the logic of cladistic analysis, the first hominins. These universal emotions are *anger*, *fear*, *sadness*, and *happiness*. Table 7.3 presents a list of additional primary emotions that various scholars have postulated but these data are drawn from studies of humans, not great apes and, thus, no speculation on the basic emotions of the first hominins.

In my view, it is likely that hominins only experienced the primary emotions of *anger*, *fear*, *sadness*, and *happiness* because we can find in sub-cortical structures connected to body systems that existed among hominins in contemporary great apes, who can serve by the logic of cladistic analysis as our proxy of what the first hominins were like. How, then, did natural selection begin to expand the emotional palette of hominins, given the existing structures that already existed. I would argue that there would be increased variation on a small set of primary emotions. Table 7.4 outlines my view of the variations on primary emotions evident in humans; and so, over the time that the brain began to grow, these may have been some of the first "new emotions" among hominins. And these probably existed after the 100cm^3 growth in the brain among later *Australopithecenes* (4 million years ago), being mostly sub-cortical with, perhaps, modest enlargement of the neo-cortex. And, once the neo-cortex began to grow after 1.6 million years ago with *Homo erectus*, much of this growth was neo-cortical which would feed back to the sub-cortex as it also continued to grow.

There is a tendency among sociologists to consider that everything about humans is cultural, including emotions. But symbolic culture only emerges with speech; and speech was a late arrival in hominin evolution. And so, many emotions were simply biologically based. The difficulties for a random process like natural selection to re-align speech centers of hominins while also creating new neo-cortical areas (e.g., *Broca's area*) allowing for the translation of gestalt-based "brain thinking" into sequential speech were formidable for a blind process like natural selection. Still, it is likely that the evolution of *more variants* on the four primary emotions began *before* the neo-cortex began to grow significantly, but once the neo-cortex grew, the like-

Table 7.3 Varying numbers of hypothesized primary emotions

Fewer primary emotions (3–6)
Sroufe (1979): pleasure, fear, anger
Kemper (1987): satisfaction, fear, anger, depression
Arnold (1960): flight, fight, defensive, aggression
Johnson-Laird/ Oatley (1992): fear, anger, sadness, disgust
Gray (1982): hope, anxiety, anger, sadness
Fehr/Russell (1984): happiness, fear, anger, sadness
Trevarthen (1984): happiness, fear, anger, sadness
Turner (1996a, b): happiness, fear, anger, sadness, surprise
Scott (1980): pleasure, love, fear, anxiety, anger, loneliness
Epstein (1984): joy, love, fear, anger, sadness
Arieti (1970): satisfaction, fear, tension, rage, unpleasure, appetite
Ekman (1984): happiness, fear, anger sadness, surprise, disgust
Moderate number of primary emotions (7–8)
Darwin (1872): pleasure, joy, affection, terror, anger, astonishment, pain
Plutchik (1980): joy, fear, anger, sadness, surprise, disgust, anticipation, acceptance
Fromme/O'Brien (1982): joy, elation, satisfaction, fear, anger, grief, resignation, shock
Panksepp (1982): fear, panic, rage, sorrow Loneliness, grief, expectancy
Malatesta/Haviland (1982): joy, fear, anger, interest, pain, brown flash, knitbrow
Higher number of primary emotions (11–12)
Emde (1980): joy, fear, anger, sadness, surprise, disgust, shame, shyness, distress, guilt, interest
Izard (1977, 1992): enjoyment, fear, anger, contempt, surprise, disgust, shame, shyness, distress, guilt, interest
Osgood (1966): joy, quiet, pleasure, fear, anxiety, anger, sorrow, amazement, disgust, interests, expectancy, boredom

Note: Virtually all scholars posit variants of h*appiness*, *sadness*, *anger*, and *fear* as primary emotions
Sources: J. Turner (2000: 68–69); Turner and Stets (2005: 14–15)

Table 7.4 Variants of primary emotions

Primary emotions	Low intensity	Moderate intensity	High intensity
Satisfaction-happiness	Content	Cheerful	Joy
	Sanguine	Buoyant	Bliss
	Serenity	Friendly	Rapture
	Gratified	Amiable	Jubilant
		Enjoyment	Gaiety
			Elation
			Delight
			Thrilled
			Exhilarated
Aversion-fear	Concern	Misgivings	Terror
	Hesitant	Trepidation	Horror
	Reluctance	Anxiety	High anxiety
	Shyness	Scared	
		Alarmed	
		Unnerved	
		Panic	
Assertion-anger	Annoyed	Displeased	Dislike
	Agitated	Frustrated	Loathing
	Irritated	Belligerent	Disgust
	Vexed	Contentious	Hate
	Perturbed	Hostility	Despise
	Nettled	Ire	Detest
	Rankled	Animosity	Hatred
	Piqued	Offended	Seething
		Consternation	Wrath
			Furious
			Inflamed
			Incensed
			Outrage
Disappointment-sadness	Discouraged	Dismayed	Sorrow
	Downcast	Disheartened	Heartsick
	Dispirited	Glum	Despondent
		Resigned	Anguished
		Gloomy	Crestfallen
		Woeful	
		Pained	
		Dejected	

Source: Data from Turner (1999a, 1999b)

lihood of further expansion of the neo-cortex would continue to grow as the sub-cortex grew, but growth would always be limited *until* speech emerged allowing humans to articulate more varied emotional and cognitive states. And here, most sociologists would jump in and pronounce that this larger brain makes cultural. This is true in the sense that humans can now denote with label with speech any emotional state, but the arousal of the emotions themselves is not cultural. It is biologically-based, even as culture sets this biology into motion. Humans cannot experience the power of emotions *without activating subcortical centers of the brain and body systems*. If, for example, I say that "I am *angry* at this very moment," when in fact I am not, does not activate real anger because the relevant subcortical areas of the brain and more general body-systems have not been fired up. The general point is this: variants and other new emotions that would evolve probably had a cultural element if hominins and humans could give them a name through speech but, unless sub-cortical centers of the brain and body systems are activated, an emotion or set of emotions is not really activated. And, as both the sub-cortex and neocortex were expanding, individuals would be "feeling" new emotional states to which they would eventually give cultural labels, but the labels are not the source of the emotions; the source lies in the neurology of humans and activation of the body systems outlined at the beginning of this chapter. The same logic applies to other ways in which palette of emotions was expanded.

A Note on the "Language of Emotions" and Hominin Survival

Chimpanzees and probably all great apes use what I term the "language of emotions" to communicate feelings as well as more instrumental intentions in interaction, mostly communicated by hand, body, and most importantly, "face" (which may be why the face of primates is not covered in fur). And, during the course of any interaction among humans (and earlier hominins as well), this "language of emotions" is in full force, even as individuals employ spoken language. Indeed, *individuals always assess implicitly the level and type of emotions accompanying speech*, and they look for confirmation in the body continence, voice inflections, and movements of body parts in seeking to know what a person is feeling.

Equally important, humans search for contradictions between what people are saying vs. what they are feeling, because it is *the feeling part of this equation that is most important in achieving attunement with others in interaction*. The famous phrase, "smile when you say that," highlights that what you are saying needs to be confirmed by body gestures that communicates emotional meanings. In this phrase, the intent of the listener using this famous phrase communicates that, if you are going to say such a disparaging thing, you had better be joking.

Indeed, the language of emotions was probably hominins *first language* carrying common meanings that allow interactions to increase solidarities. These dynamics are clearly evident among great apes whose faces and body countenance "speak volumes" in terms of achieving interpersonal alignment and even temporary solidarities. And, it is for the reason that selection hit upon expanding the palette of primary emotions as the first means for increasing social solidarity among individualistic hominins that were, as the forest began their retreat, forced to reside in more open-country habits where increased solidarity in more stable groups structures would be fitness enhancing. The fastest route to increased solidarity would be to enhance emotions, since it would be many million years—at least four to five million years—before late hominins could speak to each other via the re-aligned vocal tract, beyond discrete calls and other signs (mostly of their emotional state).

The Evolution of "First-order" Elaborations of Emotions

I am labeling the mixing of two primary emotions together, in some unknown way, as first-order emotions that are not cultural but, instead, biologically generated. As new neuro-nets connecting growing sub- and neo-cortical areas of the hominin brain evolved over the last two million years, the elaboration of culture gave labels to new, biologically-generated emotions. The reason that I can write these emotions down is because I have cultural (symbolic) labels for them, but *this obvious fact does not obviate their biology* which, is in the end, is what must be activated to experience emotions at a level of "feeling level" that activates sub-cortical areas of the brain and body systems charging up emotional states.

In Table 7.5, I suggest that combinations of two primary emotions, with one of the primary emotions being more intensely aroused biologically than the other, can be considered as "first order" elaborations beyond the expansion of primary emotions. It is possible to conceptualize fur-

Table 7.5 Combinations of primary emotions

Primary emotions		First-order elaborations
Satisfaction-happiness		
Satisfaction-happiness + aversion-fear	Generate	Wonder, hopeful, relief, gratitude, pride, reverence
Satisfaction-happiness + assertion-anger	Generate	Vengeance, appeased, calmed, soothed, relish, triumphant, bemused
Satisfaction-happiness + disappointment-sadness	Generates	Nostalgia, yearning, hope
Aversion-fear		
Aversion-fear + satisfaction-happiness	Generate	Awe, reverence, veneration
Aversion-fear + assertion-anger	Generate	Revulsed, repulsed, antagonism, dislike, envy
Aversion-fear + disappointment-sadness	Generate	Dread, wariness
Assertion-anger		
Assertion-anger + satisfaction-happiness	Generate	Condescension, mollified, rudeness, placated, righteousness
Assertion-anger + aversion-fear	Generate	Abhorrence, jealousy, suspiciousness
Assertion-anger + disappointment-sadness	Generate	Bitterness, depression, betrayed
Disappointment-sadness		
Disappointment-sadness + satisfaction-happiness	Generate	Acceptance, moroseness, solace, melancholy
Disappointment-sadness + aversion-fear	Generate	Regret, forlornness, remorseful, misery
Disappointment-sadness + assertion-anger	Generate	Aggrieved, discontent, dissatisfied, unfulfilled, boredom, grief, envy, sullenness

ther first-order states, as would be the case with a very small amount of one emotion mixed with more intense valences of another emotions, or two emotions equally mixed. Still, I doubt that this kind of fine-tuning would be fitness enhancing because a palette of too many emotions would increase the complexity of thinking, feeling, and interpreting emotions in interpersonal situations. Indeed, I think that there is ceiling on how may emotions activated by sub-cortical areas and body systems can exist, probably well under 200, although there can be many more different cultural labels for the same biologically generated emotions cross-culturally (although cross cultural studies consistently reveal that humans report the same approximate number of emotions that they can interpret). Indeed, in my travels to societies where I do not speak the language, I find it relatively easy to get help from others whose body language is more understandable than their words with the tone of my voice adding considerable emotional content that allows them to interpret by body language. Indeed, as noted earlier, the "language of emotions" was probably responsible for the initial increase in organization among hominins that allowed them to survive, and it still the way that humans today communicate with each other to form social bonds that would enable them to build new kinds of fitness-enhancing social structures.

Table 7.5 outlines my thoughts on how natural selection began to push the brain to having the capacity to "combine" in some unknown way, a greater amount of one of the four primary emotions with a less amounts of another primary emotion. Much of this could have evolved before spoken language and, as a result, increased social bonding with a more complex palette for the "language of emotions."

The Evolution of "Second Order" Elaboration of Emotions

Another way of "mixing" emotions is the combine three negative emotions. My sense is that two of the unique emotions of humans (and potentially to other high intelligent mammals and birds) arise from combining the three of the negative emotions—that is, variants of *anger*, *fear*, and *sadness*. Especially for late hominins trying increase social bonding and group solidarities (since there are no direct bio-programmers for these) combining of three negative emotions could work to

Table 7.6 The structure of shame and guilt

	Rank-ordering of constituent primary emotions		
Emotion	1	2	3
Shame	Disappointment-sadness (at self)	Assertion-anger (at self)	Aversion-fear (at consequences for self)
Guilt	Disappointment-sadness (at self)	Aversion-fear (at consequences for self)	Assertion-anger (at self)
Alienation	Disappointment-sadness (at self)	Assertion-anger (at others)	Aversion-fear (at consequences for self)

create new negative emotions that would work to increase social control and social solidarity. By shifting the order of salience or valence of the three negative primary emotions, it becomes possible to generate even more emotional states. Table 7.6 outlines what I call the "the emotions of morality" because they are experienced in relation to cultural codes and, thus, evolved as far as we know, only among humans. The negative second-order emotions of *shame* and *guilt* are the emotions of culture because they arise when a person has not met *cultural* obligations. Now, it is possible that they began to emerge before language, but they would not be fully in place until humans could articulate *cultural codes* defining moralities in diverse situations. The emotion of *alienation* is a better candidate for evolving before language because it denotes sadness and an unwillingness to commit to an ongoing set of social relations. It does not depend upon being in reference to moral codes, and might have been a useful emotion to communicate to others the need for them to respond to the negative reaction of a person (i.e., their *alienation*) from social events in emerging group structures.

Combining the three negative emotions together produced new emergent emotions that (a) draw attention to the salience of moral codes that were evolving to sustain group structures and, in the case of *alienation*, (b) lack of commitment to ongoing social relations or even the group were fitness enhancing for two reasons. The first is that they convert the negativity into a mandate for cooperative behaviors, solidarities, and group formation that *increase commitments to groups,* or in the case of *alienation*, to signal a break in commitments that needs to be attended to. Indeed, sadness as an emotion has always communicated such orientations but by adding *anger* and *fear* in the mix, a new kind of emotion emerges that make a more urgent plea to increase solidarities in a social grouping.

Virtually of the primatologists who are engaged in studies of language use among great have communicated to me that they do not think that great ape experience *shame* or *guilt*, but for evolving great apes like hominins who were seeking a way to survive in more open country and who were beginning to talk, this use of talk to create moral codes, the emotions of *shame* or *guilt* created powerful emotions facilitating social control. *Guilt* is an emotions that individuals experience when they have violated moral codings, which would increase dramatically with the evolution of speech and new cultural codes that surely began emerge once hominins could speak and create symbolic culture (values, beliefs, norms, and other codings carrying moral implications). *Shame* is the emotion experience when an individual feels like they have caused harm to others, the group, or even violated an important moral codes. Both *guilt* and *shame* are powerful emotions of social control and, in so being, push for group solidarities. Indeed, without the eventual emergence of spoken language and symbolic culture, hominins may well have gone extinct; and moreover, without *shame* or *guilt*, larger scale societies of at first thousands, and many thousands to many millions, and now in two cases billons of humans could never have evolved.

Thus, without *guilt* and *shame*, moral codes would have little power to force conformity to cultural coding; and without this emotional power, social structure among humans cannot ever be very complex. Again, emotions like *guilt* and *shame* demonstrate the interconnectedness of biology and culture because the effort to articulate moral codes, while being a product of a large neo-cortex and capacity for language and symbolic culture, depends up *sub-cortical processes generating emotions like shame and guilt*. Shame and guilt to do have their power by just the articulation of a moral code; they have their power because they activate sub-cortical emotions centers and body systems that force individuals to pay attention to cultural codes or suffer the consequences of experiencing *shame* and *guilt*, which are two the most powerful emotions that individuals can experience.

Conclusion

In this chapter, I have tried to summarize the neurology behind emotions from a sociological perspective. My review is not complete because other neurological systems, such as those creating brain waves and thus activating emotion systems are not included in my analysis. Equally important, I do not think that the neurology of emotions can be fully understood without employing an evolutionary approach. Cladistic analysis allows us to use great apes as a proxy for what early hominins were like, although obviously great apes have evolved but still within the general habit and probably niches within those habitats as well of earlier ancestors of today's great apes. Humans more direct ancestors—successive species of hominins—were being force into a very new set of habitats—secondary forests, bush lands, savannas, and even deserts—and hence had to evolve to survive in these habitats. The driving force in hominin evolution was overcoming the weak-tie structure of great ape communities in a changing climate that was pushing the forests in Africa back and increasing open-country habitats where social organization would be the only means for survival. Yet, the dilemma was that great apes do not reveal tight-knit social structures, as do monkeys and most other mammals with bio-programmers pushing for more cohesive group structures. The neurology of emotions and the evolving palette of emotions among successive species of hominin allowed them to survive forays into open-country habitats, but without a means to *forge more cohesive and permanent group structures*, it is likely that the hominin line would have gone extinct, leaving only the current refuge of today's great apes who probably where not ever very populous (because few fossils on the ancestors of today's great apes have been found, whereas hundreds of the many of ancestors of monkeys have be uncovered). Still, the few apes that were left on the planet did have a narrow niche to which they were well adapted, but when hominins split off from the last ancestor of all present-day great apes (the ancestors of chimpanzees), they were forced to become better organized or die as the forest habitats of Africa began to shrink. Without bio-programmers for cohesive groups, they were vulnerable, with this vulnerability increasing with the fact that the only route that natural selection found was to alter the neurology of hominins so that they could become more emotional and thereby achieve some degree of solidarity over what other great apes haves had revealed.

Natural selection found two routes for survival of hominins. First, and very easy, was direction selection to make hominins bipedal, which was easy since all of the present-day apes can stand up and walk around on their hind legs. This adaption probably bought some time for hominins to move in and out of forests, but did not solve the problem of becoming more social and group oriented. To increase sociality and group bases of social organization, natural selection had only one remaining route: alter the neurology of the brain and, somehow, make low-sociality hominins more group oriented so that, as groups, they could survive. Since hominins were highly intelligent for a mammal (with 350cm^3 brains), there something for selection to "select on" but how to find a way to increase group solidarities when there are no existing modules for group solidarity in great apes to select on was problematic.

The "solution" that blind natural selection eventually hit upon was to begin to enhance emotionality, selecting on sub-cortical structure buried deep in the brain. The brain is complex and hence mutations would almost always be harmful. Still, great apes have very well-develop interpersonal capacities, and they have ability to form temporary solidarities through short rituals. Moreover, they had a secret—the pre-wiring for language buried in their brain as a result of the shift of the dominant sense modality from smell to vision. Thus, there were locales in the brain that held promise for success if natural selection could blindly find them, and in its random way, select on these structures. The strategy or, less teleologically, the direction that natural selection took was to enhance all structures of the brain, target specific structures, especially those that are bilateral and on both sides of the brain and alter one side for expanded or slight altered functionality, particularly in late hominin evolution when the dormant capacity for language could be unlocked as speech around the vocal tract could be selected upon and re-aligned—which turned out to be the biggest roadblock to the evolution humans but, once overcome, it became the driving for of the evolution of human societies.

Yet, after making hominin bipedal, natural selection hit upon what would turn to be a "winning strategy": Enhance emotions because with enhanced emotion, the interpersonal skills hardwired into the neuro-anatomy of all great apes could be used to enhance the strength of social ties and, slowly, build group solidarities to the point where hominins could survive forays from the forest to predator-ridden open-country habitats—where eventually they would have to reside full time. And as emotional capacities expanded, neo-cortical centers of the brain could expand until blocked by the roadblock imposed by the difficulty of selection finding a way to re-align the complete set of structures need to produce articulated speech. Still enhanced emotionality and somewhat enhanced neo-cortex were *just enough* to allow species of hominins to survive with stronger and more enduring social ties and quasi group structures providing enough protection again predators and enough protection to protect vulnerable newborn in open-country habitats

It is clear when the breakthrough came, about 1.6 years ago, when the brain of early *Homo erectus* at 1.5 million years ago was close to size of the human brain that would eventually evolve as the roadblock of speech production would allow the hidden neurological capacity for language that had been installed 30 to 50 million years ago in great apes and, with speech, could to emerge and create symbolic culture. And with speech and symbolic culture human societies could, over the long run, become enormous and complex to the point of endangering the viability of the planet.

Neurologists, as a group tend to be overly specialized and, as result, are not that interested in the whole brain, and even less interested in how and why it evolved. Thus, it is literally up to sociology—the science of social organization—to outline the neurology of emotions and, more generally, the neurology of interpersonal behavior in humans, because otherwise humans would not exist and neurologists would not have a profession. The story of human neurology, then, is a *sociological story to tell*; and it is a story about where humans come and why we act and organize in the way we do because of our evolved neurology. Neurologists still may not be interested in the story we have to tell, but I suspect most humans would be interested because it is a very interesting story and can only be told by those interested in the neurological basis of human social organization—or by neurosociologists. Thus, for those sociologists interested the science of neurology is a viable field of study, even if many do not recognize its potential to dramatically expand knowledge about the biological basis of human social organization.

References

Arieti, S. (1970). Cognition and feeling. In M. B. Arnold (Ed.), *Feeling and emotions: The Loyola Symposium on feelings and emotions* (pp. 135–143). Academic.

Arnold, M. B. (1960). *Emotion and personality: Psychological aspects*. Columbia.

Baldwin, P. J. (1979). The natural history of the chimpanzee (Pan troglodytes verus) at Mt. Assirik, Senegal. PhD dissertation, University of Stirling, Scotland.

Baldwin, P. J., McGrew, W. C., & Tutin, C. (1982). Wide-ranging Chimpanzees at Mt. Assirik, Senegal. *International Journal of Primatology, 3*, 367–385.

Barger, N., Hanson, K. L., Teffer, K., Schenker-Ahmed, N. M., & Semendeferi, K. (2014). Evidence for evolutionary specialization in human limbic structures. *Frontiers in Human Neurosciences, 8*, 1–17. https://doi.org/10.3389/fnhum.2014.00277

Barger, N., Stefanacci, L., Schumann, C., Sherwood, C., Annese, J., Allman, J. M., Buckwalter, J. A., Hof, P. R., & Semendeferi, K. (2012). Neuronal populations in the basolateral nuclei of the amygdala are differentially increased in humans compared with apes: A stereological study. *Journal of Comparative Neurology, 520*, 3035–3054.

Barger, N., Stefanacci, L., & Semendeferi, K. (2007). A comparative volumetric analysis of the amygdaloid complex and basolateral division of the human and ape brain. *American Journal of Physical Anthropology, 134*, 392–403.

Bradshaw, J. L., & Nettleton, N. C. (1983). *Human cerebral asymmetry*. Prentice-Hall.

Collins, R. (2004). *Interaction ritual chains*. Princeton.

Damasio, A. (1994). *Descartes' error: Emotion, reason, and the human brain*. Putnam.

Darwin, D. (1872). *The expression of the emotions in man and animals*. John Murray.

Devinski, O., & Luciano, D. (1993). The contributions of the cingulate cortex to human behavior. In B. Vogt & M. Gabriel (Eds.), *Neurobiology of cingulate cortex and limbic thalamus: A comprehensive handbook*. Birkhauser.

Drevets, W. C., Price, J. R., Simpson, R. D., Reich, R., Vannier, M., & Raischie, M. E. (1997). Subenual prefrontal cortex abnormalities in mood disorders. *Nature, 386*(6627), 824–827.

Eccles, J. C. (1989). *Evolution of the brain: Creation of self*. Routledge.

Ekman, P. (1984). Expression and the nature of emotions. In K. Scherer & P. Ekman (Eds.), *Approaches to emotions*. Erlbaum.

Emde, R. N. (1980). Levels of meaning of infant emotions: A biosocial view. In W. A. Collins (Ed.), *Development of cognition, affect, and social relations*. Lawrence Erlbaum.

Epstein, S. (1984). Controversial issues in emotion theory. In P. Shaver (Ed.), *Review of personality and social psychology* (pp. 64–88). Sage.

Fehr, B., & Russell, J. A. (1984). Concept of emotion viewed from a prototype perspective. *Journal of Experimental Psychology, 113*, 464–486.

Fromme, D. K., & O'Brien, C. (1982). A dimensional approach to the circular ordering of emotions. *Motivation and Emotions, 6*, 337–363.

Goffman, E. (1959). *Presentation of self in everyday life*. Anchor.

Goffman, E. (1967). *Interaction ritual*. Anchor.

Gray, J. A. (1982). On the classification of emotions. *Behavioral and Brain Sciences, 5*(3), 431–432.

Heimer, L. (1995). *The human brain and spinal cord: Functional neouroanatomy and discretion guide*. Springer.

Izard, C. E. (1977). *Human emotions*. Plenum.

Izard, C. E. (1992). Basic emotions, relations among emotions, and emotion-cognition relations. *Psychological Review, 99*, 561–565.

Johnson-Laird, P. N., & Oatley, K. (1992). Basic emotions, rationality, and folk theory. *Cognition and Emotion, 6*(3–4), 201–223.

Kemper, T. D. (1987). How many emotions are there? Wedding the social and the autonomic components. *American Journal of Sociology, 93*, 263–268.

LeDoux, J. E. (1993a). Emotional networks in the brain. In M. Lewis & J. M. Haviland (Eds.), *Handbook of emotions*. Guilford Press.

LeDoux, J. E. (1993b). Emotional memory systems in the brain. *Behavioral Brain Research, 58*, 69–79.

LeDoux, J. E. (1996). *The emotional brain: The mysterious underpinnings of emotional life*. Simon & Schuster.

MacLean, P. D. (1990). *The triune brain in evolution: Role of paleo-cerebral functions*. Plenum Press.

MacLean, P. D. (1993). Introduction: Perspective on cingulate cortex in the limbic system. In B. A. Vogt & M. Gabriel (Eds.), *Neurobiology of the cingulate cortex and limbic thalamus: A comprehensive handbook*. Birkhäuser.

Malatesta, C. Z., & Haviland, J. M. (1982). Learning display rules: the socialization of emotion expression in intimacy. *Child Development, 53*(4), 991–1003.

Maryanski, A. (1986). African apes social structure: A comparative analysis. PhD dissertation, University of California, Irvine.

Maryanski, A. (1987). African ape social structure: Is there strength in weak ties? *Social Networks, 9*, 191–215.

Maryanski, A. (1992). The last common ancestor: An ecological-network model on the origins of human sociality. *Advances in Human Ecology, 2*, 1–32.

Maryanski, A. (1993). The elementary forms of the first proto-human society: An ecological/social network approach. *Advances in Human Evolution Ecology, 2*, 215–241.

Maryanski, A. (1995). African ape social networks: A blueprint for reconstructing early hominid social structure. In J. Steele & S. Shennan (Eds.), *The archaeology of human ancestry: Power, sex and tradition* (pp. 67–90). Routledge.

Maryanski, A. (1996). Was speech an evolutionary afterthought? In B. M. Velichkovsky & D. M. Rumbaugh (Eds.), *Communicating meaning: The evolution and development of language* (pp. 121–131). Erlbaum.

Maryanski, A. (1997). The origin of speech and its optimal size of human groups. *Critical Review, 12*, 233–249.

Maryanski, A., & Turner, J. H. (1992). *The social cage: Human nature and the evolution of society*. Stanford University Press.

McGrew, W. C. (1981). The female chimpanzee as a human evolutionary prototype. In F. Dahlberg (Ed.), *Women the gatherer* (pp. 35–73). Yale University Press.

McGrew, W. C. (1992). *Chimpanzee material culture: Implications for human evolution*. Cambridge University Press.

McGrew, W. C. (2010). In search of the last common ancestor: New findings on wild chimpanzees. *Philosophical Transaction of the Royal Society, 365*, 3265–3267.

McGrew, W. C., Baldwin, P. J., & Tutin, G. E. G. (1981). Chimpanzees in a hot, dry and open habitat: Mt. Assirik, Senegal, West Africa. *Journal of Human Evolution, 10*, 227–244.

Needham, C. (1982). *Principles of cerebral dynamics: The elementary significance of radical deduplication of the human brain*. Thomas.

Osgood, C. E. (1966). Dimensionality of the semantic space for communication via facial expressions. *Scandinavian Journal of Psychology, 7*, 1–30.

Panksepp, J. (1982). Toward a general psychobiological theory of emotions. *Behavioral and Brain Sciences, 5*, 407–467.

Plutchik, R. (1980). *Emotion: A psycho-evolutionary synthesis*. Harper and Row.

Pruetz, J. D., & Bertolani, P. (2007). Savanna chimpanzees hunt with tools. *Current Biology, 17*, 1–6.

Ruff, C. B., Trinkaus, E., & Holliday, T. W. (1997). Body mass and encephalization in Pleistocene Homo. *Nature, 387*(6629), 173–176. https://doi.org/10.1038/387173a0

Scott, J. P. (1980). The function of emotions in behavioral systems: A systems theory analysis. In Plutchik & H. Kellerman (Eds.), *Emotions: Theory, research and experience*. Academic.

Sperry, R. W. (1982). Some effects of disconnecting the cerebral hemispheres. *Science, 217*, 1223–1226.

Sroufe, A. L. (1979). Socioemotional development. In J. D. Osofsky (Ed.), *Handbook of infant development* (pp. 462–474). Wiley.

Stephan, H. (1983). Evolutionary trends in limbic structures. *Neuroscience and Biobehavioral Reviews, 7*, 367–374.

Stephan, H., & Andy, O. J. (1969). Quantitative comparative neuroanatomy of primates: An attempt at phylogenetic interpretation. *Annals of the New York Academy of Science, 167*, 370–387.

Stephan, H., & Andy, O. J. (1977). Quantitative comparison of the amygdala in insectivores and primates. *Acta Anatomica, 98*, 130–153.

Stephan, H., Baron, G., & Frahm, H. (1988). Comparative size of brains and brain components. In H. Steklis & J. Erwin (Eds.), *Comparative primate biology* (Vol. 4, pp. 1–37). Alan R. Liss.

Semendeferi, K., Lu, A., Schenker, N., & Damasio, H. (2002). Humans and great apes share a large frontal cortex. *Nature Neuroscience, 5*, 272–276.

Smith, T. S., & Stevens, G. T. (1977). Comfort regulation as a morphogenetic principle. *Advances in Group Processes, 14*, 113–155.

Trevarthen, C. (1984). Emotions in infancy: Regulators of contact and relationship with persons. In K. R. Scherer & P. Ekman (Eds.), *Approaches to emotions* (pp. 129–157). Erlbaum.

Turner, J. H. (1996a). Evolution of emotions in humans: A Darwinian–Durkheimian analysis. *Journal for Theory of Social Behaviour, 26*, 1–34.

Turner, J. H. (1996b). Cognition, emotion and interaction in the big-brained primate. In K. M. Kwan (Ed.), *Social processes and interpersonal relations* (pp. 295–315). JAI Press.

Turner, J. H. (1999a). The neurology of emotions: Implications for sociological theories of interpersonal behavior. *Social Perspectives on Emotion, 5*, 81–108.

Turner, J. H. (1999b). Toward a general sociological theory of emotions. *Journal for the Theory of Social Behaviour, 29*, 133–162.

Turner, J. H. (2000). *On the origins of human emotions: A sociological inquiry into the evolution of human affect*. Stanford University Press.

Turner, J. H. (2007). *Human emotions: A sociological theory*. Routledge.

Turner, J. H. (2021). *On human nature: The biology and sociology of what made us human*. Routledge.

Turner, J. H., & Maryanski, A. (2024). *Biosocial evolutionary analysis: Reconciling biology, psychology and sociology*. Edward Elgar.

Turner, J. H., & Stets, J. E. (2005). *Sociology of emotions*. Cambridge University Press.

Tutin, C. E. G., McGrew, W. C., & Baldwin, P. J. (1981). Responses of wild chimpanzees to potential predators. In I. B. Chiarelli & R. Corruccini (Eds.), *Primate behavior and sociobiology: Selected papers (pp. 136–141) of the VIIIth congress of the international primatological society Florence, 7–12 July 1980*. Springer.

Vogt, B. A. (1993). Structural organization of cingulate cortex. In A. Vogt & M. Gabriel (Eds.), *Neurobiology of the cingulate cortex and limbic thalamus*. Birkhauser.

8 Emotions as the Force Driving the Evolution of Hominin and Human Neurology

Jonathan H. Turner

Abstract

Much analysis of the human brain focuses on the neo-cortex, emphasizing human intelligence. Such is an appropriate topic because humans do have very large brains, relative to their body size. Still, this emphasis and indeed the celebration of human intelligence often under-emphasize the sub-cortex where emotions in mammals are activated. In this paper, the goal is argue that emotions were the *driving force behind brain evolution over the last 6 million years*. Without growth in sub-cortical areas of the brain, growth in the neo-cortex could not have occurred since all cognitions must have emotional tags to be remembered and used in thinking and decision-making. Moreover, since articulated speech and the emergence of symbolic culture depend upon enlarging the neo-cortex and other body systems, the evolution of emotions and the structure of the sub-cortex is not only critical to growth of the neo-cortex but also the capacities of humans for speech production and codification of symbolic culture. From the perspective of neuro-sociology, then, the evolution of human patterns of social organization—driven and held together by emotions—is a much better starting point in explaining how and why the hominin and eventually human brain began to grow. Such an arguments represents a quiet critique much analysis in psychology and biology. Indeed, without a sociological perspective to explain the evolution of the sub-cortex in the hominin brain, explanations of the neo-cortex, speech production, and symbolic culture are often incomplete, often relying upon "just so" stories and other more speculative endeavors. Indeed, neuro-sociology is at the center of explanations of how and why the hominin brain grew to human proportions.

Humans are the fourth great ape sharing common ancestors with present-day orangutans, gorillas, and chimpanzees. Indeed, humans and the other three great apes are so close genetically that all four great apes should belong to the genus *Homo*, despite the fact that humans tend to see themselves as distinctive and unique. Humans are, of course, unique in that they have large brains relative to their body size, but this larger brain itself is structurally very close to the smaller brains of the other three great apes. In this chapter, I will build on the discussion in the Chapters 3, 7, and 12 in developing a simple theme: hominin evolution after the splits with the ancestors of orangutans, gorillas, and chimpanzees has been driven by alterations in *the*

J. H. Turner (✉)
Department of Sociology, University of California, Riverside, Riverside, CA, USA
e-mail: jonathan.turner@ucr.edu

W. Kalkhoff et al. (eds.), *Handbook of Neurosociology*, Handbooks of Sociology and Social Research, https://doi.org/10.1007/978-3-031-95615-7_8

sub-cortical areas of the hominin brain where emotions are generated in all mammals. The eventual enlargement of the hominin brain to human proportions, the evolution of language, and the capacity for symbolic culture have all been driven by sub-cortical areas of the brain. If natural selection had not begun selecting on the sub-cortex of the first hominins that emerged some six-to seven-million years ago, humans would never have evolved; and for better or worse, the planet earth would not be over-inhabited by the fourth great ape, humans.

I employ two basic strategies for understanding great ape evolution: (1) *cladistic analysis* as developed by Alexandra Maryanski and (2) *comparative neuro-anatomy* which is my alternative to the excesses of sociobiology and evolutionary psychology (Turner, 2000, 2007; and Turner & Maryanski, 2024). Let me begin with a brief review of cladistic analysis as critical to understanding the evolution of humans, particularly the neuro-anatomy of humans.

A Very Brief Summary of Gladistic Analysis

Cladistic analysis is the label among biologists for a mode of analysis that originated in tracing the origins of languages (Jeffers & Lechiste, 1979; Foley & Gamble, 2009). The basic logic is that end points of a number of languages are believed to have evolve from a common "root" language, it is useful to examine these languages in terms of what they have in common in words, grammatical structures, and other properties of a language. By examining what languages have in common, it then is often possible to gain insights into the root language from which they all evolved. Cladistic analysis in biology uses this same logic: If various species of animals share a high proportion of their genes, then it reasonable to view them as evolving from a common ancestor.

Cladistic analysis thus allows analysts to look back in time, millions of years, to see what the *Last Common Ancestors* (LCA) of the three other great apes were like. Since humans share 99+ percent of their genes with chimpanzees, 98+% of their genes with gorillas, and 96–97% with orangutans, cladistic analysis can provide insights into the nature of humans' hominin ancestors. Thus, we can learn a great deal about the ancestors of humans, especially with respect to their neurology, by using cladistic analysis on the great apes. Figure 8.1 outlines a simple *cladogram* on the evolution of the four great apes (and several subspecies among orangutans, gorillas, and great apes).

Figure 8.1 denotes the approximate time over the last 16 million years the ancestors of the hominin line split off the ancestors of orangutans, gorillas, and chimpanzees. As is evident in Fig. 8.1, orangutans split off between 16 to 12 million years ago from what would become the ancestors of the other three great apes, with gorillas splitting off the ancestors of the hominin line around 10 to 8 million years ago, and chimpanzees splitting off the hominin line some 8 to 6 million years ago. The hominin line thus carried forward many of the traits of the orangutans, gorillas, and chimpanzees with, for our purposes, the end point of hominin evolution occurring with the emergence of various subspecies of early humans. Prominent hominins along this more recent line of evolution is a number of species, such as *Australopithecines*, *Homo habilis*, early *Homo erectus* and *Homo ergaster*, and *late Homo erectus*, can tell us a great deal about the evolution of early humans, including *Neanderthals*, *Denisovans*, and *Homo sapiens*. The fossil remains of hominins after the final split with the ancestors of great apes (chimpanzees) are an important source of knowledge about the biology of humans' ancestors, especially the neurology of these ancestors. And, coupled with even more knowledge about the biology, behavior, and neurology of extant great apes—i.e., orangutans, gorillas, and chimpanzees—it is possible to find reasonable proxies of the species of primates, especially their respective neurologies, from which humans eventually evolved. Thus, to

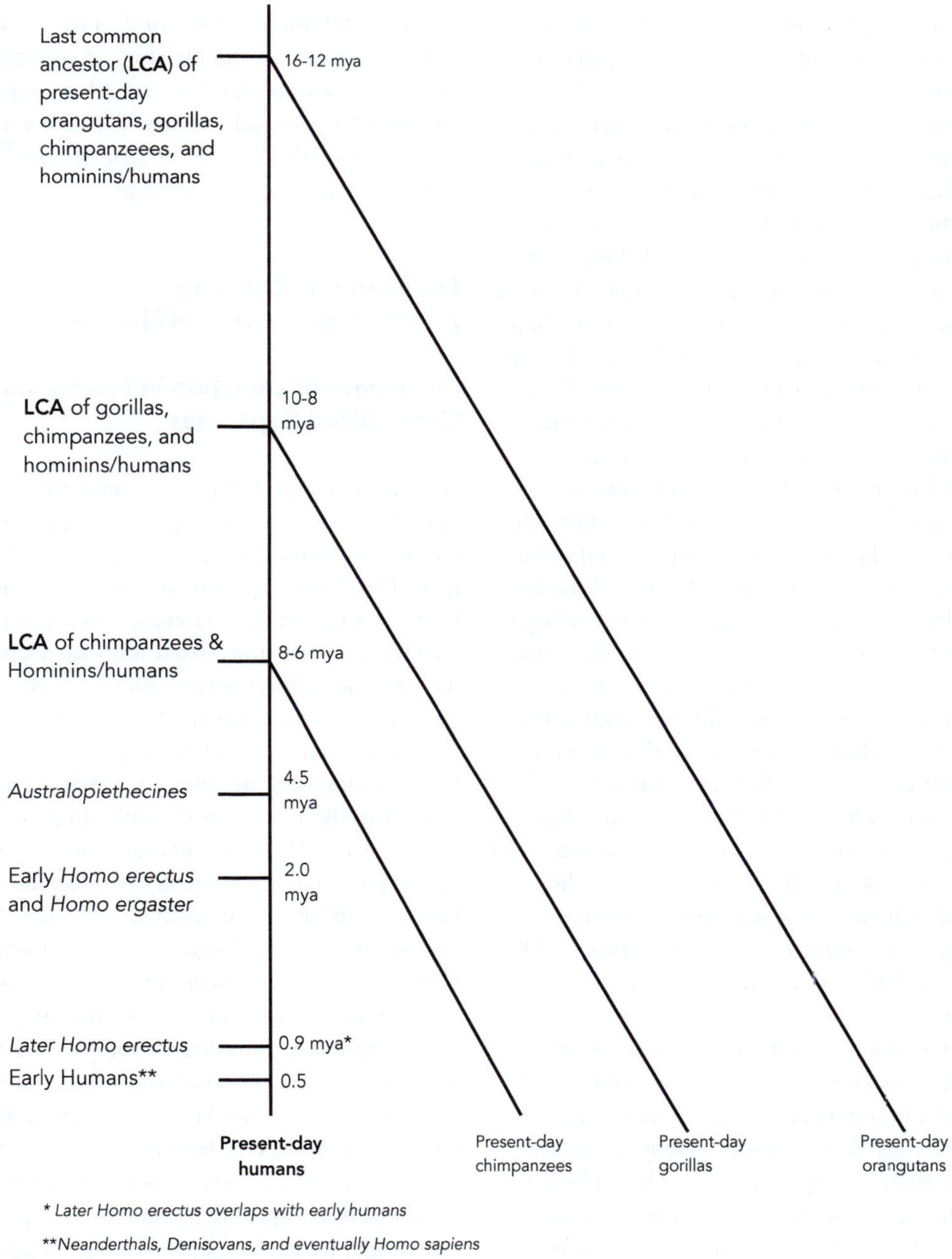

Fig. 8.1 Simplified cladogram of ancestors of hominins and humans

understanding human evolution, especially of the hominin brain, *cladistic analysis* combined with *comparative neuro-anatomy* can be viewed as an important methodology for a viable *neuro-sociology*.

Comparative Neuro-Anatomy

By comparing brains of humans, on the one side, and the other three great apes, on the other, it is possible to determine where natural selection was working on the neurology of hominins. At one time, evolutionary psychologists (e.g., Barkow et al., 1992) emphasized the notion of "modules" of the brain, suggesting that discovering new modular structures in humans compared to other primates would reveal the direction of selection along the hominin line over the last several million. My approach does not make assumptions about discrete modules of the brain as the source of specific behaviors, but instead, simply compares key structures of the brains of orangutans, gorillas, and chimpanzees with those of humans, with differences in particular structures *marking the handy work of selection as it reworked the brain inherited from the last common ancestors of humans and the other three great apes*. In Chapter 11, I summarized some of the early efforts along these lines, and the basic conclusion is that natural selection was enlarging the brain of hominins and increasing the connectivity of substructures. While there were a handful of new structures added to the human brain, most of the change involved *simply enlarging the sub-cortex where emotions are generated and the neo-cortex* where cognitions are generated and stored. Yet, as is now clear, cognitions and decision-making cannot occur unless they are tagged with emotions generated in the sub-cortex of the brain (Damasio, 1994; LeDoux, 1991, 1993, 1996) and activating body systems of humans.

Thus, selection was not so much adding new modules to the brain but, instead, increasing the size of *existing* neurological structures and their connectivity. Selection was thereby taking a more generalized strategy than implied by evolutionary psychology, working on *existing structures* by enlarging them and, adding few new structures along the way, as the sub-cortex and neo-cortex grew. Indeed, this more conservative approach should not be surprising because, to alter one structure in something as complex as large mammalian brain, creates pressures to alter other structures, and many of these needed alternations would work against fitness. Indeed, one of the reasons that the genomes of humans and the other three great apes *are so similar* is that their brains and most of their *neuro- and body- anatomy remain very similar*. So the story of comparative anatomy does not revolve around adding new modules but, instead, *enlarging existing structures of the brain whose basic structure and function existed millions of years ago*.

Evolution of Anatomy and Neuroanatomy of Hominins

The Power of Selection in Producing Generalized Capacities

As selection took the route of increasing brain size, it was operating to increase highly generalized cognitive and emotional capacities of hominins. This "strategy"[1] would enable hominins to begin to adapt to new habitats, especially open-country habitats, away from arboreal niches of the great apes (orangutans, gorillas, and chimpanzees). Selection was allowing for the possibility that hominins would be able to adapt to the changes that were occurring in Africa over the last ten million years as episodic drops in temperature led to (1) the shrinking of the forests and (2) the spread of open-country habitats like bush lands, secondary forest, savanna, and even desert habitats. As selection began to work on the brain, particularly its sub-cortical areas where emotions are generated, it also altered hominin body anatomy—arms, legs, hips, hands, and feet—in order to make hominins bi-pedal and, hence, more fit open-country habitats. These generalized alterations of the bodies of hominins along with the slower evolving changes in their neurology were all directed at expanding generalized capacities that would, in the end, allow humans to adapt to virtually any habitat on earth. Thus, hominins were walking upright among *australopithecines*

[1] Biology and evolutionary psychology often use the word "strategy" which is misleading because natural selection is *not* teleological, as the terms like "strategy" imply. Rather natural selection is working by randomly selecting on variations of structures with those increasing fitness being retained in the phenotypes and underlying genotypes of a species.

four to five million years ago; and so, while their brains had not grown much beyond those of the size of present-day great apes, their bodies had become more generalized which, in the long run, would allow them to survive outside the retreating forest canopy.

Key Pre-adaptations Evident in All Great Apes

Pre-adaptions are traits of a species that have evolved as a consequence of selection pressures on other structures. Those new traits that have little or no effects of fitness when they first emerged, but later in the evolutionary history of species, have large effects and promote fitness, are called *pre-adaptations* because they already existed as an unintended byproduct of earlier selection. Some of the "pre-adaptations"[2] were critical in this process. One early pre-adaptation was the *shift from olfactory to visual dominance* which, in turn, created as a byproduct the neurological capacity for language (Geschwind, 1965a, 1965b, 1985) which was under-utilized for millions of years and is still evident in the neurology of all great apes. In fact, when living in human contact, great apes can learn to understand human languages and to "speak" via hand signals and computer-assisted pictograms with their human trainers and caretakers. If this neurological shift had not occurred as small mammals ascended the threes after the decline of dinosaurs some 60 million years ago, hominins including humans *would never have been able to evolve*. Another critical pre-adaptation was the *grasping hands and feet* that allowed for land mammals millions of years ago to adapt to arboreal habitats; and this adaptation was to be critical in more open-country habitats where primates began to make and use of simple tools to survive and, eventually, would allow humans to build new ecological niches to which they could adapt.

Another set of pre-adaptations revolved around *sophisticated sensory systems*, ranging from capacities to gain information from complex environments to ascertain high-energy foods, hear low frequency signs necessary for spoken language, experience the world with color vision, and utilize sensitive hands and fingers in food collection and building of simple structures.

Yet another pre-adaptation was a *comparative large brain* at 375 to 400 cm, which is only a third as large as a human brain but nonetheless still very large for a mammal of the size of a great ape. Without this neurological base-line in size and complexity, it is unlikely that the hominin brain could have ever evolved.

Another important pre-adaptation was *low levels of grooming* by great apes compared to most primates (monkeys and pro-simians who constitute 98% of all primates species on earth). Rather than groom, much communication among great apes occurs through eyes, facial expressions, body countenance, and vocalizations carrying meanings. And, coupled with the pre-wiring of great apes' neurology for language, *an entirely new set of mechanisms for developing more complex interpersonal relationships* was potentially available for creating the first social structures and cultures of hominins.

The core structure of great apes compared to monkeys and most other primates can be termed *community complex* (Turner, 2021). Great apes do *not* form permanent group structures but, instead, larger communities whose members can wander around a defined territory, cognitively mapping boundaries of this territory and the demography of its members. This community complex did *not involve strong bio-programmers for permanent group formations* typical of most primates and, indeed, most mammals as well. And this larger, more flexible community structure without permanent group structures would eventually allow humans to build up large and complex societies. Indeed, most mammals have bio-programmers in their neurology to form

[2] I prefer the label, *pre-adaptation,* over current labels, such as *expectation*, because it captures the sense that traits can late dormant for millions of years and, then, suddenly become adaptive under different environmental conditions.

group structures; but hominins did not because. In the short run during early hominin evolution, this lack of group structures was a liability, but in the end it generated selection pressures to find an alternative and more flexible route to group formation built around enhanced emotions and the high level of interpersonal skills evident among all great apes.

Still another critical pre-adaptation was the long *life history characteristics* of great apes whereby great apes are slow to mature and, hence, must have mother-child bonds for many years before adulthood. This pre-adaptation would eventually allow neurologically immature offspring to pass through the cervix of females, thus allowing for *the growth of larger brains outside the womb* after childbirth. Without these life history characteristics, the hominin brain would not have been able to evolve to its present size without mothers who are genetically programmed to nurse and care for their offspring *many years after birth* (fathers are unknown in all species of great ape because of promiscuity of sexual relations.

In sum, without these pre-adaptations, a big-brained primate living in open-country habitats filled with predators but without strong bio-programmers for group formation could not have survived. Yet, other behavioral propensities of great apes would, under further selection, allow hominins to get better organized. Especially significant are the interpersonal capacities evident among all great apes that would, with additional changes to the hominin brain, allow them to develop *flexible patterns of social organizations*. For unlike most primates such as monkeys who are genetically locked into tight-knit group structures, great apes are weak-tied animals but with highly tuned interpersonal skills in reading the emotions and behavioral propensities of conspecifics.

The Interpersonal Basis of Great Ape Organization

Great apes have a large repertoire of interpersonal behavioral capacities, as summarized in Table 8.1 (Turner & Maryanski, 2024: 71–73). These capacities are the basis for the complex ways in which humans interact, as was also outlined in Chapter 12. Yet, these sophisticated interpersonal skills alone would not have allowed early hominins to survive movement into completely open-country habitats, away from the protection of the forest habitat where all great apes avoid dangers. Something more was needed: *the enhancement of emotions*. All of these interpersonal capacities, when enhanced by emotions, would allow vulnerable, slow bi-pedal evolving great apes, not prone like monkeys to form tight-knit groups, to survive in predator-ridden habitats of Africa. Indeed, if natural selection had not hit upon growing the sub-cortical areas of the hominin brain, all of the pre-adaptations and behavioral capacities of the great-ape line could not have overcome the vulnerability of a species not prone for strong ties and not possessing genetically controlled propensities (like most primates) for group formations.

Indeed, without emotions to forge bonds that are constructed by agency rather than by genetic bio-programmers, early hominins would not have been able to survive and, hence, humans would never have evolved. The driving force of hominin survival was thus the *result of intensifying the emotional capacities of hominins, thereby increasing their ability to forge stronger bonds in group-like structures* (Turner, 2000, 2002, and 2007). Once selection had hit upon this solution to the early problems of hominin survival, it would begin to allow the neo-cortex to grow and, in so doing, increase intelligence of hominins. But this was not a processes that was inevitable; and as I will emphasize, there were some key *roadblocks* along the way in using emotions as the basis for group organization. Nonetheless, what was driving neurological evolution among hominins over the last six-to-seven million years

Table 8.1 Inherited behavioral capacities from Last Common Ancestors (LCA) of hominins[a]

1.	***Reading faces and eyes of conspecifics.*** Visual dominance over haptic and auditory sense modalities, thereby subordinating other sense modalities to vision great apes can communicate common meanings through visual cues about their intended lines of action.[1]
2.	***Following eye movements to determine likely interests of conspecifics.*** Behavioral propensity to follow the gaze and eye movements of others, thus increasingly propensity to interpret gestures of eyes and face. In, fact, great apes follow gaze and eye movements to determine what another is observing and thinking the result is a high level of inter-subjective understandings of intended courses of actions among great apes.[2]
3.	***Ability of infants to imitate Orofacial movements of caretakers within weeks of birth***, thereby indicating a genetically driven propensity to seek to understand emotions of conspecifics, especially during face-to-face interactions. This capacity emerges within two weeks of birth, as it does for humans.[3]
4.	***Empathy and role-taking ("theory of mind).*** Propensity for face-monitoring for signs of action by conspecifics particularly for emotional content, thereby increasing capacity to role-take with other and to read emotional dispositions in conspecifics. What G. H. Mead (1934) termed "role-taking" of putting oneself in the positions of others, and vice versa, was thus inherited by humans as a hard-wired propensity.[4]
5.	***Body language and coordinated actions.*** Capacity to communicate meanings and coordinate actions through non-verbal signals, especially by eyes and face, and using this capacity to coordinate interaction among conspecifics as well as to coordinate collective action among conspecifics.[5]
6.	***Enhanced propensity to use imitation to learn appropriate signals and behaviors.*** Hence, the hard-wired propensity to acquire the means of symbolic communication at birth for facilitating interactions and, when needed, coordination of action among conspecifics.[6]
7.	***Rhythmic synchronization of communication.*** Field studies indicate great apes, especially chimpanzees, fall into synchronization of auditory and bodies during interaction. Rhythmic synchronization, especially of emotions, via mirror neurons that are a necessary pre-adaptation for role taking and emotion-generating interactions.[7]
8.	***Collective emotional effervescence.*** This synchronization can escalate in occasional gathering of larger numbers of conspecifics, often escalating into arousal and synchronization of emotions in what appear to be "ritual-like" interactions, very much like humans when emotions are activity in collective actions, such as festivals like Mardi Gras.[8]
9.	***Increased cortical control of sub-cortical areas of the brain through projections of axons.*** While great apes do not have the same level of cortical control over emotions as humans (although this is often tenuous), natural selection appears to have begun this process of developing neo-cortical control over sub-cortical emotion centers to reduce the chances of disrupting interactions through excessive emotional arousal.[9]
10.	***Enlarged decision-making prefrontal cortex compared to most other mammals, controlling for body size.*** While the human prefrontal cortex for decisions making by tagging cognitions with emotions is far beyond that of great apes, this structure was under some selection and thus represented a pathway for blind selection to grow further once the neo-cortex and surrounding cortices of late hominins began to grow rapidly in late hominin evolution.[10]
11.	***The ability among the great apes to learn and use language at the level of a three-year-old human child.*** This capacity follows from the precursors for language facility in Table 4.3. This capacity gave natural selection something to select upon if more refined verbal communication would be fitness enhancing. Without this pre-existing capacity brought out in great apes in laboratory studies, spoken language would never have evolved along the hominin line.[11]
12.	***The capacity for self awareness.*** The ability to recognize an image in a mirror as a reflection of self as an object in the environment, thereby creating the necessary pre-adaptation for the formation of identities and, thus, self-monitoring, control, and evaluation. Very few animals have this capacity, and for humans it became a key force in establishing strong so relations where individuals work actively to verify each others' sense of self.[12]
13.	***Reciprocity in the give and take (exchange) of resources.*** This hard-wired capacity for reciprocity exists among many higher mammals but it is particularly pronounced among great apes; and with further selection, it would become a strong bonding mechanism in animals like hominins when needed for group structures.[13]
14.	***Calculations of fairness in exchanges of resources.*** Great apes compare shares of resources with others in making judgments of fairness and justice of exchanges. This propensity would be critical in bonding among low-sociality, and it is no doubt the hard-wired basis for morality among what became culture-using animals like humans.[14]

(*continued*)

Table 8.1 (continued)

15.	***Capacity to experience variants among primary emotions, particularly among chimpanzees.*** Great apes have the same primary emotions as humans, as do almost all mammals. But, hominoisa appear to have more variants of these emotions, which again, gave blind natural selection a path to enhance and thereby increase the use of emotions as the fundamental force allowing hominins to forge stronger group ties.[15]
16.	***Capacity to experience empathy with conspecifics, thus facilitating in-depth role-taking for emotions experienced by conspecifics.*** Empathy is a critical cognitive and emotional capacity for generating stronger social bonds; and its presence among low-sociality great apes provided yet another "hook" for natural selection to enhance as a means for creating stronger social bonds along the hominin line.[16]

Citations are to data demonstrating the existence of these refined behavioral propensities in great apes and humans, as these provided a neurological base on which natural selection could "select" upon to enhance stronger social bonds among hominins and then humans. These propensities of current higher-order apes go a long way to understanding *what brain structures natural selection worked on to produce the more extensive abilities of humans* to be more social (Passingham, 1973, 1975; Sherwood et al., 2008). Many of the references cited for the 16 behavioral capacities listed above document at least some of the brain structures involved Still there is much more detailed work needed to understand the extent of great apes' capacities for inter-subjective empathy and other emotional and cognitive states

Sources: 1. Maryanski and Turner 1992; Ettlinger 1977; Jarvis and Ettlinger 1977; & Passingham 1982: 51–55; Osgood, 1966; Menzel 1971; **2.** Hare, et al. 2001, 2006; Povinelli and Eddy 1997, 2000; Itakura 1996; Baizer et al. 2007; Tomasello et al. 2001; Okamoto et al. 2002) (Hare & Tomassello, 2001; Povinelli 2000; Povinelli and Eddy 1997; **3.** Emde 1962; Ekman 1984; Sherwood et al. 2005, Tomonaga 1999; Subiaul 2007; Horowtiz 2003; Gergely & Csibra 2006; **4.** Leslie, et al. 2004, Gazzaniga & Smylie 1990; Parr et al. 2005; Tsao et al. 2003; **5.** Menzel 1971; Turner and Maryanski 2008; **5.** Tomonaga 1999; Subiaul 2007; Horowitz 2003; Gergely & Csibra 2006); **6.** Schutz-Bosbach and Prinz 2007; Rizzolatti, et al. 2002; Rizzolatti and Sinigaglia 2008; **8.** Spencer and Gillen 1899; Durkheim 1912/1947; **7.** Goffman 1967; Collins 2004; Rizzolatti et al. 2002; **9.** Raghanti et al. 2000; Sherwood 2007; Sherwood et al. 2008; Semendeferi et al. 2002; **10.** Stephan et al. 1969, 1970, 1977, 1986, 1988; **11.** Geschwind 1965a, 1965b; Damasio and Geschwind 1984; Rumbaugh and Savage-Rumbaugh 1990; Savage-Rumbaugh and Lewin 1994; Savage-Rumbaugh et al. 1978, 1988, 1993; & Bickerton 2003; **12.** Gallup 1970, 1979, 1982; **13.** Cosmides 1989; de Waal 1989, 1991, 1996, 2006. **14.** Brosnan et al. 2005; Brosnan and de Waal 2004, 2014; **15.** Turner 2000, 2002, 2007; Turner and Maryanski, 2008; **16.** Call and Tomasello 1998; Cooley 1902; Damasio 2000a, 2000b; Emde 1962; Rizzolatti & Sinigaglia 2008; Scheff 1988; Turner 2000, 2007

[Adapted from Turner and Maryanski 2024]

was emotions; and if natural selection had randomly not hit upon this solution early in hominin evolution after the split from the LCAs of chimpanzees (and earlier of gorillas and orangutans), humans would not have been able to organize themselves into the simple hunting and gathering bands, nor would humans for better or worse been able to eventually create the colossal socio-cultural structures and infra-structures of contemporary human societies over the last 10,000 years. Thus, now and in the evolutionary past, *the neurology of emotions was driving all aspect of human behavior and socio-cultural organization*. Indeed, the genetic basis for human survival is due to the effects on enhancing emotions on the pre-adaptations and behavioral capacities inherited by the LACs of orangutans, gorillas, and chimpanzees.

The Emotional Basis of Human Intelligence, Language, and Culture

If only by self-reflection, we know what natural selection and then socio-cultural selection achieved. A large brained great ape, with a high level of intelligence, a capacity for language, and an ability to create symbolic culture. In the second half of the twentieth Century, the neurological data made clear that cognitions cannot be remembered without having emotions attached to these cognitions (Damasio, 1994; LeDoux, 1996). This tagging process occurs via the *hippocampus* and related *transition cortices* (see Fig. 8.2 on p. __ for the locations of these and other critical neurological structures). For a cognition to be remembered it must be tagged with emotions which are then activated in ways that

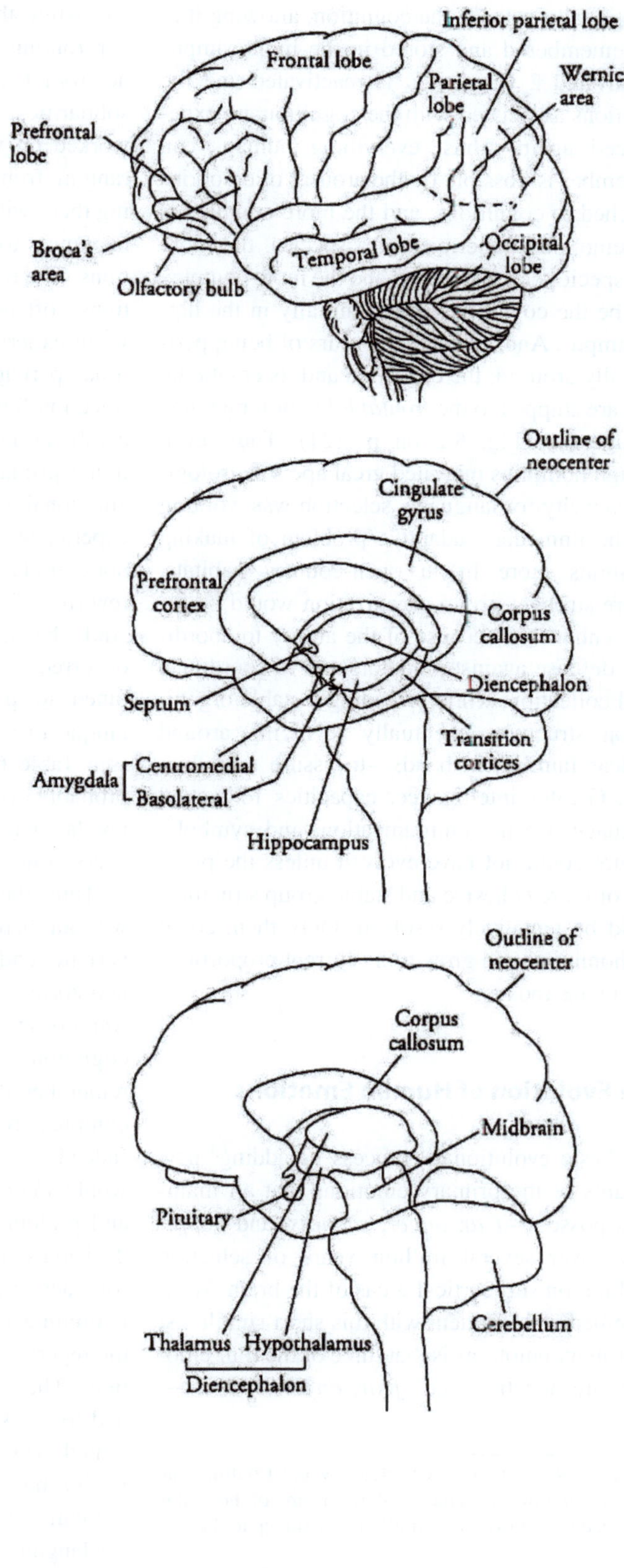

Fig. 8.2 Structures of the human brain

emotionally valence the cognition, allowing it to be remembered and stored in the hippocampus for around 2 years, if it is reactivated and the emotions associated with the cognition are experienced again. Thus, everything humans can remember is possible by the arousal of emotions attached to cognitions; and the more complex is the emotional repertoire of a species, the more this species can remember, and the more complex can be the cognitions stored initially in the hippocampus. And, then after 2 years of being periodically aroused, the cognition and its emotional tags are shipped to the *frontal lobe* for long range storage (see Fig. 8.2 on p. 121). Thus, even though hominins inherited great ape's neurological capacity for language, selection was working on the immediate adaptive problem of making hominins more fit in open-country habitats, where stronger group organization would be fitness enhancing because of the ability to coordinate defense against predators and to coordinate food collection, while providing a stable organization structure—eventually revolving around nuclear families in bands—to assure reproduction. Greater intelligence, capacities for actual language use in communication, and symbolic culture could not have evolved unless the problem of more cohesive and stable group structures could be tentatively resolved. Only then, could the hominin brain grow to its current proportions in humans today.

The Evolution of Human Emotions

The basic evolutionary process as adding[3] new variants of the primary emotions that all mammals possess—*fear, anger*, *sadness*, and *happiness*—over several million years of selection working on sub-cortical areas of the brain. What is immediately evident with this short simple list of primary emotions is that three of the four emotions are negative—i.e., *fear*, *anger*, *sadness*—and while they may be have help fitness when confronting danger in the environment, they do no promote interpersonal cohesiveness and group solidarities. My view is that natural selection worked on variants of these negative emotions ranging from low-to-higher intensity, while mixing them with variants of the positive emotion of happiness to reduce the power of negative emotions. The resulting, expanded the range of emotions softened some of the negative emotions when experienced at a low-intensity level. Over time, perhaps as much as two million years, selection found a way to simultaneously activate combinations of these four and perhaps additional primary emotions to expand the array of emotional valences that evolving hominins could experience. Just how emotions are combined is not entirely clear but it obviously involved some rewiring of the emotion centers in the sub-cortex. And the final stage of emotional evolution occurred with three of the negative were combined to produce emotions that are perhaps unique to humans: *shame*, *guilt*, and *alienation* (see Table 6). As I will argue, these emotions probably evolved with the growth of the neo-cortex, language, and culture during the last million years of hominin evolution.

Thus, the neo-cortex could not become larger without being preceded by an increase in the varieties and valences of emotions. To enlarge the neo-cortex *per se* would simply create a large warehouse without cognitions to fill it because cognitions *must be tagged by emotions* to be remembered; and complex cognitions require complex repertoires of associated emotions. Indeed, a large but largely empty neo-cortex would *decrease* fitness by consuming calories and protein *without* increasing intelligence and decision-making capacities of hominins. Thus, for each elaboration of the neo-cortex during hominin evolution, there had to be an increase in the repertoire of available emotions to tag cognitions. The sub-cortical areas of the hominin brain had to grow and become more complexly integrated to provide the needed range of emotions to justify the costs of growth in the neo-cortex. For, without a larger neo-cortex, the pre-adaptation for language, especially spoken language, could

[3]In Tables 4, 5, and 6 in Chapter 7 where I outline one scenario of how the emotional repertoire of hominins expanded over a five to six million year run up to the first humans.

not have ever evolved; and if spoken language could not evolve, neither could symbolic culture.

Thus, emotions were driving almost every aspect of neurological evolution among hominins; and in the end, the sub-cortical areas of the hominin brain became twice as large as the equivalent areas of the three great apes, controlling for body size. This growth in emotions, in turn, allowed the neo-cortex to grow to three times the size of the neo-cortex of extant non-human great apes. Thus, comparing great ape brains with human brains, just in terms of relative size of sub-cortex and neo-cortex, reveals *how* natural selection was making hominins more social and able to form more stable groups through arousing the range of positive emotions, with negative emotions used as sanctions to support emerging symbolic culture regulating behavior in groups, with *guilt*, *shame*, and *alienation* being the last emotional capacities in the evolution of late hominin and early humans.

There were, then, a series of sequences during hominin evolution when sub-cortical areas grew to enhance interpersonal bonds among low sociality and non-group forming hominins. And, with each of these changes, modest expansion of the neo-cortex could occur. Thus, the evolution of the brain was a series of feed-forward and feedback relations among sub-cortical emotion centers and the neo-cortex; and over time these cycles of feed-forward and feedback between emotion and cognitive centers began to push for spoken language and then symbolic culture. However, as the data on brain growth to be presented shortly suggests, there were clear *roadblocks* for selection is creating the capacity for spoken language; and as noted earlier, without spoken language symbolic culture could not evolve to a high degree.

Nonetheless, it was the relentless push to grow the sub-cortex that was driving everything that would allow humans to evolve as the last hominin but it was a push that was often stalled because the over-all size of the hominin brain did not grow dramatically for at least 4 to 5 million years of hominin evolution. And so, despite the roadblocks along the way to the current brain of humans, every aspect of human social organization and of growth in the brain was and is still driven by emotions. Cognitions cannot be remembered without being tagged by emotions (Domasio, 1994; LeDoux, 1991, 1993, 1996); spoken language cannot evolve without emotions pushing to expand the neo-cortex; and without spoken language the complexity of cognitions is limited and, even more significantly, symbolic culture cannot evolve. But once the neo-cortex begins to grow, as the roadblocks to spoken language are overcome, then symbolic culture can begin to evolve. And the feed backs and feed-forwards on each other and on the emotion ultimately make the growth of the neo-cortex, spoken language, and symbolic culture possible, thereby increasing the fitness of late hominins and early humans. Indeed, not until a species can experience emotions like *shame* and *guilt* can symbolic culture regulate social relations because these emotions are the emotions of social control.

Emotions, Social Solidarity, and Social Control

For millions of years, the hominin brain did not grow very much beyond what is evident among great apes who, by the logic of cladistic analysis, are our best proxy for the last common ancestor of all hominins, including humans. Thus, if hominins were about the size of chimpanzees, their brains were in the 375 to 424 cm^3. If there was modest growth in sub-cortical areas of the hominin brain, these would not be so evident because the sub-cortex is buried within the neo-cortex, separated by the *corpus callosum* (where neurons connect both sides of the brain) and the *cingulate gyrus* (which is probably like the first mammalian brains). I would argue that the early growth of the sub-cortex was, to a degree, hidden by the fact that the neo-cortex surrounds sub-cortical areas and largely determines the size of the skullcap of early fossils hominins. For the neo-cortex to grow, the sub-cortex would have to growth first, and so early increases in the size of the skullcap were modest because they reflect growth of the sub-cortex increasing the array of emotions experienced by early hominins, with growth pushing outward to grow the skullcap to a modest degree. And as hominins continued to evolve, the skullcap would continue to grow as the neo-cortex began to grow; and yet, it was not until 1.6 million years ago that larger increases in overall brain size become evident among hominins, thus marking the approximate time that the sub-cortex had grown sufficiently large to make growth of the neo-cortex fitness enhancing.

My view is that as the sub-cortex began to grow, it grew slowly with this initial growth in the sub-cortex providing *just enough* extra emotion on early hominins to increase their social bonds and capacity for more stable groups that would allow hominins to begin venturing outside the forests of Africa, at least some of the time (and then retreat back to the safety of the trees at night away from nocturnal predators). The interpersonal capacities of great apes listed in Table 8.1, when intensified by emotions, were thus enough to allow hominins with relatively small brains to survive if selection could begin to enhance their emotional repertoire the point where group solidarities could sustain more cohesive and permanent group formations.

The Growth of the Hominin Brain

As the forest began to recede in Africa, especially over the last five million years, some hominins had to spend more time in open-country habitats; and they could only do with high levels of emotional solidarity and capacity to form more stable groups. And, once emotions had reached a critical threshold, they could begin to allow for growth of the neo-cortex which, *if* roadblocks to speech could be overcome and unlock the potential for language and symbolic culture in all great apes. The "take off" point for substantial brain growth is best revealed, I think, by consulting the data on the size of the over-all brain in key hominin fossils. Table 8.2 assembles representative data on hominin brain growth.

Table 8.2 records the estimated brain volume (cm^3) of hominins and early humans as reveal by archeological data on derived from measurements on skullcaps. *Australopithecus afarensis*—the famous "Lucy"—at 3.2 millions years ago had a brain volume between 343 and 375 cm.^{3} Thus, at 3.2 million years, hominins have been evolving for at least 3 million years after the split with the LCAs of present-day chimpanzees and, perhaps surprisingly, their brains were no *larger than those of chimpanzees*, suggesting that they still must have spent considerable time in the forest canopy. The fact that they were fully bi-pedal suggests that they were foraging in more open-country habitats—bush lands, secondary forests, and even savannas—but probably returning to the protection of the forests at night. Yet, somewhat later between 3.0 and 2.5, *Australopithecus africanus* revealed growth in the brain of at least 100cm^3 which suggests that the skull cap encasing the brain had grown. I would hypothesize that most of this growth was from the sub-cortex pushing the unchanged neo-cortex outward, thus creating a larger brain but one increasing the size of the emotions centers of the brain. The resulting solidarities at the group level probably had increased to the point where, as other data suggest, *africanus* was spending more time in open country because they revealed higher levels of solidarity (Ardesch et al., 2019; Holloway, 2015; Passingham, 1973, 1975, 1982). The same was true of *Paranthropus boisei* and *aethiopicus* where data indicate that they were spending much more time in open country. Still the brain was only 50 to 100cm^3 larger than *afarensis* and present-day chimpanzees as late as 1.7 million years ago.

Then, suddenly, the brain more than doubled with early *Homo erectus* to over 900cm^3 and then with later *Homo erectus*, the brain reached the lower ranger of the human brain by (1050 cm^3) at .5 million years ago. Indeed, later *Homo erectus* overlapped with species of the first humans for some time but, in the end, *Homo erectus* was replaced by early humans. There were early signs that the brain was starting to grow rapidly because fossils of *Homo habalis* (not listed in Table 8.2) revealed a brain larger than 500cm^3 at 2.0 million years ago, although many now see H. *habilis* as an early form of *Homo erectus*.

Homo erectus and *Homo ergaster* were the precursor to humans; and so their brain had to be getting larger if humans were to evolve. This

Table 8.2 Brain size during hominin evolution

Species	Estimated date (mya[a])	Brain volume	Name of specimen
A. afarensis	3.2	343 cm^3	AL 333–105[1]
A. afarensis	3.2	375 cm^3	AL 162–28[2]
A. Africanus	3.0	425 cm^3	JLD 37/38[3]
A. Africanus	2.5	515 cm^3	Stw 505[4]
P. Aethiopicus	2.5	410 cm^3	KNM-WT 17000[5]
P. boisei	2.4	427 cm^3	Omo L399y-6[6]
P. boisei	1.7	466 cm^3	KNM-ER 732[7]
Homo erectus	1.6	932 cm^3	Java (Sangiran)[8]
Homo erectus	1.5	909 cm^3	KNM-WT 1500[9]
Homo erectus	0.9	940 cm^3	Java (Trinil 2)[10]
Homo erectus	.585	1020–1090 cm^3	China (Beijing)[11]
Early humans			
Archaic	.5–.025	880–1367 cm^3	Africa, Europe, M. East[12]
Homo Neanderhalenis	.09–.03	1200–1740 cm^3	Europe
Homo sapiens Sapiens	.4-present	1250–1600 cm^3	World

[a] Millions of years ago. *Sources*: [1] Falk (1987); [2] Falk (2015); [3] Conroy (1990); [4] Holloway (1972, 1998), Holloway et al. (2004); [5] Walker et al. (1986); [6] Holloway et al. (2004); [7] Falk (2000); [8] Holloway et al. (2004); [9] Walker and Leakey (1993); [10] Holloway et al. (2004); [11] Weidenreich (1943); and [12] Holloway (1998, 2015)

sudden growth was initially both in sub-cortical emotion centers that now worked to make the enlargement of the neo-cortex fitness enhancing; and, in all likelihood, could speak and develop symbolic culture with brains over 1000 cm.[3] Clearly *Homo erectus* was well organized, perhaps with the evolution of the nuclear family and band of hunter-gatherers which is something that does not exist in contemporary great apes and, hence, the first hominins. Further evidence of their cohesiveness and social organization is that *Homo erectus* migrated to many diverse parts of the globe long before humans did.

At the bottom of Table 8.2 are measurements on the volume of the brain of early subspecies of humans. At .5 million years ago, *archaic humans* had brain sizes ranging from 880 to 1367cm^3, *Homo neanderthalensis* at 1200 to 1740cm^3, and *Homo sapiens* at 1250 to 1600cm^3. Thus, over the last 1.6 million years, the hominin brain suddenly began to grow, with some early humans having much larger brains that *Homo sapiens sapiens* who, in all likelihood, interbred with other subspecies of humans (since extinct subspecies of humans' genes are part of the contemporary human genome).

Overcoming Roadblocks to the Growth of the Hominin Brain

As emphasized earlier, there were several roadblocks in this cycle of feed forward by emotions creating growth in the neo-cortex, which then cold feedback and encouraged further elaboration of emotions. Fortunately, the palette of emotions available to early hominins was just enough to increase attachments and group formation for them to survive. Using emotions to increase fitness was, however, a double-edged sword for several reasons. First, because the majority of primary emotions are negative, enhancing emotions could potentially work against increases social solidarity. And second, expanding the size of the brain would eventually bump up against the problem of passing through the female cervix at birth, a problem that was mitigated by *life history characteristics* that had always dictated (genetically) long prolong periods of infancy and juvenile growth during which females were highly attached to their offspring (eventually increasing when the nuclear family evolved, probably with late *Homo erectus*, brought fathers or at least a male into what among all great apes is the female-offspring bond). Contrary to much folklore, the nuclear family is not genetically driven but is the result of emotionally charged relations between males and females who, probably around two million years ago, began to form attachments based upon emotions such as *love* and *carrying* among adults (something that great apes do not engaged in, with a slight exception of female gorillas bonding with a lead silverback male (for baby sitting, in essence, and protection for predators more than intimacy and caring). But the formation of a family could not be only emotional; there also had to be cognitive and perhaps even normative forces in play, which suggests that the selection had found a route around the biggest roadblock to hominin evolution: *overcoming the problems in activating the pre-adaptation for language*, already wired in the great ape genome, as a byproduct of the early rewiring of the primate brain for visual over olfactory sensory dominance.

The Biggest Neurological Roadblock to Hominin Evolution: Speech Production

Until hominins could speak, growth of the neocortex would not be fitness enhancing because, as noted earlier, it would be an empty warehouse consuming calories and protein. All great apes can understand human languages at about the level of a three- year-old child, *if* they have been raised in a language-using environment. The neurological wiring of the association cortices in and around the *inferior parietal lobe* where the hapic, auditory, and visual sense modalities meet is the location for this pre-adaptation for language (Geschwind, 1965a, 1965b). All great apes can literally "speak" because they do not possess in and around the vocal track, lips, tongue, larynx, and other structures necessary for speech production. Bu great apes can, when exposed to a human linguistic environment, symbolically communicate with humans via other sense modalities, such as visual reading and using the hand signals of the death or by haptic manipulation of keys on computers using pictographs that can be ordered in sequences communicating meanings from great apes to humans (see Rumbaugh & Washburn, 2003; Savage et al., 1978, 1988, 1993). Thus, in my view, the slow rate of growth in the size of hominins' neo-cortex was a result of this one roadblock that would take millions of years for natural selection to finally alter key structures so as to allow articulated speech.

In addition, great apes and hence early hominins do not have a fully developed *Broca's area* located in and around the division between the frontal and temporal lobes. Broca's "hump" (Falk, 2007; Enard, 2007; Taglialatela et al., 2008) existed and could be selected upon, but without a fully-developed *Broca's area*, it would be difficult to translate brain thinking, which is more gestalt-like, into *sequences* words organized into sentences. The converse of Broca's area—that is, an area uploading information from sense modalities into the brain's mode of processing information—already existed in much more developed state than Broca's hump. Thus, this analogue of what would become *Werneike's area*, uploading sensory information into the gestalt-base of brain thinking, existed in much more developed form, which is why great apes can "understand" human speech, even though they cannot speak.

Selection on *Broca's area*, which is bilateral, could occur once speech demonstrated its fitness-enhancing consequences, but to do so, it was first necessary to overcome the problems in the many structures, regulated by many genes, that needed to be re-wired and coordinated in complex ways if human speech was to evolve. This roadblock, I would hypothesize, stood firm for millions of years, perhaps as many as five million years. But, over time, the complex pattern of selection on many structures need for fine-grained speech began at some point, probably just before the sudden doubling of the hominin brain from 400+ cm3 to 900+ cm3 at 1.6 million years ago with the emergence of *Homo erectus* (and perhaps earlier at 2.0 million years ago with *Homo habilis*, if this was a distinct species).

Earlier hominins has survived by using emotions and, equally important, they survived by "body language" in which emotions and cognitions are read visually to communicate common meanings among hominins. Since all great apes employ body language and its signaling of cognitive and emotional states, this capacity was available to hominins five to six million years ago. Indeed, great apes can clearly engage in complex communications of meanings, intentions, and emotions via the gestalts of "body language" perhaps punctuated by verbal calls in ways that could increase social and group solidarities of those early hominins that ventured out into open-country filled with predators.

Still, without speech, symbolic culture cannot emerge. Cultural "elements" can emerge and be copied among primates, such as knowledge about how to capture termites or to fend off predators with calls and throwing objects, but these are not *symbolic* in the same sense as language which communicates meanings via speech which eventually become codified into values, beliefs, plans, norms, technologies, and other symbolic formations that organize late hominin and then human societies. Still, there was a "language of emotions" generated in the mutual reading of states of affect among great apes, and their hominin descendants, which allowed them to survive in more open-country habitats. And this language of emotions is still the principle means by which humans communicate affect (Turner, 2000, 2007; Turner & Maryanski, 2024).

And, such may have also been the case, to a degree, among early hominins. For example, if selection had been working on sub-cortical emotion centers of the great ape brain, it clearly hit upon the *septum*, which is the area and surrounding neurons that generate the pleasure associated with sex. The *septum* of humans is twice the size of what it is in great apes. Since great apes are sexually promiscuous, why would the size of the more than double during hominin evolution? One answer is that selection began early in enlarging this area to carry new associative emotions *associated* with sex and the outcome of sex, offspring. Indeed, it is conceivable that emotions like "love" or "strong attachments" revolving around emotions generating "loyalty" and "caring" could have evolved very early if, indeed, the *septum* was targeted, which we know it was because of its disproportionate size in humans compared to great apes. But, if this one structure was an early target of selection, creating the emotions of male-female attachment around sexual relations, it may have been the key force in initiating attachments among sexual partners that drove the formation of early forms of the nuclear family, which is *not hard-wired in great apes or humans* as a bio-programmer. Rather, it is emotions associated with sex that could have evolved long before speech or enlargement of the neo-cortex, and the capacity for speech, or symbolic culture. If such was the case, this may have been how early hominins survived for four to five million years before the brain began to grow rapidly around 1.6 million years ago. And so, as has was the case for the septum, much of hominin evolution did not revolve around creation of new modules, although a few appear to have evolved (to be discussed later); rather, selection *works on what it is given* and alters and sometimes even repurposes an existing structure.

Such was the case for speech. Great apes can emit calls carrying meanings, but their vocals tract and surrounding structures are not organized for speech production; and as a result, it took a long time for selection to realign the many structures involved in speech production. And, if natural selection had failed to randomly find the route for such a complex realignment, humans and human culture would never have evolved. It now appears that the big breakthrough in this realignment of speech structures occurred by the emergence, as far back as two million years ago, of a series of proteins, labeled FOX, which constitute a large helix-like transcription that tag specific genes. Of particularly interest to the evolution of speech and auditory language are FOXP1 and, more significantly, FOXF2 genes. FOXP1 and FOXP2 proteins have clearly targeted the genes creating the structures that allow for speech in-and-around the vocal tract and associated structures (Crespi et al., 2017; Schulze et al., 2018; Wohgelmthuh, 2014). In particular, FOXP2 were involved in developing and coordinating the vocal tract and the muscles affecting mouth and lip for speech, with damage to these areas causing speech impairments. Originally, the discovery of these proteins was thought to have emerged only 200,000 years ago, which is long after the evolution of human speech (e.g. Enard et al., 2002a, 2002b), but more recent data reveal that FOXP2 was operating 1.8 million years ago (Enard, 2007, 2016), which provides a sufficient time frame for spoken language to emerge before the sudden growth in the size of the brain at 1.6 million years ago with early *Homo erectus* (and *Homo habalis*, if considered a distinctive species). Thus, before FOXP2, the hominin brain was small, 500cm^3 at best. Of course, FOXP2 may have been operating earlier, but what is clear is that selection was altering speech structures and had begun to get them realigned in ways that allowed for a sudden increase, indeed a doubling, of brain size in a very short time. I would hypothesize that with the beginning of articulated speech, the most difficult hurdle to spoken language was obviated which, in turn, would allowed for the beginnings of symbolic culture. And once these began to emerge, the roadblock to enlarging the neo-cortex was eliminated. As new levels of emotional complexity were evolving, more nuanced and wider variety of emotions could be attached to cognitions, thereby expanding memories and decision making that, in turn, would lead to the production of cultural codes regulating late hominin conduct.

It is at this point, as symbolic culture was emerging and regulating behavior and patterns of social organization, that the unique emotions of *guilt* and *shame* may have evolved by combining the negative emotions of *fear*, *anger*, and *sadness* in way that produce *guilt* and *shame*. These are the emotion of morality that must be *linguistically articulated*; and they are what give moral cultural codes their power because *guilt* and *shame* are two of the most painful emotions that human can experience. These emotions do not exist among great apes, and hence, did not exist among early hominins (see Boehm, 2012; Turner, 2000, 2021 for more on the biology of morality). Thus, without *guilt* and *shame*, cultural codes have no "teeth" but, with them, cultural codes are more likely to be followed for fear of negative sanctions generating *shame* and/or *guilt*. Thus, once gene enhancers went to work, selection was able to eliminate the roadblock to speech production and the building up of symbolic culture.

As I have emphasized, it is emotions that were driving even this complex process because speech can only produce symbolic culture if the codes of this culture are followed and facilitate social organization. It thus appears that *guilt* and *shame* evolved along with speech and the first cultural codes and, thereby, giving these codes regulatory power. All human societies, of any size or complexity, are dependent on the feeling of positive emotions when cultural codes are followed, and/or *guilt* and *shame* when they are not. Speech and culture exert their power because they *moralize social relations*, which depends upon emotions used in sanctioning conformity to the mandates of cultural codes. This morality is driven by emotions, with *guilt* and *shame* consolidating the three negative primary emotions into highly moralized emotions that have power to regulate human thoughts and actions.[4] Moreover, with the evolution of guilt and shame, I think that it is unlikely that speech could have produced the moralized symbolic codes necessary enduring patterns of social organization.

The Neurology of the 'Elaboration Machine'

Figure 8.3 outlines the dynamics of brain evolution based upon cladistic analysis and comparative neuroanatomy. The pre-adaptations discussed earlier and behavioral propensities inherited from the LCAs of present-day great apes outlined in: Table 8.1 pp. 119–20 are the starting point of hominin evolution, as is seen on the left side of Fig. 8.3. These inherited characteristics all have a biological basis and allowed early hominins to survive as long as they stayed near the forest canopy. But, selection pressures were also at work, given the receding forests over the last ten million years in Africa; and so, the pre-adaptations and behavioral propensities inherited from the LCAs would inevitably change and, in most cases, enhanced by selection working on enlarging the brain of hominins, very slowly at first and then about 1.6 million years ago, very rapidly. As

[4] See Table 7.6 on p. 108.

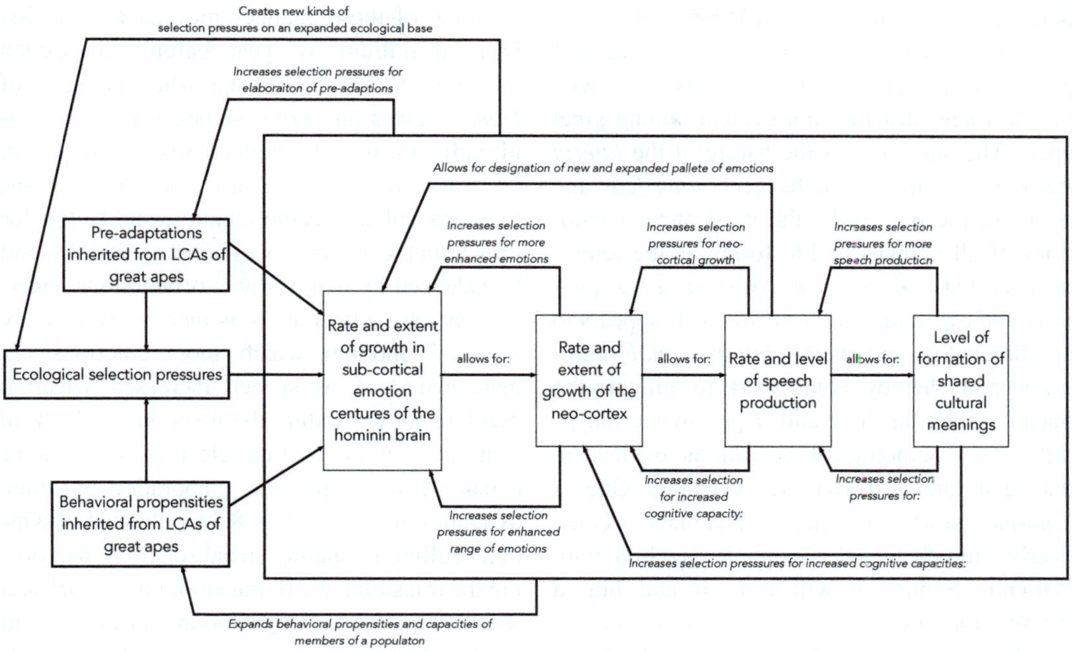

Fig. 8.3 Inherited pre-adaptations, behavioral propensities, and ecological selection pressures driving emotions and the "elaboration machine"

I have emphasized, the key to this evolutionary pattern depended upon enlarging sub-cortical emotion centers diagramed in Fig. 8.2. Since the neo-cortex is often emphasized more than the sub-cortex, it is important to emphasize again that *the evolution of the neo-cortex was driven by what was occurring in the sub-cortex* since cognitions cannot be remembered or used in decision-making without being valenced by emotions. So, the original growth of the hominin brain during the first four-to-five million years after hominins split from the LCAs of chimpanzees (and earlier, gorillas and orangutans) was not great at approximately 100cm^3 over that of the great apes, it was probably growth in the sub-cortex, pushing out the still relatively small neo-cortex outward, thereby accommodating critical growth in the sub-cortex that would eventually make fitness-enhancing the growth of the neo-cortex possible; and once the neo-cortex grew, these increased capacities of the neo-cortex would feedback and generate selection pressures for the enhancement of emotions. This feed forward from sub-cortical growth and feedback from originally small increases in the neo-cortex was *the initial engine driving evolution of hominin and then human neurology* which, despite its larger scale and connectivity, was still the basic neurology of the great apes.

More recent data using different methodologies revolving around allometric measurements confirm growth of sub-cortical areas of the brain by 50% (see Barger et al., 2007, 2012, 2014). These data confirm data gathered decades earlier[5] that the sub-cortex was growing and, moreover, that at least one new "module" (in the sense once argued by evolutionary psychology) evolved in the human *amygdala*, the ancient are generating *anger* and *fear* in different nuclei.[6] It can be asked why would natural selection enlarged the *amygdala* and installed an entirely new structure, the *lateral nucleus*, in the center of negative emotions? Selection was working to enhance positive emotions in order to increase sociality and soli-

[5]See, for example, Stephan (1972, 1983); Stephan and Andy (1969, 1977); Stephan et al. (1970); Stephan et al. (1981); Stephan et al. (1988); Eccles (1989).

[6]See also: Semendeferi et al. (2000); Semendeferi et al. (2002); Sherwood (2007); Sherwood et al. (2005, 2008).

darity among hominins so that they could form more permanent groups as a defense against predators in open-country habitats. So, why install a new structure not evident among great apes? The answer is in the nature of the *lateral nucleus*. Among the nuclei generating *fear* and *anger* in the *amygdala*, the most ancient emotions of all animals of life forms, is the *lateral nucleus* that is devoted *to assessing social contexts and expectations for behavior*. It appears to be directed at *enhancing interpersonal attunement* and, thereby, could work to mitigate the nuclei generating *fear* and *anger* in the *amygdala*. Two pathological conditions evident in damaged lateral nuclei are revealing. One is *Williams Syndrome* where individuals become overly empathetic. For example, a child with Williams Syndrome will run up and hug a stranger that think child perceives to be in emotional distress. The other pathology is the opposite of the Williams syndrome: *Autism* or the lack of a capacity for interpersonal empathy. Thus, selection was working to enhance capacities to read emotions of that could enhance interpersonal attunement among hominins. Along with the existing capacities for interpersonal attunement among great apes (see Table 8.1) inherited from the LCAs, selection was working relations to increase bonding and attunement in the midst of the centers that can disrupt social relations. Indeed, selection did something that is actually very rare human neurology: create a new module. Most change in the neurology of humans compared to great apes revolves around *increasing the size and connectivity* of existing structures, as was the case in the growth of the *septum*, *hippocampus*, *transition cortices* to the hippocampus, and other sub-cortical structures. And since the enlargement of the *amygdala* is mostly the result of the *lateral nucleus*, selection was not enlarging centers for *fear* and *anger* but, instead, creating a new module devoted to interpersonal attunement and, thereby, increasing the possibility for more stable group structures by dropping it into the most *ancient centers of sub-cortical areas* of mammals and reptiles (including birds) where anger and fear can be aroused to disrupt interpersonal attunement.

Some of these changes may have been first initiated millions of year before the sudden increase 1.6 million years ago when the brains of *Homo erectus* and *ergaster* began to grow rapidly, after the breakthrough for articulated speech. Once the roadblock for articulated speech was removed, enhanced emotionality would allow for growth in the neo-cortex as new cognitions could be valenced by new variants of emotions generated in sub-cortical areas as they activate "body systems" through which most emotions are generated.[7]And, as speech increased selection pressure for expanding the neo-cortex which, in turn, fed back to generate selection pressures for enhanced emotions, the elaboration machine dynamics outlined in Fig. 8.3 would lead to symbolic culture regulating social relations, and during the transition, the moral emotions of *guilt* and *shame* evolved in ways linking powerful and potentially painful emotions to cognitions articulating morality codified in values, beliefs, and normative systems regulating hominin and eventually fully human hunting and gathering bands, composed of nuclear families.

There were, however, several points in human evolution where the human population was declining and potentially facing extinction (see Amos & Hoffman, 2010), but the elaboration machine had generated the capacity for high- solidarity families and band structures and their cultures that humans could sustain sufficient organization and, eventually, find routes out of Africa and thereby reverse rapid drops in the number of humans on earth and, in the longer run, to over-populate earth with humans. Moreover, once the brain had reached it full size by virtue of being flexible as it passed through the female cervix at birth, humans with their now elaborate legacy from the LCAs and with the capacity to codify culture, including technologies and strategies for adaptation to new ecologies; and with small but cohesive kin units in bands coordinating hunting and gathering, late hominins and early humans were able to adjust behaviors and patterns of social organization to survive

[7]See Figs. 7.2, 7.3, 7.4, 7.5, and 7.6 in Chap. 7 for models of these "body systems."

and, eventually, prosper. Earth is literally "the planet of the fourth great ape," although overpopulation, human-induced global warming, potential nuclear war, and other challenges—all created by the neurology of the brain *driven by emotions*—may, in the end, lead to de-evolution of the fourth great ape and, no doubt, the extinction of the other three great apes and many other species on the planet.

The vulnerability of humans resides in their neurology where a low-sociality species used enhanced emotions—a good many of which are negative—to increase sociality and organizational solidarities; and then, with the emergence of speech and symbolic culture, humans could construct ever-more deadly technologies and institutional systems that arouse negative emotions and mobilize populations for activities—from warfare and environmental degradation to overpopulation and many other potential disasters—to destroy not only the remaining three great apes on the planet of the apes but also the fourth great ape. Emotions will drive these destructive actions, even though these same emotional capacities are what made it possible for the neurology of humans to create mega-societies which, ironically, often give rise to the large repertoire of negative emotions that are part of the elaboration machine from its very beginnings to the present.

Conclusion

My argument about emotions as the driving force of brain evolution is, of course, somewhat speculative. Yet, it fits both the biology and sociology of hominin evolution, without making assertions often made by sociobiologists and evolutionary psychologists. And, it uses two methodologies that, surprisingly, are hardly ever used by sociobiologists and evolutionary psychologists. Indeed, cladistic analysis and comparative neuroanatomy (among species of apes and hominins) allow access to data that, again, are hardly consulted by many engaged in evolutionary analysis. Too often, explanations are exercises in "reverse engineering" and "just so stories" explaining why in the distant past this or that behavior evolved by some change (often not specified) in the brain of the ancestors of humans. Using the brains of the three great apes as a reasonable proxy for what the brain of the first hominins was like and then making comparisons of the brains of great apes and humans, we can get a clear sense for what selection was doing to the brains of hominins, and while a few "new" structures can be found, most of the selection on the brain revolved around *increasing the size and connectivity of key brain structures*, with the sub-cortex being the starting point and then the driving force of this evolutionary process. Very few unambiguously "new" structures were involved in this six to seven million period, which makes sense because most mutations have negative effects of fitness (Fisher, 1930), and especially so in the complex structure of the brain. Dramatic changes by mutations are likely to be especially disruptive to brain neurology and, hence, to fitness of a species. A more generalized and, in many ways, more conservative approach is to alter existing structures in terms of size and, in some cases, connectivity to other structures over a long period of time. Coupled with the pre-adaptations and behavioral capacities inherited by hominins from the LACs of the great apes, we can get a much better pictures how the brain evolved, without having to make "just so" arguments but rather arguments that are compatible with the data on brain evolution.

Once we recognize that despite the relatively large brain of the first hominins, coupled with the legacy of pre-adaptations and behavioral capacities, inherited from the LACs of the three great apes (as a proxy for the first hominins), the big adaptive problem for early hominins was sociological in nature: *how to get re-organized* in a changing ecology over the last seven million years. This sociological view of the adaptive problem facing early and later hominins then guides neurological analysis as we look for neurology of what increases social bonding by animals that do not have strong social ties or patterns of group organization. This kind of analysis is just one important way that neuro-sociology can make a large contribution to biology, psychology, and evolutionary analysis in general.

References

Amos, W., & Hoffman, J. I. (2010). Evidence that two main bottleneck events shaped modern human genetic diversity. *Proceedings of the Royal Society, 277*, 131–137.

Ardesch, D. J., Scholtens, H. L. H., Li, L., Preuss, T. M., Rilling, J. K., & van den Heuvel, M. P. (2019). Evolutionary expansion of connectivity between multimodal association areas in the human brain compared with chimpanzees. *Proceedings of the National Academy of Sciences, 116*(14), 7101–7106.

Baizer, J. S., Baker, J. F., Haas, K., & Lima, R. (2007). Neurochemical organization of the nucleus paramedinaus dorsalis in the human. *Brain Research, 1176*, 45–52.

Barger, N., Hanson, K., Teffer, K., Schenker-Ahmed, N. M., & Semendeferi, K. (2014). Evidence for evolutionary specialization in human limbic structures. *Frontiers in Human Neurosciences, 8*, 277. 1–17. https://doi.org/10.3389/fnhum.2014.00277

Barger, N., Stefanacci, L., Schumann, C., Sherwood, C. C., Annese, J., Allman, J. M., Buckwalter, J. A., Hof, P. R., & Semendeferi, K. (2012). Neuronal populations in the basolateral nuclei of the amygdala are differentially increased in humans compared with apes: A stereological study. *Journal of Comparative Neurology, 520*, 3035–3054.

Barger, N., Stefanacci, L., & Semendeferi, K. (2007). A comparative volumetric analysis of the amygdaloid complex and basolateral division of the human and ape brain. *American Journal of Physical Anthropology, 134*, 392–403.

Barkow, J. H., Cosmides, L., & Tooby, J. (Eds.). (1992). *The adapted mind: Evolutionary psychology and the generation of culture*. Oxford University Press.

Bickerton, D. (2003). Symbol and structure: A comprehensive framework for language evolution. In M. S. Christiansen & L. Kirby (Eds.), *Language evolution* (pp. 77–93). Oxford University Press.

Boehm, C. (2012). *Moral origins: The evolution of virtue, altruism, and shame*. Basic Books.

Brosnan, S. F., Schiff, H. C., & de Waal, F. B. M. (2005). Tolerance for inequity may increase with social closeness in chimpanzees. *Proceedings of the Royal Society of London* (Vol. 272, pp. 253–258).

Brosnan, S. F., & de Waal, F. B. M. (2004). Fair refusal by capuchin monkeys. *Nature, 428*, 128–140.

Brosnan, S. F., & de Waal, F. B. M. (2014). Evolution of responses to (un)fairness. *Science, 346*(6207), 125776.

Call, J., & Tomasello, M. (1998). Does a chimpanzee have a theory of mind? 30 years later. *Trends in Cognitive Science, 12*, 187–192.

Collins, R. (2004). *Interaction ritual chains*. Princeton University Press.

Conroy, G. (1990). *Primate evolution*. Norton.

Cooley, C. H. (1902). *Human nature and the social order*. Princeton University Press.

Cosmides, L. (1989). The logic of social exchange: Has natural selection shaped how humans reason? Studies with the Wason selection task. *Cognition, 31*, 187–176.

Crespi, B., Read, S., & Hurd, P. (2017). Segregating polymorphisms of FOXP2 are associated with measures of inner speech, speech fluency and strength of handedness in a healthy population. *Brain and Language, 173*, 33–44.

Damasio, A. (1994). *Descartes' error: Emotion, reason, and the human brain*. Putnam.

Damasio, A. (2000a). *The felling of what happens: Body, emotions and the makings of consciousness*. New American Library.

Damasio, A. (2000b). *The feeling of what happens: Body, emotion and the making of consciousness*. Heinemann.

Damasio, A., & Geschwind, N. (1984). The neural basis of language. *Annual Review of Neuroscience, 7*, 127–147.

de Waal, F. B. M. (1989). Food sharing and reciprocal obligations among chimpanzees. *Journal of Human Evolution, 18*, 433–459.

de Waal, F. B. M. (1991). The Chimpanzee's sense of social regularity and its relation to the human sense of justice. *American Behavioral Scientist, 34*, 335–349.

de Waal, F. B. M. (1996). *Good natured: The origins of right and wrong in humans and other animals*. Harvard University Press.

de Waal, F. B. M., & Brosnan, S. F. (2006). Simple and complex reciprocity in primates. In P. K. C. P. van Schaik (Ed.), *Cooperation in primates and humans: Mechanisms and evolution* (pp. *85–106*). Springer.

Durkheim, É. (1912/1947). *The elementary forms of the religious life: A study in religious sociology*. The Free Press.

Eccles, J. C. (1989). *Evolution of the brain: Creation of self*. Routledge.

Ekman, P. (1984). Expression and the nature of emotion. In K. Scherer & P. Ekman (Eds.), *Approaches to emotion* (pp. 319–343). Erlbaum.

Emde, R. N. (1962). Level and meaning for infant emotions: A biosocial view. In W. A. Collins (Ed.), *Developments of cognition, affect, and social relations* (pp. 1–37). NJP Lawrence Erlbaum.

Enard, W. (2007). The evolution of Broca's area. IBRO history of neuroscience. https://ibro.org/wp-content/uploads/2018/07/The-Evolution-of-Brocas-Area.pdf

Enard, W. (2016). The molecular basis of human brain revolution. *Current Biology, 26*(20), R1109–R1117. https://www.ncbi.nlm.nih.gov/nlmcatalog?term=%22Curr+Biol%22%5BTitle+Abbreviation%5D

Enard, W., Khaitovich, P., Klose, J., Zollner, S., et al. (2002a). Intra- and interspecific variation in primate gene expression patterns. *Science, 296*, 340–342.

Enard, W., Przeworski, M., Fisher, S. E., Lai, C. S. L., Wiebe, V., Kitano, T., Monaco, A. P., & Paabo, S. (2002b). Molecular evolution of FOXP2: A gene involved in speech and language. *Nature, 418*, 869–872.

Ettlinger, G. (1977). Cross-modal equivalence in non-human primates. In A. M. Schiriver (Ed.), *Behavioral primaltology* (Vol. 1). Earlbaum.

Falk, D. (1987). Hominid paleoneurology. *Yearbook of Physical Anthropology, 16*, 13–30.

Falk, D. (2000). *Primate diversity*. Norton.

Falk, D. (2007). The evolution of broca's area. IBRO history of neuroscience. http://www.ibro.info/Pub/Pub_Main_Display.asp?LC_Docs_ID=3145

Falk, D. (2015). Evolution of the primate brain. In W. Henke & I. Tattersall (Eds.), *Handbook of paleoanthropology* (pp. 1495–1513). Springer.

Foley, R., & Gamble, C. (2009). The ecology of social transitions in human evolution. *Philosophical Transactions of the Royal Society of London Series B: Biological Sciences, 364*(1533), 3267–3279.

Gallup, G. G., Jr. (1970). Chimpanzees: Self-recognition. *Science, 167*, 88–87.

Gallup, G. G., Jr. (1979). *Self-recognition in chimpanzees and man: A developmental and comparative perspective*. Plenum.

Gallup, G. G., Jr. (1982). Self-awareness and the emergence of mind in primates. *American Journal of Primatology, 2*, 237–248.

Gazzaniga, M. S., & Smylie, C. S. (1990). Hemisphere mechanisms controlling voluntary and spontaneous mechanisms. *Annual Review of Neurology, 13*, 536–540.

Gergely, G., & Csibra, G. (2006). Sylvia's recipe: The role of imitation and pedagogy. In N. J. Enfield & S. C. Levinson (Eds.), *The transmission of cultural knowledge*. Oxford University Press.

Geschwind, N. (1965a). Disconnection syndromes in animals and man. Part I. *Brain, 88*, 237–294.

Geschwind, N. (1965b). Disconnection syndromes in animals and man, Part II. *Brain, 88*, 585–644.

Geschwind, N. (1985). Implications for evolution, genetics, and clinical syndromes. In S. D. Glick (Ed.), *Cerebral lateralization in nonhuman species* (pp. 247–278). Academic Press.

Goffman, E. (1967). *Interaction ritual: Essays in face-to-face behavior*. Anchor Books.

Hare, B., Call, J., & Tomasello, M. (2001). Do chimpanzees know what conspecifics know? *Animal Behavior, 61*, 139–151.

Hare, B., Call, J., & Tomasello, M. (2006). Chimpanzees deceive a human competitor by hiding. *Cognition, 101*, 495–514.

Holloway, R. L. (1972). Australopithecine endocasts, brain evolution in the hominoidea and a model of hominid evolution. In R. Tuttle (Ed.), *The functional and evolutionary biology of primates* (pp. 185–204). Aldine-Atherton.

Holloway, R. L. (1998). Relative size of the human corpus callosum redux: Statistical smoke and mirrors? *Behavioral and Brain Sciences, 21*, 333–335.

Holloway, R. L. (2015). The evolution of the hominid brain. In W. Henke & I. Tattersall (Eds.), *Handbook of paleoanthropology* (pp. 1961–1981). Springer.

Holloway, R. L., Broadfield, D. C., & and. Yuan, M. (2004). *The human fossil record, volume three: Brain endocasts—The paleoneurological evidence*. Wiley-Liss.

Horowitz, A. C. (2003). Do humans ape? Or do apes human? Imitation and intention in humans (*Homo sapiens*) and other animals. *Journal of Comparative Psychology, 117*, 325–336.

Itakura, S. (1996). An exploratory study of gaze-monitoring in non-human primates. *Japanese Psychological Research, 38*, 174–180.

Jarvis, M. J., & Ettlinger, G. (1977). Cross-modal recognition in chimpanzees and monkeys. *Neuropsychologia, 15*(4–5), 499–506.

Jeffers, R., & Lechiste, I. (1979). *Principles and methods for historical linguistics*. MIT Press.

LeDoux, J. E. (1991). Neuroscience commentary: The emotional brain. *Journal of NIH Research, 3*, 49–51.

LeDoux, J. E. (1993). Emotional memory systems in the brain. *Behavioral Brain Research, 58*, 69–79.

LeDoux, J. E. (1996). *The emotional brain: The mysterious underpinnings of emotional life*. Simon & Schuster.

Leslie, K. R., Johnson-Frey, S. H., & Grafton, S. T. (2004). Functional imaging of face and hand imitations: Towards a motor theory of empathy. *NeuroImage, 21*, 601–607.

Maryanski, A., & Turner, J. H. (1992). *The social cage: Human nature and the evolution of society*. Stanford University Press.

Mead, G. H. (1934). *Mind, self and society*. Chicago.

Menzel, E. W. (1971). Communications about the environment in a group of young chimpanzees. *Folia Primatologica, 15*, 220–230.

Okamoto, S., Tomonaga, M., Ishii, K., Kawai, N., Tanaka, M., & Matsuzawa, T. (2002). An infant chimpanzee (*Pan troglodytes*) follows human gaze. *Animal Cognition, 5*, 107–114.

Osgood, C. E. (1966). Dimensionality of the semantic space for communication via facial expressions. *Scandinavian Journal of Psychology, 7*, 1–30.

Parr, L., Waller, B., & Fugate, J. (2005). Emotional communication in primates: Implications for neurology. *Current Opinion in Neurology, 15*, 716–720.

Passingham, R. E. (1973). Anatomical differences between the neocortex of man and the other primates. *Brain Behavioral Evolution, 7*, 337–359.

Passingham, R. E. (1975). Changes in the size and organisation of the brain in man and his ancestors. *Brain and Behavior Evolution, 11*, 73–90.

Passingham, R. E. (1982). *The human primate*. Freeman.

Povinelli, D. J. (2000). *Folk physics for apes: The chimpanzee's theory of how the world works*. Oxford University Press.

Povinelli, D. J., & Eddy, T. J. (1997). Specificity of gaze-following in young chimpanzees. *British Journal of Developmental Psychology, 15*, 213–222.

Raghanti, M. A., Stimpson, J. L., Marcinkiewicz, J. L., et al. (2000). Differences in cortical serotonergic innervation among humans, chimpanzees, and macaque monkeys: A comparative study. *Cerebral Cortex, 18*, 584–597.

Rizzolatti, G., & Sinigaglia, C. (2008). *Mirrors in the brain: How our minds share actions, emotions, and experience*. Oxford University Press.

Rizzolatti, G., Fadiga, L., Fogassi, L., & Gallese, V. (2002). From mirror neurons to imitation: Facts and speculations. In W. Prinz & A. N. Meltzoff (Eds.), *The imitative mind: Development, evolution and brain bases* (pp. 247–266). Cambridge University Press.

Rumbaugh, D., & Savage-Rumbaugh, E. S. (1990). Chimpanzees: Competencies for language and numbers. In W. Stebbins & M. Berkley (Eds.), *Comparative perception, volume 2: Complex signals*. Wiley.

Rumbaugh, D., & Washburn, D. (2003). *Intelligence of apes and other rational beings*. Cambridge University Press.

Savage-Rumbaugh, E. S., & Lewin, R. (1994). *Kanzi: The ape at the brink of the human mind*. Wiley.

Savage-Rumbaugh, E. S., Murphy, J., Sevcik, R. A., Brakke, K. E., Williams, S. L., & Rumbaugh, D. M. (1993). *Language comprehension in the ape and child*. University of Chicago Press.

Savage-Rumbaugh, E. S., Rumbaugh, D., & Boysen, S. (1978). Symbolic communication between two chimpanzees (Pan troglodytes). *Science, 201*(4356), 641–644.

Savage-Rumbaugh, S., Sevcik, R. A., & Hopkins, W. D. (1988). Symbolic cross-modal transfer in two species of chimpanzees. *Child Development, 59*, 617–625.

Scheff, T. (1988). Shame and conformity: The deference-emotion system. *American Sociological Review, 53*, 395–406.

Schulze, K., Varga-Kadem, F., & Mishkin, M. (2018). Phonological working memory and FOXP. *Neuropsychologia, 108*, 147–152.

Schutz-Bosbach, S., & Prinz, W. (2007). Perceptional resonance: Action-induced modulation of perception. *Trends in Cognitive Sciences, 11*, 349–355.

Semendeferi, K., Lu, A., Schenker, N., & Damasio, H. (2000). Humans and great apes share a large frontal vortex. *Nature Neuroscience, 5*, 272–276.

Semendeferi, K., Lu, A., Schenker, N., & Damasio, H. (2002). Humans and great apes share a large frontal cortex. *Nature Neuroscience, 5*, 272–276.

Sherwood, C. C. (2007). The evolution of neuron types and cortical histology in apes and humans. In T. M. Preuss & J. H. Kaas (Eds.), *Evolution of nervous systems, Volume 4: Primates* (pp. 355–378). Academic Press.

Sherwood, C. C., Holloway, R. L., Semendeferi, K., & Hof, P. R. (2005). Is prefrontal white matter enlargement a human evolutionary specialization? *Nature Neuroscience, 8*, 537–538.

Sherwood, C. C., Subiaul, F., & Zawidzki, T. W. (2008). A natural history of the human mind: Tracing evolutionary changes in brain and cognition. *Journal of Anatomy, 212*, 426–454.

Spencer, B., & Gillen, F. J. (1899). *The native tribes of Central Australia*. Macmillan.

Stephan, H. (1972). Evolution of primate brains: A comparative anatomical investigation. In R. Tuttle (Ed.), *The functional and evolutionary biology of primates*. Aldine-Atherton.

Stephan, H. (1983). Evolutionary trends in limbic structures. *Neuroscience and Biobehavioral Reviews, 7*, 367–374.

Stephan, H., & Andy, O. J. (1969). Quantitative comparative neuroanatomy of primates: An attempt at phylogenetic interpretation. *Annals of the New York Academy of Science, 167*, 370–387.

Stephan, H., & Andy, O. J. (1977). Quantitative comparison of the amygdala in insectivores and primates. *Acta Anatomica, 98*, 130–153.

Stephan, H., Baron, G., & Frahm, H. (1988). Comparative size of brains and brain components. In H. Steklis & J. Erwin (Eds.), *Comparative primate biology, Volume 4* (pp. 1–37). Alan R. Liss.

Stephan, H., Bauchot, R., & Andy, O. J. (1970). Data size of the brain and the various brain parts in insectivores and primates. In R. Noback & W. Montagna (Eds.), *Advances in primatology, Volume 1: The primate brain*. Appleton.

Stephan, H., Frahm, H., & Baron, G. (1981). New and revised data on volumes of brain structures in insectivores and primates. *Folia Primatologica, 35*, 1–29.

Subiaul, F. (2007). The imitation faculty in monkeys: Evaluating its features, distribution, and evolution. *Journal of Anthropological Science, 85*, 35–62.

Taglialatela, J. P., Russell, J. L., Schaeffer, J. A., & Hopkins, W. D. (2008). Communicative signaling activates 'Broca's' homolog in chimpanzees. *Current Biology, 18*, 343–348.

Tomasello, M., Hare, B., & Fogleman, T. (2001). The ontogeny of gaze following in chimpanzees, *pan troglodytes*, and rhesus macaques, *Macaca mulatta*. *Animal Behavior, 61*, 335–343.

Tomonaga, M. (1999). Attending to the others' attention in macaques' joint attention or not? *Primate Research, 15*, 425.

Tsao, D., Freiwald, W. T., Mandeville, K. J., & Tootell, R. (2003). Faces and objects in macaque cerebral cortex. *Natural Neuroscience, 6*, 989–995.

Turner, J. H. (2000). *On the origins of human emotions: A sociological inquiry into the evolution of human affect. Stanford*. Stanford University Press.

Turner, J. H. (2002). *Face to face: Toward a theory of interpersonal behavior*. Stanford University Press.

Turner, J. H. (2007). *Human emotions: A sociological theory*. Routledge.

Turner, J. H. (2021). *On human nature: The biology and sociology of what made us human*. Routledge.

Turner, J. H., & Maryanski, A. (2008). The limitations of evolutionary theory from biology in explaining sociocultural evolution. *Sociologica, 3*, 1–38.

Turner, J. H., & Maryanski, A. (2024). *Biosocial evolutionary analysis: Reconciling biology, psychology, and sociology*. Edgar Elgar.

Walker, A., & Leakey, R. (Eds.). (1993). *The Nariokotome homo erectus skeleton*. Harvard University Press.

Walker, A. C., Leaky, R. E., Harris, J. M., & Brown, F. H. (1986). 2.5-myr *australopithecus boisei* from west of lake turkana, Kenya. *Nature, 322*, 517–522.

Weidenreich, F. (1943). *The skull of sinanthropus pekinensis: A comparative study on a primitive hominid skull*. Stechert.

Emotion, Brain Lateralization, and Social Relations: A Neurosociological Theory of the Basic Emotion

9

Warren D. TenHouten

Abstract

Longstanding neuroscientific debate concerns the hemispheric lateralization of emotions. Four competing views prevail. The Right-Hemisphere Model holds that all emotions are right lateralized; the Valence Model, that negatively- and positively-valenced emotions are right- and left-lateralized, respectively; the Approach-Avoidance Model, that approach- and avoidant-type emotions are left- and right-lateralized, respectively; and the Emotion-Type Hypothesis, that Ekman's six basic emotions are right-lateralized, and all other emotions are left-lateralized and socially constructed. One limitation of these models is that they do not clearly distinguish the inner mental experience of the emotions from the perception of emotions expressed by others. An emotion–lateralization model focused on the individual's inner experience of basic emotions, the Feeling–Function Model, is presented. A nearly universal ethogram that identifies the four most fundamental problems of life is described; it is proposed that the prototypical adaptive reactions to negative and positive experiences of these life-problems comprise the basic emotions. A focused review of the neuroscientific evidence supports the theoretical proposition that four basic emotions—sadness, fear, anticipation, and acceptance—are predominantly right-lateralized, and their four opposites—happiness, anger, surprise, and disgust, respectively—are left-lateralized. This emotion-lateralization model theorizes that experience of the right-lateralized emotions significantly involves global feeling-states, or states of awareness of somatic, physiological changes in mind, body, and self that precede and stimulate the right-hemisphere-based emotions. In contrast, cognitive interpretation of a situation stimulates the function-oriented left-hemisphere-based basic emotions.

W. D. TenHouten (✉)
Department of Sociology, University of California, Los Angeles, Los Angeles, CA, USA
e-mail: wtenhout@g.ucla.edu

This chapter presents a theory of the neural and sociorelational foundations of emotion that interrelates four archetypal social-relations, hemispheric brain lateralization, and the basic emotions. Long-standing neuroscientific debates concern the left and right hemispheric lateralization of emotion (Demaree et al., 2005; Wyczesany et al., 2018). Four competing views prevail: (i) the Right-Hemisphere Model (Hughlings-Jackson, 1866; Mills, 1912; Gainotti, 1972, 2012; Levy et al., 1983) claims that all emotions are right lateralized; (ii) the Valence Model (Silberman & Weingartner, 1986; Davidson et al.,

W. Kalkhoff et al. (eds.), *Handbook of Neurosociology*, Handbooks of Sociology and Social Research, https://doi.org/10.1007/978-3-031-95615-7_9

1987; Ehrlichman, 1987) postulates that the right and left hemispheres are specialized for negative and positive emotions, respectively; (iii) the Approach–Avoidance Model (Davidson, 1995; Harmon-Jones & Gable, 2018;) states that avoidance- and approach-oriented emotions are right and left lateralized, respectively; and (iv) the Emotion–Type Model (Ross et al., 1994; Ross, 2021) holds that all primary emotions are right-lateralized, and all other emotions are socially-constructed and left lateralized.

We use Fiske's (1991, 2004) Relational Models Theory and Plutchik's (1958, 1962/1991, 1980a, 1980b) psychoevolutionary theory of the emotions to critique these four models, and advance a Feeling–Function Model of the hemispheric lateralization of eight basic emotions; this model hypothesizes that (i) the inner experience of the right lateralized emotions significantly involves global feeling states, or states of awareness of physiological changes in mind, body, and self, that are partially unconscious primary processes (Weinberg, 2000); and (ii), the left hemisphere, in contrast, is specialized for goal-oriented and functional emotions that accompany plans and intentions and enable the individual to address problems and opportunities in the world. A brief, focused review of emotion lateralization research shows that, on the level of inner experience, sadness, fear, anticipation, and acceptance are predominantly right-lateralized, while the opposite set of four emotions (joy-happiness, anger, surprise, and disgust) are predominantly left-lateralize

The Archetypes of Sociality

We begin with Fiske's social-relations model of sharing, ranking, matching, and exchanging. These four dimensions of sociality have been found in very different cultures, exist to some extent in primates, and are hypothesized to have emerged out of a behavioral-action profile, or ethogram, developed in herpetology. As elaborated by Plutchik (1980a, 1980b), this fourfold ethogram asserts that temporality, hierarchy, territoriality, and identity are the four most fundamental problems of life.

Social scientists have modeled the sociocognitive structures and processes that purportedly undergird human social interactions. These vary in their dimensionality, but fourfold conceptualizations have prevailed. Several are summarized in Table 9.1. We focus on undoubtedly the most influential, general, and well-developed of these fourfold models, Relational Models Theory (Fiske, 1991, 2004; Haslam, 1995, 1997; Bolender, 2010). We use Fiske's terminology for communal-sharing (CS) and equality-matching (EM) but adopt the more general terms hierarchical-ranking (HR) and socioeconomic-exchanging (SE) for authority-ranking and market-pricing. These sociocognitive relational models can be positively or negatively valenced, depending on whether they represent opportunity or threat, gain or loss.[1]

The Four Social-Relations Models

Communally-shared social relations are close, personal, and intrinsically rewarding (Blau, 1964: 19–22). They involve institutions of family and kinship, with connection to the past through tradition, ceremony, rite, and ritual. Positive communally-shared social relations denote "duties and sentiments generating kindness and generosity among individuals perceived to be of the same kind, especially kin," without expectation of reciprocity (Fiske, 1991: 255–288). Individuals exhibit solidarity, unity, belonging, social cohesion, friendship, homophily, and a group entitativity, or a 'we-mode' (Gallotti &

[1] Relational-Models Theory has been criticized, with justification, for ignoring situations of conflict and contention, that is, negative experiences of social relations. In two scaling instruments, for example, Likert-type items present only positive experiences of the four relational models (Haslam, 1995; Haslam & Fiske, 1999). As an obvious example, individuals have negative experiences of community (CS–), for temporality involves not just birth but also death.

Table 9.1 Fourfold models of social relations theorist(s)

	Conceptualizations	Sharing	Ranking	Matching	Exchanging
Georg Simmel (1907–1910)	Forms of social interaction	Sociability	Domination	Unity and conflict	Exchange
Max Weber (1921/1978)	Social relations	Communal or associative	Authority based	Mutually responsible	Market participation
W. I. Thomas (1923/1966)	Four wishes	For response	For security	For recognition	For new experiences
Talcott Parsons (1951)	Social system imperatives	Latent-receptive	Adaptation manipulation	Integration	Goal attainment
Peter Blau (1964)	Social interaction models	Social attraction	Social power	Balanced reciprocity	Economic exchanging
M. Thompson, R. Ellis, & A. Wildavsky (1990)	Cultural theory of risk	Fatalist	Hierarchy	Egalitarianism	Individualism
Alan Page Fiske (1991)	Relational models theory	Communal sharing	Authority ranking	Equality matching	Market pricing

Frith, 2013), based on "a subjective feeling of the parties, whether affectual or traditional, that they belong together" (Weber, 1921/1978: 40).

Hierarchical ranking refers to individuals' unequal amounts of influence, authority, status, privilege, position, or rank within systems of social stratification. High social ranking can confer first access to food, shelter, comfort, and reproductive opportunities.

Socioeconomic-exchanging is a relationship mediated by values determined by a market system. It involves the commodification of resources, objects, and social positions (Weber, 1921/1978). Individuals "denominate value in a single metric, typically price" (Fiske, 1991: 15–16).

The positive experience of equality-matching obtains when participants share a common social identity and/or are functionally equivalent. Individuals "conceive of each other—or the rights, duties or actions involved in the relationship—as distinct, but as balancing each other, aligning or matching, so they are interchangeable" (Fiske, 1991: 15). Inequality-attenuating protocols range from informal turn-taking and in-kind reciprocity, to formally codified distributive justice. In decision-making, everyone has one vote; likemindedness and information sharing are prioritized (TenHouten, 2005: 121–134).

Evolutionary History of the Models

Social relations involve distinctively human cognitive capabilities including language, abstract reasoning, and the attribution of mental states to others (theory of mind[2]). Indeed, it has been argued that the relational models have arisen *sui generis* in humans (see, e.g., Tomasello, 1994). Evidence shows that they characterize disparate human cultures including various contemporary Western societies, the Moose people of Burkina Faso (Fiske, 1991), the Vai (from Liberia and Sierra Leone), the Bengali, Koreans and Chinese (Fiske, 1993), and Australian Aborigines (TenHouten, 2005). These findings strongly suggest, but do not prove, that the four relational models are culturally universal.

Compelling evidence also shows that all four relational models exist in non-human primates (Haslam, 1997). In his theory of inclusive fitness, Hamilton (1964) argued that distinguishing between kin and nonkin, and giving preference to

[2]In a survey of studies of imitation, self-recognition, deception, role-taking, and perspective-taking, in every case where non-human primate behavior has been interpreted as a sign of theory of mind, the results could be attributed to non-mentalistic processes such as associative learning or inference-based on nonmental categories (Heyes, 1998; for a review, see Call, 2007).

kin—a key feature of communal sharing—applies not only to primates, but to all social creatures, including social insects. Various species of nonhuman primates make categorical distinctions between kin and other conspecifics (Cheney & Seyfarth, 1990) and use distinct calls to communicate referentially to each (Gouzoules et al., 1984). Among chimpanzees, strong social ties—the hallmark of communal sharing—enable group-level cooperation in necessary activities such as boundary defense and hunting. In chimpanzee communities that hunt, the sharing of meat can be a social tool for nurturing and reinforcing social bonds (Hockings et al., 2007).

Social hierarchies are ubiquitous, varying in rigidity and sphere of operation across primate species (Majolo et al., 2012). Chimpanzees' agonistic exchanges typically display political complexity (de Waal, 1989). Monkeys' hierarchically-ranked social relations are distinguishable from other social interactions by distinct referential calls that communicate superordination or submission (Cheney & Seyfarth, 1990). High status is inherently unstable in primate social life and competition can be widespread and fierce. Dominant female cercopithecine monkeys, for example, launch randomly timed attacks in order to induce stress and uncertainty among subordinates (Silk, 2002).

Across primate species, grooming, for example, is commonly exchanged for support within an alliance, suggesting a shared cognitive currency in which favors and obligations can be traded. Coalitions and alliances among baboons can be compared to a marketplace, where supply and demand determine which deals are made and how profits are allocated; the main currency is strength and fighting ability (Nöe, 1992). Experimental studies of choice behavior show that, in order to decide which of two or more incommensurable goals to pursue, monkeys appear to reduce the factors favoring one goal over another to a common valuation on a single scale of subjective utility.[3]

Equality Matching, or balanced reciprocity, is exhibited by long-lived, frequently interacting organisms residing in large groups, including many primate species (Schino & Aureli, 2009). Inequality-attenuating dynamics include control of the strong by the weak. Chimpanzee societies have dominant males and alpha females, but the community as a whole typically supports the less powerful in their frequent dominance encounters, to end hostilities. Beneath the most dominant individuals, the presence of high-ranking chimps who are "effective at heading off, stopping, or ameliorating conflicts"—often together with the help of numerous lower-status individuals—represents a socioevolutionary step toward the human concept of an egalitarian society (de Waal, 2005: 82; see also Boehm, 1999).[4] The more peaceful bonobos exhibit egalitarianism, with approximate equality of males and females and only moderate competition among males for access to sexually receptive females (de Waal, 2005).

[3]In studies using a 'pay per view' task, thirsty male rhesus macaques' gazing behavior was observed in order to measure their preference for one of two visual target, fruit juice or fruit juice and a picture of a familiar monkey (Deaner et al., 2005). Orienting to one target yielded fruit juice; orienting to the other, fruit juice and a picture of a familiar monkey. Systematically varying the juice amounts and the pictures enabled estimate of the value of different types of reproductive and social stimuli in a liquid currency. The monkeys would forego juice rewards in order to view pictures of females communicating sexual signals or faces of high-ranking males; they required larger juice rewards to choose to view faces of low-ranking males and females. These and related results indicate that decisions based on value operate on a common currency that is independent of the modality of the goods under consideration.

[4]We see in this behavior, and in human hunting-and-gathering societies as well, that HR-bases relations precede EM-based relations, as equality is attained only through vigilant collective control of the selfish, bullying, aggressive, or otherwise improper or self-interested behavior of the powerful by those who are less powerful. As de Waal (2005: 78) put it, in this negating of the negative, "Egalitarianism is not based on mutual love and even less on passivity. It's an actively maintained condition that recognizes the universal human desire to control and dominate."

The Universal Ethogram

The presence of social-relational models across human cultures and various primate species suggests that these four dimensions of sociality likely evolved out of prototypical behavioral patterns that address basic problems of life (TenHouten, 2021a). Contemporary ethologists use a variety of terms to refer to a 'behavioral profile', including 'display action patterns', 'biogram', and ethogram' (Tinbergen, 1951: 7). Studies of various terrestrial animals led to the surprising discovery of how "relatively few kinds of behavior one can identify in each and how most of these are common to all" (MacLean, 1990: 100). While researching blue spiny lizards, iguanas, and chameleons, Greenberg (1977; see also Greenberg and MacLean, 1978) identified four display-action patterns necessary for social communication: an assertive *signature* display (identified by Carpenter, 1961), *courtship* displays, *submission* (and *dominance*) displays, and *territorial* displays. MacLean (1990: 112–134) utilized, and elaborated, this fourfold model to describe the daily lives of rainbow lizards and giant Komodo dragons. A crucial inference is that much of animal behavior represents adaptive reactions to an irreducible set of fundamental life problems; each can be positively or negatively valenced. The individual's potential mental responses to the resulting eight situations, we argue, comprise the basic emotions.

The Basic Emotions

As biologically programmed, goal-directed, motivational systems, emotions are expressive, adaptive reactions that occur rapidly in response to important environmental stimuli. Emotions signal problematic situations, connect goals to motives, lead behavior, and create meaning in the everyday world. Emotions are crucial to the process of sharing important information with conspecifics for assessing, then adapting to, pressing problems of life.

The most basic of emotions are widespread in the animal kingdom (Darwin, 1872/1965; Panksepp, 1998; Kremer et al., 2020). Proto-emotions have been found in Invertebrate species, including sea slugs, bees, crayfish, snails, crabs, ants, and flies; all have been shown to display cognitive, physiological, and behavioral indications of internal states "reminiscent of what we consider emotions" (Perry & Baciadonna, 2017). Emotions exist in vertebrate species, including fish (Cerqueira et al., 2017), amphibians (Laberge et al., 2000), reptiles (Novejarque et al., 2004), birds (Marzluff & Angell, 2013), and mammals—from dogs (Albuquerque et al., 2016) to humans (Darwin, 1872/1965).

The key criterion for regarding an emotion as basic, or primary, is that it has a deep evolutionary history. Darwin (1872/1965) realized that evolution governs not only anatomy and morphology but also the mind and its expressive behavior. These are as reliably characteristic of species as bodily systems and structures. Darwin adumbrated three principles regarding the emotions.[5] Particularly useful is his 'principle of antithesis', which holds that once a state of mind is linked to an associated habit, an opposite state of mind evokes an opposite reaction. Anger and fear, for example, while closely related and alike in some ways, evoke opposite responses: moving-toward or moving-away-from a source of anger or fear, respectively. This suggests that basic emotions might well exist as pairs of opposites. We use this principle to identify the basic emotions, a project that remains highly contested in emotions research.

Ekman and colleagues (Ekman et al., 1969; Ekman & Friesen, 1986) conducted cross-cultural studies of the perception of facial expressions and found six emotions that are recognized across very different cultures and therefore might well be primary: joy-happiness, anger, surprise, disgust, sadness, and fear. The emotions anger/fear and joy-happiness/sadness meet Darwin's

[5]The principles of "serviceable associated "habits and of "direct action in the nervous system" are of no value in contemporary emotion theory (Fridlund, 1992) and "antithesis" was stated in an obtuse and confusing manner (Ghiselin, 1969: 206).

criterion, but no opposites for surprise and disgust are identified. We can enlarge Ekman's 'big six' basic emotions by including the opposite of surprise (anticipation–expectation), and the opposite of disgust–rejection (acceptance). Anticipation and acceptance, however, are seldom included in hypothesized lists of basic emotions. In an inventory of twenty proposed lists of primary emotions (Turner & Stets, 2005: 14–15), anticipation–expectation is included in three lists (called 'expectancy' by Osgood, 1966 and Panksepp, 1982, 'anticipation' by Plutchik 1980a). 'Acceptance' is only listed by Plutchik (1958, 1962).

If we add anticipation and acceptance to Ekman's six, we arrive at four pairs of opposite basic emotions, eight in all, exactly as Plutchik first proposed in 1958. Over two decades later, Plutchik's (1980a: 27–30; 1980b: 146–149) model of the emotions was expanded to substantively interpret his four existential problems of life and refine the Carpenter–Greenberg–MacLean ethogram. Plutchik called these four problems of life *temporality* (the cycle of life and death), *hierarchy*, *territoriality*, and *identity*. These generalize or correspond to the ethogram behaviors of courtship, submission, territoriality, and signature, respectively. Each of Plutchik's life problems can involve either an opportunity, or a danger or threat. The potential resulting eight socio-relational situations can each motivate a distinct subjective state of mind which activates an intention, together with a motor plan, to carry out an adaptive behavioral reaction. Two decades before first *naming* these four existential life problems, Plutchik (1958) had identified eight adaptive reactions to four bivalent dimensions. These pairings of existential problems and emotions (and associated behaviors) are: (i) *joy-happiness* (e.g., as a result of advancing toward a goal) and *sadness–grief* (e.g., experiencing loss of a group member; (ii) for hierarchy, *anger* (destruction) and *fear* (protection); (iii) for territoriality, *anticipation* (exploration) and *surprise* (boundary-defense); and (iv), for identity, *acceptance* (incorporation) and *disgust* (rejection). Plutchik's psychoevolutionary model of the primary emotions is summarized in Fig. 9.1.[6] If we add the valenced relational models to this conceptualization to expand the circumplex model of Fig. 9.1b (known as Plutchik's 'wheel' of the primary emotions), we generate the causal model shown in Fig. 9.2.[7]

Efforts to classify the basic emotions, an enterprise that began in the ancient world and continues today, have three possible outcomes: (i) emotions are not natural kinds, so that all efforts at classification are fruitless endeavors, and all emotions, rather, exist *sui generis* (as suggested by Barrett, 2006; see also Brothers, 1997: 111–125); (ii) basic emotions *are* natural kinds, but their identities remain unknown; or, (iii) basic emotions are natural kinds, and one scholar has already correctly identified them (meaning that all other inventories are incorrect). We claim that the latter is what has happened: Plutchik correctly identified the basic emotions: sadness and joy-happiness, fear and anger, anticipation and surprise, and acceptance and disgust.

The presence of proto-emotions in invertebrates and emotions in vertebrates suggests that emotions are based upon a general principle of equally primordial brain organization. We propose that this principle is the hemispheric specialization of function; we next describe and then link this to the basic emotions.

---

[6]This circular arrangement of primary emotions differs from Plutchik (1962/1991: 113) in that his territoriality dimension is flipped. Later, Plutchik (1980b: 164) swapped the hierarch and territoriality axes. The new arrangement, shown in Figs. 9.1b and 9.2, is preferred here, for two reasons: first, it enables the four emotions that are right (and left) lateralized to be grouped together (from northwest to south) and also enables the positively-valenced relational models to be grouped together, as shown in figure 2. The rationale for this ordering will be presented elsewhere.

[7]In an ongoing analysis of 730 transcripts of life-historical interviews with Euro-Australians and Australian Aborigines, in which the relational models act as predictors of primary emotions, all eight path coefficients were, as hypothesized, positive and significant. These results were CS+ → joy-happiness .57, CS– → sadness .45, HR+ → anger .62, HR– → fear .92, SE+ → anticipation .70, SE– → surprise .46 (all $p < .0001$), EM+ → acceptance .70, and EM– → disgust .37.

A

Life problems	Feelings functions	Adaptive reactions	Basic emotions
temporality	feelings	reintegration	sadness
	functions	reproduction	happiness
hierarchy	feelings	protection	fear
	functions	destruction	anger
territoriality	feelings	exploration	anticipation
	functions	boundary defense	surprise
identity	feelings	incorporation	acceptance
	functions	rejection	disgust

B

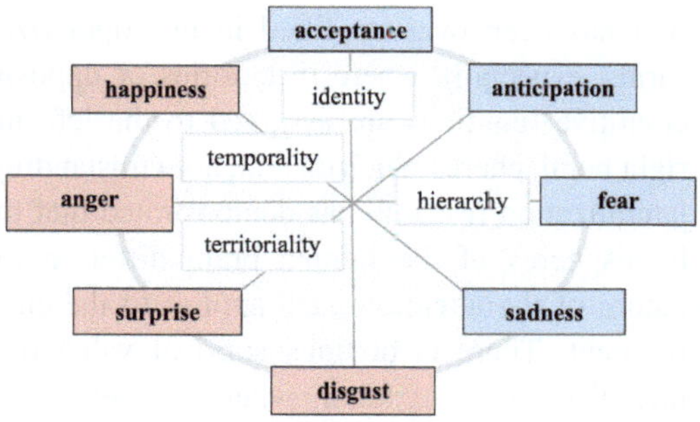

Fig. 9.1 At left, models of social relations and primary emotions. At right, circumplex or 'wheel' of the eight primary emotions. The four emotions that are predominantly lateralized to the left side of the brain are shown in red; those that draw on the right hemisphere, in blue

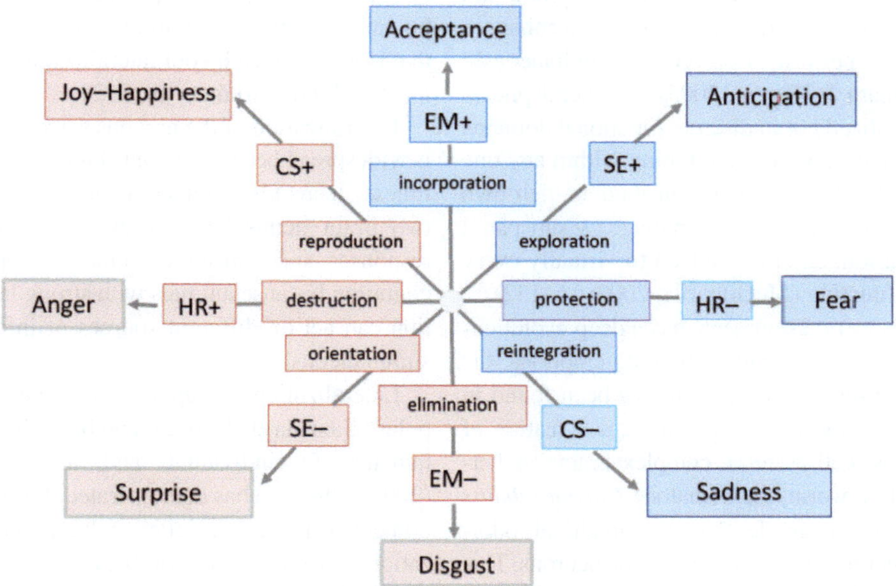

Fig. 9.2 Causal model: The eight valenced sociorelational models emerge as adaptive reactions to the prototypical problems of life, and the primary emotions in turn occur in reaction to these sociorelational situations. The four emotion functions are shown in red; the four emotion feelings, in blue

Brain Lateralization

The left and right hemispheres of the brain—for humans and much of the animal kingdom—are specialized for opposite, yet complementary, modes of information processing. Hemispheric specialization is topical in neuroscience journals (e.g., in *Laterality: Asymmetries of Brain, Behaviour, and Cognition* and *Neuropsychologia*), but it is often ignored, dismissed as a scientific paradigm (e.g., by Corballis, 1991 and by Efron, 1990; cf. TenHouten, 1992), or seen by many as "somewhat falling into disrepute" as a topic for neuroscience (a tendency described, but not endorsed, by McGilchrist (2020: 504). It has even been described as a 'myth' (Kosslyn &

Miller, 2015). The concept of cerebral lateralization has been oversimplified in its popularized forms, especially where inventories of opposite cognitive functions are assigned to the left and right hemispheres. Such disdain notwithstanding, lateralization research has demonstrated that the hemispheres of the human brain differ in the nature of the attention each applies to the environment. There is obvious survival value in a mental division of labor, where the left hemisphere narrowly focuses on grasping food or prey, while the right hemisphere enables "broad, open, and uncommitted attention, needed to watch out for predators and to interpret the intentions of conspecifics" (McGilchrist, 2020: 503). At the simplest level, as animals seek food and avoid becoming food, a lateralized brain enhances ability to perform two tasks simultaneously (Vallortigara & Rogers, 2005). The hemispheres are specialized not in discrete functional domains as such, but rather in functioning within any one domain. The hemispheres contribute in their own way to the domains of language, visuospatial skills, reason, emotion, indeed to virtually every cortical function (McGilchrist, 2009: 16–132).[8]

Hemispheric asymmetry has a deep evolutionary history. Lateralization of proto-emotions, the nervous system, and the brain has been found in invertebrate species, suggesting it is a feature of simpler as well as more complex brains. A hermaphroditic worm, the nematode *Caenorhabditis elegans*, for example, has asymmetrical odor-sensing neurons and taste perception neurons. This left/right difference enables its ability to discriminate between distinct kinds of salt ions. In the fruit fly, *Drosophila melanogaster*, the presence of asymmetry of an important brain structure enables the formation and retrieval of long-term memory (Frasnelli, 2013). Some invertebrates show lateralization of motor control (ants, spiders, cockroaches, giant water bugs), of vision or olfaction (fruit flies, red wood ants), and of behavior (slugs, pond snails, bees) (Frasnelli et al., 2012).

Brain lateralization is found across vertebrate species, including in fish, amphibians, reptiles, and birds (Stancher et al., 2018). The concept has been elaborated in mammals (Leliveld et al., 2013), and it is most developed in primates. Evidence of manual laterality in the hominid lineage has been traced to the evolutionary split with the other great apes around 6–8 million years ago. Living humans display the greatest levels of right-handedness (and inferred left hemisphere dominance) and have a greater degree of lateralization of functional specialization than any other species (Uomini & Ruck, 2018). Indeed, "there is practically no realm of human perception, cognition, emotion, or action that is not affected by our lateralized neural organization" (Güntürkün et al., 2020).

Lateralization of the nervous system and brain is widespread because it contributes to biological fitness. It avoids duplication of function of the two brain hemispheres, prevents incompatible responses, and contributes to the development of cognition because animals with strong lateralization can act on different sources of information simultaneously.

Lateralization is important not only for tadpoles (Gazzola et al., 2023) and birds, but also for humans—for individuals and, as McGilchrist (2009: 239–462) has demonstrated, for the development of human societies, cultures, and civilizations. Lateralization of behavior is adaptive and enables individual asymmetrical organisms to coordinate their behavior with others (Vallortigara & Rogers, 2020). Social species have developed lateralization at the population level, whereas solitary species develop lateralization only at the individual level (Frasnelli, 2013: 4). Consistent with Frasnelli, it follows that humans, with the most complex social life of all animals, have developed the highest level of brain lateralization.

Based on studies of corpus callosotomy (split-brain) patients, Bogen (1973) proposed that, for the left hemisphere, the self is mapped as a subset of the world, but for the right hemisphere, the world is mapped as a subset of the self. The right

[8]Mixed, ambiguous, and left handedness of course complicates generalizing about lateralization of mental functions. Without interference to the arguments made here, we will assume the individuals referred to are right handed and without a family history of sinistrality.

hemisphere is specialized for the world acting in the self, with movement from without, inward. It bears primary responsibility for viewing the state of the world as a totality, including processing information received from one's own body. The left hemisphere is specialized for the self acting in the world, with movement from within, outward. The left hemisphere's agentic, spontaneous self acts to express its will and achieve its goals. In humans, it operates in a logical, analytic, linear, and propositional mode that enables mathematical calculation and quantitative reasoning and is further specialized for tasks requiring analysis of the interior structure of stimuli. It makes sequentially organized temporal discriminations enabling linear conceptualization and measurement of a single dimension along which events are categorizable as past, present, and future.

Emotion Lateralization

With this preparation, we next consider the four pairs of emotions associated with the four problems of life.

Sadness and Joy-Happiness

Sadness

The primary emotion sadness is elicited by the loss of valued family, friends, relationships, resources, status, even metaphorically of the self. It is also triggered by setbacks related to goal-attainment, personal failure, and rejection. In humans, the experience of sadness can range from a short-lived and temporary feeling of disappointment, dejection, or downheartedness, to a longer-term sense of despondency, potentially escalating into grief and depression. While largely an affective experience, sadness involves physiological and cognitive components, including changes in heart rate, skin conductance levels, and blood pressure (Shirai & Suzuku, 2017: 288), as well as contemplation of and reflection upon the triggering events. Behaviorally, sadness may be accompanied by temporary social withdrawal. Sadness has adaptive value: it provides a coping mechanism for life's inevitable losses and setbacks, and it signals need for goal reevaluation and increased perceptual accuracy (Tobore, 2023: 3).

Studies suggest that joy-happiness and sadness–grief are strongly lateralized, with sadness primarily experienced by the right hemisphere and joy-happiness by the left hemisphere. Following damage to the left hemisphere, and with the right hemisphere in general control, an individual is apt to develop a gloomy outlook on life, perhaps bursting into tears for no reason. When the right hemisphere is injured, or suffers an epileptic attack, a patient might become cheerful, elated, and even manic. If the left hemisphere is temporally deactivated by injection of a barbiturate, the now-right-brain-dominant patient suddenly feels sad (Loring et al., 1992). The left hemisphere engages in goal-seeking, delving into the outer world, while the right hemisphere empathically connects with others. The right hemisphere is the site of our social self. In the absence of a functioning right hemisphere, our world becomes emotionally impoverished. Indeed, the right frontal lobe plays a central role in personality, in who we fundamentally are (Stuss et al., 1992). A reduction of activity in left frontal cortex—which places greater reliance on the right hemisphere—contributes to sadness. Indeed, stimulation of the left dorsal prefrontal cortex has become an antidepression target (Schutter, 2008). Numerous additional findings point to the same conclusion: sadness is strongly right lateralized.

Joy-Happiness

Joy is a foreground acute emotion, usually related to present experience. It is a pleasant feeling state characterized by outward signs of gratification. Happiness, in contrast, is a more general background sentiment. Joy and happiness thus differ in the extent to which they are experienced as temporally compressed or stretched, but they belong to the same basic emotion (Layard, 2005). Happiness can result from rewarding progress toward, or the attainment of, a desired objective (Davidson & Irwin, 1999) together with an absence of threat. It clearly meets the general criterion of movement of the self from within, outward, engaging in close or rewarding participation in the social world. From this perspective, happiness is a mixture of what Seligman (2011) in his tripartite conceptualization calls meaning, engagement, and pleasure. Excluding pleasure, we regard happiness as an emotional concomitant of meaningfully rewarding engagement in a task, project, purpose, or social relationship.

Happiness is a positive, approach-oriented emotion that primarily involves structures and networks of the left hemisphere's behavioral activation system (Gray, 1990; Demaree et al., 2005: 11–12). For right-handed adults, the experience of happiness is accompanied by brain activation in the left side of the prefrontal cortex, above and in front of the ear, while negative feelings are correlated with activation in the corresponding site in the right hemisphere (Davidson et al., 2000). Studies of patterns of regional electroencephalographic (EEG) activity induced by musical excerpts found greater left frontal activation for joyful and happy music, but greater right frontal activation for sad and fearful musical excerpts (Schmidt and Trainor, 2001). Individuals whose left hemispheres are especially active—i.e., those who manifest 'left hemisphericity' in comparison to those with 'right hemisphericity' (Bogen et al., 1972)—report more positive feelings, smile more, and are rated as happier by friends. Studies based on EEG measurement of cortical activation in newborns (Davidson et al., 2000) indicate that, when given something sweet and tasteful, the left forebrain is activated, while sour taste stimulates the right hemisphere. Significant decrease in right superior prefrontal cortex blood flow accompanies experimentally stimulated happiness. Also, visually induced recall of an aversive emotional episode is associated with significant reduction in the left-sided prefrontal cortex (Broca's area) and operculum (Davidson & Irwin, 1999). These findings reinforce the view of left-hemisphere happiness.

Happiness engages a wide network of brain regions. A fMRI study of 11 adult subjects showed that, compared to a neutral control, happiness was associated with activation of the left dorsal prefrontal cortex (Fig. 9.3, lateral view)

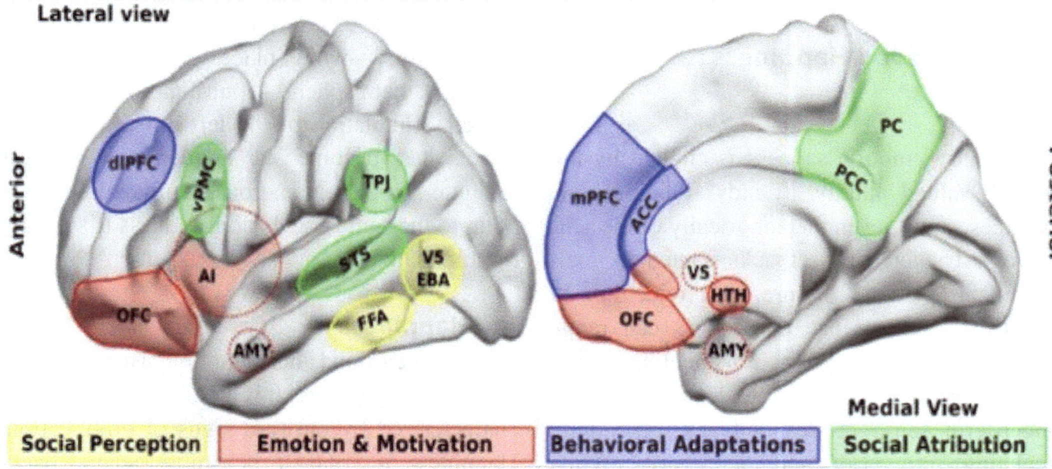

Fig. 9.3 Brain areas involved in emotion-related processes. (1) Perception of others, such as biological motion (middle temporal visual areas, V5), parts of the body (extra-striatal Body area (EBA) and faces (fusiform face area, FFA). (2) Emotional appraisal involving the amygdala (AMY), the anterior insula (AI), the anterior cingulate cortex (ACC), and the orbitofrontal cortex (OFC), which interacts with subcortical structures, including the ventral striatum (VS) and hypothalamus (HTH). Goal-directed behavior involving the dorsolateral prefrontal cortex (dlPFC) and medial prefrontal cortex (mPFC), superior temporal sulcus (STS), posterior cingulate cortex (PCC), and precuneus (PC) participate in automatic, bottom-up inferences of others' mental states, whereas structures such as the mPFC and temporoparietal junction (TPJ) are involved in theory of mind skills. Source: Billeke and Aboitiz (2013: 146)

and anterior insula, the left hypothalamus (shown in Fig. 9.3, medial view), the left anterior cingulate gyrus, midportions of the left middle temporal gyrus, and the thalamus bilaterally (C. Cerqueira et al., 2008).

Fear and Anger

Fear

The basic emotion fear has crucial survival value as a direct response to a physical danger or a threat to a valued or needed social bond. The behaviors of fear are withdrawal, avoidance, flight, and hiding. The capacity to avoid, escape, and learn from threat is adaptive for invertebrates (e.g., ants, fruit flies, moths, nautiluses, and sea slugs (Pribadi et al., 2022). As a general pattern, in response to fear-inducing situations such as predator awareness and agonistic interactions with conspecifics, fear is predominantly processed by the right hemispheres of vertebrate species, including chickens, pigeons, rodents, horses, dogs, cats, sheep, and cattle (Leliveld et al., 2013). In a study of rhesus monkeys, Kalin and colleagues (Kalin et al., 1998) compared those with stable right as opposed to left frontal EEG activation, and found the right-hemisphere activated monkeys had higher cortisol levels, were more fearful, and more defensive, suggesting fear is generated in right frontal cortex.

That rapid detection and response to threat is adaptive for survival led LeDoux (2014) to posit the existence of a vigilance system in the human brain that stimulates rapid (Öhman & Mineka, 2001) and accurate (Anderson, 2005) physiological and neural reactions to threats, even those outside of conscious awareness (Vuilleumier, 2005). Fear can be experienced either implicitly—as a creepy feeling of trepidation, as an internal 'alarm' system, or explicitly and consciously—as one is threatened by aggressive conspecifics, natural disasters, or dangerous situations. This 'two system' framework (LeDoux & Pine, 2016) suggests that the two levels of fear might differ in their underlying brainwork. A large-scale meta-analysis of neuroimaging studies of fear in 2046 subjects (Tao et al., 2021) showed that the brainwork underlying the two kinds of fear largely overlap. The strongest finding was that the key limbic structures of fear processing—regardless of level of awareness, or type of stimulus—are the left and right amygdala. The amygdala, a subcortical structure (shown in Fig. 9.3) is key for responding to a threat with fear (Öhman, 2005). It is strongly connected to visual, prefrontal, and subcortical areas of the brain. The left amygdala responds most strongly to implicit fear, while the right amygdala responds most strongly to explicit fear (Baker & Kim, 2004).

The brain's defensive brain circuit notices and then responds to threats by generating multidimensional responses, including behavioral and autonomic responses. LeDoux (2014) links the amygdala—which is key for responding to ambiguous and uncertain but salient environmental stimuli—to unconscious emotional processing. In monkeys and humans, there has evolved a direct pathway (bypassing primary sensory cortices) from the superior colliculus (located at the roof of the midbrain, which receives visual information directly from the retina), to the thalamic pulvinar (activating visual attention), to the amygdala, which mediates a rapid but coarse and unconscious visual analysis of a potential threat (Làdavas & Bertini, 2021). Evidence from unilaterally brain damaged individuals shows that this subcortical pathway is more relevant to implicit fear processing in the right hemisphere than in the same pathway of the left hemisphere (Làdavas & Bertini, 2021).

Neuroimaging research has shown that significant neural responses are elicited in the right but not the left amygdala by unconscious fear-conditioned stimuli (Morris et al., 1998). Unattended fearful facial expressions induced rapid activation of the right amygdala (as fast as 100 ms after presentation), while activation of the left amygdala occurred at later stages (Hung et al., 2010), indicating that it is the right amygdala (via the right superior colliculus and the pulvinar) most involved in fear. The left-lateralized, later-stage experience of fear is fully conscious and involves assessment of the threatening situation and formulation of an adaptive response. It is the earlier, right-lateralized, immediate, and often unconscious level of fear that can be interpreted as the basic emotion of fear. Clearly, fear is the emotion that has evolved as the key right-hemisphere-dependent mechanism involved in threat detection, avoidance, and self-defense.

Anger

Anger's classification as a basic emotion is strongly suggested by its presence in a wide variety of animal species (Darwin, 1872/1965; Panksepp, 1998: 187–205) and by its early emergence in the developing human infant (He et al., 2010). Plutchik (1962/1991: 118; 1980a, 1980b: 146–147, 154) identified the function of anger as 'destruction'. "The aim of anger," Panksepp (1998: 189) asserted, "is to increase the probability of success in the pursuit of one's ongoing desires and competition for resources." Expression of anger disturbs the offender's sense of ontological security (Garfinkel, 1967/1984: 44) and functions to downwardly recalibrate and restructure the offender's perception of their personal efficacy and/or strength of bargaining position (Sell, 2017). An individual's inner experience or sensation of anger is typically elicited by an adverse social situation concerning one's position, place, or rank in a social hierarchy. Carver and Harmon-Jones (2009) see anger as a response to goal-blockage: the angered individual confronts the anger trigger and re-establishes goal attainment. Anger's outward expression involves a decision to act, followed by the behavioral implementation of that decision. Anger is an 'approach-motivated' emotion that prioritizes the attainment of favorable outcomes.

Anger is robustly connected to the left hemisphere. This partly stems from the stronger expression of dopaminergic D_2 receptors and their reward sensitivity (Tomer et al., 2014). A meta-analysis of 61 neuroimaging studies. In a meta-analysis, Sorella and colleagues (Sorella et al., 2021) showed that the subjective experience of anger was most closely linked to activation in (i) the left insula and the left ventrolateral prefrontal cortex; (ii) Broca's left-hemisphere area for semantic tasks and speech comprehension; (iii) the left pars orbitalis; and, (iv) part of the left inferior frontal gyrus, which is also involved in language processing and comprehension. This review adds specificity to the description of anger as lateralized to the left frontal areas.

In the experience of anger there is activation of the left frontal cortex, especially the left dorsolateral prefrontal cortex, which is implicated in implementation of control and goal-attainment. The left hemisphere's frontal lobes are richly connected to areas of speech and reason. Anger involves brain areas of speech processing, including the left inferior frontal gyrus (and pars operculum), Broca's area, and the left insula (Sorella et al., 2021).

Anger also has right hemisphere involvement, as the right inferior frontal gyrus is activated in both anger perception and anger experience (Sorella et al., 2021). Anger typically arises in complex conflict situations that require self-control and situation management, tasks which requires involvement of the right hemisphere. More specifically, the right anterior cingulate cortex becomes involved in the management, regulation, and inhibition of anger once it arises (Dougherty et al., 1999). While anger directed outward involves left prefrontal activation, the suppression and inhibition of anger (Grecucci et al., 2023), and associated rumination about one's ongoing anger (Kelley et al., 2013), involves right prefrontal activation. Anger has also been linked to the right parahippocampal gyrus, which can retrieve memories of past aggravation performed by the current anger target (Vytal & Hamann, 2010).

While the personal experience of anger is left-lateralized, the processing of facial recognition of others' anger is, in comparison, right lateralized. In a meta-analysis of 5549 data sets, including 1489 healthy subjects, conducted from 2009–2021, Lukito and colleagues (Lukito et al., 2023) found that, for the perception of angry as opposed to neutral faces, the primary result was activation in the right cerebellum and right middle temporal gyrus. Anger, and other negative emotional faces—fearful, disgusted, and sad—were associated with activation of the right amygdala, consistent with other meta analyses

(Xu et al., 2021; García-Garcia et al., 2016). Activation of the ventromedial prefrontal cortex (vmPFC) (Fig. 9.3, lateral view) was predominantly right-lateralized for facial recognition of the anger expression (also fear and disgust) (Xu et al., 2021; Lukito et al., 2023). Sudden attention to an anger-infused social situation involves the right-lateralized ventral attention system, including the temporoparietal junction (Fig. 9.3, lateral view) (Downar et al., 2000). Once this external anger is perceived and attended to, the right hemisphere sends information, likely via the corpus callosum,[9] to the left hemisphere; there it can be verbally expressed and action taken. Overall, we find the personal experience of anger to be formed in the left frontal cortex, whereas the management, control, and inhibition of such anger involves the right hemisphere.

Anticipation and Surprise

Anticipation

Anticipation is important for animal and human self-regulation of simple organismic processes and complex behaviors, including planning and decision-making. One meaning of anticipation is prediction of what is apt to happen. This is essentially a cognitive process involving analysis of events using stored representations, importantly including models of the world articulated in language. From these, we draw inferences and carry out adaptive behavioral reactions to likely future events and situations. Anticipation, and its close synonyms—expectation and prediction (and more indirect kinds of anticipation)—are not intrinsically affect-laden. This sense of 'anticipation' refers not to a feeling-state or an emotion but rather to a cognitive process that is a prototypical activity of the left cerebral hemisphere and its logical–analytic mode of reasoning, functioning to attain ends and goals.

Stefanova and colleagues (Stefanova et al., 2020: 308) define an "anticipatory feeling" as "a state of awareness of physiological and neurocognitive changes that occur within an organism during a specific process of adapting to future events." An anticipatory feeling, which we consider the basic emotion 'anticipation', can be either negative or positive, as we see impending events as good or bad, gains or losses. This feeling of anticipation arises through interaction of the brain's stimulus-induced bottom-up and top-down-controlled neural processes to enable the flexible and proactive organization of behavior. Bodily changes elicit feelings of an anticipatory nature that act as interoceptive and proprioceptive physiological signals, or somatic markers (Damasio et al., 1991), that influence our behavioral choices and decisions. Drawing inferences from anticipatory feelings, individuals adjust behavior "in accordance with their intentions and future goals" (Stefanova et al., 2020: 309). An anticipatory feeling thus emerges as an emotion; it can contribute to the formulation of plans, stimulate cognitive processing including the selection and effective ordering of necessary actions, and focus attention on desired outcomes.

Anticipatory feelings are produced by a set of large-scale networks predominantly lateralized to the right hemisphere. Included is the saliency network,[10] based on brain structures that monitor the body's homeostatic state. This network includes the right dorsal and right ventral anterior insula (shown in Fig. 9.3, lateral view), whose intrinsic connectivity contributes to its performance level and the processing speed of salient information. The right insula responds actively to emotion-inducing stimuli and also to visceral

[9]The interhemispheric transfer-deficit theory of alexithymia proposes that, without a corpus callosum, split-brain patients were unable to express their feelings in words, suggesting that speech rich in metaphor, symbolization, and color (provided in part by adjectives and adverbs), requires input from the right hemisphere (Hoppe & Bogen, 1976; TenHouten et al., 1986).

[10]The saliency network (SN) is involved in detecting and filtering multimodal salient stimuli, that is, feelings, and mediates the 'switch' between the default mode network (active when not involved in a task, often involving musings about social relations) and the central executive network. The SN is largely right-hemisphere dominant, and connectivity between two of its core hubs—the frontoinsular cortex and the dorsal insular cortex, is also dominant in the right hemisphere (Zhang et al., 2019).

sensations that comprise the physiological basis of 'feelings' (Craig, 2009). The saliency network also features the right midcingulate area, which interacts with the right insula and is located behind the anterior cingulate cortex, ACC (Fig. 9.3, medial view). The saliency network connects bottom-up and top-down signal processing, central to the experience of anticipation.

Neuroimaging studies of neural activity in the interoceptive cortex (in the right posterior insula) show that consciously perceived and unconsciously processed physiological cues serve as substrates for anticipatory feelings (Critchley, 2009). These feelings address a basic problem of life: exploration of the world with expectation of finding and utilizing resources and social opportunities, while maintaining awareness of potential harm. In a dynamically changing world, the individual must continually update stored representations/models of the natural and social world with ongoing sensory inputs (Stefanova et al., 2020: 309).

Surprise

While anticipation involves voluntary orientation toward one's milieu, surprise is the prototypical adaptive reaction to unexpected and possibly unwanted penetration of one's territorial boundaries with potential consequent loss of control. As a basic emotion, surprise involves an unplanned shift of attention to a potentially problematic stimulus or situation. It is a rapid, involuntary orienting response to an unexpected (out of context) or unpredicted stimulus or event that triggers conscious attention, evaluation, and, if necessary, appropriate behavioral response (Ranganath & Rainer, 2003).

Using EEG and functional magnetic resonance imaging (fMRI) methodologies, researchers have increased knowledge of the brain infrastructures included in the orienting-response system. Unexpected stimuli are associated with left-hemisphere activations in the inferior frontal gyrus (Friedman et al., 2009). This region is implicated in the semantic analysis of a wide variety of stimuli during different task paradigms, including the extraction of meaning from environmental sounds (Lewis et al., 2004). It enables assessment of the salience of the environmental perturbation and appropriate goal-directed action. Studies investigating subjects' ability to identify 1–2 s environmental sounds show hemodynamic activation in the left supramarginal gyrus and bilateral activation in the posterior middle temporal gyrus (Lewis et al., 2004). Consistent with Friedman and colleagues' (Friedman et al., 2009) study, correct identification of sounds was correlated with activation of the left inferior frontal cortex. The results indicate that environmental sounds are stored, along with words and pictures, in left-hemisphere semantic networks. These investigations show that the primary emotional responses to surprising environmental perturbations, based on the brain's orienting-response network, are largely left lateralized.

Acceptance and Disgust

Acceptance

Plutchik (1980a, 1980b) proposed that the primary emotion acceptance addresses the fundamental life problem of establishing an 'identity', a social self. He interpreted acceptance, the positive pole of the 'identity' dimension, as the emotion of 'incorporation.' Sociologically, this means openness and/or receptivity to individuals and groups, and to a group's meaning systems, beliefs, values, and collective representations. Acceptance can manifest as a diffuse or global feeling about another or oneself, or as an acute emotional experience. It can include recognition, for example, that one wishes to incorporate another as a social intimate (TenHouten, 2021b). Theorists of human motivation have postulated, and empirically demonstrated, that human beings "have an intrinsic need for affiliation—being connected with, and accepted by, other people" (Hecht, 2014: 1; see also Baumeister & Leary, 1995; Cacioppo & Patrick, 2008).

Plutchik's conceptualization of acceptance as a primary emotion refers both to positive self-identity and identification with others. Empathic identification with others largely draws on resources of the right hemisphere (Shamay et al., 2003), particularly the right inferior parietal lobe together with the right lateral prefrontal cortex. These interacting structures inhibit propensity to impose one's own view on others, and to be more accepting of positions one has not previously taken (Decety & Chaminade, 2003). Acceptance is linked to the sociorelational phenomenon of equality-matching, which involves the attainment of likemindedness. Using Fiske's term, we see Hecht's "egalitarianism" as synonymous with the archetypal social-relations model 'equality matching' This social relationship requires a capability to see and understand from another's standpoint. The requisite 'theory of mind' (ToM) is also largely a capacity of the right hemisphere (Griffin et al., 2006).

The right hemisphere has a relative advantage over the left hemisphere in understanding and controlling pro-social dynamics of social interactions; these require self-control of one's immediate desires and consideration of others' needs. Emotional states of vulnerability trigger the intrinsic human need for empathic sociality and "being connected with and accepting other people" (Hecht, 2014: 1). This need for affiliation, for mutual acceptance, "promotes an overall pro-social mindset and corresponding emotions and behaviors: egalitarianism, closeness between people, feelings of togetherness" (Hecht, 2014: 2). Following Plutchik, we identify the key emotion of pro-sociality as 'acceptance'.

Studies of the neural basis of acceptance concur that "the human need for social connection and the desire for mutual acceptance are associated with neurophysiological activity in the right hemisphere" (Hecht, 2014: 3).[11],[12] The right hemisphere is characterized by its "openness" to the interconnectedness of things" and "is interested in others as individuals and how we relate to them. It is the mediator of empathic identification" (McGilchrist, 2009: 57; see also Shamay et al., 2003). EEG studies show that a low need for social connection is correlated with low neural activity in the right prefrontal cortex (Quirin et al., 2013).

Several regions of the right hemisphere are involved in pro-social personality traits, including affiliativeness, or the desire for closeness (and therefore of mutual acceptance), with others. Affiliativeness is associated with large size of the right anterior cingulate cortex (Whittle et al., 2008; Mortin, 2013). Agreeableness, being cooperative and pleasant in social situation, which is synonymous with acceptance, is positively correlated with gray matter volume in the right middle orbital gyrus (Nestor et al., 2013) and in the right orbitofrontal cortex (Fig. 9.3, lateral view) (Kapogiannis et al., 2013). The tendency to trust others—to accept their talk as truthful and to accept their intentions as their intentions as honest[13]—is correlated with activation of the right ventrolateral prefrontal cortex (Yanagisawa et al., 2011). After an in-depth review of the existing research, Hecht (2014: 3) declares that, taken together, the evidence suggests that "the human need for social connection and the desire to be accepted are associated with neurophysiological activity in the [right hemisphere]."

Disgust

For Plutchik, the primary emotion 'disgust' corresponds to what Gan & colleagues (Gan et al., 2022) call "core disgust." It is an evolutionarily adaptive avoidance–defense reaction. On the physiological level, core disgust facilitates rejection and avoidance of, and revulsion toward, per-

[11] Acceptance in neurobiology has largely been seen not as a basic emotion but as a strategy for regulation of problematic emotions by refraining from trying to control, suppress, or avoid them, but rather being curious about, open to, and interested in these emotions (e.g., Gross, 2015; Messina et al., 2021).

[12] Acceptance and rejection are also topical in neuroeconomics, as offers in games can be accepted or rejected. While unfair offers trigger emotional reactions such as anger, an emotional reaction to an unfair can involves heart rate deceleration as a somatic marker and right insula activation as a sociomoral judgment of unfairness, a form of moral disgust (Ohira, 2010; TenHouten, 2016).

[13] Individuals whose personality emphasizes acceptance are prone to an absorption tendency, which manifests in a naïve predisposition to believe what they are being told, to being perpetually trusting. Uninhibited levels of acceptance are demonstrated in taking in information from the other uncritically, being suggestable, even gullible. There is an underlying tendency to idealize others, so that new acquaintances are all-too-quickly embraced (Kellerman, 1979; TenHouten, 2023).

ceptibly toxic and pathogenic materials (Darwin, 1872/1965), horrid sights, disease, injury or mutilation (Olatunji & Sawchuk, 2005), and potentially harmful social interactions (Vicario et al., 2017). The behavioral response to what is deemed disgusting is a rapid distancing from the eliciting stimulus.

The term "social disgust" denotes emotion-laden reactions to behavior judged unacceptable or immoral (Gan et al., 2022). These negative sociomoral evaluations are distinct from core disgust and the visceral emotion Plutchik identified as a primary emotion. Contemporary neuroscience recognizes that the right hemisphere is disproportionately involved in the networks and structures that enable moral judgments. In an exhaustive meta-analytic review of fMRI studies of disgust, Gan and colleagues (Gan et al., 2022) found a difference in brainwork for social and core disgust. Social disgust is right-hemisphere dependent, involving a superior-temporal-frontal network. Core disgust rather involves bilateral activation of the amygdala (spreading into the right globus pallidus), insula, and inferior occipital gyrus. Besides these bilateral regions, core disgust also involves seven structures of the left hemisphere: the left inferior frontal gyrus, left declive, left fusiform gyrus (Fig. 9.3, lateral view), left inferior parietal lobule, left middle occipital gyrus, left inferior temporal gyrus, and the left superior frontal gyrus.

Neuroimaging research suggests that the insular cortex and the basal ganglia form a system that is involved in the experience, expression, and recognition of disgust (Chapman & Anderson, 2012). Disgust processing deficits are frequently seen in patients suffering from insula and basal ganglia dysfunction (reviewed by Vicario et al., 2017). In a study of patients with left or right hemisphere damage to the insula–basal ganglia system, Holtmann and colleagues (Holtmann et al., 2020) found that four left-lesioned patients had significantly lower composite disgust scores than did four right-lesioned patients and 19 normal controls. Right-lesioned patients also experienced significantly higher levels of disgust than controls. While a single test did not clearly separate the patient groups, composite measures did. Consistent with previous single-case studies in which disgust deficits result from left, but not right, insula-cortex–basal-ganglia damage, Holtmann and colleagues (Holtmann et al., 2020) suggest that a left-hemisphere basis of disgust processing is likely due to asymmetric representations of peripheral autonomic nervous activities in the human forebrain. While sociomoral disgust might involve feelings about an event or experience as a whole, the experience of core disgust is a rapid and focused functional response motivating avoidance or expulsion of what appears toxic or harmful. This study, along with neuroimaging and other kinds of data, support the hypothesis that simple, core disgust, as a primary emotion, is predominantly left lateralized.

We have found four of Plutchik's basic emotions to be primarily right-lateralized: sadness, fear, anticipation, and acceptance. These emotions' four opposites—joy-happiness, anger, surprise, and disgust—are primarily left-lateralized. Based on this analysis of the eight basic emotions, we next assess models of emotion lateralization, then present a new conceptualization, the Feeling–Function Model.

Models of Emotion Lateralization

A succession of formulations have aspired to explain the relationship between brain laterality and the emotions. We briefly critically summarize these models, then present a new formulation—the fourfold Feeling–Function Model. This model classifies the basic emotions into two groups according to their reliance on the specialized processes of the left and right sides of the brain.

The Right Hemisphere Model

The Right-Hemisphere Model posits that the right hemisphere is centrally involved in the expression and experience of emotion, particularly in perceiving the emotions of others. This view emerged over a century ago following observations of patients with right-sided damage and associated deficit (Hughlings-Jackson, 1866; Mills, 1912). The right hemisphere governs a wide array of emotional processes, including the interface between body and emotion. Compared to the left hemisphere, the right hemisphere is superior in discriminating facial expressions and in recognizing prosody (vocal intonation), humor, symbolic meaning, metaphor, and emotional content (for reviews, see Demaree et al., 2005: 2–6, McGilchrist, 2009: 58–64).

The right orbital frontal cortex is key to emotional regulation and understanding (Rolls, 2000, 2019). This structure is expanded in the right hemisphere and acts as the "senior executive" of the social-emotional right brain (Schore, 1994: 280). Self-recognition and sensory inputs concerning the self are mainly processed in the right hemisphere (Keenan et al., 2001), which is largely responsible for maintaining a coherent, continuous, and unified sense of self (Devinsky, 2000). The right hemisphere's additional ability to understand others' emotional states is a key ingredient of empathy. This emerges in early attachment behavior and directly affects the neurobiological substrate of moral development (Schore, 1994: 280). The right hemisphere's orbital-frontal areas—the "frontal moral guidance system"—mediate empathy (Bigler, 2001), while the right dorsolateral prefrontal cortex mediates our sense of moral judgment (Keenan et al., 2003).

The Right-Hemisphere Model correctly notes that the right hemisphere is involved in emotions and 'emotionality'. Its major shortcoming, however, is that the personal experience of four specific, irreducible, basic emotions—joy-happiness, anger, surprise, and disgust—are primarily, though not entirely, left-lateralized.

The Valence Model

The Valence Model developed out of clinical observations, for example, that patient with right-sided brain injuries have difficulty perceiving negative emotions), but their perception of happy faces was preserved (Adolphs et al., 1996; Borod et al., 1998). Moreover, studies of patients following amytal injections anesthetizing the right or left hemisphere produce globally positive and negative moods: the right hemisphere, on its own, produced crying, worry, a generally catastrophic reaction; the left hemisphere, resulted in smiling, laughing, a generally ebullient mood (Silberman & Weingartner, 1986). These and related finding led to the Valence Model, which postulates that the right and left hemispheres are specialized for negative and positive emotions, respectively.

A variation of the Valence Model contends that, while the experience and expression of negative and positive emotions are right- and left-lateralized, respectively, the *perception* of others' negative and positive emotions are generally right lateralized. This model additionally proposes that the experience and expression of emotions involves anterior cortical areas, whereas perception of emotions occurs with greater involvement of posterior regions (Borod, 1993; for reviews of the evidence on the Valence Model, see Demaree and colleagues (Demaree et al., 2005: 6–8) and Wyczesany (2018).

The Valence Model was undermined by research on the emotion, anger, which had been hypothesized to be a negative emotion, on the implicit ground that it is unpleasant for all involved. Harmon-Jones and Seligman (2001) found that, when subjects' control over an anger-inducing situation was manipulated (under the condition that the source of anger could be acted against and possibly controlled), there was left-hemisphere midfrontal activity. This and related studies of the lateralization of anger invalidated the Valence Model. The very idea that emotions can be usefully described as valenced was undermined.

The Approach–Avoidance Model

Following the collapse of the Valence Model, it came to be recognized that emotions are better described by their motivational content, with a left-hemisphere superiority for emotions that are approach oriented (happiness, anger) and right hemisphere superiority for avoidance-inducing emotions such as fear and disgust (Carver &

Harmon-Jones, 2009; Poole & Gable, 2014). The Approach–Avoidance Model holds that the right and left hemispheres' frontal brain areas differ not by valence but by the behavioral reactions of avoidance and approach, respectively. The left hemisphere's approach reaction aligns with Luria's (1973) description of the left-frontal cortex as involved in intention, planning, and acting in the world. A program of research followed showing that a number of approach-oriented emotions are left lateralized (described in Demaree et al., 2005: 8–11).

While left frontal activation characterized approach-oriented emotions, right frontal activation, in contrast, has been linked to depressive symptomology and the experience of withdrawal-oriented emotions, particularly sadness and fear (reviewed by Demaree et al., 2005). Demeree and colleagues (2005: 12) embrace the Approach–Avoidance Model, assessing it as compelling "not only for its excellent fit to the available data but also for its theoretical importance (i.e., motivational direction is of utmost importance to species survival and procreation)."[14]

Our model of eight basic emotions shows, however, that only half of these emotions fit the approach–avoidance/left-hemisphere–right hemisphere formulation. Of the four primary emotions we have linked to the right hemisphere, fear and sadness are avoidant, but acceptance and anticipation—the two other right lateralized emotions, which Demaree and colleagues do not mention—do not involve avoidance or withdrawal. In acceptance, we observe neither withdrawal from, nor approach to, the world. Rather, the individual endeavors to incorporate what is outside the self, to bring what is without, within. Anticipatory feelings pertain to feeling states related to one's body, mind, and social involvements. Rather than signaling withdrawal, anticipatory feelings involve exploration of the world with the prospect of discovering, and approaching, something, or someone, of interest or value.

[14]Another model, the Behavioral Activation and Behavioral Inhibition, is virtually identical to the Approach-Avoidance Model, except that it focuses on emotional traits rather than states (Demaree et al., 2005).

Of the left-hemisphere-based emotions, joy-happiness and anger are approach-oriented, but the other two left-lateralized emotions, surprise and disgust, are not approach emotions. In surprise, we pay sudden attention to some unanticipated, novel, and potentially intrusive, event in the world, and seek to take in information. This orienting response is neither withdrawal nor approach. Disgust is an emotional alarm signalling something profane or poisonous that must be expelled or kept at a distance.

The Emotion–Type Model

Elliott Ross (Ross et al., 1994; Ross, 2021) has proposed an "emotion–type hypothesis," which partitions the emotions into two groups, the 'primary' and the 'social'. By primary, Ross refers to Ekman and colleagues' six basic emotions—sadness, fear, joy-happiness, anger, surprise, and disgust. Ross claims that these primary emotions and their associated behaviors are modulated by the right hemisphere, largely based on evidence presented in Buck and Duffy (1980) and in Ross and colleagues (Ross et al., 1994). Our focused review, however, sees only two of these emotions, sadness and fear, to be right lateralized; the other four are left-lateralized.

While Ross sees the primary emotions as right-lateralized, all other emotions (i.e., the 'social' emotions) are seen as left-lateralized. These six emotions, Ross contends, are not biologically based but are rather acquired through childhood socialization, specifically by learning to "behave in a manner that is socially acceptable" according to display rules which are internalized as children "engage in various social interactions during play, school, and religious-cultural activities" (Ross, 2021: 10). The social emotions, in other words, are "culturally dependent" and socially constructed. The position we take here, however, is that all emotions are to some extent socially constructed, but they are not constructed *de novo*, for they are either primary emotions with a long evolutionary history, or are secondary- or tertiary-level mixtures of the primary emotions (TenHouten, 2007, 2013, 2021a).

The Four Emotion–Lateralization Models

Assessment of the four models is difficult for two reasons. First, the distinction between the personal experience of emotions and the perception of emotion in the communicative behavior of others is largely ignored. And second, the models tend to focus on emotionality in general, or to appraise just a few specific emotions. If we evaluate these models using the eight basic emotions proposed here, we can quantitatively judge these models in a two-step procedure. First, we exclude emotions that cannot be classified by the models' criteria. Then, we determine the proportion of the remaining emotions that permit correct lateralization predictions.

The Right-Hemisphere Model applies to all emotions, so all eight of our basic emotion are included. Erroneous predictions are made for joy-happiness, anger, surprise, and disgust, which are left lateralized, so the measure of model fit is 4/8.

While present theory assigns valence to the social-relations models that trigger the eight basic emotions, we will use 'valence' in an intuitive, informal way. The Valence Model does not account for anticipation and surprise, which are intrinsically unvalenced. Of the rest, joy-happiness and acceptance are psychologically positive, and the other four—sadness, fear, anger, and disgust, can be seen as negative. The model suggests that anger and disgust, being 'negative', are right-lateralized; this, according to our review, is mistaken, so the Valance Model's score is 4/6.

The Approach–Avoidance Model is not easily applied to acceptance and surprise. Of the rest, three—anticipation (which Plutchik links to exploration), joy-happiness, and anger—are approach oriented, while sadness, fear, and disgust are avoidant. Just one 'prediction' is wrong, as disgust is negative but left-lateralized. The model's score is 5/8.

The Emotion-Type Model sees Ekman's 'big six' as right-lateralized and all other emotions (which include acceptance and anticipation) as left lateralized. However, our review suggests that four of the six Ekman-type basic emotions—joy-happiness, anger, surprise, and disgust—are left-lateralized, so the score for this model is 2/8

The Feeling–Function Model

Given that none of the four emotion–lateralization models classifies the eight basic emotions, developed above, a new model might prove useful. We next present the Feeling–Function Model of emotion lateralization. This model (i) does not apply to emotions in general, or to 'emotionality', but is rather restricted to the eight basic emotions; and (ii), it considers these emotions only as they are personally experienced, and not as they are involved in perceiving the emotions of others.

Emotional Feelings

Given the qualitatively different ways in which the two sides of the brain process information and orient to the world, we might expect that they have qualitatively different ways of *experiencing* emotions. We propose that four *feeling*-based primary emotions—sadness, fear, anticipation, and acceptance—are right-lateralized, and their four opposites—joy-happiness, anger, surprise, and disgust—are *function*-based and left-lateralized.

The English word *feeling* means the experience of physical or emotional states, and all experiences of inner drives such as hunger, pain, and fatigue. A feeling is "a self-contained phenomenal experience" that is "subjective, evaluative, and independent of the sensations, thoughts, or images evoking them" (*APA Dictionary of Psychology* 2024). The right hemisphere experiences intense and global feeling states that are partly unconscious primary processes (Weinberg, 2000). One might thus have a global feeling of being accepted by another, or of feeling a desire to incorporate another in one's social world. No emotion is more globally linked to an individual's overall feeling state than sadness, which is right-lateralized.

As we saw in the above consideration of anticipatory feelings, feelings are states of awareness of physiological changes in the body. Feelings are generally right lateralized, as the right hemisphere is concerned with the state of the body, using somesthetic and tactile-spatial-positional information. The right hemisphere is more strongly connected to the limbic system and other subcortical structures than is the left. While the right hemisphere is to some extent involved in all emotional experience, it is particularly specialized for emotions that reflect the overall state of mind, body, and self (Joseph, 1996).

Emotional Functions

The left hemisphere is specialized for sociocognitive emotions that are generated on the basis of plans and intentions using language-based models of the world. These emotions involve actions carried out in order to accomplish the interests, needs, and welfare of the self. While emotions of the right hemisphere involve our deepest feelings, the left-hemisphere-based emotions serve necessary functions, as circumstances, planned or unplanned, require that specific actions be carried out. The word 'function', from the Latin *fungi*, *fungor*, means to perform. The performance of a function is an action, mental or physical, which enables an individual "to achieve autonomy and freedom, ...to master and influence their environments and social relationships" (Hecht, 2014: 10). The left-lateralized basic emotions are adaptive reactions that motivate empowering actions in the world. Thus, happiness motivates and rewards progress toward a goal. Anger conveys to potential competitors that an individual will act with intentionality, even fight, to attain a contested goal. Surprise mobilizes defense of territory and resources. Disgust motivates distancing from, what is toxic or harmful. These reactions involve performing actions that function to support the welfare, interests, and goals of the ego-complex.

Feelings, Functions, and Hemispheric Interactions

The neural processes that generate the experience of an emotion arise in two classes of bodily changes, which Damasio (1999: 281) identifies as "changes related to body state and changes related to cognitive state." While an emotion involves both feeling and cognition, we propose that our two sets of four basic emotions differ in their order: the right-hemisphere emotions begin with feeling states, while the left-hemisphere emotions begin with cognition. Damasio sees changes in body state as achieved by two mechanisms. First, the "body loop" uses both humeral signals (chemical messages carried in the bloodstream) and neural signals (EEG signals carried by nerve pathways) that become represented in "somatosensory structures of the central nervous system, from the brainstem on up." These result in a feeling state (Damasio, 1999: 281). Second, the "as if body loop" captures the fact that the "body loop" can take on an "as if" character. That is, the "as if body loop" mechanism bypasses the body proper, partially or entirely, which can save time and energy (Damasio, 1999: 281).

Consider anger. This emotion might begin with recognition that another's remark in conversation is insulting, hostile, demeaning, or threatening. The initial cognition in the experience of anger occurs in the frontal lobes of the left hemisphere, which are richly connected to areas of speech and reason. The meta-analysis of Sorella and colleagues (Sorella et al., 2021) shows that anger involves brain areas of speech processing, including the left inferior frontal gyrus (and pars operculum), Broca's area, and the left insula. The experience of anger typically begins with cognition in the form of inner speech, which can be quickly followed up by a retaliatory speech act. But whence come the *feelings* of anger? Recognition that one has been insulted, for example, can be communicated, in part via the corpus callosum, to the right hemisphere. Given that anger bilaterally activates insula, amygdala (threat detection and assessment), and inferior frontal gyrus (Sorella et al., 2021), we see that the right hemisphere quickly becomes involved in the anger process. A reasonable hypothesis is that these and related structures of the right hemisphere contribute to anger becoming infused with feelings, possibly manifested as 'hot under the collar" or 'seeing red'. Importantly, the right parahippocampal gyrus becomes involved in anger, as it retrieves memories of past insults from the offending other (including how one felt about these). As this right-hemisphere work is done, the assessment of the situation as a whole is reconveyed to the left hemisphere. The cognition of anger is now reinforced by a broader feeling of anger. This mental processing can occur rapidly, as information can travel 200 miles an hour in the brain. After an insult, one might have the phenomenological experience of hesitating only for a 'moment' before retaliating with words, postures, gestures, even attack behavior.

Conclusions

A focused review of emotion lateralization research shows that, of Plutchik's eight basic emotions, sadness, fear, anticipation, and acceptance are right-lateralized; their opposites—the other four basic emotions (joy-happiness, anger, surprise, and disgust)—are predominantly left-lateralized. This generalization applies to emotions as personally experienced, not as perceived in others' facial expressions, talk, and related communicative actions. We elaborate this distinction by focusing on anger: while anger is directly experienced by the left hemisphere, the right hemisphere detects anger in the speech, gestures, postures, and facial expression of others; it prioritizes attention to the situation, then forwards information to the left hemisphere, where an adaptive response is formulated and acted upon.

We argue that the emotions lateralized to the brain's right and left hemispheres can be described as emotion-feelings and emotion-functions, respectively. More specifically, the Feeling–Function Model of the hemispheric lateralization of eight basic emotions holds that right-lateralized emotions significantly involve feelings, or partially unconscious primary processes. These global feeling states represent awareness of physiological changes in mind, body, and self. The left hemisphere, in contrast, is the locus of more cognitively-infused emotions which enable the individual to detect or infer threats, problems, and opportunities. It is specialized for goal-oriented and functional emotions that accompany outward-directed plans and intentions.

From its beginning, the science of sociology is a two-dimensional enterprise, a study of mind and society. Neurosociology is widely viewed as tangential to sociology as a scientific discipline; it is considered an interdisciplinary initiative seeking an interface between social, cognitive, and affective neuroscience and sociology. However, an opposite view holds that, beyond the specific topic of emotions, the more general object of sociology—the social world—exists as a three-level phenomenon spanning mind, society, and brain. Social interaction is not possible without interactions between brains (TenHouten et al., 2022). Insofar as this view is valid, the present science of sociology remains conceptually underspecified, and therefore unable to fully address its topic, theoretically or methodologically. From this perspective, the brain provides an essential level of analysis: it is the key missing ingredient that enables advancement from the existing two-level sociology—the science of mind and society, to a new and broader paradigm, a three-level science of mind, society, and brain, of Sociology 2.0.

References

Adolphs, R., Damasio, H., Tranel, D., & Damasio, A. R. (1996). Cortical systems for the recognition of emotion in facial expressions. *The Journal of Neuroscience, 16*(23), 7678–7687.

Albuquerque, N., Guo, K., Wilkinson, A., et al. (2016). Dogs recognize dog and human emotions. *Biology Letters, 12*(1), 2015088.

Anderson, A. K. (2005). Affective influences on the attentional dynamics supporting awareness. *Journal of Experimental Psychology: General, 134*(2), 258–281.

Baker, K. B., & Kim, J. J. (2004). Amygdala lateralization in fear conditioning: Evidence for greater involvement of the right amygdala. *Behavioral Neuroscience, 118*(1), 15–23.

Barrett, L. F. (2006). Are emotions natural kinds? *Perspectives on Psychological Science, 1*(1), 28–56.

Baumeister, R. F., & Leary, M. R. (1995). The need to belong: Desire for interpersonal attachment as a fundamental human motivation. *Psychological Bulletin, 117*(3), 497–529.

Bigler, E. D. (2001). Frontal lobe pathology and antisocial personality disorder. *Archives of General Psychiatry, 58*(6), 609–611.

Billeke, P., & Aboitiz, F. (2013). Social cognition in schizophrenia: From social stimuli processing to social engagement. *Frontiers in Psychiatry, 4*(4), 358076.

Blau, P. M. (1964). *Exchange and power in social life*. John Wiley & Sons.

Boehm, C. (1999). *Hierarchy in the forest: The evolution of egalitarian behavior*. Harvard University Press.

Bogen, J. E. (1973). Hemispheric specificity, complementarity, and self-referential mapping. *Proceedings of the Society for Neuroscience, 3*, 413.

Bogen, J. E., DeZure, R., TenHouten, W. D., & Marsh, J. F. (1972). The other side of the brain. IV. The a/P ration. *Bulletin of the Los Angeles Neurological Society, 37*(2), 49–61.

Bolender, J. (2010). *The self-organizing social mind*. The MIT Press.

Borod, J. C. (1993). Cerebral mechanisms underlying facial, prosodic, and lexical emotional expression: A review of neuropsychological studies and methodological issues. *Neuropsychology, 7*(4), 445–463.

Borod, J. C., Cicero, B. A., Obler, I. K., et al. (1998). Right hemisphere emotional perception: Evidence across multiple channels. *Neuropsychology, 12*(3), 446–458.

Brothers, L. (1997). *Friday's footprint: How society shapes the human mind*. Oxford University Press.

Buck, R., & Duffy, R. J. (1980). Nonverbal communication of affect in brain-damaged patients. *Cortex, 16*(3), 351–362.

Cacioppo, J. T., & Patrick, W. (2008). *Loneliness: Human nature and the need for social connection*. W. W. Norton.

Call, J. (2007). Past and present challenges in theory of mind research in nonhuman primates. In I. C. von Hofsten & K. Rosander (Eds.), *Progress in brain research: From action to cognition* (pp. 341–353). Elsevier.

Carpenter, C. C. (1961). Patterns of social behavior in the desert iguana, *Dipsosaurus dorsalis*. *Copeia, 4*(December), 396–405.

Carver, C. S., & Harmon-Jones, E. (2009). Anger is an approach-related affect: Evidence and implications. *Psychological Bulletin, 135*(2), 183–204.

Cerqueira, C. T., Almeida, J. R., Gorenstein, C., et al. (2008). Engagement of multifocal neural circuits during recall of autobiographical happy events. *Brazilian Journal of Medical and Biological Research, 41*(12), 1076–1085.

Cerqueira, M., Millot, S., Castanheira, M. F., et al. (2017). Cognitive appraisal of environmental stimuli induces emotion-like states in fish. *Scientific Reports, 7*(1), 13181.

Chapman, H. A., & Anderson, A. K. (2012). Understanding disgust. *Annals of the New York Academy of Sciences, 1251*(1), 62–76.

Cheney, D. L., & Seyfarth, R. M. (1990). *How monkeys see the world: Inside the mind of another species*. University of Chicago Press.

Corballis, M. C. (1991). *The lopsided ape: Evolution of the generative mind*. Oxford University Press.

Craig, A. D. (2009). How do you feel—Now? The anterior insula and human awareness. *Nature Reviews Neuroscience, 10*(January), 59–70.

Critchley, H. D. (2009). Psychophysiology of neural, cognitive and affective integration: fMRI and autonomic indicants. *International Journal of Psychophysiology, 73*(2), 88–94.

Damasio, A. (1999). *The feeling of what happens: Body and emotion in the making of consciousness*. Harcourt.

Damasio, A. R., Tranel, D., & Damasio, H. C. (1991). Somatic markers and the guidance of behavior: Theory and preliminary testing. In H. S. Levin, H. M. Eisenberg, & A. L. Benton (Eds.), *Frontal lobe function and dysfunction* (pp. 217–229). Oxford University Press.

Darwin, C. (1872/1965). *The expression of the emotions in man and animals*. The University of Chicago Press.

Davidson, R. J. (1995). Cerebral asymmetry, emotion, and affective style. In R. J. Davidson & K. Hugdahl (Eds.), *Brain asymmetry* (pp. 361–387). MIT Press.

Davidson, R. J., & Irwin, W. (1999). The functional neuroanatomy of emotion and affective style. *Trends in Cognitive Science, 3*(1), 11–21.

Davidson, R. J., Jackson, D. C., & Kalin, N. H. (2000). Emotion, plasticity, context and regulation: Perspectives from affective neuroscience. *Psychological Bulletin, 126*(6), 890–909.

Davidson, R. J., Mednick, D., Moss, E., et al. (1987). Ratings of emotions in faces are influenced by the visual field to which stimuli are presented. *Brain and Cognition, 6*(4), 403–441.

de Waal, F. (1989). *Peacemaking among primates*. Harvard University Press.

de Waal, F. (2005). *Our inner ape*. Riverhead Press.

Deaner, R. O., Khera, A. V., & Platt, M. C. (2005). Monkeys pay per view: Adaptive valuation of social images by rhesus macaques. *Current Biology, 15*(6), 543–548.

Decety, J., & Chaminade, T. (2003). When the self represents the other. *Consciousness and Cognition, 12*(4), 577–596.

Demaree, H. A., Everhart, D. E., Youngstrom, E. A., & Harris, D. W. (2005). Brain lateralization of emotional processing: Historical roots and a future incorporating dominance. *Behavioral and Cognitive Neuroscience Reviews, 4*(1), 3–20.

Devinsky, O. (2000). Right cerebral hemisphere dominance for a sense of corporeal and emotional self. *Epilepsy & Behavior, 1*(1), 60–73.

Dougherty, D. D., Shin, L. M., & Alpert, M. N. (1999). Anger in healthy men: A PET study using script-driven imagery. *Biological Psychology, 46*(4), 466–472.

Downar, J., Crawley, A. P., & Mikulis, D. J. (2000). A. Multimodal cortical network for the detection of changes in the sensory environment. *Nature Neuroscience, 3*(3), 277–283.

Efron, R. (1990). *The decline and fall of hemispheric specialization*. Lawrence Erlbaum Associates.

Ehrlichman, H. (1987). Hemispheric asymmetry and positive-negative affect. In D. Ottoson (Ed.), *Duality and unity of the brain* (pp. 194–206). Macmillan.

Ekman, P., & Friesen, W. V. (1986). A new pan-cultural expression of emotion. *Motivation and Emotion, 10*(2), 159–168.

Ekman, P., Sorenson, E. R., & Friesen, W. V. (1969). Pan-cultural elements in facial displays of emotions. *Science, 164*(3875), 86–88.

Fiske, A. P. (1991). *Structures of social life: The four elementary forms of human relations*. The Free Press.

Fiske, A. P. (1993). Social errors in four cultures: Evidence about the elementary forms of social relations. *Journal of Cross-Cultural Psychology, 24*(4), 67–94.

Fiske, A. P. (2004). Relational models theory 2.0. In N. Haslam (Ed.), *Relational models theory: A contemporary overview* (pp. 3–25). Lawrence Erlbaum.

Frasnelli, E. (2013). Brain and behavioral lateralization in invertebrates. *Frontiers in Psychology, 4*, 939.

Frasnelli, E., Vallortigara, G., & Rogers, L. J. (2012). Left-right asymmetries of behavioural and nervous system in invertebrates. *Neuroscience and Biobehavioral Reviews, 36*(4), 1273–1291.

Fridlund, A. J. (1992). Darwin's anti-Darwinism in 'the expression of the emotions in man and animals'. In K. T. Strongman (Ed.), *International review of studies of emotion* (Vol. 2, pp. 13–27). John Wiley & Sons.

Friedman, D., Goldman, R., Stern, Y., & Brown, T. R. (2009). The brain's orienting response: An event-related functional magnetic resonance imaging investigation. *Human Brain Mapping, 30*(4), 1144–1154.

Gainotti, C. (1972). Emotional behavior and hemispheric side of the lesion. *Cortex, 8*(1), 41–55.

Gainotti, C. (2012). Unconscious processing of emotions and the right hemisphere. *Neuropsychologia, 50*(2), 205–218.

Gallotti, M., & Frith, C. D. (2013). Social cognition in the we-mode. *Trends in Cognitive Science, 17*(4), 160–165.

Gan, X., Zhau, X., Li, J., et al. (2022). Common and distinct neurofunctional representatives of core and social disgust in the brain: Coordinate-based and network meta-analyses. *Neuroscience and Biobehavioral Reviews, 135*(April), 104553.

García-Garcia, I., Kube, J., Gaebler, M., et al. (2016). Neural processing of emotional stimuli and the influence of age, sex and task-related characteristics. *Neuroscience & Biobehavioral Reviews, 6*(September), 773–793.

Garfinkel, H. (1967/1984). *Studies in ethnomethodology*. Polity Press.

Gazzola, A., Guadin, B., Balestrieri, A., & Pellitteri-Rosa, D. (2023). Effects of predation risk on the sensory asymmetries and defensive strategies *of Bufotes balearicus* tadpoles. *Animal Cognition, 26*(2), 491–501.

Ghiselin, M. T. (1969). *The triumph of the Darwinian method*. University of California Press.

Gouzoules, S., Gouzoules, H., & Marler, P. (1984). Rhesus monkey (*Macaca mulatta*) screams: Representational signalling in the recruitment of agonistic aid. *Animal Behaviour, 32*(1), 182–193.

Gray, J. A. (1990). Brain systems that mediate both emotion and cognition. *Cognition and Emotion, 4*(3), 269–288.

Grecucci, A., Sorella, S., & Consolini, J. (2023). Decoding individual differences in expressing and suppressing anger from structural brain networks: A supervised machine learning approach. *Behavioural Brain Research, 439*, 114245.

Greenberg, N. (1977). An ethogram of the blue spiny lizard, *Sceloporus cyanogenys* (Reptilia, Lacertilia, Iguanidae). *Journal of Herpetology, 11*(2), 177–195.

Greenberg, N., & MacLean, P. D. (1978). Ethological considerations in the experimental study of lizard behavior. In N. Greenberg & P. D. MacLean (Eds.), *The behavior and neurology of lizards* (pp. 203–224). DHEW Publication (ADM) 77–491.). U.S. Government Printing Office.

Griffin, R., Friedman, O., & Ween, J. (2006). Theory of mind and the right cerebral hemisphere. *Laterality: Asymmetries of Brain, Behaviour, and Cognition, 11*(3), 195–225.

Gross, J. J. (2015). Emotion regulation: Current status and future prospects. *Psychological Inquiry, 26*(1), 1–26.

Güntürkün, O., Ströckens, F., & Ocklenburg, S. (2020). Brain lateralization: A comparative perspective. *Physiological Reviews, 100*(3), 1019–1063.

Hamilton, W. D. (1964). The genetical theory of social behaviour. *Journal of Theoretical Biology, 7*(1), 1–52.

Harmon-Jones, E., & Gable, P. A. (2018). On the role of asymmetrical frontal cortical activity in approach and withdrawal motivation: An updated review of the evidence. *Psychophysiology, 55*(1), e12879.

Harmon-Jones, E., & Seligman, J. (2001). State anger and prefrontal brain activity: Evidence that insult-related relative left-prefrontal activation is associated with experienced anger and aggression. *Journal of Personality and Social Psychology, 80*(5), 797–803.

Haslam, N. (1995). Factor structure of social relationships. *Journal of Social and Personal Relationships, 12*(2), 217–229.

Haslam, N. (1997). Four grammars for primate social relations. In J. A. Simpson & D. T. Kenrick (Eds.), *Evolutionary social psychology* (pp. 297–316). Lawrence Erlbaum Associates.

Haslam, N., & Fiske, A. P. (1999). Relational models theory: A confirmatory factor analysis. *Personal Relationships, 6*(2), 241–250.

He, J., Degnan, K. A., McDermott, J. M., et al. (2010). Anger and approach motivation in infancy: Relations to early childhood inhibitory control and behavior problems. *Infancy, 15*(3), 246–269.

Hecht, D. (2014). Cerebral lateralization of pro- an anti-social tendencies. *Experimental Neurobiology, 23*(1), 1–27.

Heyes, C. M. (1998). Theory of mind in nonhuman primates. *Behavioral and Brain Sciences, 21*(1), 101–134.

Hockings, K. J., Humle, T., Anderson, J. R., et al. (2007). Chimpanzees share forbidden fruit. *PLoS ONE, 2*(9), e886.

Holtmann, O., Bruchmann, M., Mönig, C., et al. (2020). Lateralized deficits of disgust processing after insula-basal ganglia damage. *Frontiers in Psychology, 11*, 1429.

Hoppe, K. D., & Bogen, J. E. (1976). Alexithymia in twelve commissurotomized patients. *Psychotherapy and Psychosomatics, 28*(1–4), 148–155.

Hughlings-Jackson, J. (1866). Remarks on those cases of diseases of the nervous system in which defect of expressive is the most striking symptom. *The Medical Times and Gazette, 1*, 569–662.

Hung, Y., Smith, M. L., Bayle, D. J., et al. (2010). Unattended emotional faces elicit early lateralized amygdala–frontal and fusiform activation. *NeuroImage, 50*(2), 727–733.

Joseph, R. (1996). *Neuropsychiatry, neuropsychology, and neuroscience* (2nd ed.). Williams and Wilkins.

Kalin, N. H., Larson, C., Shelton, S. E., et al. (1998). Asymmetric frontal brain activity, cortisol, and behavior associated with fearful temperament in rhesus monkeys. *Behavioral Neuroscience, 112*(2), 286–292.

Kapogiannis, D., Sutin, A., Davatzikos, A., et al. (2013). The five factors of personality and regional cortical variability in the Baltimore longitudinal study of aging. *Human Brain Mapping, 34*(11), 2829–2840.

Keenan, J. P., Nelson, A., O'Connor, M. O., & Pascual-Leone, A. (2001). Self-recognition and the right hemisphere. *Nature, 409*(6818), 305.

Keenan, J., Wheeler, M., Platek, S., et al. (2003). Self-face processing in a callosotomy patient. *European Journal of Neuroscience, 18*(8), 2391–2395.

Kellerman, H. (1979). *Group psychotherapy and personality: Intersecting structures*. Grune & Stratton.

Kelley, N. J., Hortensius, R., & Harmon-Jones, E. (2013). When anger leads to rumination: Induction of relative right frontal cortical activity with transcranial direct current stimulation increases anger-related rumination. *Psychological Science, 24*(4), 475–481.

Kosslyn, S. M., & Miller, G. W. (2015). *Top brain, bottom brain* (2nd ed.). Simon & Schuster.

Kremer, L., Klein Holkenborg, S. E. J., Reimert, I., et al. (2020). The nuts and bolts of animal emotion. *Neuroscience & Biobehavioral Reviews, 113*(June), 273–286.

Laberge, F., Mühlenbrock-Lenter, S., Grunwald, W., & Roth, G. (2000). Evolution of the amygdala: New insights from studies of amphibians. *Brain, Behavior and Evolution, 67*(4), 177–187.

Ládavas, E., & Bertini, C. (2021). Right hemisphere dominance for unconscious emotionally salient stimuli. *Brain Sciences, 11*(7), 823.

Layard, R. (2005). *Happiness: Lessons from a new science*. Penguin Books.

LeDoux, J. E. (2014). Coming to terms with fear. *Proceedings of the National Academy of Sciences of the United States of America, 111*(8), 2871–2878.

LeDoux, J. E., & Pine, D. S. (2016). Using neuroscience to understand fear and anxiety: A two-system framework. *American Journal of Psychiatry, 173*(11), 1083–1093.

Leliveld, L. M. C., Langbein, J., & Puppe, B. (2013). The emergence of emotional lateralization: Evidence in non-human vertebrates and implications for farm animals. *Applied Animal Behaviour Science, 145*(1–2), 1–14.

Levy, J., Heller, W., Banich, M. T., & Burton, L. A. (1983). Asymmetry of perception in free viewing of chimeric faces. *Brain and Cognition, 2*(4), 404–429.

Lewis, J. W., Wightman, F. L., Brefczynski, J. A., et al. (2004). Human brain regions involved in recognizing environmental sounds. *Cerebral Cortex, 14*(9), 1008–1021.

Loring, D. W., Mead, K. J., Lee, G. P., & King, D. W. (1992). *Amobarbital effects and lateralized brain function: The Wada test*. Springer-Verlag.

Lukito, S., Fortea, L., Groppi, F., et al. (2023). Should perception of emotions be classified according to threat detection rather than emotional valence? An updated meta-analysis for whole-brain atlas of emotional faces processing. *Journal of Psychiatry & Neuroscience, 48*(5), E376–E389.

Luria, A. R. (1973). *The working brain*. Basic Books.

MacLean, P. D. (1990). *The triune brain in evolution: Role in Paleocerebral functions*. Plenum Press.

Majolo, B., Lehmann, J., de Bartoli Vizioli, A., & Schino, J. (2012). Fitness-related benefits of dominance in primates. *American Journal of Physical Anthropology, 147*(4), 652–660.

Marzluff, J. M., & Angell, T. (2013). *Gifts of the crow: How perception, emotion, and thought allow small birds to behave like humans*. Simon & Schuster.

McGilchrist, I. (2009). *The master and his emissary: The divided brain and the making of the modern world* (pp. 51–72). Yale University Press.

McGilchrist, I. (2020). Reciprocal organization of the cerebral hemispheres. *Dialogues in Clinical Neuroscience, 12*(4), 503–515.

Messina, I., Grecucci, A., & Viviana, R. (2021). Neurobiological models of emotion regulation: A meta-analysis of neuroimaging studies of acceptance as an emotion regulation strategy. *Social Cognitive and Affective Neuroscience, 16*(3), 257–267.

Mills, C. K. (1912). The cerebral mechanisms of emotional experiences. *Transactions of the College of Physicians of Philadelphia, 34*, 381–390.

Morris, J. S., Öhman, A., & Dolan, R. J. (1998). Conscious and unconscious emotional learning in the human amygdala. *Nature, 393*(6684), 467–470.

Mortin, B. E. (2013). Behavioral laterality of the brain: Support for the binary construct of hemisity. *Frontiers in Psychology, 4*, 683.

Nestor, P. G., Nakamura, M., Niznikiewicz, M., et al. (2013). In search of the functional neuroanatomy of sociality: MRI subdivisions of the orbital frontal cortex and social cognition. *Social Cognitive and Affective Neuroscience, 8*(4), 460–467.

Nöe, R. (1992). Alliance formation among male baboons: Shopping for profitable partners. In A. H. Harcourt & F. B. M. de Waal (Eds.), *Coalitions and alliances in humans and other animals* (pp. 285–321). Oxford University Press.

Novejarque, A., Lanuza, E., & Martinez-Garcia, F. (2004). Amygdalostriatal projections in reptiles: A tract-tracing study in the lizard Podarcis hispanica. *Journal of Comparative Neurology, 479*(3), 287–308.

Ohira, H. (2010). The somatic marker revisited: Brain and body in decision making. *Emotion Review, 2*(3), 245–249.

Öhman, A. (2005). The role of the amygdala in human fear: Automatic detection of threat. *Psychoneuroendocrinology, 30*(10), 953–958.

Öhman, A., & Mineka, S. (2001). Fears, phobias, and preparedness: Toward an evolved module of fear and fear learning. *Psychological Review, 108*(3), 483–522.

Olatunji, B. O., & Sawchuk, C. (2005). Disgust: Characteristic features, social manifestations, and clinical implications. *Journal of Social and Clinical Psychology, 24*(7), 932–962.

Osgood, C. E. (1966). Dimensionality of the semantic space for communication via facial expressions. *Scandinavian Journal of Psychology, 7*(1), 1–30.

Panksepp, J. (1982). Toward a general psychobiological theory of emotions. *Behavioral and Brain Sciences, 5*(3), 407–467.

Panksepp, J. (1998). *Affective neuroscience: The foundations of human and animal emotions*. Oxford University Press.

Parsons, T. (1951). *The social system: Glencoe*. Free Press.

Perry, C. J., & Baciadonna, L. (2017). Studying emotion in invertebrates: What has been done, what can be measured and what they can provide. *Journal of Experimental Biology, 220*(21), 3856–3868.

Plutchik, R. (1958). Outline of a new theory of emotion. *Transactions of the New York Academy of Sciences, 20*(5), 394–403.

Plutchik, R. (1962/1991). *The emotions: Facts, theories, and a new model, rev. ed.* University Press of America.

Plutchik, R. (1980a). A general psychoevolutionary theory of emotion. In R. Plutchik & H. Kellerman (Eds.), *Emotion: Theory, research, and experience* (*Theories of emotion*) (Vol. 1, pp. 3–34). Academic Press.

Plutchik, R. (1980b). *Emotion: A Psychoevolutionary synthesis*. Harper & Row.

Poole, B. D., & Gable, P. A. (2014). Affective motivational direction drives asymmetric frontal hemisphere activation. *Experimental Brain Research, 232*(7), 2121–2130.

Pribadi, A. K., Chalasini, S., & H. (2022). Fear conditioning in invertebrates. *Frontiers in Behavioral Neuroscience, 16*, 1008818.

Quirin, M., Meyer, F., Heise, N., et al. (2013). Neural correlates of social motivation: An fMRI study of power versus affiliation. *International Journal of Psychophysiology, 88*(3), 289–295.

Ranganath, C., & Rainer, G. (2003). Neural mechanisms for detecting and remembering novel events. *Nature Reviews Neuroscience, 4*(3), 193–202.

Rolls, E. T. (2000). The orbitofrontal cortex and reward. *Cerebral Cortex, 10*(3), 284–294.

Rolls, E. R. (2019). *The orbital frontal cortex.* Oxford University Press.

Ross, E. D. (2021). Differential hemispheric lateralization of emotions and related display behaviors: Emotion-type hypothesis. *Brain Sciences, 11*(8), 1034.

Ross, E. D., Homan, R. W., & Buck, R. D. (1994). Differential hemispheric lateralization of primary and social emotions: Implications for developing a comprehensive neurology for emotion repression, and the subconscious. *Neuropsychiatry, Neuropsychology, and Behavioral Neurology, 7*(1), 1–19.

Schino, G., & Aureli, F. (2009). Reciprocal altruism in primates: Partner choice, cognition, and emotions. *Advances in the Study of Behavior, 39*, 45–69.

Schmidt, L. A., & Trainor, L. J. (2001). Frontal brain electrical activity (EEG) distinguishes valence and intensity of musical emotions. *Cognition and Emotion, 15*(4), 487–500.

Schore, A. N. (1994). *Affect regulation and the origin of the self: The neurobiology of emotional development.* Lawrence Erlbaum Associates.

Schutter, D. J. L. G. (2008). Antidepressant efficacy of high-frequency transcranial magnetic stimulation over the left dorsolateral prefrontal cortex in double-blind controlled designs: A meta-analysis. *Psychological Medicine, 39*(1), 65–75.

Seligman, M. E. P. (2011). *Flourish: A visionary new understanding of happiness and Well-being*. Atria Books.

Sell, A. (2017). Recalibration theory of anger. In T. K. Shackelford & V. A. Weekes-Shackelford (Eds.), *Encyclopedia of evolutionary psychological science* (pp. 1–15). Springer.

Shamay, S. G., Tomer, R., Berger, B. D., & Aharon-Peretz, J. (2003). Characterization of empathy deficits following prefrontal bran damage: The role of the right ventromedial prefrontal cortex. *Journal of Cognitive Neuroscience, 15*(3), 324–337.

Shirai, M., & Suzuku, N. (2017). Is sadness only one emotion? Psychological and physiological responses induced by two different situations: "Loss of someone" and "failure to achieve a goal". *Frontiers in Psychology, 8*, 288.

Silberman, E., & Weingartner, H. (1986). Hemispheric lateralization of functions related to emotion. *Brain and Cognition, 5*(3), 322–353.

Silk, J. B. (2002). Practice random acts of aggression and senseless acts of intimidation: The logic of status contests in social groups. *Evolutionary Anthropology, 11*(6), 221–225.

Simmel, G. (1907–1910/1971). *On individuality and social forms* (D. N. Levine, Ed.). The University of Chicago Press.

Sorella, S., Grecucci, A., Piretti, L., & Job, R. (2021). Do anger perception and the experience of anger share common neural mechanisms? Coordinate-based meta-analytic evidence of similar and different mechanisms from functional neuroimaging studies. *NeuroImage, 230*, 117777.

Stancher, G., Sovrano, V. A., & Vallortigara, G. (2018). Motor asymmetries in fish, amphibians, and reptiles. *Progress in Brain Research, 238*, 33–56.

Stefanova, E., Dubljević, O., Herbert, C., et al. (2020). Anticipatory feelings: Neural correlates and linguistic markers. *Neuroscience & Biobehavioral Reviews, 113*(June), 308–324.

Stuss, D. T., Gow, C. A., & Heatherington, C. R. (1992). "No longer gage": Frontal lobe dysfunction and emotional change. *Journal of Consulting and Clinical Psychology, 60*(3), 349359.

Tao, D., He, Z., Lin, Y., et al. (2021). Where does fear originate in the brain? A coordinate-based meta-analysis of explicit and implicit fear processing. *NeuroImage, 227*, 117686.

TenHouten, W. D. (1992). Cerebral lateralization: A scientific paradigm in crisis? Critique of Efron and Corballis. *Journal and Social and Evolutionary Systems, 15*(3), 319–326.

TenHouten, W. D. (2005). *Time and society*. Harper & Row.

TenHouten, W. D. (2007). *A general theory of emotions and social life*. Routledge.

TenHouten, W. D. (2013). *Emotion and reason: Mind, brain, and the social domains of work and love*. Routledge.

TenHouten, W. D. (2016). Embodied feeling and reason in decision-making: Assessing the somatic marker hypothesis. *REPLACES, Revista Latinoamericana de Estudios sobre Cuerpos, Emociones y Sociedad [Latin American Journal of Studies on Bodies, Emotions, and Society], 8*(20), 87–97.

TenHouten, W. D. (2021a). Basic emotion theory, social constructionism, and the universal ethogram. *Social Science Information, 60*(4), 610–630.

TenHouten, W. D. (2021b). Climbing, and falling off, Plato's ladder of love: The emotions of love and love's undoing. In E. Vanderheiden & C.-H. Mayer (Eds.), *International handbook of love: Transcultural and transdisciplinary perspectives* (pp. 101–121). Springer.

TenHouten, W. (2023). Love's place in the spectrum of affect, one of 24 secondary emotions: Implications for psychiatric theory. *International Review of Psychiatry, 35*(1), 97–112.

TenHouten, W. D., Hoppe, K. D., Bogen, J. E., & Walter, D. O. (1986). Alexithymia: An experimental study of cerebral commissurotomy patients and normal controls. *American Journal of Psychiatry, 143*(3), 312–316.

TenHouten, W. D., Schussel, L., Gritsch, M. F., & Kaplan, C. D. (2022). Hyperscanning and the future of neurosociology. *Sociological Methodology, 53*(1), 139–157.

Thomas, W. I. (1923/1966). Motivation: The wishes. In M. Janowitz (Ed.), *On social organization and social personality: Selected papers* (pp. 117–139). The University of Chicago Press.

Thompson, M., Ellia, R., & Wildavsky, A. (1990). *Cultural theory*. Westview Press.

Tinbergen, N. (1951). *The study of instinct*. Oxford University Press.

Tobore, O. T. (2023). On the beauty of sadness: It's okay to say I am sad, thank you. *Communicative & Integrative Biology, 16*(1), 2211424.

Tomasello, M. (1994). The question of chimpanzee culture. In R. W. Wrangham, W. C. McGrew, F. B. M. de Waal, & P. G. Heltne (Eds.), *Chimpanzee cultures* (pp. 301–317). Harvard University Press.

Tomer, R., Slagter, H. A., Christian, B. T., et al. (2014). Love to win or hate to lose? Asymmetry of dopamine D2 receptor binding predicts sensitivity to reward vs. punishment. *Journal of Cognitive Neuroscience, 26*(5), 1039–1048.

Turner, J. H., & Stets, J. E. (2005). *The sociology of emotions*. Cambridge University Press.

Uomini, N. T., & Ruck, L. (2018). Manual laterality and cognition through evolution: An archeological perspective. *Progress in Brain Research, 238*, 295–323.

Vallortigara, G., & Rogers, L. J. (2005). Survival with an asymmetrical brain: Advantages and disadvantages of cerebral lateralization. *Behavioral and Brain Sciences, 28*(4), 575–633.

Vallortigara, G., & Rogers, L. J. (2020). A function for the bicameral mind. *Cortex, 124*(March), 274–285.

Vicario, C. M., Rafel, R. D., Martino, D., & Avenanti, A. (2017). Core, social and moral disgust are bounded: A review on neural bases of repugnance in clinical disorder. *Neuroscience & Biobehavioral Reviews, 80*(Sept), 185–200.

Vuilleumier, P. (2005). How brains beware: Neural mechanisms of emotional attention. *Trends in Cognitive Science, 9*(12), 585–594.

Vytal, K., & Hamann, S. (2010). Neuroimaging support for discrete neural correlates of basic emotions: A voxel-based meta-analysis. *Journal of Cognitive Neuroscience, 22*(12), 2864–2885.

Weber, M. (1921/1978). *Economy and society: An outline of interpretive sociology* (Vol. 1., G. Roth & C. Widdich, Trans., Eds.). University of California Press.

Weinberg, I. (2000). The prisoners of despair: Right hemisphere deficiency and suicide. *Neuroscience & Biobehavioral Reviews, 24*(8), 799–815.

Whittle, S., Yücel, M., Fornito, A., et al. (2008). Neuroanatomical correlates of temperament in early adolescence. *Journal of the American Academy of Child & Adolescent Psychiatry, 47*(6), 682–693.

Wyczesany, M., Capotosto, P., Zappasodi, F., & Prete, G. (2018). Hemispheric asymmetries and emotions: Evidence from effective connectivity. *Neuropsychologia, 121*(December), 98–105.

Xu, P., Peng, S., Luo, Y. J., et al. (2021). Facial expression recognition: A meta-analytic review of theoretical models and neuroimaging evidence. *Neuroscience & Biobehavioral Reviews, 127*(August), 820–836.

Yanagisawa, K., Masui, K., Furutani, K., et al. (2011). Does higher general trust serve as a psychosocial buffer against social pain? An NIRS study of social exclusion. *Social Neuroscience, 6*(2), 190–197.

Zhang, Y., Suo, X., Ding, H., et al. (2019). Structural connectivity profile supports laterality of the saliency network. *Human Brain Mapping, 40*(18), 5242–5255.

Part III

Mind, Self, Interaction, and Sociality

10 Relationships Between Neurosociology, Foundational Social Behaviorism, and Currents in Symbolic Interaction

David D. Franks

Abstract

This chapter discusses the contributions of neurosociology to what symbolic interactionists refer to as the foundational social behaviorism of G.H. Mead. First, the varieties in symbolic interaction are described. A theme of this chapter is that if a field so different from symbolic interaction as neuroscience can contribute to it, then neuroscience can certainly contribute to all of sociology. This chapter starts with describing symbolic interaction's early interest in aphasia, then how it supports accounts through evidence given by split-brain research. Then memory is shown to be highly unreliable and interpretive. Much of this chapter describes the issues that created pragmatism and closes with how mirror neurons support the philosophy of pragmatism and contribute to our ability to speak, another area critical to symbolic interaction.

Of the three categories above, the most difficult to define is symbolic interaction. Easiest in a sense is social behaviorism in which George Herbert Mead did the defining for those willing to read his various lectures as in *Mind, Self and Society* (1934), *The Philosophy of the Act* (1938), as well as *The Philosophy of the Present* (1959), which contains the Carus Lectures presented at Berkeley the year before his death in 1931. Even though it was not highly developed in his time, he insisted that the physiological mechanisms underlying the social were indispensable to understanding minds, and in several places he made this point. For example,

> …This experienced world does not appear except when the various excitements reach certain points in the central nervous system; it is also true that if you cut off any of those channels you wipe out so much of that world.
> What the behaviorist... ought to do is to take the complete act… the whole process of conduct…as his unit in his analysis. In doing that he has to take into account, not simply the nervous system, but also the rest of the organism, for the nervous system is only a specialized part of the entire organism. (1934: 111)

For further examples, on pages 111–115 in *Mind Self and Society*, he mentions the central nervous system at least 17 times. As Clark McPhail[1] points out, Mead worked in what is now called neuroscience before joining Dewey at the University of Chicago. This is an important point

D. D. Franks (✉)
Department of Sociology, Virginia Commonwealth University, Richmond, VA, USA

[1]Private communication. Prof. McPhail was influential in SI and was its distinguished lecturer in 2005. He brought Mead's initial experience as a neurological researcher at the University of Chicago to my attention.

W. Kalkhoff et al. (eds.), *Handbook of Neurosociology*, Handbooks of Sociology and Social Research, https://doi.org/10.1007/978-3-031-95615-7_10

because the vast majority of current symbolic interactionists today are not aware of this and even may or may not be interested in Mead's works.

As this author remembers, symbolic interaction (SI) was promulgated as a society in the late 1960s and was born in large part as a reaction to Talcott Parsons and traditional structural sociology. For a significant part of the membership since the 1980s, contemplating what Mead said was a matter of unproductive interpretation. They rejected the third person, scientific voices espousing "truth statements" like Mead's.

Instead there was an emphasis on the new and the untraditional. At times, annual sessions took the form of theatrical performances and especially ethnographies. David Maines had to urge his colleagues away from the extremes of the postmodernist "linguistic turn" wherein "everything was within the text." "New readings," wrote Denzin (1992: 513), "are justified not because they yield more truth, but because they are new." Maines (1995) countered that "... if the new is good because it's new and if the old is bad because it's boring, then what do we do if we come up with something old that works?" As we shall see, this is especially true in regard to the pragmatic tradition of Dewey and Mead.

Herbert Blumer, Mead's teaching assistant, became known as his "spokesman" and coined the term symbolic interaction. In doing so, he moved SI significantly far from the social behaviorism of Mead and the early American pragmatism. (See McPhail and Rexroat 1979.) Because of Blumer's influence, the interpretive symbol as an intangible word became the emphasis, and the pragmatic notion that words subserve action was lost.

The diversity within the SI society should not be ignored though that is not the point of this essay. A few were even interested in neuroscience. There was always a friendly tension between Denzin espousing ethnographic interpretations of movies and Carl Couch, who was committed to developing general principles of group processes and traditional hypothesis testing. Sheldon Stryker was another adherent of Mead's who diverged from the members of the society by espousing a "social structuralist position" (see Stryker 1985). For many SI members, the objectivity of neuroscience, if they attended to it at all, was really a construction of male domination. Auto-ethnologies came into vogue, and for many an exclusive reliance on qualitative methods replaced the impersonal character of traditional science. There was a decided move away from Mead's behaviorist pragmatism and toward a complete relativization of Mead. This was in spite of his clear insistence on maintaining the "possibility of error" which extreme relativism does not allow because it implies a truth from which that error diverges. The possibility of error is also an *a prior* assumption of hypothesis-testing. The new view implied that traditional science was just another story.

It is important to keep these two versions of symbolic interaction separate so one does not confuse Mead's foundational efforts with current tendencies. There is also significant variation among current symbolic interactionists. For example, Clark McPhail was another SSSI member who took a more traditional methodological position in studying collective behavior and was very interested in Mead's interest in the neuroscience of his day. A third, less visible group teach and use foundational concepts but are not members of the association. Examples of these are Victor Gecas and Andrew Weigert as well as Darwin Thomas.

Symbolic interaction would seem to be as different from neuroscience as any field in sociology; if brain science can be seen as playing a relevant role in symbolic interaction, it has passed the most difficult test of its relevancy to all of sociology.

This chapter will demonstrate that neurosociology can confirm important concepts in foundational symbolic interaction and more recent SI and also broaden its range. I will discuss neurological insights and interests by early symbolic interactionists and their recognition of correlations between thinking and movements of bodily parts as used in speech. The very name symbolic interaction implies a stress on the intangible and symbolic nature of its referent and public constructions. The tangible resists our push.

Abstractions like space, time, and grammar do not; to think of holding them physically is an absurdity. In this sense, the symbol is extrasensory—a perfect illustration of what we call mind as an emergent from the physical brain. Since the symbol's link to its referent is customary and not fixed causally, it follows that there will be a major stress on the interpretive nature of symbolic life. We shall see how brain science is largely consistent with this because "the brain actively selects, interprets, edits and changes the very quality of incoming information to fit its own requirements and limitations." The brain projects its own character on the world just as symbols do (Franks 2007).

Another issue given attention in this chapter deals with the unconscious and recent neuroscience research. This material broadens the range of foundational symbolic interaction. Neuroscientists and cognitive psychologists have collected a large amount of very credible evidence for the importance of the unconscious and symbolic interaction, with its emphasis on the self-conscious control of behavior, ignores the unconscious at its own peril. Gazzaniga writes that we are only conscious of two percent of what the brain does. Thought and language, as important as they are, do not characterize all of the brain's thought. Rather than thinking in words, the right side of the brain thinks in some type of gestalt-based mode (Gazzaniga 1998: 21). According to Jonathan Turner, SI tends to overemphasize thought and language as important as they are. "If all of thought was internalized speech we would think very slowly because speech is lineal and must draw out one phoneme at a time. Our brains work much faster than that, and this is why we have Broca and Wernicke's area to down load and up load speech to the brain's way of thinking." Scheff (1990: 51) also emphasizes the impressive speed of social interaction. He tells us that "Because of the ambiguity of expression it seems necessary that internal searches must be so rapid that they must occur as many different parallel processes." He quotes Anderson (1983) that internal problem-solving events take only 1 ms (one thousandth of a second).

In spite of their differences, social behaviorism and research on mirror neurons are intimately based on a similar epistemology—that of Mead's pragmatism including as it does the metatheory of *transaction*.

After a presentation of the social construction of memory, this chapter deals with the curbs that pragmatism puts on social construction, and Gazzaniga's findings from his split-brain research will be described to show how it gives robust empirical evidence to Lyman and Scott's theory of accounts.

Early Symbolic Interactionists' Interests in the Neurological Processes

Aphasia

According to the notable SI leaders Lindesmith et al. (1988), the work on aphasia by neurologists Henry Head (1926), Kurt Goldstein (1948), and A.R. Luria (1976) was of great interest because it gave strong evidence for the important causal relationship between human speech and thought. To symbolic interactionists as well as neuroscientists among others, these two processes are critical for being human. Clearly these cognitive abilities were seriously disabled in the syndrome of aphasia that attacked Broca's area and Wernicke's area of the temporal lobes. The findings of these neurologists provided important evidence for the influence of language on all things human. For Lindesmith et al., human thought and language are intimately related, and they chided the early neuroscientists above for reifying them into separate processes. Such reification is now justified by current neuroscience and Jonathan Turner's remarks above. It followed that thinking and speaking were also intricately related since making "symbolic indications" to one's self was a key process in this perspective. In severe aphasia, intelligence and personality are disorganized, as well as memory of words. They add that "The scaffolding on which speech is developed is built up in relation to hearing, vision, and the sensory motor skills that were involved in uttering words" (Lindesmith et al. 1988).

The loss of capacities in aphasia may include an inability to read, write, or name familiar objects (Lindesmith et al. 1988). Out of sight, out of mind characterizes their mind-frames, and they are stuck in the concrete, having difficulty with abstractions. These neurologists also note that the focus on the concrete meant that the continuous flow of experience is disrupted and the patient's images of things not present were unstable and disorganized.

The Revisionist Nature of Perception

Another early agreement between neuroscience and Meadian theory is the revisionist nature of perception. In terms of its connection with the world, the source of perception can be seen as coming from two directions. One is the sensed world of stimuli that indeed, we may interpret, but which we do not initiate. A loud noise may be interpreted any number of ways, but the fact that it happened in the first place is not such a matter. Such stimuli come to us as passive observers and can be seen as "read out *from*" the world. The old Enlightenment British empiricists saw this as the main avenue of knowledge since to them it was the major link between the knower and the known. This epistemology is variously known as "copy" or "correspondence theory" since the human brain was seen as a blank, impressionable *Tabula Rasa* filled in by sensed experience much like the impression on malleable dough being formed by the "cookie cutter" of experience. Learning theory is consistent with this since the person is seen as passive and knowledge is mainly "read out *from*" the world separate from our will. This goes against the voluntarism so important to Mead and symbolic interaction theorists.

More compatible with SI theory and its focus on conscious interpretation underlying social constructionism is what we project *onto* the world. Here the focus of causation comes from us with the help of the interpretations and definitions of the situation stemming from our social groups. This theory of perception was typical of British idealists. A prime example of this position can be seen clearly in our perceptions of gestalt pictures that lend themselves to equally plausible interpretations even though the sensory drawings and dots originally drawn and comprising the possible perceptions remain stable. The fact that sensation by itself gives only sensation and nothing else was not recognized by the British empiricists.

> As Albert Einstein ...clearly recognized, the belief in an objective world with scientific objects in it the same for all observers, is a theoretically inferred world, not a purely empirically given knowledge. (N.S.F. Northrop 1948: 43)

The Chicago pragmatists saw these positions as dualistic and the source of never-ending debates that assured a lack of intellectual progress. In place of thinking of the link between the knower and the known world as the rationalist's reason or the empiricist's bodily-given senses, Mead and Dewey saw the link as stemming from human earthly *behavio*r. We become conscious of the world through the way it responds to our actions on it. In neurosociological terms, we only act successfully to that which the object allows. Researchers working with mirror neurons use the concept of "affordances" that an object makes possible. It is difficult, indeed, to use the stem of a daisy as a nail or to bounce a solid lead square.

In order to avoid the pitfall of dualism which places the mind and body, the individual and society, the knower and the known, and the organism and environment in antithetical and contradictory realms, pragmatism looks at these contrasts in a *transactional* manner.

> If interaction assumes the organism and the environmental objects to be present separate forms of existence...prior to their entry into joint as investigation, then transaction assumes no pre-knowledge of either organism or environment alone as common system... (Dewey and Bentley 1949: 123)

In neurosociological terms, pertaining to the dualism between concrete body and intangible mind, how do we get from the electrochemical, synaptic connections within the brain to the direct but intangible experience we are conscious of here and now?

Neuroscience has produced strong evidence that we cannot copy the world the way it is. First, our five senses cannot possibly register all of the multitudinous possible stimuli in the "obdurate world" to use the phrase of symbolic interactionist's major theorist, Herbert Blumer. Buckminster Fuller warned that what humans can sense is "less than a millionth of reality" (Christian 1977).

But more than that, even if we could sense these things, our brain-given senses are *transducers*. In terms of sight, Blumer's obdurate world turns out to be made up of electronic waves of different lengths traveling at a speed of 186,000 miles per hour (Christian 1977: 193). The eyes of the brain change electronic waves into a form to which the brain can accommodate. For example, the computer program that changed WordPerfect to Microsoft Word was a transducer. The brain is highly revisionist because we can never see the world simply as light waves, i.e., as the world is. What we can do is accurately perceive the obdurate world as the consequence of this brain/environment relationship. Thus, while we may have a *relational*, transactional view, we avoid the kind of relativism where, as Mead says, "anything goes." Arendt (1958: 236) stresses this revisionism in her quotation of Eddington that "whatever we perceive is as different from the incoming data as a telephone number is to its subscriber." Likewise, Mead spoke of this as the "objective reality of perspectives." Here the perspective is made up of what an animal brings to the environment in terms of sensory transducers (read *into*) and what the environment offers in terms of what can be successfully transduced (read out *from*). Since this "transactional slab" is objectively there in nature, and although it is relational, it still has a relational deterministic character that is not relative. This places important limits on constructionism that some interactionists ignore. They often see color as solely determined by the fact that different societies socialize their members to see different colors and to ignore others. But color is not completely socially constructed. We can readily learn to see other colors that our linguistic categories ignore. (See also Lakoff and Johnson 1999: 23–26.)

The revisionist character of the perception of color may be clearly seen when we consider that the electric spikes which different senses send to the brain are all the same. The experiential difference for the individual is the brain area to which the stimulus is sent (Franks 2003). A similar argument can be made for all the senses other than color. When I smell an odor, gaseous molecules are sent to the lining of the olfactory membranes of my nose. Here the same kind of argument must be made as that of color. There is no smell in the rose per se; fragrances are a transaction of odor messages and how they stimulate the lining of my nose. To say it is in the rose is a "semantic fallacy."

Another reason that a non-revisionist, correspondence theory fails is that each of our senses must provide an extreme simplification of what is "out there" independent of our will. The limited size of the brain only allows reception of a small portion of what is "out there." Each sensed image must be reduced by a factor of 100. Gazzaniga (1985) tells us that rather than incorporating each microscopic flash, the eye constructs a gestalt or a pattern of the "shape of things."

An excellent statement on the revisionist statement of perception is written by Wentworth and Ryan (1992: 29) about the vast number of processes the brain goes through in constructing the sight of a doll:

> Once in the brain, "feature extraction" takes place. Edges, horizontal and vertical lines, mappings of textures and shadings are abstracted from the impulse-coded doll and scattered across the visual cortex. This twice shattered doll has connections among several specialized areas and seconds, with the experience of the doll antedated so that no processing lag time is noticeable.

Consistent with the above, they state that "the brain is neither the passive mirror of the world nor its camera…. Like a master carpenter, the brain takes the blue print, tools and materials and brings them together into a dwelling." Then they add, "What a wonderful physiological foundation for social constructionism" (Wentworth and Ryan 1992: 30).

The Revisionist Nature of Memory

Daniel Schacter's revisionist characterization of memory is set forth in his classic, The Seven Sins of Memory (2001). We do not remember the actual act in long-term memory. What we remember is the last time we remembered something. Furthermore, every long-term past is a reconstructed past. Therefore we do not have direct memories of the original acts. Lindesmith et al. (1988: 137).have focused on the social construction of collective memory following Mead's theory of the symbolic nature of the past.

Our memory is very often tied to the self and/or our own interests. Again, Turner states that memory emotions tagged to cognition are stored in the hippocampus for 2 years and then shipped up to the frontal lobes where they are activated. He says that self is a proxy for emotions since cognitions involving self will almost always arouse them.

In principle, collective memories are no different when they construct self-serving memories of the group. Individually we remember those events that confirm or threaten the self or our groups. We remember events that produce emotion, and one of the things we are most emotional about is ourselves. We remember best those things that interest us. Thus, the seven sins are really about self-maintenance rather than sins. The past is remembered in self-enhancing ways.

In the late 1990s, considerable attention was drawn to the possibility of making persons believe they had been abused when they were not, or even thinking they had been perpetrators of abuse when they were not. These instances were caused by the inadvertent suggestions of therapists or other interrogators (See Loftus 1999). However one looks at memory, it is interpretive to the core.

Split-Brain Research as Evidence for the Validity of the Symbolic Interaction Theory of Accounts

This section explores one way that neuroscience can give empirical confirmation of SI hypotheses that would otherwise be impossible in that framework. The theory of accounts was developed in 1968 by Scott and Lyman. Here they diverged from the psychological notion of a motive as the "wellspring" of action and shifted emphasis to the more sociological notion of motives as words that we direct to others when we have done something untoward and need to "save face." Motives here are *imputed* by one's audience and *avowed* by the actor. They have little to do with the truth about one's private motives. Actors can, or cannot, believe their own accounts; they are part of facework to make their audiences accept our behavior.

They depend on the social context. Men will talk to other men about their girl friends in ways they would never do when talking to their girls. They are presenting different selves, one that is "on the make" and tough and one that is "in love, thoughtful, and tender".

Closely associated with accounts is *Techniques of Neutralization* by Sykes and Matza (1957). These techniques allow delinquents to "neutralize or justify their activities, which reduces conventional social controls and can allow their behavior to continue". Once again, they have little to do with the truth.

Neuroscientific Supports for Accounts

Michael Gazzaniga (1985) began his split-brain research by operating on patients suffering from extreme epilepsy. This involved severing the corpus callosum in half. This is a mass of 200 million cables in the middle of the brain that enables the two halves of the brain to communicate with each other. Such communication is needed because the two halves are capable of different things and thus complement each other. The left brain communicates linguistically in right-handed people, and the right brain is mute, communicating only by electronic means. The differences in the two hemispheres have been exaggerated in the past, but the fact remains that both are needed. When a written message is presented to the right hemisphere, the severed right side can no longer inform the conscious left side what it is doing. What is remarkable is that when patients whose corpus callosums have been severed are asked to draw a horse, they will start drawing one without any recognition of what they are doing until the final form of the horse emerges.

When patients were asked why they drew what they did, they always came up with a quick answer that made sense to them and that they thoroughly believed. Here the action occurred first, and then the verbal explanation followed. Of course, they were the only one who is deceived. Even more relevant to Scott and Lyman's sociology of accounts are two other experiments. In the first, the experimenters sent a message to the right brain of the patients that they should rise and leave the room. When they did so, they were asked why they left. The answers came quickly that they were going to get a coke or get some fresh air. They never said that they did not know why they left. In the second case, the right brain was told to laugh. When the subjects did so, they were asked why they were laughing, and again, they never said, "I don't know," but a reason is forthcoming that only they believe. The verbalizations the sociologists term "accounts" or "vocabulary of motives" are to neuroscientists "confabulations." The only recourse the sociologist has available to get evidence about the true reasons for this behavior would be to ask the subject, but this would be patently absurd since the subjects believed their own accounts. In the cases above, it was clear to the researchers that they did these things because they were asked. In this research, neuroscience provided empirical evidence for the existence of accounts that sociologists could not have obtained by their own methods.

Mirror Neurons and a Confirmation of the Pragmatic Priority of Action

Because the brain is the organ by which we know, anyone studying its implications unwittingly becomes an epistemologist. This is especially true if we seek, in Lakoff and Johnson's words, "an empirically responsible philosophy." In 1999 in their *Philosophy and the Flesh*, they argued that human language was largely metaphorical and that these metaphors involved our experience with bodily actions…. "Do you grasp what I mean?" "I hammered that home to them." "You are talking over my head or beneath me"—I won't bore you with other thousands of examples. But they also insisted that philosophy be informed by what we know of the mind and its embodiment through neuroscience. Because we all share bodily experiences, this increases intersubjectivity. Society is dependent on this for its very existence.

Mirror Neurons

As discussed in Chap. 14 by Liza Azziz-Zadeh and Sook-lei Liew, a mirror neuron fires when we watch others perform a certain action. But it does more than that. Unconsciously it actively simulates, or copies, other's actions on the observer's motor cortex. In an important sense, we do behaviorally what we watch. These neurons are inherently social and directed toward the actions of others; they also actively combine motor activity and perception. Furthermore, and most important, mirror neurons fire when the action observed can be seen as *intentional on the part of the observed actor*, thus preparing us for the lightening quick responses which are made necessary by the speed of social interaction. (Some reservations of how this reflects intentionality can be found in Churchland 2011: 139–143.)

For researchers in mirror neurons, the relation between our motor intentions and the way we perceive the world is an "affordance." An affordance is something that the environment offers to an animal's behavior such as surfaces that provide support, objects to be manipulated, and substances that can be eaten. Interestingly, the old Chicago pragmatists used the very same word. Researchers Rizzolatti and Singigaglia (2008) argue that the visual perception of an object implies the immediate and automatic selection of those properties that facilitate our behavioral interaction with it. For these neuroscientists, we perceive the world not so much as "it is" in some final sense but in terms of how it answers to or facilitates our actions.

This means once again that what we sense and reason about is limited. The environment becomes constructed and objectified only in relation to the individual's limited motor capacities, senses, and brains. The German term *Umwelt* captures this organism/environment relation very well. It refers to the world carved out for our attention by our own capacities, sensitivities, and motor repertoires (Becker 1964).

The research on mirror neurons implies the existence and relevance of just this social behaviorist view. The mirror neurons of dancers responded most strongly to observations of familiar dance steps in line with the style of dancing the observers routinely used. More pointedly, when human subjects were shown videos of motor behaviors of different animals communicating, like monkeys lip-smacking, dogs barking, and humans talking, mirror neurons became engaged only at the observation of humans talking, even though visually it was obvious that these all were communications with moving mouths. Since we have no spaces in our motor cortex for barking and lip-smacking, we cannot simulate them.

Mirror Neurons and Speech

Here again we find supportive evidence for the early pragmatist's program. As we have seen, the *products* of speech and human symbolic thought are *extra*sensory, like time and space, or the notion of redness in and of itself, independent of any actual red object. But the *process* of speech is behavioral. An important tenet of social behaviorism, and thus foundational symbolic interaction, is that as humans solve problems they are really talking to themselves, making subliminal movements with those muscles associated with speech like the tongue, lips, and larynx. Deaf children privately thinking to themselves likewise make subliminal hand and body gestures of sign language that underlie their symbolic thought. The congenitally blind, who have never seen gestures involved in sign language, nonetheless make motor movements when they talk. Gestures are not ancillary to talk but an inherent part of it just as grammar is. We even gesture on the phone with no one to see us.

How do mirror neurons enter the picture at this point? Once again, we do not just passively hear talk—we actively *do* the talk, in this case, the talk of other people. This is because talk is behavioral. We perceive talk not so much as sounds directly through our ears, but we *comprehend* it more indirectly through making the same speech movements ourselves in the motor cortex (Iacoboni 2008: 103). We simultaneously enact the same speech movements others are making. Their talk literally becomes our talk with the provision that it is automatic and unconscious.

As seen in this volume, more than several social neuroscientists see mirror neurons as capturing the social intent of others. This position holds that concepts are represented in the brain by the same sensory–motor circuitry on which the concept relies. The general *concept* of grasping would be represented by increased activity in the sensory motor areas devoted to grasping *actions*. Some of these neuroscientists include Aziz-Zadeh and Damasio (2008), and Gibbs (2006). They also describe the last 20 years of research that gives evidence that language is biologically linked to action through mirror neurons in a way congruent with the priority given to action by the pragmatist social behaviorists. They insist that language is not separate from motor behavior but rather inherently grounded in actions. This is an important part of Mead's social behaviorism. Neuroscientific research provides convergent validity here and is a powerful tool to strengthen evidence for the validity of the perspective of social behaviorism or the "embodied semantic hypothesis."

Mead's Four Stages of the Act

Mead's four stages of the act are more in line with mirror neuron research. First, comes action as some impulse to behave toward an object, and second, what we perceive most clearly is what answers to or facilitates our interests and intentions to act. As Damasio wrote in his *Descartes' Error* (1994), "Perception is more involved with action than we think. Perceiving is as much acting on the environment as it is about receiving signals from it." All perception is selective and our behavioral intentions do the selecting. Books on the shelf that we have ignored for years come immediately and clearly into sight as we reach for them.

The next stage of the act is manipulation—*doing* something with the object or, if the act involves people, taking the role of the other in social interaction. Last, we have consummations. Here consummation reaches back teleologically to pull the different stages into being. This is teleological because the later stages are present in the first stages and control the others. The whole framework is voluntaristic because in the first stage, the image of consummation is symbolic and open to the actor's change. This also contrasts with a version of reinforcement theory wherein the stimulus comes first, we then perceive it, we next think (at least sometimes), and last, we act in a way dictated by our conscious thought.

Conclusion

In this chapter, I have shown how neuroscience can be helpful to perspectives or fields that some might (wrongly) consider diametrically opposed to it—foundational social behaviorism and even symbolic interaction in some cases. I have also shown how the neuroscience of mirror neurons confirms the pragmatic metatheories of Dewey and Mead. This is a prime example of David Maines' rebuttal to Denzin that "if the new is good because it's new and if the old is bad because its boring, then what do we do if we come up with something old that works?" Hopefully I have given enough evidence to illustrate that pragmatism works as far as mirror neurons are concerned. Neuroscience additionally can provide helpful critiques of the perspective of symbolic interaction, in this case, creating a balanced approach to social constructionism and relativism.

References

Anderson, J. R. (1983). *The architecture of cognition*. Harvard University Press.

Arendt, H. (1958). *The human condition*. Double Day Anchor Books.

Aziz-Zadeh, L., & Damasio, A. (2008). Embodied semantics for actions: Findings from functional Bain imaging. *The Journal of Physiology, 102*(1–3), 335–376.

Becker, E. (1964). *The revolution in psychiatry: The new understanding of man*. The Free Press of Glencoe.

Christian, J. (1977). *Philosophy an introduction to the art of wondering* (2nd ed.). Holt, Rhineheart and Winston.

Churchland, P. (2011). *Braintrust: What neuroscience tells us about morality*. Princeton University Press.

Damasio, A. (1994). *Descartes' error: Emotion, reason and the human brain*. Avon books NY.

Denzin, N. (1992). *Symbolic interaction and cultural studies: The politics of interpretation*. Blackwell.

Dewey, J., & Bentley, A. (1949). *The knower and the known*. Beacon.

Franks, D. (2003). Mutual interests, different lenses: Current neuroscience and symbolic interaction. *Symbolic Interaction, 26*(4), 613–630.

Franks, D. (2007). Neurosociology. In G. Ritzer (Ed.), *Blackwell encyclopedia of sociology* (p. 3185). Blackwell Publishing.

Gazzaniga, M. (1985). *The social brain: Discovering the networks of the mind*. Basic Books.

Gazzaniga, M. (1998). *The Mind's Past*. University of California Press.

Gibbs, R. (2006). *Embodiment and cognitive science*. Cambridge University Press.

Goldstein, K. (1948). *Language and language disturbances: Aphasic symptom complexes and their significance for medicine and theory of language*. Grune and Sutton.

Head, H. (1926). *Aphasia and kindred disorders of speech* (Vol. 1). Macmillian.

Iacoboni, M. (2008). *Mirroring people: The new science of how we connect with others*. Farrar, Straus and Giroux.

Lakoff, G., & Johnson, M. (1999). *The philosophy of the flesh: The embodied mind and its challenge to western thought*. Basic Books.

Lindesmith, A., Strauss, A., & Denzin, N. (1988). *Social Psychology* (7th ed.). Prentice Hall. Inc..

Loftus, E. (1999). Creating false memories. In T. S. American (Ed.), *The scientific American book of the brain*. Lyons Press.

Luria, A. R. (1976). *Cognitive development: Its culture and social foundations* (M. Martin-Morillas & L. Solotaroff, Trans., M. Cole Ed.). .: Harvard University Press.

Maines, D. (1995). *Some Comments on postmodernism: The so-called new interpretive turn*. Paper presented at the annual meeting of the Midwest Sociological Society.

McPhail, C., & Rexroat, C. (1979). Mead VS Blumer: The divergent methodological perspectives of social behaviorism and symbolic interactionism. *American Sociological Review, 44*(June), 449–467.

Mead, G. H. (1934). *Mind, self and society: From the standpoint of a social behaviorist*. University of Chicago Press.

Mead, G. H. (1938). *The philosophy of the act*. University of Chicago Press.

Mead, G. H. (1959). *The philosophy of the present: The Paul Carus lectures*. The Open Court Publishing Co.

Northrop, F. S. C. (1948). *The logic of the sciences and the humanities*. Macmillan.

Rizzolatti, G., & Singigalia, C. (2008). *Mirrors in the brain: How our minds share actions and emotions*. Oxford University Press.

Scheff, T. J. (1990). *Microsociology: Discourse, emotion and social structure*. The University of Chicago Press.

Stryker, S. (1985). Symbolic interaction and role theory. In G. Lindzey & E. Aronson (Eds.), *The handbook of social psychology* (pp. 311–378). Random House.

Sykes, G., and Matza D. (1957). "Techniques of Neutralization: A Theory of Delinquency." *American Sociological Review, 22*, 664–670

Wentworth, W. M., & Ryan, J. (1992). Balancing mind body and culture: The place of emotion in social life. In D. D. Franks & V. Gecas (Eds.), *Social perspectives on emotion* (pp. 25–46). JAI Press.

The Origin of the Human Mind: Tracing the Evolutionary Steps

11

Alexandra Maryanski

As natural selection acts solely by accumulating slight, successive, favorable variations, it can produce no great or sudden modifications; it can act only by very short and slow steps.

Charles Darwin, 1859, *On the Origin of Species.*

Abstract

Why is the human brain so large and complex? The brain's wrinkled folding does not easily reveal its secrets, but it carries its development in its anatomy and little of it involved adaptations to the modern world. In the light of evolution, what was put together and when? Large brains are metabolically costly, so its subcortical and neocortical enhancements must reflect formidable challenges in past habitats. What, then, were the challenges that selection had to overcome? What pressures created a rare self- identity, and a social mind with a cooperative intelligence that makes large-scale societies filled with millions of strangers possible? The evolutionary record tells us that the human brain was configured in a variegated mosaic pattern by a series of far-reaching evolutionary novelties. Hence, notions that the human brain is a "gift of the gods," a result of large mutations, or that "human nature" is a product of mental molecules or mechanisms that evolved during the Pleistocene are misplaced. In reality, as this essay will document, our brain and its inherent proclivities are the outcome of over *65 million years* of primate evolution.

A. Maryanski (✉)
Department of Sociology, University of California, Riverside, Riverside, CA, USA
e-mail: Alexandra.maryanski@ucr.edu

Introduction

Why do humans have such large and complex brains? Our brain does not easily reveal its secrets, for unlike other body parts, we cannot feel our brains, nor easily see through its wrinkled folds. Humans have bigger brains than most mammals— on average about 1300 cubic centimeters (cc's) along with nearly 90 billion nerve cells or neurons. All our bodily processes and behaviors stem from this neural activity. Our brain weights about 3 pounds (1.4 kg or about 3% of average body weight), never shuts down, relies on glucose as fuel, and uses about 20% of metabolic energy daily. Imaging technology and neuroscientists have yielded profound insights, but the brain remains a mystery beneath the surface (DeCasien et al., 2022).

W. Kalkhoff et al. (eds.), *Handbook of Neurosociology*, Handbooks of Sociology and Social Research, https://doi.org/10.1007/978-3-031-95615-7_11

What of the mind? For George Herbert Mead ([1934]1974), the mind is not an entity but a process, a brain in action or a behavioral capacity (e.g. perception, reasoning). Natural selection is a conservative and directed force, so the encephalization[1] of our brain, and what Mead (p. 149) called a "mental content" or a *self* must reflect intense selection pressures in past environments where survival and reproductive success were problematic.[2] What, then, were the challenges that selection had to overcome? What pressures created a *social mind* that can engage in both strong and *weak* tie social relations, thereby generating a cooperative intelligence that makes large-scale societies filled with millions of strangers possible?[3]

The goal of this chapter is to broaden the scope of neurosociology[4] by *not* addressing how the brain interprets or shapes outcomes, or what it entails, but by demystifying more of its secrets by asking what made it all possible in the first place. What foundational milestones account for its remarkable complexity and disproportionately large neocortex? What proclivities are hard-wired? When did a self-identity arise along with other facets of human nature? In the light of evolution, *what* was put together and *when?*

As brain anatomy reflects brain evolution, little of which evolved as adaptations to the modern world, one portal into its many secrets is to trace its evolutionary history. As Sergio Almécija et al. (2021: 1) underscored, the majority of "human features" are primitive traits inherited from primate…ancestors. Even humans' distinctively large brains and delayed maturation are framed within a primate trend of increased encephalization and slower life history…."

As Old-World monkeys and apes share most of our phyletic history, with chimpanzees sharing 98% of human DNA (and nearly as much with gorillas and orangutans), answers can be found in primate evolution. The basic argument can be summarized as follows: The fossil and molecular records tell us that the human brain was assembled in a variegated mosaic pattern during the course of 65 years of primate evolution. And the evidence points to a series of watershed events that ultimately shaped our cognitive software: (a) a sensory revolution to visual dominance, (b) a legacy of life history characteristics, (c) a catastrophic hominoid extinction, (d) a shift in trans-

[1]Cranial capacity refers to the volume of the inside of the skull, an indicator of brain size. The unit of measure is cubic centimeters (or CC's).The relationship between brain size and body size for most species are closely correlated, so we would expect larger animals to have bigger brains than smaller ones. The encephalization quotient (EQ) is the ratio of observed brain size to the predicated size given the overall body size of a species.

[2]It would be a misconception to assume that all human traits are beneficial or adaptive. Yet, as Darwin and Wallace discovered, selection is a directed (but mindless) force that acts on traits with adaptive value *in a given environment*. Selection is the architect of evolution, but it can only act on the genetic variation in a gene pool generated by the *random* forces of evolution— mutation, genetic drift, and gene flow. The primate, hominoid, and hominin traits sketched in this essay likely stem largely from selection forces, but also by tradeoffs and even traits not driven by selection (e.g. the flux of genetic drift), so that some adaptations are just barely good enough for survival and reproductive success. While paleoanthropology and paleoneurology entertain competing hypotheses, lots of phylogenetic uncertainties, and longstanding controversies, this chapter rests on the best supported knowledge to date.

[3]Most social mammals use genetic-based or kinship ties for interpersonal cooperation that restricts organizational size. Macro-societies with millions of members and a complex division of labor are rare and limited to two taxonomic orders—insects and human primates. Insect societies (e.g., ants) use tactile and olfactory chemical cues for cooperative interactions among different caste members—a kind of "personal anonymity." But humans can organize on many levels— from small kinship-based societies to large-scale societies with some strong ties but mostly weak ties. For an insightful analysis, see Richard Machalek (1992).

[4]Pioneer scholars in this new subfield are Warren TenHoughten and Charles Kaplin (1973). Neurosociology became a formal subfield in the 1990s (see Franks & Smith, 1999). Inherently interdisciplinary, neurosociology addresses such abiding questions as to how the brain shapes perception and behavior, the functions of various brain structures, the emergent self, the social nature of the brain, human nature, how culture and social structure shape the brain, and how the brain shapes reality. Key objectives are to create a unifying research program by advancing sociological theory, as well as developing useful practical applications (for discussions see (Franks, 2010; Hopcroft, 2013; Hammond, 2013; Kalkhoff et al., 2016; Turner & Maryanski, 2024; Brashears, 2013).

port, (e) and an open-country niche. To appreciate their foundational implications for human neural development, we need to begin with early primate evolution because *most* facets of human neocortex are not special to humans at all but *hallmarks* of primate evolution in general. As Walter Hartwig (2007: 31) phrased it,

> [P]rimates are a radiation of mammals in which growth of structures related to the brain has taken precedence over growth of structures related to the body. Comparatively speaking, other lineages of mammals have "spent" their genetic variation on more specialized biological adaptations…For primates, the evolutionary path has stayed a general course in which neural development, rather than somatic, has characterized the emergence of new taxa.

Hence, notions that the human brain was a "gift of the gods," created by large random mutations, or that "human nature" was configured by mental molecules or mechanisms that evolved during the Pleistocene, as argued by some evolutionary psychologists, are misguided. In reality, *the human brain and its inherent proclivities are the product of 65 million years of primate evolution.* But before venturing into a "stranger than fiction" history that calls up images of the *Rise of the Planet of the Apes,"* let me begin with a little background on living primates for readers unfamiliar with the primate order.

Living Primates and the Early Evolution of Primates

The Present-Day Primate Order

The primate order has about 400+ living species divided up into two suborders: *Prosimii* or prosimians (descendants of early primates— the lemurs and lorises or *Strepsirrhines*), and *Anthropoidea* or anthropoids (monkeys, apes and humans or *Haplorhines).* The anthropoids comprise three superfamilies (1) *Ceboidea* or New World monkeys (in Central and South America), (2) *Cercopithecoidea* or Old-World monkeys (in Africa, Asia, and Europe) and, (3) *Hominoidea* or apes and humans (in Asia, Africa, and global for humans). New World monkeys and Old-World monkeys are identified mostly by their location and distinctive differences in nostrils, whereas apes and humans (or hominoids) are usually identified by their large body size (compared to monkeys) lack of tails, and a novel skeletal structure. Most living primate *species* are monkeys. Ape and human *species* are 5% of primate *species*, a tiny hominoid family that includes the little gibbons and siamangs (*Hylobatidae),* the three great apes—orangutans (*Pongo*), gorillas (*Gorilla*), chimpanzees (*Pan*) and humans (*Homo*).

Most primates live in year-round societies. For highly intelligent animals that are socially inclined, mature slowly, and live a long time, such permanent living arrangements require the coordination and cooperation of adult males, females, and varying ages of infants, juveniles, and adolescents. Females have a long gestation for mammals, give birth to one infant at a time (multiple births are rare), followed by a long socialization process where youth learn to (a) control their emotions, (b) follow societal rules, (c) and elicit proper social behaviors. As Daniel Gebo (2014: 181) remarked, "Living primates represent a diverse group of mammals…*As a group they represent the ancestral foundation for all human biology* (emphasis added).

Early Primate Evolutionary Trends

The primate order was launched when a shrew-like animal took to the trees in the late Cretaceous (circa 65 Mya) (Pozzi et al., 2014). Selection forces than acted to convert a mammal physically and cognitively adapted to ground-living to the ultra-extreme challenges of tree-living. One early tack was paring a dependence on olfaction, which is a superior sense for locating objects in a ground-living niche, but in a tree-living niche invites a misstep and death by gravity. Hence, selection acted early in primate evolution to enhance vision by converting light sensitive photoreceptor neurons from a mostly two-dimensional world of length and breadth into a three-dimensional world of air-space and depth. And it is this early swing away from smell to vision that would prove foundational for the later expansion of the primate neocortex.

Proto-primates or *Plesiadapiforms* appear in the fossil record in the late Cretaceous during a warm period in Europe and North America (especially in Montana and Wyoming (Manitilla et al., 2021). These small-bodied proto-primates (with about 11 diverse families and 150 + species) had bushy tails, short arms and legs, and a quadrupedal gait. Cranial materials represented by *Carpolestes simpsoni* show they had small brains, eyes to the sides of the skull, and an imposing snout, but the olfactory bulb was slightly smaller with an uptick in cortical tissue over other Paleocene mammals. *Carpolestes* also possessed a primate-like grasping foot and a nail (instead of a claw) on its big toe (for discussions on proto-primates see Granotosky & Young, 2023; Godinot, 2015; Manitilla et al., 2021; Bloch & Boyer, 2007; Ross & Martin, 2007).

True primates, or ancestral prosimians appear in China, Europe, and North America at the Paleocene-Eocene boundary (circa 56–34 Mya) when the earth was covered in vast forests. Two well-represented superfamilies are *Adapoidea* with 46 genera and 105 species, who resemble living lemurs, and *Omomyoidea* with 120 species who may be ancestral to monkeys, apes and humans (Godinot, 2015). At this stage, both superfamilies had evolved a bony rim to protect the eyes, and neocortical enhancements that included tissue expansions in the occipital area (visual) and parietal era (general somatic and motor) for more adept grasping, balance, and hand-to-mouth feeding guided by vision. By the Eocene epoch, primate eyes had moved to a forward position for three-dimensional vision, nails replaced claws on most digits, and sensitive soft tactile pads had evolved to put prosimians more in touch (pun intended) with their environment. Skeletal remains show they leaped and hopped, a locomotor pattern that would require visual acuity for propelled movement forward and upward. Thus, near the beginning of primate evolution, the olfactory bulb had decreased a little with a corresponding increase in vision, with enhancements in motor and tactile skills for survival in a precarious three-dimensional world of length, breadth, and depth. (Granotosky & Young, 2023; Godinot, 2015; Fleagle, 2013; Silcox & Lopez-Tórres, 2023; Kemp, 2024).

Anthropoid Origins: A New Kind of Primate

By the end of the Eocene (circa 34 Mya), the earth had cooled, and by the Oligocene (circa 34–23 Mya) the lush rainforests had transitioned into dry grasslands in the Northern regions, and many prosimian species disappeared. The Southern regions, however, stayed warm and ushered in new primates—basal anthropoids (i.e., ancestral to monkeys, apes, and humans).

The Fayum Depression near Egypt is a desert today, but it was once a jungly rainforest with meandering streams and an abundance of plants and animals (Harrison, 2013). Prosimians are in the Fayum, but most primates are basal anthropoids represented by two families: the *Parapithecidae* and the *Propliopithecidae.* The *Parapithecidae* typified by the genus *Apidium*, is likely ancestral to New World monkeys because their dentition resembles *both* prosimians and New World monkeys. *The Propliopithecidae* represented *by* the genus *Aegyptopithecus (*circa 34 Mya*) is likely ancestral to Old World monkeys, apes, and humans*. In looks, *Aegyptopithecus* is a hybrid with a jumble of prosimian and monkey-ape features. It was tree-living, weighted about 13 pounds, and a quadrupedal gait. Brain volume is relatively small at 32 cc's but the frontal lobes are heightened and suggestive of more cognitive complexity (Radinsky, 1973). The olfactory bulb is reduced and the visual cortex is clearly enhanced. *Aegyptopithecus* also had the dental formula (number and types of teeth) of all living Old-World monkeys, apes, and humans (indicating their phyletic closeness). And, significant for tracing phyletic evolution, it had five cone-shaped cusps on its lower molars—a unique character trait found only in ancestral hominoids, living apes, and humans[5] (Begun, 2023—also see footnote 12).

Aegytopithecus is a pivotal early anthropoid because it signals a qualitative shift towards more cortical-centered behavior with decreased olfactory receptors that project to limbic zones, and increased visual receptors that project to neocortical zones, a structure geared towards understanding a variable and changing external environment. And this evolutionary trend was to culminate in the capacity of *Homo sapiens* to represent the world symbolically (for discussions see Falk, 2015; Harrison, 2013; Begun, 2023).

Cortical Trends in Primate Evolution

Prosimians reigned supreme for millions of years until the Oligocene epoch (circa 34–23 Ma). Global cooling and vanishing forests account for their extinction in Northern zones, but in Southern tropical zones the forests did not disappear but became concentrated near the equator. Today, prosimian descendants are confined to Madagascar or occupy nocturnal niches in Asia and Africa. Many reasons account for why prosimians lost their competitive edge but niche overlap with anthropoids surely played a key role.

The *Competitive Exclusion Principle* states that two species cannot coexist if they rely on the same resources, so even a slight uptick in sensory equipment for procuring resources would constitute a competitive edge. Cranial endocasts show that ancestral prosimians had much smaller frontal lobes (for higher cognitive functions) than early anthropoids like *Aegyptopithecus*. While the cortical networks of living prosimians (i.e. lemurs and lorises) are expanded over their ancestors, their wet nose (similar to a dog) and specialized glands used to directly deposit odors or "scent signals" for marking territory and conspecific communication are unmistakable signs that their ancestors used olfaction to some degree to perceive the world and communicate (Gebo, 2014; Napier & Napier, 1985; Begun, 2023).

While the sense modalities are crafted to work together and to respond to particular stimuli, it is physically impossible for even the most complex and sophisticated sensory equipment to signal all possible environmental stimuli. This means that an organism's "software" and, notably, the modality that takes the leading role reflects the adaptive strategies of its ancestors. For example, most mammals cannot detect ultraviolet light. An exception are Arctic reindeer (Rangifer) who live in a relatively rich UV light environment with rods and cones sensitive to ultraviolet light. With selection pressure that included short-wavelength vision, Arctic reindeer were able to adapt to the extreme seasonal differences in levels of light (Hogg et al., 2011). So, the sensory equipment installed by selection is all about adaptation or die.

In a comparison of smell receptors (with chemical properties) versus visual receptors (with physical properties as they capture light off objects), they are organized differently and project differently, with olfaction housed near the limbic lobe of subcortical tissue and vision housed in the occipital lobe of neocortical tissue.[6] Hence, the primate shift to visual dominance is

[5] Teeth are of crucial importance as they are purely under genetic control and help to identify who's related to whom in primate evolution. All mammals have incisors, canines, premolars and molars. A dental formula is an abbreviated way of denoting the number and types of teeth in each quadrant of the mouth. A primate with a 2.1.3.3. dental formula — prosimians and New World monkeys— means they have 2 incisors, 1 canine, 3 premolars and 3 molars in *each* quadrant or 36 teeth. One with a 2.1.2.3. dental formula has 32 teeth as do Old World monkeys, apes and humans. Dentition is also used to identify habitat, body size, diet, and life History characteristics (to be discussed) and even migration patterns. Also see footnote 12.

[6] All mammals have a neocortex, a part of the cerebral cortex that appeared on top of the early mammalian forebrain. Four sections or lobes became distinguished—frontal, temporal, parietal, and occipital, with tracts of cortical tissue given over for sensory perception. The visual, auditory, and general somatic zones are lodged respectively in the occipital, temporal, and parietal lobes, with each specialized to extract particular environmental features when activated by visual (light waves), acoustic (sound waves) or tactile (touch). Small- brained mammals have only a little neocortex, but in primates the neocortex expanded and in humans it takes up about 80% of brain volume (Kaas, 2011).

far more than a sensory shakeup. It represented an extraordinary leap away from subcortical olfactory fields, involved more with survival and the emotional centers of the brain, toward neocortical fields involved more with understanding a variable and changing world with much higher cognitive functioning. No direct evidence is available to tell us when this revolutionary change occurred, but an approximate timeline is important and can be estimated by employing the historical comparative method or cladistics, a standard tool used in textual criticism, historical linguistics, and comparative biology.

The Sensory Revolution to Visual Dominance: A Cladistic Analysis

Molecular data indicate that the *Last Common Ancestor* (LCA) of Old-World monkeys, apes, and humans lived during the late Oligocene or Early Miocene (circa 30–24 Mya). Using cladistics, we can bring this ancestor back to life, or at least its sensory system. The logic behind this remarkable discovery procedure is that an "original" or last common ancestor (LCA) of a closely related set of entities can be reconstructed by identifying "evolutionary novelties" or clusters of character traits that are uniquely shared by its descendants. For example, in historical linguistics this method is used to identify cognates shared by a set of related languages with the goal of reconstructing the mother "speech forms" from which the daughter languages were derived. Or in textual criticism, alternative forms of an original manuscript (e.g., the Bible) are the related entities, and the goal is to reconstruct the original text from which the later texts were copied. In biogeography and evolutionary biology, the procedure is essentially the same but is called cladistics (for discussions see Jeffers & Lehiste, 1970; Platnick & Cameron, 1977; Forey et al., 1992, and Maas, 1958).

To assure objectivity, cladistics relies on two sources of independent confirmation: a *relatedness hypothesis* to ascertain that the shared "evolutionary novelties" are not due to chance but the outcome of descent from a common ancestor, and a *regularity hypothesis* to ascertain that the modifications from ancestor to descendants follow a regularity or a systemic bias to link the related groups to each other and to their last common ancestor (Jeffers & Lehiste, 1970: 17–25). As our goal is to date past events and put them in chronological order, the timeline for when vision replaced smell is critical as it was the driving force for further cortical expansion. So, we will need extant prosimians and New World monkeys to serve as our control groups because they are closely related sister taxa and make ideal benchmarks for identifying what distinctive "evolutionary novelties" were inherited by Old-World monkeys and apes/humans *before they split from the mother population into two lineages* in the late Oligocene or early Miocene.[7]

A truly surprising discovery is the sheer sophistication of the LCA sensory equipment so early in anthropoid evolution. Another is that it stayed nearly intact in living Old World monkeys, apes and humans *after* 25 million years of anthropoid evolution. Since sensory inputs effect the brain's "functional architecture" and serve as the gateway to awareness, let me provide a sample of selection's remarkable handiwork for each major sense modality, and especially for how selection revolutionized the visual system (Kirk, 2006: 76; and Purvis et al., 1996).

The Olfactory and Gustatory Modalities

For most mammals, the olfactory modality is dominant for locating objects in space because it is self-alerting, transmits information actively or

[7]The phyletic relationship between New World monkeys (the *Platyrrhines*) and Old-World monkeys, apes and humans (the *Catarrhines*) is an enigma. Did New World monkeys evolve from ancestral prosimians that once lived in North America? Or did they evolve from ancestral Old-World primates? If so, how did they get to Central and South America? Geological Data now support the hypothesis that a landmass broke away from the African continent and the monkeys rafted on "floating islands" to South America sometime during the Oligocene (see Seiffert et al., 2020). Given their African origin, it was prudent to include them.

passively, and has superior long-distance properties for detecting the lingering odors of predators and prey (Hoover, 2010). But in anthropoids it plays a minor role when compared to prosimians.[8] For example, the vomeronasal organ (VNO), a chemoreceptor accessory to the olfactory that is used to detect pheromones and other chemosensory functions, is well developed in extant prosimians and to a lesser degree in New World monkeys but is merely a vestige organ in Old World monkeys, apes, and humans. In timing the loss of the VNO, Timothy Smith et al. (2011, 2168) concluded that the "VNO and other olfactory structures indicate a complex pattern of reduction…Taken together, anatomical and genetic data [support] a hypothesis that the catarrhine VNS became non-functional prior to the divergence of the cercopithecoid [Old World Monkey] and hominoid lineages." However, Old World anthropoids (including humans) still use olfaction for sexually related functions, avoiding toxic chemicals, judging the tastefulness of foods, and for humans to relight old memories. (for discussions see Stein & Rowe, 2011; Rouquier & Giorgi, 2007; Gilad et al., 2003).

The gustatory organ located on the tongue and palate often works with smell and is essential for controlling and maintaining food intake, triggering taste-induced sensations of pleasure or displeasure, and for assuring nutritional energy. What effect might a reduced olfactory gland have on the gustatory system? Fossil data are thin (as the gustatory is a soft tissue structure), but a pivotal change is an uptick in sweetness receptors. In short, simple sugary substances that tasted sweet to most primates were activated by *ancestral receptors.* In contrast, complex sugary substances were detected *only* by Old World anthropoids (Nofre et al., 1996: 761). For example, only Old-World monkeys, apes, and humans can perceive Aspartame as tasting sweet (Xia et al., 2011). Nofre et al. (1996, 761) concluded that Old World anthropoids comprise an evolutionary "compact group" and that they inherited this novelty during the Oligocene from the "ancestral stock of the Old-World simians" or early anthropoids (Glaser, 2007: 128; Nofre et al., 1996: 761).) They also propose that enhanced sweetness receptors were a major transition in Old World anthropoid evolution because it improved "food search efficiency and dietary choice…especially fruit, which could have favoured…mental development, and, later, the emergence of humans."[9]

The Auditory Modality

The auditory cortex in housed in the temporal lobes and functions in cortical and subcortical regions (Hackett, 2003, 2008; King & Nelken, 2009). Three properties of acoustic sounds are *wavelength, frequency,* and *amplitude.* Frequency effects pitch and tone quality; a higher frequency has a higher pitch than a lower frequency, and pitch aids in distinguishing between low and high sounds and sound quality (e.g., a pleasant or unpleasant tone). Early mammals evolved a sensitivity to high frequency sound waves, which are strongly directional and easily reflected, and it afforded a competitive edge for mammals (Heffner & Heffner, 2008; Heffner, 2004). Primates, however, underwent a steady *decrease* in high frequency sensitivity and a corresponding *increase* in low frequency sensitivity. Today, all living primates have low-frequency hearing but threshold varies along phylogenetic lines. Prosimians have much better high-frequency sensitivity and less low-frequency sensitivity than Old World monkeys, apes, and humans. New World monkeys have a low-frequency sensitivity that falls middle-range between prosimians on one side and Old-World monkeys, apes, and

[8] As a subcortical sensory organ, the olfactory bulb is located far from cortical regions (e.g., Broca's area) for language. This probably accounts for why humans have difficulty classifying and naming customary smells (Falk, 2000: 68). It is also why the aromas of familiar smells will evoke at times strong emotional sensations linked with subcortical zones.

[9] We can thank (or curse) our Oligocene ancestor for our novel sweetness receptors. It surely explains our fondness for rich fat-laden desserts and "junk foods" where sucrose or other refined sugars are the usual ingredient.

humans on the other, a significant variation that cannot be ascribed to body size differences.

Overall, Old-World monkeys, apes, and humans have similar acoustic profiles so they all hear a sound much in the same way (Coleman, 2009: 88). Given the small differences in their auditory sensitivity, it was surely in place before the split between Old World monkeys and apes/humans, with selection later favoring for the most part only slight auditory modifications. For as Hackett (2008: 775) relates, "there is both direct and indirect evidence that the major features of nonhuman primate auditory cortex organization are conserved in humans" (for discussions see Coleman & Colbert, 2010; Heffner, 2004; Coleman et al., 2010).

Why would selection favor a sensitivity to low frequency sound waves? In a rainforest with dense year-round foliage, heavy rainfall and thick humidity, higher frequencies are less valuable because they are strongly directional and fade quickly. In contrast, low frequency waves in a cluttered multi-canopied forest travel better around objects. As sounds in the natural world are mostly made by animals, low frequency sounds (like the slithering noise made by a snake) can be heard from longer distances. The primate auditory system is also automatically alerting so it can serve as a wide-ranging early detection system (see Masterton & Diamond, 1973: 431).

The auditory neural circuitry in anthropoids is also an efficient receptor for close range and long-distance communication. Of *great significance* is that in monkeys, apes, and humans *sounds* heard and *perceived* are fully under neocortical (or voluntary control) for higher cognitive processing. By contrast, in Old World monkeys and apes, *sounds made* by individual vocal tracts are largely (but not fully) emotionally based and controlled by subcortical structures. In short, monkeys and apes have cortically based hearing and perception, but relatively fixed *species-specific vocal calls.* Only in *hominin* evolution did selection modify the auditory cortex for voluntary individualized vocalizations—a *key preadaptation for speech in the Homo lineage.* (for discussion see Snowdon, 1990; Cheney & Seyfarth, 1980; and Maryanski, 1997).

The Haptic or Touch Modality

The somatic modality is a massive network of body receptors located in the primate skin (cutaneous inputs) and embedded in muscles, tendons, and joints (kinesthetic inputs) that transmit information to the parietal lobe for fine sensorimotor coordination. Haptic refers to "active" touch using the prehensile hands and feet to manipulate stationary objects for tactile impressions. All primates actively touch objects by using sensitive fingertip receptors and most have an opposable thumb and big toe for grasping. The human foot, however, no longer has a grasping toe because of the conversion to bipedal locomotion (for discussions see Ito et al., 2021; Gebo, 2014; Gibson, 1962, Lederman & Klatzky, 2009).

The primate line began with committed arborealists so grasping hands and feet were a crucial adaptation (Kaas, 2011; Wise, 2007). The fossil record documents these adaptations, but living primates can best interpret these evolutionary trends. Present-day prosimians show a heightening of motor and sensory zones that is reflected in their excellent grasping and climbing abilities and the use of prehensile hands for putting food into their mouth. However, anthropoid evolution involved an elaborate expansion of the posterior parietal cortex, which is reflected in their superior prehensile hands for detecting what lies under the surface of objects and exploring their environment. And, unlike prosimians who groom with a dental comb and toilet claw,[10] anthropoids use their prehensile hands (Napier & Napier, 1985: 78; Preuss, 2007: 18–19).

Tactile perception in primates is under cortical control for learned, voluntary, and intentional responses to environmental stimuli (Preuschoft & Preuschoft, 1994).[11] New World monkeys also

[10]A dental comb is composed of incisors and canines in the lower jaw of prosimians and compressed to form a scraper or comb. A toilet claw is a specialized claw or nail on the prosimian foot.

[11]Tactile perception in humans is the "reality sense" with vibration powers so acute that for deaf and blind individuals it can serve as a surrogate for many hearing functions (Keller, 1904). Haptic or "active touch" also has an inherent ability to recognize objects by processing temporal

have enhanced grasping skills compared to prosimians, but limited finger flexibility. By comparison, Old-World monkeys, apes, and humans inherited finely tuned prehensile hands and independent control of their fingers for delicate motions.

The Visual Modality

Vision is located mostly in the occipital lobe. Early in primate evolution, selection equipped prosimian orbits with a bony rim for protection and moved the eyes to a forward-facing position for a wider binocular affect and greater depth perception. In anthropoid evolution, selection enhanced the orbits with a postorbital closure (a complete bony cup) and heighted binocular vision to stereoscopic vision. Visual association areas were also expanded to retain information on the stability and distance of objects. One evolutionary novelty (absent in prosimians) is a depression in the macula of the retina for a finely tuned fovea for sharp, high-acuity vision (Ross & Martin, 2007). Another is augmented color vision. While the primate retina is activated by light-waves, the image formed depends on the density of rod and cone photoreceptors. Prosimians and New World monkeys have a dichromatic or two-cone color system, whereas Old World monkeys, apes, and humans have a trichromatic or three-cone system (Moritz et al., 2014; Jacobs, 2004). A three-cone system is far more sensitive to dim light and to contrasts in brightness for the detection of even slight color variations and the fine particulars of objects and scenes (Gebo, 2014; Campbell & Maffer, 1976; Preuss, 2007: Jacobs, 2007; Lucas et al., 2003).

sequential patterns. Individuals blind from birth can also rely on their prehensile hands to guide their movements in a bounded space by stringing together a succession of independent chucks of information to perceive a pattern or a structure of the whole by connecting the individual parts in a linear frame rather than a three-dimensional spatial frame (Von Senden, 1932/1960 and see Keller, 1904: 135 who wrote of this phenomenon). Yet, this remarkable capacity is rarely used unless vision is deficient (Freides, 1973: 302; Preuss, 2007).

Enhanced color vision is also useful for locating predators, especially at long distances (Dominy et al., 2004). It also introduced into the visual spectrum a "continuum of hues" for distinguishing, for example, young leaves with optimal nutrition from more mature leaves, or for distinguishing unripe fruits from ripe fruits. Vision is also central for interpreting the meaning of hand gestures, "body language," and for facial displays such as the "play face" (a precursor to human laughter) and, of course, by direct eye-to-eye contract. It may have also aided anthropoid social communication by adding a powerful perceptual dimension for the gathering and reading of conspecific facial cues, which are critical to "body language" (Hiramatsu et al., 2017; Vorobyev, 2004; Lucas et al., 2003).

Overall, vision is paramount for primate spatial information and is the major integrator for most environmental stimuli. It also guides the information pickup of the auditory and tactile senses, for in cases of intersensory conflict, visual inputs will usually determine what is being perceived (Conroy, 1990; Martin, 1990; Radinsky, 1973). Visual perception is also paramount for monkey and apes communication, and despite the important role played by speech, it is primary for humans as well. As Kass and Balaram (2014: 1) noted, "All primates, including humans, are highly visual creatures…the number, size, and complexity of brain structures involved in visual processing has expanded dramatically in the primate order, far more than those of any other species in the mammalian lineage."

The Relatedness and Regularity Hypothesis

To corroborate this cladistic analysis, we need to first entertain the null hypothesis that the Old World anthropoid sensory equipment was independently acquired by each monkey, ape, and human species after splitting from the LCA mother population. The null hypothesis cannot be ruled out, but improbable. The *Relatedness Hypothesis* is strongly supported by molecular and fossil data documenting that Old World monkeys/apes/humans shared a relatively recent LCA (in geological time). Second, distinctive differences exist in the sensory equipment of Old-World anthropoids when compared to our benchmark prosimians and New-World monkeys. Third, the sheer number of heritable traits or "evolutionary novelties" compared to our control groups, and the overall small variance in Old World anthropoid sensory equipment makes the null hypothesis of independent acquisition nearly impossible. As Jon Kass (2008: 385) expressed it: "if a feature or trait is present in all members of a group of phylogenetically related mammals, then it is parsimonious to infer that the common ancestor of the group also had that trait, and that it was retained in all the preserved lines of evolution from that ancestor."

The *Regularity hypothesis* requires that modifications *after* Old-World monkeys and apes/humans split into two lineages evidence a systemic regularity that is congruent with the traits of the Oligocene mother population. One difficulty is that, overall, Old-World anthropoids have minimal or no changes in their sensory equipment because of stable selection pressures on these traits. Apes and humans, however, have enrichments. And, in particular, humans have two major modifications with revolutionary consequences. One is the human foot with its partial loss of prehensile function (i.e., the opposable toe for grasping objects). On this, as Ito et al. (2021: 1–2) relate: "To adapt to habitual bipedal locomotion, the human foot has evolved significantly…The fact that the human foot is morphologically distinct from that of other primates indicates that the modification of the foot structure is vital to the evolution of human bipedal locomotion."

Another modification is the conversion of the human vocal channel from a mostly fixed species-specific call system under subcortical control to an open-ended vocal channel under cortical control. Simply put, this recasting was foundational for the evolution of human speech. To be sure, these refinements for humans are monumental (and discussed later), but in terms of strictly fulfilling the requirements of the regularity hypothesis, they all evidence a systemic regularity because their underlying elements are clearly traceable back to a shared anthropoid adaptation. We can therefore confidently conclude that the sensory architecture of Old-World monkeys, apes and humans was *not* acquired independently by each taxon but stems from an ancestral Oligocene mother population. When viewed in the light of evolution, then, the shift from smell to visual dominance took place early in anthropoid evolution and served as a *catalyst for further cortical expansion in the hominoid (ape and human) lineage.*

It is tempting to ask: who was this elusive LCA? We don't know yet but *Saadanius hijazensis*, an anthropoid that lived in Afro-Arabia between 28 and 30 million years ago offers clues into the LCA identity and the origins of Old-World monkeys, apes and humans. Substantial craniodental materials show *Saadanius* was intermediate between *Aegyptopithecus* and early Miocene monkeys and apes as its features are within the range of the predicated traits of the LCA. This confirms that the split between Old World monkeys and apes/humans took place between 29 and 27 Mya, a date in agreement with the molecular clock (Zalmout et al., 2010; Begun, 2023; Harrison, 2013; Fleagle et al., 2026: 345).

Although the mother population with its "avant-garde" sensory equipment withered away, it did so after spinning off two daughter lineages: Old World monkeys (*Cercopithecoidea*) and apes/humans (*Hominoidea*). What happened next is a classic "who-done-it" mystery with apes starting out as evolutionary blockbusters and ending up as evolutionary failures.

The Origin of Old-World Monkeys and Apes and the Legacy of Life History Characteristics

The early Miocene (circa 23–16 Ma) opened with rising temperatures and the arrival of new players on the evolutionary playing field—Early Old-World monkeys (*Cercopithecoidea)* represented by *Victoriapithecidae,* a small family of monkeys, and early hominoids (Proconsuloidea), represented by *Proconsul and Ekembo.*

Victoriapithecus is the likely ancestor of living Old World monkeys. In looks, this monkey weighted about 10 pounds, had a tail, a cranium size of about 54 cc's, a bilophodont cusp pattern that is a signature feature of living Old-World monkeys, and a quadrupedal gait.[12] *Victoriapithecus* is special because monkey species are rare during the early and middle Miocene. (Miller et al., 2009; Benefit, 1999; Dean & Leakey, 2004: 406; Fleagle et al., 2026: 379).

Proconsul and Ekembo were large-bodied tailless apes, with a Y-5 cusp pattern, the signature molar pattern of all ancestral apes, living apes and humans (see footnote 12). *Proconsul* had a cranial capacity of 160 cc's with an enhanced visual cortex and frontal lobe, whereas *Ekembo* had cognitive capacities comparable to the most intelligent present-day monkeys (Begun, 2023). Early apes were arboreal, had a dietary dependence on soft, ripe fruits, and during the "golden years" of the Early and Middle Miocene, so reproductively successful that they filled most niches in the vast African forests. Strangely enough, and important is that most early apes had a skeletal structure that resembled present-day monkeys with long flexible backs, quadrupedal gaits, and forelimbs positioned beneath their bodies.

However, co-existing with *Ekembo and Proconsul* were rare apes with a more modern ape-like anatomy, represented by *Morotopithecus bishop. Morotopithecus* had a short stiff back, a suspensory-like posture, and flexible forelimbs that allowed use of the prehensile hands for vertical climbing and below-branch arboreality (Walker et al., 1983; Ward, 2015; McNulty, 2015; Andrews, 2020; Urciuoli & Alba, 2023).

Early ape species varied in body size, but most were large-bodied when compared to monkeys. What advantage is there to a larger body size? For openers, bigger animals are less victimized by predators, especially in an arboreal habitat. And, with lots of food and little competition animals can grow larger and reach maturity later. Today, significant differences exist between Old World monkeys, apes and humans in such *Life History* Characteristics as gestation, growth, maturation, reproduction and longevity.[13] For example, a 50-pound adult baboon (a very large monkey) has a gestation phrase of 175 days, a nursing phase of 420 days, an infant phase of 1.6 years, a juvenile phase of 4.4 years and an adult phase of 23 years, with intervals of 1.7 years between single births. By comparison, a 115-pound adult chimpanzee has a gestation phase of 228 days, a nursing phase of 1460 days, an infant phase of 3 years, a juvenile phase of 7.0, and an adult phase of 34 years, with 5.6 years between single births. And this disparity is not purely a reflection of body size as the little gibbon ape at 15 pounds (who is smaller than many monkeys), has a gestation phrase of 205 days, a nursing phrase of 730 days, an infant phrase of 2 years, a juvenile phase of 6.5 years, and an adult phase of 23 years with 2.7 years between births (Wolpoff, 1999; Falk, 2000).[14]

These *Life History* differences must have evolved *after* monkeys and apes split from their anthropoid common ancestor. But how far back?

[12] I am emphasizing cusp patterns under genetic control because they are sometimes the only way to identify whether a fossil is a monkey or an ape. As discussed, *all* Old-World monkeys, apes and humans have a 2.1.2.3 dental formula or 32 teeth (see footnote 4). But Old-World monkeys have 4 cusps on the surface of their molars, each linked by a loph so they have a *bilophodont* molar pattern. In contrast, ape and human lower molars have 5-cusps on their molars. Recall that the hybrid *Aegyptopithecus* had a 5-cusp pattern or Y-5 pattern which is the more primitive molar configuration. The bilophodont configuration of Old-World monkeys is the more recent one and it may have originated with *Victoriapithecus.*

[13] The comparison here is only between Old World monkeys and apes. When compared with most mammals, monkeys have extended life history characteristics and live relativity long lives.

[14] In mammals, phylogeny and ecology seemingly play a larger role than body mass for extended life history characters. Following Healy et al., 2019, prolonged traits may evolve in an environment with a slower pace of life. For example, arboreal mammals live longer than those on the ground (also see Leigh & Blomquist, 2011). Of interest, the fact that little gibbons share extended life histories with apes and humans buttresses the finding that extended life history traits originated with early apes because gibbons branched away from a LCA with great apes/hominins between 18 and 20 million years ago.

We can answer this question by examining the teeth of young Miocene hominoids who died when their first molar cusps were emerging. Then, once the age of each ape is established, their fossilized molars can serve as a proxy for other traits because life history characters are systemic such that a long gestation period is coupled with prolonged time in the womb and nursing, an extended infant and juvenile phase, longer maturation, and a longer life. This procedure was done on a number of hominoid species including— *Afropithecus turkanensis* (circa 17.5 million years) and *Sivapithecus parvade* (circa 10 million years ago) with impressive results. The molars eruptions of Miocene apes all fall outside the range of living Old World monkeys and well within (or above) the mean of chimpanzees. As Kelley and Smith (2003, 326) relate: "This… finding…is compatible with the hypothesis that there was a shift to the prolonged life histories that characterize extant apes [and humans] early in the evolution of the Hominoidea."

This means that early in ape/proto-hominin evolution, extended *Life History Traits* were baked into hominoid phylogeny, probably as a consequence of a stable rain forest environment with little predation (Maclatchy, 2004; Healy et al., 2019). And prolonged life history traits are correlated with an uptick in brain size—and size matters. A larger brain with sophisticated cortically-based modalities guided by vision has more cortical tissue and can process more sensory information for greater perceptual awareness. In fact, scholars have argued that increased input in visual receptors has been a prime factor in the evolution of the primate brain (Kirk, 2006; Kass, 2007; Barton, 1998). In addition, slower maturation entails a much longer parental investment, a prolonged socialization period, and a longer life (Healy et al., 2019). Later, in *hominin* (human ancestor) evolution, these characteristics could be subjected to directional selection for as Potts (2004: 211) highlighted, they could serve as "a precondition for selection leading to brain enlargement and enhanced cognition."

Thus, early in ape/proto-hominin evolution, precursors for further encephalization of the neocortex were already in place: larger-sized bodies that correlate with larger brain size, slower gestation, longer nursing, and longer infancy and juvenile phases that would, if selected on, prepare the way for the evolution of the hominin mind.[15]

The Great Ape Die-off And A Rare Mutation with Fitness Value

The Downfall of the Apes

By the Middle Miocene (circa 15–11 Mya) the idyllic paradisial world of hominoids took a negative turn when Africa became cooler, leading to a decline in ape populations. But their fortunes greatly improved when a land bridge opened, allowing for a migration to the warmer, southern parts of Europe and Asia, which led to a large radiation and diversity of apes (Andrews & Kelley, 2007). Thus, hominoids enjoyed a second "golden age" and the forests literally became the "lands of the apes."

But over time, as the earth continued to cool, apes faced a catastrophic decline in numbers and species. By contrast, Old World monkeys, who were rare during most of the Miocene began to proliferate, gradually moving into the vacated niches of the dying apes (Fortelius & Hokkanen, 2001; Keller & Barron, 1987: 147; Pilbeam & Young, 2004; Potts, 2004).

Why did Old-World monkeys flourish and apes nearly die out? A detailed account is not possible here (but see Turner & Maryanski, 2008, 2024), but large-bodied apes with prolonged life history traits would be at a disadvantage in a colder climate with less resources. A long-standing hypothesis is that monkeys also gained a competitive dietary edge over apes by evolving a specialized digestive system that could effectively neutralize the toxic compounds in unripe fruits instead of waiting for fruits to ripen, as apes and humans must do today (Andrews, 1981;

[15]Hominin is the taxonomic term that includes all living humans, past humans, and all extinct relatives that walked upright (e.g., Australopithecus). It replaced the term hominid, which is now more inclusive and includes our closest living and extinct ape relatives.

Temerin & Cant, 1983; Hunt, 2016; Maclatchy et al., 2019).[16] Apes are frugivores, so a dietary disruption would have dire consequences. And so it came to be that by the late Miocene (11 to 7 Mya) most apes died en masse after millions of years of survival and great reproductive success (Kay & Ungar, 1997; Potts, 2004; Pilbeam & Young, 2004).

The Last Common Ancestor of Great Apes and Humans

How did the LCA of apes and humans survive this hominoid wipeout? Addressing this question is crucial because it altered the course of hominoid evolution. As mentioned earlier, most early apes like *Proconsul* embodied a mosaic mixture of characteristic ape features that included a lack of tail and a hominoid cusp pattern but had "monkey-like" limb proportions, a pronograde posture, and a four-limbed locomotor pattern similar to monkeys today. See Fig. 11.1 for a skeleton of a present-day Old World monkey to visually appreciate in a general sense the morphology of early apes. In contrast, rare early apes like *Morotopithecus* shared a loss of tail and a hominoid cusp pattern with their *Proconsul* cousins but they also had an "ape-like" post-cranial skeleton, an orthograde posture, and a locomotor pattern that allowed for some degree of suspensory arm-hanging for two-limbed travel. See Fig. 11.2 for a drawing of present-day great apes and humans that highlights their novel skeletal structure and, notably, an upper anatomy that evolved originally for two limbed arboreal travel. So, present-day apes and humans are the living descendants of a Miocene ancestor with an orthograde posture and arm-hanging skills. As Young et al. (2023: 1438) expressed it, "Humans can… brachiate and seem to have retained the locomotor capabilities of their arboreal ancestors."

What makes suspensory travel so strikingly original (and peculiar) it that it uses only the hands, arms and shoulders for support and propulsion. This requires (a) a strong finger flexion, for if the hand is to suspend the body like hanging fruit, it must be formed into a kind of hook, (b) a hand that can propel the body through space, and (c) an extreme range of supination made possible by a specialized wrist joint. Apes and humans have about 180 degrees of supination so that a body held by the prehensile hands can pivot at about the needed half circle for suspension (for discussions see Lewis 1974; Maclatchy et al., 2019; Oxnard, 1963: 166; Gregory, 1916, Nakatsukasa, 2008; Urciuoli & Alba, 2023; Maclatchy, 2004; Young et al., 2023).

Suspensory travel is not just noteworthy, it very likely contributed to the expansion of the ape/proto-hominin neocortex. To perform arm-rotational travel requires precise hand-eye coordination as the motor cortex must react quickly to what the visual cortex "sees" in sending signals to the brain. It also required finer cortical motor control for precise delicate movements compared to the stereotyped four-footed locomotion of *Proconsul* or *Victoriapithecus*. Moreover, it called for mastering individually acquired navigational skills, especially for the proficient calculation of egocentric distance requiring, in turn, greater memory for learning secondary depth cues. A reliance on two limbs for travel (without a tail for balance) would also promote greater cortical proficiency to detect cues to surface features (e.g. smoothness, roughness) and what lies under the surface of objects, and to precisely gauge the support capacity of a tree branch.

We can also logically infer that using the hands alone for suspension with the arms hanging above the head for feeding or travel would also intensify selection for expansion of association cortices for coordinating haptic (parietal lobe) and visual responses (occipital lobe). Although hominoid vision is dominant for locating objects in space, it is insufficient for vertical climbing and deficient for gripping tree limbs above the head. For unlike quadrupedal monkeys

[16] Present-day Colobine monkeys have digestive systems with multichambered stomachs and a chemistry that enables them to ingest and effectively neutralize the toxic compounds in plants, leaves, seeds and unripe fruits. The appearance of *Microcolobus*, an ancestral colobine about 10 million years ago nicely fits the unripe-fruit hypothesis because apes began their accelerated decline during the late Miocene (Raaum et al., 2005).

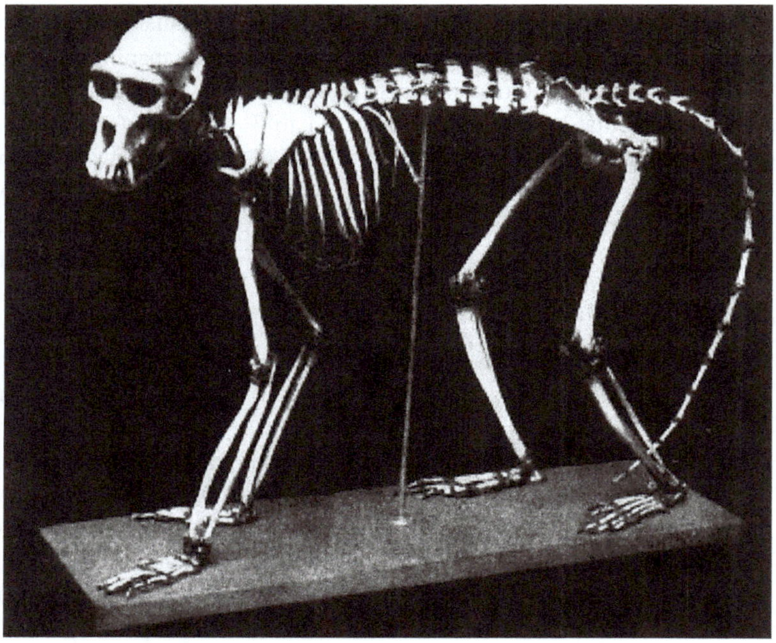

Fig. 11.1 Skeleton of an Old World Papionin, a highly successful group of living Old World monkeys that include mandrills, macaques, baboons, geladas, etc., displayed on a stand against a dark background. The anatomical diagram shows the morphological structure of this monkey, and notably its curved spinal posture and quadrupedal, four-footed gait (Swedwell, 2011)

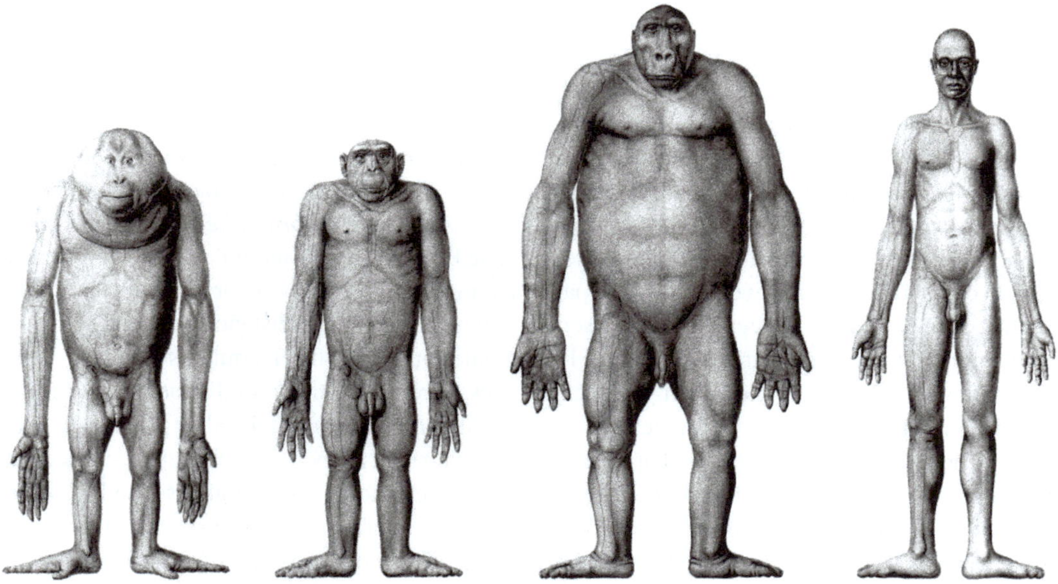

Fig. 11.2 Great ape males and a human male. Each hominoid is set to the same scale in an erect position without body hair and with the lower limbs straightened for comparison. Their unique upper anatomy leaves no doubt that they shared a common ancestor with a suspensory like posture. Along with the little gibbon ape, hominoids make up only a tiny handful of primate species today. From left to right: Pongo (orangutan), Pan (chimpanzee), Gorilla, and Homo. Drawn by Professor Adolph Schultz (1933). Courtsey of E. Schweizerbart'sche Verlagsbuchhandlung

and bipedal humans who use vision almost entirely for guiding locomotion, the quaint nature of forelimb locomotion requires assistance from other modalities, which, in turn, would lead to greater cross-modal association of, for example, the somosensory and visual cortices to prevent sensory disharmony. Thus, for the rare suspensory apes, the *hominoid* brain was evolving for foraging fitness in marginal niches in the receding forests of Africa.

But how did the LCA of apes/early hominins escape extinction? An arboreal quadrupedal adaptation means that most apes were "top-branch" crown feeders, while the suspensory apes were "under- branch" borderline feeders. So, given the ripe/unripe hypothesis that monkeys outcompeted apes by evolving a capacity to consume unripe fruits, it would leave the quadrupedal apes literally dying of hunger. But neither a quadrupedal ape nor any monkey could step on the outermost ends of fragile, swaying branches to pick fruits. This gymnastic feat required the skillful interplay of a trapeze artist, a hominoid that could distribute its weight among several branches or suspend its body below or above the delicate thin branches (Fleagle, 2013). As natural selection is not governed by chance, perfection, or progress, we can assume a "survival of the fitness," despite the fact that an arm-swinging adaptation may have been just good enough for survival.

The social structure of suspensory apes also likely contributed to their survival and we can use their acrobatic habits to infer organizational patterns. A cladistic analysis previously done on present-day great ape societies (to reconstruct the LCA social structure of apes and humans), revealed that all contemporary apes live in fission-fusion societies with social structures built from a few strong ties and mostly weak ties. Great apes also share a rare migration pattern of female dispersal at puberty (and for orangutans and gorillas male dispersal as well) (see Maryanski & Turner, 1992; Maryanski, 1992; Turner & Maryanski, 2024). In contrast, in Old World monkey societies males leave at puberty and females stay in their natal group, enabling them to forge strong-tied genetic or kin-based matrilines (Rhine & Maryanski, 1996).

Now if we assume that the surviving Miocene apes exploited a niche with scattered fruits at the margins, a social structure of strong-tied related females that foraged together (as female monkeys do) would be a maladaptive strategy as it could never support large or even stable tight-knit groups. But a society with promiscuous mating and a female dispersal pattern blocks the creation of matrilines—and maximizes foraging fitness and hence survival.

Thus, for contemporary apes (and perhaps for the LCA) there is a strength in weak ties because individuals are not embedded in strong-tie cliques, giving them the freedom to be self-reliant individualists and to forage independently in space and, except for mother and dependent offspring, to be relatively unencumbered with kinship obligations. A weakly tied social structure also calls for a more inclusive mind-set, a special cognition with the deft and versatile capacity to remember dispersed members, who may not see each other for weeks or more, but still easily reconnect on an occasional basis with community members.

Peter Andrews (2015: 135) also reports on another possible survival edge for the LCA of apes and humans—a rare genetic mutation that raised uric acid levels. Nearly all extant primates (and all mammals) have low uric acid levels because they possess an active enzyme called uricase (*urate oxidase*) that breaks down uric acid so it can easily be removed in urine (Bannasch et al., 2008). The LCA had an inactive uricase enzyme that blocked this decomposition. Following Andrews (2015: 135), the Miocene apes relied heavily on ripe fruits with high sugar levels. But in the presence of fructose, uric acid facilitates the accumulation and storage of triglycerides, a type of fat. Andrew suggests that in lean times the storage of body fat facilitated by the uricase mutation may have enabled the LCA and its descendants to survive the Miocene environment and, later, glacial conditions during the Pleistocene epoch. Today, only apes and humans carry this rare mutation and, thanks to our LCA, elevated uric acid levels. This is not usually a problem, although an excess of uric acid (from refined sugars, organ meats, etc.) can lead to gout, obesity, diabetes, and cardiovascular diseases because fructose is directly metabolized into uric acid (for discussions, see Andrews, 2015; Maliuolo et al., 2016; Álvarez-Lario & Macarró-Vicente, 2010; Li et al., 2021).

Great Ape Cognition and the Leap to a Personal Self

Molecular clock estimates indicate that the LCA ancestor of great apes and humans lived about 15 million years ago. This pivotal ancestor is unknown, but *Nyanzapithecus*, a 13-million-year-old ape recently unearthed provides tantalizing clues to the LCA profile because its morphology is close to living apes (and humans) and its projected brain volume is substantially larger than Old World monkey brains from the same period. As Isaiah Nengo et al. (2017: 173) relates, "Phylogenetic, cranial, and dental analyses [of *Nyanzapithecus]* ...offer compelling insight into the ancestral morphology that gave rise to the clade containing extant apes and humans."

By the late Miocene preserved ape fossils are rare, but one well represented genus is the Asian ape *Sivapithecus.* For nearly 5 million years (from 12 to 8 million Mya), this large-bodied hominoid foraged in dense forests (especially in India and Pakistan) and subsisted on a plant and fruit diet. *Sivapithecus* has a cranium size within the range of present-day apes, who average about 400 cc's, an extended *Life History Profile (*essential for letting the brain grow larger), and facial features that closely resemble present-day orangutans (Begun, 2007, 2015, 2023; Sherwood et al., 2008; Urciuoli & Alba, 2023; Andrews, 2020; Puch, 2022). While endocranial casts of fossilized skulls like Sivapithecus can accurately reproduce intriguing bulges, fissures, and depressions for determining the brain's size and shape, to get insights into the cognitive traits of ancestral hominoids and what is truly distinct about human neuroanatomy, we need to turn to living apes.

The present-day hominoid brain and the Old-World monkey brain are similar except for an increased convoluted surface, which buries the surface of the neocortex in the fissures (Semendeferi & Damasio, 2000; Campbell & Maffer, 1976). Cortical folding harbors substantially more neuron densities in a limited space, which allows for more intricate circuity. This trend is apparent in great ape and human motor cortex, as it is more refined than that of monkeys, perhaps because of a suspensory adaptation. Hominoid prehensile hands also evidence more cortically-based precision for gripping objects while the soft fingertip pads have enhanced raised ridges for enriched tactile perception. (Holloway, 1968; Napier & Napier, 1985). During hominoid evolution, cortical tissue was also augmented in specialized association zones with complex multimodal regions for sensory convergence. And novel changes in neurochemistry and circuitry enhanced cell activity for memory, learning, and other higher cognitive functions (for discussions see Sherwood & Hof, 2007).

An emergent property of the hominoid cognitive leap was the appearance of a *social mind* with revolutionary facets. One component is *a sense of self.* The touchstone for self-recognition is the "looking glass" test of identifying "who is in the mirror." Monkeys, apes, and humans have virtually the same visual equipment so all "see" a mirror as a physical object, but they differ in perceiving who *is* in the mirror. Monkeys "see" another monkey, whereas great apes and humans "see" *self* and respond to it with playful gestures. When an animal can recognize itself as an object in the environment, to use G.H. Mead's words, it has the potential to have an abstract sense of self (Mead, 1934).

A personal self is exceedingly rare in the animal kingdom and, thus far, only great apes, cetaceans, elephants, and humans are known to have a self-identify (Gallup, 1998; Gallup et al., 2014; Walraven et al., 1995). Indeed, Suddenforf and Collier-Baker (2009: 1676) contend that in the primate world it is "only the descendants of a hominoid that probably lived between 13.8 and 18 Mya ago [that] have so far reliably demonstrated that they know who it is that… looks back at them when they look in the mirror." And if this cognitive capacity to recognize self in a mirror can enhance fitness, it can be selected upon to produce what humans reveal today: a series of identities that direct and guide their behaviors and emotional responses to others and social structures.

A self-identity opens up a broader range of cognitive boundaries as it is an enormous liberating force, promoting cognitive flexibility and intentional purposeful behavior. As Mead (1934)) emphasized, a self- consciousness entails an ability to be both subject and object to yourself, and this reflexivity is a special form of social consciousness because if you can take yourself as an object, you can reflect that others do as well; and this enables you to assume the role of the other. Following Tomasello and Herrmann (2010: 3–8), chimpanzees live in a social world where they can engage in various types of voluntary social relationships, including ones based on *"friendship."* They can identify and engage in third party relationships, and they have the social intelligence to interpret the acts and goals of their social partners, form alliances, and even manipulate social maneuvers by using a kind of "*Machiavellian* intelligence." They can also express empathy for a grieving member and even act like they understand what caused the sorrowful emotions expressed by the other (Biro et al., 2010). As Tomasello and Herrmann (2010) relate: "Great apes…operate on their cognitive worlds in ways very similar to humans."

Another upshot of the hominoid mind was a cognitive leap to a new order of things—from visual representations of physical objects to their symbolic representations. Apes interact routinely with conspecifics by using symbolic gestures that are channeled through their visual-touch modalities (which are under cortical control). According to Tomasello and Call (2007: 226), the "apes are learning or inventing many of their gestures, and they are choosing particular gestures for particular contexts, following up with other gestures if the first one does not work. Apes have social/communicative goals, and in their gestural communication… they are pursuing them flexibly." In chimpanzee communities, for example, the gestural vocabulary is familiar to all age and sex classes. However, juveniles and young adults favor particular gestures not used by adults (much like human teenagers). According to Jane Goodall, 1986, chimpanzees simply invent new gestures when needed.

Abundant evidence also exists that linguistically-trained apes can use symbols referentially to represent three-dimensional objects in space or to anticipate future events using human sign-language or a computer board of abstract lexigrams. One ape, for example, signed "candy tree?" to a caregiver in anticipation of an upcoming Christmas holiday when a tree is decorated with colorful sugar treats for the chimpanzees (Fouts & Fouts, 1993). Yet, beyond everything, great apes inherited from their LCA with humans an auditory cortex primed to link a linguistic phonemic code with visual images and sounds. Apes cannot "talk" as their *vocal responses* are still mostly under subcortical control. But since their *auditory perception is* cortically-based, they have the latent capacity when raised in a linguistic environment to *spontaneously* learn the grammar rules of human speech, and to comprehend complete sentences at about the level of a 3-year-old human child. In response, they use either gesture or a computer board to punch out the correct series of abstract lexigrams for shared symbolic communication (for discussions, see Savage-Rumbaugh et al., 1993; Savage-Rumbaugh & Lewin, 1994; Maryanski, 1996, 1997; Benson & Greaves, 2005; Pär et al., 2005). While great apes and humans share a treasure chest of these inherited pre-adaptations, only the human mind was augmented for a much higher-order of cognition.

Hominin Evolution and a Terrestrial Niche

Chimpanzees and proto-hominins split from their LCA between 5 and 8 million years ago. An assortment of late Miocene taxa are candidates for the earliest basal hominin, but their hominin status remains controversial (Fleagle et al., 2026). *Sahelanthropus tchadensis* (circa 7 Mya) of Chad in Central Africa, the only candidate with a preserved cranium, is estimated to have an endocranial volume of 360–370 cc's (Zollikofer et al., 2005).

In the early Pliocene (circa 5 Mya) a number of unambiguous hominin species appear in the African fossil record, with most species placed in the genus *Australopithecus*. As cognates, ancestral chimpanzees and early australopithecines were likely similar in appearance but the austrapiths had two standout features: non-projecting canines and a habitual bipedal gait— *the* defining trait of a hominin).[17] *Australopithecus afarensis* (nicknamed Lucy's species) is the best-known early hominin species, and lived from approximately 3.8 to 2.9 Million years ago. The cranial proportions of *afarensis* indicate a mean volume of 446 cc's, which is in line with present-day chimpanzees (405 cc's). Overall, despite a timespan of millions of years of survival and reproductive success, the brains of the many species in the Australopithecine lineage remained in the chimpanzee range (although there is a very modest enlargement trend).

During the late Pliocene and early Pleistocene (circa 2.3–1.8 million years ago) the *Homo* lineage appeared in Africa. Early *Homo* species represented by *Homo habilis* were not much taller or heavier than australopithecines, but they had an altered shaped skull with more modern frontal lobes and a jump in cranial capacity between 752 and 687 cc's (the human range is 1200 to 1600 cc's) (Tattersall, 2023).

The next hominin grade represented by *Homo erectus* appeared between 2.1 to 1.8 years ago. Here we have an ancestor with smaller jaws and teeth, larger body size (some adult individuals were in the range of 5.9—5.11 (176–180 cm) along with a more human-like morphology from the neck down. All signs point to this hominin as on the road to humanity, especially in brain size with an average brain volume of 970 cc's. *Erectus* was a true open-country dweller. Indeed, this lineage survived for nearly 2 million years and its geographic distribution after leaving Africa covers the scope of Europe and Asia (Fleagle, 2013). So, while the brain size of Austrapithecines barely changed over millions of years, the migrations of *Homo* populations called for novel adaptations that would lead to early modern *Homo sapiens* who first appeared in Africa about 300,000 years ago (for detailed discussions see Tattersall, 2023; Fleagle, 2013; Antón & Middleton, 2023).

A new ecology always triggers selection pressures and the more at odds from the previous one, the more pronounced the pressures for change. Yet, adaptive responses to change are always guided by traits already present in a species' taxonomic repertoire. As the legendary ethologist Hans Kummer (1971, 129) explained, "New behavioral adaptations are possible only if the necessary dispositions are within the scope of the behavioral heritage and if they can be accommodated within the existing social system." So, when hominins migrated to diverse terrestrial zones, they carried with them the hominoid legacy of a forest adaptation that would limit the kinds of structural changes and elaborations possible.

What obstacles did they face? A dominant visual system in open country with scattered trees and roaming predators was surely a stumbling block because it requires active attention, is inadequate in low visibility, and useless after dark. Hominins also lacked projecting canines or any built-in-defense weapons—even claws. Yet, it would take an act of evolutionary gymnastics at this late stage of *hominoid* evolution to change a visually oriented primate into an olfactory oriented primate. H*ominoid* proclivities for weak ties, individualism, and self-reliance would also place constraints on the kinds of organizational arrangements possible. However, once *Homo* left the security of a leafy canopy, selection pressures were surely triggered for tighter-knit groups.

But what kinds of social forms were possible? One organizational model has early *Homo* populations organized like baboon troops (Papio) because baboons originally lived in propinquity with early hominins in open terrain and faced similar selection pressures for food, sleeping sites, and protection from predators. If early *Homo* were organized like contemporary baboons, group living would consist of hierarchical matrilineages with adult males positioned in military formation at the front and rear flanks of the troop as they moved across open terrain.

[17]To date, we have no fossils that are directly ancestral to living chimpanzees. So, their evolutionary history after the split with hominins is virtually undocumented and the same is true for gorillas. In contrast, orangutans (who shared a LCA with proto-hominins, and the African apes [circa 15–16 Mya] have relatives. For example, the Asian ape *Sivapithecus* discussed above is considered directly ancestral to the extant orangutans of Sumatra and Borneo.

Yet, the cladistic analysis done on contemporary ape social networks mentioned earlier documented that female-biased dispersal after puberty is a deeply-rooted hominoid trait and would restrict matrilineal formations (see Maryanski & Turner, 2024; Turner & Maryanski, 2008; Turner & Maryanski, 2024). Support for this hypothesis also comes from the chemistry of early *Austrapithecine* dentition indicating that females grew up in one locality and moved to another locality after puberty (Callaway, 2011 and Copeland et al., 2011). Genetic Research on European Neanderthal populations also support the dispersal of adolescent females at puberty (Skov et al., 2022: 524; Lalueza-Fox et al., 2011: 252). If female migration after puberty is shared by great apes, *Austrapithecines* and *Neanderthal* populations and is also congruent with findings from ethnographic research that hunter-gatherer populations overwhelmingly practiced patrilocality, it is reasonable to conclude that early hominin females also dispersed from their natal unit after puberty. It also offers clues into the social structure of early hominin societies, and the origin of the human family[18] (for discussions see Ember & Ember, 1983; Murdock, 1949; Service, 1965; Maryanski, 2021).

As species usually start out with the organizational structure they inherit, a second organizational model is a blueprint of chimpanzee societies because it is far more likely that early *Homo* would start out with a social formation somewhat similar to their closest living relative. Indeed, as Gagneux and Varki (2001: 2) related, "if taxonomic classification were based solely upon genomic DNA sequence similarity, the nearly 99% identity of human, chimpanzee, and bonobo genomes would require a reclassification of the latter two into the genus *Homo*."

As discussed, common chimpanzees live in a fission-fusion community organization where females after puberty move to another community. Hence, chimpanzee society is anchored by natal males who are self-reliant individualists, often forge alone, hang out with a preferred friend, or join short impromptu gatherings to socialize. Mating is promiscuous, so incoming females usually forage alone and raise their offspring alone. Thus, unlike Old World monkeys who are organized into stable groups and extended kinship matrilines, chimpanzees have no extended kinship obligations, so they are free to move about their macro-level community and interact with a greater number of individuals. Indeed, in chimpanzee societies weak ties serve to link a large, fluid population into a community form of organization, and in this sense provides support for Mark Granovetter's famous "strength of weak ties" thesis (Granovetter, 1973).

Now if living great apes have social networks with weak tie proclivities, early hominins likely did as well. How then might selection have modified and expanded on such a fluid, relational structure? A detailed discussion is not possible (see Turner & Maryanski, 2008, 2024, Turner, 2021) but as hominins ventured further away from their forest home, a more tightly-knit social structure would be crucial for survival. Yet, if we reject the baboon model and if the chimpanzee community model is not sufficiently structured, what alternatives are possible?

The best clues are found in studied hunter-gatherers. For at least 99% of human history (or until garden farming was adopted about 10,000 years ago), humans were food collectors. So, it is not surprising that hunting and gathering societies share with chimpanzees many organizational features such as a wide- ranging area with band members moving about in a yearly seasonal round. Both societal types are also "fission-fusion" societies because residents within a "big band" region can freely join or leave a local band depending on individual preferences and procurable resources. However, a hunter-gatherer society is anchored by self-sufficient nuclear families

[18] A human proclivity, of course, is a tendency to behave in a certain way and in most instances can be overridden by ecological factors, demographic constraints, subsistence activities, external or internal warfare, and other sociological forces. Yet, matrilocal residence (even when combined with avunculocal residence) is limited to a small number of societies. For example, in the Ethnographic Atlas of 858 societies (Murdock, 1949, 1967), the residence rules are the following: Matrilocality 13.1%; Avunculocality 4.3%; Duolocality 0.9%; Bilocality 8.5%; Neolocality 4.7% and, *Patrilocality 68.5%* (Pasternak, 1976: 44; Ember & Ember, 1983; Maryanski, 2021).

who can forage apart from other families when resources are scarce, and come together when resources are plentiful. Another similarity discussed earlier is that the majority of hunter-gather societies (and most traditional societies) favor *female-biased dispersal*, although exogamy is dressed up with invented marriage rules, which are the touchstone for the creation of human kinship networks.

Thus, the hallmark of a hunter-gatherer society is a loose-knit social structure and not a tight-knit group. Even kinship and marriage norms (e.g., bilateral kinship, easy divorce) are designed to accommodate a shifting collection of individuals. And, like chimpanzees, emphasis is placed on individual mobility, personal autonomy, and fluid networks (Maryanski, 2018, 2021). In short, hunter-gatherers do not display the high collectivism that is often imputed to human nature. They do, however, have a pivotal institution — stable kinship groupings which likely allowed hominins to be sufficiently organized to survive in open-country conditions (see Bailey & Aunger, 1990 for a comparative relational analysis, and Turner & Maryanski, 2008 for a detailed discussion).

It is meaningful to now ask: What neurological changes were possible for a hominin that migrated to an open country niche? For the most part, even the most intense selection forces could not have uprooted the hominoid cognitive legacy. Early *Homo* had not only benefited from nearly 65 million years of primate brain evolution, but humans are the end product of over 20 million years of ape evolution, making it near impossible for large, radical mutations to have taken hold. As R.A. Fisher noted long ago: "the probability that individual mutations will contribute to evolution is in inverse correlation to the intensity of their effect on the developing phenotype." Indeed, Fisher made it crystal clear that species are not "passively awaiting the next favourable mutation" (quoted in Stebbins, 1969: 104; Fisher, 1959: 16–17). Of course, many *small* mutations (or point mutations) occurred during hominin evolution, but much of the burden on natural selection was to extend, elaborate, combine, or alter *already existing characteristics* of the hominin neuroanatomy. Selection could work successively on the tail ends of bell-shaped distributions of existing traits that enhanced fitness and, in this way, alter the hominin genome.

For example, as uncovered earlier by cladistic analysis, the LCA of Old-World anthropoids bequeathed to its monkey and ape descendants an auditory cortex under cortical control for *incoming sounds*. By contrast, the vocal channel for individual *sound production* remained a relatively fixed species-specific call system. But during hominin evolution, selection placed the vocal channel under cortical control to allow for the creation of purely voluntary sounds (perhaps to assist the visual modality as sounds are a self-alerting detection system). Then, over time, the hominin vocal anatomy also underwent modifications that involved the tongue, larynx, pharynx, and related structures, so that more and more finely tuned and articulated sounds could be made. And, eventually, the cultural/social creation of a system of linguistic units.

Hominin neuroanatomy also underwent extensive modifications in the augmentation of association tissues, greater complexity in multimodal regions, and more intricate interconnections among brain parts. Selection also usurped already existing zones on the left side of the brain, such as from *Broca's hump* to *Broca's area* in the frontal lobe, (which is functionally unique in humans but cytoarchitecturally homologous in higher primates) to control the production of linguistic vocal utterances as speech is downloaded from the way the brain thinks, which for the most part, is in gestalts rather than words. Then, in addition to other neurological refinements, such as linking up an already in place capacity among the great apes for symbolic representations, much of the neurological work for speech had already been done.[19]

Thus, spoken language is not the result of dramatic mutations nor a mysterious property of humans. It results from modifications and elaborations to an *already existing anthropoid sensory*

[19]The realigning of all the structures for articulated speech took a long time and was, in many ways, the last "roadblock" to the evolution of the hominin brain.

system and the extension of symbolic and other cognitive abilities *already present* in our hominoid legacy. As Deacon (2007: 531) phrased it: "Generations of comparative neuroanatomists.

have failed to identify even one major novel brain structure in humans. This suggests that our special adaptations for language are the result of using previously evolved primate brain structures in new ways and in new combinations." That selection acted to continuously promote brain development in human ancestors means it enhanced survival and reproductive success, despite the fact that a big brain is exceedingly costly to maintain.

Finally, we should ask: What dramatic events occurred to spark the cognitive genus of *Homo Sapiens?* One factor is the invention of standardized tools. Crude stone tools first appear about 2.6 Mya with *Austrapithecines*, but during the reign of *Homo erectus* (from 1.8 to about 250,000 years ago) more specialized tools were invented, including hand axes, scrapers and cleavers—signaling the first cognitive leap in technological evolution. About 800,000 years ago, hominins also began to make effective use of fire for staying warm, seeing at night, and warding off predators. Fire technology led to cooking foods— another revolutionary technology that, in turn, ushered in such physiological changes as reduced tooth size and reduction in jawbone morphology. Richard Wrangham (2009, Wrangham et al., 1999) hypothesized that cooking foods allowed hominins to consume more calories and, in turn, helped to grow the brain. Thus, the material traces that *Homo erectus* left behind tell us this hominin had undergone cognitive changes tied to the rewiring of the brain. And, it is likely that *erectus* served as the ancestral stock for the rise of later species of *Homo* (i.e. Neanderthals, modern humans, and probably Denisovans).

Fully modern looking humans with big brains, rounded skulls, and true chins appear in the fossil record about 300,000 years ago. Yet, despite the eight billion of us alive today, the human species has less genetic variation than any other living primate, including chimpanzees (Barbujani et al., 2013). This enigma has been recently linked to a series of bottlenecks events, one apparently between 930,000 and 813, 000 years ago when a severe bottleneck event seemingly wiped out 98.7% of human ancestors (Hu et al., 2023). Whatever the bottleneck, we reveal so little variability in our genome, we should not be surprised when confronted with problems of survival, humans work out relatively similar solutions. It also puts to rest notions that *Homo sapiens* can be divided up into different races on the basis of illusionary qualitative distinctions among populations (Barbujani, 2015; Barbuyani et al., 2013).

Thus, goes the strange tale of how the hominin mind evolved in a mosaic pattern, although it reveals a unity of, and continuity in structure and function. First, an early anthropoid ancestor with visual dominance set the stage for a primate tuned to the outside world and to voluntary, cortically-based behavior. Then, after the split between monkeys and apes, their subsequent evolution took different paths, with apes dominating the forests for nearly 15 million years until monkeys gained the upper hand, and still do today. The extinction of all quadrupedal apes followed but the arm-swinging apes survived; and one of those surviving lineages became the LCA of the great apes and humans. By the time of *Homo*, hominins were already decked out with all kinds of preadaptations on which selection could go to work. And, the hominin/human mind eventually emerged.

Yet, the genius of the human social mind rests not only on its amazing cognitive and intellectual potential but on its amazing capacity to cooperate with others—even strangers. As Tomasello and Herrmann (2010: 5) highlighted, "Virtually all of human's highest cognitive achievements are not the work of individuals acting alone but rather of individuals collaborating in groups…ending up with such things as complex social institutions structured by joint goals, division of labor, and communicative symbols." Thus, selection built on the elementary collaboration of great ape weak tie networks to fashion a strikingly novel *social mind,* a sociality that allows for the integration of millions of individuals and, in some cases, billions of personal strangers.

Yet, the human mind has an evolutionary novelty truly all its own—*selection favoring change itself*. Human evolution took place during great climatic change, from warm periods through long glacial periods of cold and ice. The hominin brain faced such dramatic changes and endured such constant selection pressures that the long-term outcome was a structure built around creativity, variation in socialization, and the ability to adapt to a sweeping variety of new challenges. Thus, the human mind is wired for adaptability; it is not locked into many hard-wired predispositions. It is specialized, and, yet, still highly generalized. The integration of culture with biology is what allows for this capacity—a truly unique capacity that along with our one-of-a-kind sociality is the true "secret of the human mind."

References

Almécija, S., Hammond, A. S., Thompson, N. E., Pugh, K. D., Moyà-Solà, S., & Alba, D. M. (2021). Fossil apes and human evolution. *Science, 372*, 587.

Álvarez-Lario, B., & Macarrón-Vicente, J. (2010). Uric acid and evolution. *Rheumatology (Oxford), 49*(11), 2010–2015.

Andrews, P. (1981). Species diversity and diet in monkeys and apes during the Miocene. In C. B. Stringer (Ed.), *Aspects of human evolution*. Taylor and Francis.

Andrews, P. (2015). *An ape's view of human evolution*. Cambridge University Press.

Andrews, P. (2020). Last common ancestor of apes and humans: Morphology and environment. *Folia Primatologica, 91*, 122–148.

Andrews, P., & Kelley, J. (2007). Middle Miocene dispersals of apes. *Folia Primatologica, 78*, 328–343.

Antón, S. C., & Middleton, E. R. (2023). Making meaning from Fragmenary fossils: Early *homo* in the early to early middle Pleistocene. *Journal of Human Evolution, 179*, 103307.

Bailey, R., & Aunger, R. (1990). Humans as primates. The social relationships of Efe pygmy men in comparative perspective. *International Journal of Primatology, 11*, 127–146.

Bannasch, D., Safra, N., Young, A., Karmi, N., Schaible, R., & Ling, G. (2008). Mutations in the SLC2A9 gene cause Hyperuricosuria and Hyperuricemia in the dog. *PLoS Genetics, 4*(11), e1000246.

Barbujani, G. (2015). Race: Genetic aspects. *International Encyclopedia of the Social and Behavioral Sciences, 19*, 825–832.

Barbujani, G., Ghirotto, S., & Tassi, F. (2013). Nine things to remember about human genome diversity. *Tissue Antigens, 82*, 155–164.

Barton, R. A. (1998). Visual specialization and brain evolution in primates. *Proceedings of the Royal Society of London. Series B: Biological Sciences, 265*, 1933–1937.

Begun, D. (2007). Fossil record of Miocene hominoids. In W. Henke & T. Hardt (Eds.), *Handbook of Paleoanthropology 1* (pp. 921–977). Springer-Verlag.

Begun, D. (2015). Fossil Record of Miocene Hominoids. In W. Henke & I. Tattersall (Eds.), *Handbook of Paleoanthropology 2* (pp. 1261–1332). Springer-Verlag.

Begun, D. (2023). Catarrhine origins and evolution. In C. S. Larsen (Ed.), *A companion in biological anthropology* (pp. 381–399). John Wiley & Sons.

Benefit, B. R. (1999). *Victoriapithecus:* The key to Old World monkey and Catarrhine origins. *Evolutionary Anthropology, 7*, 155–174.

Benson, J., & Greaves, W. (2005). *Functional dimensions of ape-language discourse*. Equinox.

Biro, D., Humle, T., Koops, K., Sousa, C., Hayashi, M., & Matsuzawa, T. (2010). Chimpanzee mothers at Bossou, Guinea carry the mummified remains of their dead infants. *Current Biology, 20*, 351–352.

Bloch, J., & Boyer, D. (2007). New skeletons of Paleocene-Eocene Plesiadapiformes. In M. J. Ravosa & M. Dagosto (Eds.), *Primate origins: Adaptations and evolution* (pp. 535–581). Springer.

Brashears, M. (2013). Humans use compression heuristics to improve the recall of social networks. *Scientific Reports, 3*, 1513.

Callaway, E. (2011). Female Australopiths seek brave new world. *Nature, 338*.

Campbell, F., & Maffer, L. (1976). Contrast and spatial frequency. In R. Held & W. Richards (Eds.), *Recent progress in perception* (*Scientific American*). W.H. Freeman and Company.

Cheney, D. L., & Seyfarth, R. M. (1980). Vocal recognition in free-ranging vervet monkeys. *Animal Behavior, 28*, 739–751.

Coleman, M. (2009). What do primates hear? A meta-analysis of all known nonhuman primate Behavioral audiograms. *International Journal of Primatology, 30*, 55–91.

Coleman, M., & Colbert, M. (2010). Correlation between auditory structures and hearing in non-human primates. *Journal of Morphology, 27*, 511–532.

Coleman, M., Kay, R., & Colbert, M. (2010). Auditory morphology and hearing sensitivity in fossil new world monkeys. *The Anatomical Record, 293*, 1711–1721.

Conroy, G. (1990). *Primate evolution*. W.W. Norton & Company.

Copeland, S., Sponheimer, M., de Ruiter, D., Lee-Thorp, J., Codron, D., le Roux, P., Grimes, V., & Richards, M. (2011). Strontium isotope evidence for landscape use by early Hominins. *Nature, 474*, 76–78.

Deacon, T. W. (2007). The evolution of language systems in the human brain. In J. Kaas & T. Preuss (Eds.), *Evolution of nervous systems: A comprehensive reference, V4* (pp. 529–547). Elsevier.

Dean, M. C., & Leakey, M. (2004). Enamel and dentine development and the life history profile of

Victoriapithecus macinnesi from Maboko Island, Kenya. *Annals of Anatomy, 186*, 405–412.
Decasdien, A., Barton, R., & Higham, J. (2022). Understanding the human brain: Insights from comparative biology. *Trends in Cognitive Sciences, 26*, 432–445.
Dominy, N., Ross, C., & Smith, T. (2004). Evolution of the special senses in primates: Past, present and future. *The Anatomical Record Part A, 281*, 1078–1082.
Ember, M., & Ember, C. (1983). *Marriage, family and kinship: Comparative studies of social organization*. HRAF Press.
Falk, D. (2000). *Primate diversity*. W.W. Norton & Company.
Falk, D. (2015). Evolution of the primate brain. In W. Henke & I. Tattersall (Eds.), *Handbook of Paleoanthropology* (pp. 1496–1525). Springer Heidelberg.
Fisher, R. A. (1959). Natural selection from the genetical standpoint. *Australian Journal of Science, 22*, 16–17.
Fleagle, J. G. (2013). *Primate adaptation and evolution*. Academic Press.
Fleagle, J., Baden, A., & Gilbert. C. (2026). *Primate adaptation and evolution. Elsevier*.
Forey, P. L., Humphries, C. J., Kitching, I. L., Scotland, R. W., Siebert, D. J., & Williams, D. M. (1992). *Cladistics: A practical course in systematics*. Clarendon Press.
Fortelius, M., & Hokkanen, A. (2001). The trophic context of hominoid occurrence in the later Miocene of Western Eurasia: A primate-free view. In L. de bonis, G. Koufos, & P. Andrews (Eds.), *Hominoid evolution and climatic change in Europe* (pp. 19–47). Cambridge University Press.
Fouts, R. S., & Fouts, D. H. (1993). Chimpanzees' use of sign language. In P. Cavalieri & P. Singer (Eds.), *The great ape project: Equality beyond humanity* (pp. 28–41). St. Martin's Press.
Franks, David. 2010. Neurosociology.
Franks, D., & Smith, T. (1999). *Mind, brains and society*. Emerald Group Publishing.
Freides, D. (1973). Human information processing and sense modality: Cross-modal functions, information complexity, memory, and deficit. *Psychological Bulletin, 81*(5), 284–310.
Gagneux, P., & Varki, A. (2001). Genetic differences between humans and great apes. *Molecular Phylogenetics and Evolution, 18*, 2–13.
Gallup, G. G. (1998). Self-awareness and the evolution of social intelligence. *Behavioural Processes, 42*, 239–247.
Gallup, G. J., Platek, S., & Spaulding, K. (2014). The nature of recognition revisited. *Trends in Cognitive Science, 18*, 57–58.
Gebo, D. (2014). *Primate comparative anatomy*. Johns Hopkins University Press.
Gibson, J. J. (1962). Observations on active touch. *Psychological Review, 69*, 477–491.
Gilad, Y., Man, O., Pääbo, S., & Lancet, D. (2003). Human specific loss of olfactory receptor genes. *Proceedings of the National Academy of Sciences, 100*(6), 3324–3327.
Glaser, D. (2007). The evolution of the sweetness receptor in primates. In J. Kaas & T. Preuss (Eds.), *Evolution of nervous systems: A comprehensive reference, V4* (pp. 121–128). Elsevier.
Godinot, M. (2015). Fossil record of the primates from the Paleocene to the Oligocene. In W. Henke & I. Tattersall (Eds.), *Handbook of Paleoanthropology* (pp. 1137–1260). Springer.
Goodall, J. (1986). *The chimpanzees of Gombe: Patterns of behavior*. Harvard University Press.
Granotosky, M., & Young, J. (2023). Primate locomotion: A comparative and developmental perspective. In C. S. Larsen (Ed.), *A companion to biological anthropology* (pp. 587–602). Wiley Blackwell.
Granovetter, M. (1973). The strength of weak ties. *American Journal of Sociology, 78*, 1360–1380.
Gregory, W. K. (1916). Studies of the evolution of the primates. *Bulletin of the American Museum of Natural History, 35*, 239–355.
Hackett, T. (2003). The comparative anatomy of the primate auditory cortex. In A. Ghazanfar (Ed.), *Primate audition: Behavior and neurobiology* (pp. 199–226). CRC Press.
Hackett, T. (2008). Anatomical organization of the auditory cortex. *Journal of the American Academy of Audiology, 19*, 774–779.
Hammond, M. (2013). The neurosociology of reward release, repetition, and social emergence. In D. D. Franks & J. H. Turner (Eds.), *Handbook of Neurosociology* (pp. 311–329). Springer.
Harrison, T. (2013). Catarrhine origins. In D. Begun (Ed.), *A companion to Paleoanthropology* (pp. 376–381). John Wiley & Sons.
Hartwig, W. (2007). Primate evolution. In C. Campbell, A. Fuentes, K. MacKinnon, M. Panter, & S. Bearder (Eds.), *Primates in perspective* (pp. 19–31). Oxford University.
Healy, K., Ezard, T., Jones, O., Salguero-Gómex, R., & Buckley, Y. (2019). Animal life history is shaped by the pace of life and the distribution of age-specific mortality and reproduction. *Nature Ecology & Evolution, 3*(8), 1217–1224.
Heffner, R. (2004). Primate hearing from a mammalian perspective. *The Anatomical Record, 281A*, 1111–1122.
Heffner, H., & Heffner, R. (2008). High frequency hearing. In P. Dallos, D. Oertel, & R. Hoy (Eds.), *Handbook of the senses: Audition* (pp. 55–60). Elsevier.
Hiramatsu, C., Melin, A., Allen, W., Dubuc, C., & Higham, J. (2017). Experimental evidence that primate trichromacy is well suited for detecting primate social colour signals. *Proceedings of the Royal Society B, 284*, 20162458.
Hogg, C., Neveu, M., Stokkan, K.-A., Folkow, L., Cottrill, P., Douglas, R., Hunt, D., & Jeffery, G. (2011). Arctic reindeer extend their visual range into the ultraviolet. *The Journal of Experimental Biology, 214*, 2014–2019.

Holloway, R. (1968). The evolution of the primate brain: Some aspects of quantitative relations. *Brain Research, 7*, 121–172.

Hoover, K. (2010). Smell with inspiration: The evolutionary significance of olfaction. *Yearbook of Physical Anthropology, 143*, 63–74.

Hopcroft, R. (2013). Neurosociology and theory of mind (ToM). In D. D. Franks & J. H. Turner (Eds.), *Handbook of Neurosociology* (pp. 231–254). Springer.

Hu, W., Hao, Z., Pengyuan, D., Vincenzo, F., Manzi, G., Cui, J., Fu, Y., Pan, Y., & Li, H. (2023). Genomic inference of a severe human bottleneck during the early to middle Pleistocene transition. *Science, 381*, 979–984.

Hunt, K. (2016). Why are there apes? Evidence for the co-evolution of ape and monkey ecomorphology. *Journal of Anatomy, 228*, 630–685.

Ito, K., Nakamura, T., Suzuki, R., Negishi, T., Oishi, M., Nagura, T., Jinzaki, M., & Ogihara, N. (2021). Comparative functional morphology of human and chimpanzee feet based on three-dimensional finite element analysis. *Bioengineering and Biotechnology, 9*, 760486.

Jacobs, G. H. (2004). Photopigment variations and the evolution of anthropoid vision. In C. F. Ross & R. F. Kay (Eds.), *Anthropoid origins* (pp. 645–664). Kluwer/Plenum.

Jacobs, G. H. (2007). The comparative biology of photopigments and color vision in primates. In J. Kaas & T. Preuss (Eds.), *Evolution of nervous systems: A comprehensive reference, V4* (pp. 80–85). Elsevier.

Jeffers, R., & Lehiste, I. (1970). *Principles and methods for historical linguistics*. MIT Press.

Kaas, J. (2011). Neocortex in early mammals and its subsequent variations. *Annuals of the New York Academy of Sciences, 1225*, 28–36.

Kaas, J., & Balaram, P. (2014). Current research on the organization and function of the visual system in primates. *Eye and Brain, 6*(Suppl 1), 1–4.

Kalkhoff, W., Thye, S., & Pollock, J. (2016). Developments in Neurosociology. *Sociology Compass, 10*(3), 242–258.

Kass, J. (2007). The evolution of sensory and motor systems in primates. In J. Kaas & T. Preuss (Eds.), *Evolution of nervous systems: A comprehensive reference, V4* (pp. 35–57). Elsevier.

Kass, J. (2008). The evolution of the complex sensory and motor system of the human brain. *Brain Research Bulletin, 75*, 384–390.

Kay, R., & Ungar, P. (1997). Dental Evidence for Diet in some Miocene Catarrhines with Comments on the Effects on Phylogeny on the Interpretation of Adaptation. In D. Begun, C. Ward, & M. Rose (Eds.), *Function, phylogeny, and fossils: Miocene Hominoid evolution of adaptation*. Plenum.

Keller, H. (1904). *The world I live in*. Hodder and Stoughton.

Keller, G., & Barron, J. (1987). Paleodepht distribution of Neogene hiatuses. *Paleoceanography, 2*, 67–713.

Kelley, J., & Smith, T. M. (2003). Age at first molar emergence in early Miocene *Afropithecus turkannsis* and life-history evolution in the *Hominoidea*. *Journal of Human Evolution, 44*, 307–329.

Kemp, A. (2024). Effect of binocular visual cue availability on fruit and insect grasping performance in two Cheirogaleids: Implications for primate origins hypotheses. *Journal of Human Evolution, 188*, 103456.

King, A., & Nelken, I. (2009). Unraveling the principles of auditory cortical processing: Can we learn from the visual system? *Nature Neuroscience, 12*, 698–701.

Kirk, C. (2006). Visual influences on primate encephalization. *Journal of Human Evolution, 51*, 76–90.

Kummer, H. (1971). *Primate societies*. Aldine.

Lalueza-Fox, C., Rosas, A., Estalrrich, A., Gigli, E., Campos, P., et al. (2011). Genetic evidence for Patrilocal mating behavior among neandertal groups. *Proceedings of the National Academy of Sciences, 108*(1), 250–253.

Lederman, S. J., & Klatzky, R. L. (2009). Haptic perception: A tutorial. *Attention, Perception and Psychophysics, 71*, 1439–1459.

Leigh, S., & Blomquist, G. (2011). Life history. In C. Campbell, A. Fuentes, K. MacKinnon, S. Bearder, & R. Stumpf (Eds.), *Primates in perspective* (pp. 418–428). Oxford University Press.

Lewis, O. J. (1974). The wrist articulations of the Anthropoidea. In F. Jenkins (Ed.), *Primate locomotion*. Academic Press.

Li, Z., Hoshino, Y., Tran, L., & Gaucher, E. (2021). Phylogenetic articulation of uric acid evolution in mammals and how it informs a therapeutic uricase. *Molecular Biology and Evolution, 39*(1), msab312.

Lucas, P., Dominy, N., Riba-Hernandex, P., Stoner, K., Yamashita, N., LorÍa-Calder6n, E., Petersen-Pereira, W., Rojas-Durán, Y., Salas-Pena, R., Solis-Madrigal, S., Osorio, D., & Darvell, B. (2003). Evolution and function of routine trichromatic vision. *Evolution, 57*, 2636–2643.

Maas, P. (1958). *Textual criticism*. Oxford University Press.

Machalek, R. (1992). Why are large societies rare. In L. Freese (Ed.), *Advances in human ecology* (Vol. 1, pp. 33–64). Jai Press.

Maclatchy, L. (2004). The oldest ape. *Evolutionary Anthropology, 13*, 90–103.

Maclatchy, L. M., Rossie, J., Houssaye, A., Olejniczak, A. J., & Smith, T. M. (2019). New hominoid fossils from Moroto II, Uganda and their bearing on the taxonomic and adaptive status of *Morotopithecus bishop*. *Journal of Human Evolution, 132*, 227–246.

Maiuolo, J., Oppedisano, F., Gratteri, S., Muscoli, C., & Mollace, V. (2016). Regulation of uric acid metabolism and excretion. *International Journal of Cardiology, 213*, 8–14.

Manitilla, G., Chester, S. G. B., Clemens, W. A., Moore, J. R., Sprain, C. J., Hovatter, B. T., Mitchell, W. S., Mans, W., Mundil, R., & Renne, P. R. (2021). Earliest Paleocene Purgatoriids and the initial radiation of stem primates. *Royal Society Open Science, 8*, 2100050.

Martin, R. D. (1990). *Primate origins and evolution: A phylogenetic reconstruction*. Chapman & Hall.

Maryanski, A. (1992). The last common ancestor: An ecological-network model on the origin of human sociality. *Advances in Human Ecology, 2*, 1–32.

Maryanski, A. (1996). Was speech an evolutionary afterthought? In B. Velichkovsky & D. Rumbaugh (Eds.), *Communicating meaning: The evolution and development of language* (pp. 79–97). Lawrence Erlbaum Associates.

Maryanski, A. (1997). Primate communication and the ecology of a language niche. In U. Segerstråle & P. Molnár (Eds.), *Nonverbal communication: Where nature meets culture* (pp. 191–209). Lawrence Erlbaum Associates.

Maryanski, A. (2018). *Émile Durkheim and the birth of the gods: Clans, incest, totems, Phratries, hordes, Mana, taboos, corroborees, Sodaities, menstrual blood, apes, Churingas, Cairns, and other mysterious things*. Routledge.

Maryanski, A. (2021). The sociology of kinship: A case for looking Back to the future. In S. Abruytn & O. Lizardo (Eds.), *Handbook of classical sociological theory* (pp. 267–299). Springer.

Maryanski, A. & Turner, J. (2024). An approach to evolutionary sociology and its implications for theorizing on socio-cultural evolution. *Köln Z Soziol, 76*, 351–389.

Maryanski, A., & Turner, J. H. (1992). *The social cage: Human nature and the evolution of society*. Stanford University Press.

Masterton, B., & Diamond, I. (1973). Hearing: Central neural mechanisms. In E. Carterette & M. Friedman (Eds.), *Handbook of Perception no. 3, biology of perceptual systems* (pp. 408–448). Academic Press.

McNulty, K., Begun, D., Kelley, J., Manthi, F., & Mbua, E. (2015). A systematic revision of proconsul with a description of a new genus of early Miocene hominoid. *Journal of Human Evolution, 84*, 42–61.

Mead, G. H. (1934). *Mind, self and society*. University of Chicago Press.

Miller, E. R., Benefit, B. R., McCrossin, M. L., Plavan, J., Leakey, M., El-Barkooky, A., Hamdan, M., Gawad, M., Hassan, S., & Simons, E. (2009). Systematics of early and middle Miocene old world world monkeys. *Journal of Human Evolution, 57*, 195–211.

Moritz, G., Melin, A., Fred, Y., Bernard, H., Ong, P., & Dominy, N. (2014). Niche convergence suggests functionality of the nocturnal fovea. *Frontiers in Integrative Neuroscience, 8*, 61.

Murdock, G. (1949). *Social structure*. Macmillan.

Murdock, G. (1967). *Ethnographic atlas*. University of Pittsburgh.

Nakatsukasa, M. (2008). Comparative study of Moroto vertebra specimens. *Journal of Human Evolution, 55*, 581–588.

Napier, J. R., & Napier, P. H. (1985). *The natural history of the primates*. MIT Press.

Nengo, I., Tafforeau, P., Christopher, C., Fleagle, J., Miller, E., Fribel, C., Fox, D , Feinbert, J., Pugh, K., Berruyer, C., Mana, S., Engle, Z., & Spoor, F. (2017). New infant cranium from the African Miocene sheds on ape evolution. *Nature, 548*, 171.

Nofre, C., Tinti, J. M., & Glaser, D. (1996). Evolution of the sweetness receptor in primates. 11. Gustatory responses of non-human primates to nine compounds known to be sweet in Man. *Chemical Senses, 21*, 747–762.

Oxnard, C. E. (1963). Locomotor adaptation in the primates. In J. Napier & N. A. Barnicot (Eds.), *Locomotor adaptations in the primates* (Vol. 10). Symposia of the Zoological Society in London.

Pär, S., Fields, W., & Savage-Rumbaugh, S. (2005). *Kanzi's primal language: The cultural initiation of primates into language*. Palgrave Macmillan.

Pasternak, B. (1976). *Introduction to kinship and social organization*. Prentice-Hall.

Pilbeam, D., & Young, N. (2004). Hominoid evolution: Synthesizing disparate data. *Comptes Rendus Palevol, 3*, 305–321.

Platnick, N., & Cameron, H. D. (1977). Cladistic methods in textual, linguistic, and phylogenetic analysis. *Systematic Zoology, 26*, 380–385.

Potts, R. (2004). Paleoenvironmental basis of cognitive evolution in great apes. *American Journal of Primatology, 62*, 209–228.

Pozzi, L., Hodgson, J., Burrell, A., Sterner, K., Raaum, R., & Disotell, T. (2014). Primate phyletic relationships and divergence dates inferred from complete mitochondrial genomes. *Molecular Phylogenetic and Evolution, 75*, 165–183.

Preuschoft, S., & Preuschoft, H. (1994). Primate nonverbal communication: Our communication heritage. In W. Nöth (Ed.), *Origins of semiosis* (pp. 66–100). Mouton de Gruyter.

Preuss, T. M. (2007). Primate brain evolution in phylogenetic context. In J. Kaas & T. Preuss (Eds.), *Evolution of nervous systems: A comprehensive reference, V4* (pp. 1–34). Elsevier.

Puch, K. (2022). Psylogenetic analysis of middle-late Miocene Apes. *Journal of Human Evolution, 165*, 103140.

Purvis, D., White, L., Zheng, D., Andrews, T., & Riddle, D. (1996). Brain size, behavior and the allocation of neural space. In D. Magnusson (Ed.), *The lifespan development of individuals: Behavioral, neurobiological, and psychosocial perspectives* (pp. 162–178). Cambridge University Press.

Raaum, R., Kirstin, L., Sterner, N., Noviello, C., Stewart, C., & Disotell, T. (2005). Catarrhine primate divergence dates estimated from complete mitochondrial genomes: Concordance with fossil and nuclear DNA evidence. *Journal of Human Evolution, 48*, 237–257.

Radinsky, L. (1973). Aegyptopithecus endocasts: Oldest record of a Pongid brain. *American Journal of Physical Anthropology, 39*, 239–248.

Rhine, R., & Maryanski, A. (1996). A twenty-one year history of a dominant stumptail matriline. In J. E. Fa & D. G. Lindburg (Eds.), *Evolutionary biology and behavior of macaques evolution and ecology of macaque societies* (pp. 473–499). Cambridge University Press.

Ross, C. F., & Martin, R. D. (2007). The role of vision in the origin and evolution of primates. In J. Kaas & T. Preuss (Eds.), *Evolution of nervous systems: A comprehensive reference* (Vol. 4, pp. 59–76). Elsevier.

Rouquier, S., & Giorgi, D. (2007). The loss of olfactory receptor genes in human evolution. In J. Kaas & T. Preuss (Eds.), *Evolution of nervous systems: A comprehensive reference, V4* (pp. 129–139). Elsevier.

Savage-Rumbaugh, S., & Lewin, R. (1994). *Kanzi: The ape at the brink of the human mind*. John Wiley.

Savage-Rumbaugh, S. J., Murphy, J., Seveik, K., Brakke, S. L. W., & Rumbaugh, D. (1993). Language comprehension in the ape and child. In *Monographs of the Society for Research in Child Development 58*. University of Chicago Press.

Seiffert, E., Tejedor, M., Fleagle, J., Nove, N., Cornejo, F., Bond, M., de Vries, D., & Campbell, K. (2020). A Parapithecid stem anthropoid of African origin in the Paleogene of South America. *Science, 368*, 194–197.

Semendeferi, K., & Damasio, H. (2000). The brain and its Main anatomical subdivisions in living hominoids using magnetic resonance imaging. *Journal of Human Evolution, 38*, 317–332.

Service, E. (1965). *Primitive social organization*. Random House.

Sherwood, C., Subiaul, F., & Zawidzki, T. (2008). A natural history of the human mind: Tracing evolutionary changes in brain and cognition. *Journal of Anatomy, 212*, 426–454.

Sherwood, C. C. & Hof, P. R. (2007). The evolution of neuron types and cortical histology in apes and humans. In *Evolution of Nervous Systems: A Comprehensive Reference V4 (pp 355–378)*. Elsevier.

Silcox, M., & López-Torres, S. (2023). Primate origins: The earliest primates and Euprimates and their role in the evolution of the order. In C. S. Larsen (Ed.), *A companion to biological anthropology* (pp. 365–379). Wiley Blackwell.

Skov, L., Peyégne, S., Popli, D., Lasi, L., Devièse, T., et al. (2022). Genetic insights into the social organization of Neanderthals. *Nature, 610*, 519–525.

Smith, T., Garrett, E., Bhatnagar, K., Bonar, C., Bruening, A., Dennis, J., Kinznger, J., Johnson, E., & Morrison, E. (2011). The Vomeronasal organ of new world monkeys (Platyrrhini). *The Anatomical Record, 294*, 2158–2178.

Snowdon, C. (1990). Language capacities of non-human animals. *Yearbook of Physical Anthropology, 33*, 215–243.

Stebbins, G. L. (1969). *The basis of progressive evolution*. University of North Carolina Press.

Stein, P., & Rowe, B. (2011). *Physical anthropology*. McGraw Hill.

Suddenforf, T., & Collier-Baker, E. (2009). The evolution of primate visual self-recognition: Evidence of absence in lesser apes. *Proceedings of the Royal Society, 276*, 1671–1677.

Swedwell, L. (2011). African Papionins. In C. Campbell, A. Fuentes, K. Mackinnon, S. Bearder, & R. Stumpf (Eds.), *Primates in perspective* (pp. 241–277). Oxford University Press.

Tattersall, I. (2023). Endocranial volumes and human evolution. *F1000Research, 12*, 565.

Temerin, L., & Cant, J. (1983). The evolutionary divergence of old-world monkeys and apes. *American Naturalist, 122*, 335–351.

TenHouten, W., & Kaplin, C. (1973). *Science and its mirror image: A theory of inquiry*. Harper and Row.

Tomasello, M., & Call, J. (2007). Ape gestures and the origins of language. In J. Call & M. Tomasell (Eds.), *The gestural communication of apes and monkeys* (pp. 221–239). Mahwah.

Tomasello, M., & Herrmann, E. (2010). Ape and human cognition: What's the difference? *Current Directions in Psychological Science, 19*, 3–8.

Turner, J. (2021). *On human nature: The biology and sociology that makes us human*. Routledge.

Turner, J., & Maryanski, A. (2008). *On the origin of societies by natural selection*. Paradigm Publishers.

Turner, J., & Maryanski, A. (2024). *Biosocial evolutionary analysis: Reconciling biology, psychology and sociology*. Edward Elgar Publishing.

Urciuoli, A., & Alba, D. (2023). Systematic of Miocene apes: State of the art of a Never ending controversy. *Journal of Human Evolution, 175*, 103309.

Von Senden. M. (1932/1960). *Space and sight. The perception of space and shape in the congenitally blind before and after operation* (P. Heath, Trans.). Methucn.

Vorobyev, M. (2004). Ecology and evolution of primate color vision. *Clinical and Experimental Optometry, 87*, 231–238.

Walker, A., Falk, D., Smith, R., & Pickford, M. (1983). The skull of proconsul Africanus: Reconstruction and cranial capacity. *Nature, 305*, 525–527.

Walraven, V., Elsacker, L., & Verheyen, R. (1995). Reactions of a group of pygmy Chimpanzees (*Pan paniscus)* to their mirror-images: Evidence of self-recognition. *Primates, 36*, 145–150.

Ward, C. (2015). Postcranial locomotor adaptations of hominoids. In C. S. Larsen (Ed.), *A companion in biological anthropology* (pp. 1364–1380). John Wiley & Sons.

Wise, S. P. (2007). Evolution of ventral premotor cortex and the primate way of reaching. In J. Kaas & T. Preuss (Eds.), *Evolution of nervous systems: A comprehensive reference 4* (pp. 157–165). Elsevier.

Wolpoff, M. (1999). *Paleoanthropology*. McGraw-Hill.

Wrangham, R. (2009). *Catching fire: How cooking made us human*. Profile Books.

Wrangham, R., Jones, J., Laden, G., Pilbeam, D., & Conklin-Brittain, N. L. (1999). The raw and the stolen: Cooking and the ecology of human origins. *Current Anthropology, 40*, 567–594.

Xia, L., Bachmanox, A., Maehashi, K., Li, W., Lim, R., Brand, J., Beauchamp, G., Reed, D., Thai, C., & Floriano, W. (2011). Sweet taste receptor gene variation and aspartame taste in primates and other species. *Chemical Senses, 36*, 453–475.

Young, M., Virga, J., Kantounis, S., Lynch, S., Chernik, N., Gustafson, J., Cannata, M., Flaim, N., & Granatosky, M. (2023). How pendular is human brachiation? When form does not follow function. *Animals, 13*, 1438.

Zalmout, L., Sanders, W., Latchy, L. M., Gunnell, G., Al-Mufarreh, Y., Ali, M., Nasser, A.-A., Al-Masari, A., Al-Sobhi, S., Nadhra, A., Matari, A., Wilson, J., & Gingerich, P. (2010). New Oligocene primate from Saudi Arabia and the divergence of apes and old-world monkeys. *Nature, 466*, 360–365.

Zollikofer, C. P., Ponce de Leon, M., Lieberman, D., et al. (2005). Virtual cranial reconstruction of *Sahelanthropus tchadensis*. *Nature, 434*, 755–759.

The Neuro-anatomy of Humans' Complex Interpersonal Capacities

12

Jonathan H. Turner

Abstract

Humans are, in essence, evolved great apes, sharing a very high percentage of their genes with the three great apes: chimpanzees, gorillas, and orangutans. The neurology of all great apes allows them to communicate in episodes of interaction; and indeed, the genetic wiring for language comprehension exists among all great apes. The interpersonal capacities of great apes are outlined with an eye to understanding the neurological basis of human social interaction. Interaction among humans is very much the same as that among great apes but, at the same time, is very unique because of a much larger brain, both sub-cortically where emotions are generated, and neo-cortically where cognitions are formed and stored. Moreover, while great apes can understand human languages, they cannot speak because they lack the requisite vocal tract and key neurological structures, such as Broca's area, but they can communicate via sign languages and computers using pictograms that can be arrayed to produce sentences carrying common meanings. By comparing great ape and human brains, it is possible to understand not just speech and symbolic culture that differentiates humans from great apes, but also the elaboration of the dimensions along which human interaction occurs. These dimensions make human interaction more complex than among great apes, but the neurology involved is not all that unique, except for a few key structures and, more importantly, the size and connectivity within and among sub-cortical and neo-cortical areas of the human brain. An effort is made to conceptualize these interpersonal dynamics in an expanded view of the approach employed by George Herbert Mead but with an emphasis on the neurological basis of human interaction.

Social structures and culture are created and sustained by interaction among individuals. This emphasis social structure and culture differentiates sociologists from biologists and psychologist who, for the most part, study behavior, *per se*, leading to many useful insights, but at the same time, the narrow focus on behavior alone tends to under-emphasized the effects of interpersonal behavior in creating, sustaining, or even tearing down the socio-cultural structures from

J. H. Turner (✉)
Department of Sociology, University of California, Riverside, Riverside, CA, USA
e-mail: jonathan.turner@ucr.edu

W. Kalkhoff et al. (eds.), *Handbook of Neurosociology*, Handbooks of Sociology and Social Research, https://doi.org/10.1007/978-3-031-95615-7_12

which human societies are built. Moreover, social structures and their cultures are ultimately built from the dynamics inhering in interaction among individuals in encounters. Indeed, humans' rather remarkable inter-personal capacities, along with their enlarged sub-cortical areas generating emotions and their neo-cortical areas allowing for intelligence and enhance decision making are what allow humans to have such complex and nuanced interpersonal abilities and, moreover, to build up larger-scale social structures and their cultures.

One useful approach to understanding how and why humans have these capacities requires that we engage in evolutionary analysis to discover the origins of the special neurology making humans so adept at reading each others emotional states when interacting. In this way, we can make connections among human biology and behavior, on the one side, and on the other side, with the building up, maintaining, and even tearing down socio-cultural formations that are the building blocks of human societies. The approach for pursuing this agenda is what can be termed comparative neuro-anatomy.

Comparative Neuro-Anatomy

Almost 25 years ago, as someone trained in symbolic interactions and interaction more generally, I recognized that it was important to understand the neurology of inter-personal behavior. I realized that the work with Alexandra Maryanski on great ape behaviors could be extended to comparative neuro-anatomy (Turner, 2021; Turner & Maryanski, 2024), whereby it would be possible to compare human brains with those of humans' closest biological relatives, the great apes (chimpanzees, gorillas, and orangutans). Since humans and great apes share a very high percent of their genes (99+ percent with chimpanzees, 98+ percent with gorillas, and 96–97 percent with orangutans), it would be possible to conduct what is known as *cladistic analysis* on great apes and humans (See Turner & Maryanski, 2024; Andrews, 1981, 2015, 2020; Andrews & Martin, 1987; McGrew, 2010).

Cladistic Analysis: Looking Back in Time and Hard-wired Behavior Propensities

When present-day life forms share a high degree of their genes, it is a reasonable hypothesis that they once shared a common ancestor. The very high percentage of shared genes among chimpanzees, gorillas, orangutans, and humans means that these primates share common ancestors around 12 million years ago with the ancestors of present-day orangutans, 8–9 million years ago with the ancestors of current subspecies of gorillas, and 6–7 million years ago with the subspecies of common chimpanzees.[1] Around 6–7 million years ago, *hominins* split off from the ancestors of current great apes to initiate a line of evolution among new great ape species, termed *hominins*, that would eventually lead to the emergence of humans.[2]

At the behavioral and interpersonal level, great apes have well developed interpersonal skills that sustain great ape communities which, however, *do not* reveal permanent group structures—the basic organizational platform of most mammals. Rather, great apes live in communities or home ranges (that can be many square miles) in which many temporary interpersonal "hook-ups" occur. But on the whole, apes have *very few strong social ties.* Among orangutans, there are virtually *no* enduring strong ties, except mother-offspring ties that are broken forever when male and female offspring leave their material community, never to return; the same is true among gorillas. Among chimpanzees, all female offspring break the strong bond with mothers at puberty and move to other communities. In contrast to orangutans and gorillas, however, chimpanzee males remain in their mother's

[1] *Bonobo chimpanzees* are not included, even though they share as many common genes with humans as common chimpanzees, but they evolved later in all likelihood from common chimpanzees around 2.0 to 2.5 million years ago.

[2] For more detailed analysis of the evolution of great apes and hominins, see: (Maryanski & Turner, 1992; Turner & Maryanski, 2005, 2024; Turner, 2000, 2002, 2005, 2007, 2021).

community and visit them but do not form a permanent group with their mothers. Rather, males periodically visit their mothers for brief visits, while forming more stable friendships with other males and their brothers but even these stronger relations *do not* lead to permanent day-to-day group formations

Yet despite the lack of permanent groups, all great apes possess well develop interpersonal capacities, which may seem strange for such weak-tie, non-group forming animals. A moment's reflection, however, makes clear why they have these *exceptionally refined interpersonal abilities*. In a weak-tie social structure where individuals wander around the community, hook up with fellow community member for a brief time, and then move on requires a high degree of intelligence, which great apes have with brains of at least 350cm^3 or a bit more, allowing them to remember previous interactions with other members of a community. And, when two great apes meet and interact, they engage in greeting rituals, seemingly pick up from their last meeting, and then close interactions with ritualized departures—much like humans do. Thus, in a week tie-social structure, it is *interpersonal acuity that makes communities viable* and, moreover, that make large societies with few strong ties possible.

In Alexandra Maryanski's (1986, 1987, 1992, 1993, 1995, 2013) pioneering *cladistic analysis* of great social structures, she coded all of the field studies over the last 100 years of primate research and coded tie relations among adult females and males, among adult females, among adult males, and females and their offspring.[3] Males and their offspring cannot be coded because great apes do not organize into families, and hence, paternity is never known because all great apes are *highly promiscuous* sexually. And, as noted above, the only strong ties are between females and their offspring, and these are broken at puberty when all females migrate away from their community, never to return, and when both males *and* females among gorillas and orangutans leave their natal community at puberty, with only male chimpanzee offspring remaining in their community for a lifetime.

Since the ancestors of present-day great apes are the ancestors of *hominins* on the line to humans, it is reasonable to assume by the logic of cladistic analysis that humans must have these same hard-wired interpersonal capacities as great apes. Thus, *much inter-personal behavior is biologically and, more specifically, neurological based*—with an inherited neurology being central to understanding human interpersonal behavior. Without going into detail at this point, let me list the behavioral and interpersonal and capacities of great apes that are relevant to understanding their interpersonal facility:

1. *Reading Faces and Eyes of Conspecifics*. With visual dominance over haptic and auditory sense modalities, thereby subordinating other sense modalities to vision, great apes can communicate common meanings through visual cues about their intended lines of action (Maryanski & Turner, 1992; Jarvis & Ettlinger, 1977; Menzel, 1971).
2. *Following Eye Movements to Determine Likely Interests of Conspecifics*. Great ape behavioral propensities to follow the gaze and eye movements of others leads to the propensity to interpret gestures of eyes and face. In, fact, great apes follow gaze and eye movements to determine what another is observing and thinking The result is a high level of inter-subjective understandings of the intended courses of actions among great apes (Itakura, 1996; Baizer et al., 2007; Tomasello et al., 2001; Hare et al., 2001; Povinelli, 2000; Povinelli & Eddy, 1997).
3. *Ability of Infants to Imitate Orofacial Movements of Caretakers Within Weeks of Birth*, thereby indicating a genetically driven propensity to seek to understand the emotions of conspecifics, especially during face-to-face interactions. This capacity emerges within two weeks of birth, as it does for humans (Ekman, 1984; Tomonaga, 1999; Subiaul, 2007).

[3]These data have been updated over the last 40 years, again supporting the portrayal outlined above about great ape social relations.

4. *Empathy and Role-Taking ("theory of mind).* Great apes engage in intense face-monitoring for signs of action by conspecifics particularly for emotional content, thereby increasing capacity to read emotional dispositions of conspecifics. What G. H. Mead (1934) termed "role-taking" of putting oneself in the positions of others, and *vice versa*, was thus inherited from great apes by humans as a hard-wired propensity (Leslie et al., 2004, Gazzaniga & Symlie, 1990; Parr et al., 2005; Tsao et al., 2003).
5. *Capacity to communicate meanings and coordinate actions through non-verbal signals*, especially by eyes and face, allows great apes to coordinate interaction among conspecifics as well as to coordinate collective action among conspecifics (Menzel, 1971; Turner & Maryanski, 2008).
6. *Enhanced Propensity to Use Imitation to Learn Appropriate Signals and Behaviors.* Hence, great apes are hard to acquire the means of symbolic communication at birth for facilitating interactions and, when needed, is used to coordinate action among conspecifics (Tomonaga, 1999; Subiaul, 2007; Horowitz, 2003; Gregely & Csibra, 2006).
7. *Rhythmic Synchronization of Communication.* Field studies indicate great apes, especially chimpanzees, document their propensity to fall into synchronization of auditory sounds and body movements, and presentations of self during interaction. Rhythmic synchronization, especially of emotions via mirror neurons, were thus a necessary pre-adaptation for establishing capacities for a smooth flow to interaction (Schutz-Bosbach & Prinz, 2007; Rizzolatti et al., 2002; Rizzolatti & Sinigaglia, 2008).
8. *Collective Emotional Effervescence.* This synchronization of gestures and emotions can escalate in occasional gatherings of larger numbers of conspecifics, often escalating into arousal and synchronization of emotions in what appear to be "ritual-like" interactions, very much like humans when emotions are activity in collective actions, such as festivals like Mardi Gras (Spencer & Gillian, 1899; Durkheim, 1912/1947; Goffman, 1967; Collins, 2004; Rizzolatti et al., 2002; Rizzolatti & Sinigaglia, 2008).
9. *Increased Cortical Control of Sub-cortical Areas of The Brain Through Projections of Axons.* While great apes do not have the same level of cortical control over emotions as humans (although this is often tenuous), natural selection appears to have begun this process of developing neo-cortical control over sub-cortical emotion centers to reduce the chances of disrupting interactions through excessive emotional arousal (Raghanti et al., 2000; Sherwood, 2007; Sherwood et al., 2005; Semendeferi et al., 2002).
10. *Enlarged Decision-making Prefrontal Cortex Compared to Most Other Mammals, controlling for body size.* While the human prefrontal cortex for decisions making by tagging cognitions with emotions is far beyond that of great apes, this structure was under some selection and thus represented a way for blind selection to grow further once the neocortex and surrounding cortices of late hominins began to grow rapidly in late hominin evolution (Stephan, 1983; Stephan & Andy, 1969; Stephan et al., 1970, 1981, 1988).
11. *The Ability Among The Great Apes to Learn And Use Language at The Level of a Three-year-old Human Child.* This capacity represents a hard-wired the precursors for language facility. This capacity gave natural selection something to select upon if more refined verbal communication would be fitness enhancing. Without this pre-existing capacity brought out in great apes in laboratory studies of great apes communicating with humans, spoken language would never have evolved along the hominin line (Geschwind, 1965a, 1965b; Rumbaugh & Savage-Rumbaugh, 1990; Savage-Rumbaugh & Lewin, 1994; Savage-Rumbaugh et al., 1993; Bickerton, 2003).

12. *The Capacity for Self Awareness*. The ability to recognize an image in a mirror as a reflection of self as an object in the environment, thereby creating the necessary pre-adaptation for the formation of identities and, thus, self-monitoring self and others in interaction (Gallup, 1970, 1979, 1982; Gallup et al., 2014).
13. *Reciprocity in The Give And Take (Exchange) of Resources.* This hard-wired capacity for reciprocity exists among many higher mammals but it is particularly pronounced among great apes; and with further selection, it would become a strong bonding mechanism in animals like hominins when needed for more stable group structures (Cosmides, 1989; de Waal, 1989, 1991; de Waal & Brosnon, 2006).
14. 14. *Calculations of Fairness in Exchanges of Resources*. Great apes compare shares of resources with others in making judgments of fairness and justice of exchanges. This propensity would be critical in bonding among low-sociality animals like great apes, and it is no doubt the hard-wired basis for morality among what became culture-using animals like humans (Bronsnan, 2006; Bronsnan & de Waal, 2004, 2014).
15. *Capacity to Experience Variants Among Primary Emotions*, particularly among chimpanzees. Great apes have the same primary emotions as humans, as do almost all mammals. But, great apes appear to have more variants of these emotions, which again, gave blind natural selection a path to enhance and thereby increase the use of emotions as the fundamental force allowing hominins to forge stronger group ties (Turner, 2000, 2002, 2007, 2021; Turner & Maryanski, 2008, 2024).
16. *Capacity to Experience Empathy with Conspecifics,* thus facilitating in-depth role-taking for emotions experienced by conspecifics. Empathy is a critical cognitive and emotional capacity for generating stronger social bonds; and its presence among low-sociality great apes provided yet another "hook" for natural selection to enhance as a means for creating stronger social bonds along the hominin line (Call & Tomasello, 1998; Cooley, 1902; Damasio, 2000; Emde, 1962, 1980; Rizzolatti & Sinigaglia, 2008; Scheff, 1988).[4]

As is evident in this list, great apes have many of the same neurological wiring for interpersonal capacities as humans, in more rudimentary form to be sure, but the key point is that, once these behavioral and interpersonal capacities existed, they would be available for natural election to enhance and elaborate these capacities. And, this is the important point: Neurological capacities outlined above had probably existed for millions of years before hominins split off from the ancestors of today's great apes. So, in this sense, looking at great apes is looking back in time, seeing humans distant ancestors through the "telescope" provided by cladistic analysis. And, what is evident is this: Humans' hominin ancestors as they completed splitting off the great-ape evolutionary line some 6–8 millions ago possessed many of the capacities for sophisticated interpersonal behavior that humans reveal today. Most of these have a neurological basis, and so, enlarging the sub-cortical areas where emotions are generated would, in turn, allow for the neo-cortex to grow and, once the roadblocks to speech began to be overcome some 1.5 millions, spoken language would soon produce symbolic culture, *all built upon existing interpersonal capacities* that had existed along the great ape evolutionary line for millions of years. So, much of the heavy lifting

[4]Citations are to data demonstrating the existence of these refined behavioral propensities in great apes and humans, as these provided a neurological base on which natural selection could "select" upon to enhance stronger social bonds among hominins and then humans. These propensities of current higher-order apes go a long way to understanding *what brain structures natural selection worked on to produce the more extensive abilities of humans* to be more social (Passingham, 1973, 1975; Sherwood et al., 2008). Still there is much more detailed work needed to understand the extent of great apes' capacities for intersubjective empathy and other emotional and cognitive states.

for enhanced interpersonal capacity had been done *over a 20 to 30 million years ago* as the great ape clade evolved, providing a basis for subsequent selection to rework the brain for enhanced interpersonal behaviors in interpersonal relations among humans.

Coupled to what used to be termed "pre-adaptations," which is the label that I still prefer over the new term, *exaptations*, that is currently used, the existing hard-wired neurology for language that *had been wired into the brain of some apes at the very beginning of primate evolution*. There is sat in the association areas in and around the *inferior parietal lobe* where the *occipital, temporal*, and *haptic lobes* meet pushing for vision over olfaction (which the dominant sense modality of most mammals but not humans). It is unlikely that spoken language could have evoluted in hominins almost 60 million years later if this random byproduct of selection for visual dominance had not occurred (because, obviously, vision is more fitness- enhancing that olfaction in living in the tree dimension world of the arboreal habitat). Even so, it took millions of years of hominin evolution to take advantage of this pre-adaption for language as random selection slowly realigned of the vocal tract and surrounding structures of hominins who, as evolving great apes, initially had little capacity for fine-tuned vocalization necessary for spoken language.

The 'Language of Emotions' and Interpersonal Relations

The evolution of the neurology of the hominin brain was driven by selection enlarging key sub-cortical structures of the hominin brain and the body-systems generating emotions (see Chapters 3, 7, 8). As a larger palette and range of emotions evolved, the neo-cortex could grow; and after roadblocks to spoken language were finally overcome, speech could then be used to create symbolic culture—*always* valenced by emotions. Great apes can read emotions with a high degree of acuity via the visual modality but cannot speak because of their compressed vocal tracts. But their neurology allows them to communicate through sign language. Perhaps more importantly, they can like humans today communicate emotional states through face and body "languages." This capacity for a "language of emotions" was where natural selection *began to alter the neurology of hominins*. With a capacity to communicate meaning to others about their feelings, they had the rudiments of a language of emotions, an adaptive capacity that allowed hominins to survive in open-country habitats in Africa by bonding through an enhanced repertoire of emotions to forge strong bonds and group solidarities. Indeed, most mammals, birds, and reptiles have similar capacities, with more intelligent mammals being more emotional because *intelligence is based upon the capacity to tag cognitions with emotions* (Damasio, 1994; LeDoux, 1991, 1993, 1996).

Thus, most intelligent mammals have an implicit "language of emotions" communicated via vocal calls, movements and juxtaposition of bodies, and movements their hands, arms, heads, beaks, trunks, wings, tails, flippers, etc. This language of emotions dominates interpersonal relations and collective organization among all intelligent species. So, it is not surprising that natural selection worked on emotions *if* changes in behavior, social relations, or group organization were necessary in new ecological habitats requiring changes in phenotypes and behaviors. Such was the case for hominins as selection first made them bi-pedal and then began to increase the size of sub-cortical emotion centers to make them more emotional and able to form, over millions of years, more stable group structures which would be fitness-enhancing in the open-country habitats that were emerging over the last ten million years as the forests in Africa began to retreat.

And, among all intelligent animals, interactions among conspecifics generally involve emotion displays and interpretations of these displays for discerning how a conspecific will act. As is apparent in the above list of behavior capacities and propensities of great apes, and hence hominins' common ancestor with the ancestors of current great apes, enhanced emotions increased fitness by allowing for more permanent bonds among emotionally "attached" conspecifics.

What is surprising, in some ways, is that it was not until about 40 years ago that the "Theory of Mind" was proposed by D. G. Premack and G. Woodruff (1978) with the question: "Do chimpanzees have a theory of mind?" That is, can the gestures emitted by one chimp allow others to put themselves "in the mind" of the chimp emitting emotions and, thereby, respond accordingly. To a sociologists, this would be obvious, but in the 1970s there was a great deal of resistance by some scholars to what was obvious (but still not accepted in some circles): great apes can learn sign language and thus communicate with humans, thereby blowing up the myth that language capacities are what distinguish humans from their "lesser" evolutionary cousins (i.e., the great apes), even though it became obvious when it became possible to map the genomes of great apes and humans revealing that they share a very high proportion of genes.

This debate among primatologists, neurologists, and psychologists led to neuroscience studies to do what G. H. Mead (a century ago) had termed *role-taking* or the mutual reading of the "mind" of others. Mead[5] developed these ideas at least 70 years before Premack and Woodruff (1978); and if they had known about Mead and the accumulated theoretical and empirical work in sociology about interaction, they might have used a less awkward label than "Theory of Mind." At any rate, their question led to important studies in neurology and psychology (e.g., Call & Tomasello, 1998; Saxe & Kansisher, 2003) recognizing that the *prefrontal cortex*, *posterior temporal sulcus*, *temporo-parietal junction*, and other structures facilitated the capacity for successful role-taking—that is, the process of "getting reciprocally inside each others' brains" so as to understand the feelings and dispositions to act in certain ways of each individual in an interaction.

Using G. H. Mead's Theory to Expand the Conceptualization of Interaction Dynamics

George Herbert Mead (1934) was part of the pragmatist movement in philosophy, but he became an icon in early American sociology because of his insights into the dynamics of interaction. Surprisingly, he did not analyze emotions in any detail but instead focused on the fundamental mechanism by which humans interact and, thereby, derive a sense of self and create the structures that hold societies together. He saw interaction revolving around the process of "role-taking" whereby individuals read each others' gesturers—both verbal and non-verbal—to determine their dispositions to act in a particular manner. In so doing, they could mutually communicate their respective self-conceptions to others and their likely courses action. In fact, Mead used the label *role-taking* to encompass much more that mutual reading by individuals the self and roles to be enacted. The notion of role-taking was, in fact, not about only roles, but any dimension of interaction. Yet, it was until the early 1960s that scholars began to expand upon Mead's fundamental insight. Ralph H. Turner (1962) was took the first conceptual step in expanding the notion of role-taking by introducing the notion of *role-making*. For R. H. Turner,[6] much more than self and roles being are communicated in mutual talk and body language. At a minimum, expectations, cultural norms, values, beliefs, emotions, and much more through are reciprocally communicated during any interaction. And, since that time, scholars expanded what became known as *symbolic interaction,* a loose confederation of theorists and researchers in sociology addressing many more dimensions of interaction. In the list above of the interpersonal capacities of great apes, it is evident that a chimpanzees or any great ape with comparative small brain compared to

[5]However, Mead tended to think that language was acquired socially and downplayed its genetic basis, which is perhaps understandable in an era when genetics was just emerging as a driving force of evolution.

[6]I am often asked, as J. H. Turner, if R. H. Turner was my father (two sociologists living in southern California). He was not (he was too young) but my exposure to his work as an undergraduate still influences my work 60 years later.

humans can do everything that Mead outlined, even though he tended to see the capacities for self and role-taking as unique to humans. But great apes and few other animals possess a sense of self (Gallup, 1970, 1979, 1982; Gallup et al., 2014), and all animals with or without a sense of self can communicate in vocalizations and expressive gestures a great deal of information to each other, which in turn, facilitates the flow of interaction and formation of groups.

The neurology of hominins and then the last hominin, humans, allows for a much more robust mutual signaling of self, intentions, and expectations of persons as they interact. I argue that Mead's initial insight on role-taking, or more recently, the notion of "Theory of Mind" are much more complex than originally outlined, once the capacities of humans in interaction are more fully documented (see Turner, 2002, 2021). And where possible, I will at least mention some of the neurology involved, but more importantly, suggest lines of research on the neurology of these "making" and "taking" of self, situation, roles, status, and other dimensions of social interaction.

Figure 12.1 outlines a simple model of some of the important dimensions along which the "*making*" (though talk and non-verbal gestures) dimensions of self and situation relevant to the interaction. The reciprocal dynamic is "*taking*" (interpreting the talk and non-verbal gestures) that others see as salient dimensions. In fact, much mutual *taking* and *making* communicates *individuals' respective perceptions of the salient portions of the social structure and cultural context of the interaction*, above and beyond presentation of their self conceptions of themselves and expectations for how self should be treated. Figure 12.1 lists many of the most important dimensions that are communicated in human interactions.

The *hippocampus* and *frontal lobe* store memories that, in turn, are made available via the *prefronal cortex*, and given emotional valences by sub-cortical areas of the brain and activation of body-systems (see Chapter 7 on the *neurotransmitter system*, *the peptide and hormonal system*, *the autonomic system*, and *musculoskeletal system*). Alfred Schutz (1932) termed these emotionally valenced cognitions, *stocks of knowledge at hand,* that are communicated in the mutual exchanges of talk and *body-language*. The seeming rather long lists of "making" and "taking" in Fig. 12.1 is a more elaborated view of role-taking and making along *additional dimensions that human neurology makes possible*. These are, in essence, what people pay attention to in face-to-face interactions; and while they overlap, they each represent a dimension of interpersonal behavior that can come into play as individuals listen to talk and read body language. Some degree of consensus among those dimensions' salience is necessary in a given situation *if* interactions are to flow and become synchronized. And as individuals interact, they generate feedback through human neurology, both the neo-cortex and sub-cortex. The much larger size of both the neo-cortex and sub-cortex in humans expands the complexity of the process of interaction, especially as, first of all, spoken language and symbolic culture evolved along the late hominin line and, then, as the complexity social structures eventually began to evolve during societal evolution. Each of these dimensions are summarize below

Role-Taking and Role-Making

As mentioned above, Ralph H. Turner (1962) proposed that the reciprocal of Mead's notion of *role-taking* should be *role-making.* When individuals role-take, they read the gestures of others, while at the same time, present gestures to others in an effort to *make* roles for themselves *vis-à-vis* others. With this explicit and implicit presentation of gestures, or what Irving Goffman (1959) termed the "presentation of self," individuals provide the information necessary for others to role-take. Sometimes, this information is deliberately orchestrated but, equally important, are expressive signals that simply slip out unconsciously. In either case, humans can be seen as not only role-taking but also as *role-making* animals; and the more individuals role-make, the easier will the role-taking efforts of others become.

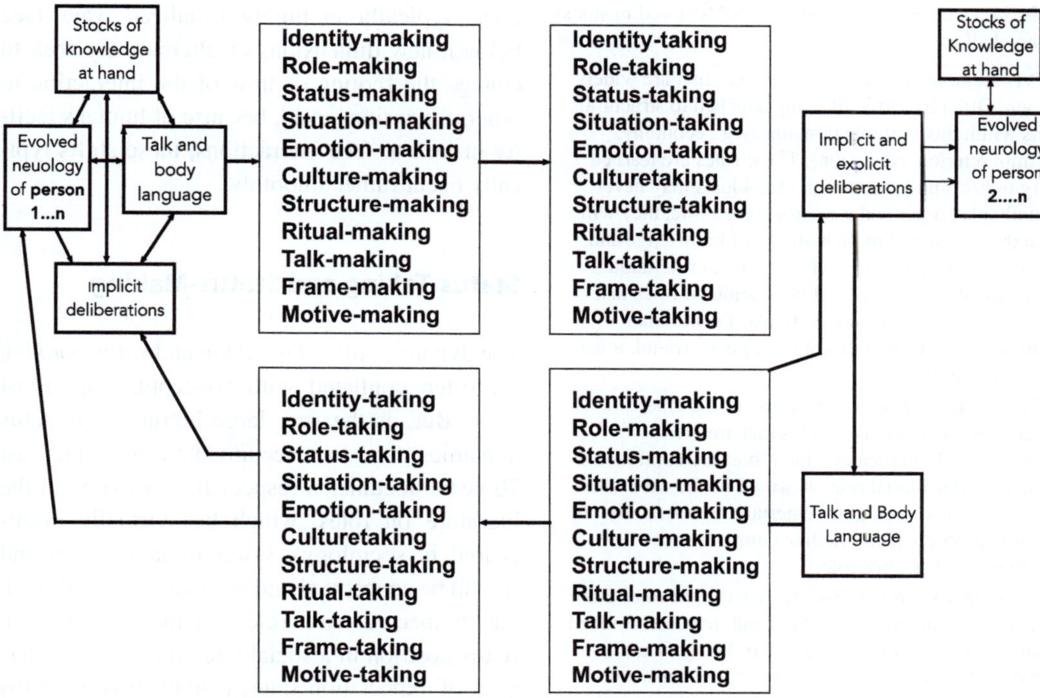

Fig. 12.1 Model of Fundamental Components Communicated During Human Interaction

Role-making and role-taking both depend upon individuals possessing "stocks of knowledge at hand"—to use Schutz's (1932/1967) adaptation of Edmund Husserl's vocabulary. Humans carry in their *frontal cortex* inventories of roles for most situations, and they draw upon these inventories when making a role for themselves and when trying to determine the role being asserted by others. Table 12.1 outlines some of the basic types of roles that humans store in their long-term memory banks in *the frontal cortex*. As an individual role-makes and role-takes, this person implicitly assembles a role for self, while seeking to understand the role that others are making for themselves. These respective roles will also carry moral evaluations and be valenced with emotions pulled up from sub-cortical areas of the brain, via the neuro-nets connecting the *prefrontal cortex to emotion modules and nuclei in the sub-cortex* of the human brain. What is needed from neuroscience is a better understanding of how these inventories of roles are stored, how they are retrieved, how they become emotionally valenced, and how they are assembled along the patterns outlined in Table 12.1.

Some time ago (Turner, 1997, 1998, 2000,2002), I began to recognize that G. H. Mead and R. H. Turner (1962) had produced a useful way to conceptualize other key interpersonal dynamics. In fact, too much is packed into Mead's and R. H. Turner's notion of roles; and by unpacking the dimension as is outlined in Fig. 12.1 a much better conceptualization of interaction is possible. As Fig. 12.1 emphasizes, the imagery of "taking" (through reading situational gestures and cues) and "making" (by deliberately and unconsciously orchestrating presentations of gestures) is useful when isolating other interpersonal processes beyond roles, and so, as will become evident, I will conceptualize *all interpersonal dynamics* through the prism provided by the notions of *taking* and *making*. Moreover, by unpacking these dimensions along which interaction occurs, we are also in a better position to discovery the neurology underlying any of these dimensions (Turner, 2011, 2019, 2021).

Table 12.1 Inventories of types of roles stored in stocks of knowledge

1. ***Preassembled roles***. These are roles that are widely known. Often by virtue of being attached to structural units within institutional domains (e.g., economy, kinship, religion, education). These roles are learned early in life, and even though individuals may never actually play a particular preassembled role, they will have the role stored in their stocks of knowledge and, as a result, be able to identify the role in role-taking. There are also more general preassembled roles that are widely known outside particular institutional domains, such as the roles of host, guest, friend, lover, and companion.
2. ***Combinational roles***. These are, as the label suggests, combinations of roles that are well understood. Often preassembled roles are combined to form a combinational role, as would be the case with a host who is also a daughter entertaining her family kin unit. Or, a generalized role like companion can be combined with a work role.
3. ***Generalized roles***. These are generalized roles that individuals generally understand and that can be added to almost any other role. The role of being assertive, social, upbeat, gracious, shy, reserved, serious, and diligent, for example, are all stored in people's stocks of knowledge, and they can recognize variants of these role alone or in combination with other roles.
4. ***Trans-situational roles***. These are roles that people carry with them from situation to situation. They are typically associated with membership in particular social categories or categoric -units. For instance, gender, class, and ethnicity generate expectations for how individuals will play roles, and typically, individuals carry these roles and play them out attached to another role.

Note: These four types of roles are only examples of how roles are cognitively stored in stocks of knowledge. There are certainly hundreds and perhaps thousands in a complex society of roles stored in these stocks. There are, no doubt, other dimensions along which roles are organized in the frontal cortex, but the key point is that role-taking and role making are made much easier because individuals already have the building blocks of a role stored in their stocks of knowledge, thereby making role-taking and role-making considerable easier and faster

The order of presentation of these *making* and taking processes is somewhat arbitrary since, in any given interaction, some of these dimensions become more important than others. Moreover, as each interaction unfolds the initial dimensions of making and taking suggest other dimensions which become part of a *re-making* and *re-taking* process. Furthermore, and at any given movement, participants in an interaction may introduce, typically in highly ritualized ways (see below), new dimensions or alternatively seek to change the footing or flow of the interaction to other dimensions. And, because of humans facility in face-to-face interactions, these shifts typically occur rather smoothly.

Status-Taking and Status-Making

The dynamics of *status-taking* and *status-making* are often conflated with conceptualizations of roles. But, as the very large literature on status dynamics that has accumulated over the last 70 years documents (especially compared to the literature on roles, which has virtually disappeared in sociology), status dynamics can and should be conceptualized as distinct from, though interconnected with, role dynamics. True, each status position in a social structure generally has a set of expectation states (established by status beliefs and norms) that constrain the roles that individual's play. Moreover, role-taking will often rely upon *status cues* as a source of useful information in role-taking. The converse can also be true: roles can be used to determine the status location of others when social structures are more fluid and ambiguous. In such cases, roles become cues for status-taking, or the process whereby individuals determine the status of others in a situation *vis-a-vis* self. Moreover, as researchers, such as Joseph Berger and colleagues (e.g., Berger et al., 1977; Berger & Zelditch, 1985) have clearly documented in the large literature of status organizing processes, interaction revolves around a process of not only determining the status of others (the *status-taking* dynamic), but these research projects have demonstrated that status dynamics revolve around games of *micro-politics* and *micro-economics* in which individuals seek to assert a given status *vis-a-vis* others and the expectations states on self and others that this status carries.

Some neurological data document that the *amygdala*—the center for *fear* and *anger* in all mammals and reptiles—is involved in status dynamics. This finding alone signals that the neurology of status is somewhat different than that for roles. True, much like roles, people store inventories of status information in their *frontal cortex* and use these stocks of knowledge in status-taking and status-making, but status dynamics arouse more intense emotions than role dynamics. The more individuals *status-make* and seek to establish their superiority over others, the more will *anger* and *fear* be activated in the *amygdala*, probably because of the micro-politics of rank and power that flow in many interactions. These involve implicit and explicit aggression and fear responses in the interaction between a person and others. The more successful are persons in establishing their higher rank, the more sub-cortical areas generating, at a minimum, *satisfaction* or, at a maximum, *happiness* will also be activated—above and beyond what is evident for successful role-making. As Theodore Kemper and Randall Collins (1990) have theorized, individuals experience positive emotional energy when they have successfully established higher status than others, whereas those who they have failed to make for themselves the status that they had expected to establish in a situation will experience negative emotions, and the more individuals lose status during the course of interaction, the more will the core centers for *anger, fear,* and *sadness* as well as *shame* and *alienation* be activated. To a lesser extent, the same dynamics hold for role-making, but I would argue that status dynamics activate more areas of the brain because they are explicitly tied to people's ability to sustain their sense of self-worth and power *vis-a-vis* others in social contexts.

Structure-Taking and Structure-Making

Interactions generally occur within a social structure, but individuals in interaction often signal to others the relevant aspects of social structures that are to guide the interaction, which can be different than the structure in which individuals are actually incumbent. Once the relevant structure is implicitly established, many other dimensions of "making" and "taking" fall into place, such as that status of individuals in the present structure in which individuals are seeing themselves as co-present, or an alternative structure along with other diffuse status considerations like gender and age. The roles that individuals can appropriately play also become clear, as do the cultural dimensions (e. g., norms, beliefs, values) that are to guide the interaction. And as these are implicitly accepted, the rituals appropriate for opening, forming, and closing the interaction become clear, as do the rules of talk and the kinds of frames appropriate for the interaction. Moreover, the emotions appropriate for the interaction also become easier to calibrate. Indeed, the nature of virtually all of the elements of making and taking become easier to understand once participants in the interaction fails to fall into line with the structural context that has been established.

As a general rule, any of the making and taking dimensions outlined in Fig. 12.1 can operate this way, if they are highly salient to participants in an interaction. Establishment of one highly salient dimension will constrain which other dimensions are to be relevant in an interaction. Some are more likely than others, but once a subset of the dimensions is implicitly agreed upon by individuals in interaction, this subset guides the rest of the making and taking dynamics. And, if a shift is to be made, it is almost always done by *ritual-making* and *ritual-taking* that announcing a shift in *frames*, and whatever else is going to be relevant to the interaction.

This ability to set up a set of the dimensions outlined in Fig. 12.1, and to alter them without too much disruption in the interaction is possible because of humans high level of acuity in reading not only verbal gestures but also body-languages communicating emotions. Moreover, the substance and the mode of speech in *talk-making* and *talk-taking* is often a signal of a shift in relevant dimensions guiding the interaction. All of these dynamics are possible by, of course, speech, but *body language* is often just as important. And, these capacities are the result of natural selection expanding the emotional repertoire of humans and, of course, the early rewiring of the brain for speech (in and around the *inferior parietal lobe*) that millions of years later led to reworking of the elements of the vocal tract and surrounding structures for speech production allowing for symbolic culture.

Moreover, as noted above, humans can be physically present on one structure but, in fact, invoking the salience of another social structure. And, humans can make shifts in the salient social structures because of the capacity to announce the shift and then emit those gestures relevant to the shift. For example, in one structure people may find themselves in conversations that have nothing to do with the status locations and culture of the structure in which they are physically co-present. This capacity to invoke a *different structural locations* in the "mind" is normally done effortlessly because of the large neo-cortex of humans and the ability to arouse appropriate affect, rituals, styles of speech, culture, and other dimensions listed in Fig. 12.1. This ability allows humans to have complexly layered interactions that can "jump structural or situation venues" by simply changing roles, status, social structures, frames, cultural referents, and other dimensions along which interpersonal relations flow.

Thus, the reason that the dimensions in Fig. 12.1 are not listed in sense a hierarchal order is because a shift in any of them will require a shift in other related dimensions. And this dynamic occurs constantly in human interaction because the evolved neurology allowing for storage and categorization of cognitions in memory (a *large frontal lobe* and *hippocampus*) and the capacity to retrieve rapidly and reshuffle salient cognitions (*prefrontal cortex*) and arouse new emotion states (*sub-cortex* and *body-systems* (neurotransmitter system, autonomic nervous system, musculoskeletal system, and peptide-hormonal system outlined in Chapter 7).

Situation-Taking and Situation-Making

Structure-taking and structure-making are often mediated by *situation-taking* and *situation-making*. In particular, the nature of the situation can be of three types (Collins, 1975): (1) *work-practical*, (2) *social*, and (3) *ceremonial*. These are generalized and can occur in many different types of situations and social structures, but whatever the nature of the more inclusive situation and social structure where this situation occurs, humans are attuned to *which of these types of situations* is more salient, which in turn provides general guidance for other types of "taking" and "making". Also, it is rare that a situation are purely one type, but rather, most situations involve a mix of these three types; and so given the relative salience of each, individuals have a means for adjusting behaviors and for determining other types of relevant "taking and making" outlined in Fig. 12.1. At a minimum, understanding the normative structure of a situation allows individuals to "*structure-take and structure-make*" as well as "*cultural-take and make*" which, in turn, may provide guidance on other dimensions outlined in Fig. 12.1. Thus, the structures in which individuals find themselves force re-adjustments in the relative amounts of *work-practical*, *social*, and *ritual* behaviors to be in play, as well what other dimensions listed in Fig. 12.1 are also in play. For example, going to work and place of religious worship generate situations within social structures and their cultures that guide the relative amounts of work-practical, social, and ritual-making (through *ritual-taking*).

These kinds of adjustments as they stimulate additional types of "making and taking" soon establish a firm bases for how individuals are to interact. And, in this way what is obviously a long list of "making and taking" dynamics listed in Fig. 12.1 becomes quickly more focused. Thus, whatever dimension of "making and taking" are initially salient sets off a rapid, and generally, a rather effortless assessment of the salience of "other dimensions of making and taking." I think that such is the case because individuals initially tag cognitions with emotions in the *hippocampus*, and then when remembered and re-arouse the emotional tags are eventually shipped up to the *frontal cortex* for longer-term memory. Moreover, these emotionally tagged cognitions are catalogued (in some unknown way) to be easily retrieved and blended by the *prefrontal cortex* with other dimensions that seem relevant in a given situation. This process occurs very rapidly and can alter the mix of dimensions in play.

Culture-Taking and Culture-Making

Mead borrowed from W. Wundt's (1912/1916) "folk psychology" the notion that individuals not only role-take with specific others but also with "generalized others." The generalized other is a "community of attitudes," and there can be multiple communities at varying levels of generality. If we unpack this notion of role-taking with generalized others, we can re-conceptualize the process *as culture-taking* or reading situational cues, scanning stocks of knowledge, and assembling the relevant cultural elements guiding the behaviors of individuals in a situation. The converse of this process of culture taking is *culture-making* where individuals orchestrate gestures, speech, and use of situational props to assert *what* cultural elements they think *should be relevant* in a situation. From people's mutual efforts at culture-taking and culture-making, a set of situational expectations is generated, and these expectations revolve, I believe, around the dimensions of situational culture summarized in Table 12.2.

Individuals carry cultural codes in their stocks of knowledge, stored in the *frontal cortex* and tagged with emotions from *sub-cortical areas* of the brain. Part of these stocks of knowledge are inventories of understandings as to when and where particular types of cultural codes can be invoked and used and when they cannot. Table 12.2 simply emphasizes the dimensions of a process that I have, at times, termed *normatization* or the assembling of expectations for episodes of interaction (Turner, 2002, 2011). These expectations revolve around *categorizing* (persons and situations) in an appropriate manner, establishing the relevant *frames* (Goffman, 1974), using the appropriate *forms of talk* and *body language*, invoking the appropriate *rituals* (Goffman, 1959,1967, 1981, 1983), and experiencing and displaying the proper *emotions and feelings* (Hochschild, 1979, 1983).

At a more general level, these situational expectations are constrained by beliefs and ideologies, institutional norms, and cultural values; and thus, as people scan stocks of knowledge, they first determine the relevant values, beliefs, and broader norms before they fine-tune these during the process of *normatization*. Individuals generally know that episodes of interaction, or *encounters* to use Erving Goffman's (1983) term, are embedded in larger-scale social structures that evidence an ideology (moral codes about proper conduct within an institutional

Table 12.2 Culture-taking and culture-making

Culture-taking and culture-making operate to establish expectations for how individuals should interact during the course of an encounter. These expectations revolve around the following axes:
1. ***Micro-framing of the encounter:*** The process of culture-taking and culture-making that imposes expectations for what can be included and, conversely, what is to be excluded as subjects of talk and nonverbal behaviors.
2. ***Categorizing the encounter:*** The process of culture-taking and culture-making in which individuals typify: (a) The categoric-unit memberships of participants in the encounter; (b) The relative amounts of work-practical, social, and ceremonial activity to be conducted in the encounter; (c) The degree of intimacy to be achieved with others along a continuum of treating others as personages (people as only representatives of categoric units or as incumbent in positions of corporate units), persons (with some knowledge of others as individuals), and intimates (with more in-depth knowledge of others); and (d) the relative authority/power of self and others, and on the basis of these nodes of categorization. Expectations for behaviors of self and others are developed.
3. ***Forming communication in the encounter.*** The process of culture-taking and culture-making by which expectations for the proper modes of (a) talk and conversation as well as (b) expressions of body language and demeanor.
4. ***Ritualizing the encounter:*** The process of culture-taking and culture-making in which expectations are developed for the appropriate rituals to (a) open and close interaction, (b) form and structure the flow of interaction, (c) symbolize the significance of the interaction, and (d) repair breaches to the interaction.
5. ***Emotionally energizing the encounter:*** The process of culture-taking and culture-making whereby expectations for the nature and valence of (a) emotions to be felt by a person and (b) emotions to be displayed to others are established.

domain) and broad institutional norms, and they understand the societal value premises on which these ideologies and institutional norms are based. These become, in essence, a *general or macro-frame* that delimits the options of individuals as they *culture-take* and *culture-make* within an encounter. They will also seek to set up situational expectations in the encounter with a more micro-frame that allows them to (a) establish categories for self, others, and situations; (b) settle on forms of talk; (c) enact rituals needed to open, close, track, and repair the interpersonal flow; and (d) emit emotions that are acceptable within the general frame provided by values, ideologies, and institutional norms. Yet, since there are still many options, even within this more delimited micro-frame, a considerable amount of ongoing negotiation during the processes of culture-taking and culture-making will occur in most encounters.

I have, of course, unpacked the notion of "generalized other" considerably beyond Mead's formulation, but this is necessary in order to have a more robust conception of culture and how it is assembled "on the ground" during the course of interaction. Since culture is almost always *moralized*, it reveals both cognitive and emotional dimension. Emotions moralize cognitions and give them the power to constrain interaction by allowing others to sanction those who deviate from cultural prescriptions and proscriptions or, even more effectively, by making deviants feel *shamed* and *guilty* for their conduct (thus leading them to sanction themselves and become motivated to make amends to others).

Frame-Making and Frame-Taking

The neurology of culture-taking and culture-making revolves around stores of emotionally charged cognitions about how people *should behave* in situations and the mechanisms by which individuals can scan relevant stores of "knowledgeability" (held in the *frontal lobe* and *hippocampus,* and retrieved by the *prefrontal cortex*) to assemble mutually acceptable cultural expectations. This process is facilitated by two *levels of framing* (Goffman, 1974), first at the level of society and macrostructure and, second, at the level of the encounter embedded in macro structures and meso-structures. If the embedding is unclear, then individuals must work very hard to figure out how to *normatize* an episode of interaction because they do not have the first-level frame. As far as I know, there is no great understanding of how framing dynamics are generated in the human brain, although emotionally tagged cognitions in the *frontal lobe*, scanned by the *prefrontal cortex* are certainly involved. The *prefrontal cortex* is probably involved in both levels of framing, rapidly invoking the relevant macro-level frames that, in turn, direct the process of micro-framing at the level of the encounter. And, once framed at the micro level, categorization, talk and body language, feelings, and rituals become clear—thereby allowing the encounter to be *normatized.*

Identity-Making and Identity-Taking

Higher primates are one of the few mammals that can identify their reflection in a mirror as their own image, and as I will argue later, this neurological capacity was selected upon and blown into the behavioral ability to possess multiple types of identities, including those listed in Table 12.3. Since identities are emotionally laced cognitions about self, they are a powerful motive force in human interaction. Again, the *prefrontal cortex* connects the cognitive part of an identity to sub-cortically generated emotions, but again, it is unclear just how this is done. It may be, as Burke and Stets (2009) have emphasized, that there is an *identity standard* stored in the *frontal cortex*, and this standard is evoked by the *prefrontal cortex* in making decisions about whether or not an identity has been verified by others. And, depending upon whether or not the person perceives the identity standard to have been met, the *prefrontal cortex* sends to *sub-cortical* emotion centers signals activating an array of positive or negative emotions. Emotions are

Table 12.3 Types of identities involved in identity-making and identity-taking

1. ***Core-identities***: The sum total of identities that individuals combine into a general core identity, which is not fully conscious and yet highly emotionally charged. When individuals cannot successfully identity-make at the level of their core identity, they will experience more intense emotions.
2. ***Categoric-unit identities***: These are identities attached to membership in various social categories—e.g., gender, age, sexual preference, ethnicity, religious affiliation, etc.—that are salient in a society and situations where individuals interaction. When such identities are affirmed, individuals experience positive emotions, and when they are not affirmed, they are likely to experience negative emotions.
3. ***Corporate-unit identities***. These are identities attached to incumbency in various types of corporate units—groups, organizations, and communities. These identities are often linked to the status locations and roles that individuals play in corporate units, with confirmation or non-confirmation of the identity attached to a role generating either positive or negative emotions.
4. ***Role-identities***. These are identities that can be attached to corporate units or more generalized roles—friend, host, emotional moods (happy, sad, playful), femininity, masculinity, etc.—that are played along with roles attached to positions in corporate units or membership in categoric units.

Note: These identities are arrayed in a rough hierarchy, with core-identities at the top, and role identities at the bottom. Emotions are potentially the strongest at core and categoric identities, with corporate unit and roles identities carrying less powerful emotional responses. Higher identities are often more complex and not fully consciously understood, whereas lower level identities are more clearly understand and easily articulated

positive when these outputs to the identity standard are congruent, whereas the arousal of negative emotions signals that there is disjuncture between the identity standard, on the one side, and the behavioral outputs and responses of others to these outputs, on the other. Again, the neurology here is not clear because sociologists have not designed studies using imaging technologies to determine how the brain is operating during the self-verification process. Psychologists and economists, among the social scientists, have been far more active in imaging studies than sociologists, but the questions posed by sociologists tend to be less interesting to researchers in these disciplines.

Motive-Taking and Motive Making

I see these two motive states as the most important in that individuals will devote most of the motive-taking and motive-making efforts to determining if these needs are being realized. Mead implied and most subsequent "symbolic interactionists" have emphasized that *verification of self or identity* is the most powerful motive state directing a person's behavior (Stryker, 1980; McCall & Simmons, 1978; Burke & Stets, 2009; Turner, 2002, 2011, 2021). Individuals are constantly identity-taking to determine which of many identities that a person has is most salient in a situation, and if it is possible, they will seek to verify this identity. The second most salient motive state is to *realize a profit in the exchange* of resources and especially a profit that meets normative standards of what would be considered "fair" and "just" in a given culture (as determined by *culture-taking*). Thus, in individuals motive-taking, they seek information about what resources are being exchanged, at what ratio, and at what level of fair exchange; reciprocally, individuals in their motive-making are signaling through their emotional arousal on a positive to negative scale whether or not the exchange of resources is fair and just.

There is clearly some hard-wiring for these motive states, since calculations of fairness and justice run through the primate line and perhaps the higher mammalian line as well (Brosnan & de Waal, 2004; Brosnan et al., 2005; de Waal, 1991, 1996). The *amygdala* is clearly the key sub-cortical region of the brain in calculations of justice, providing the emotional (moral) outrage for injustices, whereas the *anterior cingulate* and *neurotransmitter system* provide much of the positive emotions when justice and fairness are realized. The *frontal cortex* stores stocks of knowledge about justice norms and criteria for calculating justice, whereas the *prefrontal* cortex connects these stocks of knowledge to the relevant *sub-cortical emotion centers* in ways that are not fully understood but critical to understanding human behavior.

The process of calculating justice and fairness in exchanges is complicated by the dynamics briefly outlined in Table 12.4 where comparison points are mentioned. Resource payoffs relative to costs and investments in receiving these payoffs are always made relative to a comparison point or points. Depending upon *which* comparison point or set of points listed in Table 12.4 is invoked, the sense of justice or injustice will vary; and so, the key to understanding individuals and collective senses of fairness revolves around the neurology of invoking comparison points-clearly, something done in the prefrontal cortex. Just how this is done is unclear, even after a considerable amount of research by neuro-economics on decision-making dynamics.

The other motive states listed in Table 12.4 are also important but not as powerful as verification of self or identity and fairness in exchange payoffs. People need to feel that they are part of the ongoing interpersonal flow, but this need for *group inclusion* does not necessarily mean high solidarity with others. Individuals always want to feel themselves to be part of an interaction, but their need for solidarity or simple inclusion will vary. Individuals also need to feel that others are *trustworthy,* sincere, and respectful of self. And they have needs for *facticity* or the sense that things are as they seem and that, for the purposes of a given interaction, self and others are experiencing the same obdurate world.

When any of the need states listed in Table 12.4 is not realized, negative emotions are aroused to the point of breaching the encounter. The more powerful the need state and the greater the failure to meet these need, the greater will be the negative emotional reaction, and the more likely will the encounter be disrupted. Because of this potential for highly intense negative emotional arousal, individuals motive-take to be sure that they understand which needs among others are most salient and whether or not these needs are being met. They rely mostly on emotional cues to make this determination because negative emotions demand attention because of their disruptive potential. As long as emotions of others remain on the positive side, individuals can relax their motive-taking, but part of the reciprocal of *motive-taking*—i.e., *motive-making*—is to signal not only one's needs but whether or not these needs are being realized in an interaction. Again, the emission of negatively valenced emotional cues is critical to keeping others informed during the process of *motive-making*.

Table 12.4 Universal need states driving motive-taking and motive-making

1. ***Verification of identities:*** Needs to verify one or more of the four basic identities that individuals present in all encounters: (a) *Core identity:* The conceptions and emotions that individuals have about themselves as persons that they carry to most encounters. (b) *Categoric-unit identity:* The conception that individuals have of themselves by virtue of their membership in categoric units which, depending upon the situation, will vary in salience to self and others; when salient, individuals seek to have others verify this identity. (c) *Corporate-unit identity:* The conception that individuals have about their incumbency in corporate units (groups, organizations, and communities) and/or their identification with the members, structure, and culture of a corporate unit; when individuals have identity. (d) *Role identity:* The conception that individuals have about themselves as role players, particularly roles embedded in corporate units nested in institutional domains; the more a role identity is lodged in a domain, the more likely will individuals seek to have others verify this identity.
2. ***Making a profit the exchange of resources:*** Needs to feel that the receipt of resources by persons in encounters exceeds their costs and investments in securing these resources and that their shares of resources are "just" and "fair" compared to (a) the shares that others receive in the situation and (b) reference points that are used to establish what is a just share. Reference points can include others·in a situation, alternative situations where exchange can occur, norms specifying just rates of exchange, or abstracted distributions of what people in general receive in exchanges.
3. ***Group inclusion:*** Needs to feel that one is a part of the ongoing flow of interaction in an encounter, and the more focused is the encounter, the more powerful is this need.
4. *Trust:* Needs to feel that others' are predictable, sincere, respective of self, and capable of sustaining rhythmic synchronization through talk and body language.
5. ***Facticity:*** Needs to feel that, for the purposes of the present interaction, individuals share a common inter-subjectivity, that the situation is indeed as it seems, and that the situation has an obdurate character.

When individuals have more idiosyncratic need states, they will have to be more demonstrative in their *motive-making*, if they expect others to motive-take with them successfully. And the more idiosyncratic are need states of individuals, the more likely will something go wrong in the situation because signaling unique motives is difficult, particularly since individuals may not be fully conscious of their needs until it is too late. They may only come to recognize that they had a need after it has gone unfulfilled, thereby *creating frustration* to go along with *anger, fear, hurt,* and other negative emotions. For those motive-taking, they will initially focus on the more universal need states listed in Table 12.4 to be sure that these needs are covered, and the result may be a failure to recognize additional need states that others may bring to an encounter. Thus, failure in motive-taking and motive-making is endemic to human social interaction—once again emphasizing the fragility of sociality and group solidarity among humans

Emotion-Taking and Emotion-Making

Human interaction is not possible without individuals' capacities to signal and read emotional phonemes strung together by syntax. The first language of hominins and later humans was *the language of emotions*. This language is visually based because humans, as an evolved primate, are visually dominant and thus rely primarily on vision for sensory inputs from the environment. *Emotion-taking* is primarily a process of reading gestures visually, particularly of face but also body movements, positioning, and countenance. Individuals' sources of information on emotions can come through voice inflections via the auditory sense modality (*Wernicke's area* on the *posterial temporal lobe* near the *inferior parietal lobe*) or acts of touching through the haptic sense modality (*parietal lobe*), but emotion-taking will typically involve efforts to visually verify emotional inputs via other sense modalities. *Emotion-making* is, to some degree, outside of volitional control because emotions emanate from sub-cortical areas of the brain—for example, *amygdala, hippocampus, peptide and neurotransmitter body systems*—and cannot be completely controlled by the neurons from the frontal lobe and prefrontal cortices, by *Broca's area* in the temporal area controlling speech production, or by the *Sylvian fissure* (separating the parietal from frontal lobes) regulating many muscles and the *musculoskeletal body system*. Emotions can be partially filtered through the cortical censors, and humans are such adept *emotion-takers* that they can generally recognize this distorting effect of the filters and make the appropriate assessments of "the real" emotions that others are experiencing. At times, people reveal considerable expressive control and are able to regulate and orchestrate their emotional displays in accordance with cultural expectations, despite their "real" emotions. Still, if emotions are strong and especially if they are negative, this control is always tenuous, even when negative emotions have been repressed and are out of conscious awareness (see discussion below). Nonetheless, because culture-taking and culture-making inevitably pull emotion ideologies, feeling rules, and display rules from culture (Hochschild, 1979, 1983), culture-taking and culture-making inevitably force some efforts at control of emotion-making.

Emotion-taking and emotion-making are complicated by the fact that humans are capable of activating *defense mechanisms* to control their emotional experiences and their displays of emotions to others. The most important source of information of many interactions can be distorted either at the behavioral output stage or at the reflected appraisals stage of emotion-taking. Humans have very large neurons extending down from the *prefrontal cortex* through the *amygdala* to other emotion-generating sub-cortical modules that can push emotional memories from conscious awareness, storing them in the *hippocampus*, it appears. The result is that people are often unaware of powerful emotions that may affect their behavioral outputs. For example, individuals may be shocked when others tell them (from their emotion-taking efforts) that they seem *sad, mad, anxious, fearful*, or *angry*. Or, a person can often be truly surprised by their sudden emotional reactions to certain events because the emotions had been part of a repressed memory, which was stimulated by the current situation. Such is the power of repression and other defense mechanisms. At times, the repression is so complete that the repressed emotions are not even evident at the behavioral output end of interactions, with the result that others cannot effectively emotion-take with a person. Still, repression is never complete, at least in the long run, and over time, those who consistently emotion-take with a person will begin to make attributions about the sources of this person's emotional responses and dispositions.

When emotions are repressed, they will generally reemerge but often in transmuted forms. For example, as many clinicians would argue, repressed *shame* will often come out as aggression and *anger*. Moreover, repressed emotions will generally increase in intensity over time, and as they do so, they will eventually break through the *neo-cortical* and/or *hippocampal censors* and become an inadvertent part of emotion making that others pick up in their *emotion-taking*. Because individuals are implicitly aware of repression of emotions, they typically redouble their efforts at emotion-taking to be sure that they understand another's feelings, and in cultures like those in western societies that have an influential psychiatric subculture culture, people are consciously looking for signs of repression. A lot is at stake in the dynamics revolving around repression because, if people cannot honestly emotion-make (that is, "give off their true feelings) and accurately emotion-take ("read" emotions in self and others), the most important bonding force in human interaction—the arousal of emotions—can remain ambiguous, thus creating tensions and breaches in ongoing interactions.

If an episodes of interaction are embedded in more inclusive socio-cultural formations (e.g., groups, organizations, communities, institutional domains), then expectations for emotions will be clearer than when interactions are not embedded because the expectations flow from culture and are activated in roles and status positions that further specify what emotions should be felt and expressed. Thus, the more structured is an encounter, the more individuals can make assumptions about the emotions of others in their emotion-taking efforts, and the more they can orchestrate emotional displays to conform to expectations tied to culture and social structure. As Hochschild (1983) has argued, individuals must often engage in "emotion work" under conditions where they are expected to feel and display particular emotions that they do not feel. And thus, the more emotion work by individuals, the more likely are unacceptable emotions to be held in check, but like repression, it is even more difficult to hold back emotions that a person is consciously feeling but is forbidden to display. At some point, these emotions will come out and be picked up by others in emotion-taking. And so, like repression, emotion-taking by individuals will often implicitly seek to determine if there are other emotions being held back consciously or unconsciously by individuals. And the longer is an episode of interaction, the more likely are these less acceptable emotions to become part of emotion-making by a person and, hence, part of the emotional information gleaned by emotion-taking by others.The neurology of these dynamics is not well understood, but the large increase in the size of the *hippocampus* in humans compared to other higher primates (Turner, 2000; Stephan, 1983; Stephan & Andy, 1969, 1977; Stephan et al., 1988; Eccles, 1989; see also Barger et al., 2007, 2012, 2014). Controlling for body size which is correlated with brain size is one indicator that natural selection on humans and their hominin ancestors created a sub-cortical area to store unconscious memories, but this area is also part of the normal memory dynamics of all mammals. On the one side through the *transition cortices*, working memory is sustained, with short-term memories being shuttled to the *hippocampus* for storage. If these memories are activated again, especially with their original emotional tags (what Damasio, 1994, termed the "somatic marker hypothesis"), then they are more likely to be remembered and after two years shipped up to the *frontal lobe* for longer term memory and often marked by cultural symbols. From the other side, unpleasant and painful emotions tied to particular situations can be shipped down from the *prefrontal lobe* to the *hippocampus* to remove them from conscious awareness, and they can be kept there for long periods of time. Yet, since the *hippocampus* is designed to ship memories back to the frontal lobe, it is often difficult to hold them from consciousness. Still, if the cortical blockers are powerful and the memory stays in the hippocampus for long periods of time, a person may remain unaware of certain painful memories. Yet, this person may still reveal some of these emotions in emotion-making and, as noted above, may remain unaware that these emotions are "leaking" out and becoming part of others emotion-taking.

Ritual-Making and Ritual-Taking

Virtually every interaction among humans begins with (1) greeting rituals to initiate interaction, (2) forming rituals that space and distribute talk and body language during the course of interaction, and (3) closing or exit rituals to mark the termination of an interaction, with indications of how the interaction will be picked up again. Rituals are emitted through both speech and body language that set an interaction on a *footing*, to use Goffman's label, and they vary considerably in terms of where an encounter is embedded—local situation, social structure, culture and potentially any locale of an interaction.

Opening rituals set the emotional mood, the form of talk, the frame of what is to be talked about, and virtually any aspect of "making" and "taking"; and as the interaction proceeds, rituals smooth the flow of interaction and are often invoked in turn-taking among individuals in determining who can speak and hold the conversational floor. Rituals are always used to shift frames of talk and behavior to punctuate key points of emphasis. Rituals can also be used to sanction those who violate norms of conversation, to establish the proper mix of work-practical, social, and the nature of the rituals are appropriate, to relative status and the roles that people play.

Rituals thus designate the dimensions outlined in Fig. 12.1 as they are introduced or discarded from the interaction and, in essence, every shift in the talk and expressive behaviors of those in an interaction. Rituals are, then, critical to alerting individuals of what is important and how to behave, while also being markers for changes in the flow of the interaction. In this way, individuals are alerted to changes in the dimensions that are *most salient* as any proceeds from its beginning to end. In so doing, rituals alert the *prefrontal cortex* to search memories in the *neo-cortex* cortex or even those still in the *hippocampus* about what is important and salient, how the interaction is to proceed, and when changes in some aspect of the interaction are about to occur. Rituals are thus a kind of early "warning signal" that a shift is occurring, thereby activating memory centers (*frontal lobe* and *hippocampus*) and the *prefrontal cortex* to pull new cognitions from the memory banks of the brain and, most importantly, alert the *sub-cortical emotions centers* (and related *body-systems*) for changes in the level and nature of affect emitted in an interaction. When *ritual-making* and *taking-fails*, it is immediately evident as individuals are unsure as to how to proceed, often to the point of breaching the interaction. When rituals are not adequately introduced into the flow of a conversation, individuals are usually labeled as "rude" and "difficult" or some other label suggesting a lack of tact that, in reality, is a failure to execute a ritual when necessary to alter the flow of an interaction. Indeed, without rituals, interactions can simply fall apart from breaching, which often sets into motions repair rituals to close the breach. Indeed, if interpersonal sanctions are employed, these too are often couched in highly ritualized ways (in the flow of words and affect) as sanctions that, if successful, avoid breach in the interaction and, if unsuccessful, also operate to breach the interactions. Given that humans sub-cortical areas are capable of generating many more negative and positive emotions, the tendency for humans to ritualize interactions (as do great apes, thereby suggesting that "ritualization" of encounters has a biological basis), it should not be surprising that natural selection intensified this biologically-based propensity evident in most higher mammals as a means for mitigating the arousal of negative emotions. Rituals are thus also linked to cultural prescriptions and proscriptions that are part of the cultural basis of any interaction.

Early founders of sociology like Herbert Spencer (1874–1896) and, more prominently, Emile Durkheim (1912/1947), recognized the significance or rituals to sustaining the social order. Social structures and their cultures, as well as the dynamics of encounters of individuals in interaction, are not viable without the extensive use of appropriate rituals as a means of navigating the beginning, flow, and termination of an interaction. Just where in the human brain, if this is even the right way to phrase the question, does this capacity reside?

I doubt if there is a clear module, like there is for language and speech; it may be spread across many cortices and nuclei in the brain, linked by neuro-networks to both speech, memory, and body movements. Still, discovering the locations in the brain where rituals are activated would uncover one of the most important forces in creating and sustain social relations and social structures

Talk-Making and Talk-Taking

I have saved for the last dimension a dynamics of interaction that occurs early on, but in most interactions, the form of talk is understood by memories of past interactions that become expectations during iterations of an encounter. Moreover, in their stocks of knowledge (*frontal lobe*) individuals generally know how to talk in most situations and are motivated to do so because, to not talk in the proper form, volume, and mood is to cause breaches in the interaction. Erving Goffman's early book on *Forms of Talk* (1983) was one of this first in sociology to address this dynamic in detail. From a neurological perspective the capacity for language was a by product of the rewiring of the great ape brains for speech production, evident in centers around the *inferior parietal lobe* in great apes.[7] Thus, spoken language in the human measure only occurs among great apes, but non-human great apes lack the necessary structure in their vocal tracts and, thus, can only "speak" and manifest the neurological capacity for language through the visual (sign language) or visual-haptic (typing messages in a machine with icons). Ironically, monkeys are much for proficient in the emitting sounds necessary for language (in the structure of their vocal tracts and surrounding structures), but lack a sufficiently large neo-cortex to take advantage of a neurology that contains the pre-adapatation that allows great apes and humans to "speak" because of their greater intelligence.

With talk can come symbolic culture, and as these fed off each other (as well as the *neo-cortex* and *sub-cortex*) during late hominin evolution, speech is fully intertwined with cultures and social structures at all levels social organization, as well as the roles and status positions from which these structures are constructed. Thus, talk is constrained by virtually all of the dimensions listed in Table 12.1. And the result is that forms and styles of talk became an important part of any interaction, with different forms of talk being invoked in virtually all encounters, depending upon the nature of the encounter as constrained by other elements "making" and "taking" listed in Fig. 12.1. Thus, different configurations of these dimension in any encounter also set up expectations for the form of talk, and if the appropriate form is not used in an encounter, it can breach the flow of interaction.

Forms of talk vary along many dimensions, including the emotional valences of talk, the volume of talk, the grammar of talk, the pace of talk, and many other variable forms of articulation. And, humans generally know which configuration of these variations is appropriate and normatively acceptable, and which are not. Thus, an encounter and all of the other dimensions that can come into play in the "making and taking" constrain forms of talk and body-language as well. To talk in the wrong form is, at a minimum, considered rude and, at a maximum, a breach of the encounter. And yet, most of the time, individuals "know" the right form, and given cues from the situation, structure, culture, frame, roles, and status, they can produce "the proper form of talk" is employed. A large *neo-cortex* can store the needed information; a develop *prefrontal cortex* can "pick" the proper form of talk.

[7]Monkeys have the same association cortices as great apes but do allow for speech because they do not have the requisite volume of the neo-cortex, as do other apes not in the great-ape family.

Conclusion

In this chapter, my goal has been to outline how the neurology of humans is still much the same as it was for millions of years during great ape evolution. There are relatively few new brain structures because introducing a new structure into the complex neurology of the hominin brain would be almost harmful in structures as complex of a large mammalian brain. Rather, natural selection relied upon *selection on existing structures*, and by selecting on larger structures on one side of the Bell Curve describing the distribution from smaller to larger, natural selection could keep the complexity of interconnections in tact while gaining increased cognitive and emotional power by simply enlarging the brain.

Human interaction could become more complex, operating on many dimensions, but still within the basic framework of the neurology inherited from the ancestors to present-day great apes. Size and perhaps some increased connectivity allows humans to interact on many different levels and with complex dimensions in play; and indeed, most interactions proceed smoothly and, if breached, they can generally be repaired. As is evident, I can only offer speculations about how the neurology is involved in interpersonal behavior, but the goal of neuro-sociology is to discover how the neurology of humans allows from complex patterns of interaction among humans to occur. Biology and psychology do not tend to look at interaction, *per se*, but rather the analysis behavior(s) dominate their literatures. For sociologists, the goal is to understand the robust nature of interaction and how it operates, creates, or alters not just behavior, and not just interaction, per se, but *also* social structure and culture. To understand these processes neuro-sociology needs to be more robust and more oriented to the evolution of hominin and then human neurology to understand how the brain, working through interpersonal behaviors, makes society and culture possible. Thus, neuro-sociology should not overspecialize as is the case in biology, psychology, and economics; it should remain broad and robust in re-interpreting findings in other disciplines studying neurology but, even more importantly, in creating a new kind of neurology—*neuro-sociology*.

References

Andrews, P. (1981). Species diversity and diet in monkeys and apes during the Miocene. In C. Stringer (Ed.), *Aspects of human evolution*. Taylor & Francis.

Andrews, P. (2015). *An ape's view of human evolution*. Cambridge University Press.

Andrews, P. (2020). Last common ancestor of apes and humans: Morphology and environment. *Folia Primatology, 91*, 122–148.

Andrews, P., & Martin, L. (1987). Cladistic relationships of exant and fossil hominoids. *Journal of Human Evolution, 16*, 101–118.

Baizer, J. S., Baker, J. F., Haas, K., & Lima, R. (2007). Neurochemical organization of the nucleus *Paramedinaus dorsalis* in the human brain. *Brain Research, 1176*, 45–52.

Barger, N., Hanson, K., Teffer, K., Schenker-Ahmed, N. M., & Semendeferi, K. (2014). Evidence for evolutionary specialization in human limbic structures. *Frontiers in Human Neurosciences, 8*(277), 1–17. https://doi.org/10.3389/fnhum.2014.00277

Barger, N., Stefanacci, L., Schumann, C., Sherwood, C. C., Annese, J., Allman, J. M., Buckwalter, J. A., Hof, P. R., & Semendeferi, K. (2012). Neuronal populations in the basolateral nuclei of the amygdala are differentially increased in humans compared with apes: A stereological study. *Journal of Comparative Neurology, 520*, 3035–3054.

Barger, N., Stefanacci, L., & Semendeferi, K. (2007). A comparative volumetric analysis of the amygdaloid complex and basolateral division of the human and ape brain. *American Journal of Physical Anthropology, 134*, 392–403.

Berger, J., Fisek, M. H., Norman, R. Z., & Zelditch, M. (1977). *Status characteristics in social interaction*. Elsevier.

Berger, J., & Zelditch, M. (1985). *Status, rewards, and influence*. Jossey-Bass.

Bickerton, D. (2003). Symbol and structure: A comprehensive framework for language evolution. In M. S. Christiansen & L. Kirby (Eds.), *Language evolution* (pp. 77–93). Oxford University Press.

Brosnan, S. F. (2006). Nonhuman species' reactions to inequity and their implications for fairness. *Social Justice Research, 19*, 153–185.

Brosnan, S. F., & de Waal, F. B. M. (2004). Fair refusal by capuchin monkeys. *Nature, 428*, 128–140.

Brosnan, S. F., & de Waal, F. B. M. (2014). Evolution of responses to (un)fairness. *Science, 346*(6207), 125776.

Brosnan, S. F., Hillary C., Schiff, C., & de Waal, F. B. M. (2005). Tolerance for inequity may increase with social social closeness in Chimpanzees. *Proceedings. Biological Sciences, 272*, 253–280.

Burke, P., & Stets, J. E. (2009). *Identity theory*. Oxford University Press.
Call, J., & Tomasello, M. (1998). Does a chimpanzee have a theory of mind? 30 years later. *Trends in Cognitive Science, 12*, 187–192.
Collins, R. (1975). *Conflict sociology*. Academic Press.
Collins, R. (2004). *Interaction ritual chains*. Princeton University Press.
Cooley, C. H. (1902). *Human nature and the social order*. Princeton University Press.
Cosmides, L. (1989). The logic of social exchange: Has natural selection shaped how humans reason? Studies with the Wason selection task. *Cognition, 31*, 187–176.
Damasio, A. (1994). *Descartes' error: Emotion, reason, and the human brain*. Putnam.
Damasio, A. (2000). *The felling of what happens: Body, emotions and the makings of consciousness*. New American Library.
de Waal, F. B. M. (1989). Food sharing and reciprocal obligations among chimpanzees. *Journal of Human Evolution, 18*, 433–459.
de Waal, F. B. M. (1991). The chimpanzee's sense of social reality and its relation to the human sense of justice. *American Behavioral Scientist, 34*, 335–349.
de Waal, F. B. M. (1996). *Good natured. The origins of right and wrong*. Harvard.
de Waal, F. B., & Bronson, S. F. (2006). Simple and complex reciprocity in primates. In P. Kappeler & C. P. van Schaik (Eds.), *Cooperation in primates and humans: Mechanisms and evolution* (pp. 85–106). Springer.
Durkheim, É. (1912/1947). *The elementary forms of the religious life: A study in religious sociology*. The Free Press.
Eccles, J. C. (1989). *Evolution of the brain: Creation of self*. Routledge.
Ekman, P. (1984). Expression and the nature of emotion. In K. Scherer & P. Ekman (Eds.), *Approaches to emotion* (pp. 319–343). Erlbaum.
Emde, R. N. (1962). Level and meaning for infant emotions: A biosocial view. In W. A. Collins (Ed.), *Developments of cognition, affect, and social relations* (pp. 1–37). NJP Lawrence Erlbaum.
Emde, R. N. (1980). Levels of meaning of infant emotions: A biosocial view. In W. A. Collins (Ed.), *Development of cognition, affect, and social relations*. Lawrence Erlbaum.
Gallup, G. G., Jr. (1970). Chimpanzees: Self-recognition. *Science, 167*, 88–87.
Gallup, G. G., Jr. (1979). *Self-recognition in chimpanzees and man: A developmental and comparative perspective*. Plenum.
Gallup, G. G., Jr. (1982). Self-awareness and the emergence of mind in primates. *American Journal of Primatology, 2*, 237–248.
Gallup, G. G., Jr., Platek, S., & Spaulding, K. (2014). The nature of recognition revisited. *Trends in Cognitive Science, 18*, 57–58.
Gazzaniga, M. S., & Smylie, C. S. (1990). Hemisphere mechanisms controlling voluntary and spontaneous mechanisms. *Annual Review of Neurology, 13*, 536–540.
Gergely, G., & Csibra, G. (2006). Sylvia's recipe: The role of imitation and pedagogy. In N. J. Enfield & S. C. Levinson (Eds.), *The transmission of cultural knowledge*. Oxford University Press.
Geschwind, N. (1965a). Disconnection syndromes in animals and man, Part I. *Brain, 88*, 237–294.
Geschwind, N. (1965b). Disconnection syndromes in animals and man, Part II. *Brain, 88*, 585–644.
Goffman, E. (1959). *The presentation of self in everyday life*. Anchor Books.
Goffman, E. (1967). *Interaction ritual: Essays in face-to-face behavior*. Anchor Books.
Goffman, E. (1974). *Frame analysis: An essay on the organization of experience*. Harper & Row.
Goffman, E. (1981). *Interaction ritual*. Pantheon.
Goffman, E. (1983). *Behavior in public spaces*. The Free Press.
Hare, B., Call, J., & Tomasello, M. (2001). Do chimpanzees know what conspecifics know? *Animal Behavior, 61*, 139–151.
Hochschild, A. R. (1979). Emotion work, feeling rules and social structure. *American Journal of Sociology, 85*, 551–575.
Hochschild, A. R. (1983). *The managed heart: Commercialization of human feeling*. University of California press.
Horowitz, A. C. (2003). Do humans ape? Or do apes human? Imitation and intention in humans (homo sapiens) and other animals. *Journal of Comparative Psychology, 117*, 325–336.
Itakura, S. (1996). An exploratory study of gaze-monitoring in non-human primates. *Japanese Psychological Research, 38*, 174–180.
Jarvis, M. J., & Ettlinger, G. (1977). Cross-modal recognition in chimpanzees and monkeys. *Neuropsychologia, 15*(4–5), 499–506.
Kemper, T. D., & Collins, R. (1990). Dimensions of micro-interaction. *American Journal of Sociology, 96*, 32–68.
LeDoux, J. E. (1991). Neuroscience commentary: The emotional brain. *Journal of NIH Research, 3*, 49–51.
LeDoux, J. E. (1993). Emotional memory systems in the brain. *Behavioral Brain Research, 58*, 69–79.
LeDoux, J. E. (1996). *The emotional brain: The mysterious underpinnings of emotional life*. Simon & Schuster.
Leslie, K. R., Johnson-Frey, S. H., & Grafton, S. T. (2004). Functional imaging of face and hand imitation: Towards a motor theory of empathy. *NeuroImage, 21*, 601–607.
Maryanski, A. (1986). African ape social structure: A comparative analysis. PhD dissertation, University of California, Irvine.
Maryanski, A. (1987). African ape social structure: Is there strength in weak ties? *Social Networks, 9*, 191–215.
Maryanski, A. (1992). The last common ancestor: An ecological-network model on the origins of human sociality. *Advances in Human Ecology, 2*, 1–32.

Maryanski, A. (1993). The elementary forms of the first proto-human society: An ecological/social network approach. *Advances in Human Evolution Ecology, 2*, 215–241.

Maryanski, A. (1995). African ape social networks: A blueprint for reconstructing early hominid social structure. In J. Steele & S. Shennan (Eds.), *The archaeology of human ancestry: Power, sex and tradition* (pp. 67–90). Routledge.

Maryanski, A. (2013). The origin of speech and its implication for optimal size of groups. *Critical Review, 12*, 233–249.

Maryanski, A., & Turner, J. H. (1992). *The social cage: Human nature and the evolution of society*. Stanford University Press.

McCall, G. P., & Simmons, J. L. (1978). *Identities and interactions*. Basic books.

McGrew, W. C. (2010). In search of the last common ancestor: New findings on wild chimpanzees. *Philosophical Transaction of the Royal Society, 365*, 3265–3267.

Mead, G. H. (1934). *Mind, self and society*. University of Chicago Press.

Menzel, E. W. (1971). Communications about the environment in a group of young chimpanzees. *Folia Primatologica, 15*, 220–230.

Parr, L., Waller, B., & Fugate, J. (2005). Emotional communication in primates: Implications for neurology. *Current Opinion in Neurology, 15*, 716–720.

Passingham, R. E. (1973). Anatomical differences between the neocortex of man and the other primates. *Brain Behavioral Evolution, 7*, 337–359.

Passingham, R. E. (1975). Changes in the size and organization of the brain in man and his ancestors. *Brain and Behavior Evolution, 11*, 73–90.

Povinelli, D. J. (2000). *Folk physics for apes: The chimpanzee's theory of how the world works*. Oxford University Press.

Povinelli, D. J., & Eddy, T. J. (1997). Specificity of gaze-following in young chimpanzees. *British Journal of Developmental Psychology, 15*, 213–222.

Premack, D., & Woodruff, G. (1978). Does the chimpanzee have a theory of mind? *Behavior and Brain Science, 1*, 515–526.

Raghanti, M. A., Stimpson, J. L., Marcinkiewicz, J. L., et al. (2000). Differences in cortical serotonergic innervation among humans, chimpanzees, and macaque monkeys: A comparative study. *Cerebral Cortex, 18*, 584–597.

Rizzolalatti, G., & Sinigaglia, C. (2008). *Mirrors in the brain: How our minds share actions, emotions, and experience*. Oxford University Press.

Rizzolatti, G. and Sinigaglia, C. (2008). *Mirrrors in the braiin: How our minds share actions, emotions, and experience*. Oxford. Oxford

Rizzolattti, G., Fadiga, L., Fogassi, L., & Gallese, V. (2002). From mirror neurons to imitation: Facts and speculations. In W. Prinz & A. N. Meltzoff (Eds.), *The imitative mind: Development, evolution and brain bases* (pp. 247–266). Cambridge University Press.

Rumbaugh, D., & Savage-Rumbaugh, E. S. (1990). Chimpanzees: Competencies for language and numbers. In W. Stebbins & M. Berkley (Eds.), *Comparative perception, volume 2: complex signals*. Wiley.

Savage-Rumbaugh, E. S., & Lewin, R. (1994). *Kanzi: The ape at the brink of the human mind*. Wiley.

Savage-Rumbaugh, E. S., Murphy, J., Sevcik, R. A., Brakke, K. E., Williams, S. L., & Rumbaugh, D. M. (1993). *Language comprehension in the ape and child*. University of Chicago Press.

Saxe, R., & Kansisher, N. (2003). People thinking about people thinking: The role of the temporo-parietal junction in theory of mind. *NeuroImage, 19*, 1835142.

Scheff, T. (1988). Shame and conformity: The deference-emotion system. *American Sociological Review, 53*, 395–406.

Schutz, A. (1932/1967). *The phenomenology of the social world*. Northwestern University Press.

Schutz-Bosbach, S., & Prinz, W. (2007). Perceptional resonance: Action-induced modulation of perception. *Trends in Cognitive Sciences, 11*, 349–355.

Semendeferi, K., Lu, A., Schenker, N., & Damasio, H. (2002). Humans and great apes share a large frontal cortex. *Nature Neuroscience, 5*, 272–276.

Sherwood, C. C. (2007). The evolution of neuron types and cortical histology in apes and humans. In T. M. Preuss & J. H. Kaas (Eds.), *Evolution of nervous systems, volume 4: Primates* (pp. 355–378). Academic Press.

Sherwood, C. C., Holloway, R. L., Semendeferi, K., & Hof, P. R. (2005). Is prefrontal white matter enlargement a human evolutionary specialization? *Nature Neuroscience, 8*, 537–538.

Sherwood, C. C., Subiaul, F., & Zawidzki, W. (2008). A natural history of the human mind: Tracing evolutionary changes in brain and cognition. *Journal of Anatomy, 212*, 426–454.

Spencer, H. (1874–1896/1898). *The principles of sociology, 3 volumes*. Appleton.

Spencer, B., & Gillen, F. J. (1899). *The native tribes of Central Australia*. Macmillan.

Stephan, H. (1983). Evolutionary trends in limbic structures. *Neuroscience and Biobehavioral Reviews, 7*, 367–374.

Stephan, H., & Andy, O. J. (1969). Quantitative comparative neuroanatomy of primates: An attempt at phylogenetic interpretation. *Annals of the New York Academy of Science, 167*, 370–387.

Stephan, H., & Andy, O. J. (1977). Quantitative comparison of the amygdala in insectivores and primates. *Acta Anatomica, 98*, 130–153.

Stephan, H., Baron, G., & Frahm, H. (1988). Comparative size of brains and brain components. In H. Steklis & J. Erwin (Eds.), *Comparative primate biology, volume 4* (pp. 1–37). Alan R. Liss.

Stephan, H., Bauchot, R., & Andy, O. J. (1970). Data size of the brain and the various brain parts. In C. R.

Noback & W. Montagna (Eds.), *Advances in primatology, volume 1: The primate brain*. Appleton.
Stephan, H., Frahm, H., & Baron, G. (1981). New and revised data on volumes of brain structures in insectivores and primates. *Folia Primatologica, 35*, 1–29.
Stryker, S. (1980). *Symbolic interactionism: A structural version*. Benjamin/Cummings.
Subiaul, F. (2007). The imitation faculty in monkeys: Evaluating its features, distribution, and evolution. *Journal of Anthropological Science, 85*, 35–62.
Tomasello, M., Hare, B., & Fogleman, T. (2001). The ontogeny of gaze following in chimpanzees, *pan troglodytes*, and rhesus macaques, *macaca mulatta*. *Animal Behavior, 61*, 335–343.
Tomonaga, M. (1999). Attending to the others' attention in macaques' joint attention or not? *Primate Research, 15*, 425.
Tsao, D., Freiwald, W., Knutsen, T., Mandeville, J., & Tootell, R. (2003). Faces and objects in macaque cerebral cortex. *Natural Neuroscience, 6*, 989–995.
Turner, R. H. (1962). Role-taking: Process versus conformity. In A. Rose (Ed.), *Human behavior and social processes* (pp. 20–40). Houghton Mifflin.
Turner, J. H. (1997). The evolution of emotions: The basis of human social organization. In U. Segerstråle & P. Molnár (Eds.), *Nonverbal communication: Where nature meets culture* (pp. 211–228). Erlbaum.
Turner, J. H. (1998). The evolution of moral systems. *Critical Review, 11*, 211–232.
Turner, J. H. (2000). *On the origins of human emotions: A sociological inquiry into the evolution of human affect*. Stanford University Press.
Turner, J. H. (2002). *Face to face: Toward a theory of interpersonal behavior*. Stanford University Press.
Turner, J. H. (2005). Social control and emotions. *Symbolic Interaction, 28*, 475–485.
Turner, J. H. (2007). *Human emotions: A sociological theory*. Routledge.
Turner, J. H. (2011). Extending the symbolic interactionist theory of interaction processes: A conceptual outline. *Symbolic Interaction, 34*(3), 330–339.
Turner, J. H. (2019). The effects of cultural, structural, and interpersonal dynamics on interaction rituals. In E. B. Weininger, A. Lareau, & O. Lizardo (Eds.), *Ritual, emotion, violence: Studies on the micro-sociology of Randall Collins* (pp. 153–178). Routledge.
Turner, J. H. (2021). *On human nature: The biology and sociology of what made us human*. Routledge.
Turner, J. H., & Maryanski, A. (2005). *Incest: Origins of the taboo*. Routledge/Paradigm.
Turner, J. H., & Maryanski, A. (2008). The limitations of evolutionary theory from biology in explaining sociocultural evolution. *Sociologica, 3*, 1–38.
Turner, J. H., & Maryanski, A. (2024). *Bio-social evolutionary analysis: Reconciling biology, psychology, and sociology*. Edgar Elgar.
Wundt, W. (1912/1916). *Elements of folk psychology*. Macmillan.

Neurosociology of Social Roles

13

Alexander Shkurko

Abstract

Social roles are the interface between the macro-level organization of society and individual behavior and interaction. The chapter explores how the recent emergence of neurosociology can contribute to a better understanding of social roles. It identifies three ways to approach the problem of roles from the perspective of neurosociology. First, it begins with the intersection between a sociological notion of role taking and the Theory of Mind, which is widely studied in social neuroscience. Second, it explores how the dual-process approach in cognitive psychology and neuroscience can be used to understand the interplay between the automated and controlled processes underlying social roles. Third, it shows how recent advances in neuroscience revealing the modular organization of the neural system can contribute to new ideas and hypotheses in the neurosociology of roles.

A. Shkurko (✉)
Independent scholar, Ulyanovsk, Russia

The Notion of Role in Social Sciences

The notion of role is one of the most useful in the social sciences. Broadly speaking, it refers to a set of expectations associated with a specific position in a social structure or a situation, be that a large-scale stratification system or a contextual structure of dyadic interaction. The idea of "role" is equally successfully used by the different paradigms in sociology as well as in other disciplines, from social psychology to international relations. For sociology, it links micro- and macro-levels of analysis and reveals important aspects of the specific 'social' nature of human life (Turner, 2006).

The core of the role idea is that it specifies cognitions, emotions, attitudes, behaviors, rights and duties, and/or other socially meaningful attributes of an actor depending on *where* in the social structure this actor is situated. Different people (or other actors) can hold the same role and follow the same set of expectations, and the same actor can switch from one set of role expectations to another. What matters is that roles are pre-existent and form shared knowledge. Breaking role expectations may be perceived as a norm violation and lead to social control. Importantly, a role is not simply a behavioral model. Rather, it encompasses the whole range of actor attributes needed to maintain the interaction and underlying social structure. The following examples

W. Kalkhoff et al. (eds.), *Handbook of Neurosociology*, Handbooks of Sociology and Social Research, https://doi.org/10.1007/978-3-031-95615-7_13

illustrate the variety and complexity of roles across different levels of social organizations.

Any type of organization is based on a division of tasks, and any job position is a role, with formally and/or informally described rights, duties, available resources, and relations to others. An employee is expected to know what to do in their position, whom to report to and whom to command, how to cooperate with colleagues, where to find necessary information, when to leave the workplace, and so on. Importantly, job-related roles within organizations are related to a large-scale division of labor in societies. A medical specialist may have a specific role in a medical organization, but they may also be a member of a broader category of "doctors." From the point of view of other members of society, they have to "know something about health and drugs" and "must cure diseases."

Some roles are diffuse and universal, whereas others are specific and contextual. Gender roles or roles associated with a social class are universal and encompass a broad range of behaviors and attitudes, from daily routines and interactions to life strategies. By contrast, the role of a "train passenger" is restricted to one context and one type of behavior. Roles regulate behaviors as well as emotions and attitudes. Individuals are expected to *act* in certain ways in the "parent" role but also to *feel* love and empathy toward their children. Many roles are also relational, and actors must switch from one role in the system to another. For example, a person is a child to their parents and is a parent to their own children. In the same way, a "speaker" at a conference may take on the role of "audience member" at the end of their presentation. Even an ordinary conversation requires the interchanging roles of "speaker" and "listener" to flow smoothly.

At the highest levels, roles can be associated with collective and institutional actors, as in the case of the "leading company" in the market or the "middle power" in international relations. Despite a seemingly different nature, the notion of role in such cases generally remains the same, as the community expects specific actions from each actor. For example, a leading company is expected to innovate regularly or to lobby its interest in the government, whereas the middle power is expected to support human rights and maintain the rule-based international order. The role of collective actors is related to the corresponding roles of individual actors (e.g., the roles of ministers in a national government).

Thus, the notion of role is a useful tool for describing the social world at different levels. It functions as an interface between the micro- and macro levels, between an individual and social structure or culture. Focusing on roles raises many questions. How does a particular role come to have specific content? How do people become role-holders? How do different roles coexist in the human mind? How do people switch, often so effortlessly, from one role to another? How are different roles related to each other? What determines how a person (or another type of actor) performs a role and what happens when they don't meet role expectations? What happens when a role is perceived differently by various actors? How are individual roles related to the roles of collective actors? And so on.

In what follows, I explore whether and how progress in the neurosciences can help shed light on at least some of these questions and improve our understanding of roles. In Sect. "Theories of Social Roles", I briefly describe some key theoretical approaches to roles in the social sciences. Section "Neurosociological Perspective on Roles" introduces the general perspective of neurosociology and a neurosociological approach to roles. In Sect. "Role-Taking and the Theory of Mind", I take and extend the example of role-taking (Franks, 2013) to show what relevant studies in social neuroscience can and cannot do for questions of sociological interest. In Sect. "Should Roles Be Explicit?—The Dual-Process Approach to Roles", building on the role of deliberate and automatic processes in role-taking, I address how a "dual-process" approach—one that distinguishes between automated and controlled processes—can advance theory and research on social roles. Section "Complexity of Roles, Multiple Roles, and Modularity of the Neural System: Beyond the Dual Approach" goes beyond the dual approach and explores recent advancements in our understanding of the modular organization of the human brain and how these may contribute to the study of roles. Section "Final Remarks" contains some final, brief remarks about the future of the neurosociology of roles.

Theories of Social Roles

Even though role-related terminology is widespread in sociology and other social sciences, it is rarely elaborated theoretically and is seldom the focus of theoretical discussions. For example, George Ritzer's *Sociological Theory* explicitly discusses "role" only once in the section on structural functionalism, and the concept is mentioned in passing several times in the sections on symbolic interactionism and a few other approaches (Ritzer & Stepnisky, 2018).

The conventional view of social roles was introduced by Linton in the 1930s (Linton, 1936) and elaborated by Parsons in his theory of social systems (Parsons, 1951). By necessity, roles are associated with statuses and implement a shared set of value standards, which regulate social interaction within a particular context. Roles, as well as statuses, are relational, (i.e., there must be a complementary and interdependent set of status-roles, and their content is defined irrespective of who holds the role). Within this relational structure, expectations connected to particular role are accompanied by norms and sanctions. Expectations for ego are at the same time sanctions to alter within an interaction situation (Ibid., p. 40). Roles vary in terms of the degree of institutionalization. A complex of fully institutionalized interdependent roles comprises a social institution. Institutionalized roles are supported by a shared culture (value standards) and personal commitments to role expectations. Thus, roles in Parsons' theory are not a marginal concept. Rather, they are core elements of institutions and the social order more generally.

Further elaboration of the functionalist approach involved switching from a higher-order view of roles as elements of stable institutions and social order to a lower-order view. Along these lines, Merton (1957) introduced role-set theory as an example of "middle range" theorizing in sociology. Role-set theory implies that an actor situated in a particular position takes several different and potentially conflicting roles. Since roles are relational by nature, then the content of a given role should be different depending on these specific relations. For example, the role of a student differs when dealing with teachers, administrators, or other students. Importantly, such a role-set is not the same as multiple roles played by an individual in various social contexts and is integral to specific institutional settings.

An alternative symbolic interactionist approach to roles developed alongside the functionalist approach. In the interactionist approach, the main focus is on the flexible and emergent nature of roles during social interaction. One of the key concepts in G.H. Mead's interactionism is "role-taking" (Mead, 1972). In contrast to the functional perspective, the concept of "role" in interactionism is not related to an abstract, functional position within the broader institutional organization. Instead, what matters most is the human ability to understand others' thoughts and actions and to mirror other's thoughts and action in self. This, however, does not mean that role-taking is simply intention recognition or the reconstruction of another's internal states, which is called Theory of Mind (ToM) in contemporary psychology. First, the process outlined in Mead's work leads to behavioral control. In a given situation, actors must imagine different courses of action and how others in the situation may respond to each one, and then the actor must make a choice about what action to take. Second, role-taking is a continuous process wherein individuals switch between different others' perspectives as they emerge in different situations. This is essentially a developmental process where a child acquires social intelligence and personality, and this ultimately culminates in the adoption of the perspective of "generalized other." Here, Mead's approach brushes up against Parsons' functionalism, as the perspective of the generalized other allows one to understand specific role *relations* (e.g. between students and teachers, children and parents, different members of a workgroup, etc.) in a depersonalized manner, i.e., it allows to understand what anybody taking this specific place, should do to make the whole process possible.

Dynamic and interactional aspects of roles were even more important in Goffman's (1959) dramaturgical approach. Here, roles are expectations regarding behavior in specific situations, but they are not imposed normative restrictions on individuals. Instead, they are the subject of negotiation and strategic manipulation. People literally play roles in interaction to present themselves in accordance with others' expectations while still pursuing their own interests, and this role performance implies a social-cognitive process of understanding others' views and expectations. From this perspective, although individuals are seen as more active role *makers*, in highly institutionalized contexts the role expectations and relations between different roles are highly structured. Ultimately, Goffman developed a more restricted view of individual role performances, which he came to see as significantly constrained and determined by shared frames of perception and implicit regulation of behavior (Goffman, 1974).

An attempt to integrate and critically analyze sociological and social-psychological perspectives on roles was advanced by Biddle in the late 1970s. Appreciating the growing popularity of role theory, he scrutinized the key questions regarding what roles are and how they are related to behavior (Biddle, 1979). Biddle argued that the concept of "role" was too vague, but he also identified several characteristics of roles shared by most scholars: they pattern specific behaviors; they are related to a set of people sharing a similar position; they generally carry explicit expectations; they have functional consequences; people socialize others by teaching roles (Ibid., p. 8). However, none of these axioms were self-evident, nor were they widely recognized by social scientists. Biddle also problematized the common view of roles as regulating behavior through expectations. He stressed that by "expectations" scholars and researchers often mean different things, including anticipated reactions, norms, values, thoughts, or feelings. He also pointed out that roles are typically defined within a context, and this context, rather than the role itself, determines behavior in many different ways depending on available resources and other structural factors.

Despite conceptual ambiguities and theoretical problems, Biddle remained a proponent of role theory. He attempted to develop the theory by focusing on the limitations of behaviorism; by seeing the most interesting aspects of human social life in symbols and meaning. In this vein, Biddle viewed roles as dealing with both behavioral *and* symbolic aspects of human life. Roles are contextualized behaviors, rather than expectations. More specifically, roles are specific behaviors characteristic of various persons in similar contexts (Ibid., p. 58). However, the idea of roles implies our ability to compare and generalize across situations and persons and extract specific behavioral patterns. From this starting point, Biddle provides a detailed description of various types of roles (personal, contextualized, functional), their characteristics (generality, complexity, uniqueness), their differentiation and integration, and several related concepts (e.g. identity and position) to combine existing theories and approaches into a broader conceptual framework for further analysis and research. Still, his approach remains behavioral, in which observing characteristic behavior is the main way to study roles empirically. All internal phenomena associated with roles, such as attitudes, emotions, and cognitions, are of secondary importance.

After Biddle, developments in role theory were modest. Despite Biddle's ambitions and attempts at conceptual synthesis and theoretical integration, work on roles remained rather fragmentary and conceptually diverse. For example, in developmental psychology and educational research, role-taking (i.e., understanding others' thoughts, feelings, and actions) is seen as useful for explaining educational success (e.g. Enright & Lapsley, 1980). In social psychology, social roles are often seen as stereotypes and sets of beliefs that explain important differences between people with common identities, as in the case of sex differences (Eagly & Wood, 2012). In international relations, roles are associated with states' positions vis-à-vis other states on various political issues, as well as with the social learning processes underlying role-playing and role-changing (Harnisch et al., 2011).

In sociology, role theory is still mainly understood in terms of the two general perspectives outlined above (structural functionalism and symbolic interactionism) with relatively modest modifications (Stryker, 2006; Turner, 2006). In terms of more significant developments, Stryker (2006) proposed a "structural symbolic interactionist frame" that treats humans as active actors involved in symbolically-mediated social interactions. However, actors are not completely free in their actions and pursuit of their interests. Rather, they are constrained by the learned understanding of stable social structures and their elements as well as by roles, which are shared behavioral expectations. In this approach, roles are cognitive structures that result from the categorization and classification of the social world. Role expectations are the direct consequence of the fact that people perceive others as members of a particular category (i.e., social position) with associated behavioral expectations (Ibid., p. 226). These definitions of the situation emerge and change as the interaction flows. They are negotiated and socially constructed, they are sometimes a matter of conflicting interpretations, and they vary in terms of their stability and strength. At the behavioral level, roles are probed via the role-making process, and they are "corrected" in line with feedback from social counterparts.

Lack of conceptual clarity and theoretical unification is associated with several "problems" for role theory. For one thing, the structural-functional perspective has been criticized for its rather static view of roles and for downplaying human freedom and creativity. Along these lines, Cicourel (1972) and Collins (1986) criticize the idea of roles as epitomizing the rationalistic cognitive view that human behavior is guided by non-problematic, explicit, and clearly defined ideas, whereas actual social life is generally organized around tacit knowledge, routines, and implicitly *negotiated* behaviors. Furthermore, some critics point to the diminishing scope and significance of the associated notions of "status" or "position" in sociology. These notions are largely restricted to treatments of inequality and stratification where interests and resources are more important than roles (Barbalet, 2006).

As a result of existing uncertainties and modest theoretical developments, some critics conclude that social role is a vanishing concept in sociology (Barbalet, 2006). I would argue that this conclusion is hardly true. The study of social roles and related phenomena remains widespread in many disciplines. According to the Dimensions database (dimensions.ai), there were more than 15 thousand scientific publications with the phrase "social roles" in the title or abstract, as of September 2023. The majority of publications are in two subfields: "Sociology" (more than one thousand) and "Gender Studies" (slightly less than one thousand). Also, there are no obvious signs of declining interest in the study of social roles. In fact, the number of publications in the area of "Human Society" is steadily growing.

In sum, it would be fair to say that at present, role theory in the social sciences remains underdeveloped and full of uncertainties, ranging from conceptual confusion to theoretical and deeper metatheoretical disagreements. Despite this, roles and many related constructs (e.g., role performance, role-taking, role-making, role differentiation, role change, role set, role conflict, role model, etc.) are part of a common vocabulary and are ongoing objects of serious and legitimate research in social sciences. Pervasive use of the "role" concept shows that social scientists appreciate the range of phenomena associated with it. This brings us to ask whether the emergent field of neurosociology has anything to unique contribute to advancing our understanding of roles and related phenomena.

Neurosociological Perspective on Roles

Neurosociology seeks to benefit from the remarkable recent progress in brain research and to explore how theories, methods, and findings from neuroscience can advance our understanding of social action, social relations, and social organization. As an emergent field of study, neurosociology has no solidified agenda. Progress in the field so far has been exploratory, and the area is quite diverse in terms of approaches and the range of topics of interest (Franks & Turner, 2013). One of the most significant barriers to the development of neurosociology is the lack of original empirical research in the area. The study of the neural bases and mechanisms of social phenomena—including social cognition, attitudes, intergroup relations, social interaction, etc.—is currently the mainstay of a more well-established interdisciplinary synthesis of neuroscience and social psychology: social neuroscience.

In sociology, by contrast, attempts to explore the nexus of social and neural questions, specifically, and social and biological questions, more generally, have been comparatively timid. On the bright side, this means that there are open opportunities to investigate how sociologically relevant issues and problems can benefit from neuroscience. This is true for virtually *any* area of sociological interest, including the study of social roles.

In general, there are several ways in which sociology can benefit from the study of neural processes. First, neuroscience provides new methods for studying subjective phenomena, and to the extent that sociology relies on them, new methods provide additional information about such phenomena and/or help validate traditional methods.

Second, the study of neural processes can be used to clarify and conceptualize important theoretical constructs of sociology, especially those that refer to subjective states and processes, such as attitudes, intentions, or expectations. If we accept the assumption that specific subjective phenomena should be reliably associated with a specific neural process, then the inability to associate a sociological construct with a specific neural process or mechanism can be treated as a reason for reconceptualization. In the case of social roles, we can ask whether role expectations related to normative beliefs (what role holders should do) and those related to anticipated behavior (what we think they will do) have identifiable and interrelated neural signatures. The absence of such persistent relations may indicate that normative and anticipatory aspects of expectations should be treated as different constructs rather than parts of one construct.

Third, the study of neural mechanisms can contribute to the understanding of social mechanisms. For example, Kalkhoff and others (2020) used EEG to investigate how social status affects a group task performance and showed that this relation is mediated by performance expectations, in particular. Of particular sociological importance are such comparative investigations of the different neural mechanisms that may underlie similar observable behaviors, beliefs, emotions, norms, or other social phenomena. The specification of such mechanisms can then be used to understand and predict differences and changes in relevant outcomes in other social contexts. Knowledge of the neural mechanisms underlying different aspects of roles or different types of roles would be useful when it comes to advancing our understanding of how and why role performance succeeds of fails.

Fourth, neurosociology can seek to understand what is possible and what is essentially impossible in society, taking into account the possibilities and limitations of brain structure and function. In terms of affordance theory, we can consider the brain as a kind of "affordance," on which social systems depend for their constitution and functioning. In terms of roles, this means addressing questions such as: What kind of neural organization is necessary to perform a social role with certain characteristics (e.g., being an objective and impartial judge)? Is the human brain capable of playing such a role? How is role performance affected by such and such characteristics of the brain? Or, in a radically transformative way: How can the brain be tuned or enhanced to ensure that it meets the requirements of a particular role, say a politician committed to the public good?

To pursue these goals, there are several strategies for neurosociological study. One strategy is to review and synthesize findings from neuroscience that are relevant for a given sociological problem to determine whether and how they advance our understanding. However, even this relatively simple task requires some familiarity with the neurosciences. Being able to understand and appreciate differences in concepts, theories, and methods across disciplines is needed to identify and incorporate relevant findings. A more advanced strategy is to use social neuroscience findings or even available raw data (e.g., from open databases such as Neurosynth) to pose and test sociological arguments. The next level is to develop research projects and conduct neurosociological research using methods from neuroscience (e.g., Kalkhoff et al., 2020; Firat, 2021). A good level of understanding of both neuroscientific and sociological theories and methods can lead to the development of "radical" new ideas that shed light on the brain-society relationship. In what follows, I employ the first, most basic strategy by reviewing some insights and findings from social neuroscience and related fields and show how they contribute to our understanding of social roles. By necessity, the review is selective and is not intended to cover all relevant literatures and issues.

Role-Taking and the Theory of Mind

Taking the perspective of other people is crucial for predicting their behavior as well as for establishing social relations and creating social cohesion. This ability is described as "role-taking" in sociology and as "Theory of Mind" (ToM) in psychology and social neuroscience. Role-taking is a good starting point for the neurosociology of roles because there are several well-developed research lines in social neuroscience that explore how people perceive and understand others' emotions, intentions, and actions. However, these advancements cannot be directly interpreted in "role-taking" terms. As Franks (2013) has argued, there are important differences between role-taking and Theory of Mind. Although at first glance they both deal with taking the perspective of other people, role-taking is a broader, more complex process. For one thing, role-taking is not limited to the mere reconstruction of alter's thoughts and feelings. Rather, it takes a stance on alter's view of self, incorporates this view into the self-concept, and converts it into behavior by ego deemed necessary for the cooperative process of social interaction (Franks, 2013). Moreover, the notion of role-taking has sociological importance because it transcends the dyadic relations in face-to-face interaction, and it generalizes many perspectives into an overall attitude of a social group. Also, through role-taking, an individual has to see him/herself through the eyes of particular role holders (e.g., a boss, a doctor, a spouse, etc.)

The differences between role-taking and ToM are important, and they reveal the challenges of directly importing ideas from social neuroscience into (neuro-)sociology. At the same time, such ideas and findings still have some theoretical utility. Mead insightfully assumed that the human central nervous system is somehow prepared to take into account both the attitudes of specific individuals as well as the attitudes of the entire social group at large, or the ingroup (Mead, 1972, p. 255). As of now, the mechanisms by which this occurs remain only partially understood, but they can help us better understand the process of role-taking.

One important possibility is that role-taking does not depend on one key neural system but more likely involves several systems and functions within the brain. Of particular relevance here are specific functions that differentiate brain activity in mentalizing versus non-mentalizing tasks. Studies in social neuroscience identify several brain areas and mechanisms involved in ToM processing, and two of them are of particular importance. According to van Overwalle (2009), there are two main brain areas that subserve ToM systematically and across different tasks and contexts. One is the medial prefrontal cortex (mPFC), a neocortical (evolutionarily more recent) region that is typically associated with higher-level cognitive functions, including strategic decision-making, the long-term learning and storage of information, as well as the processing of norms and values. The second area is the temporoparietal junction (TPJ), which is located at the intersection of temporal and parietal lobes. The TPJ is directly linked with the brain's more primitive limbic system, and it is involved in the processing of sensory input from both the external environment and the body.

According to van Overwalle (2009), the TPJ plays a key role in inferences about others' goals and intentions in specific situations, whereas mPFC is involved in the development of more explicit representations of broader social contexts and others' stable traits. This is partially because the TPJ integrates visual social information from the human mirror neuron system (hMNS). Mirror neurons are neurons that are activated both when we perform some action and also when we observe the same action performed by others (Gallese et al., 2004). The discovery of mirror neurons is among the most important findings in neuroscience, providing a very simple explanation of how we perceive and understand others. Though views on the role of the mirror neuron system in social cognition have become more skeptical over time, the hMNS is still considered part of the ToM neural network, namely the low-level basis of action understanding (Heyes & Catmur, 2022). The TPJ does not rely solely on mirror neurons but, according to van Overwalle (2009), it uses information from them as "input" while at the same time providing a higher-order representation, including the important distinction between "self" and "other." It is worth mentioning that more recent research reveals that another brain area adjacent to the TPJ, the posterior superior temporal sulcus (pSTS), may actually play a more significant role in the integration of several types of ToM-related information (Yang et al., 2015). In turn, the mPFC, which itself is a heterogeneous and multifunctional region, is involved in tasks that require understanding the stable traits of others, guiding behavior across many situations, understanding the general rules of appropriate action (social scripts), and social categorization (i.e., differential thinking about the members of various social groups).

Although debates about the neural basis of ToM and related processes continue, the anatomical and functional differences between relevant, distinct brain areas are of special interest to the theory of role-taking. Taking into account the conceptual differences between role-taking and ToM (see above), the temptation arises to suppose that role-taking should be more associated with the mPFC than the TPJ because it involves more abstract social knowledge representation and regulation of social behavior. By comparison, the TPJ/pSTS is more involved with the processing of immediately available social information and perspective taking. As such, it depends on processing that is essentially automatic, such as face recognition or action observation via mirror neurons. This raises an important question regarding the role of explicit and implicit processes in role-taking and ToM.

A typical design of a ToM experiment uses the so-called "false-belief task." In this task, an observer sees a scene where one character places a ball in a basket and leaves the scene. Then, another character removes the ball from the basket and places it in a box. Then, the first character returns, and the participant has to indicate whether they will seek the ball in the basket or the box. This design requires participants to explicitly reason about others' mental states. However, when participants are not directly instructed to infer others' reasoning, ToM construction can rely on spontaneous, automatic processes. Several studies investigated this issue by focusing on spontaneous ToM construction. Boccadoro and others (2019) reported on a series of experiments where participants observed video scenes of characters developing either true or false beliefs. Two of the three experiments also included a standard (explicit) version of the false belief task. Findings indicated that spontaneous ToM construction (at the belief formation phase) activated a large cluster in the temporoparietal cortex including the TPJ, but it did not activate the mPFC. Moreover, a comparison of the imaging data with activation maps in the Neurosynth online database revealed that the patterns of activation during the observation of false belief formation were associated with regions involved in visuospatial and working memory functions rather than ToM. They concluded that spontaneous ToM is a general 'submentalizing' cognitive function related to the identification of discrepancies between contradictory information—in this case between ego's true and other's false beliefs. The process is fast and spontaneous, and it does not require inhibition of ego's perspective.

In the end, what do these studies of the neural basis of ToM say about role-taking? The sociological notion of role-taking is assumed to be an explicit and effortful process, one that requires the integration of information about others' thoughts and feelings, prediction of their behavior, and taking the perspective of others as part of the self and with consequences for one's own behavior. ToM was also traditionally supposed to be explicit and associated with the deliberate reconstruction of others' mental states. However, neural evidence shows that it can rely on low-level, implicit processes. Both the mPFC and the TPJ—the two most frequently mentioned parts of the 'ToM neural network'—are involved in explicit mentalizing, but the TPJ also seems to be the basis of spontaneous mentalizing, whereas the mPFC is not.

At first glance, it seems logical to think of role-taking as based on, though not reduced to, the "conscious" part of the ToM (i.e., mPFC-related). However, three points challenge this view. First, if ToM is partially based on spontaneous, 'submental' processes and if ToM contributes to role-taking, then role-taking should also depend on implicit processes (i.e., processes that an actor is not, or cannot be, aware of). This should be particularly true in direct social interaction, when actors have no overt need to think about the intentions and mental states of the others. Even if role-taking is a complex mental representation, its inputs come from the unconscious, bottom-up processes such as mirror neurons. Second, the sociological function of role-taking is explained in terms of social structure, inequality, and other macro-level phenomena (Franks, 2013; Love & Davis, 2021). Many findings within social neuroscience and neurosociology show that perceptions of social relations and social inequality, as well emotional and behavioral predispositions to respond differently to actors occupying different positions in the social structure, are also automatic and unconscious (e.g., Cunningham et al., 2004; Kalkhoff et al., 2020; Koski et al., 2015; Lieberman, 2007). Thus, there is good reason to believe that the relationships been role-taking and social structure also rely on low-level social cognitive functioning. Third, it is precisely the explicit interpretation of roles (i.e., that actors are aware of their own and others' roles and are able to describe and evaluate them) that leads many theorists to criticize the construct and find it inadequate for explaining social phenomena (e.g., Collins, 1986). In light of these challenges, it may be useful to include *both* explicit (conscious, recognizable, and often symbolically represented) and implicit (outside of conscious control) processes in the conceptualization and study of social roles.

Should Roles Be Explicit?—The Dual-Process Approach to Roles

The two theoretical perspectives on roles outlined above—namely the structural-functional and interactionist perspectives—differ in how they treat human autonomy and agency. In their extreme and caricatured forms, they either deny human autonomy and see people as strictly conforming to abstract social roles, or they see roles as freely chosen and strategically manipulated by the actor. However, in both cases, roles are assumed to be explicitly recognized by both the role-holder and other actors. Irrespective of their theoretical approach, most scholars agree on the explicit nature of roles (Biddle, 1979, p. 8). However, rejecting this view would arguably lead to an improved understanding of roles—one that benefits from advances in cognitive and neurosciences.

In the evolution of sociology, acceptance of implicit/unconscious aspects of human cognition and behavior remains a matter of notable dissensus. Generally, sociological theories reluctantly incorporate implicit or unconscious processes and mechanisms of social cognition and behavior into their explanatory models. However, to some extent, these processes are recognized as potentially important and, in some theories, play a key role. For example, in ethnomethodology, behavioral regulation in typical social contexts is based on implicit expectations (Garfinkel, 1967).

Giddens (1984) relies upon an outdated Freudian theory of psychics to accept the role of unconscious motivations. Goffman's theory of frames points to implicit mechanisms underlying the organization of actions and the automatic way by which people switch between frames of understanding (Goffman, 1974). Habermas (1984) recognized that both communicative and strategic actions are at least partially implicit. Collins' (1986) theory of interaction ritual chains highlights the role of implicit processes that underlie shared social beliefs and structures.

In the neurosciences, implicit processes and their relationship to reflexive mind have enjoyed more mainstream acceptance. The "dual-process" or "dual-systems" theories (Frankish, 2010; Lieberman, 2007; Sherman et al., 2014) may have special utility in sociology, especially with respect to the study of roles. Despite important differences and debates within the perspective, its core idea is simple: in the human mind, there are two qualitatively different types of processes supported by two types of cognitive and neural systems. The Type 1 system (or System I) is fast, automatic, effortless, and unconscious, whereas the Type 2 system (or System II) is slow, controllable, effortful, and conscious. System I is evolutionarily older than System II, and it is supported by a different set of neural mechanisms. However, the dual-process approach is not simply that there are two types of mental processes, but that they provide two alternative paths for behavioral response or action. Both systems can respond to a situation or a task in their own way, and these responses can be identical or different.

There is no unified dual theory. Instead, there are a lot of theories, models, and applications of the general principle mentioned above. Furthermore, there are important differences concerning the content of attributes associated with each type of process. Approaches also differ in terms of the phenomena to which the dual principle is applied. These range from the specific (e.g., decision-making or attitudes) to the more general (i.e., the whole organization of the human mind). In addition, there is an ongoing theoretical debate regarding the interaction between the two systems. Some theories see them as acting consequently (i.e., the default mode of response to environmental cues is automatic, but the controlled system can intervene and change the course of action), while other theories see the two systems as acting in parallel or in a more complex and multilayered way (e.g. Strack & Deutsch, 2014). Finally, as discussed below, the most recent approaches take a more complex and intricate view of the human mind—one that goes far beyond the binary distinction in the dual process model. Even so, the dual process approach turned out to be a very useful and successful research program that stimulated a lot of important findings in cognitive science and social psychology (Evans, 2018).

The implications of the dual-process/system approach for sociology are straightforward. If each system is a potential source of a behavioral response or subjective state, then focusing solely on the conscious and controlled system will result in an inability to identify the role of the second system and, consequently, will fail to provide a proper explanation. If a behavioral outcome, such as a decision, is generated by implicit attitudes (System I), and System II merely rationalizes it, then accepting this rationalization as a cause of action is not convincing as an explanation and is not effective as a prediction. If the automatic system is capable of producing attitudes or making decisions, then it is difficult to ignore it in a sociological study. Since the two systems require different resources and rely on different mechanisms, it is reasonable to expect that different tasks, environments, social contexts, or individual abilities will affect how and to what extent each system contributes to the final outcome.

There is a growing interest in incorporating the dual-process approach into sociological studies. Among the most notable theoretical advances, there have been several attempts to incorporate the dual-process approach in cultural sociology (DiMaggio, 1997; Lizardo, 2017; Lizardo et al., 2016), economic sociology (Vila-Henninger, 2021), and political science (Rathburn, 2020; Vila-Henninger, 2020). On the empirical side, Vaisey (2009) used a secondary analysis of panel data from the US National Study of Youth and

Religion to clarify the role of cultural beliefs in regulating behavior: is it a tool for post hoc rationalization and justification, or a source of motivation? More recently, Brett and Miles (2021) conducted a meta-analysis of psychological studies to investigate the distribution of preferences for automatic and deliberate thinking across various social populations, thereby directly incorporating the dual-process model into the sociological agenda. Chernozub (2023) developed a graphical associative test as a proxy measure for studying implicit attitudes and combined it with measures of explicit attitudes in traditional mass surveys to describe and predict electoral behavior.

None of these studies, however, capitalized on neuroscientific contributions that provide a much more detailed, accurate, and reliable description of the two systems. Fortunately, several researchers adopted the methods of brain research and conducted the first neurosociological studies in the strict sense. Gregory et al. (2009) developed a research program to study unconscious processes in direct social interaction. In particular, they showed that manipulating the paraverbal auditory signal processed by the right hemisphere of the brain during communication affects social cohesion and task group performance. In a more recent study, Kalkhoff et al. (2020) used EEG to empirically investigate unconscious performance expectations as a mediator between social status and behavior. Given that expectations are among the most important sociological concepts, yet are typically hidden from consciousness and reflection, operationalizing them using information about the brain activity is particularly important (see below). Finally, Firat (2021) used the dual-process approach in a direct sense, to argue that two types of values, communal and agentic, may perform a similar stress-buffering function by adopting different neural mechanisms associated with implicit and explicit processes, respectively.

The use of the dual-process model to incorporate implicit processes into sociological study is justified because traditional sociological methodologies do not provide such opportunities, not because the theories are ignorant of them. It would be wrong, however, to take interest in the unconscious as an argument for devaluing deliberate reasoning and intentional action and their role in social organization. System II is at least as interesting to neurosociology as System I, largely because the core elements of the constitution of society are explicit. These include systems of beliefs, legal order, major social exchanges, policies, goals and, among other, roles. In fact, there are many relatively simple ways by which the dual process/systems approach, *strengthened by brain research*, might improve the sociological understanding of roles and related phenomena. Along these lines, Fig. 13.1 proposes a general framework for advancing this agenda.

The similar "dual" logic can be applied to other role-related issues. For example, not only does expected behavior rely more or less on automated processes, but role expectations themselves may be more or less explicit. The social dynamics of modern societies are accompanied by a tendency to make role expectations explicitly stated and formalized. Choosing a role might also be seen as a purely explicit and deliberate strategic decision, but because decision-making is subject to dual-process interpretation, automatic processes likely play a role. The figure does not specify the nature of both types of processes, nor does it spell out how they interact. However, when it comes to our understanding of roles, the model illustrates the heuristic potential of the dual process approach, and it proposes some promising avenues of neurosociological research.

One straightforward strategy is to investigate the structure and interaction of automatic and controlled processes in role processes, both in terms of the expected and performed content of roles as well as other role-related processes (e.g., role evaluation, transition, etc.). Even if roles are explicit and formalized—as is often in the case of job-related roles, for example—their routinized implementation is largely based on automatic processes. The process of role internalization can be seen as a shift from deliberate to automatic processing of roles through learning and rewards. It also speaks to the development of neural pathways that integrate the sensory and motor functions necessary to perform a relevant action

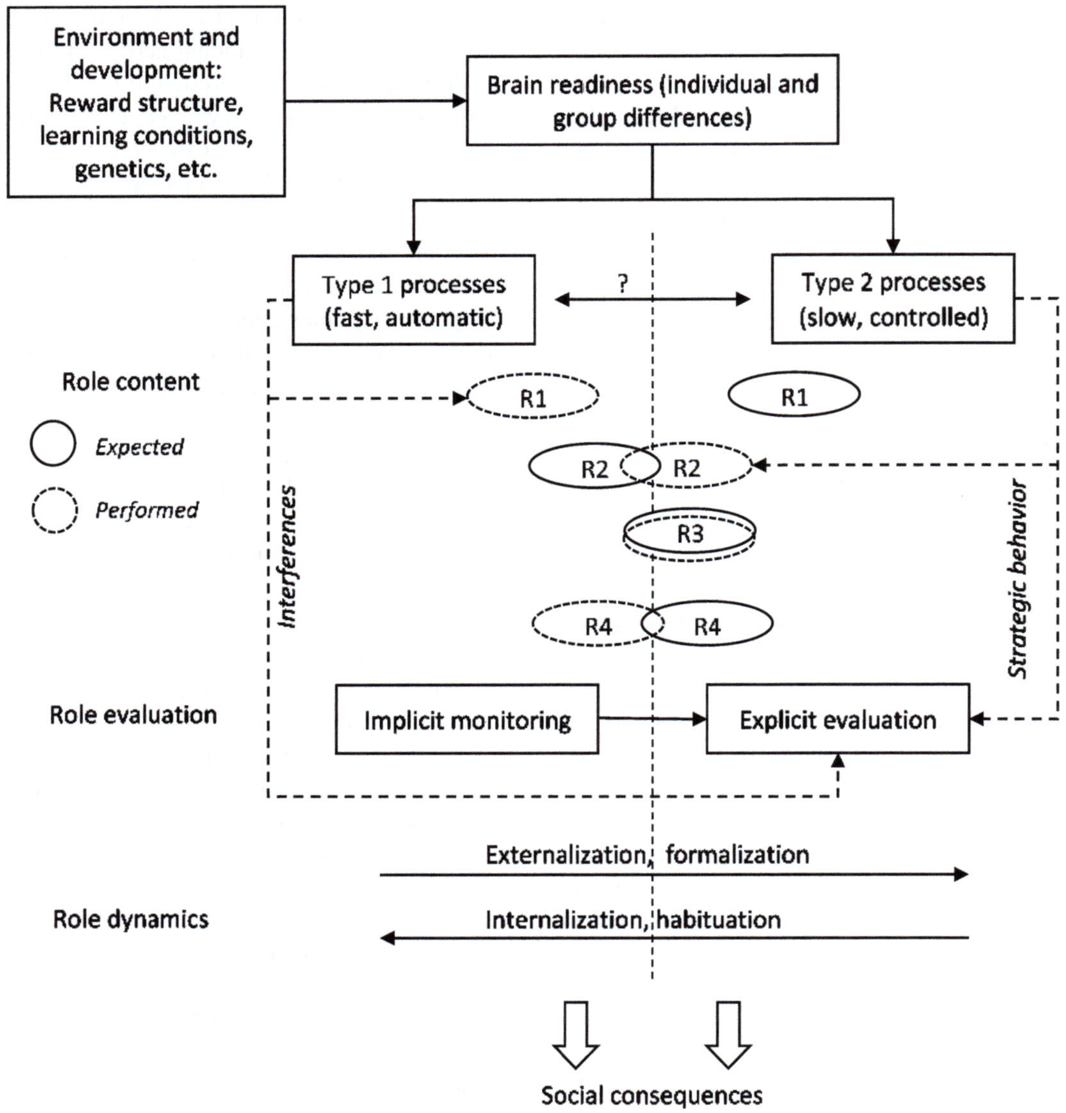

Fig. 13.1 The dual-process approach to social roles

without the involvement of deliberate reasoning. Both expected and performed roles may differ in terms of the extent to which deliberate (versus more automatic) processing is involved. Tensions may arise when the two systems create a difference between role expectations and actual role performances—i.e., when a role requires a highly automated behavior but is performed deliberately (e.g., a poorly trained firefighter), or vice versa (e.g., a recruiter who does not bother to fairly and carefully screen an applicant and makes decisions based on implicit attitudes).

Neurocognitive research can provide a detailed and solid empirical basis for the understanding of role processing and internalization. Toward this end, electroencephalography (EEG) and similar methods can be especially useful. Together with low cost, a key advantage of EEG is high sensitivity to the temporal dynamic of neural processes, which facilitates identification and disentanglement of the two types of processes within the dual systems model (Banks, 2018). In neurosociological research on roles, EEG can be used to identify the involvement of automatic processes in various situations, such as (1) whether a role-relevant response to a task is automatic or controlled, (2) how fast and effortlessly a person enacts a role or switches to a new one, or (3) whether evaluation of others' role performance fits background expectancies or not.

Another promising avenue of research would be to compare different social groups/populations in terms of their "brain-role match," or the degree to which the architecture and functioning of the cognitive system is "ready" to perform a role. In line with Brett and Miles (2021), it is reasonable to suggest that social populations differ in their general preferences for deliberate or automatic processes, and this observation can be extended to the recognition of various degrees of dependence of role-related phenomena on each type of process. Many professional roles require deliberate reasoning, while relational social roles (e.g. parent) require emotional attachment (love and care), which is typically associated with System I and may rely on evolutionary/ancient neural mechanisms such as oxytocin secretion. Repeated inability to invoke the appropriate neurocognitive and neuroaffective systems is likely to have dysfunctional consequences at the macro-level.

It is worth noting that increasing rationalization in modern societies leads to the development of explicitly stated, even formalized, role expectations, and these impose requirements on emotions, cognitions, and behaviors that presumably rely on automated processes. This can give rise to a different kind of role-brain matching problem. For example, the organizers of a recent neuroscience conference sent out a code of conduct to attendees detailing the "expectations for conduct." The code described what participants could and could not do during the event. Prohibited behaviors included unwanted sexual behavior, misgendering, bathroom policing, or violations of personal space. This is an interesting case that illustrates both (1) the implicit assumption that scientists are suspicious and untrustworthy animals unable to control themselves, and (2) the belief that imposing explicit regulations will successfully control actions that are to a significant extent automatic.

The brain's preparedness for various aspects of automatic and deliberate processes depends on developmental, social, and cultural factors. For example, the ability for cognitive control (associated with lateral parts of PFC) develops steadily and gradually, whereas sensitivity to reward (associated with ventral striatum) develops in an inverted U-shape and peaks in adolescence, which explains greater risk-seeking behavior during that phase of life (Shulman et al., 2016). Education is known to foster analytical thinking and Type 2 (deliberate) processes (Brett & Miles, 2021). Different cultural settings lead to differences in the way values are processed in the brain as well as the degree of automated processing of protected values (Kaplan et al., 2017). In each case, the group-level differences in brain functioning are associated with different degrees of preparedness for particular role demands. If people's brains develop in a collectivist society that facilitates the inclusion of close others in the self-concept (Ng et al., 2010), it is reasonable to expect that an official who tries to avoid the conflict of interest when dealing with a relative (e.g., distributing grants or filling vacancies) will have more difficulty following the rules than an official socialized in an individualistic society. This is because the brain will literally perceive the losses of others' as their own.

A third avenue of research using the dual approach is to examine the automated processes, which can intervene in role-related cognitions and behaviors. Social and cognitive neuroscience have identified various aspects of the "social brain," which may affect how roles are perceived, performed, or negotiated. The evolution of the human mind required the development of social cognitive abilities for dealing with specific problems of living in a complex social environment: detecting social threats, managing group relations, navigating social hierarchies, or mating (Kaschak & Maner, 2009). The result is that the brain functions in a manner that allows fast and easy monitoring of social relations and responses to social situations. Social neuroscience has revealed many specific neural mechanisms responsible for automatic face recognition, threat cues detection, social categorization and selective response to ingroup and outgroup members, selective attention to status symbols, evaluation of others' physical attractiveness, etc. Of particular importance are neural systems supporting the representation of affiliation structures and social ranks (Beasley et al., 2012; Dwortz et al., 2022; Shkurko, 2013; Stolier & Freeman, 2017). There are many ways these processes can be triggered automatically by relevant social cues and intervene in various aspects of role-related processes. Two examples illustrate the issue.

In the aforementioned neurosociological experiment by Kalkhoff et al. (2020), participants were status differentiated based on task performance and then performed a similar joint task. In theory, the established performance-based status differences should lead to aligned expectations regarding performance at the joint task. In this experimental context, we can interpret the status-related expectations as specific functional roles: both participants *could* influence each other's task-related behaviors, but expected contributions are asymmetrical. That is, one partner is supposed to play a "leader" role (i.e., due to greater competence), and the other is supposed to play a "follower" role. The problem is that such expectations are often "hidden," not only from an observer, but also from the participants themselves. Mirroring Garfinkel's ethnomethodological experiments, the researchers used EEG to detect the brain's response to violations of expectations, or expected role behavior. That such expectations actually exist and function as a mediator of behavior is supported by the revealed differences in electrophysiological responses associated with choice uncertainty (i.e., subjective uncertainty about one's own task performance).

Using a different analytical approach, Kalkhoff et al. (2020) more directly measured expectation violations with a specific EEG event-related potential (ERP) component, the Feedback-Related Negativity (FRN). In general, FRN amplitudes are higher in response to unexpected stimuli. In the study by Kalkhoff et al. (2020), the FRN amplitude was greater for higher status participants whose influence attempts were repeatedly rejected by a lower status partner than for lower status participants whose influence was repeatedly accepted by a higher status partner, which is also expectation-violating. The asymmetry here is likely attributable to status loss aversion. That is, differences in status produce aligned (though often unconscious) performance expectations; therefore, correcting one's own solution to a problem after receiving feedback from a partner may be the most rational thing to do. However, status-based differences in the *reaction* to expectation violations introduce a source of bias into role performances that can interfere with the process of rational, informed correction. There may be a pronounced (irrational) tendency for higher status actors to dismiss conflicting ideas from lower status actors without fully weighing the value of those ideas simply because lower-status people are not supposed to disagree.

In a different study, Love and Davis (2021) conducted two experiments to reveal how women's race and status affect role-taking *accuracy*, operationalized as the ability to correctly predict the behavior of a partner in dyadic interaction. They showed that under normal social circumstances, black women are more accurate than white women (presumably due to greater role-taking effort by lower-status individuals), but after imposing an artificial status manipulation that gave black female participants a status advantage, the discrepancy was reversed (i.e., the lower-status white female participants became more accurate).

Although the experiment did not employ any implicit measures, there are reasons to believe that incorporating them might enhance the analysis. Previous research reveals that social signals trigger activity in neural networks associated with attention and other functions necessary for role-taking (Dwortz et al., 2022). Since "race" is the available social information in the situation examined, the question arises: how is the conflict between manipulated status cues and highly internalized and implicit race-based status representations resolved? Does the newly introduced and contradictory information inhibit the neural mechanisms of salience detection (particularly associated with the amygdala), or does it invoke executive control (associated with prefrontal regions)? Does it weaken the associative chains linking the representations of race and status, and if so, to what extent? Or, perhaps, reversing positions within a structure only reinforces the representation of the abstract hierarchical social structure, preventing a more egalitarian approach to task-related roles. Moreover, if low-status black women rely on Type 1 social cognition in perspective-taking, at least under "normal" circumstances, this may call into question the authors' idea that they are more engaged in "expressive labor." Labor is effortful, but the Type 1 processes are effortless. If the attention of lower status individuals is automatically modulated toward higher-status individuals, this may be seen as a useful adaptation. Status inequalities are inevitable, for example in organizations, and since managers have more power, employees must monitor managers carefully. If such monitoring is automatic, it helps to reduce the cognitive effort necessary to perform a role-taking task. The same logic applies to more persistent forms of inequality, including those related to race. The social structure may require an unequal distribution of "expressive labor," but reliance on the implicit processes helps to reduce or buffer such inequalities. Due to incongruent status representations, expressive labor (i.e. attentional and role taking effort) might be greater for white women in the "reverse" condition of Love and Davis' experiment compared to black women in the "normal" condition. Probing such questions seems useful for the sociological study of roles, yet it is difficult to answer them without considering the interplay between the automatic and controlled processes.

In sum, the dual-process approach provides a clear and relatively simple research program for the sociological study of roles and role-related issues based on methodological, theoretical, and empirical progress in cognitive psychology and neuroscience. Future research will reveal how controlled versus automatic processes are involved in role behavior, role evaluation, role transition, role exit, role-taking, role making, role conflicts, and other related issues across different types of roles and various social populations. In addition, future research will show how these two types of processes interact with one another, what types of structural, institutional or cultural settings favor their development, and it will identify the social consequences of these interactions. This potentially fruitful agenda of neurosociological research brings the field of sociology, more generally, up to speed with ongoing, contemporary debates within the neurosciences.

Complexity of Roles, Multiple Roles, and Modularity of the Neural System: Beyond the Dual Approach

A widespread assumption in the social sciences, as well as in common sense, is that individuals are best thought of as *unitary* actors. Even if an actor has an internal structure with different goals or motives, they are still a holistic and well-integrated entity, which is the basic unit of social ontology, even in system theories. The brain is recognized as the material receptacle of this social homunculus, no matter whether it is accepted as a determining essence or simply as a condition of existence. The dual system approach switches to a different view: there are two homunculuses in the brain, not one. Robert Sheckley would say that "you still haven't got much to work with." In all likelihood, though, the situation is much more complex.

One reason why many proponents of the dual approach prefer the term dual "process" over dual "systems" is that the automatic system does not work as a single system, but rather is a set of different systems, or modules (Frankish, 2010; Evans, 2018). Moreover, there is emerging evidence that the conscious system is not a single module that can only focus on one object, but rather it can be viewed as composed of multiple streams focused on multiple internal and external objects (Fernyhough et al., 2018). Further, the fact that the dual approach often associates the two types of processes with very different sets of attributes and numerous dichotomies (e.g. unconscious/conscious, automatic/controlled, associative/propositional, affective/cognitive, implicit/

explicit, and so on) led researchers to seek more nuanced models that go beyond the more basic distinction between two (and only two) systems (e.g. Evans, 2009; Amodio, 2014; Moors, 2014). In cognitive neuroscience, the alternative theoretical description of brain-to-behavior relations is based on a more complex, *modular* view of the neural system.

The modularity of complex systems means that they are subdivided into interconnected but relatively independent subsystems, with more intensive connections within modules than between them. This provides flexibility and allows for complex behaviors. Importantly, modules are not simply different functional parts of the united system, but rather are clusters in a distributed network. Different modules can evolve independently and operate in parallel, while at the same time contributing to the overall network's functionality. The modular view of complex systems was initially developed by Simon (1962), whose idea of "nearly decomposable systems" is now frequently used in organizational science. Nearly decomposable systems are complex systems in which the components operate almost independently in the short run, but not in the long run; that is, their inter-part connections are "weak but not negligible" (Ibid., p. 474). Later, Fodor (1983) developed a modular view of the human mind, defining modular cognitive systems as domain-specific, innate, hardwired in the neural system, computationally autonomous (i.e. not relying on shared computational resources), and not assembled from a general stock of simpler processes (ibid., p. 37).

In social and cognitive neuroscience, the modular view is widely accepted, but it has evolved over time. Initially, cognitive and social neuroscience relied upon the identification of task-dependent activations in specific brain areas or neuronal populations. The goal here was to identify the neural systems responsible for specific functions (e.g. the role of mirror neurons or TPJ in Theory of Mind). An important limitation of this "phrenological" approach is that most brain areas are involved in multiple tasks and functions. Later, then, the focus shifted to connections *between* brain areas in connection with psychological functions, even without significant changes in regional activation. Even more recently, methodological and technical advances, as well as the availability of neuroimaging big data, gave rise to "network neuroscience," which describes the brain in terms of network analysis. Here, the brain is understood as a modular network system with nodes and edges as well as specific topological features, such as the number of modules and degree of modularity. Developments in network neuroscience go beyond earlier attempts to determine what brain areas "light up" with particular tasks or behavioral measures. Instead, it additionally focuses on connections between brain regions and seeks to reveal large-scale neural systems that recruit different brain areas and show patterns of coherent activations (Mišić & Sporns, 2016; Sporns & Betzel, 2016). The modules in this view are spatially and temporally overlapping, and, even more importantly, they are flexible and subject to change over time. Furthermore, there are individual differences in the modular organization of neural systems leading to differences in cognitive plasticity and sensitivity to cognitive training and other interventions (Gallen & D'Esposito, 2019). There are also important age-related differences in the modular architecture of the brain. Changes in the degree of modularity have an inverted U-shape, with a decreasing number of interhemispheric connections and greater hemispheric-related modularity earlier and later in the lifespan (Puxeddu et al., 2020). Finally, the use of multivariate analytic techniques allows us to associate flexible modules with sets of behavioral phenotypes (i.e., multiple behavioral measures), thereby providing a more realistic picture of brain-behavior relationships (Mišić & Sporns, 2016).

How does the emerging work on the modular organization of the brain contribute to the sociological study of roles? First, it is worth noting that sociology has some advantages over psychology in dealing with networks and network-based complex systems. Network analysis has established traditions in the social sciences. Moreover, social network analysis already incorporates the notion of role as pre-defined or emerging patterns of behavior (Forestier et al., 2012). Yet this does not explain how social roles in their conventional meaning can be understood vis-à-vis the modular organization of the brain.

What is more important is that the social roles are complex, variated, and multiple, and individual-level processing of role-related phenomena (e.g. role-making, role performance, role evaluation, etc.) recruits various underlying neural modules and networks. As a theoretical construct, roles can include behaviors, attitudes, emotions, skills, cognitive abilities, etc. Further, there are different roles performed by an individual, and different sets of role attributes within one role (role-set). At present, neuroscientific research does not look at social roles as a measurable correlate of brain activity. However, there is no reason to expect that social roles are somehow embedded in specialized modules, or that each role is performed by a separate module or the like. Rather, we might expect that different roles (operationalized by any sociologically appropriate measure) are associated with different neural modules (revealed by a data-driven study of neural activity). In a very simplified and schematic way, various types of correspondence can be depicted in the following way (Fig. 13.2).

The left part of this simplified conceptual scheme describes the conventional sociological view of roles at the individual level. The right part describes the modular view of associated brain functioning.

Each role can be described as a set of attributes (i.e., characteristic behaviors, emotions, abilities, etc.). Roles differ in the number of their attributes, ranging from very narrow and specific (e.g. a situation where passengers simply must sit in place and not disturb others) to broad and diffuse (e.g., parents' roles involve a broad and indefinite set of activities and reactions, from daily caring to empathy and strategic investments decisions). If a role implies various sets of attributes in different role-related contexts, these constitute role sets.

Each role or role-related process is represented by several modules in the central nervous system (and, probably, in the peripheral system as well), organized in a specific manner. It is reasonable to suppose that some roles rely upon a larger number of modules than others. It is unlikely that roles can be represented by single modules. However, stable and well-internalized roles may be supported by identifiable modular networks with the characteristic spatial and temporal pattern of neural activity. This is a matter of future neurosociological research. Modules that are not directly associated with role-specific attributes can be active for various reasons (e.g. triggered by role-irrelevant stimuli) and interfere with role performance, role representation, etc.

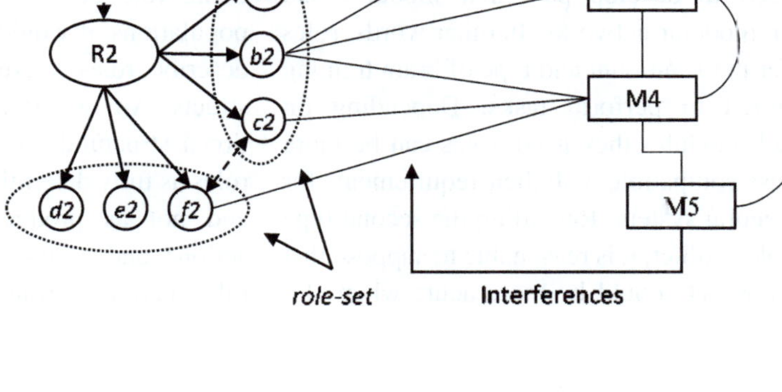

Fig. 13.2 Types of correspondence between roles and neural modules. R refers to roles; a, b, c, etc. refer to role attributes; M refers to modules

Despite its simplicity, the modular view suggests some ways to pose neurosociological problems, formulate testable hypotheses, and address questions of more general sociological interest. First of all, there is the basic descriptive task of identifying the neural basis of role-related phenomena. This task does not differ from routine research objectives in social and cognitive neuroscience, but it would incorporate behavioral or other observable measures to operationalize sociological (rather than psychological) constructs. Even such descriptive research will be useful in revealing whether roles (as theoretical constructs) have an identifiable neural basis and whether the neural signatures can be used as reliable measures of roles, their components, and characteristics, e.g. degree of internalization.

Second, the modular view on brain functioning helps to formulate specific hypotheses on brain-role relations. Here are some examples:

- By definition, modules are supposed to be relatively independent of each other. Thus, if a particular role is associated with several modules, they should contribute to role performance or role evaluation unevenly. This means that in different contexts, some modules (e.g. related to automatic motor reactions) become more actively involved than others (e.g. deliberate decisions).
- There are two types of role conflict depicted in the scheme in Fig. 13.2. One is between two different roles, R1 and R2 (e.g. conflict between work and family roles), and the second is between homologous attributes within a role-set, *c2* and *f2* (e.g. loyalty to a boss vs. loyalty to colleagues). The modular organization has a direct impact on both types of conflicts. Different roles compete for the time needed to develop particular modules and form modular networks. In other words, roles differ in the amount and type of brain training required to perform tasks. Depending on which modules they need, roles can be more or less compatible with their requirements for the neural system. Regarding the second type of role conflict, it is reasonable to suppose that the conflict should be more acute when elements of a role-set rely on the same specific module and compete for its resources.
- A larger degree of modularity should lead to a more difficult role transition, especially when roles rely on different modules and modular networks. Also, age-related role transitions may depend on the degree to which roles depend on brain areas with various levels of flexibility and inter-cluster connectivity over the lifespan (Puxeddu et al., 2020).
- Roles are associated with stable modular networks with identifiable characteristics. Therefore, roles can be a source of role-induced interferences in *irrelevant* contexts. After the end of role-playing or role exit (e.g. after retirement), people often have difficulties with "switching," or the transition to other roles. From a functional sociological point of view, this effect is hard to explain: the switch to a different context or institutional setting should lead to acceptance of relevant roles and corresponding role behavior or attitudes. However, if we see roles as hardwired in the brain, such effects are inescapable: the more fixed and developed neural network is necessary to perform a role, the more difficult it is to erase such an established network.

Finally, the modular basis of role-related processes has consequences for the macro-level sociological perspective. Individual and group (e.g. age-related) differences in the modular organization of the brain allow us to investigate the causes and consequences of role-brain match/mismatch at a microscopic level unavailable to standard sociological methods. However, there are important macro-level applications of the modular neurosociological perspective. By comparing the modular organization underlying particular roles across various individuals and social populations, it would be possible to identify and describe roles as such—as shared social constructs. An important criticism of the conventional structural theory of roles is that it treats roles as fully determined by system requirements and not as products of individual interests, actions, and strategies. Identification of the neural signatures of roles would provide a clear measure of the intersubjectivity of roles. If a role is represented similarly in the brains of many individuals—with a similar topology of the modular network—then it would make sense to view that particular role as a well-institutionalized social construct with shared social meaning. By contrast, significant differences in the modular architecture associated with a given role (e.g., a gender role) would signal its emergent or contested status and could illuminate the detailed nature of these ambiguities.

Final Remarks

Neurosociology and social neuroscience have overlapping research interests and focus on common problems, mirroring the overlapping interests and problems of sociology and social psychology. Nevertheless, the barrier preventing neurosociology from capitalizing on empirical brain research in the neurosciences is significant. Indeed, the lack of empirical research programs in neurosociology (rather than exceptional but singular studies) remains the main roadblock to the development of the field. At the same time, there are relatively simple and clear objectives for such empirical programs. First, no fundamental barriers outright prevent the use of methods from neuroscience to identify the neural foundations of sociological constructs. For example, the study of the neural basis of social roles is not inherently more difficult than the study of the self-concept or theory of mind. Second, the systematic investigation of socially relevant brain functioning across various social populations is a clear and reasonable objective, yet this remains largely in the wheelhouse of cultural neuroscience. Whether findings from such empirical research will be useful or groundbreaking within sociological paradigms is currently not clear. Yet in neurosociology they have value in their own right, representing a useful attempt to describe society as the community of brains or as the community of (sub-individual) modular networks.

Roles are a particularly good target for neurosociological investigation. Despite some conceptual and theoretical ambiguity, "role" has a conventional meaning shared by most social scientists. The key point is that the concept interlinks micro- and macro levels of social reality. In any meaningful sense, roles capture shared beliefs about what people should do (and feel) in various contexts. As I hope to have shown, there are enormous opportunities for the inclusion of cognitive and neuroscience findings into the (neuro-)sociological study of social roles. As this perspective advances, we will likely come to understand roles as stable patterns of neural activity—patterns governed by complexity and associated with distinctive social effects emergent and identifiable at the macro-level of social organization.

Acknowledgments I thank the anonymous reviewer for comments and suggestions, which were invaluable to both the form and content of the chapter.

References

Amodio, D. M. (2014). Dual experiences, multiple processes: Looking beyond dualities for mechanisms of the mind. In W. Sherman, B. Gawronski, & Y. Trope (Eds.), *Dual process theories of social mind* (pp. 560–577). The Guilford Press.

Banks, A. P. (2018). Comparing dual process theories: Evidence from event-related potentials. In W. De Neys (Ed.), *Dual process theory 2.0* (pp. 66–81). Routledge.

Barbalet, J. (2006). Social role. In B. S. Turner (Ed.), *The Cambridge dictionary of sociology* (pp. 583–584). Cambridge University Press.

Beasley, M., Sabatinelli, D., & Obasi, E. (2012). Neuroimaging evidence for social rank theory. *Frontiers in Human Neuroscience, 6*, 123. https://doi.org/10.3389/fnhum.2012.00123

Biddle, B. (1979). *Role theory: Expectations, identities, and Behaviors*. Academic Press.

Boccadoro, S., Craco, E., Hudson, A. R., Bardi, L., Nijhof, A. D., Wiersema, J. R., Brass, M., & Mueller, S. C. (2019). Defining the neural correlates of spontaneous theory of mind (ToM): An fMRI multi-study investigation. *NeuroImage, 203*, 116193. https://doi.org/10.1016/j.neuroimage.2019.116193

Brett, G., & Miles, A. (2021). Who thinks how? Social patterns in reliance on automatic and deliberate cognition. *Sociological Science, 8*, 96–118. https://doi.org/10.15195/v8.a6

Chernozub, O. L. (2023). Graphic associative test of attitudes as a convenient implicit measurement tool for mass polls. *RUDN Journal of Sociology, 23*(1), 122–141. https://doi.org/10.22363/2313-2272-2023-23-1-122-141

Cicourel, A. (1972). *Cognitive sociology*. Penguin.

Collins, R. (1986). On the microfoundations of macrosociology. *The American Journal of Sociology, 86*(5), 984–1014.

Cunningham, W. A., Johnson, M. K., Raye, C. L., Gatenby, J. C., Gore, J. C., & Banaji, M. R. (2004). Separable neural components in the processing of black and white faces. *Psychological Science, 15*, 806–813. https://doi.org/10.1111/j.0956-7976.2004.00760.x

DiMaggio, P. (1997). Culture and cognition. *Annual Review of Sociology, 23*, 263–287. https://doi.org/10.1146/annurev.soc.23.1.263

Dwortz, M. F., Curley, J. P., Tye, K. M., & Padilla-Coreano, N. (2022). Neural systems that facilitate the representation of social rank. *Philosophical Transactions of the Royal Society B, 377*(1845), 20200444. https://doi.org/10.1098/rstb.2020.0444

Eagly, A. H., & Wood, W. (2012). Social role theory. In P. A. M. VanLange, T. E. Higgins, & A. W. Kruglanski (Eds.), *Handbook of theories of social psychology* (Vol. 2, pp. 458–476). Sage Publications.

Enright, R. D., & Lapsley, D. K. (1980). Social role-taking: A review of constructs, measures, and measurement properties. *Review of Educational Research, 50*(4), 647–674.

Evans, J. S. B. T. (2009). How many dual-process theories do we need? One, two, or many? In J. Evans & K. Frankish (Eds.), *In two minds: Dual processes and beyond* (pp. 33–54). Oxford University Press.

Evans, J. S. B. T. (2018). Dual process theory: Perspectives and problems. In W. De Neys (Ed.), *Dual process theory 2.0* (pp. 137–156). Routledge.

Fernyhough, C., Alderson-Day, B., Huriburt, R. T., & Kühn, S. (2018). Investigating multiple streams of consciousness: Using descriptive experience sampling to explore internally and externally directed streams of thought. *Frontiers in Neuroscience, 12*, 494. https://doi.org/10.3389/fnhum.2018.00494

Firat, R. B. (2021). A neurosociological theory of culturally and structurally situated cognition and ethnoracial stress. *Frontiers in Sociology, 6*, 695042. https://doi.org/10.3389/fsoc.2021.695042

Fodor, J. A. (1983). *The modularity of mind*. The MIT Press.

Forestier, M., Stavrianou, A., Velcin, J., & Zighed, D. A. (2012). Roles in social networks: Methodologies and research issues. *Web Intelligence and Agent Systems: An International Journal, 10*(1), 117–133. https://doi.org/10.3233/WIA-2012-0236

Frankish, K. (2010). Dual-process and dual-system theories of reasoning. *Philosophy Compass, 5*(10), 914–926. https://doi.org/10.1111/j.1747-9991.2010.00330.x

Franks, D. D. (2013). Why we need Neurosociology as well as social neuroscience: Or—Why role-taking and theory of mind are different concepts. In D. D. Franks & J. H. Turner (Eds.), *Handbook of Neurosociology* (pp. 27–32). Springer.

Franks, D. D., & Turner, J. H. (Eds.). (2013). *Handbook of Neurosociology*. Springer.

Gallen, C. L., & D'Esposito, M. (2019). Brain modularity: A biomarker of intervention-related plasticity. *Trends in Cognitive Sciences, 23*(4), 293–304. https://doi.org/10.1016/j.tics.2019.01.014

Gallese, V., Keysers, C., & Rizzolatti, G. (2004). A unifying view of the basis of social cognition. *Trends in Cognitive Science, 8*, 396–403. https://doi.org/10.1016/j.tics.2004.07.002

Garfinkel, H. (1967). *Studies in ethnomethodology*. Prentice-Hall.

Giddens, A. (1984). *The constitution of society: Outline of the theory of structuration*. University of California Press.

Goffman, E. (1959). *Presentation of self in everyday life*. Anchor.

Goffman, E. (1974). *Frame analysis: An essay on the organization of experience*. Harper Colophon.

Gregory, S. W., Kalkhoff, W., Harkness, S. K., & Paull, J. L. (2009). Targeted high and low speech frequency bands to right and left ears respectively improve task performance and perceived sociability in dyadic conversations. *Laterality, 14*(4), 423–440. https://doi.org/10.1080/13576500802598181

Habermas, J. (1984). *The theory of communicative action* (Vol. 1, 2). Beacon Press.

Harnisch, S., Frank, C., & Maull, H. W. (Eds.). (2011). *Role theory in international relations. Approaches and Analyses. Approaches and analyses*. Routledge.

Heyes, C., & Catmur, C. (2022). What happened to mirror neurons? *Perspectives on Psychological Science, 17*(1), 153–168. https://doi.org/10.1177/1745691621990638

Kalkhoff, W., Melamed, D., Pollock, J., Miller, B., Overton, J., & Pfeiffer, M. (2020). Cracking the black box: Capturing the role of expectation states in status processes. *Social Psychology Quarterly, 83*(1), 26–48. https://doi.org/10.1177/0190272519868988

Kaplan, J. T., Gimbel, S. I., Dehghani, M., Immordino-Yang, M. H., Sagae, K., Wong, J. D., et al. (2017). Processing narratives concerning protected values: A cross-cultural investigation of neural correlates. *Cerebral Cortex, 27*(2), 1428–1438. https://doi.org/10.1093/cercor/bhv325

Kaschak, M. P., & Maner, J. L. (2009). Embodiment, evolution, and social cognition: An integrative framework. *European Journal of Social Psychology, 39*, 1236–1244. https://doi.org/10.1002/ejsp.664

Koski, J. E., Xie, H., & Olson, I. R. (2015). Understanding social hierarchies: The neural and psychological foundations of status perception. *Social Neuroscience, 10*(5), 527–550. https://doi.org/10.1080/17470919.2015.1013223

Lieberman, M. D. (2007). Social cognitive neuroscience: A review of core processes. *Annual Review of Psychology, 58*, 259–289. https://doi.org/10.1146/annurev.psych.58.110405.085654

Linton, R. (1936). *The study of man*. Appleton-Century.

Lizardo, O. (2017). Improving cultural analysis: Considering personal culture in its declarative and nondeclarative modes. *American Sociological Review, 82*(1), 88–115. https://doi.org/10.1177/0003122416675175

Lizardo, O., Mowry, R., Sepulvado, B., Stoltz, D. S., Taylor, M. A., Van Ness, M. A., & Wood, M. (2016). What are dual process models? Implications for cultural analysis in sociology. *Sociological Theory, 34*(4), 287–310. https://doi.org/10.1177/0735275116675900

Love, T. P., & Davis, J. L. (2021). Racial differences in women's role-taking accuracy: How status matters. *Sociological Status, 8*, 150–169. https://doi.org/10.15195/v8.a8

Mead, G. H. (1972 [1934]). *Mind, self, and society. From a standpoint of social behaviorist*. University of Chicago Press.

Merton, R. K. (1957). The role-set: Problems in sociological theory. *The British Journal of Sociology, 8*(2), 106–120.

Mišić, B., & Sporns, O. (2016). From regions to connections to networks: New bridges between brain and behavior. *Current Opinion in Neurobiology, 40*, 1–7. https://doi.org/10.1016/j.conb.2016.05.003

Moors, A. (2014). Examining the mapping problem in dual-process models. In W. Sherman, B. Gawronski, & Y. Trope (Eds.), *Dual process theories of social mind* (pp. 20–34). The Guilford Press.

Ng, S. H., Han, S., Mao, L., & Lai, J. C. L. (2010). Dynamic bicultural brains: fMRI study of their flexible neural representation of self and significant others in response to culture primes. *Asian Journal of Social Psychology, 13*(2), 83–91. https://doi.org/10.1111/j.1467-839X.2010.01303.x

Parsons, T. (1951). *The social system*. The Free Press of Glencoe.

Puxeddu, M. G., Faskowitz, J., Betzel, R. F., Petti, M., Astolfi, L., & Sporns, O. (2020). The modular organization of brain cortical connectivity across the human lifespan. *NeuroImage, 218*, 116974. https://doi.org/10.1016/j.neuroimage.2020.116974

Rathburn, B. (2020). Towards a dual process model of foreign policy ideology. *Current Opinion in Behavioral Sciences, 34*, 211–216. https://doi.org/10.1016/j.cobeha.2020.04.005

Ritzer, G., & Stepnisky, J. (2018). *Sociological theory* (10th ed.). Sage Publications.

Sherman, J. W., Gawronski, B., & Trope, Y. (Eds.). (2014). *Dual process theories of social mind*. The Guilford Press.

Shkurko, A. (2013). Is social categorization based on relational ingroup/outgroup distinction? A meta-analysis. *SCAN, 8*, 870–877. https://doi.org/10.1093/scan/nss085

Shulman, E. P., Smith, A. R., Silva, K., Icenogle, G., Duell, N., Chein, J., & Steinberg, L. (2016). The dual system model: Review, reappraisal, and reaffirmation. *Developmental Cognitive Neuroscience, 17*, 103–117. https://doi.org/10.1016/j.dcn.2015.12.010

Simon, H. A. (1962). The architecture of complexity. *Proceedings of the American Philosophical Society, 106*(6), 467–482.

Sporns, O., & Betzel, R. F. (2016). Modular brain networks. *Annual Review of Psychology, 67*, 613–640. https://doi.org/10.1146/annurev-psych-122414-033634

Stolier, R. M., & Freeman, J. B. (2017). A neural mechanism of social categorization. *Journal of Neuroscience, 37*(23), 5711–5721. https://doi.org/10.1523/JNEUROSCI.3334-16.2017

Strack, F., & Deutsch, R. (2014). The reflective-impulsive model. In J. W. Sherman, B. Gawronski, & Y. Trope (Eds.), *Dual process theories of social mind* (pp. 92–104). The Guilford Press.

Stryker, S. (2006). Traditional symbolic interactionism, role theory, and structural symbolic interactionism: The road to identity theory. In J. H. Turner (Ed.), *Handbook of sociological theory* (pp. 211–231). Springer.

Turner, R. H. (2006). Role theory. In J. H. Turner (Ed.), *Handbook of sociological theory* (pp. 233–254). Springer.

Vaisey, S. (2009). Motivation and justification: A dual-process model of culture in action. *American Journal of Sociology, 114*(6), 1675–1715. https://doi.org/10.1086/597179

van Overwalle, F. (2009). Social cognition and the brain: A meta-analysis. *Human Brain Mapping, 30*, 829–858. https://doi.org/10.1002/hbm.20547

Vila-Henninger, L. A. (2020). A theory of 'popular political legitimation': A dual-process model approach to legitimation and political socialization. *Journal for the Theory of Social Behaviour, 50*(4), 490–515. https://doi.org/10.1111/jtsb.12255

Vila-Henninger, L. A. (2021). A dual-process model of economic behavior: Using culture and cognition, economic sociology, behavioral economics, and neuroscience to reconcile moral and self-interested economic action. *Sociological Forum, 36*, 1271–1296. https://doi.org/10.1111/socf.12763

Yang, D. Y.-J., Rosenblau, G., Keifer, C., & Pelphery, K. A. (2015). An integrative neural model of social perception, action observation, and theory of mind. *Neuroscience and Biobehavioral Reviews, 51*, 263–275. https://doi.org/10.1016/j.neubiorev.2015.01.020

What Are the Neurosociological Foundations of Identities and Identity-Related Processes?

14

Richard E. Niemeyer

Abstract

This chapter bridges the gap between the fields of neuroscience and micro-sociological by examining the relevance of research on the default mode network (DMN) to identity theory. I posit that the neurological activity patterns characterizing the DMN constitute the biological mechanisms underlying identity theory's core concepts and processes. Incorporating these biological insights provides empirical support for identity theory's core concepts and arguments, deepens micro-sociologists understanding of the relationship between the mind, brain, and social structures, and introduces many exciting avenues for further research.

Author's Note. This work was supported by the Air Force Office of Scientific Research, funding award FA9550-23-1-0453.

This work contains no conflicts of interest for any contributing author.

R. E. Niemeyer (✉)
Warfighter Effectiveness Research Center,
Department of Behavioral Sciences & Leadership,
United States Air Force Academy, USAF Academy,
El Paso County, CO, USA
e-mail: Richard.niemeyer@afacademy.af.edu

Introduction

Sociologists have long speculated how biology, social structure, and culture influence human cognition and behavior. However, it is only within the last few decades that the necessary research technologies, background knowledge, and philosophical frameworks have sufficiently developed to advance sociological thinking beyond mere speculation. In this chapter, I draw upon these developments to examine the neurological mechanisms underlying an essential part of sociological social psychology—identity theory (Burke & Stets, 2022). Through this examination, I aim to refine identity theory's accuracy, precision, and explanatory power and offer a blueprint for similar advancements in other areas of micro-sociological research.

Neuroscientists have spent the past 25 years identifying the structural and functional characteristics of the brain's intrinsic connectivity networks (ICN) (Biswal et al., 2010; Corbetta & Shulman, 2002; Fox et al., 2006; Gusnard et al., 2001; Hagmann et al., 2008; Raichle & Gusnard, 2005; Raichle et al., 2001). These networks consist of interconnected brain regions that work together to generate specific mental states or perform various cognitive and behavioral acts. One ICN, the *default mode network* (DMN), appears to generate the cognitive processes that identity theorists consider essential for social interaction (Andrews-Hanna et al., 2010, 2010; Buckner

W. Kalkhoff et al. (eds.), *Handbook of Neurosociology*, Handbooks of Sociology and Social Research, https://doi.org/10.1007/978-3-031-95615-7_14

et al., 2008; Leech et al., 2011; Spreng & Grady, 2010). These processes include possessing a sense of self, engaging in self-reflection, interpreting the meaning of physical and social experiences, taking the perspective of others, and covertly preparing for potential actions (Burke & Stets, 2022). However, the precise relationship between DMN-related research and identity theory remains to be determined, as well as its relevance.

My goal in this chapter is to demonstrate how research on the DMN can be causally linked to the identity theory research program without succumbing to biological reductionism. Specifically, I will argue that the patterns of neuronal activity generated by the DMN constitute the biological mechanisms underlying identity theory's socio-cognitive concepts and processes. I will argue that research on the DMN uniquely substantiates identity theory's claims by validating its concepts and processes anatomically and physiologically. Additionally, I will argue that a deeper understanding of the neurosociological foundations of identity-related processes can expand the explanatory scope of identity theory into new research domains.

To this end, I divide this chapter into four parts. Part 1 briefly describes identity theory's core concepts and processes. In Part 2, I explore how Mead's concept of meaning can bridge contemporary neuroscience and identity theory. In Part 3, I expand upon this initial integration to propose several neurosociological mechanisms responsible for identity theory's central concepts and processes. Finally, in Part 4, I briefly discuss the benefits of my proposed neurosociological translation of identity theory.

Overview of Identity Theory

An identity is "the set of meanings that define individuals as occupants of roles in society" (Burke & Stets, 2022:1). Identity *theory* is a micro-sociological research program that examines how social structures, social interactions, and individuals' identities mutually influence each other (Burke & Stets, 2022).[1] The explanatory goal of identity theorists is to explicate how an individual's identities initially form; once formed, how identities organize an individual's behavior in predictable ways; and finally, how the meaning of an identity and its associated cognitive and behavioral processes potentially change in response to repeated interactions with others.

Identity theorists recognize four bases of identities: person identities, role identities, group identities, and categorical identities (Burke & Stets, 2022:166–175). Person identities are the set of meanings that define the characteristics of a person *as an individual*. Role identities define individuals in terms of the *roles they perform*, and group identities define individuals as *members of a social group (e.g., clubs, organizations, etc.)*. Categorical identities are meanings that define individuals as *members of social categories like race, class, and gender.* Two dimensions further characterize each of these identities. The first is the conventional dimension or shared expectations, which reflects the sociocultural definitions and behavioral expectations associated with an identity (Serpe & Stryker, 2011; Stets & Serpe, 2013). The second dimension is the idiosyncratic dimension or shared meanings (Serpe & Stryker, 2011; Stets & Serpe, 2013). It consists of the individual's distinctive interpretation of the meanings and responsibilities associated with an identity and her definition of the kinds of behaviors that she can appropriately engage in. For example, Burke and Stets (2022) note that the conventional dimension that defines the role-responsibilities of a professor includes teaching courses, grading papers, and holding

[1] Like most sociological research programs, identity theory has evolved over time. In the past, it was possible to distinguish three separate lines of research carried out by George McCall and J.L. Simmons; Sheldon Stryker and colleagues; and Peter J. Burke, Jan E. Stets, and colleagues. Recently, Burke and Stets (2022) articulated a version of identity theory that incorporates these distinct research programs into a single framework, as do Serpe and Stryker (2011) and Stets and Serpe (2013). Following this synthesis, this chapter will treat the concepts and theories offered by each of the three programs as elements of a single framework.

office hours. However, an individual may inject into their professor role-performance an idiosyncratic dimension that defines the role's responsibilities, including mentor, confidant, or friend. Similarly, the conventional dimension of an individual's moral identity may state that "a good person does not steal" (Stets, 2010). Nonetheless, the idiosyncratic dimension of the identity may qualify the conventional definition as "a good person does not steal from members of their social network; outsiders, though, are fair game."

Identity theorists posit that individuals learn their identity dimensions through socialization and others' reflected appraisals (Burke & Stets, 2022). Following the structural symbolic interaction tradition (Stryker, 2002), socialization is defined here as the process by which individuals are formally taught by the members of their society the meanings of the signs, symbols, and behaviors that characterize their culture. Through this socialization process, individuals learn the conventional dimensions of their identities. The set of meanings that eventually define an individual's identity is called the identity standard. Following Cooley (1902), a reflected appraisal is the information people convey during a social interaction that indicates how each is defining and interpreting the meaning of the other. For example, Christy's approving look during Conrad's lecture may indicate to Conrad that he is being perceived as a competent professor. If Conrad continues to experience this identity verification from Christy and his other students, the standard for his professor identity will be "competent."

The perceptual control process is a theoretical model of how the identity standard dynamically influences individuals' behavior throughout their daily tasks (Burke & Stets, 2022). The model postulates that the meanings stored in the identity standard guide an individual's behavior (e.g., a competent professor lectures this way and not that way). Others then interpret the meaning of this behavior and provide feedback in the form of actual appraisals. Once perceived, a person compares the meaning of this feedback with the meanings stored in his or her identity standard. An individual's identity is said to be verified if the meaning of the reflected appraisal matches the meaning stored in his or her identity standard. Positive emotions are the phenomenological consequence of identity verification. If the meaning of the reflected appraisal does not match the identity standard, then an error signal is generated; generally speaking, this error signal is experienced as a negative emotion. This negative emotion, in turn, motivates the individual to engage with the stimulus until the perceived meanings match those stored in the identity standard. Importantly, identity theory argues that this process can occur consciously or subconsciously (Burke & Stets, 2022). In short, this means that the perception and interpretation of the meaning of self-relevant stimuli, the comparison of the meaning with the meanings stored in the identity standard, and the motivation to respond to the failure to verify an identity can occur as an automatic or deliberately controlled cognitive process.

Further, identity theorists argue that the perceptual control process can occur even when other individuals are not directly present to provide a reflected appraisal (Burke & Stets, 2022). For example, when anticipating a social interaction, an individual will draw upon the meanings stored in his or her identity standard to construct a course of action that reflects these meanings. Identity theorists maintain that this is done by the individual mentally rehearsing potential lines of behavior and imagining how others may react. Importantly, identity theorists claim that these imagined, reflected appraisals generate the same positive and negative emotions as those actual appraisals experienced during face-to-face interaction. Beyond organizing future behaviors into a coherent plan of action, the anticipatory feelings of approval partially legitimize an individual's cultural understanding and personal definitions of an identity. In short, this is because emotions generally reinforce or dissuade behaviors regardless of their source (i.e., imagined vs. actual). Once this set of behaviors is performed in front of others, the emotions generated by the actual appraisals fully legitimize or delegitimize these understandings (Burke & Stets, 2022:42).

A Prelude to the Neurosociology of Identity Theory

Integrating micro-sociological theory and neuroscience can be challenging because the two fields employ very different data, theoretical frameworks, and research methods. However, there are some similarities between the two fields that can serve as a starting point for integration, provided one knows where to look. For example, recall that identity theorists define an identity as the conventional and idiosyncratic meanings that are self- and socially ascribed to an individual. As noted by Burke and Stets (2022), this definition is significantly influenced by the work of the classical sociologist George H. Mead (1934). Interestingly, Mead frequently referenced neurons and the nervous system when defining concepts in *Mind, Self, and Society,* particularly when discussing his concept of meaning. This connection between biology, Mead's ideas, and identity theory provides a crucial starting point for establishing the neurological foundation of identity theory concepts, which can then be expanded to other aspects of the framework.

A Neurocognitive Translation of Mead's Concept of Meaning

According to Mead (1934), the meaning of an object is determined by how an individual is prepared to act toward it. Regarding the meaning of a telescope, Mead says:

> If we want to trace the responses of the astronomer, we have to go back into his *central nervous system*, back to a whole *series of neurons*; and we find something there that answers to the exact way in which the astronomer approaches the instrument under certain conditions....the values which we say the instrument has are values through the relationship of the object to the person who has that sort of attitude. If a person did not have that *particular nervous system*, the instrument would be of no value. *It would not be a telescope*. (Mead, 1934:29; emphasis added)

Notice here the degree to which Mead binds this concept of meaning to the existence of some neurological substrate; indeed, in the last line of the quote, he explicitly notes that the neurological processes underlying the behavioral attitude toward the telescope are precisely what makes it a telescope. Without this biological substrate, the telescope as such would not exist.

In the neuroscience literature, Mead's (1934:29) "whole series of neurons" is more precisely described as a cell assembly. A cell assembly is a collection of neurons that coordinate their patterns of neural activity to represent information about the environment (Hebb, 2002). To borrow Mead's example, this means that a discrete collection of neurons responds in a deliberate and coordinated fashion whenever an individual perceives a telescope, while another cell assembly responds whenever he or she perceives a horse. Memory engrams are cell assemblies that store perceptions, other experiences, and information in the brain for later recall (Takamiya et al., 2020). Given this function, I argue that conventional and idiosyncratic meanings constituting an individual's identity standard are represented in his or her brain by memory engrams.

Types of Declarative Knowledge

According to neuroscience research, memory can be divided into distinct types (Squire & Kandel, 2009). Memory that an individual can consciously recollect is called declarative knowledge or explicit memory. Semantic memory is a subtype of declarative memory and is defined as an individual's knowledge of general facts, e.g., the capitals of major countries, the names of famous sports teams, the grammar and vocabulary of their language, etc. Episodic memory is a second subtype of declarative memory that is defined as the knowledge an individual has about specific events that they experienced from the first-person perspective, e.g., his memory of the time he traveled to Paris after graduating from high school or the memory of the time he watched his favorite sports team win the championship game.

A third type of declarative memory is called semantic autobiographical (D'Argembeau & Salmon, 2011; Levine et al., 2004). Like episodic memory, semantic autobiographical memory is a form of self-relevant knowledge; like semantic memory, though, this knowledge about the self is not tied to a specific "episode" in an individual's autobiographical history. In other words, semantic autobiographical memory is knowledge about oneself that has been abstracted from specific experiences. For example, an individual's memory of her mood yesterday during lunch is episodic knowledge; an individual's knowledge about her general disposition is a semantic autobiographical memory. Importantly, the critical difference between semantic, episodic, and semantic autobiographical memory is not so much the content of the knowledge *per se* but the degree to which the memory is tied to a specific moment or abstracted from numerous individual events.

Although the identity theory concepts of conventional and idiosyncratic dimensions are defined in terms of "sociocultural" versus "person meanings," and the neuroscience concepts of semantic and episodic memory are defined in terms of "general" versus "particular" experiences, we argue nonetheless that these two sets of concepts are interchangeable. The following example illustrates the reason for this. A professor is defined as an individual who teaches at the university or college level; the primary responsibilities of a professor include designing and teaching courses, performing research, and grading assignments. This general knowledge about the meaning of the word professor is semantic knowledge. In the above summary of identity theory, I noted that Burke and Stets (2022) cite teaching courses, grading papers, and holding office hours as examples of the conventional dimension of the role-responsibilities of a professor. As seen here, the phenomena captured by the neuroscience concept of semantic knowledge and the identity theory concept of conventional meanings are essentially the same.

Burke and Stets (2022) further note that the idiosyncratic dimension of the professor identity can include being a friend or mentor. It is reasonable to assume that an individual could develop this idiosyncratic definition after specific interactions with "mentor-type" or "friend-type" professors during graduate school. It is also possible that a professor could learn these meanings through a process of reflexivity after experiencing specific instances of positive reinforcement from her students as she engaged in mentor-like or friend-like behaviors and reflected upon them. Because individuals can deliberately recall specific examples of these experiences from the first-person perspective, they are episodic memories. The general idiosyncratic meaning of "professor as friend" that is abstracted from these specific episodic instances, though, is semantic autobiographical knowledge. Based upon these similarities, the neurological concepts of episodic and semantic autobiographical memory are interchangeable with the identity theory concept of the idiosyncratic dimension of meaning.

Measuring Meaning in the Brain

Researchers have recently begun to investigate the neurological foundations of how individuals recall semantic, episodic, and semantic autobiographical knowledge using a noninvasive neuroimaging technology called functional magnetic resonance imaging (fMRI) (Buckner et al., 2008; Burianova et al., 2010; D'Argembeau & Salmon, 2011; Kompus et al., 2009; Levine et al., 2004; Spreng & Grady, 2010; Spreng et al., 2009). The correspondence between Mead's concept of meaning, contemporary understandings of memory, and the concept of meaning in identity theory means these technologies can also be applied to research on identity-related processes. Given the importance of fMRI to both lines of research, it is necessary to review how this technology works.

fMRI measures the neural activity underlying a cognitive process by taking advantage of the blood oxygen level-dependent (BOLD) effect (Cacioppo & Berntson, 2005). Like all cells, neurons must draw oxygen from the blood to function. Because of this requirement, the body constantly delivers to the brain a steady flow of blood to maintain its resting state or baseline levels of cellular activity, i.e., the level of neural activity that occurs when an individual is not actively engaged in a cognitive or behavioral act. Also, like all other cells in the body, greater levels of cellular activity require greater amounts of oxygen. To meet this greater demand, the body increases the amount of blood delivered to a brain region whenever it is "activated" to perform some cognitive or behavioral function.

Because the oxygen demands of neurons are relatively low when a brain region is "at rest," the ratio of the amount of oxygenated to deoxygenated hemoglobin in the blood is high. This is because the body can deliver new oxygenated blood to the region faster than the associated neurons can consume it. When neurons suddenly increase their activity level at the onset of a cognitive or behavioral act, they briefly begin to consume oxygen from the blood faster than the body can replace it. As a result, the ratio of oxygenated to deoxygenated hemoglobin decreases; an increase immediately follows this transient state in the amount of blood sent to the active brain region, which, in turn, increases the ratio of oxygenated to deoxygenated hemoglobin. To summarize, fMRI can measure changes in neural activity by measuring changes in the ratio of oxygenated to deoxygenated hemoglobin. Further, fMRI allows researchers to determine where a specific cognitive act occurs in the brain by identifying which brain regions demonstrate a change in blood oxygen level at both the onset and during the entire performance of the cognitive activity.

Considering how fMRI operates, neuroscientists examining the neurological foundations of declarative knowledge generally employ two types of research designs (Cacioppo & Berntson, 2005; D'Argembeau & Salmon, 2011). The first involves using fMRI to record changes in neural activity while an individual deliberately recalls a semantic, episodic, or semantic autobiographical memory. For example, Wiggs and colleagues (Wiggs et al., 1998) examined the neural correlates of semantic memory by showing participants a black-and-white line drawing of common objects (such as a thimble, an American football, or an igloo) and then asking them to either recall the object's name or specify the color typically associated with the object (e.g., thimbles are generally silver and American footballs are generally brown). Similarly, the neurological correlates of episodic memory are generally examined by recording changes in brain activity as individuals recall very specific moments in their lives (e.g., what they did on their last vacation, etc.) (D'Argembeau & Salmon, 2011). Semantic autobiographical memory, in turn, is examined by asking individuals to evaluate if a series of descriptive statements accurately describe them; specifically, such statements include the following: I am a shy person. I am a father. I am a good person. Etc. (D'Argembeau & Salmon, 2011; Ochsner et al., 2005; Rameson et al., 2010). The recorded levels of "task-induced" neural activity are then compared to those recorded before the onset of the task to determine which regions demonstrated increased oxygen consumption levels. Areas of the brain that demonstrate increased oxygen consumption are then identified as playing a functional role in enabling memory recall.

The second research design involves a series of tasks that sequentially require the recollection of semantic, episodic, and semantic autobiographical knowledge, and the patterns of increased neural activity are then compared across tasks. Brain regions that only show increased activation levels during specific kinds of memory recall are said to be functionally discrete; brain regions that show increased activation across multiple tasks are said to serve as a common foundation for multiple types of recollection.

The Neurological Foundations of Declarative Knowledge

More than a decade of extensive research employing the above empirical designs has consistently identified the following brain regions as responsible for enabling the recollection of semantic, episodic, and semantic autobiographical knowledge (see Fig. 14.1). First, portions of the lateral temporal and the posterior inferior parietal cortices have routinely been demonstrated to be functionally involved in the recollection of semantic memory (Binder et al., 2009; Prince et al., 2007; Wiggs et al., 1998). Second, portions of the medial prefrontal cortex, posterior cingulate cortex/precuneus, hippocampus, and anterior cingulate cortex have all been reliably shown to play a functional role in recalling episodic memory (see Fig. 14.1) (Addis et al., 2009; D'Argembeau & Salmon, 2011; Kwan et al., 2010; Levine et al., 2004; Leyhe et al., 2009; Prince et al., 2007). Third, a comparison of the brain regions activated to generate semantic autobiographical knowledge reveals significant overlap with those that generate general semantic knowledge; i.e., both semantic and semantic autobiographical knowledge are generated by regions of the brain that are responsible for recalling knowledge that is not tied to a specific event. The key difference though is that the recollection of semantic autobiographical knowledge involves "episodic" areas of the brain that process self-related information and enable deliberate cognitive control (Levine et al., 2004). This means semantic autobiographical knowledge is neurologically distinct from general semantic knowledge to the degree that the former requires a greater amount of cognitive effort to recall, as well as by the fact that it contains a self-relevant dimension.

To summarize our argument thus far, the identity theory concepts of conventional and idiosyncratic meanings are conceptually identical to semantic, episodic, and semantic autobiographical memory. Further, because an individual's identity standard is defined as the complete stock of meanings a person has for a given identity, the concept of an identity standard is co-terminate with an individual's complete stock of episodic, semantic, and semantic autobiographical knowledge. Recognizing this correspondence is essential for three reasons. First, it provides an important clue for identifying which neuroscience research programs may apply to identity theory, e.g., those research programs that are related to the three forms of memory. Second, the fact that physiologically distinct mechanisms generate all three forms of memory indicates that the difference between idiosyncratic and conventional meanings is more than simply a heuristic distinction; instead, it is an accurate representation of empirical reality. Finally, the fact that fMRI can be used to observe the neurological processes underlying declarative knowledge means that this same neuroimaging technology can be used to study how the conventional and idiosyncratic meanings may or may not affect an individual's behavior. The larger implications of this last point will be discussed in greater detail in the next section.

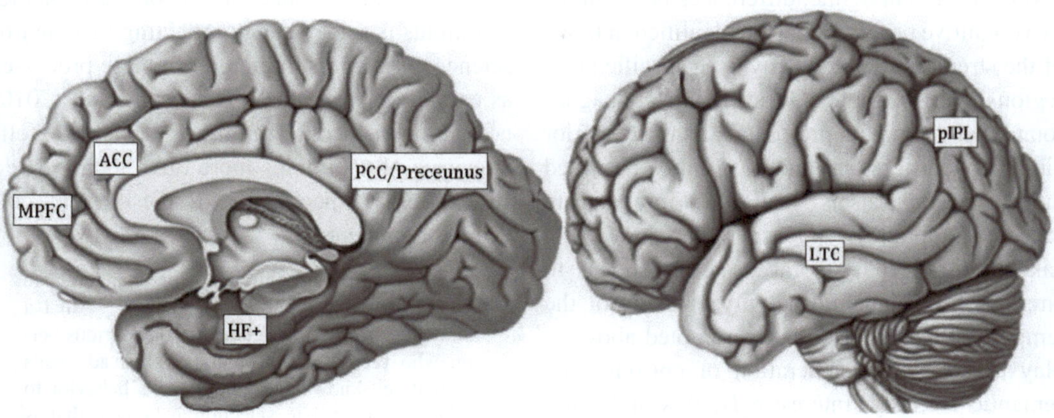

Fig. 14.1 Brain regions responsible for generating episodic memory (**left**) involve the medial prefrontal cortex (MPFC), anterior cingulate cortex (ACC), hippocampal formation (HF+) and posterior cingulate cortex/Preceunus (PCC/preceunus). Brain regions responsible for generating semantic memory (**right**) include lateral temporal cortex (LTC) and posterior inferior parietal lobe (PIPL). Views of the brain are taken from midsagittal, dissected down the midline of the brain (**Left**) and an exterior view (**Right**)

The Neurosociology of Identity Theory

The Neuroscience of Self- and Reflected Appraisals

As was noted in the overview above, an important assumption of identity theory is that individuals can engage in self- and reflected appraisals (Burke & Stets, 2022). Using fMRI in the manner discussed above, neuroscientists have investigated three research questions related to these two processes: First, what are the neurological processes responsible for generating self- and reflected appraisals? Second, to what degree do the neurological foundations of self- and reflected appraisals overlap? Third, what does a shared neurological substrate tell us about how individuals can accurately assess how others are interpreting them?

A remarkable finding that is consistently confirmed across studies shows that both self- and reflected appraisals are largely generated by the same set of brain regions (see Fig. 14.2); specifically, these include the medial prefrontal cortex, the posterior cingulate/precuneus, and multiple regions of the temporal lobe (for original evidence of this correlation, as well as an extensive summary of similar findings, see Ochsner et al., 2005). At the same time, differences between the two cognitive processes can be identified in terms of the strength of neural activity in specific brain regions. For example, Ochsner and colleagues found that the medial prefrontal and the posterior cingulate cortices are more strongly activated during self-appraisals than they are during reflected appraisals (Ochsner et al., 2005). Likewise, reflected appraisals were shown to employ stronger neural activity in many of the temporal brain regions that were noted above to play a role in the generation of episodic and semantic memory. Interestingly, this and other studies have found that reflected appraisals from the perspective of close others tend to demonstrate stronger activity in the brain regions associated with episodic memory and emotion, while reflected appraisals from the perspective of unfamiliar others tend to show stronger neural activity in the brain regions responsible for generating semantic knowledge (Ochsner et al., 2005).

These patterns of neural activation relate to identity theory in the following ways. First, the fact that a distinct pattern of neural activity can characterize each type of appraisal empirically validates identity theory's taken-for-granted assumption that self- and reflected appraisals are distinct cognitive processes. Second, the involvement of brain regions associated with declarative knowledge during the performance of self- and reflected appraisals empirically validates identity theory's argument that individuals draw upon their identity standard when defining themselves or interpreting how others are defining them. At the same time, though, the indication that reflected appraisals from the perspective of close others are neurologically distinct from those performed from the perspective of unfamiliar others may suggest the need for identity theory to rethink how a person's level of familiarity with others may modify identity-related processes.

The Neuroscience of the Implicit Perception of Self-Relevant Meanings

A distinguishing characteristic of the self- and reflected appraisals discussed above is the degree to which an individual explicitly decides whether a stimulus is self-relevant; reflecting this, neuroscience researchers often refer to these processes as explicit appraisals. As Rameson et al. (2010) note, though, people implicitly detect the self-relevance of a stimulus more often than they do so explicitly; specifically, Rameson and colleagues comment:

> …when flipping through television channels, one usually does not explicitly ask oneself: "Am I a funny person who likes comedies or a serious person who prefers documentaries?" Instead, one's implicit self-knowledge guides one's behavior to make an appropriate selection in a way that is adaptive, automatic and below the level of awareness. (Rameson et al., 2010:701)

As noted in the identity theory overview above, the perceptual control process also assumes that individuals can consciously and subconsciously

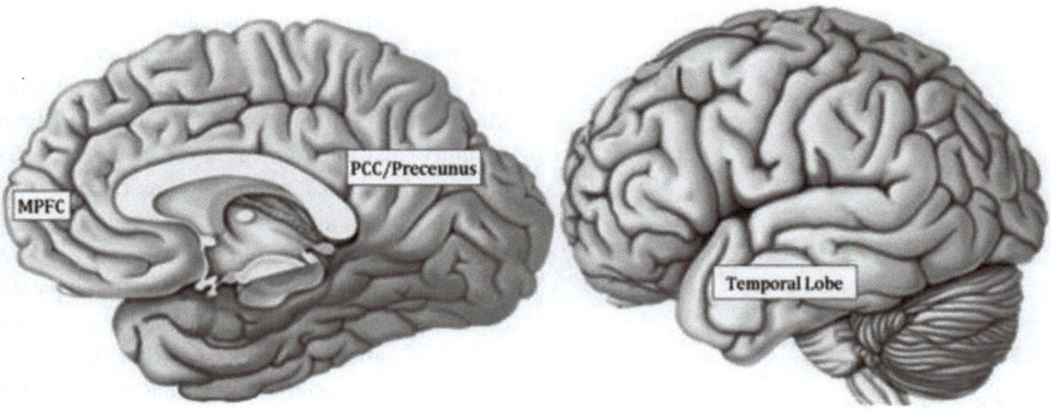

Fig. 14.2 Brain regions involved in self and reflected-appraisals. Medial prefrontal cortex (MPFC) and posterior cingulate cortex (PCC) as seen from the midsagittal view (**Left**) and Temporal lobe as seen from the exterior view of the brain (**Right**). Locations of brain regions are approximate

perceive self-relevant stimuli. Unlike neuroscience research, identity theorists have never empirically specified *how* the implicit detection of self-relevant stimuli is possible. Or, in other words, how can a person automatically distinguish the small subset of stimuli that represents a reflected appraisal from the totality of stimuli that constantly bombards their sensory system? In this section, I review examples of how modern neuroscience research has identified and described some of the mechanisms underlying this ability.

Recently, Rameson and colleagues used fMRI to compare the patterns of neural activity associated with the implicit and explicit perception of self-relevant stimuli (Rameson et al., 2010). In this study, a group of participants were asked to rate how well the adjectives "athletic" and "scientific" applied to their self-concept and how important each concept was to their sense of self. Researchers coded participants as possessing an athletic identity if they saw themselves as "athletic" and considered it a key aspect of their self-concept. Likewise, participants who indicated that "scientific" was an accurate and important description of their self-concept were coded as possessing a scientific identity.

Following this, each participant was shown a stream of 88 emotionally neutral images of athletic- and scientific-related situations and asked to identify for each one whether a human being was present in the scene. The requirement to detect a person in the image was used to divert the participant's conscious attention away from the "athletic" or "scientific" theme underlining each scene. This task aimed to determine if nonconsciously processed stimuli are "tagged" by the brain as being self-relevant and, if so, which areas of the brain are responsible. After performing this task, the participants were shown 40 athletic- and 40 scientific-related adjectives and asked to respond quickly with "me" or "not-me." Here, the goal of the task was to determine which brain regions were involved in the explicit evaluation of the self-relevance of a stimulus.

According to Rameson et al.'s (2010) results, the implicit and explicit detection of self-relevant stimuli are largely processed by the same brain regions. Specifically, these regions include the medial prefrontal and ventral medial prefrontal cortices, the amygdala, the subgenual anterior cingulate cortex, the posterior cingulate/precuneus, and the ventral striatum (see Fig. 14.3). Further, the strength of neural activity generated by a brain region during the explicit performance task is significantly correlated with the strength of the neural activity generated during the performance of the implicit task. For this reason, Rameson et al. (2010) conclude that processing self-relevant information engages roughly the same neural structures regardless of whether it is done implicitly or explicitly.

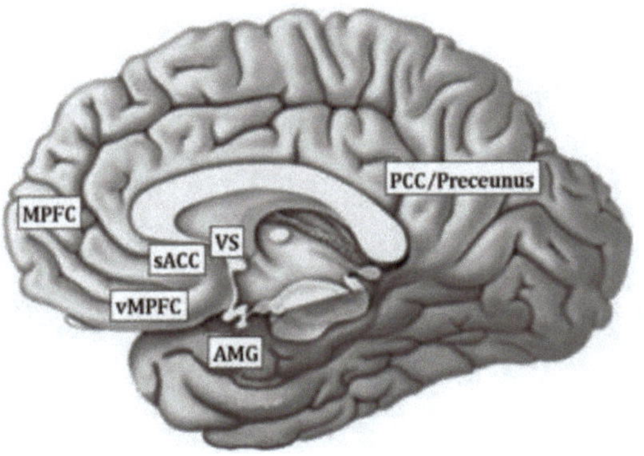

Fig. 14.3 Brain regions involved in implicit and explicit detection of self-relevant stimuli. Locations of brain regions are approximate as seen in the midsagittal view. Brian regions include medial prefrontal cortex (MPFC), ventral medial prefrontal cortex (vMPFC), amygdala (AMG), ventral striatum (VS), posterior cingulate cortex (PCC), subgenual anterior cingulate cortex (sACC)

Importantly though, Rameson and colleagues are careful to note that a brain region called the dorsal medial prefrontal cortex was observed to be uniquely involved in the explicit perception of self-relevant stimuli (Rameson et al., 2010). Following a theoretical model proposed by Northoff and Bermpohl (2004), Rameson and colleagues interpret this pattern of neural activity to mean that the ventral medial prefrontal cortex is functionally responsible for tagging and representing incoming stimuli as self-relevant at the preconscious level; the dorsal medial prefrontal cortex, in turn, is functionally responsible for consciously evaluating this relevance. Accordingly, this is why the ventral medial prefrontal cortex is involved in implicit and explicit tasks; i.e., regardless of the task type, incoming stimuli must be tagged and internally represented before they are consciously reflected upon. Interestingly, Northoff and Bermpohl (2004) reference a study by Kawasaki and colleagues in which the ventral medial prefrontal cortex began to process self-relevant stimuli 200 ms after it was presented, while increased activity in the dorsal medial prefrontal cortex did not occur until 300–800 ms later (Kawasaki et al., 2001). They suggest that this difference in activation time is indicative of functional specialization of the ventral and dorsal medial prefrontal cortices.

Beyond empirically validating identity theory's assumption that the mind can consciously and subconsciously perceive self-relevant stimuli, the above studies suggest that altered functionality in the dorsal and ventral medial prefrontal cortices commonly associated with mental illness may act to modulate identity-related processes in significant ways (Northoff & Bermpohl, 2004). For example, an altered pattern of neural activity in the dorsal medial prefrontal cortex of patients diagnosed with either severe depression or anxiety is significantly related to several cognitive symptoms that are indicative of these illnesses (Kim et al., 2011; Lemogne et al., 2009; Northoff, 2007; Sheline et al., 2010).

Considering the role that the dorsal medial prefrontal cortex plays in the various identity processes discussed thus far, the relationship between this brain area and these cognitive symptoms suggests that depression and anxiety negatively influence the self- and reflected appraisal, perception, and role-rehearsal processes, as well as negatively modulating the positive emotions experienced when an identity is confirmed. Given this, identity theory would predict that individuals diagnosed with depression or anxiety should have difficulty establishing and maintaining a positive identity. Interestingly, qualitative interviews with young adults diagnosed with bipolar disorder report experiencing this exact problem (Inder et al., 2008).

The Neuroscience of the Perceptual Control Process

An important argument made by identity theorists is that the conventional and idiosyncratic meanings of an identity, the ability to engage in self- and reflected appraisals, and the ability to perceive and interpret self-relevant information are not individual processes that occur in isolation; instead, they are all a part of the larger, continuously ongoing, identity control process (Burke & Stets, 2022). Until now, we have largely investigated the neurological foundations of the individual components of the perceptual control process; in this section, we will illustrate how research on the default mode network can be used to integrate these different patterns into a single neurological system.

The default mode network is comprised of 11 brain regions that are structurally organized into two subnetworks connected by two central hubs (see Fig. 14.4) (Andrews-Hanna et al., 2010; Buckner & Carroll, 2007; Buckner et al., 2009; Hagmann et al., 2008; Sporns, 2011). One of the central hubs is defined by an area of the brain commonly called the anterior medial prefrontal cortex; the other is the posterior cingulate cortex discussed previously. The first of the subnetworks is called the dorsal medial prefrontal cortex system. As its name suggests, this subsystem contains the dorsal medial prefrontal cortex discussed above and a small number of other brain regions. The second subnetwork is called the medial temporal system; this network consists of the medial temporal lobe, the ventral medial prefrontal cortex, and the hippocampus (all of which were already discussed above), as well as a few other brain regions (Leech et al., 2011).

The default mode network has been empirically demonstrated to be functionally responsible for generating and controlling three broadly defined cognitive states and processes (Stawarczyk et al., 2011). The first of these cognitive processes is called endogenously focused attention (Buckner et al., 2008; Buckner & Carroll, 2007; Fox et al., 2005). This pattern of neural activity is associated with the deliberate generation and effortful control of internal cognitions; phenomenologically, these internal cognitions are experienced whenever an individual covertly thinks, deliberates, ruminates, recalls, considers, or fantasizes about something in their "mind."

The internal cognitions generated by the default mode network can be divided into two subtypes based upon their content and structural-functional connectivity patterns (Leech et al., 2011). The medial temporal subsystem largely performs the first subtype of internal cognitions and consists of episodic/autobiographical memory, imaginary scenario construction, and prospection (Leech et al., 2011). Episodic/autobiographical memory was already defined above. Imaginary scenario construction, in turn, is defined as the ability to imagine with detail fictitious or hypothetical events (Buckner, 2010; Buckner & Carroll, 2007). Finally, prospection is defined as the ability to view oneself from a future perspective, e.g., thinking about oneself

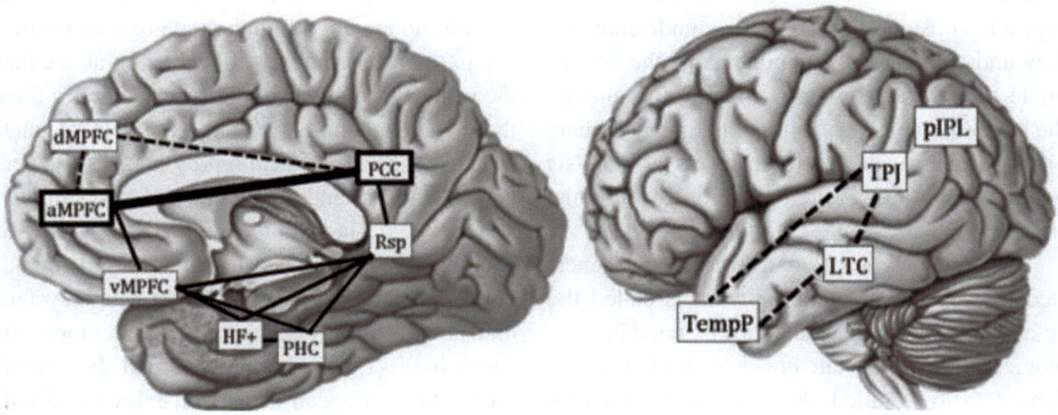

Fig. 14.4 Brain regions comprising the default mode network. The thick solid line represents the connection of the two main hubs of the default mode network; the posterior cingulate cortex (PCC) and anterior medial prefrontal cortex (aMPFC). The posterior inferior parietal lobe (pIPL) and the brain regions connected by thin solid lines represent the medial temporal lobe (MTL) subsystem which includes ventral medial prefrontal cortex (vMPFC), retrosplenial cortex (Rsp), hippocampal formation (HF+) and the parahippocampal cortex (PHC). Brain regions connected by thin dashed lines represent the dorsal medial prefrontal conrtex (dMPFC) subsystem and include temporal pole (TempP), lateral temporal cortex (LTC), and temporal parietal junction (TPJ). Locations of brain regions are approximate as seen in the mid-sagittal view (**Left**) and exterior view (**Right**) of the brain

making dinner later that evening or taking a trip the following weekend (Buckner, 2010; Buckner & Carroll, 2007). According to identity theory, these cognitive processes are nearly identical with the mental activity that they argue an individual engages in when they rehearse potential lines of future action. The dorsal medial prefrontal cortex subsystem predominantly generates the second set of internal cognitions and involves the processes of self-reflection, as well as self- and reflected appraisals that were also discussed above (Corbetta et al., 2008; Fox et al., 2005; Leech et al., 2011).

The fact that the totality of the individual brain regions discussed above is part of a single structural-functional network explains why the same brain regions are routinely activated in so many different cognitive processes. In short, each individual mental act is simply a specific instance of a more general cognitive process (e.g., deliberately generated and controlled internal mentation). This is an important insight because it indicates that the various component parts of the identity control process can only be fully understood as a coherent and dynamic whole. At the same time, though, the fact that the various subprocesses all possess clearly defined empirical referents indicates that each dimension of the identity control process can be isolated for explicit consideration. Finally, understanding how and why the subcomponents of the default mode network dynamically interact to generate the larger whole makes it possible for researchers to move empirically between levels of analysis without committing ecological or reductionist errors.

The second type of cognitive activity generated by the default mode network is called the default state (Raichle et al., 2001). Stated in sociological terms, this state consists of what Alfred Schutz (1967) called duration: a "continuous coming-to-be and passing-away" of undirected thoughts, which is phenomenologically experienced as the constant transition from a "now-thus" to a "new now-thus." The reason that this mental state is referred to as "default" is because the associated pattern of neural activity is what the brain "defaults" to whenever not engaged in a cognitively effortful task (Preminger et al., 2011; Raichle & Gusnard, 2005). Finally, the third process related to the default mode network is sometimes called the sentinel hypothesis (Buckner et al., 2008; Stawarczyk et al., 2011). This specific pattern of activity is believed to represent a state of "watchfulness" (Gilbert et al., 2007), whereby the brain engages in the general and unfocused monitoring of an individual's internal and external environment for salient or potentially relevant stimuli.

Researchers have repeatedly shown that the cognitive and neurological content of default and sentinel processes is heavily imbued with self-relevant information (Buckner et al., 2008; Christoff et al., 2009; Smallwood & Schooler, 2006; Spreng et al., 2010). This means that when the mind wanders, it tends to think about autobiographical memories, future plans and goals, and other self-related thoughts. Likewise, the sentinel state of activity involves constantly cycling through episodic and semantic knowledge that may need to be recalled to address a situation in the environment. In a certain sense, the sentinel state of activity represents a constant simulation of the external environment based on the brain's previous experiences.

Considering a proposed theoretical model by Northoff (2011), these two dimensions of the default mode network may relate to identity theory in the following way. Recall from above that the ventral medial prefrontal cortex acts as a filter that tags incoming stimuli as self-relevant. Because the neurological representation of a perceived stimulus and the memory of said stimulus are identical (Iacoboni et al., 2005; Squire & Kandel, 2009), and because the ventral medial prefrontal cortex is part of a larger neural system that is constantly cycling through memories of past stimuli, it may be the case that the ventral prefrontal cortex can detect self-relevant stimuli by matching the incoming signal to the corresponding pattern of neural activity stored in memory. If this hypothesis is correct, the model would not only explain how the perception component of the identity control process occurs, but it would also empirically verify that the identity control process is constant and ongoing and occurring at both the subconscious and conscious levels.

To summarize, the default mode network is a large network of structurally and functionally connected brain regions that engage in three types of neural activity. These three types of neural activity underlie the conscious and subconscious enactment of the perceptual control process.

The Importance of Neuroscience Research to Identity Theory

My arguments in Parts 2 and 3 suggest that incorporating neuroscience insights strengthens identity theory in four ways. First, it provides identity theory the capacity to define its concepts with greater precision. Because concepts are merely symbolic constructs created by the researcher, their ability to accurately reflect the phenomena they represent is limited by the observational power of the research technologies available (Collins, 1994). As such, a single concept can erroneously be used to refer to two sets of unrelated phenomena simply because the characteristics that distinguish them occur at an inaccessible level of analysis. A similar problem can occur with hypothesized theoretical statements when a single process is argued to describe two empirically distinct phenomena—called a lumping error (Proctor & Niemeyer, 2019). For example, the neuroscience research discussed in Part 3 indicates that the idiographic dimension of an identity indiscriminately refers to two distinct empirical processes, i.e., semantic autobiographical and episodic memory. The ability to specify the neurological correlates of an identity enables identity theorists to formulate a more precise conceptual definition of the idiographic dimension.

Second, neuroscience can identify elements of an identity process that may occur below the social, behavioral, and cognitive levels. This is an important contribution because a failure to account for these influences will increase the probability that a theory will fail to predict or accurately explain the outcomes of social interactions correctly. To illustrate this point, consider the following proposition: A change in cognitive variable X is positively related to a change in environmental variable Y. According to this proposition, a researcher should expect to see an increase in variable X whenever they observe an increase in the value of variable Y.

Let us assume that environmental variable Y is only positively related to cognitive variable X if neurological variable Z is above some threshold k. Under these circumstances, the proposition would only appear correct if neurological variable Z happened by chance to be above threshold k when the theory was applied. In short, a failure to consider the input of the neurological level of analysis may result in an accurate theory *by chance*, but its rightful incorporation can result in a correct theory *by design*. Because variations in hormone levels and functional-connectivity patterns within and between individuals have been shown to impact the neurological mechanisms underlying identity-related processes, it is essential to understand how the psychological dimensions of identity theory are influenced by their biological substrate (Burnett & Blakemore, 2009; Cacioppo & Patrick, 2008; Fair et al., 2008).

Third, understanding the biological foundations of social behavior can help identity theorists specify the scope conditions of their theories with greater accuracy. Scope conditions are the circumstances under which a theoretical claim is expected to hold (Cohen, 1989). Accurately specified scope conditions are essential because they can aid researchers in diagnosing why a proposition failed to explain or predict a phenomenon of interest. On the one hand, said failure can stem from the relationships between the variables specified in the model not accurately reflecting the "real" nature of the processes they are meant to represent. On the other hand, it could also be that the "correct" theoretical explanation was applied to the "correct" set of phenomena but not under the "correct" conditions. Knowing the relevant scope conditions of a theory can help discern between these two possibilities by allowing researchers to determine if the observed deviations from the expected outcome are attributable to conditions exogenous to the model. For this reason, more precisely stated scope conditions are better than their less precise counterparts because they provide greater accuracy when diagnosing the sources of "theory failure."

Finally, neuroscience research can help "grow" identity theory in new directions. According to Berger et al. (1985; Berger & Zelditch, 1993; Wagner, 1984, 2007; Wagner & Berger, 1985), sociological theories grow in at least five ways, one of which is called proliferation. Formally stated, theoretical proliferation is defined as the following: "We say of two theories, T1 and T2, that T2 is a proliferation of T1 if T2 enlarges the range of application of the concepts and principles in T1 to social phenomena beyond the original domain or the original set of problems within a domain addressed by T1 (Berger & Zelditch, 1993)." As noted above, recognizing the neurological foundations of identity-related processes provides essential insights into how mental illness can negatively impact the development and maintenance of identities. Given this, the proliferation of identity theory into either the sociology of mental illness or clinical psychology may provide critical insights into how mental illness, the identity control process, and social structures interact to influence the quality of life of pathological populations. These insights, in turn, may suggest new holistic interventions that consider a larger domain of potential solutions for assisting those diagnosed with mental illness.

Summary and Conclusion

This chapter aimed to illustrate the empirical and theoretical relevance of neuroscience research on the default mode network to identity theory. Toward this end, I examined how specific identity theory concepts and processes can be empirically described in neurological terms. I then demonstrated how the neurological foundations for identity theory's concepts and processes are components in a more extensive and dynamic system of ongoing neurological activity called the default mode network. This realization is significant because it anatomically and physiologically validates identity theory's argument that identity-related processes are complex integrated systems and must be analyzed as such.

Disclaimer The views expressed in this publication are those of the author and do not reflect the official policy or position of the US Air Force, Department of Defense, or the U.S. Government. PA#: USAFA-DF-2024-453.

References

Addis, D. R., Sacchetti, D. C., Ally, B. A., Budson, A. E., & Schacter, D. L. (2009). Episodic simulation of future events is impaired in mild Alzheimer's disease. *Neuropsychologia, 47*(12), 2660–2671. https://doi.org/10.1016/j.neuropsychologia.2009.05.018

Andrews-Hanna, J. R., Reidler, J. S., Huang, C., & Buckner, R. L. (2010). Evidence for the default network's role in spontaneous cognition. *Journal of Neurophysiology, 104*(1), 322–335. https://doi.org/10.1152/jn.00830.2009

Andrews-Hanna, J. R., Reidler, J. S., Sepulcre, J., Poulin, R., & Buckner, R. L. (2010). Functional-anatomic fractionation of the brain's default network. *Neuron, 65*(4), 550–562. https://doi.org/10.1016/j.neuron.2010.02.005

Berger, J., Wagner, D. G., & Zelditch, M., Jr. (1985). Theory growth, social processes, and metatheory. In J. H. Turner (Ed.), *Theory building in sociology: Assessing theoretical accumulation* (pp. 19–42). Sage.

Berger, J., & Zelditch, M. (1993). *Theoretical research programs: Studies in the growth of theory*. Stanford University Press.

Binder, J. R., Desai, R. H., Graves, W. W., & Conant, L. L. (2009). Where is the semantic system? A critical review and meta-analysis of 120 functional neuroimaging studies. *Cerebral Cortex, 19*(12), 2767. https://doi.org/10.1093/cercor/bhp055

Biswal, B. B., Mennes, M., Zuo, X. N., Gohel, S., Kelly, C., Smith, S. M., Beckmann, C. F., Adelstein, J. S., Buckner, R. L., Colcombe, S., Dogonowski, A. M., Ernst, M., Fair, D., Hampson, M., Hoptman, M. J., Hyde, J. S., Kiviniemi, V. J., Kotter, R., Li, S. J., Lin, C. P., Lowe, M. J., Mackay, C., Madden, D. J., Madsen, K. H., Margulies, D. S., Mayberg, H. S., McMahon, K., Monk, C. S., Mostofsky, S. H., Nagel, B. J., Pekar, J. J., Peltier, S. J., Petersen, S. E., Riedl, V., Rombouts, S., Rypma, B., Schlaggar, B. L., Schmidt, S., Seidler, R. D., Siegle, G. J., Sorg, C., Teng, G. J., Veijola, J., Villringer, A., Walter, M., Wang, L. H., Weng, X. C., Whitfield-Gabrieli, S., Williamson, P., Windischberger, C., Zang, Y. F., Zhang, H. Y., Castellanos, F. X., & Milham, M. P. (2010). Toward discovery science of human brain function. *Proceedings of the National Academy of Sciences of the United States of America, 107*(10), 4734–4739. https://doi.org/10.1073/pnas.0911855107

Buckner, R. L. (2010). The role of the hippocampus in prediction and imagination. *Annual Review of Psychology, 61*, 27–48. https://doi.org/10.1146/annurev.psych.60.110707.163508

Buckner, R. L., Andrews-Hanna, J. R., & Schacter, D. L. (2008). The brain's default network—Anatomy, function, and relevance to disease. In *Year in cognitive neuroscience 2008* (pp. 1–38). Blackwell Publishing. https://doi.org/10.1196/annals.1440.011

Buckner, R. L., & Carroll, D. C. (2007). Self-projection and the brain. *Trends in Cognitive Sciences, 11*(2), 49–57. https://doi.org/10.1016/j.tics.2006.11.004

Buckner, R. L., Sepulcre, J., Talukdar, T., Krienen, F. M., Liu, H. S., Hedden, T., Andrews-Hanna, J. R., Sperling, R. A., & Johnson, K. A. (2009). Cortical hubs revealed by intrinsic functional connectivity: Mapping, assessment of stability, and relation to Alzheimer's disease. *Journal of Neuroscience, 29*(6), 1860–1873. https://doi.org/10.1523/JNEUROSCI.5062-08.2009

Burianova, H., McIntosh, A. R., & Grady, C. L. (2010). A common functional brain network for autobiographical, episodic, and semantic memory retrieval. *NeuroImage, 49*(1), 865–874. https://doi.org/10.1016/j.neuroimage.2009.08.066

Burke, P. J., & Stets, J. E. (2022). *Identity theory: Revised and expanded* (2nd ed.). Oxford University Press. https://doi.org/10.1093/oso/9780197617182.001.0001

Burnett, S., & Blakemore, S. J. (2009). Functional connectivity during a social emotion task in adolescents and in adults. *European Journal of Neuroscience, 29*(6), 1294–1301. https://doi.org/10.1111/j.1460-9568.2009.06674.x

Cacioppo, J. T., & Berntson, G. G. (2005). *Social neuroscience: Key readings*. Psychology Press.

Cacioppo, J. T., & Patrick, W. (2008). *Loneliness: Human nature and the need for social connection*. WW Norton & Company.

Christoff, K., Gordon, A. M., Smallwood, J., Smith, R., & Schooler, J. W. (2009). Experience sampling during fMRI reveals default network and executive system contributions to mind wandering. *Proceedings of the National Academy of Sciences, 106*(21), 8719. https://doi.org/10.1073/pnas.0900234106

Cohen, B. (1989). *Developing sociological knowledge*. Nelson-Hall.

Collins, R. (1994). Why the social sciences won't become high-consensus, rapid-discovery science. *Sociological Forum, 9*(2), 155–177.

Cooley, C. H. (1902). *Human nature and the social order*. Charles Scribner's Sons.

Corbetta, M., Patel, G., & Shulman, G. L. (2008). The reorienting system of the human brain: From environment to theory of mind. *Neuron, 58*(3), 306–324. https://doi.org/10.1016/j.neuron.2008.04.017

Corbetta, M., & Shulman, G. L. (2002). Control of goal-directed and stimulus-driven attention in the brain. *Nature Reviews Neuroscience, 3*(3), 201–215. https://doi.org/10.1038/nrn755

D'Argembeau, A., & Salmon, E. (2011). The neural basis of semantic and episodic forms of self-knowledge: Insights from functional neuroimaging. In C. López-Larrea (Ed.), *Sensing systems in nature*. Landes Biosciences. https://doi.org/10.1007/978-1-4614-1704-0_18

Fair, D. A., Cohen, A. L., Dosenbach, N. U. F., Church, J. A., Miezin, F. M., Barch, D. M., Raichle, M. E., Petersen, S. E., & Schlaggar, B. L. (2008). The maturing architecture of the brain's default network. *Proceedings of the National Academy of Sciences of the United States of America, 105*(10), 4028–4032. https://doi.org/10.1073/pnas.0800376105

Fox, M. D., Corbetta, M., Snyder, A. Z., Vincent, J. L., & Raichle, M. E. (2006). Spontaneous neuronal activity distinguishes human dorsal and ventral attention systems. *Proceedings of the National Academy of Sciences of the United States of America, 103*(26), 10046–10051. https://doi.org/10.1073/pnas.0604187103

Fox, M. D., Snyder, A. Z., Vincent, J. L., Corbetta, M., Van Essen, D. C., & Raichle, M. E. (2005). The human brain is intrinsically organized into dynamic, anticorrelated functional networks. *Proceedings of the National Academy of Sciences of the United States of America, 102*(27), 9673–9678. https://doi.org/10.1073/pnas.0504136102

Gilbert, S. J., Dumontheil, I., Simons, J. S., Frith, C. D., & Burgess, P. W. (2007). Comment on "Wandering minds: The default network and stimulus-independent thought". *Science, 317*, 43b. https://doi.org/10.1126/science.1140801

Gusnard, D., Akbudak, E., Shulman, G., & Raichle, M. E. (2001). Role of medial prefrontal cortex in a default mode of brain function. *NeuroImage, 13*(6), S414–S414. https://doi.org/10.1073/pnas.071043098

Hagmann, P., Cammoun, L., Gigandet, X., Meuli, R., Honey, C. J., Wedeen, V., & Sporns, O. (2008). Mapping the structural core of human cerebral cortex. *PLoS Biology, 6*(7), 1479–1493. https://doi.org/10.1371/journal.pbio.0060159

Hebb, D. O. (2002). *The organization of behavior: A neuropsychological theory*. L. Erlbaum Associates.

Iacoboni, M., Molnar-Szakacs, I., Gallese, V., Buccino, G., Mazziotta, J. C., & Rizzolatti, G. (2005). Grasping the intentions of others with one's own mirror neuron system. *PLoS Biology, 3*(3), e79. https://doi.org/10.1371/journal.pbio.0030079

Inder, M. L., Crowe, M. T., Moor, S., Luty, S. E., Carter, J. D., & Joyce, P. R. (2008). "I actually don't know who I am": The impact of bipolar disorder on the development of self. *Psychiatry: Interpersonal and Biological Processes, 71*(2), 123–133. https://doi.org/10.1521/psyc.2008.71.2.123

Kawasaki, H., Adolphs, R., Kaufman, O., Damasio, H., Damasio, A. R., Granner, M., Bakken, H., Hori, T., & Howard, M. A. (2001). Single-neuron responses to emotional visual stimuli recorded in human ventral prefrontal cortex. *Nature Neuroscience, 4*(1), 15–16. https://doi.org/10.1038/82850

Kim, M. J., Gee, D. G., Loucks, R. A., Davis, F. C., & Whalen, P. J. (2011). Anxiety dissociates dorsal and ventral medial prefrontal cortex functional connectivity with the amygdala at rest. *Cerebral Cortex, 21*, 1667–1673. https://doi.org/10.1093/cercor/bhq237

Kompus, K., Olsson, C. J., Larsson, A., & Nyberg, L. (2009). Dynamic switching between semantic and episodic memory systems. *Neuropsychologia, 47*(11), 2252–2260. https://doi.org/10.1016/j.neuropsychologia.2008.11.031

Kwan, D., Carson, N., Addis, D. R., & Rosenbaum, R. S. (2010). Deficits in past remembering extend to future imagining in a case of developmental amnesia. *Neuropsychologia, 48*(11), 3179–3186. https://doi.org/10.1016/j.neuropsychologia.2010.06.011

Leech, R., Kamourieh, S., Beckmann, C. F., & Sharp, D. J. (2011). Fractionating the default mode network: Distinct contributions of the ventral and dorsal posterior cingulate cortex to cognitive control. *Journal of Neuroscience, 31*(9), 3217. https://doi.org/10.1523/JNEUROSCI.5626-10.2011

Lemogne, C., le Bastard, G., Mayberg, H., Volle, E., Bergouignan, L., Lehericy, S., Allilaire, J. F., & Fossati, P. (2009). In search of the depressive self: Extended medial prefrontal network during self-referential processing in major depression. *Social Cognitive and Affective Neuroscience, 4*(3), 305–312. https://doi.org/10.1093/scan/nsp008

Levine, B., Turner, G. R., Tisserand, D., Hevenor, S. J., Graham, S. J., & McIntosh, A. R. (2004). The functional neuroanatomy of episodic and semantic autobiographical remembering: A prospective functional MRI study. *Journal of Cognitive Neuroscience, 16*(9), 1633–1646. https://doi.org/10.1162/0898929042568587

Leyhe, T., Muller, S., Milian, M., Eschweiler, G. W., & Saur, R. (2009). Impairment of episodic and semantic autobiographical memory in patients with mild cognitive impairment and early Alzheimer's disease. *Neuropsychologia, 47*(12), 2464–2469. https://doi.org/10.1016/j.neuropsychologia.2009.04.018

Mead, G. H. (1934). *Mind, self and society*. University of Chicago Press.

Northoff, G. (2007). Psychopathology and pathophysiology of the self in depression—Neuropsychiatric hypothesis. *Journal of Affective Disorders, 104*(1–3), 1–14. https://doi.org/10.1016/j.jad.2007.02.012

Northoff, G. (2011). The brain and its self: Concepts of self and the cortical midline structures. In S. Han & E. Pöppel (Eds.), *Culture and neural frames of cognition and communication* (pp. 41–63). Springer. https://doi.org/10.1007/978-3-642-15423-2_3

Northoff, G., & Bermpohl, F. (2004). Cortical midline structures and the self. *Trends in Cognitive Sciences, 8*(3), 102–107. https://doi.org/10.1016/j.tics.2004.01.004

Ochsner, K. N., Beer, J. S., Robertson, E. R., Cooper, J. C., Gabrieli, J. D. E., Kihsltrom, J. F., & D'Esposito, M. (2005). The neural correlates of direct and reflected self-knowledge. *NeuroImage, 28*(4), 797–814. https://doi.org/10.1016/j.neuroimage.2005.06.069

Preminger, S., Harmelech, T., & Malach, R. (2011). Stimulus-free thoughts induce differential activation in the human default network. *NeuroImage, 54*(2), 1692–1702. https://doi.org/10.1016/j.neuroimage.2010.08.036

Prince, S. E., Tsukiura, T., & Cabeza, R. (2007). Distinguishing the neural correlates of episodic memory encoding and semantic memory retrieval. *Psychological Science, 18*(2), 144. https://doi.org/10.1111/j.1467-9280.2007.01864.x

Proctor, K. R., & Niemeyer, R. E. (2019). *Mechanistic criminology*. Routledge.

Raichle, M. E., & Gusnard, D. A. (2005). Intrinsic brain activity sets the stage for expression of motivated behavior. *The Journal of Comparative Neurology, 493*(1), 167–176. https://doi.org/10.1002/cne.20752

Raichle, M. E., MacLeod, A. M., Snyder, A. Z., Powers, W. J., Gusnard, D. A., & Shulman, G. L. (2001). A default mode of brain function. *Proceedings of the National Academy of Sciences of the United States of America, 98*(2), 676–682. https://doi.org/10.1073/pnas.98.2.676

Rameson, L. T., Satpute, A. B., & Lieberman, M. D. (2010). The neural correlates of implicit and explicit self-relevant processing. *NeuroImage, 50*(2), 701–708. https://doi.org/10.1016/j.neuroimage.2009.12.098

Schutz, A. (1967). *The phenomenology of the social world*. Northwestern University Press.

Serpe, R. T., & Stryker, S. (2011). The symbolic interactionist perspective and identity theory. In *Handbook of identity theory and research* (pp. 225–248). Springer. https://doi.org/10.1007/978-1-4419-7988-9_10

Sheline, Y. I., Price, J. L., Yan, Z., & Mintun, M. A. (2010). Resting-state functional MRI in depression unmasks increased connectivity between networks via the dorsal nexus. *Proceedings of the National Academy of Sciences, 107*(24), 11020. https://doi.org/10.1073/pnas.1000446107

Smallwood, J., & Schooler, J. W. (2006). The restless mind. *Psychological Bulletin, 132*(6), 946. https://doi.org/10.1037/0033-2909.132.6.946

Sporns, O. (2011). *Networks of the brain*. MIT Press. https://doi.org/10.7551/mitpress/8476.001.0001

Spreng, R. N., & Grady, C. L. (2010). Patterns of brain activity supporting autobiographical memory, prospection, and theory of mind, and their relationship to the default mode network. *Journal of Cognitive Neuroscience, 22*(6), 1112–1123. https://doi.org/10.1162/jocn.2009.21282

Spreng, R. N., Mar, R. A., & Kim, A. S. N. (2009). The common neural basis of autobiographical memory, prospection, navigation, theory of mind, and the default mode: A quantitative meta-analysis. *Journal of Cognitive Neuroscience, 21*(3), 489–510. https://doi.org/10.1162/jocn.2008.21029

Spreng, R. N., Stevens, W. D., Chamberlain, J. P., Gilmore, A. W., & Schacter, D. L. (2010). Default network activity, coupled with the frontoparietal control network, supports goal-directed cognition. *NeuroImage, 53*(1), 303–317. https://doi.org/10.1016/j.neuroimage.2010.06.016

Squire, L. R., & Kandel, E. R. (2009). *Memory: From mind to molecules*. Roberts & Co.

Stawarczyk, D., Majerus, S., Maquet, P., & D'Argembeau, A. (2011). Neural correlates of ongoing conscious experience: Both task-unrelatedness and stimulus-independence are related to default network activity. *PLoS One, 6*(2), 14. https://doi.org/10.1371/journal.pone.0016997

Stets, J. E. (2010). The social psychology of the moral identity. In S. Hitlin & S. Vaisey (Eds.), *Handbook of the sociology of morality* (pp. 385–409). Springer. https://doi.org/10.1007/978-1-4419-6896-8_20

Stets, J. E., & Serpe, R. T. (2013). Identity theory. In *Handbook of social psychology* (pp. 31–60). Springer Netherlands. https://doi.org/10.1007/978-94-007-6772-0_2

Stryker, S. (2002). *Symbolic interactionism: A social structural version*. Blackburn Press.

Takamiya, S., Yuki, S., Hirokawa, J., Manabe, H., & Sakurai, Y. (2020). Dynamics of memory engrams. *Neuroscience Research, 153*, 22–26. https://doi.org/10.1016/j.neures.2019.03.005

Wagner, D. G. (1984). *The growth of sociological theories*. Sage.

Wagner, D. G. (2007). The limits of theoretical integration. *Social Justice Research, 20*(3), 270–287. https://doi.org/10.1007/s11211-007-0045-9

Wagner, D. G., & Berger, J. (1985). Do sociological theories grow? *American Journal of Sociology, 90*, 697–728. https://doi.org/10.1086/228142

Wiggs, C. L., Weisberg, J., & Martin, A. (1998). Neural correlates of semantic and episodic memory retrieval. *Neuropsychologia, 37*(1), 103–111. https://doi.org/10.1016/S0028-3932(98)00044-X

15 Neurosociology and Theory of Mind (ToM)

Rosemary L. Hopcroft

Abstract

Theory of mind (ToM) is a term used to describe a person's ability to understand another person's mental states. It overlaps with, but is not the same as, empathy for another person. The most common test of ToM is the false belief task that determines whether a person can understand that another person holds a false belief. Children can usually pass this test by the time they are about 4 years old, although many autistic children can never pass the false belief test. The development of ToM usually improves as a child grows older, parallel to language development. It has been argued that the mirror neuron system is involved with both the development of ToM and language learning. There are several theories of ToM and growing development of computer models of ToM, as they are useful to the development of artificial intelligence. For sociologists, ToM is an important cognitive development that underlies many of the social processes that are the topic of sociological study.

R. L. Hopcroft (✉)
Department of Sociology, University of North Carolina at Charlotte, Charlotte, NC, USA
e-mail: rlhopcro@charlotte.edu

Theory of Mind (henceforth ToM) is the term used to describe a person's ability to understand another person's mental states. It is somewhat cryptically called "ToM" because it shares some features with the explanations scientists call theories. That is, ToM refers to the ability to use the observed behavior of another person to develop an explanation for the behavior involving that person's mental processes, make predictions about the person's future behavior, and use those predictions to explain new observations (Malle, 2005, p. 225). For example—"She chose only the chocolate candy because she likes chocolate best. In the future, if she has a choice of candy, she will choose chocolate candy again." "She took the strawberry candy because there was no chocolate candy."

ToM overlaps with, but is not the same as, empathy (Schurz et al., 2021). Definitions of ToM include the ability to make inferences about others' beliefs and emotions, with the latter sometimes referred to as "affective ToM." Empathy is usually defined as being able to infer and understand others' emotional states, but it also sometimes involves being able to cognitively reason about those states. The latter is often referred to as "cognitive empathy" (Walter, 2012). In a review of the neuroimaging literature on ToM and empathy, Schurz et al. (2021) found evidence that ToM and empathy share certain types of brain activity across different tasks and stimuli, suggesting that cognitive processes

W. Kalkhoff et al. (eds.), *Handbook of Neurosociology*, Handbooks of Sociology and Social Research, https://doi.org/10.1007/978-3-031-95615-7_15

involved in ToM and empathy are overlapping, but not the same.

ToM forms an important bridge between those sciences focusing on the individual (e.g., neuroscience and psychology) and those focusing on the group (e.g., sociology). For neuroscientists and cognitive psychologists, development of ToM allows the individual to be fully social and function normally in the social world. Individuals who have trouble understanding other peoples' mental states have social difficulties, while severe lack of understanding of others' mental states becomes pathological. For sociologists and other social scientists, ToM may be regarded as the psychological underpinning of the social nature of individuals. Microsociology (and social psychology more generally) is grounded in the ability of individuals to think about and anticipate the behaviors of others, and hence in ToM.

In this chapter, I first discuss one of the standard tests used to gauge a child's ToM, the false belief test. Then I trace the typical development of ToM in young children as revealed by the work of neuroscientists and cognitive psychologists. I also discuss ToM's association with language learning, the brain bases of ToM, the non-typical development of ToM in autistic children, and theories of ToM (including computer models of ToM). The last area is of vital importance to the continued development of artificial intelligence. Last, I discuss the relevance of the development of ToM in children for sociological work on the emergence of the self, as well as other sociological perspectives.

I discuss the work of social neuroscientists and cognitive psychologists first because it deals with a lower level of analysis (the individual) while sociology deals with a higher level of analysis (the group). Given this, sociology both subsumes and transcends the more micro-oriented disciplines, and thus logically is discussed last. An important subtheme of this chapter, as that of this whole volume, is that sociological work builds on, but does not contradict, the findings of neuroscience, psychology and the other life sciences. In this view, ToM is a cornerstone of the sociological edifice.

The False Belief Task

One way to assess a child's ToM is the "false belief task" used for preschoolers, although there are many other tests that gauge children's theory of mind beyond and before preschool, for example the "reading the mind in the eyes test" for older children (Greenberg et al., 2023; Wellman, 2018; Wimmer & Perner, 1983), and tests that extend the concept of theory of mind, for example the "accidental transgressor" task to measure a morally-relevant theory of mind (Killen et al., 2011).

The false belief task can take many forms, but all seek to determine whether the child can understand that another person may be motivated by a false belief. For example, the task may be as follows: The child is told that Max puts his candy in a green cabinet, then he goes outside to play. While Max is outside, his mother moves the candy into a blue cabinet. The child is asked: "When Max comes back in, where will he look for his candy?" A child who answers correctly that Max will look for his candy in the wrong, i.e., green, cabinet is said to have passed the false belief test. Although the age of success differs somewhat depending on the particular version of the task used, generally children can pass this test by the time they are 4–5 years old (Saxe, 2006; Wellman, 2018). Ability to pass the false belief test occurs at about the same age in all countries (Lillard & Skibbe, 2005, p. 281; Rakoczy, 2022). Three-year olds generally do not pass the test and have trouble understanding that Max will not know everything that they themselves know (i.e., the true location of the candy). To them, all others see the world like they do, and there is no notion of other people having another perspective.

Children younger than 4 years old can pass an implicit version of this task, however. Garnham and Perner (2001) use an ingenious method of finding out whether children aged on average about 3 and a half can understand another's false belief, even if they cannot express it to someone else. In this version of the false belief test, children are confronted with a model containing two slides reached by one stair. On the side of each slide is a box. The child is told the following story, which is also acted out by two small figurines and a toy ball: Alan plays with a ball and places it in one of the boxes by one of the slides, then goes away to climb the stair to the slides. While he is doing this, Rebecca comes and switches the ball to the box by the other slide, but Alan does not see this. Meanwhile, Alan decides he wants to play with the ball again. Some of the children were asked "Which slide will Alan come down?" Others were told: "Quick! Move the mat to catch Alan as he comes down the slide!" Children were much more likely to move the mat to the correct slide than to answer the question correctly and explicitly, showing an implicit understanding of Alan's false belief. That is, children can understand that another may have a false belief before they can verbalize this understanding. There have been a variety of other implicit false belief tests that suggest that children as young as 15-months old can form some sort of implicit understanding that another's behavior will be influenced by their false beliefs (Onishi & Baillargeon, 2005).

Development of ToM in Typical Children

The fact that children can pass an implicit version of the false belief test before they can verbally give the correct answer suggests that the child's ToM is acquired slowly over early childhood. A great deal of research in neuroscience and psychology tracks this development in young children.

Before a child can understand the content of another person's mental states, they have to understand that others have mental states. Babies are very interested in other people and quickly learn to differentiate them from objects (Legerstee, 1992). Even newborn babies imitate the actions of other people, such as sticking out a tongue (Meltzoff & Moore, 1977, for recent controversy and support for the original finding see Davis et al., 2021). Babies aged between 5 and 8 weeks imitate a person sticking out a tongue, but not similar behaviors produced by an object, suggesting that they recognize the difference (Legerstee, 1992). Babies also understand that some behaviors that people perform have motivations, and some behaviors don't (i.e., they are accidental). Children can differentiate between intentional and accidental behavior by the age of 9 months (Wellman, 2018; Woodward, 1999).

Many authors argue that gaze following is an important precursor to ToM as this suggests that children know that the gazer is intentionally paying attention to something. It suggests some understanding of the intentional states of others (Baron-Cohen, 1995). In humans, gaze-following behavior emerges between 3 and 18 months of age (Butterworth & Cochran, 1980; Corkum & Moore, 1995; Scaife & Bruner, 1975). By 14–18 months of age children can follow adult eye gaze by following the eyes only (Brooks & Meltzoff, 2002; Moore & Corkum, 1998). By 18 months, children start to call adults' attention to objects by pointing to them or holding them up (Golinkoff, 1986).

At approximately the same age (18–24 months), infants develop the capacity to recognize themselves in mirrors (Brüne & Brüne-Cohrs, 2006; Gallup et al., 2002; Reddy et al., 2010). This is demonstrated by the mirror test, where a researcher applies a red mark to the child's nose and then puts the child in front of a mirror. Children show self-recognition when they investigate the mark on their nose in the mirror, and most children can do this by the age of two. Before the age of two, children react socially to the reflection as if it were another person, smiling, making noises, etc. (Gallup et al., 2002). The ability to pass the mirror test is not limited to humans, since all of the great apes, as well as dolphins and elephants, can pass this test and recognize themselves in a mirror. It is very likely that whales can do so as well (de Waal, 2009).

Self-recognition is the beginning of the development of a sense of self. Embarrassment and the use of personal pronouns also occur at about this age, suggesting that children are developing self-awareness and some understanding of how others see them (Rochat, 2003). Being aware of oneself is an important step in being aware of your own thinking. Being aware of your own thoughts also means you are more likely to be able to infer what others are likely to think in similar situations (Gallup et al., 2002). This is also the time that pretend play begins (Lewis & Ramsay, 2004; Stone, 2006). Pretend play often involves pretense about the desires, beliefs, and intentions of others, and so is an important step in understanding the minds of others. In pretend play, however, children at this age often make mistakes about what others can see and think (Flavell et al., 1981).

Pretend play often involves role playing, where children practice being another person in a different role with a different repertoire of behaviors. This not only allows children to practice seeing things from another person's point of view (the beginning of role taking), but it allows children to practice the socially acceptable behaviors of people in their social group. The sociologist George Herbert Mead thought this role playing was vital to the full development of sense of self (Turner et al., 1989), as discussed below.

Between 2 and 3 years, children develop their understandings of the desires and emotions of others (Saxe et al., 2004). Researchers have found that children first understand that others can have different desires to themselves, next they understand that others can have different beliefs, next they can understand that others can have differential knowledge, then the understanding that others can have false beliefs, and last they can understand that a person may feel one way but display a different emotion (Wellman, 2018).

The development of ToM appears to be similar across cultures, with some slight differences in timing. There are also differences between cultures in adults' ToM (Flavell, 1999). For example, Americans are more likely to give trait reasons for why a person behaved a certain way, while people from other cultures are more likely to give situational reasons (Lillard & Skibbe, 2005, p. 285). Liu et al. (2008) found similar patterns of development of ToM as measured by passing the false belief test in China, Hong Kong, Canada, and the United States, although the timing did differ across regions. The timing of development for children in mainland China and children in the United States was very similar, while Canadian children developed earlier, and Hong Kong Chinese children developed later. The factors responsible for the differences in timing were not clear, however. Wellman (2018) has found that Chinese students in Beijing learn things in a slightly different order to Western children, in that they first understand that others may have different access to knowledge before they understand that others can have different beliefs to themselves. Other contextual factors besides culture also matter for the timing of the development of ToM. For example, development of ToM is often slower than typical in the deaf children of hearing parents (but not in deaf children of deaf parents), there is also evidence that the development of ToM can be significantly speeded up (Wellman, 2018).

ToM also improves with age. As children age, they are more likely to explain actions with reference to the actor's goals. In particular, between middle childhood (ages about 6–12)and adulthood, the recognition of psychological goals of others increases greatly (Lovett & Pillow, 2010; Greenberg et al., 2023). Part of the brain called the "default mode network" that is implicated with ToM and empathy (among other things) (Smallwood et al., 2021) becomes significantly more integrated as children get older, suggesting that the ability to understand others continues to improve with age (Fair et al., 2008). Greenberg et al. (2023) found a peak in ToM as measured by the Eyes Test for both females and males at age 20. For females, there was a decline from age 20–50, followed by a further decline after 50. For males there was also a decline from about age 20–58, and then a steep decline after age 58. There are also individual and sex differences in Theory of Mind. Using the "reading the mind in the eyes" test for Theory of Mind, Greenberg et al. (2023) find an on-average female advantage in the eyes test across all ages and in 36 out of 57 countries studied.

ToM: Association with Language

Development of the ability to pass the false belief test is also associated with word learning (Malle, 2005, p. 229) and verbal ability (Wellman, 2018). After all, the primary difference between the explicit and implicit versions of the false belief task is that the explicit version requires children to verbalize what they know. There is other evidence that developing the ability to pass the false belief test is not entirely dependent on learning the ability to represent one's thoughts verbally.

For example, a study by Andrew S. Gordon (2006) plotted the frequency of usage of words related to knowledge and beliefs by age of children. He assumed that if the ability to represent others' thoughts in words was necessary for the child to understand the mind states of others, then before a child was able to pass the false belief test there would be a large increase in knowledge and use of words related to knowledge and beliefs. He found that children use an increased number of words related to knowledge and beliefs from the age of two, but it is a steady increase indicative of developing linguistic competency. There is no qualitative shift right before children can pass the false belief test. Nor do the type of concepts expressed by children change dramatically at any time. This suggests that the ability to pass a false belief test is not because of a large improvement in the ability to express ones understanding of the mental states of others.

Others have argued that it is children's developing ToM that aids in their developing linguistic competency. Bloom (2000, p. 61) writes "children use their naïve psychology or ToM to figure out what people are referring to when they use words." This means that children do not simply take literal interpretations of statements such as "Would you mind telling me what time it is?" and respond with "I don't mind at all." Instead, they understand the intent of the speaker, and so respond appropriately with the time. A variety of research shows that children use the eye gaze, body language and facial expressions of the speaker to understand the meaning and intentions of the speaker, and this helps them learn new words (Tomasello & Barton, 1994).

Continued improvement of language abilities fuels the interactional process involved in role-playing, role-taking, socialization, and the emergence of the self, as language is the most effective symbol system available to humans. The full development of language capabilities is also necessary for adult role performance and presentation of self as described by Goffman and others (see below).

Brain Bases of ToM

In typically developing adults, ToM is supported by regions including the temporoparietal junction (TPJ), precuneus (PreC), anterior temporal lobes (ATL), dorsomedial prefrontal cortex (dmPFC), and ventromedial prefrontal cortex (vmPFC). (Wang et al., 2021). These regions are often collectively called the ToM (or mentalizing) system (Bedny et al., 2009; Smallwood et al., 2021; Vistoli et al., 2011).

Mirror Neurons and ToM

It has been argued that mirror neurons play an important role in ToM (although for a critique see Caramazza et al., 2014; Heyes & Catmur, 2022). Mirror neurons are neurons that fire when a person performs an action and also fire when a person sees another person performing the same action (See Chap. 12 of this volume and Rizzolatti & Craighero, 2004; Montgomery et al., 2007). Although it was originally thought that mirror neurons were concentrated in the inferior parietal lobule, the lower part of the precentral gyrus and the posterior part of the inferior frontal gyrus (Rizzolatti & Craighero, 2004; Vistoli et al., 2011), subsequent research shows that mirror neurons or cells with mirroring properties are distributed in many motor, sensory and emotional brain areas (Bonini, 2017; Heyes & Catmur, 2022).

Rizzolatti et al. (2001) suggested that by mirroring observed actions, we can understand another's actions from our own point of view, and we can also better predict the mental state of the person performing the action. Thus, it was proposed that mirror neurons may be directly involved in ToM (Iacoboni et al., 2005; Oberman et al., 2005; Oberman & Ramachandran, 2007; Pineda & Hecht, 2009; Rizzolatti et al., 2001). There is evidence that ToM does use some of the regions belonging to the mirror neuron system in addition to the mentalizing system (Vistoli et al., 2011), suggesting that cognitive processes involved with Theory of Mind overlap with those of the mirror neuron system. Other scholars suggest that mirror neurons are involved in the lower level processing of observed actions rather than understanding the actors' mental states, so cannot be involved in ToM. (Heyes & Catmur, 2022; Hickok, 2014). However, a recent study in mice shows that mirror neurons in the hypothalamus are involved in the perception and response to aggression from others, suggesting that mirror neurons are involved in understanding others' intent in social interactions (Yang et al., 2023).

Mirror neurons have been shown to be involved with empathy (Schulte-Rüther et al., 2007). There does appear to be complementarity between the two brain systems (Geiger et al., 2019). A recent metanalysis found that people high in psychopathic traits (a lack of empathy for others) had less activation in regions of the brain associated with mirror neurons than others during emotional processing tasks (Penagos-Corzo et al., 2022), suggesting that the mirror neuron network is involved with empathy.

Mirror neurons may also be involved in language learning (Bel et al., 2009; Oberman & Ramachandran, 2007; Ramachandran, 2011, p. 144). These authors suggest that mirror neurons help an infant repeat a word or words he or she has just heard, as mirror neurons in the baby's brain echo the neurons of the speaker. As evidence of the role of mirror neurons in language learning, Stephens et al. (2010) found that there is similar neural activity in the same regions of both the speaker's and listener's brains. Although on average this mirroring was delayed, there was also a predictive, anticipatory response in the listener's brain regions. The more this anticipatory response occurred, the more the communication was successful. The regions where the neural coupling was observed were those that have been implicated in the production and comprehension of language: early auditory areas (A1+), superior temporal gyrus, angular gyrus, temporoparietal junction (these areas are also known as Wernicke's area), parietal lobule, inferior frontal gyrus (also known as Broca's area), and the insula. Two of these areas, the parietal lobule and the inferior frontal gyrus have also been associated with the mirror neuron system (Rizzolatti & Craighero, 2004). The role of mirror neurons in understanding speech has been debated by Hickok (2014).

Lack of Development of ToM in Autistic Children

Autism has been described as a problem of the atypical development of ToM (Andreou & Skrimpa, 2020; Baron-Cohen, 1995; Moran et al., 2011; Schroeder et al., 2010). Autistic children have great difficulty in understanding the mental states of others and, with the exception of some, cannot pass the false belief test. Although they can recognize themselves in a mirror at about the same age as neurotypical children, they may have a variety of problems regarding sense of self and tend to treat the self as a thing rather than a person (Reddy et al., 2010). Some might find the most basic social interactions difficult and stressful. Unlike neurotypical children, some autistic children do not engage in pretend play. This deficit in autistic individuals seems to be independent of other cognitive abilities and general intelligence.

Poor ToM skills in autistic children are also associated with impaired language use. About 30 percent of autistic individuals have no language at all (Bloom, 2000, p. 78). Individuals diagnosed with milder forms of autism, formerly known as Asperger's syndrome, often talk like "little professors." Their language is highly literal. So, they will answer a question put by a telephone caller "Is your mother available?" by saying "yes" and hanging up. Some autistic individuals are almost neurotypical in their language, and these individuals tend to have better skills at understanding the thoughts of others (Frith & Happe, 1994).

A variety of evidence suggests that autism involves deficits in the mirror neuron system (Andreou & Skrimpa, 2020; Bel et al., 2009; Cattaneo et al., 2007; Schroeder et al., 2010; Williams et al., 2001). The neuroscientist V. S. Ramachandran and his colleagues showed this as follows (Ramachandran, 2011). As the brain works, it emits various waves, including a mu wave. Whenever a person engages in a voluntary action, the mu wave is suppressed. It is also suppressed when a person watches another person engage in a voluntary action—so it appears to reflect mirror neuron activity. Ramachandran used EEG (electroencephalography) to measure the mu waves. A medium functioning autistic child who was willing to have electrodes placed on his scalp was the subject. As in neurotypical individuals, the child's mu wave was suppressed when he engaged in any simple voluntary action. Yet, when he watched someone else perform the same action, the mu wave was not suppressed, suggesting that his mirror neurons were not firing. Deficits in the mirror neuron system may also help account to the difficulties autistic children have with language (Oberman & Ramachandran, 2007). Other studies of mu wave suppression in autistic children have found no difference to neurotypical controls (Heyes & Catmur, 2022). However, a review of recent studies (Andreou & Skrimpa, 2020) found a clear pattern of aberrant mu suppression in individuals with autism.

Theories of ToM

There are currently three major theories of the development of ToM from psychology and neuroscience (Flavell, 1999). The first is the "Theory theory" (Gopnick & Wellman, 1994; Wellman & Gelman, 1998). This is the idea that ToM is the development of an informal theory of other minds and is developed much as a scientist develops a theory, through prediction and testing. According to this view, children revise and improve their ToM in response to experience. Sometimes such theories are referred to as Child Scientist theories (Gurney et al., 2021).

The second theory of development of ToM is the Modular theory (Fodor, 1992; Leslie, 1994; Baron-Cohen, 1995). Modular theory posits that ToM is supported by a distinct, separate, brain module that matures along with other brain modules. The development of ToM is thus overwhelmingly a process of physical maturation. Experience may trigger of the ToM mechanism but does not determine the mechanism itself (Brüne & Brüne-Cohrs, 2006). Sometimes modular theory is treated as a subset of Theory theory. In this view, the developing ToM in a child is seen as the maturation of a domain-specific module rather than the refining of a general-purpose algorithm (Gallese & Goldman, 1998).

The last theory is "Simulation theory" (Harris, 1992). This is the theory that children use their own mental states to simulate the mental states of others. Thus, the ability to pass the false belief test is an extension of self-understanding. Experience is crucial in this development. It is especially through role taking in pretend play that children learn to simulate the experiences of others.

It is likely that all three theories are correct to a certain extent. The existence of specific regions of the brain associated with ToM tasks supports the modular theory, while the developmental timetable for ToM supports the theories that stress the role of experiential learning (theory theory and simulation theory). The fact that thinking about the self uses the same brain regions as thinking about others, particularly the mPFC, supports the simulation theory of ToM (Mitchell et al., 2006). Others have argued that the role of mirror neurons in ToM reasoning also supports the simulation theory (Gallese & Goldman, 1998).

Computer Models of Theory of Mind

Theories of Theory of Mind have become important with the development of artificial intelligence, and the task of imbuing artificial intelligence with "human-like" cognition (Gurney et al., 2021). Models of human cognition (including Theory of Mind) are currently being developed in cognitive neuroscience, computational neuroscience, and artificial intelligence research (Kriegeskorte & Douglas, 2018). Models include those based on reinforcement learning, partially observable Markov decision processes, utility maximization, and Bayesian inference (for details and a review see Gurney et al., 2021).

ToM and the Emergence of the Sociological Self

Development of a ToM is key to developing a sense of self. Here, work on ToM links directly with sociological understandings of the emergence of the mind and self, particularly the work of the early sociologists Charles Horton Cooley (1864–1929) and George Herbert Mead (1863–1931). Understanding the symbolic minds of others is central to Charles Horton Cooley's and George Herbert Mead's sociological theories of the self. Through understanding other people's mental states, we make self-appraisals, e.g., we see ourselves as others see us if we think what they think is important. This shapes both how we understand ourselves and what we choose to do. Cooley (1902) referred to this as the "looking glass self." Understanding the minds of others also helps us learn the rules and culture of our society and become fully socialized selves.

Cooley (1902) thought that members of our primary group were particularly important in constructing his looking glass self. That is, it is our understanding of their views of us that do most to influence our self-understandings. George Herbert Mead (1934) incorporated much of Cooley's thinking into his own understanding of the emergence of the mind and self. By understanding the minds of others and their views of us, we develop a sense of ourselves as an individual with a mind of our own. Mead stressed that this was a maturational, interactional and adaptational process. An individual acts, sees the response of others, then adjusts future behaviors to elicit desired responses (Turner et al., 1989, pp. 440–441). With time, a stable repertoire of behavior and sense of self emerges. Both Cooley and Mead saw this feedback process as symbolic and linguistic. The discovery of the role of mirror neurons in this process and their role in helping individuals understand the minds of others provides an additional biological basis for the process (Franks, 2010, p. 89).

For Mead and Cooley, our understanding of other people's minds is also an important source of social control, as we usually, but not always, want others' views of us to be favorable. Initially, the primary group is the most important group of others, but as an individual grows older the relevant groups become larger and more diverse until an individual understands the larger "community of attitudes" among the wider social group. Thus, an individual can take the role of the other, but also the larger community—what Mead called "the generalized other." At this stage, the self is fully formed and relatively stable. This self, in turn, becomes an important determinant of what a person chooses to do.

For Mead, role playing was central to this process of the emergence of the self (Turner et al., 1989, p. 448). Understanding the minds of others helps us play roles, because we know what others expect us to do. This is first manifested in pretend play, which occurs as noted above about 18–24 months of age. Thus, children literally practice the behaviors of people in their group in various roles. A child at this age may pretend to be a mother or father, baby or child, teacher or waiter. It is by practicing roles that one learns them and also learns to put oneself in the place of another in different roles, and thus can more readily anticipate the behaviors of others in other roles. This helps us adjust our own behaviors in the various roles we play and helps us anticipate the behaviors of others. Mead thought this is what ultimately made a large society possible.

Both Cooley, and to a lesser extent Mead, avoid an oversocialized view of the self in which the self is entirely a construction of the social environment. A largely forgotten qualification of his "looking glass self" made by Cooley is that a self that is constructed entirely from the looking glass self is weak and incomplete (Franks & Gecas, 1992). The initial sense of self for Cooley is an inborn sense of aggression and assertiveness associated with the more aggressive words like "mine" and "my" instead of reflective appraisals. Mead's concept of role-taking as critical to self-development is a more social process, placing greater emphasis on a prior learning of what he called a "universe of discourse" or significant symbols. Thus, for Mead in contrast to Cooley, role-taking was the process of responding to one's self as the actor thinks others would, and using this anticipated response to control ones' future behavior. This fosters flexible behavior necessary in a social animal. Without sharing significant symbols, this would be impossible (See Chap. 4 above).

However, Mead did allow for some autonomous agency in his work separating the "I" from the "me" (Turner et al., 1989, pp. 444–445). The "me" is our understanding of how others see us, the "I" is the actor capable of spontaneous, individual behavior. It is through the behavior of the actor "I" that individual differences and individual drives are manifest. Mead thought that high status, better integrated members of a social group have a stronger sense of "me", while low status, more poorly integrated members of a social group have a stronger sense of "I" (Turner et al., 1989, p. 445).

ToM and Sociological Perspectives

Understanding the minds of others (and hence ToM) is crucially involved in a variety of other sociological perspectives. Most of these are microsociological. Space precludes covering all sociological theories that involve understanding the minds of others, however a few are discussed here. First is the symbolic interactionist school closely associated with Cooley and Mead. This perspective examines the use of symbols in interaction: how actors use symbols to obtain desired results, and how they read and react to others use of symbols. Symbols include words, but also non-verbal signs such as physical behavior, facial expressions, styles of dress and self-presentations, and so on. Erving Goffman (1959) particularly studied role performance, and how individuals use a variety of symbols and props in their role performances.

The exchange theory tradition within sociology sees all social behavior as exchange of valued goods. For example, a friendship may be seen as the exchange of liking and support between individuals, a marriage may be seen as an exchange of love and fidelity between husband and wife. Such exchanges depend on the anticipation of the behavior of the exchange partner—a friend offers support because she wants to be supported in the future, and she knows that if she doesn't offer her support this may not happen. This involves an understanding of the other's mind and that person's reaction if support is withheld. Similarly, perspectives such as game theory and rational choice theory, although not including an explicit ToM, all involve understandings of how individuals (or corporate entities) make decisions about behavior, and how those decisions are likely to be influenced by the behavior of another person or corporate entity.

Expectations states theory likewise is grounded in our assumptions about others' mental processes and behavior given their social characteristics (Berger et al., 1977). High status individuals, for example, are expected to think and behave in a competent manner, lower status individuals are expected to think and behave less competently. There is also research within the expectations tradition that examines how our understanding of other peoples' expectations of us, given our various characteristics, shapes our subsequent behavior (Webster & Whitmeyer, 2002). ToM is also implicated in a variety of other important sociological processes, for example, the formation of self concept (e.g. Burke, 1991), judgments about morality (Guglielmo et al., 2009), normative processes, and many others.

Conclusion

The development of ToM or the understanding of the mental states of others is the end product of the development of cognitive prosocial skills by infants and toddlers that all neurotypical children undergo. ToM overlaps with, but is not the same as, empathy. An important marker in the development of ToM occurs in children between 3 and 4, in that they learn that other people have different beliefs compared to their own and that these beliefs may be false. It is associated with the development of language abilities, but it is not the same thing. Just as the child's language abilities improve with age, evidence suggests that children become progressively better at understanding the mental states of others as they grow older.

Neuroimaging studies show that certain regions of the brain are involved in ToM tasks. These overlap with the mirror neuron system, the system of neurons in the brain that mirror the neurons of others performing actions, and the functioning of the mirror neuron system seems to support ToM tasks. However, the mirror neuron system complements, but does not replace, the ToM system in the brain. Problems with ToM abilities are notable in individuals with autism, who also show deficits in mirror neuron systems. Autistic children also have problems with language. As a result, severely autistic children cannot function normally in the social world.

Theories and computational models of ToM are currently important in the further development of "human-like" artificial intelligence. For sociology, ToM is an important cognitive development that underlies many of the social processes theorized about and researched by sociologists, including emergence of the mind and self as discussed by Cooley and Mead. Any sociological or other social scientific perspective that involves actors' anticipations of the behavior and response of others is of necessity grounded in this cognitive development. ToM thus links sociology and other social sciences to the life sciences. At the same time, cognitive psychology and neuroscience can learn much about group dynamics from sociologists.

References

Andreou, M., & Skrimpa, V. (2020). Theory of mind deficits and neurophysiological operations in autism spectrum disorders: a review. *Brain Sciences, 10*(6), 393.

Baron-Cohen, S. (1995). *Mindblindness: An essay on autism and ToM*. MIT Press.

Bedny, M., Pascual-Leone, A., & Saxe, R. (2009). Growing up blind does not change the neural bases of ToM. *Proceedings of the National Academy of Sciences, 106*(27), 11312–11317.

Bel, L., Ronald, M., Pineda, J. A., & Sharma, A. (2009). Motor–auditory–visual integration: The role of the human mirror neuron system in communication and communication disorders. *Journal of Communication Disorders, 42*, 299–304.

Berger, J., Hamit Fisek, M., Norman, R. Z., & Zelditch, M., Jr. (1977). *Status characteristics and social interaction: An expectations states approach*. Elsevier.

Bloom, P. (2000). *How children learn the meaning of words*. MIT Press.

Bonini, L. (2017). The extended mirror neuron network: Anatomy, origin, and functions. *The Neuroscientist, 23*(1), 56–67.

Brüne, M., & Brüne-Cohrs, U. (2006). Theory of mind—evolution, ontogeny, brain mechanisms and psychopathology. *Neuroscience & Biobehavioral Reviews, 30*(4), 437–455.

Brooks, R., & Meltzoff, A. N. (2002). The importance of eyes: How infants interpret adult looking behavior. *Developmental Psychology, 38*(6), 958–966.

Burke, P. J. (1991). Identity processes and social stress. *American Sociological Review, 56*(6), 836–849.

Butterworth, G., & Cochran, E. (1980). Towards a mechanism of joint visual attention in human infancy. *International Journal of Behavioral Development, 3*(3), 253–272.

Caramazza, A., Anzellotti, S., Strnad, L., & Lingnau, A. (2014). Embodied cognition and mirror neurons: A critical assessment. *Annual Review of Neuroscience, 37*, 1–15.

Cattaneo, L., Fabbri-Destro, M., Boria, S., Pieraccini, C., Monti, A., Cossu, G., & Rizzolatti, G. (2007). Impairment of actions chains in autism and its possible role in intention understanding. *Proceedings of the National Academy of Sciences, 104*(45), 17825–17830.

Cooley, C. H. (1902). *Human nature and the social order*. Charles Scribner's Sons.

Corkum, V., & Moore, C. (1995). Development of joint visual attention in infants. In C. Moore & P. J. Dunham (Eds.), *Joint attention: Its origins and role in development* (pp. 61–83). Lawrence Erlbaum Associates, Inc.

Davis, J., Redshaw, J., Suddendorf, T., Nielsen, M., Kennedy-Costantini, S., Oostenbroek, J., & Slaughter, V. (2021). Does neonatal imitation exist? Insights from a meta-analysis of 336 effect sizes. *Perspectives on Psychological Science, 16*(6), 1373–1397.

de Waal, F. (2009). *The age of empathy*. Crown Publishing.

Fair, D. A., Cohen, A. L., Dosenbach, N. U. F., Church, J. A., Miezin, F. M., Barch, D. M., Raichle, M. E., Petersen, S. E., & Schlaggar, B. L. (2008). The maturing architecture of the brain's default network. *Proceedings of the National Academy of Sciences, 105*(10), 4028–4032.

Flavell, J. H. (1999). Cognitive development: Children's knowledge about the mind. *Annual Review of Psychology, 50*, 21–45.

Flavell, J. H. B. A., Everett, K. C., & Flavell, E. (1981). Young children's knowledge about visual perception- Further evidence for the level 1-level 2 distinction. *Developmental Psychology, 17*, 99–103.

Fodor, J. A. (1992). A theory of the child's theory of mind. *Cognition, 44*(3), 283–296.

Franks, D. D. (2010). *Neurosociology*. Springer.

Franks, D. D., & Gecas, V. (1992). Autonomy and conformity in cooley's self theory: The looking glass self and beyond. *Symbolic Interaction, 15*, 49–68.

Frith, U., & Happé, F. (1994). Language and Communication in Autistic Disorders. *Philosophical Transactions: Biological Sciences, 346*(1315), 97–104. http://www.jstor.org/stable/56024

Gallese, V., & Goldman, A. (1998). Mirror neurons and the simulation ToM-reading. *Trends in Cognitive Sciences, 2*(12), 493–501.

Gallup, G. G., Jr., Anderson, J. R., & Shillito, D. J. (2002). The mirror test. In M. Bekoff, C. Allen, & G. M. Burghardt (Eds.), *The cognitive animal* (pp. 325–334). MIT Press.

Garnham, W. A., & Perner, J. (2001). Actions really do speak louder than words—but only implicitly: Young children's understanding of false belief in action. *British Journal of Developmental Psychology, 19*, 413–432.

Geiger, A., Bente, G., Lammers, S., Tepest, R., Roth, D., Bzdok, D., & Vogeley, K. (2019). Distinct functional roles of the mirror neuron system and the mentalizing system. *NeuroImage, 202*, 116102.

Goffman, E. (1959). *The presentation of self in everyday life*. Double Day.

Golinkoff, R. M. (1986). I beg your pardon?: The preverbal negotiation of failed messages. *Journal of Child Language, 13*(3), 455–476.

Gopnick, A., & Wellman, H. M. (1994). The 'theory' theory. In L. A. Hirschfield & S. A. Gelman (Eds.), *Mapping the mind: Domain specificity in cognition and culture* (pp. 257–293). Cambridge University Press.

Gordon, A. S. (2006). In M. A. Arbib (Ed.), *Language evidence for changes in a theory of mind. Action to language via the mirror neuron system*. Cambridge University Press.

Greenberg, D. M., Warrier, V., Abu-Akel, A., Allison, C., Gajos, K. Z., Reinecke, K., Rentfrow, P. J., Radecki, M. A., & Baron-Cohen, S. (2023). Sex and age differences in "theory of mind" across 57 countries using the English version of the "Reading the Mind in the Eyes" Test. *Proceedings of the National Academy of Sciences, 120*(1), e2022385119.

Guglielmo, S., Monroe, A., & Malle, B. F. (2009). At the heart of morality lies folk psychology. *Inquiry, 52*(5), 449–466.

Gurney, N., Marsella, S., Ustun, V., & Pynadath, D. V. (2021). Operationalizing theories of theory of mind: A survey. In *AAAI fall symposium* (pp. 3–20). Springer Nature Switzerland.

Harris, P. L. (1992). From simulation to folk psychology: The case for development. *Mind Language, 7*, 120–144.

Heyes, C., & Catmur, C. (2022). What happened to mirror neurons? *Perspectives on Psychological Science, 17*(1), 153–168.

Hickok, G. (2014). *The myth of mirror neurons: The real neuroscience of communication and cognition*. WW Norton & Company.

Iacoboni, M., Molnar-Szakacs, I., Gallese, V., Buccino, G., Mazziotta, J. C., & Rizzolatti, G. (2005). Grasping the intentions of others with one's own mirror neuron system. *PLoS Biology, 3*(3), e79.

Killen, M., Mulvey, K. L., Richardson, C., Jampol, N., & Woodward, A. (2011). The accidental transgressor: Morally-relevant theory of mind. *Cognition, 119*(2), 197–215.

Kriegeskorte, N., & Douglas, P. K. (2018). Cognitive computational neuroscience. *Nature Neuroscience, 21*(9), 1148–1160.

Legerstee, M. (1992). A review of the animate-inanimate distinction in infancy: implications for models of social and cognitive knowing. *Early Development Parenting, 1*, 59–67.

Leslie, A. M. (1994). ToMM, ToBy and Agency: Core architecture and domain specificity. In L. A. Hirschfield & S. A. Gelman (Eds.), *Mapping the mind: Domain specificity in cognition and culture* (pp. 119–148). CUP.

Lewis, M., & Ramsay, D. (2004). Development of self-recognition, personal pronoun use, and pretend play during the 2nd year. *Child Development, 75*, 1821–1831.

Lillard, A. S., & Skibbe, L. (2005). ToM: Conscious attribution and spontaneous trait inference. In R. R. Hassin, J. S. Uleman, & J. A. Bargh (Eds.), *The new unconscious* (pp. 277–305). Oxford University Press.

Liu, D., Wellman, H. M., Tardif, T., & Sabbagh, M. A. (2008). ToM development in Chinese children: A meta-analysis of false-belief understanding across cultures and languages. *Developmental Psychology, 44*(2), 523–531.

Lovett, S. B., & Pillow, B. H. (2010). Age-Related Changes in Children's and Adults' Explanations of Interpersonal Actions. *The Journal of Genetic Psychology, 171*(2), 139–167.

Malle, B. F. (2005). Folk ToM: Conceptual foundations of human social cognition. In R. R. Hassin, J. S. Uleman, & J. A. Bargh (Eds.), *The new unconscious* (pp. 225–255). New York, Oxford University Press.

Mead, G. H. (1934). *Mind, self and society*. University of Chicago Press.

Meltzoff, A. N., & Keith Moore, M. (1977). Facial imitation in infants. *Science, 198*, 75–78.

Mitchell, J. P., Malia Mason, C., Macrae, N., & Banaji, M. R. (2006). Thinking about others: The neural substrates of social cognition. In J. T. Cacioppo, P. S. Visser, & C. L. Pickett (Eds.), *Social neuroscience* (pp. 63–82). MIT Press.

Montgomery, K. J., Isenberg, N., & Haxby, J. V. (2007). Communicative hand gestures and object-directed hand movements activated the mirror neuron system. *Social Cognitive and Affective Neuroscience, 2*(2), 114–122.

Moore, C., & Corkum, V. (1998). Infant gaze following based on eye direction. *British Journal of Developmental Psychology, 16*(4), 495–503.

Moran, J. M., Young, L. L., Saxe, R., Lee, S. M., O'Young, D., Mavros, P. L., & Gabrieli, J. D. (2011). Impaired ToM for moral judgment in high-functioning autism. *Proceedings of the National Academy of Sciences, 108*, 2688.

Oberman, L. M., Hubbard, E. M., McCleery, J. P., Altschulera, E. L., Ramachandrana, V. S., & Pineda, J. A. (2005). EEG evidence for mirror neuron dysfunction in autism spectrum disorders. *Cognitive Brain Research, 24*, 190–198.

Oberman, L. M., & Ramachandran, V. S. (2007). The simulating social mind: The role of the mirror neuron system and simulation in the social and communicative deficits of autism spectrum disorders. *Psychological Bulletin, 133*(2), 310–327.

Onishi, K. H., & Baillargeon, R. (2005). Do 15-month-old infants understand false beliefs? *Science, 308*(5719), 255–258.

Penagos-Corzo, J. C., Cosio van-Hasselt, M., Escobar, D., Vázquez-Roque, R. A., & Flores, G. (2022). Mirror neurons and empathy-related regions in psychopathy: Systematic review, meta-analysis, and a working model. *Social Neuroscience, 17*(5), 462–479.

Pineda, J. A., & Hecht, E. (2009). Mirroring and mu rhythm involvement in social cognition: Are there dissociable subcomponents of ToM? *Biological Psychology, 80*(3), 306–314.

Rakoczy, H. (2022). Foundations of theory of mind and its development in early childhood. *Nature Reviews Psychology, 1*(4), 223–235.

Ramachandran, V. S. (2011). *The tell-tale brain*. Norton Company.

Reddy, V., Williams, E., Costantini, C., & Lang, B. (2010). Engaging with the self: Mirror behaviour in autism, Down syndrome and typical development. *Autism, 14*(5), 531–546.

Rizzolatti, G., & Craighero, L. L. (2004). The mirror-neuron system. *Annual Review of Neuroscience, 27*, 169–192.

Rizzolatti Giacomo, L., Fogassi, L., & Gallese, V. (2001). Neurophysiological mechanisms underlying the understanding and imitation of action. *Nature Review of Neurosocience, 2*, 661–670.

Rochat, P. (2003). Five levels of self-awareness as they unfold early in life. *Consciousness and Cognition, 12*(4), 717–731.

Saxe, R. (2006). Four brain regions for one ToM? In J. T. Cacioppo, P. S. Visser, & C. L. Pickett (Eds.), *Social neuroscience* (pp. 83–102). MIT Press.

Saxe, R., Carey, S., & Kanwisher, N. (2004). Effect of subjective perspective taking during simulation of action: A PET investigation of agency. *Nature Neuroscience, 4*(5), 546–550.

Scaife, M., & Bruner, J. S. (1975). The capacity for joint visual attention in the infant. *Nature (London), 253*(5489), 265–266.

Schroeder, J. H., Desrocher, M., Bebko, J. M., & Catherine Cappadocia, M. (2010). The neurobiology of autism: Theoretical applications. *Research in Autism Spectrum Disorders, 4*, 555–564.

Schulte-Rüther, M., Markowitsch, H. J., Fink, G. R., & Piefke, M. (2007). Mirror neuron and theory of mind mechanisms involved in face-to-face interactions: a functional magnetic resonance imaging approach to empathy. *Journal of Cognitive Neuroscience, 19*(8), 1354–1372.

Schurz, M., Radua, J., Tholen, M. G., Maliske, L., Margulies, D. S., Mars, R. B., Sallet, J., & Kanske, P. (2021). Toward a hierarchical model of social cognition: A neuroimaging meta-analysis and integrative review of empathy and theory of mind. *Psychological Bulletin, 147*(3), 293.

Smallwood, J., Bernhardt, B. C., Leech, R., Bzdok, D., Jefferies, E., & Margulies, D. S. (2021). The default mode network in cognition: A topographical perspective. *Nature Reviews Neuroscience, 22*(8), 503–513.

Stephens, G. J., Silbert, L. J., & Hasson, U. (2010). Speaker–listener neural coupling underlies successful communication. *Proceedings of the National Academy of Sciences, 107*(32), 14425–14430.

Stone, V. E. (2006). ToM and the evolution of social intelligence. In J. T. Cacioppo, P. S. Visser, & C. L. Pickett (Eds.), *Social neuroscience* (pp. 103–129).

Tomasello, M., & Barton, M. (1994). Learning words in nonostensive contexts. *Developmental Psychology, 30*(5), 639–650.

Turner, J. H., Beeghley, L., & Powers, C. H. (1989). *The emergence of sociological theory*. Wadsworth.

Vistoli, D., Brunet-Gouet, E., Baup-Bobin, E., Hardy-Bayle, M.-C., & Passerieux, C. (2011). Anatomical and temporal architecture of ToM: A MEG insight into the early stages. *NeuroImage, 54*, 1406–1414.

Walter, H. (2012). Social cognitive neuroscience of empathy: Concepts, circuits, and genes. *Emotion Review, 4*(1), 9–17.

Wang, Y., Metoki, A., Xia, Y., Zang, Y., He, Y., & Olson, I. R. (2021). A large-scale structural and functional connectome of social mentalizing. *NeuroImage, 236*, 118115.

Webster, M., & Whitmeyer, J. M. (2002). Modeling second-order expectations. *Sociological Theory, 3*, 306–327.

Wellman, H. M. (2018). Theory of mind: The state of the art. *European Journal of Developmental Psychology, 15*(6), 728–755.

Wellman, H. M., & Gelman, S. A. (1998). Knowledge acquisition in foundation domains. In D. Kuhn & R. S. Siegler (Eds.), *Handbook of child psychology* (Cognition, perception and language) (Vol. 2, pp. 523–573). Wiley.

Williams, J. H. G., Whiten, A., Suddendorf, T., & Perrett, D. I. (2001). Imitation, mirror neurons and autism. *Neuroscience and Biobehaviour Review, 25*, 287–295.

Wimmer, H., & Perner, J. (1983). Beliefs about beliefs: Representation and constraining function of wrong beliefs in young children's understanding of deception. *Cognition, 13*, 103–128.

Woodward, A. L. (1999). Infants' ability to distinguish between purposeful and non-purposeful behaviors. *Infant Behavior & Development, 22*(2), 145–160.

Yang, T., Bayless, D. W., Wei, Y., Landayan, D., Marcelo, I. M., Wang, Y., DeNardo, L. A., Luo, L., Druckmann, S., & Shah, N. M. (2023). Hypothalamic neurons that mirror aggression. *Cell, 186*(6), 1195–1211.

Brains, Bodies, and Their Worlds: Towards a Model of Predictive Social Actors

16

Jacob E. Cheadle, K. J. Davidson-Turner, and Bridget J. Goosby

Abstract

Despite the growing confidence in biosocial research in sociology, integrating the brain into sociological theory has progressed slowly. Traditionally seen as a "bottom-up" processor converting sensory signals into cognitive and emotional representations, the brain is now increasingly viewed as a predictive Bayesian organ. This perspective suggests that the brain develops error-minimizing probabilistic models through active engagement with its environment's "generative processes." It embodies "top-down" "generative models," tested and refined against "bottom-up" sensory errors. Such brains regulate their bodies to actively engage with the world, altering perceptions or the environment to conform to predictions. Perception provides sensory errors needed to evaluate and potentially change models (i.e., learning), helping the brain embody the statistical regularities of its existence. Brains learn about the world through action-perception cycles, optimizing models and reducing uncertainty. This chapter explores sociological concerns such as mirror neurons, role-taking/making, and social action, aligning with microsociological insights that view the situation as the primary scene of action. Emphasizing evolving cognitive and emotional trajectories, these insights highlight the importance of socializing and enculturating experiences scaffolded by patterns of shared attention in social situations. This framework lays the foundation for understanding how predictive neurosociological actors shape and are shaped by their sociocultural environments.

Introduction

This chapter presents the brain as embodying *predictive models of its body in its world* (Barrett, 2020; Clark, 2015; Parr et al., 2022), underscoring intrinsic interconnections among brains, bodies, broader social ecologies, and the minds that exchanges among them give rise to. Part of the appeal of the *brain as a predictive modeler* is a theoretical grounding in well-defined assumptions and principles by which brains attempt to make sense of the joint states of their bodies and

J. E. Cheadle (✉) · B. J. Goosby
Department of Sociology, Population Research Center, and The Center on Aging and Population Sciences, The University of Texas at Austin, Austin, TX, USA
e-mail: j.e.cheadle@utexas.edu; bgoosby@prc.utexas.edu

K. J. Davidson-Turner
Department of Sociology, The University of Texas at Austin, Austin, TX, USA
e-mail: kjdavidsonturner@utexas.edu

W. Kalkhoff et al. (eds.), *Handbook of Neurosociology*, Handbooks of Sociology and Social Research, https://doi.org/10.1007/978-3-031-95615-7_16

worlds, regulating the former so that actors may act with their bodies within and upon the latter (Friston, 2009, 2010; Clark, 2013). Elaborating upon this emerging perspective, the *active inference framework* (AIF; Friston, 2009, 2010; Friston & Kiebel, 2009) emphasizes the interdependence of prediction, action, and inference as the dynamics by which the world, others within it, and the self are disclosed (Clark, 2015). By statistically attuning to conditions and dynamically linking bodies to the microsocial and larger structural contours that define individuals' probabilistic risk and opportunity landscapes (e.g., Bonilla-Silva, 1997; Glass & McAtee, 2006; Link & Phelan, 1995), the AIF motivates insights into neurosociological and more traditional areas of sociological inquiry and theory in which actors enculturate, coordinate, learn, plan, act, and adapt (Franks, 2010, 2019; Kalkhoff et al., 2016).

We discuss two different models of the brain in this chapter. First, we explore the traditional *responsive* view, which most readers are probably familiar with. Then, we look at the developing *predictive* perspective, which may be less well-known (though see Davis, 2013). After that, we introduce the AIF in a non-technical way to further our understanding of the predictive brain by including the role of action-perception cycles in minimizing prediction errors, both so that actors may *act* and to better incorporate the *social* and the *neuro*. Finally, with the AIF underwriting predictive brains, we delve into some sociological concerns, such as mirror neurons and George Herbert Mead's notion of role-taking and theory of the act (Franks, 2010; Mead, 1934). Our goals here are programmatic and introductory, aiming to provide a foundation in important neuroscientific concepts while non-exhaustively 'seeding' sociological concerns. By emphasizing the notion of agency and the dynamic interplay between an organism and its environment, the AIF perspective reciprocates with sociological insights, grounding actors in their neuro-socio-cultural interdependencies. Moreover, this view bridges the traditional divide between biological and social constructionist perspectives by making each dependent upon the other.

A Tale of Two Models of the Brain

The "Bottom-Up" Model of the Brain

The traditional model of the brain that sociologists have "inherited" from neuroscience, psychology, and the biomedical disciplines to varying degrees, going back to Sherrington (1900), is a "bottom-up" information processor (Barbas, 2015; Bubic et al., 2010), reflecting its stimulus-response origins (Barrett, 2017b; Keller & Mrsic-Flogel, 2018). This heritage is evident in the ubiquitous language of, for example, *respond*, *react*, *trigger*, and *activate* as experience is "processed" whenever brains, bodies, actors, and organisms, more generally, are described. The modern incarnation of the brain is thus commonly portrayed in inherently retrospective ways such that the neural computations and mental representations of the world we experience are typically a bit behind whatever happens to be happening. Consequently, perceptual experiences and the mental events following a stimulus are described as *representations* of sensory stimuli--things that have already happened.

The traditional model, shown in Fig. 16.1, is "bottom-up" in the sense that the fine-grained detail of sensory signal processing at the entry points into the brain, when signals from outside are transduced into signals within (e.g., vision), flow "upwards" through feature detectors and towards more abstracted pattern matching and compressed processing layers as information is passed along for cognitive (i.e., association cortices), emotional (i.e., limbic regions), and action processing (i.e., motor commands), in service to response generation (Hutchinson & Barrett, 2019). An information processing cascade, instantiated by a visual stimulus, for example, and giving rise to a mental event, is outlined in Fig. 16.1a. A more elaborated version of this model is presented at the mind level in Fig. 16.1b and at a more general brain level in Fig. 16.1c. These depictions express common understandings about how (in a high-level sense) stimuli are processed, represented, and integrated in service to response and action generation.

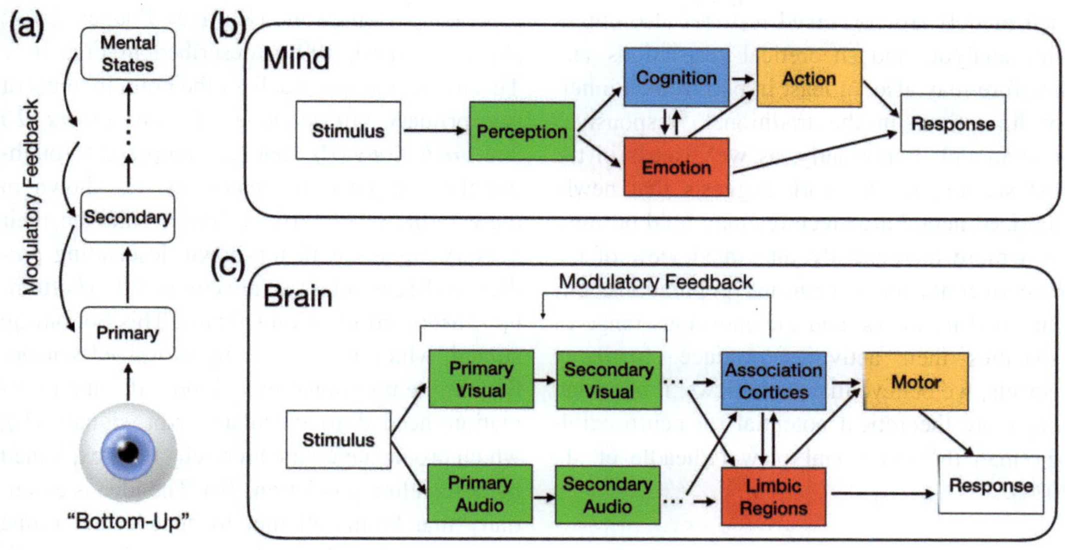

Fig. 16.1 The "bottom-up" model. (**a**) Internal representations are generated by bottom-up input and top-down signals act as modulatory signals (Keller & Mrsic-Flogel, 2018). (**b**) A mind-level depiction of the bottom-up representational framework. (**c**) A brain-level level representation of this framework. The colors characterize the mapping of mental faculties to their mapping in the brain (green: sensation and perception; blue: cognition; red: emotion; yellow: action). Adapted from Hutchinson and Barrett (2019)

One implication is that the mental events comprising thoughts, feelings, and the actions that constitute behavior follow from sensory input from the world, potentially modulated by past experiences or "top-down" control of these stimulus-response links. Because sensory information processing proceeds in a sequence of neuronal firing patterns that trigger cognitions, emotions, motor commands, and perhaps a dialogue among them, mental events can constitute their own stimuli. Consequently, external triggers can become both future internal triggers and subsequent sources of computational modulation. Modulation thus also has a kind of *retrospectivity* in the sense that it can also represent a response to ongoing computational activity that was *triggered by* a stimulus, whether internal or external. One feature of this model is that it seems to fit common cognitive and emotional experiences of the mind, where environmental deviations trigger rapid changes to physiology, thoughts, and feelings that can be engaged with, evaluated, and appraised cognitively as attention moves among bodily, emotional, cognitive, and external states.

Notably, this view of the brain has been criticized for failing to develop an integrated theory (Hawkins & Dawkins, 2021; Kotchoubey et al., 2016; Stam, 2015). Consequently, descriptive and explanatory models of neural function and activity have tended to propose processing strategies and computational heuristics that are bespoke and tailor-made for particular sensory and cognitive contexts[1] (Walsh et al., 2020). Mirroring to some extent the challenge of mid-level theories and their potential divergences versus grand theories and loss of specificity in particular applications, such theoretical fragmentation undoubtedly reflects the vast complexity of the brain in conjunction with the diversity of capabilities our brains exhibit. Reliance upon bespoke models are, at least in part, a natural consequence of this complexity. However, how

[1] In this vein, Hommel (2020:1303) argues that the current trends toward "mechanistic" explanations in cognitive neuroscience and psychology are better said to be pseudo-mechanistic. They suggest that "approaches that do not aim to explain effects of particular experimental paradigms but, rather, account for general human capacities" are needed.

such models are conceived *a priori* also comes with analytic and theoretical restrictions and therefore may also, at least in part, reflect inherent limitations in the traditional "responsive" brain model. Importantly, as we discuss in the next section, recent work suggests that newly theorized neural architectures may hold promise for a more theoretically integrated view of the brain over its many neuronal populations, circuits, and networks, and the enormous range of capacities their activities produce. In many respects, we believe that such views, if accurate, hold more theoretical potential for neurosociology than the traditional view (Cheadle et al., 2024).

The Predictive Brain

An increasingly influential revision to the traditional model that proposes more theoretical integration is the *hierarchical predictive coding/processing* framework (Knill & Pouget, 2004; Rao & Ballard, 1999), described in Fig. 16.2. This model reconceptualizes the brain in terms of two primary informational streams, *errors (E)* and *predictions (P)*, that are compared throughout the computational hierarchy. As shown in these terms in Fig. 16.2a, brains simulate their sensory signals with top-down descending prediction (P) cascades that are compared to bottom-up sensory information streams. This bottom-up stream, which reflects the hypothesized sensory flow in the traditional model, provides the information needed to calculate error signals (E), which propagate up the hierarchy until explained by descending predictions (P). The idea is essentially that brains attempt to encode and refine their simulations by optimizing models to minimize errors, updating their predictions in the moment and possibly changing their models to better account for those errors that cannot be explained by prior experience under the current set of potential hypotheses/beliefs about the

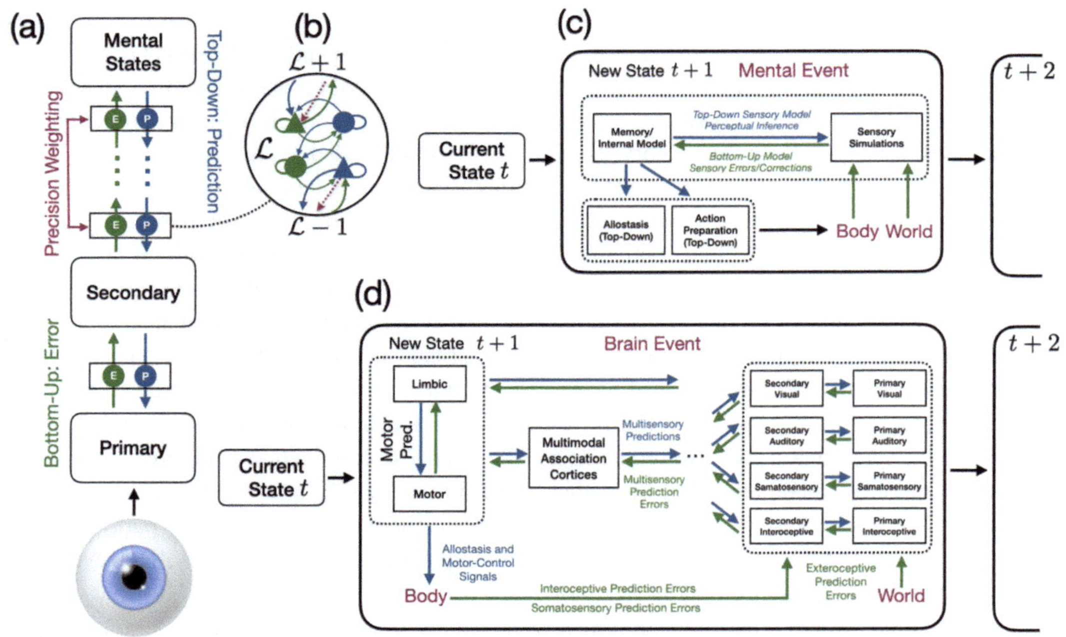

Fig. 16.2 Predictive coding model. (**a**) In a hierarchical predictive processing framework, predictions cascade "top down" the hierarchy and are compared to errors ascending the hierarchy from the "bottom-up" (Keller & Mrsic-Flogel, 2018). (**b**) A cutout depicting neural dynamics adapted from Seth (2013) and Friston and Kiebel (2009). (**c**) A mind-level depiction of the predictive processing framework with forward (predictive) and inverse (error) informational flows. (**d**) A brain-level level representation of this framework. Blue arrows represent prediction signals descending the computational hierarchy and green arrows represent ascending sensory signals against which predictions are compared. Adapted from Hutchinson and Barrett (2019)

causes of sensory signals. In a step that reverses the traditional view of the brain, when errors are adequately explained and predictions suffice, *the prediction is the signal* (Clark, 2013, 2023). The implication is that where the conventional model can be said to be *representational*, the predictive view is inherently *constructionist* (Barrett, 2017a; Hutchinson & Barrett, 2019).

Compared to the corresponding bottom-up sensory error streams from which representations are assembled in the traditional model, the core top-down prediction streams are decompressed as they descend the computational hierarchy as information flows from areas of great compression and abstraction into those responsible for ever-finer features and details. The supporting dynamics cover many scales of neuronal organization, interconnection, structure, and modulatory mechanisms (i.e., neurotransmitters, peptides, and other molecules). Supporting this computational architecture are proposed "state units," exemplified in the cutout in Fig. 16.2b, color-coded for error (green) and prediction (blue) neurons, with projection neurons as triangles and inhibitory interneurons as circles (Friston & Kiebel, 2009; Seth, 2013). Top-down predictions flow from level $\mathcal{L}+1$ to $\mathcal{L}$, while errors ascend from $\mathcal{L}-1$ to $\mathcal{L}$ so that prediction error within a state unit is a linear mixture of hierarchical and lateral connections (Friston, FitzGerald, et al., 2017; Seth, 2013). The upward flow of errors is precision modulated by the dashed downward purple arrow (e.g., dopaminergic and oxytocin modulation; Seth & Friston, 2016) so that errors viewed as imprecise can be downweighed and perhaps fail to propagate, possibly signaling a need for enhanced attention to increase the precision of new sensory data, or to reject and overwrite the sensory stream.

One of the features of this model that we believe is of value to sociologists is the inherent brain-body-world coupling it entails, which brings forward dynamics not included in traditional views relying on responsivity and homeostatic error-correcting feedback (Arnaldo et al., 2022; Barrett et al., 2016; Sterling, 2012, 2015). Figure 16.2c, d presents the more general system-level model as mental and brain events, parallel to the depictions in Fig. 16.1b, c, but emphasizing the integration and interdependence of multiple systems, including those in the body. For example, the term *allostasis* in Fig. 16.2 refers to the predictive regulation of the internal milieu of the body (Sterling, 2012), a systems concept that captures many signaling dynamics and physiological self-actions, situating physiological regulation as a set of coordinated activities across systems, organs, tissues, and thus presenting a view well beyond the more traditional notion of reactive system-specific homeostatic error-correcting feedback (see Billman, 2020 for a review of the concept of homeostasis). Brains anticipate needs before they arise, whether raising blood pressure before standing, the feeling of thirst well before actual dehydration, or increasing available energy to mobilize action when threatened (see Sterling, 2012, 2015; Theriault et al., 2021). Allostasis can, therefore, be understood as predictions about whether and how physiological states need to be changed and coordinated, including regulatory adaptations that seek to resolve prediction errors towards anticipated future states.

For example, allostasis attempts to avoid the anticipated harm of an acute stressor through coordinated and metabolic costly regulation patterns and subsequent fight-or-flight actions, which can be highly prescribed and constrained based on the social setting and cultural norms of that time and place. One implication is that regulatory allostatic state changes are themselves predictions about how states will change if directed to do so, which, like all predictions, may be more or less successful (i.e., efficient regulation in a healthy body, but imprecise and inefficient regulation for a body that has been undermined by years of chronic stress; McEwen, 1998; Bobba-Alves et al., 2022). But how does the brain know the current states of its body or whether the proposed changes had the intended consequences? The "data" acquired to these ends is obtained via *interoception*, which comprises the expectations about current states and signals ascending from throughout the body to the brain, including their interpretation and integration (Barrett & Simmons, 2015; Chen et al., 2021; Kleckner et al., 2017), a dynamic critical to modern biosocial constructionist theories of emotion (Barrett, 2017a, 2017b). Allostasis and interoception are thus part of the same regulatory monitoring system linking predictions to sensory error streams (Kleckner et al., 2017).

The motor control signal pathways in Fig. 16.2 also predict the body's physical kinematics (i.e., movement, action, and location), which are monitored via *proprioception*, a crucial part of the *somatosensory system* (e.g., touch). It is worth recognizing that motor signal pathways support all physical action here. For example, a microsocial interaction ritual such as a conversation is accomplished through sequences of precise motor commands that generate vocalizations and all the other subtleties of facial, postural, and gestural expression. Figure 16.2, therefore, brings together multiple processes needed for a brain to assemble information about what is happening in and of its body and in and of the world so that it may act upon and with them. In so doing, predictive models have been applied to understand a range of capacities[2] while multimodal and multiscalar evidence of the physical implementation of predictive processing is accumulating via diverse research streams (e.g., Chanes & Barrett, 2016; Ficco et al., 2021; Hutchinson & Barrett, 2019; Kleckner et al., 2017; Mikulasch et al., 2023; Shipp et al., 2013; Walsh et al., 2020).

[2]Some examples include cognitive effort (Parr et al., 2023), attention (Friston, 2009), value (Friston, 2010), learning (Smith et al., 2020), memory (Barron et al., 2020), rumination and imagination (Hesp et al., 2020), communication (Friston & Frith, 2015a, Friston & Frith, 2015b; Vasil et al., 2020), speach (Caucheteux et al., 2023), reading (Friston et al., 2018), curiosity and insight (Friston, Lin, et al., 2017), action understanding (Friston et al., 2011), affect and emotion (Barrett, 2017b; Hesp et al., 2021; Joffily & Coricelli, 2013; Seth & Critchley, 2013; Van de Cruys, 2017), depression and anxiety (Arnaldo et al., 2022; Barrett et al., 2016; Harshaw, 2015; Shaffer et al., 2022), socializing (Theriault et al., 2021), the mind (Badcock et al., 2019), meta-awareness (Sandved-Smith et al., 2021), and culture (Kastel et al., 2023; Veissière et al., 2020). For additional examples, see Hutchinson and Barrett (2019).

The Active Inference Framework

The predictive (Bayesian) brain has been used to motivate the Active Inference Framework (AIF; Friston, 2009, 2010; Friston & Kiebel, 2009), which emphasizes that prediction is not just in service to accuracy, but to guiding action and behavior towards preferred sensory experiences and away from those that are harmful. The scope is from thriving to death at the extremes, with various states of health/disease and (un)happiness (etc.) further in towards the middle. As Parr et al. (2022:26) write, "[t]echnically, Active Inference agents come equipped with models that assign high marginal probabilities to the states they prefer to visit or the observations they prefer to obtain. For a fish, this means a high marginal likelihood of being in water. This implies that organisms implicitly expect the observations they sample to be within their comfort zone (e.g., physiological bounds)." If the central goal of a predictive brain is to learn the statistical regularities of its body, social, and physical worlds, and then to direct action towards preferred experiences and away from those that are harmful, how does it go about doing so?

In the AIF, brains and other complex living systems aim to minimize *surprise*[3] by constantly updating their internal models to reduce the discrepancies between their predicted and actual sensory observations. Prediction here does a great deal of work, encompassing both cognitive processes and the fulfillment of basic survival needs. The minimization, more accurately an approximation via a proxy called *free energy* (Friston, 2009), is achieved via adaptive cycles of perceptual inference and action, hence *active inference*. *Perception* to update constructed representations in the moment and learning in the long-term, and *action* to change perceptions to

[3]Surprise or *surprisal* is the negative log probability of an observation, $-\ln P(o)$, and is large and positive when probabilities are low and approaches 0 as probabilities approach 1. This concept is the negative of Bayesian model evidence, also called the marginal likelihood per the quote above, and is intimately connected to Shannon Entropy, the long-term average of surprise, $H[\ln P(o)] = -E[\ln P(o)]$ (Parr et al., 2022).

confirm beliefs or provide new information. For humans, this endeavor is guided through socializing and enculturating experiences, which together constitute the *generative processes* of their embedding environments (Parr et al., 2022). Actors seek to learn the generative processes that structure their risk and opportunity landscapes so they, too, can be social participants who engage in activities that facilitate surviving and, to the extent possible, thriving.

Not surprisingly, much sociology documents, characterizes, and explains the generative social processes that probabilistically structure and organize both experiences and the actions afforded to different actors (Grusky, 2014; Grusky & Hill, 2017). From the brain's perspective, the result of this engagement with and by the world is the encoding of these statistical regularities within individuals' *probabilistic generative (predictive) models* of how sensory observations are generated, and thus beliefs about both observable and hidden variables (Parr et al., 2022). Such models, comprised of neural structures wired together through the biodynamics of experientially entrained embodiment (i.e., Hebbian plasticity, 'neurons that fire together, wire together'). Predictive processing provides a mechanistic account or process theory of the neuronal dynamics subserving the higher-order functions by which social actors with social brains make sense of a complex, uncertain, and dangerous world, but is ultimately incomplete because inference alone is insufficient. A brain is also an organ of action on its mutually constitutive body and the environments it inhabits, and so can exercise some control over the signals it sends out to engage with the world, thereby exerting some influence over the signals it receives, and thus a degree of *agency* over what and how its generative models develop (i.e., learns).

Predictive Brain, Bayesian Brain

An important proposition is that the predictive brain is (approximately) Bayesian (Friston, 2010; Knill & Pouget, 2004; Parr et al., 2022; Seth & Friston, 2016). The problem of inference about states in the body and the world is the inherent uncertainty of the situation a brain, which is trapped in a black box we call a skull (Rieke et al., 1999, cited in Clark, 2013), is in. Hidden away from direct connections to the world, it attempts to make inferences about the causes of the electrochemical and other signals received from inside and outside its body. Brains do not *see* directly; they predict the signals received from the eyes via the optic nerve, continuously correcting prediction errors and updating constructed representations of the environment. Because life flows in temporal sequence, the notion of model *updating* as new information is gathered is essential. At the momentary level, we may focus attention to see an object more clearly to update the precision of our constructed representation, while on grander temporal scales neural structure may change more dramatically and durably, as when learning why so-called "mirror neurons" provide a neurobiological foundation for Mead's concept of role-taking (e.g., Franks, 2010).

Given the existence of a prior generative model of observations and hypotheses (or beliefs) about causes (or states), the challenge is how to optimally decide when to update models: what hypotheses explain a set of observations given that observations were made under prior hypotheses about what could be observed? The problem involves updating conditional probabilities, and Bayes' Theorem provides the optimal method for updating and refining a model that generates probability distributions of prior beliefs (i.e., predictions) as new information is acquired. This *Bayesian belief updating* is key to minimizing and managing surprise as it allows the brain to optimize when to update itself and when to learn by altering predictions to better minimize errors. In other words, the idea is that the brain approximates Bayes' rule for updating prior beliefs encoded in distributions of predictions into new *posterior* beliefs about those hypotheses given the new observations. A related way to think about this is that observations collected under some model of the world (the *likelihood* or the probability of observations given prior beliefs/ hypotheses) are, in conjunction with prior beliefs, transformed into an evaluation of hypotheses or states given those observations (the *posterior*).

Posteriors match priors when there are no sensory prediction errors, indicating successful prediction (and thus providing) inference about the causes of sensory signals. Error propagation is a consequence of hypotheses or predictions at each step in the computational chain that are unlikely given what has been observed, necessitating prior-posterior updates. Underlying this perspective is the idea that the fundamental (or even mechanistic) causes of sensory experience are hidden behind the 'sensory veil' enshrouding the brain.[4] *Bayesian model inversion* thus takes observed consequences given hypotheses/beliefs and infers causes from their consequences. In other words, brains seek to infer the most probable hypotheses or explanations of the states of external and internal environments given sensory observations, to explain what one perceives via its exteroceptive, interoceptive, and somatosensory signals per Fig. 16.2. The goal is thus to form accurate hypotheses or beliefs about sensory states reflecting the *joint* probabilities of hypotheses *and* observations, a problem of revealing generative processes through embodied and embodying socializing and enculturating experiences.

AIF: Action-Perception Cycles, Free Energy, and Expected Free Energy

Whether considering humans or other complex living systems more generally (i.e., *hierarchical* AIF; Friston, 2009, 2010; Friston & Kiebel, 2009), the goal is not just to make accurate inferences about current states of the world, but to solicit observations that are consistent with a variety of goals from the most fundamentally biochemical (e.g., temperature, oxygen, etc.) to highly abstract and conceptual for humans (e.g., aesthetic needs). The AIF holds that it is prediction "all the way down" in that all life evolved to inhabit niches, albeit some more specific and others more general, and thus each embodies predictions about the world it will inhabit (Ramstead et al., 2021). To exist and persist, basic environmental parameters must not drift too far from evolved expectations; all life is a fish in water (Parr et al., 2022). There is a reason the composition of scuba diver tanks does not deviate too far and only in prescribed ways from the air of our atmosphere. To do otherwise could be catastrophic. We see here a reflection of Maslow's "Hierarchy of Needs" (Maslow, 1964), which begins with the evolved physiological requirement of cells and other bodily necessities and then becomes increasingly abstract--safety, love and belonging, esteem, cognitive needs, self-actualization, and transcendence. When viewed through the AIF lens, this hierarchy reflects minimization of *negative* prediction errors across scales, from those supporting physiological integrity moment-to-moment through life to those that are psychological, highly abstract, subjective, and have great temporal depth over the life span.

Viewed in this way, sociological insights about power, wealth, and privilege are recovered in the AIF in terms of the consequences of social conditions for basic biological and psychosocial imperatives (Parr et al., 2022) for those social groups who are statistically less likely to have their needs met and errors controlled and managed. Thus, and in parallel to the general sociological concern for inequity and the way it undermines life quality (Goosby et al., 2018; Grusky & Hill, 2017; Neckerman & Torche, 2007), the AIF brings forward the insight that the goal is not just to identify the state of things, but to contextualize and make the most of them given our own natures and, as social entities, the ways our natures intersect and intertwine with those of others. Prediction alone lacks the broader context

[4] Technically, this veil, a conditional independence structure called a Markov Blanket (Pearl, 1998), plays key roles in deeper theoretical treatments of active inference in complex systems biology (Ramstead et al., 2023). Perhaps there is potential for a *free energy sociology* beyond individual human actors and for the systems we create through our higher-order, complex collective dynamics, mirroring that life, in general, is comprised of complex systems of complex systems [of...]. This view would begin with the idea that social structures both form Markov Blankets and appear to seek to persist over time via adaptive epistemic and pragmatic acts that minimize free energy (e.g., Poirier et al., 2021).

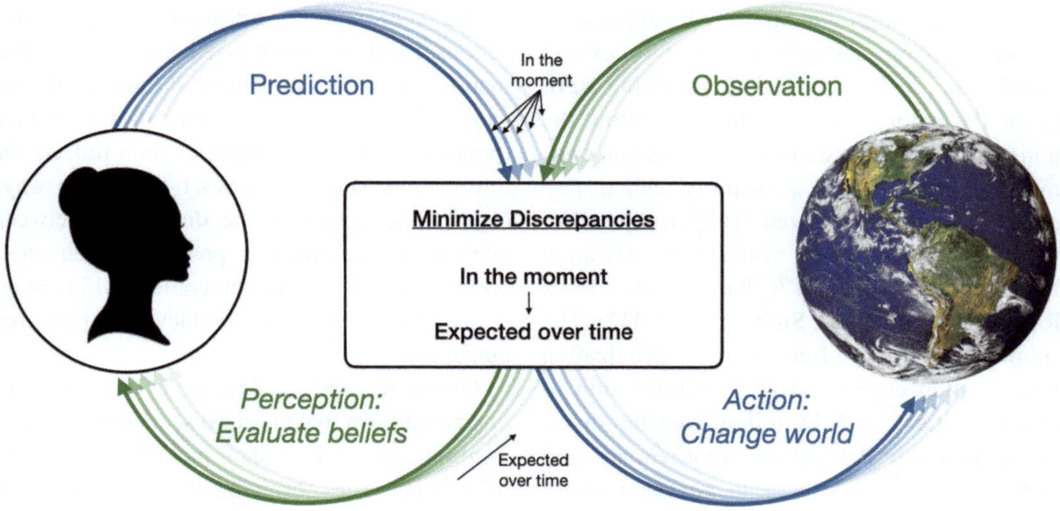

Fig. 16.3 Action-perception cycles. Perception is used to test models through sensory observations while actions change the world, both by changing observations (i.e., a change in perspective) and changing the world when it is acted upon. Each cycle represents free energy at a point in time, while those projected arbitrarily forward in time represent the expected free energy of different action policy sequences. Adapted from Parr et al. (2022)

of what it means to both require and desire certain conditions and experiences. Additionally, it does not encompass the full scope of interactions between the actor (or organism) and its environment or the nature of the ways these interactions unfold, which are profoundly interdependent with those of conspecifics sociologically, and are also fundamentally ecological as all living systems depend upon and contribute to the greater biosphere.[5]

The AIF predicates a simple dynamic subserving predictive inference and error correction from which rich complexity unfolds: brains seek to infer the most probable explanations for the states of their internal and external environments, accomplished by something with the look and feel of Bayesian updating to optimize error minimizing generative models that guide action through *action-perception cycle* sequences over time (Friston, 2009, 2010, 2013). The Fig. 16.3 depiction of the AIF (Friston, 2009, 2010, 2013) in terms of action-perception cycles is shown in two ways, the first representing what is happening *in the moment*, and the second by adding temporal depth as expectations over time and *into the future* (i.e., planning; Friston et al., 2018; Parr et al., 2022). The idea is that *action* in the moment is the means by which actors "probe" and attempt to alter the statistical regularities of the generative processes of their embedding environments, while the resulting *perceptions* optimize generative models through the state dynamics of prediction error updating to optimize models (Parr et al., 2022).

Action, whether by directly altering the external environment to conform to one's model, or by repositioning the actor within it to shift their perspective to evaluate beliefs, facilitates learning (posterior update) or validates what has been learned via the perceptual consequences (affirming priors or "self-evidencing"; Hohwy, 2016). Inference is thus posed as a process of active engagement with body and world, a kind of dialogue that unfolds through cycles of action and perception. These dynamics work together, challenging models and states of the world in cycles of action and perception, of change without and change within. Brains seek to optimize the consequences of their bodily actions and thus perceptual experiences to avoid those that are harmful, life threatening, or otherwise undermining.

[5]For an application of the AIF to the biosphere see Rubin et al. (2020).

The idea is highly temporally localized as it is about making sense of the states of body and world *now*, and managing or resolving *next,* given the actor's history. This idea turns the notion of exact Bayesian inference into an optimization problem of a quantity called *free energy,* which is minimized via approximate (or variational) Bayesian inference (Friston, FitzGerald, et al., 2017; Parr et al., 2022; Ramstead et al., 2020; Smith et al., 2022). The simpler variational free energy optimization *now* is often expressesd as a minimization of two quantities. First, by minimizing the divergence between the actor's approximate and embodied *prior* and the true *posterior* under exact Bayesian inference, and then by minimizing the surprise (commonly termed *surprisal* or *evidence*) of the observation. The first term captures the notion of perceptual inference and learning, and the second the use of action to acquire unsurprising observations or those that provide positive prediction errors (i.e., when things go better than expected; Sterling, 2012), each guiding the other by testing, validating, and/or adapting.

A unique capacity of the human brain is it's temporal depth over a range of capacities so that minds can extend both into the past and future. This idea, captured in Fig. 16.3 as *expected free energy*, extends the temporal component towards expected future states given potential action sequences. The idea is that expected free energy is an *expectation* based on *simulating* potential observations projecting into the future under different "action policies" with the goal of regulating and minimizing free energy over arbitrary periods of time. What these two concepts, free energy and expected free energy, both ways of optimizing an approximation to the Bayesian inversion, bring forward is how brains manage their prediction errors and thus regulate their energetics. For expected free energy, this is accomplished by planning and adapting action sequences that bring new knowledge, referred to as *epistemic action*, and actions toward the acquisition of desired ends and preferences, referred to as *pragmatic action.*[6] Epistemic and pragmatic actions, therefore, work together to gather information about how to achieve desired goals and ends (epistemic action) and then to realize them (pragmatic action). Epistemic action reflects the information gain that comes by acting in ways expected to maximize the divergence between one's prior and the true posterior given one's preferences for certain outcomes, while pragmatic value reflects achieving preferred outcomes.

Of course, expected free energy is an *expectation* based upon hypothesized sequences of simulated outcomes based upon models entrained by prior experiences, bringing the embodiment of experiences interdependent with encompassing generative processes within individuals' embodied subjective generative models. Prediction comes with no guarantee that an action plan is good and likely to succeed or that a good plan that is likely to succeed will be successful. If anything, life has a way of teaching hard lessons about such (un)certainties. What happens in the flow of real life, from moment to moment, is the attempt to minimize free energy. Expected free energy brings forward our capacities for thinking, planning, and coordinating with others in our attempts to manage free energy, including the willingness to tolerate increased short-run free energy risk at times for a lower average free energy over the long term. Action plans that fail or cannot be realized are likely to increase free energy, bringing along expensive metabolic costs that burden bodies and wear them down while increasing negative affective and emotional experiences (Barrett, 2017a; Barrett et al., 2016; Bobba-Alves et al., 2022). The challenge of minimizing free energy in both the short and long term is the problem of "the human condition" and is perhaps the lynchpin of health, happiness, and survival as it speaks both to our most basic survival needs now and the

[6] Parr et al. (2022) discuss these and other potential expressions for free and expected free energy, each of which provides unique intuitions, along with some ways they subsume other models (e.g., expected utility theory).

more abstract needs for value and meaning that typically unfold over longer periods of time.

Discussion

In this introductory tale of two models of the brain, the second tale of prediction subsumes the original story of response and representation, turning a one-way stream of bottom-up informational flows into a form of top-down construction optimized by minimizing error flows. Whereas the traditional model is successful to the extent it can represent environmental conditions and guide appropriate responses to them, the predictive brain is successful to the extent it can sufficiently minimize the errors it makes when predicting the joint states of its body, the world, and the consequences of the actions it takes within and upon them both. In other words, where the traditional view is concerned with veridicality or truth (Hoffman, 2019; Prakash et al., 2021), prediction emphasizes error minimization and, thus, the *statistical plausibility* of an inference, which may not provide a true account (Parr et al., 2022). These views ultimately highlight the key challenges that come with the inherent uncertainties of existence and, thus, the need to make mutually interdependent inferences about bodies, worlds, and actions as both models present different paths to *persisting as an entity* in terms of true perceptions versus those that (merely) minimize surprise[7] (Hoffman, 2019; Prakash et al., 2021).

We summarize this predictive model in terms of an actor in their world per the AIF in Fig. 16.4, bringing together the depictions in Figs. 16.2 and 16.3. Actors in Fig. 16.4 are a confluence of factors across many different intersecting layers of organization, both within and outside bodies, which are continuously engaged in action-perception cycles mediated by adaptive error dynamics throughout the neurocomputational hierarchy. The error-minimizing brain is both (approximately) Bayesian and generative (Friston, 2009, 2010, 2013) in that past experiences supply potential explanations for sensory signals, providing an inverse mapping between current beliefs, current experiences, and subsequent expectations (Knill & Pouget, 2004; Rao & Ballard, 1999). Brains in this rendition test hypotheses or beliefs about events in bodies and the world through the inferential dynamics of action and perception, i.e., change interoceptive/exteroceptive signals via allostatic/motor commands and evaluate hypothesis/predictions/beliefs via error signals in light of new data. This view on inference as *active* poses a process theory emphasizing the centrality of ongoing, dynamic interactions between actors and the generative processes of their environments (Friston, FitzGerald, et al., 2017), unfolding continuously in a multimodal dance of action-perception consequences embodied in generative models (Parr et al., 2022).

Figure 16.4 describes an ongoing dialogue in which individuals both change and are changed by the world outside, each shaping the other and becoming integral parts of each other's embodied predictive "machinery." We learn to reason about the events in our lives, offering hypotheses and explanations at every turn while evincing numerous well-documented cognitive biases[8] (e.g., Robertson & Kesselheim, 2016; Thomas, 2018). Human predictive capacities leverage the past to simulate the future, enabling the development and selection of action policies *hoped* or perhaps *believed* to be most likely to achieve desired ends, adapting them to changing conditions, new information, and errors. Recognizing that life is a balancing act of exploration and pragmatic action for all actors seeking to manage the risks that

[7]Indeed, establishing causation (i.e., veridicality) is a problem of design from the start and is not always achievable, whereas estimating correlations (i.e., error minimization) is both generally possible and much simpler.

[8]Certainly, there is much to think about here with regard to false beliefs and the current popularity of conspiracy theories. There are no guarantees our explanations will coincide with true causes, just attempts, especially when false beliefs may not have direct survival consequences (e.g., believing one can fly). Human history is a history of false beliefs about the causes of experiences and events in the world. Moreover, it is often safer to be wary of potential threats in the world than otherwise, and there are countless examples of actual conspiracies (Shermer, 2022).

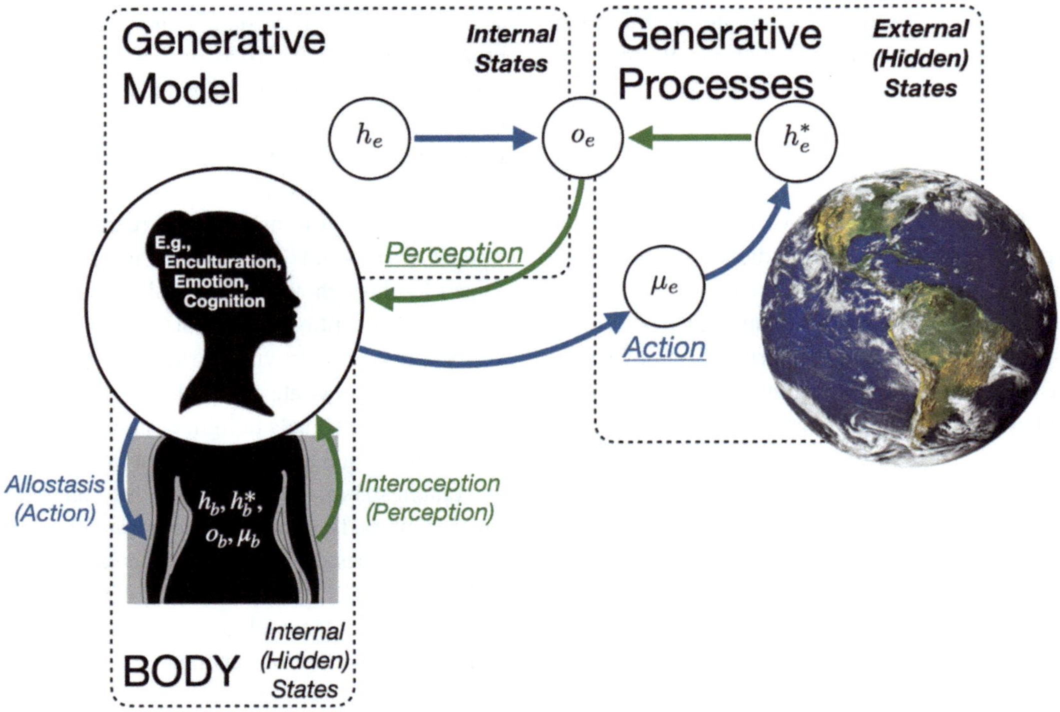

Fig. 16.4 Global description of the AIF model. A depiction of brain-mediated perception-action processes. Direct connections from the world to the brain/body (e.g., environmental toxins, viruses, etc.) are omitted. Adapted from Parr et al. (2022)

come with surprising states inimical to existence, the old fish out of water metaphor, balanced against the quest to find those experiences that resolve in the ways hoped for, hopefully even better than expected (Joffily & Coricelli, 2013; Sterling, 2012; Van de Cruys, 2017), active inference speaks to the challenge of being alive. We seek to do the things *we do* because *we are the things we are*, by which error minimization and thus confirmation of priors provide evidence for our own existence as just *that very thing* (Hohwy, 2016).

Accordingly, action consequences in Fig. 16.4 ground perceptual experiences just as perceptual consequences guide actions, all bounded and circumscribed within sequences of social interactions across different social settings through time. Building on Durkhemian and Goffmonian microsocial insights, Collins (2005) referred to these dynamics as *interaction ritual chain* trajectories in which the *situation* is both the sociological starting point and scene of action, from which individuals' motivations towards future action are encouraged or undermined (i.e., emotional energy; Collins, 1981). As elaborated in Veissière et al. (2020), settings dynamics channel and constrain attention and thus perceptual experience, including what and how to attend to the world and others with in it, from which we learn to explain both others' actions and through them the affordances accorded our own actions (Franks, 2013a). These themes speak directly to the centrality of microsocial interactional processes within which actors *take the role of the other* (role-taking; Mead, 1934) to look and see themselves through others' eyes so that they may adopt both perspectives different from their own and new self-perspectives (Franks, 2013b; Maryanski, 2013).

The role-taking theoretical touchpoint has been of considerable interest since neurosociology began gaining momentum and cohering into a relevant and important subdisciplinary area under the auspices of (e.g.) Franks (2010) and the first *Handbook of Neurosociology,* edited by Franks and Turner (2013); see also Turner, 2000; Massey, 2002). Because role-taking allows one to be both subject and object to oneself (Maryanski, 2013), to anticipate oneself as others (are expected to) do based on prior experience, it provides guides for both action and self-becoming in a truly social sense (Hopcroft, 2013). Role-taking thus plays a critical role in social interaction rituals by coordinating the reciprocal and reflexive dimensions of the mutual focus of attention within a situation (Collins, 2005), providing opportunities from which the self is internalized and in which the individual relies on others to complete their self-picture (Goffman, 1956). Because the consequences of action depend upon the positions and perspectives of other actors such that self-control echoes social control (Franks, 2013b), a core problem of humans' intense sociality is the problem of *inferring others' inferences on the self.*

The *neuro*-social dimension of role-taking is a simulation and, thus, an error-minimization problem, raising the question of how neural mechanisms support and optimize this capacity. Mirror neurons have been proposed to solve this problem of how we connect to others and put ourselves in their shoes (Lakoff, 2013). Key to the excitement around these neurons, both within the neurosciences and neurosociology (see also Lizardo, 2007; Lizardo et al., 2020; Summers-Effler et al., 2015), was that such neurons were active when monkeys both performed a goal-directed action and when they observed an experimenter or another monkey performing the same or similar action (di Pellegrino et al., 1992; Rizzolatti et al., 1996). As reviewed in Bonini et al. (2022), parallel discoveries in many different species, including humans, and various cell types distributed in motor, sensory, emotional, and other brain areas subserving social interaction quickly followed (see also Liew & Aziz-Zadeh, 2013). Consequently, mirror-neuron circuitry is the same for *perception* and *action* (Lakoff, 2013) and appears to be key to "coordinating what goes on beyond the boundary of the skin with what goes on within it" (Solymoski, 2013, p. 94). In short, mirror neurons were quickly realized to be at the essential core of human capacities for understanding and predicting the behaviors of others, but not in any sense independent of prior experience because such capacities arise from the same coordinated computations used to generate one's own behaviors (Molnar-Szakacs & Uddin, 2013).

Interestingly, based on both experimental findings and anatomofunctional architecture, much recent evidence supports the notion that mirror neurons are part of our ability to *predict and anticipate* others rather than merely to *represent* and *respond* to them (reviewed in Bonini et al., 2022). For example, predicting others' actions so that one may update their own motor plans (including speech) during interaction is more efficient than merely reacting after the fact. It also implies that the entrainments of successful interactions reflect converging predictions (i.e., generative models) as communication induces perceptual learning so that others literally change our minds (i.e., embodiment), and *vice versa* (Friston & Frith, 2015b). The neurosociological implications of mirror neurons thus quite directly apply to important interactional social processes, translating the intentions of others via our perceptions of them, providing a basis for *theory of mind* (for ToM from sociological perspectives, see Hopcroft, 2013; Turner, 2013), and role-taking, where "actors experience not only what they think the other is thinking, but experience their own selves from their construction of the other's different cognitive perspective and emotional predispositions" (Franks, 2013b, p. 28).

However, the traditional "bottom-up" view[9] that mirror neurons are *responses to*, or *representations of,* sensory information has an "inversion problem" because the generative model of observations cannot be inverted (i.e., an inference about hypotheses given data) unless each sensory input is associated uniquely with one cause (Kilner et al., 2007a, 2007b). Friston and Frith (2015a) argue that the infinite regress induced by

[9]This is more commonly referred to as "feed-forward" (as compared to "feed-back") in the neuroscience literature, but we find "bottom-up" to be more descriptive in the present context.

modeling one who is modeling you modeling them (i.e., role-taking the role-taker), is resolved when both possess the same model as evidenced in the correspondences between the predictions about sensations received and emitted among interlocutors. Critically, in the hierarchical predictive view, the model is inverted as a combination of prediction and error streams *throughout* the cortical hierarchy, with the most abstracted (and implicitly sociological) and compressed areas at the top providing contextualizing predictions within which others' action-perceptions are situated (Kilner et al., 2007b). By way of example to illustrate the importance of hierarchy in all of this, Kilner et al. (2007a) address the Dr. Jekyll and Mr. Hyde thought-experiment of Jacob and Jeannerod (2005). In this experiment, one observes a hand taking hold of a scalpel to be applied to a human body. How does one tell if Dr. Jekyll is attempting to heal or Mr. Hyde is attempting to harm?

Jacob and Jeannerod (2005) argue that it is impossible to distinguish the intent and, thus, the actor from observational motor signals alone. In other words, there is too little information in mirror neuron activity alone to invert the generative model and issue predictions that statistically discriminate intent and, therefore, plausibly recognize who wields the scalpel. Yet, when viewed as part of a larger predictive hierarchy, Kilner et al. (2007a) show that the predictive coding model expands the purview to the larger sensory context in which the observed actions take place. Such high-level sensory empirical priors provide hypotheses contextualizing the observed actions, conditionalizing beliefs about them, and reaffirming the microsociological insight that the situation is the scene of action. This view echoes the larger sociological insight that individual action can only rarely be truly individuated from social context. Thus, if we imagine our observation to be in a dimly lit operating theater with rows of medical students huddled in tiered seats encircling the surgical stage where the scalpel is wielded with clinical precision, the inference weighs quickly and powerfully toward Dr. Jekyll's attempt to heal. So much the difference if the observer catches the flash of light on a scalpel in a dim London backstreet.

This model is consistent with the idea that we understand other people by invoking those experiences that resonate with what we perceive of theirs.[10] Moreover, the mirror neuron system is active along with "mentalizing" regions supporting complex social cognition (Liew & Aziz-Zadeh, 2013), which is always multimodal and hierarchical over levels spanning from the highly specific to general and abstract (Theriault et al., 2021). The predictive Bayesian view thus provides a recontextualization of well-recognized mirror-neuron mechanisms that supply a rich source of neurosociological insight and furnish a foundation for important human capacities, such as role-taking. This capacity is a key facet of understanding the setting that resides within others, whose qualia we infer as we imagine, simulate, and predict them in their world, in which we too are a part.

Actors also adapt their self-presentations by making inferences about what others will perceive, whereby they negotiate roles for themselves in different social situations and contexts (i.e., *role-making*; Turner, 1962 cited and discussed in Turner, 2013). The difference between *role-taking* and *role-making* can therefore be seen as the distinction between *inferring how we are seen* versus *presenting as we intend to be seen*. The key difference is thus whether one attends to the sensory consequences of their own or others' actions (Kilner et al., 2007a, 2007b). The distinction, in other words, is the difference between optimizing self-actions versus inferring goals, intentions, and perceptions from other-actions. Such dynamics reflect what one is attending to in ongoing and mutually unfolding action-perception cycles entrained in the flow of time. In the case of role-making, the self-goal is known and one acts to achieve it (i.e., intended consequences). In the case of role-taking, evalua-

[10] This appears to provide a foundation for understanding the *assumed similarity bias* in the correspondence between how people describe both themselves and others (Thielmann et al., 2020).

tions of others' self-goals and inferences *vis a vis* the observer-self are optimized by minimizing perceptual prediction errors involving others. Role-taking and role-making, therefore, represent reciprocal aspects of the same action-perception cycle, each informing and shaping the other by integrating perceptual evidence from prior experiences with "you or others like you" and situational contexts "more or less like this one."

Interestingly, Theriault et al. (2021) argue that there is an important relationship between what we have written about free energy here and the challenges of mutual prediction during social interaction. That is, free energy increases (i.e., metabolic costs) when one fails to meet a conspecific's expectations, a concept they refer to as a "sense of should" to capture the idea that behavioral and other expectational conformities ease the friction of social encounters when we do what others think we ought.[11] When one's behavior does not match others' predictions, others' subsequent behaviors are less likely to match one's predictions, producing error cascades with the consequence that mutual models may fail to converge. Consequently, a shared reality may not emerge, roles may not be negotiated successfully, and the situation may be "breached" (Garfinkel, 1967). When things go wrong and errors abound, the benign result might be an unsatisfying and awkward failed social ritual, or it could be far more dramatic and even deadly if things go terribly wrong and turn violent.[12]

In contrast, positive social outcomes reflect positive prediction errors and error reduction rates, as with humor (Joffily & Coricelli, 2013; Van de Cruys, 2017), pinging our reward systems (Sterling, 2012, 2020) and increasing positive emotional energy. Role-taking and role-making thus collaborate to optimize social encounters via efficient predictive action, minimizing negative prediction errors and free energy when things go well, and producing positive prediction errors when things go better than expected. Expected free energy is also reflected in Mead's theory of the act, in which, as Franks (2013b, p. 28) notes (emphasis original), "we are *pulled* along in a teleological fashion by our anticipations of our own consummation of the act." Indeed, free energy resonates with Mead's four stages of the act (Franks, 2010) in which action begins as an impulse to behave in some way, perception is guided to facilitate our interests and intentions to act, something is done with objects or people (e.g., role-taking/making), and "where the last stage is present in the first stage" (Shook, 2013, p. 41) the act is consummated. In other words, the fulfillment of generative model predictions (i.e., error minimization) is coordinated in-the-moment to minimize free energy, but is pulled into the future by our best guesses about which action policy sequences and at what costs are most likely to fulfill our epistemic and pragmatic goals. Action-perception cycles (attempt to) minimize expected free energy for acts in the *now*, while successful action policies (attempt to) minimize the time average of free energy of *the act* for temporally deeper sequences entailing both epistemic and pragmatic engagements.

Not surprisingly, sociologists have also noted that Mead's four stages of the act are consistent with mirror neuron research generally (Franks, 2010, 2013a, 2019). These theoretical concerns represent important dynamics in microsocial situations in Fig. 16.4. The breadth of scope, however, intends to supply the potential for reciprocal neuro-sociological exchanges beyond these local matters, situating them within the broader sociological contexts and structures unique to specific times and places. This idea mirrors the need to understand the situation, which is critical for mir-

[11]This, as the authors note, assumes that both participants are motivated by others' expectations and that both feel obligated to perform (or not) some behaviors, so can be adapted to alternative situations by changing these constraints.

[12]We are intrigued by Barrett's (2017a) discussion of "emotional granularity." It is fascinating to think that the rich conceptual elaboration of *feelings* enables high degrees of emotional regulation. In contrast, a characteristic of "toxic masculinity" may be limited emotional granularity that could be consequential to physical expressions culminating in violence, a physical rather than conceptual expression of frustration, anger, distress, etc. See also Fridman et al. (2019) for an application to police violence.

ror neurons to produce inferences explaining others' action motivations and perspectives, and thus for successful role-taking/making. For example, we recently presented this model (Cheadle et al., 2024) and applied it to, or at least "seeded," theoretical concerns in sociological stress process research, the sociology of emotions, and cognitive cultural sociology. Each of these areas has had some explicit and plenty of implicit ideas about brains that this model updates in some way. We argued there that the *stress response* and *anticipatory stress* have been undertheorized in terms of neuro-bio-social dynamics, despite their importance for understanding the social contours of health. The model presented here helps to reorient *the physiological stress response* towards allostasis, and *anticipation* towards socially structured and experientially entrained predictions.

Additionally, definitions of emotions in the sociology of emotions are heterogeneous and diverse (for reviews, see Bericat, 2016; Stets, 2010, 2012; Thoits, 1989; Turner, 2009; Turner & Stets, 2006), even as affect is under-theorized (von Scheve, 2018), which may, in some cases, be the more appropriate concept. The model here both relates these factors to stress processes through allostasis and clarifies them by defining *affect* (i.e., the valence and arousal of feelings; (Russell, 1980)) as a high-level construction that indexes interoceptive sensations of allostatic states and changes (i.e., coordinated adjustment of homeostatic parameters; Schulkin & Sterling, 2019; Arnaldo et al., 2022). Affect is then proposed to provide the foundation for emotions (Barrett, 2017a, 2017b), which (largely) emerge as enculturated concepts used to construct a joint explanation of body states (i.e., allostasis and interceptive signals) and the broader situational context (i.e., exteroception). Relatedly, cognitive cultural sociologists have also adopted "dual process" models of cognition (see Lizardo et al., 2016, 2020) that are both open to considerable critique (Ignatow, 2021) and are reframed in terms of priors and precision in the AIF (Friston et al., 2016), turning metaphors about neurocognitive processes into neural activities differentiated by varying precisions and degrees of effort (i.e., a schema is not computed *fast* because of some special "Type 1" neural system [e.g., Boutyline & Soter, 2021], but because it is a highly precisely held belief; Parr et al., 2023).

Expanding the view further, *generative processes* in Fig. 16.4 are just a recasting of social structure, culture, and broader social conditions that intimate the objective probabilities of events, states, and outcomes out there in the world. *Generative processes* supply situations and, along with others, their uncertainties, which experiences entrain in *generative models* embodied in neural structure, providing individuals subjective and correlational approximations to whatever realities they happen to bump up against. Fascinatingly, Strand and Lizardo (2022a, 2022b) have written insightfully about probabilistic interpretations of Weber (Weber, 1913, 2019) and Bourdieu (e.g., Bourdieu, 1973). They argue that the *field* came in Bourdieu's later works to capture distributions of Weber's objective *Chance* (i.e., generative processes), while individuals' subjective Weberian approximations were captured in the *habitus* (i.e., individuals' statistically entrained generative models). In light of these insights, they called for a probabilistic sociology, though they did not mention Bayesian interpretations of objective and subjective probabilities, including the way objective-subjective probabilities can be combined to make sense of the reverse inference problem whereby causes are inferred from observations (i.e., the posterior; see Clayton, 2021; Gill, 2002). The view we present here satisfies their notion of a probabilistic sociology, which is not so different than a probabilistic biology (Ramstead et al., 2018, 2021, 2022), and is shown here to be an inherently neurosociological problem as it reflects the fundamental challenge of perception and inference under uncertainty that all living systems are subject to (Parr et al., 2022).

There is, admittedly, more we would like to elaborate upon here, but we can cover only so much ground. As we said at the outset, our goal has been to "seed" potential sociological concerns. By way of example, we showed how mirror neurons and Meadian insights from prior neurosociological work come together more cohesively in this framework than is supplied by traditional conceptualizations of the brain. We have argued similarly about related topics elsewhere (Cheadle et al., 2020, 2024; Goosby et al., 2018; Goosby & Cheadle, 2024), intimating, we believe, the expanding potential for reciprocal neuro-social exchanges. Our intent is not strictly to reinterpret sociological phenomena in terms of neuro-social interactions prioritizing neuroscientific theories. Rather, it has been the homologous nature of them, the *inference of perceived* mutual reinforcement, that has inspired us. There is a great deal of bio-social reciprocity in culturally encoded brains that captures, we believe, an important priority in interdisciplinary work. In fact, reciprocity, the thing that makes a friend a *good friend*, is what transforms interdisciplinarity into transdisciplinarity. Nature acknowledges our disciplinary boundaries only because we construct them thusly, and for our own convenience, at that.

Conclusion

This chapter introduced foundational concepts such as predictive processing, the Bayesian brain hypothesis, and variational Bayesian inference from the perspective of the AIF. This perspective, while still developing, has advanced enough to invite a reconsideration of the brain as a *prospective constructor*. Predictive coding models will continue to evolve, elaborate, and gain complexity, yet the foundations for deeper neurosociological integration are both in place and are likely to remain similar, if not unchanged, should the overarching normative model stay intact. This level of abstraction motivates a neurobiologically adaptive theoretical social actor, but does not necessarily oblige the methods, designs, and technologies of neuroscience practice and brain measurement *per se*.[13] Any such theoretical *neurosociological actor* should conform to sociological insights into how the social structures in which individuals are situated constrain and channel experience and, thus, the rich array of outcomes and nuances evinced within and between societies and cultures. Just as genes provide the basal probabilities of both an individual's capacities and their distributions over populations, the worlds around us shape the statistical regularities of our situated experiences and, thus, the expectations and explanations that different groups and individuals hold as they go about the tasks of living. The value of such a neurosociological model is that it is multidimensionally integrative, both bringing together insights across disciplines and across areas of sociological inquiry, at least in those situations in which minds (e.g., emotion, cognition) and bodies (e.g., health physiology, allostatic load) are at play.

Funding This research was supported by grant, P30AG066614, awarded to the Center on Aging and Population Sciences at The University of Texas at Austin by the National Institute on Aging, and by grant, P2CHD042849, awarded to the Population Research Center at The University of Texas at Austin by the Eunice Kennedy Shriver National Institute of Child Health and Human Development. The content of this article is solely the responsibility of the authors and does not necessarily represent the official views of the National Institutes of Health or the University of Texas at Austin.

References

Arnaldo, I., Corcoran, A. W., Friston, K. J., & Ramstead, M. J. D. (2022). Stress and its sequelae: An active inference account of the etiological pathway from allostatic overload to depression. *Neuroscience &*

[13]Sociologists are now using the tools of neuroscience directly (e.g., Kalkhoff et al., 2020; Kiat et al., 2016, 2017; Kiat & Cheadle, 2017, 2018; Kiat, Cheadle, & Goosby, 2018a, 2018b; Kiat, Dodd, et al., 2018) and have begun using neuro-modulated physiological correlates of social behaviors, experiences, and environments outside of the lab and in real life using wearable sensors (Cheadle et al., 2020; Jelsma et al., 2024; Jelsma et al., 2021; Zhang et al., 2021). Looking towards additional potential sociological applications, Summers-Effler et al. (2015) discuss incorporating mirror neurons into ethnographic research, while the AIF is highly mathematical and is built upon a strong computational foundation (Parr et al., 2022; Smith et al., 2022), which holds promise for applications in computational sociology (e.g., Foster, 2018). Of course, there are already good examples of theoretical applications that involve brains (e.g., Niemeyer, 2013) and sociologists may look towards measuring different aspects of expectancies, explanations, and even belief concordances in future survey research.

Biobehavioral Reviews, 135, 104590. https://doi.org/10.1016/j.neubiorev.2022.104590

Badcock, P. B., Friston, K. J., & Ramstead, M. J. D. (2019). The hierarchically mechanistic mind: A free-energy formulation of the human psyche. *Physics of Life Reviews, 31*, 104–121. https://doi.org/10.1016/j.plrev.2018.10.002

Barbas, H. (2015). General cortical and special prefrontal connections: Principles from structure to function. *Annual Review of Neuroscience, 38*(1), 269–289. https://doi.org/10.1146/annurev-neuro-071714-033936

Barrett, L. F. (2017a). *How emotions are made: The secret life of the brain*. Houghton Mifflin Harcourt.

Barrett, L. F. (2017b). The theory of constructed emotion: An active inference account of interoception and categorization. *Social Cognitive and Affective Neuroscience, 12*(1), 1–23.

Barrett, L. F. (2020). *Seven and a half lessons about the brain*. Houghton Mifflin Harcourt.

Barrett, L. F., Quigley, K. S., & Hamilton, P. (2016). An active inference theory of allostasis and interoception in depression. *Philosophical Transactions of the Royal Society B: Biological Sciences, 371*(1708), 20160011. https://doi.org/10.1098/rstb.2016.0011

Barrett, L. F., & Simmons, W. K. (2015). Interoceptive predictions in the brain. *Nature Reviews Neuroscience, 16*(7), 419–429. https://doi.org/10.1038/nrn3950

Barron, H. C., Auksztulewicz, R., & Friston, K. (2020). Prediction and memory: A predictive coding account. *Progress in Neurobiology, 192*, 101821. https://doi.org/10.1016/j.pneurobio.2020.101821

Bericat, E. (2016). The sociology of emotions: Four decades of progress. *Current Sociology, 64*(3), 491–513. https://doi.org/10.1177/0011392115588355

Billman, G. E. (2020). Homeostasis: The underappreciated and far too often ignored central organizing principle of physiology. *Frontiers in Physiology, 11*, 200. https://www.frontiersin.org/articles/10.3389/fphys.2020.00200

Bobba-Alves, N., Juster, R.-P., & Picard, M. (2022). The energetic cost of allostasis and allostatic load. *Psychoneuroendocrinology, 146*, 105951. https://doi.org/10.1016/j.psyneuen.2022.105951

Bonilla-Silva, E. (1997). Rethinking racism: Toward a structural interpretation. *American Sociological Review, 62*(3), 465–480. https://doi.org/10.2307/2657316

Bonini, L., Rotunno, C., Arcuri, E., & Gallese, V. (2022). Mirror neurons 30 years later: Implications and applications. *Trends in Cognitive Sciences, 26*(9), 767–781. https://doi.org/10.1016/j.tics.2022.06.003

Bourdieu, P. (1973). The three forms of theoretical knowledge. *Social Science Information, 12*(1), 53–80. https://doi.org/10.1177/053901847301200103

Boutyline, A., & Soter, L. K. (2021). Cultural schemas: What they are, how to find them, and what to do once you've caught one. *American Sociological Review, 86*(4), 728–758. https://doi.org/10.1177/00031224211024525

Bubic, A., Von Cramon, D. Y., & Schubotz, R. I. (2010). Prediction, cognition and the brain. *Frontiers in Human Neuroscience, 4*, 1094. https://doi.org/10.3389/fnhum.2010.00025

Caucheteux, C., Gramfort, A., & King, J.-R. (2023). Evidence of a predictive coding hierarchy in the human brain listening to speech. *Nature Human Behaviour, 7*(3), 430–441. https://doi.org/10.1038/s41562-022-01516-2

Chanes, L., & Barrett, L. F. (2016). Redefining the role of limbic areas in cortical processing. *Trends in Cognitive Sciences, 20*(2), 96–106. https://doi.org/10.1016/j.tics.2015.11.005

Cheadle, J. E., Davidson-Turner, K. J., & Goosby, B. J. (2024). Active inference and social actors: Towards a neuro-bio-social theory of brains and bodies in their worlds. *KZfSS Kölner Zeitschrift Für Soziologie Und Sozialpsychologie, 76*, 317. https://doi.org/10.1007/s11577-024-00936-4

Cheadle, J. E., Goosby, B. J., Jochman, J. C., Tomaso, C. C., Yancey, C. B. K., & Nelson, T. D. (2020). Race and ethnic variation in college students' allostatic regulation of racism-related stress. *Proceedings of the National Academy of Sciences, 117*, 31053. https://doi.org/10.1073/pnas.1922025117

Chen, W. G., Schloesser, D., Arensdorf, A. M., Simmons, J. M., Cui, C., Valentino, R., Gnadt, J. W., Nielsen, L., Hillaire-Clarke, C. S., Spruance, V., Horowitz, T. S., Vallejo, Y. F., & Langevin, H. M. (2021). The emerging science of interoception: Sensing, integrating, interpreting, and regulating signals within the self. *Trends in Neurosciences, 44*(1), 3–16. https://doi.org/10.1016/j.tins.2020.10.007

Clark, A. (2013). Whatever next? Predictive brains, situated agents, and the future of cognitive science. *Behavioral and Brain Sciences, 36*(3), 181–204. https://doi.org/10.1017/S0140525X12000477

Clark, A. (2015). *Surfing uncertainty: Prediction, action, and the embodied mind (Illustrated edition)*. Oxford University Press.

Clark, A. (2023). *The experience machine: How our minds predict and shape reality*. Pantheon.

Clayton, A. (2021). *Bernoulli's fallacy: Statistical illogic and the crisis of modern science*. Columbia University Press.

Collins, R. (1981). On the microfoundations of macrosociology. *American Journal of Sociology, 86*(5), 984–1014.

Collins, R. (2005). *Interaction ritual chains*. Princeton University Press.

Davis, J. (2013). Persistent inequality: A neurosociological perspective. In D. D. Franks & J. H. Turner (Eds.), *Handbook of neurosociology* (pp. 333–348). Springer. https://doi.org/10.1007/978-94-007-4473-8_21

di Pellegrino, G., Fadiga, L., Fogassi, L., Gallese, V., & Rizzolatti, G. (1992). Understanding motor events: A neurophysiological study. *Experimental Brain Research, 91*(1), 176–180. https://doi.org/10.1007/BF00230027

Ficco, L., Mancuso, L., Manuello, J., Teneggi, A., Liloia, D., Duca, S., Costa, T., Kovacs, G. Z., & Cauda, F.

(2021). Disentangling predictive processing in the brain: A meta-analytic study in favour of a predictive network. *Scientific Reports, 11*(1), 16258. https://doi.org/10.1038/s41598-021-95603-5

Foster, J. G. (2018). Culture and computation: Steps to a probably approximately correct theory of culture. *Poetics, 68*, 144–154. https://doi.org/10.1016/j.poetic.2018.04.007

Franks, D. D. (2010). *Neurosociology: The nexus between neuroscience and social psychology*. (1st ed. 2010 edition). Springer.

Franks, D. D. (2013a). Relationships between neurosociology, foundational social behaviorism, and currents in symbolic interaction. In D. D. Franks & J. H. Turner (Eds.), *Handbook of neurosociology* (pp. 139–148). Springer. https://doi.org/10.1007/978-94-007-4473-8_11

Franks, D. D. (2013b). Why we need neurosociology as well as social neuroscience: Or—Why role-taking and theory of mind are different concepts. In D. D. Franks & J. H. Turner (Eds.), *Handbook of neurosociology* (pp. 27–32). Springer. https://doi.org/10.1007/978-94-007-4473-8_3

Franks, D. D. (2019). *Neurosociology: Fundamentals and current findings*. Springer. https://doi.org/10.1007/978-94-024-1600-8

Franks, D. D., & Turner, J. H. (Eds.). (2013). *Handbook of neurosociology*. Springer. https://doi.org/10.1007/978-94-007-4473-8

Fridman, J., Barrett, L. F., Wormwood, J. B., & Quigley, K. S. (2019). Applying the theory of constructed emotion to police decision making. *Frontiers in Psychology, 10*, 1946. https://doi.org/10.3389/fpsyg.2019.01946

Friston, K. (2009). The free-energy principle: A rough guide to the brain? *Trends in Cognitive Sciences, 13*(7), 293–301. https://doi.org/10.1016/j.tics.2009.04.005

Friston, K. (2010). The free-energy principle: A unified brain theory? *Nature Reviews Neuroscience, 11*(2), 2. https://doi.org/10.1038/nrn2787

Friston, K. (2013). Active inference and free energy. *Behavioral and Brain Sciences, 36*(3), 212–213. https://dx.doi.org.ezproxy.lib.utexas.edu/10.1017/S0140525X12002142

Friston, K., FitzGerald, T., Rigoli, F., Schwartenbeck, P., O'Doherty, J., & Pezzulo, G. (2016). Active inference and learning. *Neuroscience & Biobehavioral Reviews, 68*, 862–879. https://doi.org/10.1016/j.neubiorev.2016.06.022

Friston, K., FitzGerald, T., Rigoli, F., Schwartenbeck, P., & Pezzulo, G. (2017). Active inference: A process theory. *Neural Computation, 29*(1), 1–49. https://doi.org/10.1162/NECO_a_00912

Friston, K., & Frith, C. (2015a). A duet for one. *Consciousness and Cognition, 36*, 390–405. https://doi.org/10.1016/j.concog.2014.12.003

Friston, K. J., & Frith, C. D. (2015b). Active inference, communication and hermeneutics. *Cortex, 68*, 129–143. https://doi.org/10.1016/j.cortex.2015.03.025

Friston, K., & Kiebel, S. (2009). Predictive coding under the free-energy principle. *Philosophical Transactions of the Royal Society B: Biological Sciences, 364*(1521), 1211–1221. https://doi.org/10.1098/rstb.2008.0300

Friston, K., Lin, M., Frith, C. D., Pezzulo, G., Hobson, J. A., & Ondobaka, S. (2017). Active inference, curiosity and insight. *Neural Computation, 29*(10), 2633–2683. https://doi.org/10.1162/NECO_a_00999

Friston, K., Mattout, J., & Kilner, J. (2011). Action understanding and active inference. *Biological Cybernetics, 104*(1), 137–160. https://doi.org/10.1007/s00422-011-0424-z

Friston, K. J., Rosch, R., Parr, T., Price, C., & Bowman, H. (2018). Deep temporal models and active inference. *Neuroscience & Biobehavioral Reviews, 90*, 486–501. https://doi.org/10.1016/j.neubiorev.2018.04.004

Garfinkel, H. (1967). *Studies in ethnomethodology*. Prentice-Hall. http://swbplus.bsz-bw.de/bsz00202960Xinh.htm

Gill, J. (2002). *Bayesian methods: A social and behavioral sciences approach/Jeff Gill*. Chapman & Hall/CRC.

Glass, T. A., & McAtee, M. J. (2006). Behavioral science at the crossroads in public health: Extending horizons, envisioning the future. *Social Science & Medicine, 62*(7), 1650–1671. https://doi.org/10.1016/j.socscimed.2005.08.044

Goffman, E. (1956). *The presentation of self in everyday life*. University of Edinburgh, Social Sciences Research Centre.

Goosby, B. J., & Cheadle, J. E. (2024). The immune system is a complex system: Inflammatory morbidity and systemic racism. *KZfSS Kölner Zeitschrift Für Soziologie Und Sozialpsychologie, 76*, 713. https://doi.org/10.1007/s11577-024-00952-4

Goosby, B. J., Cheadle, J. E., & Mitchell, C. (2018). Stress-related biosocial mechanisms of discrimination and African American health inequities. *Annual Review of Sociology, 44*(1), 319–340.

Grusky, D. B. (2014). *Social stratification: Class, race, and gender in sociological perspective* (4th ed.). Routledge.

Grusky, D. B., & Hill, J. (Eds.). (2017). *Inequality in the 21st century: A reader* (1st ed.). Routledge.

Harshaw, C. (2015). Interoceptive dysfunction: Toward an integrated framework for understanding somatic and affective disturbance in depression. *Psychological Bulletin, 141*(2), 311–363. https://doi.org/10.1037/a0038101

Hawkins, J., & Dawkins, R. (2021). *A thousand brains: A new theory of intelligence* (1st ed.). Basic Books.

Hesp, C., Smith, R., Parr, T., Allen, M., Friston, K. J., & Ramstead, M. J. D. (2021). Deeply felt affect: The emergence of valence in deep active inference. *Neural Computation, 33*(2), 398–446. https://doi.org/10.1162/neco_a_01341

Hesp, C., Tschantz, A., Millidge, B., Ramstead, M., Friston, K., & Smith, R. (2020). Sophisticated affective inference: Simulating anticipatory affective dynamics of imagining future events. In *Active inference* (pp. 179–186). Springer. https://doi.org/10.1007/978-3-030-64919-7_18

Hoffman, D. D. (2019). *The case against reality: Why evolution hid the truth from our eyes / Donald Hoffman* (1st ed.). W.W. Norton & Company.

Hohwy, J. (2016). The self-evidencing brain. *Noûs, 50*(2), 259–285. https://doi.org/10.1111/nous.12062

Hommel, B. (2020). Pseudo-mechanistic explanations in psychology and cognitive neuroscience. *Topics in Cognitive Science, 12*(4), 1294–1305. https://doi.org/10.1111/tops.12448

Hopcroft, R. L. (2013). Neurosociology and theory of mind (ToM). In D. D. Franks & J. H. Turner (Eds.), *Handbook of neurosociology* (pp. 231–241). Springer. https://doi.org/10.1007/978-94-007-4473-8_16

Hutchinson, J. B., & Barrett, L. F. (2019). The power of predictions: An emerging paradigm for psychological research. *Current Directions in Psychological Science, 28*(3), 280–291. https://doi.org/10.1177/0963721419831992

Ignatow, G. (2021). Cognitive sociology after relational biology. *Sociological Forum, 36*(S1), 1253–1270. https://doi.org/10.1111/socf.12764

Jacob, P., & Jeannerod, M. (2005). The motor theory of social cognition: A critique. *Trends in Cognitive Sciences, 9*(1), 21–25. https://doi.org/10.1016/j.tics.2004.11.003

Jelsma, E. B., Goosby, B. J., & Cheadle, J. E. (2021). Do trait psychological characteristics moderate sympathetic arousal to racial discrimination exposure in a natural setting? *Psychophysiology, 58*(4), e13763. https://doi.org/10.1111/psyp.13763

Jelsma, E., Zhang, A., Goosby, B. J., & Cheadle, J. E. (2024). Sympathetic arousal among depressed college students: Examining the interplay between psychopathology and social activity. *Psychophysiology, 61*(9), e14597. https://doi.org/10.1111/psyp.14597

Joffily, M., & Coricelli, G. (2013). Emotional valence and the free-energy principle. *PLoS Computational Biology, 9*(6), e1003094. https://doi.org/10.1371/journal.pcbi.1003094

Kalkhoff, W., Melamed, D., Pollock, J., Miller, B., Overton, J., & Pfeiffer, M. (2020). Cracking the black box: Capturing the role of expectation states in status processes. *Social Psychology Quarterly, 83*(1), 26–48. https://doi.org/10.1177/0190272519868988

Kalkhoff, W., Thye, S. R., & Pollock, J. (2016). Developments in neurosociology. *Sociology Compass, 10*(3), 242–258. https://doi.org/10.1111/soc4.12355

Kastel, N., Hesp, C., Ridderinkhof, K. R., & Friston, K. J. (2023). Small steps for mankind: Modeling the emergence of cumulative culture from joint active inference communication. *Frontiers in Neurorobotics, 16*, 944986. https://www.frontiersin.org/articles/10.3389/fnbot.2022.944986

Keller, G. B., & Mrsic-Flogel, T. D. (2018). Predictive processing: A canonical cortical computation. *Neuron, 100*(2), 424–435. https://doi.org/10.1016/j.neuron.2018.10.003

Kiat, J. E., & Cheadle, J. E. (2017). The impact of individuation on the bases of human empathic responding. *NeuroImage, 155*, 312–321. https://doi.org/10.1016/j.neuroimage.2017.05.006

Kiat, J. E., & Cheadle, J. E. (2018). Tick–tock goes the croc: A high-density EEG study of risk-reactivity and binge-drinking. *Social Cognitive and Affective Neuroscience, 13*(6), 656–663.

Kiat, J. E., Cheadle, J. E., & Goosby, B. J. (2018a). The impact of social exclusion on anticipatory attentional processing. *International Journal of Psychophysiology, 123*, 48–57. https://doi.org/10.1016/j.ijpsycho.2017.11.010

Kiat, J. E., Cheadle, J. E., & Goosby, B. J. (2018b). Wait for it: The impact of social exclusion on anticipatory attentional processing. *International Journal of Psychophysiology, 123*, 48–57.

Kiat, J. E., Dodd, M. D., Belli, R. F., & Cheadle, J. E. (2018). The signature of undetected change: An exploratory electrotomographic investigation of gradual change blindness. *Journal of Neurophysiology, 119*(5), 1629–1635.

Kiat, J. E., Straley, E., & Cheadle, J. E. (2016). Escalating risk and the moderating effect of resistance to peer influence on the P200 and feedback-related negativity. *Social Cognitive & Affective Neuroscience, 11*(3), 377–386. https://doi.org/10.1093/scan/nsv121

Kiat, J. E., Straley, E., & Cheadle, J. E. (2017). Why won't they sit with me? An exploratory investigation of stereotyped cues, social exclusion, and the P3b. *Social Neuroscience, 12*(5), 612–625. https://doi.org/10.1080/17470919.2016.1223167

Kilner, J. M., Friston, K. J., & Frith, C. D. (2007a). Predictive coding: An account of the mirror neuron system. *Cognitive Processing, 8*(3), 159–166. https://doi.org/10.1007/s10339-007-0170-2

Kilner, J. M., Friston, K. J., & Frith, C. D. (2007b). The mirror-neuron system: A Bayesian perspective. *Neuroreport, 18*(6), 619–623. https://doi.org/10.1097/WNR.0b013e3281139ed0

Kleckner, I. R., Zhang, J., Touroutoglou, A., Chanes, L., Xia, C., Simmons, W. K., Quigley, K. S., Dickerson, B. C., & Barrett, L. F. (2017). Evidence for a large-scale brain system supporting allostasis and interoception in humans. *Nature Human Behaviour, 1*, 69.

Knill, D. C., & Pouget, A. (2004). The Bayesian brain: The role of uncertainty in neural coding and computation. *Trends in Neurosciences, 27*(12), 712–719. https://doi.org/10.1016/j.tins.2004.10.007

Kotchoubey, B., Tretter, F., Braun, H. A., Buchheim, T., Draguhn, A., Fuchs, T., Hasler, F., Hastedt, H., Hinterberger, T., Northoff, G., Rentschler, I., Schleim, S., Sellmaier, S., Tebartz Van Elst, L., & Tschacher, W. (2016). Methodological problems on the way to integrative human neuroscience. *Frontiers in Integrative Neuroscience, 10*, 41. https://doi.org/10.3389/fnint.2016.00041

Lakoff, G. (2013). Neural social science. In D. D. Franks & J. H. Turner (Eds.), *Handbook of neu-*

rosociology (pp. 9–25). Springer. https://doi.org/10.1007/978-94-007-4473-8_2

Liew, S.-L., & Aziz-Zadeh, L. (2013). The human mirror neuron system, social control, and language. In D. D. Franks & J. H. Turner (Eds.), *Handbook of neurosociology* (pp. 183–205). Springer. https://doi.org/10.1007/978-94-007-4473-8_14

Link, B. G., & Phelan, J. (1995). Social conditions as fundamental causes of disease. *Journal of Health and Social Behavior*, 80–94. http://dx.doi.org.ezproxy.lib.utexas.edu/10.2307/2626958

Lizardo, O. (2007). "Mirror neurons," collective objects and the problem of transmission: Reconsidering Stephen Turner's critique of practice theory. *Journal for the Theory of Social Behaviour, 37*(3), 319–350. https://doi.org/10.1111/j.1468-5914.2007.00340.x

Lizardo, O., Mowry, R., Sepulvado, B., Stoltz, D. S., Taylor, M. A., Van Ness, J., & Wood, M. (2016). What are dual process models? Implications for cultural analysis in sociology. *Sociological Theory, 34*(4), 287–310. https://doi.org/10.1177/0735275116675900

Lizardo, O., Sepulvado, B., Stoltz, D. S., & Taylor, M. A. (2020). What can cognitive neuroscience do for cultural sociology? *American Journal of Cultural Sociology, 8*(1), 3–28. https://doi.org/10.1057/s41290-019-00077-8

Maryanski, A. (2013). The secret of the hominin mind: An evolutionary story. In D. D. Franks & J. H. Turner (Eds.), *Handbook of neurosociology* (pp. 257–287). Springer. https://doi.org/10.1007/978-94-007-4473-8_18

Maslow, A. H. (1964). *Religions, values, and peak-experiences, by Abraham H. Maslow*. Ohio State University Press.

Massey, D. S. (2002). A brief history of human society: The origin and role of emotion in social life: 2001 presidential address. *American Sociological Review, 67*(1), 1–29. https://doi.org/10.2307/3088931

McEwen, B. S. (1998). Protective and damaging effects of stress mediators. *The New England Journal of Medicine, 338*, 171. https://doi.org/10.1056/NEJM199801153380307

Mead, G. H. (1934). *Mind, self & society from the standpoint of a social behaviorist [by] George H. Mead ... Edited, with introduction, by Charles W. Morris*. The University of Chicago Press.

Mikulasch, F. A., Rudelt, L., Wibral, M., & Priesemann, V. (2023). Where is the error? Hierarchical predictive coding through dendritic error computation. *Trends in Neurosciences, 46*(1), 45–59. https://doi.org/10.1016/j.tins.2022.09.007

Molnar-Szakacs, I., & Uddin, L. Q. (2013). The emergent self: How distributed neural networks support self-representation. In D. D. Franks & J. H. Turner (Eds.), *Handbook of neurosociology* (pp. 167–182). Springer. https://doi.org/10.1007/978-94-007-4473-8_13

Neckerman, K. M., & Torche, F. (2007). Inequality: Causes and Consequences. *Annual Review of Sociology, 33*, 335–357. https://doi.org/10.1146/annurev.soc.33.040406.131755

Niemeyer, R. E. (2013). What are the neurological foundations of identities and identity-related processes? In D. D. Franks & J. H. Turner (Eds.), *Handbook of neurosociology* (pp. 149–165). Springer. https://doi.org/10.1007/978-94-007-4473-8_12

Parr, T., Holmes, E., Friston, K. J., & Pezzulo, G. (2023). Cognitive effort and active inference. *Neuropsychologia, 184*, 108562. https://doi.org/10.1016/j.neuropsychologia.2023.108562

Parr, T., Pezzulo, G., & Friston, K. J. (2022). *Active inference: The free energy principle in mind, brain, and behavior*. The MIT Press.

Pearl, J. (1998). Graphical models for probabilistic and causal reasoning. In P. Smets (Ed.), *Quantified representation of uncertainty and imprecision* (pp. 367–389). Springer. https://doi.org/10.1007/978-94-017-1735-9_12

Poirier, P., Faucher, L., & Bourdon, J.-N. (2021). Cultural blankets: Epistemological pluralism in the evolutionary epistemology of mechanisms. *Journal for General Philosophy of Science, 52*(2), 335–350. https://doi.org/10.1007/s10838-019-09472-8

Prakash, C., Stephens, K. D., Hoffman, D. D., Singh, M., & Fields, C. (2021). Fitness beats truth in the evolution of perception. *Acta Biotheoretica, 69*(3), 319–341. https://doi.org/10.1007/s10441-020-09400-0

Ramstead, M. J. D., Badcock, P. B., & Friston, K. J. (2018). Answering Schrödinger's question: A free-energy formulation. *Physics of Life Reviews, 24*, 1–16. https://doi.org/10.1016/j.plrev.2017.09.001

Ramstead, M. J. D., Kirchhoff, M., Constant, A., & Friston, K. (2021). Multiscale integration: Beyond internalism and externalism. *Synthese, 198*, 41. https://doi.org/10.1007/s11229-019-02115-x

Ramstead, M. J. D., Kirchhoff, M. D., & Friston, K. J. (2020). A tale of two densities: Active inference is enactive inference. *Adaptive Behavior, 28*(4), 225–239. https://doi.org/10.1177/1059712319862774

Ramstead, M. J. D., Sakthivadivel, D. A. R., Heins, C., Koudahl, M., Millidge, B., Da Costa, L., Klein, B., & Friston, K. J. (2023). On Bayesian mechanics: A physics of and by beliefs. *Interface Focus, 13*(3), 20220029. https://doi.org/10.1098/rsfs.2022.0029

Ramstead, M. J. D., Seth, A. K., Hesp, C., Sandved-Smith, L., Mago, J., Lifshitz, M., Pagnoni, G., Smith, R., Dumas, G., Lutz, A., Friston, K., & Constant, A. (2022). From generative models to generative passages: A computational approach to (neuro) phenomenology. *Review of Philosophy and Psychology, 13*, 829. https://doi.org/10.1007/s13164-021-00604-y

Rao, R. P. N., & Ballard, D. H. (1999). Predictive coding in the visual cortex: A functional interpretation of some extra-classical receptive-field effects. *Nature Neuroscience, 2*(1), 1. https://doi.org/10.1038/4580

Rieke, F., Warland, D., Steveninck, R. D. R. V., & Bialek, W. (1999). *Spikes: Exploring the neural code*. (Reprint edition). Bradford Books.

Rizzolatti, G., Fadiga, L., Gallese, V., & Fogassi, L. (1996). Premotor cortex and the recognition of motor actions. *Cognitive Brain Research, 3*(2), 131–141. https://doi.org/10.1016/0926-6410(95)00038-0

Robertson, C. G., & Kesselheim, A. (2016). *Blinding as a solution to bias: Strengthening biomedical science, forensic science, and law*. Academic Press.

Rubin, S., Parr, T., Da Costa, L., & Friston, K. (2020). Future climates: Markov blankets and active inference in the biosphere. *Journal of the Royal Society Interface, 17*(172), 20200503. https://doi.org/10.1098/rsif.2020.0503

Russell, J. A. (1980). A circumplex model of affect. *Journal of Personality and Social Psychology, 39*(6), 1161–1178. https://doi.org/10.1037/h0077714

Sandved-Smith, L., Hesp, C., Mattout, J., Friston, K., Lutz, A., & Ramstead, M. J. D. (2021). Towards a computational phenomenology of mental action: Modelling meta-awareness and attentional control with deep parametric active inference. *Neuroscience of Consciousness, 2021*(1), niab018. https://doi.org/10.1093/nc/niab018

Schulkin, J., & Sterling, P. (2019). Allostasis: A brain-centered, predictive mode of physiological regulation. *Trends in Neurosciences, 42*(10), 740–752. https://doi.org/10.1016/j.tins.2019.07.010

Seth, A. K. (2013). Interoceptive inference, emotion, and the embodied self. *Trends in Cognitive Sciences, 17*(11), 565–573. https://doi.org/10.1016/j.tics.2013.09.007

Seth, A. K., & Critchley, H. D. (2013). Extending predictive processing to the body: Emotion as interoceptive inference. *The Behavioral and Brain Sciences, 36*(3), 227–228. https://doi.org/10.1017/S0140525X12002270

Seth, A. K., & Friston, K. J. (2016). Active interoceptive inference and the emotional brain. *Philosophical Transactions of the Royal Society B: Biological Sciences, 371*(1708), 20160007. https://doi.org/10.1098/rstb.2016.0007

Shaffer, C., Westlin, C., Quigley, K. S., Whitfield-Gabrieli, S., & Barrett, L. F. (2022). Allostasis, action, and affect in depression: Insights from the theory of constructed emotion. *Annual Review of Clinical Psychology, 18*(1), 553–580. https://doi.org/10.1146/annurev-clinpsy-081219-115627

Shermer, M. (2022). *Conspiracy: Why the rational believe the irrational*. Johns Hopkins University Press.

Sherrington, C. S. (1900). The muscular sense. In E. A. Sharpey-Schäfer (Ed.), *Textbook of physiology*. Pentland.

Shipp, S., Adams, R. A., & Friston, K. J. (2013). Reflections on agranular architecture: Predictive coding in the motor cortex. *Trends in Neurosciences, 36*(12), 706–716. https://doi.org/10.1016/j.tins.2013.09.004

Shook, J. R. (2013). Social cognition and the problem of other minds. In D. D. Franks & J. H. Turner (Eds.), *Handbook of neurosociology* (pp. 33–46). Springer. https://doi.org/10.1007/978-94-007-4473-8_4

Smith, R., Friston, K. J., & Whyte, C. J. (2022). A step-by-step tutorial on active inference and its application to empirical data. *Journal of Mathematical Psychology, 107*, 102632. https://doi.org/10.1016/j.jmp.2021.102632

Smith, R., Schwartenbeck, P., Parr, T., & Friston, K. J. (2020). An active inference approach to modeling structure learning: Concept learning as an example case. *Frontiers in Computational Neuroscience, 14*, 41. https://www.frontiersin.org/articles/10.3389/fncom.2020.00041

Solymoski, T. (2013). Can the two cultures reconcile? Reconstruction and neuropragmatism. In D. D. Franks & J. H. Turner (Eds.), *Handbook of neurosociology* (pp. 83–97). Springer. https://doi.org/10.1007/978-94-007-4473-8_7

Stam, H. J. (2015). The neurosciences and the search for a unified psychology: The science and esthetics of a single framework. *Frontiers in Psychology, 6*, 1467. https://doi.org/10.3389/fpsyg.2015.01467

Sterling, P. (2012). Allostasis: A model of predictive regulation. *Physiology & Behavior, 106*(1), 5–15. https://doi.org/10.1016/j.physbeh.2011.06.004

Sterling, P. (2015). *Principles of neural design*. The MIT Press.

Sterling, P. (2020). *What is health?: Allostasis and the evolution of human design*. The MIT Press.

Stets, J. E. (2010). Future directions in the sociology of emotions. *Emotion Review, 2*(3), 265–268. https://doi.org/10.1177/1754073910361975

Stets, J. E. (2012). Current emotion research in sociology: Advances in the discipline. *Emotion Review, 4*(3), 326–334. https://doi.org/10.1177/1754073912439791

Strand, M., & Lizardo, O. (2022a). Chance, orientation, and interpretation: Max Weber's neglected probabilism and the future of social theory. *Sociological Theory, 40*(2), 124–150. https://doi.org/10.1177/07352751221084712

Strand, M., & Lizardo, O. (2022b). For a probabilistic sociology: A history of concept formation with Pierre Bourdieu. *Theory and Society, 51*(3), 399–434. https://doi.org/10.1007/s11186-021-09452-2

Summers-Effler, E., Van Ness, J., & Hausmann, C. (2015). Peeking in the black box: Studying, theorizing, and representing the micro-foundations of day-to-day interactions. *Journal of Contemporary Ethnography, 44*(4), 450–479. https://doi.org/10.1177/0891241614545880

Theriault, J. E., Young, L., & Barrett, L. F. (2021). The sense of should: A biologically-based framework for modeling social pressure. *Physics of Life Reviews, 36*, 100–136. https://doi.org/10.1016/j.plrev.2020.01.004

Thielmann, I., Hilbig, B. E., & Zettler, I. (2020). Seeing me, seeing you: Testing competing accounts of assumed similarity in personality judgments. *Journal of Personality and Social Psychology, 118*(1), 172–198. https://doi.org/10.1037/pspp0000222

Thoits, P. A. (1989). The sociology of emotions. *Annual Review of Sociology, 15*(1), 317. https://doi.org/10.1146/annurev.so.15.080189.001533

Thomas, O. (2018). Two decades of cognitive bias research in entrepreneurship: What do we know and where do we go from here? *Management Review*

Quarterly, 68(2), 107–143. https://doi.org/10.1007/s11301-018-0135-9

Turner, R. H. (1962). Role-taking: Process versus conformity. In *Human behavior and social processes*. Routledge.

Turner, J. H. (2000). *On the origins of human emotions: A sociological inquiry into the evolution of human affect* (1st ed.). Stanford University Press.

Turner, J. H. (2009). The sociology of emotions: Basic theoretical arguments. *Emotion Review, 1*(4), 340–354. https://doi.org/10.1177/1754073909338305

Turner, J. H. (2013). Neurology and interpersonal behavior: The basic challenge for neurosociology. In D. D. Franks & J. H. Turner (Eds.), *Handbook of neurosociology* (pp. 119–137). Springer. https://doi.org/10.1007/978-94-007-4473-8_10

Turner, J. H., & Stets, J. E. (2006). Sociological theories of human emotions. *Annual Review of Sociology, 32*, 25–52.

Van de Cruys, S. (2017). Affective value in the predictive mind. In T. K. Metzinger & W. Wanja (Eds.), *Open mind*. MIND Group. https://predictive-mind.net/papers/affective-value-in-the-predictive-mind/getAbstract

Vasil, J., Badcock, P. B., Constant, A., Friston, K., & Ramstead, M. J. D. (2020). A world unto itself: Human communication as active inference. *Frontiers in Psychology, 11*, 417. https://www.frontiersin.org/articles/10.3389/fpsyg.2020.00417

Veissière, S. P. L., Constant, A., Ramstead, M. J. D., Friston, K. J., & Kirmayer, L. J. (2020). Thinking through other minds: A variational approach to cognition and culture. *Behavioral and Brain Sciences, 43*, e90. https://doi.org/10.1017/S0140525X19001213

von Scheve, C. (2018). A social relational account of affect. *European Journal of Social Theory, 21*(1), 39–59. https://doi.org/10.1177/1368431017690007

Walsh, K. S., McGovern, D. P., Clark, A., & O'Connell, R. G. (2020). Evaluating the neurophysiological evidence for predictive processing as a model of perception. *Annals of the New York Academy of Sciences, 1464*(1), 242–268. https://doi.org/10.1111/nyas.14321

Weber, M. (1913). Some categories of interpretive sociology. *Sociological Quarterly, 22*(2), 151–180.

Weber, M. (2019). *Economy and society: A new translation/Max Weber*. Harvard University Press. https://doi.org/10.4159/9780674240827

Zhang, A., Goosby, B., & Cheadle, J. E. (2021). In the flow of life: Capturing affective socializing dynamics using a wearable sensor and intensive daily diaries. *Socius, 7*, 23780231211064009. https://doi.org/10.1177/23780231211064009

Neuroscience, Cognition, and Social Networks

17

Matthew E. Brashears and Brea Perry

Abstract

Humans are among the most intelligent species, using powerful cognitive abilities to develop sophisticated language and culture and to expand beyond their original range to all corners of the Earth. Similarly, humans are remarkably social, building elaborate social structures characterized by substantial division of labor and coordination in activities without the need for tight genetic relatedness. In this chapter, we review research that not only connects these abilities, but suggests that they are mutually dependent. We begin by reviewing the social brain hypothesis—a prominent evolutionary explanation for our high intelligence. We then discuss how the social brain manifests in specific brain structures with distinct consequences for social behavior and, by extension, social networks. This segues to current thinking about the activity at the level of minds, which extends the capabilities manifest in brains, enabling the formation and maintenance of complex social networks that might otherwise exceed humans' cognitive capacity. Finally, the chapter concludes with a consideration of what we know as well as what directions researchers should pursue in the future at the intersection of neuroscience, cognition, and networks, and why they should do so.

M. E. Brashears (✉)
Department of Sociology, University of South Carolina, Columbia, SC, USA
e-mail: Brasheam@mailbox.sc.edu

B. Perry
Department of Sociology, Indiana University, Bloomington, IN, USA
e-mail: blperry@iu.edu

Introduction

Superficially, brains and social networks seem entirely divorced from one another. The brain is a physical computational organ that supports simple and complex behaviors in a wide variety of animal species, including humans. In contrast, a social network is a set of agents (e.g., individuals) who are connected by sets of relationships (i.e., ties) that bind them together. Whereas brains can be held in one's hand and are possessed by individuals, networks are ephemeral and are possessed in common by network members. Nevertheless, processes unfolding at the individual level necessarily impact social networks, and we have good reason to think that many of the features of our brains evolved to support social activity. Indeed, a substantial body of research leads to the conclusion that human intelligence emerged primarily as a result of our social style of life (e.g., Dunbar, 1992, 1998; Dunbar & Shultz, 2017; Humphrey, 1976), suggesting that at its core, human cognition is fundamentally ori-

W. Kalkhoff et al. (eds.), *Handbook of Neurosociology*, Handbooks of Sociology and Social Research, https://doi.org/10.1007/978-3-031-95615-7_17

ented towards social activity. However, if we accept this proposition, we must confront a logical consequence: if our social activity is supported by neural features evolved over thousands of years, then logically some, or most, of our sociability would not be possible without those features. In other words, just as the presence of a given neural structure or brain function can enable a type of sociability, so too can an absence prevent it. It is only by understanding how neural structures and brain activity give rise to social network behaviors that we can fully understand human groups and, likewise, only by understanding the dynamics of human groups can we hope to grasp the forces that shaped the human brain. Moreover, if human cognition is especially sensitive to social influence, understanding connections between the brain and social networks is essential for explaining and addressing individual and group differences in cognitive development, health, and aging.

In this chapter we begin by discussing the evolution of human intelligence, which sets the stage for the connection between our brains and our social networks. The human brain is not a general purpose computational artifact, like a laptop, but is a processing organ shaped by evolution to accomplish the goals necessary for survival, among which was the creation, maintenance, and exploitation of social ties. We then introduce the crucial distinction between neural structure and function on the one hand and the algorithmic behavior of the brain on the other. Far from requiring a single response to a given stimulus, evolution has crafted an impressive degree of flexibility into the human brain. We then discuss the neural structures and functions that enable and in some ways constrain human social life. Finally, we proceed to a discussion of some of the schemas and heuristics (i.e., the algorithms) that humans "execute" on their neurology that enable performance beyond what our raw hardware might otherwise acheive. These heuristics are critical to the large and flexible social networks we presently enjoy, but are shaped by and therefore ultimately rooted in the limitations and capabilities of brain structure and function. Ultimately, we call for increased attention to the ways in which our neurology, and how we employ it, shape the structure and function of our social networks and vice versa.

The research we review in this chapter largely draws on personal social network, or egocentric network, theory and methodology. A personal social network is the set of others to whom an individual is connected and the ties between them. For example, if we were to ask an individual for those persons with whom they interact regularly and then ask about the connections between those alters, we would gain information about a network that is personal to that individual (see also Perry et al., 2018). As such, the focus is on the individual, or ego, their immediate associates, or alters, and the unique constellation of connections in which they are embedded, regardless of where they are located or how they know the focal person. In contrast, the structure of all ties between all individuals in one socially or geographically bounded group comprises a sociocentric network. For example, if we collect information on who talks to whom for all members of a particular corporate branch, the network is not personal to any one specific individual and thus we are dealing with one sociocentric network. The distinction is primarily one of data collection methodology. In the personal network case, we are not trying to map all social relations within some defined setting, but instead choose a focal person, or ego, and map the relations connecting them to specific others, unbounded by context. As such, this methodology typically produces data on many networks (i.e., one for each ego), and is ideal for examining individual variation in social connectedness. In practice, the differences between personal and sociocentric networks constitute a spectrum (e.g., within any given sociocentric network, we can define a personal network by choosing a focal individual and identifying their direct alters). But typically, personal networks are collected in the absence of a sociocentric network (as is common in survey data collection), providing greater flexibility in sampling strategies and broader population inferences.

Our focus on personal networks reflects our subject matter; to the extent that brains shape sociocentric networks, it is only via their manifestation of structures in personal networks. Thus, we focus on the forms of social network analysis that capture the linkages between the brain and the wider social network. However, while these social networks are personal, they are far from idiosyncratic. Despite substantial turnover in the individuals that make up a personal social network, the structure and composition of these networks (e.g., hierarchical organization, density of interconnections, percent kin) tend to remain fairly stable over time (Lubbers et al., 2010; Wellman et al., 1997). This is likely attributable to the persistence of social signatures, or habitual patterns of interaction with members of personal networks (e.g., frequency of contact with network members, tendency to introduce associates to each other), which contributes to a personal network structure that is stable and hierarchical (Saramaki et al., 2014). Social signatures are likely driven by humans' cognitive limitations, discussed below, making it unfeasible to maintain intimacy and frequent contact with more than some small subset of the overall network (Powell et al., 2012). The relative within-person stability of global personal network properties coupled with substantial between-person variation in them lays the foundation for a body of research examining associations between cognitive processes, brain structure and function, and features of personal social networks.

The Evolution of Brains and Minds

Humans are arguably one of, if not the, most intelligent extant species.[1] This high intelligence has allowed human societies to harness natural forces, develop powerful technologies, and to construct an elaborate physical culture. While it may appear as if the evolutionary case for high intelligence is obvious, this is not quite true. The human brain consumes approximately 20–25% of our metabolic resources (Rolfe & Brown, 1997; Sokoloff, 1960), but only accounts for roughly 3% of an adult human's body mass. In a child, brain tissue may account for as much as 50% of all metabolic activity (Kennedy & Sokoloff, 1957), thus representing a tremendous expenditure of resources. If the average adult consumes 2000 calories per day, then 400–500 of these calories are solely used to support the brain. If the rest of our bodies consumed calories at the same rate as our energy-intensive brains, the average human would need to consume a daunting 13,000 calories per day to keep up with the demands. So what function justifies the evolution of such a ferociously expensive organ?

One answer is that our powerful and expensive brains enable us to fashion tools and develop technology. While our brains do enable this activity, the rapid pace of technological advancement we now enjoy has been a recent phenomenon. This suggests that highly complex tool use emerged relatively late in the sequence that gave rise to high intelligence in humans. In contrast, simple tools fashioned from elements of the natural world are not exclusive to humans and have been observed in use by chimpanzees, baboons, and dolphins among others (Cheney & Seyfarth, 2019; Mann et al., 2012). This suggests that using tools may require some level of intelligence, but will not inevitably trigger runaway increases in intelligence. Additionally, evolution favors traits that are useful now rather than traits that will only become useful later, and therefore the eventual ability to create and deploy powerful tools in the far future does not ensure selection for greater tool use in the present.

A second argument focuses on the availability of calorie rich foods. As a changing environment made calorie dense foods more readily available, it became possible for an organism to support a powerful brain (Clutton-Brock & Harvey, 1980). Yet, calorie availability is a necessary, rather than a sufficient, condition for the evolution of power-

[1] At least as judged by humans, which may admit a degree of bias. Indeed, it is challenging to assess intelligence between species and it may not be possible to do so even in principle (Erickson, 2014). Likewise, it can be difficult to distinguish cognitive abilities like mimicry from superficially similar phenomena such as response facilitation (Byrne, 1995). Herein we primarily focus on intelligence in terms of general problem solving intelligence and view it as more crucial for our argument that humans are considered to be highly capable in this area rather than that they are uniquely superior relative to all non-human species.

ful brains. Rather than increase the calories consumed by each individual member of a species (as is needed to support larger brains), evolution could just as easily have favored increases in population size, dispersing genes into a larger number of individual carriers.

Other arguments present various physical challenges that high intelligence allows a species to overcome, such as identifying/predicting when time-limited resources (e.g., ripe fruit) or highly calorie dense foods will be available, or by enabling an organism to engage in extractive foraging (Clutton-Brock & Harvey, 1980; Melin et al., 2014; Parker, 2015). The difficulty with these physical challenges is that they are largely stable over evolutionary time scales and so can be solved without the need for high levels of general intelligence. For example, many species have evolved to migrate between seasonally available food supplies without being forced to develop significant intelligence. Consequently, the connection between extractive foraging and intelligence in primates is inconsistent (Dunbar & Shultz, 2017) and general intelligence may be unnecessary in species that are capable of learning behaviors. While mimicry is a a cognitively complex task, many skills (e.g., breaking fruit open with rocks) can spread via response facilitation (e.g., heightened tendency to experiment with rocks and fruit as a consequence of witnessing others) rather than true mimicry (Byrne, 1995). Given the caloric expense of brain tissue, solutions to these physical challenges that did not involve the expansion of the brain ought to have been, and quite likely were, favored. The upshot of the preceding is that high intelligence provides an advantage in completing a variety of important tasks, but none of those tasks appear to be sufficient to explain the emergence of high intelligence. For that, we must look to an even more significant challenge: managing social life.

First articulated as the "Machiavellian Intelligence Hypothesis" (Humphrey, 1976) the modern "Social Brain Hypothesis" (Dunbar, 1998; Dunbar & Shultz, 2017) argues that high intelligence allows humans to compete more effectively with their in-group members. Compared to solitary species, members of group living species gain many advantages but are always surrounded by others who prefer the same sorts of food, shelter, and mates as themselves. The support that social living provides in meeting environmental challenges comes at the cost of heightened competition with conspecifics. The social brain hypothesis asserts that individuals with the greatest levels of intelligence, compared to members of their own group, will be most able to pursue their own competitive interests relative to other group members, allowing them to secure better access to desirable resources without risking the integrity of the group itself. Better access to these resources should permit greater fertility and enable more of those offspring to be raised to adulthood. As a result, a disproportionate number of the next generation will be offspring of these smarter indivduals and, if intelligence is at least partly genetic, the average intelligence of the group will tend to increase. However, the smarter next generation must compete not against a stable external challenge, but against the now smarter pool of in-group members. The most intelligent members of this generation should, again, be most reproductively successful, thereby leaving a smarter subsequent generation. In short, because increases in intelligence are driven by the need to compete within one's own group, the evolution of still greater intelligence is always favored (i.e., it is always adaptive to be a bit smarter than the rest of one's own group).[2] Therefore, intelligence evolved for such a purpose should tend to continue increasing until some external limit is encountered.[3]

If the social brain hypothesis accounts for human intelligence, then we should be able to identify evidence in support of it. Substantial brain growth associated with the shift to social

[2]It should be noted that this logic would not apply to eusocial insects, despite their large group sizes, because the only way individuals can pass on their genes is by supporting the hive as a whole. Whereas humans face a tradeoff between their individual genetic legacy and the group, eusocial insects experience only congruence between the needs of the group and propograting their own genes into the future.

[3]For example, further increases in brain mass push calorie requirements unsustainably high.

diurnal living, increasing the opportunity for social interaction, is suggestive (Shultz et al., 2011), but the smoking gun would be a relationship between brain size and structure on one hand and the size/complexity of social networks on the other. Robin Dunbar (1992) evaluated this proposition by examining the relationship between primary group size and the neocortical ratio (i.e., neocortex volume divided by the remaining volume of the brain) for thirty-eight primate genuses, ranging from *Avahi* (i.e., certain lemurs) to *Pan* (i.e., chimpanzees and bonobos, our closest extant non-human relatives) and uncovered a clear positive correlation. Using these results, he predicted that human primary social groups should contain approximately 150 individuals (i.e., "Dunbar's Number"). Continued research indicates that there is a general relationship between neocortical ratio and network size in primates, including humans, even with the addition of numerous controls (e.g., there appears to be a strong link between cognitive ability and network size; (Stiller & Dunbar, 2007) as well as cognitive limits on sociability (Roberts et al., 2009)). Likewise, there is evidence that the social brain hypothesis may explain the high intelligence of cetaceans, particularly dolphins, despite their radically different brain structure (Connor & Mann, 2006) and potentially that social complexity impacts the cognitive flexibility of Poecilid fish (Triki et al., 2023). As such, ample evidence supports a link between our brains and the growth of our social networks.

While the logic of the social brain hypothesis is compelling, a number of limitations prevent it from being entirely convincing. Primarily, it is silent as to which brain areas should be "larger" and, likewise, is vague as to what makes a social network more complex. The former is a concern in that it encourages exploration of multiple measures focusing on different aspects of brain size or structure that have little functional link to social behavior or outcomes. This is precisely the approach taken in Dunbar's classic (1992) paper. The lack of specificity allows the hypothesis to be supported if any one of the measures exhibits the anticipated relationship with group size, while falsification requires failure on all measures. Given that evolution is not constrained to achieve improvements only via one pathway, this lack of specificity is somewhat unavoidable, but it still grants the hypothesis more bites at the apple than is strictly desirable. For example, the intelligence of cetaceans may well be due to social factors (Connor & Mann, 2006), supporting the social brain hypothesis, but cetaceans lack a neocortex. Thus, it is clear that the link between the neocortex and group size is not specifically predicted by theory nor uniquely necessary in order to claim support for the account.

Nevertheless, the definitional issues surrounding social network complexity are by far the more problematic ones, as there are numerous species that live in large groups without necessarily being especially intelligent individually. For example, fish sometimes live in massive schools and eusocial insects assemble into sometimes massive hives, but few would regard either as especially intelligent. This implies that some additional factor is needed in order for group size to cause increases in intelligence. Among fish schools, intelligence is likely unnecessary because coordination requires only simple reactions to the behaviors of one's immediate neighbors. Among eusocial insects, large group sizes are accompanied by physical specialization, reducing the behavioral flexibility required from different units. In contrast, among primates, the key factor linking intelligence to group size may have been the rise of fission-fusion societies in which members must often anticipate the reactions of individuals who are not currently present but who can be expected to return soon (Cheney & Seyfarth, 2019). However, each of the above are post-hoc accounts for why the social brain hypothesis applies in one instance and not in another. These accounts protect the hypothesis from falsification but do not reflect arguments or postulates embedded in the theory itself. As such, while Dunbar's research is extremely interesting, there is no theoretical reason to assume that the neocortical ratio specifically should be the key factor and the examples of fish and eusocial insects indicate that group size, absent some other factor, is a necessary but not sufficient factor to give rise to human-like levels of intelligence.

Consistent with this premise, other attempts to derive predicted group sizes based on brain volume yield highly variable estimates with large confidence intervals (e.g., Lindenfors et al., 2021), suggesting that some other factor or factors are involved. Likewise, some evidence supports the key relevance of social understanding of relations between bodies rather than relations between intentions (e.g., Barrett et al., 2022), suggesting that other features (e.g., emotional intelligence) might be more critical drivers than goal-directed planning. In short, the social brain hypothesis is presently under-specified.

Despite these drawbacks, the social brain hypothesis has produced evidence supporting the claim that intelligence is connected in some fashion to group size and complexity, at least among primates. As such, it is reasonable to adopt this as a guiding frame. Human intelligence was substantially, if not primarily, shaped by the need to manage social interaction with other in-group members, and selection favoring this capacity is likely to manifest a number of other consequences in human cognition and behavior.

Distinguishing Mind and Brain

Substantial research supports the role of social life in the evolution of human intelligence. Some of this evidence (e.g., Dunbar's research) identifies an apparent relationship between the size of certain brain structures and the size/complexity of social networks. This provides good reason to expect that our social behavior is deeply intertwined with our neural structure and function. However, it is important to be clear about what this means, and what it does not.

Concluding that social behavior depends upon neural functions does not mean that brain structure determines social behavior. This is a reductive explanation requiring that a given set of circumstances mechanically produce the same output, and is frequently, and rightly, rejected by social scientists.[4] *Homo Sapiens* enjoys impressive cognitive flexibility, with the ability to adopt new beliefs, skills, and behaviors quite rapidly, and over long periods of the life course. Indeed, this behavioral flexibility is quite likely one of the key drivers of human success, enabling us to spread across the planet long before the development of advanced technology. As a result, we should be highly skeptical that evolution would have hard-wired social responses into humans. In other words, given that humans are characterized by considerable flexibility, it should require a great deal of evidence to convince us that in some specific area we have none.

Instead, evolution has created competencies that humans are able to employ flexibly to accomplish objectives, but these competencies impose constraints even as they empower. As an analogy, evolution has equipped humans with dexterous hands and opposable thumbs that allow us to grasp objects and engage in fine manipulation. However, the structure of those hands also imposes constraints; our wrists twist only in particular ways, and to particular degrees, and our thumbs and fingers clamp only in certain configurations. Likewise, while humans can learn a wide variety of behaviors via operant conditioning, our response to various conditioning schedules (i.e., patterns of reinforcement or punishment) is essentially identical to that of other species (e.g., Domjan, 2018). While we have the ability to learn many behaviors, the manner in which that learning occurs is more strongly controlled by brain structure and function. The neural basis of our social cognitive competencies function in the same way; our brains enable us to track and manipulate social information, but at the same time the manner in which they accomplish this feat necessarily imposes constraints. While these constraints can sometimes be circumvented, the strategies we employ cannot exceed the bounds of what our brains can accomplish.

Given this reality, we use two terms—brain and mind—to refer to two related, but distinct, concepts. When we use the term "brain," we mean the physical structures and neurological processes (e.g., brain functions) that execute cognitive operations. Our neural structure and activ-

[4]For example, claims that males or females exhibit specific behaviors because they are evolved to do so.

ity is thus a property of the brain. In contrast, when we use the term "mind" we are referring to the heuristics and algorithms that are being executed by the brain. Needless to say, given that the brain stores information physically in terms of the presence/absence of synapses and thought occurs as a consequence of the physical operation of neurons, mind is always reflected in the physical configuration of brain. Yet, the heuristics that the brain is executing can be characterized independently of the brain. In effect, the brain resembles the hardware of a computer while mind is the software.[5] Certainly, software is physically present as a sequence of physical changes in storage media, and current states of electrical circuits, but the software is nonetheless distinct from the hardware. Mind is resident in, and depends upon, the brain, but the mind is distinct from the brain.

Below, we turn first to the operation of the brain, describing the neurological mechanisms that contribute to building and maintaining social networks and social network characteristics associated with brain structure and function in adults. We then turn to the operation of mind, discussing what we know about the heuristics and algorithms that allow us to make the most out of our social brains.

Brain and Social Networks

A body of research in social neuroscience, reviewed below, has increasingly documented bidirectional associations between brain structure and function and personal social network characteristics. On one hand, neurological features, such as gray matter volume or synaptic activity in specific brain areas, shape social learning and behavior in patterned ways. These regular patterns of interaction are the basis of ongoing relationships that comprise social networks. On the other hand, social context, including social, cognitive, and affective stimuli transmitted through social networks, can alter brain structure and function.

Neurological Drivers of Social Networks

Much of the neuroimaging work examining covariance of brain and social network characteristics in humans has attempted to test the social brain hypothesis (Dunbar, 1998) and thus has focused on social network size. This research suggests that social network size is positively associated with gray matter volume, brain activity, and functional connectivity in multiple regions, including the amygdala, the prefrontal cortex, the orbitofrontal cortex, and the entorhinal cortext (Bickart et al., 2011; Hampton et al., 2016; Kanai et al., 2012; Kwak et al., 2018; Lewis et al., 2011; Noonan et al., 2018; Powell et al., 2012; Schmalzle et al., 2017; Von Der Heide et al., 2014). In general, these associations are strongest and most consistent in the amygdala and orbitofrontal cortex, which are responsible for perceiving and processing social information. Variability in findings related to specific brain regions may be due to differences in the measures of social network size used across studies since distinct cognitive capabilities may be required for maintaining, for example, larger online networks versus smaller real-world networks (Kanai et al., 2012; Von Der Heide et al., 2014). Associations between brain structure and function and heterogeneity in humans' social network characteristics are likely explained in large part by variation in social information processing, or social cognitive abilities. Survival in a complex and dynamic social world necessitates frequent social interpretation and decision-making, as well as behavioral adaptation to changing social environments. Social information processing—the ability to perceive, interpret, and understand social information—underlies this adaptability and is a process distinct from general cognitive processing (Grainger et al., 2023; Stanley & Adolphs, 2013).

To date, three major neurological systems have been implicated in individual variation in social cognitive abilities (Lamblin et al., 2017).

[5]This analogy should not be pressed too far. The operation of the brain is quite distinct from that of our electronic computers in a variety of respects.

First, the affective system is associated with perceiving social and emotional significance and motivation to approach or avoid social situations, and includes the amygdala, hypothalamus, hippocampus, and other regions (Guyer et al., 2016). Second, the cognitive-regulatory system is involved in rapid evaluation of meaning and context from social stimuli and subsequent emotional and behavioral responses to it. This system includes the orbitofrontal, dorsomedial, and ventral pre-frontal cortex (Nelson et al., 2005). Third, the mentalizing system mediates the ability to perceive and assign mental states to oneself and to others (i.e., theory of mind), and incorporates the medial pre-frontal cortex, temporoparietal junction, and superior temporal sulcus (Schurz et al., 2014). Similarly, tracking popularity and social value is key for a social species and research with adolescents implicates both the fusiform face area and the dorsolateral prefrontal cortex in this function (Dai et al., 2023).

These social cognitive abilities have been widely implicated in the capacity of humans and nonhuman primates to develop and maintain social connections (Adolphs et al., 1998; Baron-Cohen, 2000; Hauck et al., 1998; Krendl et al., 2022; Krendl & Betzel, 2022). Specifically, social cognitive abilities such as remembering faces, inferring others' beliefs and intentions (i.e., cognitive theory of mind), understanding what others are feeling (i.e., mentalizing, or affective theory of mind), and sharing the emotional experience of another person (i.e., empathizing) are essential for navigating social interactions (Decety & Jackson, 2006; Henry et al., 2023; Krendl & Heatherton, 2009). In particular, cognitive and affective theory of mind have been shown to play a key role in facilitating social interactions (Bishop-Fitzpatrick et al., 2017; Watson et al., 1999) by increasing prosocial behavior (Caputi et al., 2012), reducing misunderstandings (Hughes & Leekam, 2004), and promoting social intelligence (Yeh, 2013).

Over the past few decades, research in social neuroscience has identified areas of the brain, and increasingly functional connections between them, that are associated with social information processing. These are, by and large, the same areas of the brain associated with social network differences across individuals, implicating social cognitive abilities as a likely underlying mechanism. For example, being more accurate in inferences about others' emotional states (i.e., mentalizing) is associated with having larger personal social networks (Kardos et al., 2017; Stiller & Dunbar, 2007). At the same time, humans' social network size and social bridging (i.e., having social networks that are more expansive, less affectively close, and less densely interconnected) is correlated with gray matter volume in regions of the brain associated with mentalizing and emotional processing, including the amygdala and orbitofrontal cortex, and with functional connectivity between these regions (Bickart et al., 2012; Kanai et al., 2012; Manchella et al., 2023; Powell et al., 2012; Von Der Heide et al., 2014).

Consistent with this literature, recent research employing sociocentric networks (i.e., whole social networks) suggests that a person's position in a social network may be associated with brain function. A study of entire villages found that an individual's in-degree (i.e., the number of people who reported being connected to them), but not out-degree (i.e., the number of people they reported being connected to), was associated with prefrontal cortex volume (Kwak et al., 2018). Additionally, having a higher in-degree has been associated with exhibiting neural responses to audio-visual stimuli that are more similar to peers in areas of the brain linked to social cognition (Baek et al., 2022). Because in-degree is an indicator of popularity and capacity to attract other people for friendship, these findings provide a conservative test of the underlying role of social cognitive skills in associations between social network characteristics and brain volume than is provided by people's own accounts of their personal or egocentric social networks, which tend to be over-reported (Feld & Carter, 1998; Perry et al., 2018).

Motivation to engage in social interaction and seek social stimuli may also play an important role in the link between brain structure and function and social network size. For example, functional connectivity of the amygdala with brain regions implicated in social perception, affiliation, and episodic memory is associated with larger social networks (Bickart et al., 2012; Von Der Heide et al., 2014). This suggests that greater amygdala activation may be associated with finding social stimuli more interesting and rewarding, thus motivating cultivation of larger social networks (McClure et al., 2004; Von Der Heide et al., 2014). Studies such as these provide strong indirect evidence that brain structure and function shape social cognitive processing and rewards which, in turn, are associated with social network characteristics.

Effects of Social Networks on the Brain

Relative to the inverse, there has been less research documenting how being embedded in social networks with particular characteristics shapes brain structure and function. However, it is clear that the human brain is extremely responsive to social-environmental stimuli (i.e., neuroplasticity), such as exposure to language or human touch (Dinse, 2021), perhaps stemming from intensive evolutionary pressure to achieve social fitness. As such, using social cognitive skills in the context of social interaction is a form of cognitive exercise or stimulation that may lead to greater brain volume or functional connectivity in brain regions implicated in social cognition. The clearest causal evidence comes from studies of nonhuman primates where features of social networks, including group size and an individual's position in it, can be experimentally manipulated. For example, Sallet et al. (2011) observed that macaques assigned to larger social groups experienced increases in gray matter volume and in functional connectivity in regions of the temporal and frontal cortex implicated in social cognition. These same regions were also correlated with higher social dominance, indicating that processing increasingly complex social information may alter the brain in ways that improve social fitness (Noonan et al., 2014; Sallet et al., 2011). Similar prospective and experimental evidence from other model organisms confirms these relationships and suggests that interactions in larger and more complex social groups can induce gene expression related to neuroplasticity (e.g., neuronal growth) in regions of the brain responsible for social decision-making (Cardoso et al., 2015; Fernald, 2015; Williamson et al., 2016).

A related body of research has examined the impact of lacking social connections (i.e., social isolation) and social exclusion on the brain. For example, numerous experimental and observational studies using rodent and primate models have demonstrated profound neurological consequences of chronic social isolation, finding evidence of brain plasticity and impaired neuronal activity in signaling pathways in the amygdala, hippocampus, and prefrontal cortex (Djordjevic et al., 2012; Xiong et al., 2023). These brain effects have also been documented in humans following experiences of early life social deprivation (e.g., being reared in isolation in an institution) and are similar to neurological patterns observed in adults with depression and anxiety (Mehta et al., 2009; Tottenham et al., 2010; Xiong et al., 2023). In children, social enrichment is an experience-expectant stimulus such that forming close, emotional bonds with a parent or other adult is necessary for normal brain development (Lally & Mangione, 2017). Among healthy adults, people who self report having an absence of satisfying social relationships exhibit reduced gray matter volume in the amygdala, hippocampus, and left superior temporal sulcus—regions of the brain involved in processing the actions, motivations, and emotions of others (i.e., theory of mind; Zovetti et al., 2021). Together, these studies suggest that a lack of social stimulation affects normal brain development in early life and influences neuroplasticity across the life course.

In the context of human aging, researchers have begun to identify associations between social network characteristics, neurodegeneration, and cognitive decline, particularly in the case of Alzheimer's disease and related dementias (ADRD). This body of evidence largely points to the role of social bridging, or expansive and loosely-connected networks containing affectively weaker relationships that cut across heterogeneous social groups (Perry, McConnell, et al., 2022). Social scientists have shown that access to "weak" ties provides opportunities for exposure to novel social stimuli, including a diversity of ideas, information, activities, verbal and nonverbal social cues, faces, and speech patterns; (Aral & Van Alstyne, 2011; Brashears & Quintane, 2015; Burt, 2004; Granovetter, 1977; Thoits, 2011). Bridging interactions are likely to be more cognitively enriching than familiar, repetitive, and comfortable exchanges with family and other close ties (Perry, McConnell, et al., 2022). The cognitive enrichment provided by social bridging may shape brain health directly by triggering gene expression and catalyzing functional connectivity and neuronal growth, including brain repair (Ainsworth et al., 2021; Holtmaat & Svoboda, 2009; Sale et al., 2009; Sallet et al., 2011). These effects of social bridging on the brain are consistent with recent studies indicating that greater social network complexity, measured as irregular contact with a diversity of social roles, frequent interactions with friends, and less dense networks are associated with healthy brain volume and connectivity, better cognitive functioning, and a slower trajectory of cognitive decline in older adults (Ellwardt et al., 2015; James et al., 2012; Sommerlad et al., 2019).

Social bridging may also indirectly affect trajectories of cognitive decline through lifelong accumulation of cognitive reserve. Up to one-third of older adults meeting full neuropathological criteria for ADRD (e.g., build up of amyloid and tau in the brain) exhibit no significant cognitive impairment, suggesting that neurodegeneration may be moderated by compensatory processes, including social enrichment (Bennett et al., 2006; Peng et al., 2022; Perry, McConnell, et al., 2022; Perry, Roth, et al., 2022; Sharifian et al., 2019). Those with high levels of social engagement, more friends, and a diverse set of associates are more resilient than similar peers, on average, to the cognitive effects of neurodegeneration. For example, Perry and colleagues (Perry, Roth, et al., 2022) find that among older adults with high levels of social bridging, expected adverse cognitive effects of brain atrophy in regions implicated in socioemotional learning and processing (e.g., the amygdala) are completely attenuated. Taken together, these results suggest that bridging networks may provide social cognitive stimulation that exercises the brain, making it more resilient to neuropathological changes common in aging.

In sum, the research reviewed here indicates strong and robust associations between social network characteristics and brain structure and function that are likely bidirectional. Social cognitive abilities rooted in larger brain volume and greater functional connectivity between social processing and memory centers in the brain enable the formation and maintenance of more and increasingly complex configurations of social ties. At the same time, because these features of the brain and related cognitive skills likely evolved through intense selective pressure to live harmoniously in larger social groups, the brain appears to be particularly sensitive to patterned social stimuli not only in childhood, but across the life course into old age. As a result, social networks and associated social experiences (e.g., social isolation) can influence neuronal growth, pruning, and decay, affecting the structural properties and functional connectedness of the brain.

Mind and Social Networks

Intentional social behavior requires that social networks be present as constructs in the brain (i.e., activation states of neurons) and the mind (i.e., algorithmic states). This occurs both at the level of long term memory, or storage of information that we are not currently thinking about, and at the level of working memory, or storage of information that we are actively thinking about. However, while long term memory is apparently large in total capacity,[6] working memory is quite

[6]The exact capacity of human long term storage remains a topic of research. It must be finite, given that the brain contains a finite number of neurons and synapses, but the exact upper limits are not known with certainty.

limited, and these limits create a dilemma when it comes to capturing networks in the mind.

In order for humans to reason effectively about social networks, we not only must collect and store information about our relationships with other individuals, but also about the relationships between those others. However, a group of size n includes n(n − 1) possible relations among all members.[7] In a group of five, the individual would need to track twenty relations, of which only four (20%) directly involve the individual. If we increase the size of the group modestly from five persons to ten, the total number of relationships increases to ninety, of which only nine (10%) directly involve the individual. The implications are clear: even small human groups contain potentially large numbers of relationships to be cognitively represented and small increases in group size can dramatically increase the amount of information to be stored and processed. Further enhancing these demands, it is necessary to model the knowledge and intentions of others (i.e., a theory of mind). The existence of theory of mind in our extant primate relatives is uncertain (e.g., Cheney & Seyfarth, 2019), but humans demonstrate the possession of "intentionality" or multiple nested layers of theory of mind (Stiller & Dunbar, 2007). Many forms of popular entertainment (e.g., romantic comedies) only work because the audience is capable of tracking the beliefs and intent of the characters, including their evident beliefs about each other, without being taxed to the point that the exercise becomes unenjoyable. But while humans evidently can manage to track social information with this level of richness, it nonetheless adds to the cognitive challenge of social life.

The dramatic increase in information processing as networks grow points to a potential drawback in Dunbar's Number. If a fixed amount of brain (e.g., neocortex) is needed to store each new relationship, the rapid increase in relationships to be tracked should quickly produce impractically large brains. Dunbar's model (Dunbar, 1992; eq. 1) avoids this challenge by assuming bigger neocortices are more efficient,[8] but does not explain how efficiency gains are achieved. Moreover, research (Bernard et al., 1987; DiPrete et al., 2011; Killworth et al., 1990; McCarty et al., 2001) indicates that Dunbar's number may be an underestimate, requiring human brains to achieve even larger gains than already assumed by the model.

The solution to this paradox lies not in brain, but in mind. Human social recall is often superior to recall of other types of information; social network structures (e.g., people connected through friendship ties) are recalled more accurately than identical non-social networks (e.g., locations linked by roads) (e.g., Janicik & Larrick, 2005; Simpson et al., 2011; Van Kreveld & Zajonc, 1966). Yet, paradoxically, observed and recalled interactions show mismatches at the dyadic (i.e., person-to-person; Bernard & Killworth, 1977), triadic (i.e., three-person group; (Killworth & Bernard, 1979), and clique (i.e., larger saturated group; Bernard et al., 1982) levels. Both of these findings can be correct because humans are not good at remembering the specific interactions, but appear to be quite good at remembering overall patterns of contact into which specific interactions can be fit (Freeman et al., 1987, 1988, 1989; Freeman & Romney, 1987).

Schemas, or preexisting patterns of information, are known to improve and accelerate the

[7]This assumes that relations can be asymmetric (e.g., person A considers person B to be a friend while person B does not view person A the same way). Assuming symmetry would lead to the same ultimate conclusion.

[8]According to the model, if the neocortical ratio of an organism increases from 0.96 to 3.22 (i.e., across the range of extant non-human primates from *Avahi* to *Pan*) the estimated number of direct ties that organism could maintain would increase by roughly 64 while the number of indirect ties it could track would increase by approximately 4250. A further rise in its neocortical ratio to 4.1 (i.e. equivalent to *homo sapiens*) would increase the estimated number of direct ties the organism could manage by approximately 83 and the number of indirect ties it could track by roughly 17,600. Put differently, the model indicates that the smaller increase in neocortical ratio from *Pan* to *Homo* (i.e. +0.88) would produce a more substantial gain in the ability to track and manage social relations than would the larger increase in neocortical ratio from *Avahi* to *Pan* (i.e. +2.26). Dunbar's model requires that as the neocortex gets *bigger* every cubic centimeter of it must also get *better* at processing social information.

learning process (e.g., Bartlett, 1995; Brewer & Treyens, 1981; Martin, 1993; Neisser, 1976). A simple schema might be the common practice in the United States of relaying phone numbers as two groups of three digits and one group of four digits. The total number of digits is the same, but by "chunking" the numbers together using a pre-existing framework the digits become easier to encode and recall. Likewise, schemas provide individuals with pre-existing frameworks for identifying the salient features in a situation, facilitating comprehension and recall. For example, an experienced soccer fan is better able to follow the action in a fast-paced game than a neophyte because they can rapidly categorize events and identify those that are meaningful to the situation. Prior work on networks (De Soto, 1960) found that subjects recalled hierarchical relations more quickly when the relations matched an implicit schema. For example, asymmetric ties (i.e., A selected B but B does not select A) were learned more rapidly and accurately when the relationship was described in terms of authority (e.g., "commands") rather than friendship. Individuals employ schemas based on multiple characteristics, including kinship (Brewer & Yang, 1994), as well as context (Brewer & Garrett, 2001), and geographic location (Brewer & Garrett, 2001; Killworth & Bernard, 1982) to improve recall.

More recent work shows that "compression heuristics" (rules that allow social networks to be reconstructed from partial information) improve recall by making it unnecessary to memorize many dyads (Brashears, 2013). For example, authority relations are often characterized by transitivity (i.e., A ⟶ B, B ⟶ C, A ⟶ C) and so remembering a small number of edges could hypothetically allow many others to be reconstructed. A number of characteristics function as compression heuristics, including kinship, affective balance, and triadic closure (Brashears, 2013; Brashears et al., 2016; Brashears & Brashears, 2016); indeed, it appears that triadic closure may be a default compression heuristic that must be overridden with cognitive effort when it appears to be inappropriate to the situation (Brashears & Quintane, 2015). This suggests that social network recall under conditions of cognitive load or time pressure is likely to be biased in favor of closed triads. Women also exhibit a significant advantage in social network recall relative to men (Brashears et al., 2016) though this advantage disappears when we focus specifically on recall of structural holes (i.e., gaps separating sections of the social networks from each other: see Quintane et al., 2025). In short, humans achieve high levels of recall efficiency by optimizing their "software" in the form of schemas and heuristics rather than their "hardware" in the form of more neural tissue. Observed social network recall is consistent with mechanistic models of recall that employ these heuristics (Omodei et al., 2017), as well as Large Language Models (LLM) trained using human generated data (Wang, Cui & Kleinberg, 2024), confirming that heuristic use is sufficient to account for observed social networks. However, while many compression heuristics are applicable to multiple social networks, the use of heuristics will produce characteristic errors (e.g., recalling non-kin networks as more balanced than they really are; see Brashears & Brashears, 2016). Indeed, the utility of a compression heuristic depends on the degree to which its dictates are consistent with the structure of the underlying social network. As such, use of heuristics that fit the social network well should lead on average to improved recall when compared to effortful cognition, but using an incorrect heuristic will likely lead to poorer recall of the social network relative to effortful cognition. Thus, the value of a given heuristic is situationally contingent rather than absolute. Both the accurate recall of social networks facilitated by compression heuristics, as well as the inevitable errors that they entail, should influence networking behavior and thus realized social networks. Learning of heuristics can result from instruction (e.g., cultural education about family relationships) but also emerges through abstraction from direct observation of interactions (Son et al., 2023), and thus relies both on deduction and induction.

Priming, or activation of certain cognitive states or expectations, also impacts network recall. When exposed to social exclusion, individuals tend to activate denser social networks (O'Connor & Gladstone, 2015), and behave similarly when experiencing negative emotion generally (Shea et al., 2015). Critically, this line of work suggests that when bringing certain social networks to mind (i.e., out of long term storage and into working memory) the needs of the present, and the affective and motivational characteristics of that moment, can influence the process. As a result, networking behavior is influenced at a very basic level by transitory moods and attitudes as well as cognitive acumen. As a consequence, emotional states that are typical of an individual or a situation can exert a systematic influence on cognitive social network activation and subsequent social behavior. Priming effects can be measured in ego networks via the "jump ratio" (Sun et al., 2021; Sun & Smith, 2018), which quantifies the difference between the order in which individuals are recalled and their proximity to each other in the network. Similarly, content spreading through social networks can be influenced by heuristics operating at the individual level that govern both responses to errors (Brashears & Gladstone, 2016) as well as the potential (dis)utility of deception (Money, 2023). Individuals who encounter apparently flawed messages received through networks make judgments as to whether to ask for clarification, to infer the intended meaning of the message, or to discard the signal entirely as garbled. Heuristics shaping this decision process as well as inferences used to "repair" a flawed message ultimately help shape the speed and accuracy with which a social contagion can spread via social network ties. Likewise, individuals do not invariably choose to pass on information at all, or may manipulate or falsify the information for their own benefit, the good of others, or to protect the group. Over time these content-focused heuristics shape social network evolution directly as mistakes in error correction, or poor choices to disclose (or not) sensitive information will influence behavior and impact the affective strength of the social network.

In summary, schemas, compression heuristics, and priming all reveal the importance of mental "software" for social network encoding and recall. In all cases, the same underlying neurological capability supports these learned algorithms. Moreover, these mechanisms demonstrate that accepting the evolutionary or neural roots of a phenomenon does not obligate us to accept a simplistic or rigid set of expectations for behavior. Humans evolved neurological structures that permit considerable flexibility, and it is therefore no surprise to discover that they use it.

Future Directions

Extensive research demonstrates the importance of social life for the emergence of human intelligence, uncovers the roots of sociability in brain structure and function, and lays out the specific learned capacities that support social activity. Nonetheless, we are keenly aware that we are at the very beginning of the process of understanding how our social networks depend upon, and in turn influence, brain and mind.

One key challenge is the need to connect brain activation to specific algorithmic behaviors; in other words, to directly link brain to mind. The revolution in brain imaging brought about by fMRI, PET scanning, and (to a lesser extent) high fidelity EEG has provided crucial information about which sections of the brain are active for specific purposes. This knowledge has aided in uncovering the probable evolutionary sequences through which particular functions became represented in the brain, revealing signs of exaptation (i.e., adaptation of a structure evolved for one purpose to service a new goal). But there remains in many cases a gulf between knowing where in the brain a function is supported and knowing how that function is supported. Greater availability of inexpensive brain imaging coupled with improved understanding of the functions of mind offers new research avenues. Researchers might test subjects in an fMRI using tasks developed to explore specific compression heuristics or social cognitive skills, providing insight into exactly how the execution of different mental programs or cognitive resources are reflected in brain activity. Other opportunities in this domain are legion. Success in this area would provide new understanding of how particular types of social deficits result from neurological insult, as well as suggest new therapeutic avenues to address these issues. For example, much of the social deficits that characterize autism spectrum disorder (ASD) are thought to derive from differences in autistic and neurotypical brains. By understanding how the brain supports and enables social behavior, we gain the potential to craft more effective interventions for ASD and related disorders.[9] Likewise, by connecting social "programs" to underlying neurology, we gain a stronger ability to trace particular features of social networks to specific evolutionary timeframes or forces, essentially using biology to track social development and vice versa through the historical record. Such insights would be critical to answering questions and addressing key debates about phenomena like altruism, cooperation, competition, and social bias that are fundamental concerns of sociobiology and the social sciences.

A second key complication is understanding how mental activity is translated into social network structure. It appears clear from the evidence that large, complex social networks (at least in hominids) require powerful brains. Likewise, it is logical to infer that if and when a social structure grows too complex, people are more likely to make social mistakes (e.g., faux pas) that degrade or destroy relationships, thereby bringing the network back into manageable limits. But while such inferences are reasonable, we lack an understanding of exactly how variations in brain structure or function and mental representations manifest as social behaviors that, in turn, produce realized network structure. An intriguing early effort in this regard (Smith et al., 2012) focuses on the process of drawing social information from memory. This scheme distinguishes between the "potential network" of persons who are known, the "cognitively activated network" of persons who come to mind regarding a specific situation, the "mobilized network" of persons contacted, and the "realized network" of persons who respond positively to mobilization. As such, it is the most coherent model to date of how mental activity translates into manifested networks via a set of nested stages. Nonetheless, while the steps from potential to cognitively activated to mobilized networks are partially or wholly cognitive in nature and focused on memory, a great deal of non-cognitive, non-memory issues remain to be identified. For example, when choosing to mobilize an associate, how do individuals decide what method of mobilization to employ? Do they make a direct request, manipulate the other into taking a desired action, or do something else instead? Likewise, how does the tactic chosen by the mobilizer influence (or not) the target (i.e., shape the transition from mobilized to realized networks)? Answers to these questions will likely connect to other brain areas and, in turn, evolutionary processes that will add more detail to our understanding of human biological and social development. Addressing these questions would also provide insight into individual variation in people's ability to leverage social networks for desired outcomes and increased life chances (i.e., social capital)—specifically, does this variation have neurological or cognitive origins and is this a point of intervention to reduce inequalities?

A third issue stems from the distinction between capacity and motivation. The social brain hypothesis, and much of our discussion, has centered around identifying the brain functions and mind schemas that provide us with capacity for social activity. Yet, less attention has been given to understanding the motivation for specific outcomes or for particular tactics. Beyond simple distinctions between personality traits such as extroversion and introversion (which are themselves variably connected to realized networks: see Brashears et al., 2020) many brain functions may be tied not to the capacity to take social action, but to preferences for specific actions and outcomes as well as degrees of willingness to act in pursuit of them.

[9] It should be noted that the modern neurodiversity movement argues, in part, that ASD is not necessarily a disorder, but rather should be understood as a type of diversity. We do not take a stance on this point, but note that the name of the movement is itself predicated on the notion that what happens at the level of neurons shapes events at the level of social interaction. Improving our knowledge of the links between the brain and the social will provide new opportunities for all types of thinkers.

For example, variation in amygdala activation may be associated not only with ability to cultivate large networks, but with finding social stimuli more interesting and rewarding (McClure et al., 2004; Von Der Heide et al., 2014). Ignoring motivation may result in many researchers struggling in vain to identify a capacity difference in the ability to manage, for example, brokerage positions when the difference is actually one of willingness.

Conclusion

Research examining connections between mind, brain, and social networks provides critical insight into humans' neural, cognitive, and behavioral responses to social challenges and opportunities. Integrating social network theory and methodology with social neuroscience has the potential to improve predictions of the mechanisms underlying variation in individual and group behavior. This knowledge, in turn, can help researchers identify and prevent adverse consequences of social isolation on wellbeing and behavior, and develop more effective interventions to augment or maintain social networks and social cognitive abilities for neurological, psychological, and physical health across the life course.

Beyond individual health and wellbeing, improved knowledge of the role of brain and mind in producing and supporting social networks can play an important role in improving organizations, institutions, and digital tools that support human life. Virtually everything of value that is done by humans is done by teams and in many cases these teams rely on preexisting organizational and bureaucratic structures. Such structures are usually designed either without reference to the capacities of their members, or rely on imprecise ideas about these capabilities and limitations. Improved knowledge of the role of the brain and mind in sociability opens the door to adaptation of structures to make them more human-friendly, more comprehensible, and thus more effective than might otherwise be possible. Likewise, better knowledge about how we, as humans, grasp social phenomena offers the opportunity to build this knowledge into our growing array of artificially intelligent assistants. The future will include numerous AI decision support tools and failing to build them in a socially aware manner will fundamentally limit their effectiveness.

Perhaps most importantly, research into how our brains, our minds, and our networks co-create each other provides a window into what it means to be human. The need to manage the social has defined our species and appears to have been responsible for the development of our impressive level of intelligence. Likewise, the legacy of this process likely shapes how we think about and understand problems in the present. By understanding these connections, we gain the ability to see ourselves more deeply and more clearly. Ultimately, the most fundamental value of this research may lie in its ability to help us understand ourselves.

Acknowledgements This work was funded by the National Institute on Aging (5R01AG057739; 5R01AG070931; 5P30AG010133), and by an Indiana University Collaborative Research Grant through the Vice President for Research. This project also received support from the Indiana Clinical and Translational Sciences Institute, funded in part by Award Number UL1TR002529 from the National Institutes of Health, National Center for Advancing Translational Sciences, Clinical and Translational Sciences Award. The content is solely the responsibility of the authors and does not necessarily represent the official views of the National Institutes of Health.

References

Adolphs, R., Tranel, D., & Damasio, A. R. (1998). The human amygdala in social judgment. *Nature, 393*(6684), 470–474.

Ainsworth, M., Sallet, J., Joly, O., Kyriazis, D., Kriegeskorte, N., Duncan, J., Schüffelgen, U., Rushworth, M. F., & Bell, A. H. (2021). Viewing ambiguous social interactions increases functional connectivity between frontal and temporal nodes of the social brain. *Journal of Neuroscience, 41*, 6070.

Aral, S., & Van Alstyne, M. (2011). The diversity-bandwidth trade-off. *American Journal of Sociology, 117*(1), 90–171. https://doi.org/10.1086/661238

Baek, E. C., Hyon, R., López, K., Finn, E. S., Porter, M. A., & Parkinson, C. (2022). In-degree centrality in a social network is linked to coordinated neural activity. *Nature Communications, 13*(1), 1118.

Baron-Cohen, S. (2000). Theory of mind and autism: A review. In *International review of research in mental retardation* (Vol. 23, pp. 169–184). Academic Press. https://doi.org/10.1016/S0074-7750(00)80010-5

Barrett, L., Henzi, S. P., & Barton, R. A. (2022). Experts in action: Why we need an embodied social brain hypothesis. *Philosophical Transactions of the Royal Society B: Biological Sciences, 377*(1844), 20200533. https://doi.org/10.1098/rstb.2020.0533

Bartlett, F. C. (1995). *Remembering: A study in experimental and social psychology*. Cambridge University Press.

Bennett, D. A., Schneider, J. A., Tang, Y., Arnold, S. E., & Wilson, R. S. (2006). The effect of social networks on the relation between Alzheimer's disease pathology and level of cognitive function in old people: A longitudinal cohort study. *The Lancet Neurology, 5*(5), 406–412.

Bernard, H. R., & Killworth, P. D. (1977). Informant accuracy in social network data II. *Human Communication Research, 4*(1), 3–18.

Bernard, H. R., Killworth, P. D., & Sailer, L. (1982). Informant accuracy in social-network data V. An experimental attempt to predict actual communication from recall data. *Social Science Research, 11*(1), 30–66.

Bernard, H. R., Shelley, G. A., & Killworth, P. (1987). How much of a network does the GSS and RSW dredge up? *Social Networks, 9*(1), 49–61.

Bickart, K. C., Hollenbeck, M. C., Barrett, L. F., & Dickerson, B. C. (2012). Intrinsic amygdala-cortical functional connectivity predicts social network size in humans. *Journal of Neuroscience, 32*(42), 14729–14741. https://doi.org/10.1523/JNEUROSCI.1599-12.2012

Bickart, K. C., Wright, C. I., Dautoff, R. J., Dickerson, B. C., & Barrett, L. F. (2011). Amygdala volume and social network size in humans. *Nature Neuroscience, 14*(2), 163.

Bishop-Fitzpatrick, L., Mazefsky, C. A., Eack, S. M., & Minshew, N. J. (2017). Correlates of social functioning in autism spectrum disorder: The role of social cognition. *Research in Autism Spectrum Disorders, 35*, 25–34.

Brashears, M. E. (2013). Humans use compression heuristics to improve the recall of social networks. *Scientific Reports, 3*(1), 1513.

Brashears, M. E., & Brashears, L. A. (2016). The enemy of my friend is easy to remember: Balance as a compression heuristic. In S. R. Thye & E. J. Lawler (Eds.), *Advances in group processes* (Vol. 33, pp. 1–31). Emerald Group Publishing Limited. https://doi.org/10.1108/S0882-614520160000033001

Brashears, M. E., Brashears, L. A., & Harder, N. L. (2020). Where you are, what you want, and what you can do: The role of master statuses, personality traits, and social cognition in shaping ego network size, structure, and composition. *Network Science, 8*(3), 356–380.

Brashears, M. E., & Gladstone, E. (2016). Error correction mechanisms in social networks can reduce accuracy and encourage innovation. *Social Networks, 44*, 22–35.

Brashears, M. E., Hoagland, E., & Quintane, E. (2016). Sex and network recall accuracy. *Social Networks, 44*, 74–84.

Brashears, M. E., & Quintane, E. (2015). The microstructures of network recall: How social networks are encoded and represented in human memory. *Social Networks, 41*, 113–126.

Brewer, D. D., & Garrett, S. B. (2001). Evaluation of interviewing techniques to enhance recall of sexual and drug injection partners. *Sexually Transmitted Diseases*, 666–677.

Brewer, W. F., & Treyens, J. C. (1981). Role of schemata in memory for places. *Cognitive Psychology, 13*(2), 207–230.

Brewer, D. D., & Yang, B. L. (1994). Patterns in the recall of persons in a religious community. *Social Networks, 16*(4), 347–379.

Burt, R. S. (2004). Structural holes and good ideas. *American Journal of Sociology, 110*(2), 349–399. https://doi.org/10.1086/421787

Byrne, R. W. (1995). *The thinking ape: Evolutionary origins of intelligence*. Oxford University Press.

Caputi, M., Lecce, S., Pagnin, A., & Banerjee, R. (2012). Longitudinal effects of theory of mind on later peer relations: The role of prosocial behavior. *Developmental Psychology, 48*(1), 257.

Cardoso, S. D., Teles, M. C., & Oliveira, R. F. (2015). Neurogenomic mechanisms of social plasticity. *Journal of Experimental Biology, 218*(1), 140–149.

Cheney, D. L., & Seyfarth, R. M. (2019). *Baboon metaphysics: The evolution of a social mind.* University of Chicago Press. https://www.degruyter.com/document/doi/10.7208/9780226102429/html

Clutton-Brock, T. H., & Harvey, P. H. (1980). Primates, brains and ecology. *Journal of Zoology, 190*(3), 309–323.

Connor, R. C., & Mann, J. (2006). Social cognition in the wild: Machiavellian dolphins. *Rational Animals*, 329–367.

Dai, J., Jorgensen, N. A., Duell, N., Capella, J., Maza, M. T., Kwon, S.-J., Prinstein, M. J., Lindquist, K. A., & Telzer, E. H. (2023). Neural tracking of social hierarchies in adolescents' real-world social networks. *Social Cognitive and Affective Neuroscience, 18*(1), nsad064.

De Soto, C. B. (1960). Learning a social structure. *The Journal of Abnormal and Social Psychology, 60*(3), 417.

Decety, J., & Jackson, P. L. (2006). A social-neuroscience perspective on empathy. *Current Directions in Psychological Science, 15*(2), 54–58.

Dinse, H. R. (2021). Neuroplasticity in humans. In M. L. Zeise (Ed.), *Neuroscience for psychologists* (pp. 193–230). Springer.

DiPrete, T. A., Gelman, A., McCormick, T., Teitler, J., & Zheng, T. (2011). Segregation in social networks based on acquaintanceship and trust. *American Journal of Sociology, 116*(4), 1234–1283.

Djordjevic, J., Djordjevic, A., Adzic, M., & Radojcic, M. B. (2012). Effects of chronic social isolation on Wistar rat behavior and brain plasticity markers. *Neuropsychobiology, 66*(2), 112–119.

Domjan, M. (2018). *The essentials of conditioning and learning*. American Psychological Association. https://psycnet.apa.org/record/2017-58770-000

Dunbar, R. I. (1992). Neocortex size as a constraint on group size in primates. *Journal of Human Evolution, 22*(6), 469–493.

Dunbar, R. I. (1998). The social brain hypothesis. *Evolutionary Anthropology: Issues, News, and Reviews: Issues, News, and Reviews, 6*(5), 178–190.

Dunbar, R. I. M., & Shultz, S. (2017). Why are there so many explanations for primate brain evolution? *Philosophical Transactions of the Royal Society B: Biological Sciences, 372*(1727), 20160244.

Ellwardt, L., Van Tilburg, T. G., & Aartsen, M. J. (2015). The mix matters: Complex personal networks relate to higher cognitive functioning in old age. *Social Science & Medicine, 125*, 107–115.

Erickson, R. P. (2014). Are humans the most intelligent species? *Journal of Intelligence, 2*(3), 119–121.

Feld, S., & Carter, W. C. (1998). Foci of activity as changing contexts for friendship. *Placing Friendship in Context*, 136–152.

Fernald, R. D. (2015). Social behaviour: Can it change the brain? *Animal Behaviour, 103*, 259–265.

Freeman, L. C., Freeman, S. C., & Michaelson, A. G. (1988). On human social intelligence. *Journal of Social and Biological Structures, 11*(4), 415–425.

Freeman, L. C., Freeman, S. C., & Michaelson, A. G. (1989). How humans see social groups: A test of the Sailer-Gaulin models. *Journal of Quantitative Anthropology, 1*(3), 229–238.

Freeman, L. C., & Romney, A. K. (1987). Words, deeds and social structure: A preliminary study of the reliability of informants. *Human Organization, 46*(4), 330–334.

Freeman, L. C., Romney, A. K., & Freeman, S. C. (1987). Cognitive structure and informant accuracy. *American Anthropologist, 89*(2), 310–325.

Grainger, S. A., Crawford, J. D., Riches, J. C., Kochan, N. A., Chander, R. J., Mather, K. A., Sachdev, P. S., & Henry, J. D. (2023). Aging is associated with multidirectional changes in social cognition: findings from an adult life-span sample ranging from 18 to 101 years. *The Journals of Gerontology: Series B, 78*(1), 62–72.

Granovetter, M. S. (1977). The strength of weak ties. In *Social networks* (pp. 347–367). Elsevier.

Guyer, A. E., Silk, J. S., & Nelson, E. E. (2016). The neurobiology of the emotional adolescent: From the inside out. *Neuroscience & Biobehavioral Reviews, 70*, 74–85.

Hampton, W. H., Unger, A., Von Der Heide, R. J., & Olson, I. R. (2016). Neural connections foster social connections: A diffusion-weighted imaging study of social networks. *Social Cognitive and Affective Neuroscience, 11*(5), 721–727

Hauck, M., Fein, D., Maltby, N., Waterhouse, L., & Feinstein, C. (1998). Memory for faces in children with autism. *Child Neuropsychology, 4*(3), 187–198. https://doi.org/10.1076/chin.4.3.187.3174

Henry, J. D., Coundouris, S. P., Mead, J., Thompson, B., Hubbard, R. E., & Grainger, S. A. (2023). Social frailty in late adulthood: social cognitive and psychological well-being correlates. *The Journals of Gerontology: Series B, 78*(1), 87–96.

Holtmaat, A., & Svoboda, K. (2009). Experience-dependent structural synaptic plasticity in the mammalian brain. *Nature Reviews Neuroscience, 10*(9), 647–658.

Hughes, C., & Leekam, S. (2004). What are the links between theory of mind and social relations? Review, reflections and new directions for studies of typical and atypical development. *Social Development, 13*(4), 590–619.

Humphrey, N. K. (1976). The social function of intellect. In P. P. G. Bateson & R. A. Hinde (Eds.), *Growing points in ethology*. Cambridge University Press.

James, B. D., Glass, T. A., Caffo, B., Bobb, J. F., Davatzikos, C., Yousem, D., & Schwartz, B. S. (2012). Association of social engagement with brain volumes assessed by structural MRI. *Journal of Aging Research, 2012*, 512714.

Janicik, G. A., & Larrick, R. P. (2005). Social network schemas and the learning of incomplete networks. *Journal of Personality and Social Psychology, 88*(2), 348.

Kanai, R., Bahrami, B., Roylance, R., & Rees, G. (2012). Online social network size is reflected in human brain structure. *Proceedings of the Royal Society B: Biological Sciences, 279*(1732), 1327–1334.

Kardos, P., Leidner, B., Pléh, C., Soltész, P., & Unoka, Z. (2017). Empathic people have more friends: Empathic abilities predict social network size and position in social network predicts empathic efforts. *Social Networks, 50*, 1–5.

Kennedy, C., & Sokoloff, L. (1957). An adaptation of the nitrous oxide method to the study of the cerebral circulation in children; normal values for cerebral blood flow and cerebral metabolic rate in childhood. *The Journal of Clinical Investigation, 36*(7), 1130–1137.

Killworth, P. D., & Bernard, H. R. (1979). Informant accuracy in social network data III: A comparison of triadic structure in behavioral and cognitive data. *Social Networks, 2*(1), 19–46.

Killworth, P. D., & Bernard, H. R. (1982). A technique for comparing mental maps. *Social Networks, 3*(4), 307–312.

Killworth, P. D., Johnsen, E. C., Bernard, H. R., Shelley, G. A., & McCarty, C. (1990). Estimating the size of personal networks. *Social Networks, 12*(4), 289–312.

Krendl, A. C., & Betzel, R. F. (2022). Social cognitive network neuroscience. *Social Cognitive and Affective Neuroscience, 17*(5), 510–529.

Krendl, A. C., & Heatherton, T. F. (2009). Self versus others/self-regulation. *Handbook of Neuroscience for the Behavioral Sciences, 2*, 859–878.

Krendl, A. C., Kennedy, D. P., Hugenberg, K., & Perry, B. L. (2022). Social cognitive abilities predict unique aspects of older adults' personal social networks. *The Journals of Gerontology: Series B, 77*(1), 18–28.

Kwak, S., Joo, W., Youm, Y., & Chey, J. (2018). Social brain volume is associated with in-degree social network size among older adults. *Proceedings of the Royal Society B: Biological Sciences, 285*(1871), 20172708.

Lally, J. R., & Mangione, P. (2017). Caring relationships: The heart of early brain development. *YC Young Children, 72*(2), 17–24.

Lamblin, M., Murawski, C., Whittle, S., & Fornito, A. (2017). Social connectedness, mental health and the adolescent brain. *Neuroscience & Biobehavioral Reviews, 80*, 57–68.

Lewis, P. A., Rezaie, R., Brown, R., Roberts, N., & Dunbar, R. I. (2011). Ventromedial prefrontal volume predicts understanding of others and social network size. *NeuroImage, 57*(4), 1624–1629.

Lindenfors, P., Wartel, A., & Lind, J. (2021). 'Dunbar's number' deconstructed. *Biology Letters, 17*(5), 20210158.

Lubbers, M. J., Molina, J. L., Lerner, J., Brandes, U., Ávila, J., & McCarty, C. (2010). Longitudinal analysis of personal networks. The case of Argentinean migrants in Spain. *Social Networks, 32*(1), 91–104.

Manchella, M. K., Logan, P. E., Perry, B. L., Peng, S., Risacher, S. L., Saykin, A. J., & Apostolova, L. G. (2023). Associations between social network characteristics and brain structure among older adults. *Alzheimer's & Dementia, 20*, alz.13534.

Mann, J., Stanton, M. A., Patterson, E. M., Bienenstock, E. J., & Singh, L. O. (2012). Social networks reveal cultural behaviour in tool-using dolphins. *Nature Communications, 3*(1), 980.

Martin, C. L. (1993). New directions for investigating children's gender knowledge. *Developmental Review, 13*(2), 184–204.

McCarty, C., Killworth, P. D., Bernard, H. R., Johnsen, E. C., & Shelley, G. A. (2001). Comparing two methods for estimating network size. *Human Organization, 60*(1), 28–39.

McClure, S. M., York, M. K., & Montague, P. R. (2004). The neural substrates of reward processing in humans: The modern role of fMRI. *The Neuroscientist, 10*(3), 260–268.

Mehta, M. A., Golembo, N. I., Nosarti, C., Colvert, E., Mota, A., Williams, S. C. R., Rutter, M., & Sonuga-Barke, E. J. S. (2009). Amygdala, hippocampal and corpus callosum size following severe early institutional deprivation: The English and Romanian Adoptees Study Pilot. *Journal of Child Psychology and Psychiatry, 50*(8), 943–951.

Melin, A. D., Young, H. C., Mosdossy, K. N., & Fedigan, L. M. (2014). Seasonality, extractive foraging and the evolution of primate sensorimotor intelligence. *Journal of Human Evolution, 71*, 77–86.

Money, V. (2023). Demonstrating anticipatory deflection and a preemptive measure to manage it: An extension of affect control theory. *Social Psychology Quarterly, 86*(2), 151–169.

Neisser, U. (1976). *Cognitive psychology*. Appleton-Century-Crofts.

Nelson, E. E., Leibenluft, E., McClure, E. B., & Pine, D. S. (2005). The social re-orientation of adolescence: A neuroscience perspective on the process and its relation to psychopathology. *Psychological Medicine, 35*(2), 163–174.

Noonan, M. P., Mars, R. B., Sallet, J., Dunbar, R. I. M., & Fellows, L. K. (2018). The structural and functional brain networks that support human social networks. *Behavioural Brain Research, 355*, 12–23.

Noonan, M. P., Sallet, J., Mars, R. B., Neubert, F. X., O'Reilly, J. X., Andersson, J. L., Mitchell, A. S., Bell, A. H., Miller, K. L., & Rushworth, M. F. (2014). A neural circuit covarying with social hierarchy in macaques. *PLoS Biology, 12*(9), e1001940.

O'Connor, K. M., & Gladstone, E. (2015). How social exclusion distorts social network perceptions. *Social Networks, 40*, 124–128.

Omodei, E., Brashears, M. E., & Arenas, A. (2017). A mechanistic model of human recall of social network structure and relationship affect. *Scientific Reports, 7*(1), 17133.

Parker, S. T. (2015). Re-evaluating the extractive foraging hypothesis. *New Ideas in Psychology, 37*, 1–12.

Peng, S., Roth, A. R., Apostolova, L. G., Saykin, A. J., & Perry, B. L. (2022). Cognitively stimulating environments and cognitive reserve: The case of personal social networks. *Neurobiology of Aging, 112*, 197–203.

Perry, B. L., McConnell, W. R., Coleman, M. E., Roth, A. R., Peng, S., & Apostolova, L. G. (2022). Why the cognitive "fountain of youth" may be upstream: Pathways to dementia risk and resilience through social connectedness. *Alzheimer's & Dementia, 18*(5), 934–941.

Perry, B. L., Pescosolido, B. A., & Borgatti, S. P. (2018). *Egocentric network analysis: Foundations, methods, and models*. Cambridge University Press.

Perry, B. L., Roth, A. R., Peng, S., Risacher, S. L., Saykin, A. J., & Apostolova, L. G. (2022). Social networks and cognitive reserve: Network structure moderates the association between amygdalar volume and cognitive outcomes. *The Journals of Gerontology: Series B, 77*(8), 1490–1500.

Powell, J., Lewis, P. A., Roberts, N., Garcia-Finana, M., & Dunbar, R. I. (2012). Orbital prefrontal cortex volume predicts social network size: An imaging study of individual differences in humans. *Proceedings of the Royal Society B: Biological Sciences, 279*(1736), 2157–2162.

Quintane, E., Brashears, M. E., Gonzalez-Gomez, H, & Brands, R. (2025). Gender, network recall, and structural holes. *Personnel Psychology*, 1–17.

Roberts, S. G., Dunbar, R. I., Pollet, T. V., & Kuppens, T. (2009). Exploring variation in active network size: Constraints and ego characteristics. *Social Networks, 31*(2), 138–146.

Rolfe, D. F., & Brown, G. C. (1997). Cellular energy utilization and molecular origin of standard metabolic rate in mammals. *Physiological Reviews, 77*(3), 731–758.

Sale, A., Berardi, N., & Maffei, L. (2009). Enrich the environment to empower the brain. *Trends in Neurosciences, 32*(4), 233–239.

Sallet, J., Mars, R. B., Noonan, M. P., Andersson, J. L., O'Reilly, J. X., Jbabdi, S., Croxson, P. L., Jenkinson, M., Miller, K. L., & Rushworth, M. F. (2011). Social network size affects neural circuits in macaques. *Science, 334*(6056), 697–700.

Saramaki, J., Leicht, E. A., Lopez, E., Roberts, S. G. B., Reed-Tsochas, F., & Dunbar, R. I. M. (2014). Persistence of social signatures in human communication. *Proceedings of the National Academy of Sciences, 111*(3), 942–947.

Schmalzle, R., Brook O'Donnell, M., Garcia, J. O., Cascio, C. N., Bayer, J., Bassett, D. S., Vettel, J. M., & Falk, E. B. (2017). Brain connectivity dynamics during social interaction reflect social network structure. *Proceedings of the National Academy of Sciences of the United States of America, 114*(20), 5153–5158.

Schurz, M., Radua, J., Aichhorn, M., Richlan, F., & Perner, J. (2014). Fractionating theory of mind: A meta-analysis of functional brain imaging studies. *Neuroscience & Biobehavioral Reviews, 42*, 9–34.

Sharifian, N., Manly, J. J., Brickman, A. M., & Zahodne, L. B. (2019). Social network characteristics and cognitive functioning in ethnically diverse older adults: The role of network size and composition. *Neuropsychology, 33*, 936.

Shea, C. T., Menon, Y., Smith, E. B., & Emich, K. (2015). The affective antecedents of cognitive social network activation. *Social Networks, 43*, 91–99.

Shultz, S., Opie, C., & Atkinson, Q. D. (2011). Stepwise evolution of stable sociality in primates. *Nature, 479*(7372), 219–222.

Simpson, B., Markovsky, B., & Steketee, M. (2011). Power and the perception of social networks. *Social Networks, 33*(2), 166–171.

Smith, E. B., Menon, T., & Thompson, L. (2012). Status differences in the cognitive activation of social networks. *Organization Science, 23*(1), 67–82.

Sokoloff, L. (1960). The metabolism of the central nervous system in vivo. *Handbook of Physiology, Section I, Neurophysiology, 3*, 1843–1864.

Sommerlad, A., Sabia, S., Singh-Manoux, A., Lewis, G., & Livingston, G. (2019). Association of social contact with dementia and cognition: 28-year follow-up of the Whitehall II cohort study. *PLoS Medicine, 16*(8), e1002862.

Son, J.-Y., Bhandari, A., & FeldmanHall, O. (2023). Abstract cognitive maps of social network structure aid adaptive inference. *Proceedings of the National Academy of Sciences, 120*(47), e2310801120.

Stanley, D. A., & Adolphs, R. (2013). Toward a neural basis for social behavior. *Neuron, 80*(3), 816–826.

Stiller, J., & Dunbar, R. I. (2007). Perspective-taking and memory capacity predict social network size. *Social Networks, 29*(1), 93–104.

Sun, H., Brashears, M. E., & Smith, E. N. (2021). Network representation capacity: How social relationships are represented in human mind. In M. L. Small, B. L. Perry, B. A. Pescosolido, & E. B. Smith (Eds.), *Personal networks: Classic readings and new directions in egocentric analysis* (pp. 555–572).

Sun, H., & Smith, E. B. (2018). Cognitive search and social networks. In *Academy of Management Conference, Chicago, United States*.

Thoits, P. A. (2011). Mechanisms linking social ties and support to physical and mental health. *Journal of Health and Social Behavior, 52*(2), 145–161.

Tottenham, N., Hare, T. A., Quinn, B. T., McCarry, T. W., Nurse, M., Gilhooly, T., Millner, A., Galvan, A., Davidson, M. C., Eigsti, I., Thomas, K. M., Freed, P. J., Booma, E. S., Gunnar, M. R., Altemus, M., Aronson, J., & Casey, B. J. (2010). Prolonged institutional rearing is associated with atypically large amygdala volume and difficulties in emotion regulation. *Developmental Science, 13*(1), 46–61.

Triki, Z., Zhou, T., Argyriou, E., Sousa de Novais, E., Servant, O., & Kolm, N. (2023). Social complexity affects cognitive abilities but not brain structure in a Poecilid fish. *BioRxiv*, 2023–08.

Van Kreveld, D., & Zajonc, R. B. (1966). The learning of influence structures. *Journal of Personality, 34*(2), 205–223.

Von Der Heide, R., Vyas, G., & Olson, I. R. (2014). The social network-network: Size is predicted by brain structure and function in the amygdala and paralimbic regions. *Social Cognitive and Affective Neuroscience, 9*(12), 1962–1972.

Wang, Y., Cui, H., & Kleinberg, J. (2024). Microstructures and accuracy of graph recall by large language models. *Advances in Neural Information Processing Systems (NeurIPS), 38*, 1–29.

Watson, A. C., Nixon, C. L., Wilson, A., & Capage, L. (1999). Social interaction skills and theory of mind in young children. *Developmental Psychology, 35*(2), 386.

Wellman, B., Wong, R. Y., Tindall, D., & Nazer, N. (1997). A decade of network change: Turnover, persistence and stability in personal communities. *Social Networks, 19*(1), 27–50.

Williamson, C. M., Franks, B., & Curley, J. P. (2016). Mouse social network dynamics and community structure are associated with plasticity-related brain gene expression. *Frontiers in Behavioral Neuroscience, 10*, 152.

Xiong, Y., Hong, H., Liu, C., & Zhang, Y. Q. (2023). Social isolation and the brain: Effects and mechanisms. *Molecular Psychiatry, 28*(1), 191–201.

Yeh, Z.-T. (2013). Role of theory of mind and executive function in explaining social intelligence: A structural equation modeling approach. *Aging & Mental Health, 17*(5), 527–534.

Zovetti, N., Rossetti, M. G., Perlini, C., Brambilla, P., & Bellani, M. (2021). Neuroimaging studies exploring the neural basis of social isolation. *Epidemiology and Psychiatric Sciences, 30*, e29.

The Neural Dynamics of Interpersonal Synchrony

18

Joseph Dippong and Jon Overton

Abstract

Over several decades, research on interactional synchrony has demonstrated the power of social situations for creating coordinated and rhythmic patterns of shared behavior between and within individuals and groups. More recently, dating back to the early 2010s, work within social neuroscience has revealed the extent to which such behavioral patterns are reflected in and driven by shared neural activity. While the broader literature on interactional synchrony has developed around a fairly robust set of theoretical models, research on neural synchrony and interbrain synchronization lacks a strong theoretical backbone. In this chapter, we argue in favor of developing a neurosociology of synchrony that draws on well-established sociological theories to make sense of and bring coherence to a diverse set of empirical findings within social neuroscience. We review research on behavioral, biological, and neural synchrony, calling attention both to ways that sociological theory can help explain diverse findings and to how such findings can open new avenues of theoretical inquiry for sociologists. Additionally, we provide examples of research that highlights the value that sociological theory brings to neuroscience and that neuroscience brings to sociology, identifying open questions and future directions. In doing so, we call special attention to work on vocal accommodation as an example for neurosociologists to follow and a potential starting point for developing a unique neurosociology of interpersonal synchrony.

The Neural Dynamics of Interactional Synchrony

When observing two or more people engaged in interaction, the degree to which their behaviors visibly align and mirror each other is often quite striking. For example, when a speaker leans toward their interaction partner during a particularly intimate moment of conversation, we intuitively expect that the person to whom they are speaking will reciprocate and mimic that behavior—often automatically and without thinking—bridging the physical and social space between them. Interactional synchrony like this occurs across a wide array of social contexts. Whether it is a pair of close friends hiking who find themselves walking naturally in stride, or a group of largely unacquainted baseball fans who throw their hands in the air in unison to celebrate their team's hero hitting a home run, social settings can generate an interactional rhythm that coordinates the behaviors of everyone present so that

J. Dippong (✉) · J. Overton
University of North Carolina at Charlotte, Charlotte, NC, USA
e-mail: jdippong@charlotte.edu; jovert11@uncc.edu

W. Kalkhoff et al. (eds.), *Handbook of Neurosociology*, Handbooks of Sociology and Social Research, https://doi.org/10.1007/978-3-031-95615-7_18

the group moves as a single social unit (Hall, 1976; Kalkhoff et al., 2011).

Many of the behaviors involved in interactional synchrony occur at a superficial outer level, where even the most casual observer can see them. But beyond shared gestures and speech patterns, the rhythm of interaction also coordinates a variety of inward biological processes that are not as easily observed. Such inward synchrony is evident in research that demonstrates that romantic partners' heartbeats (Shimshock et al., 2024) and electrodermal activity (Coutinho et al., 2019) synchronize during emotional discussions; that rituals facilitate heart rate synchrony between participants and observers (Konvalinka et al., 2011); that during conversation, speakers' vocal frequencies converge in a non-conscious process of social alignment (Dippong, 2020; Gregory, 1983); and that participation in a shared task can generate synchronous brain activity (Condon & Ogston, 1967; Yun et al., 2012).

Decades of research across the social sciences has focused on examining and explaining the social and biological mechanisms that drive behavioral synchrony, as well as on how synchrony shapes a variety of interactional outcomes of interest to sociologists. In general, synchrony serves to enhance and create social bonds (Cacioppo et al., 2014). Higher levels of synchrony are associated with greater solidarity (Kalkhoff et al., 2011), better communication quality (Gregory et al., 1997), and increased romantic attraction (Cohen et al., 2024), to name only a few outcomes. The effects of interactional synchrony are so beneficial that Smith (2013) goes as far as to describe synchronized action as a social "immune response" that provides relief in moments of distress.

Despite the well documented importance of interactional synchrony for a multitude of social processes, almost no sociological research has delved into its neural underpinnings. Although a good deal of work has been done within the field of social neuroscience to elucidate the neural processes that drive behavioral synchrony and inter-brain synchronization, the absence of a sociological viewpoint within the paradigm has led to a patchwork of disparate empirical findings (and a few consistent trends) with virtually no theoretical framework to explain them. Indeed, Holroyd (2022) points out that the lack of theoretical backbone, along with a bit of methodological incoherence, presents serious challenges for research on interbrain synchronization.

There are gaps, then, in the social scientific understanding of neural processes related to synchrony that neurosociology is well suited to address. Aside from providing a theoretical structure to existing patterns of findings, taking a neurosociological vantage point on interactional synchrony offers the potential to refine sociological theory using the tools and methods of neuroscience and to illuminate new approaches to tackling seemingly unanswerable questions (Kalkhoff et al., 2016).

Our aim in this chapter is to advance the idea of a neurosociology of synchrony in two ways: (1) by calling attention to opportunities to utilize sociological theories and paradigms to bring coherence to and provide context to the disparate findings that have emerged within the neuroscience of behavioral synchrony, and (2) by highlighting how existing work on the neural processes associated with interactional synchrony connects to sociological studies of vocal accommodation. We begin with an overview of interactional synchrony, focusing on the different forms that synchrony can take and how different sensory modalities affect synchrony. Next, we discuss relevant findings from social neuroscience regarding the neural processes involved in interpersonal synchrony and inter-brain synchronization. In doing so, we consider specific examples of sociological research on vocal accommodation as a mutually beneficial connecting point between biosociology, neurosociology and social neuroscience more broadly. We conclude with a summary of our argument that a neurosociology of interactional synchrony has the potential to illuminate new avenues of inquiry in sociology and provide a stronger theoretical framework for the neuroscience of synchrony.

General Processes Involved in Interactional Synchrony

Researchers have observed a multitude of ways that people engage in interactional synchrony at the behavioral, physiological, and neural levels (Nguyen, Abney, et al., 2021). Early studies by Condon and colleagues (see, e.g., Condon & Ogston, 1967; Condon & Sander, 1974) identified many of the different forms of synchrony that occur during interaction and laid the groundwork for extensive research on interactional synchrony. Analyzing film footage of parents interacting with their children, Condon and Ogston (1967) observed two forms of outward synchrony. First, people synchronize their actions with themselves. Self-synchrony occurs when, for example, a speaker coordinates their arm and hand gestures with the rhythm of their own speech or when they lean forward and widen their eyes while revealing a juicy bit of gossip (Dippong & Kalkhoff, 2018). Second, people synchronize their behaviors with those of their interaction partner(s). Interactional synchrony involves a variety of forms of nonconscious behavioral co-adjustment, including aligning speech patterns and vocal pitch, as well as aligning movements, such as when two people spontaneously and unintentionally stop conversation to take a drink at the same time. While the overall interaction reveals a broad and observable guiding rhythm, it is often composed of dozens of micro-rhythms that occur below the level of actors' awareness (Condon & Ogston, 1967).

Neural Synchrony and Physiological Synchrony

Along with outward forms of synchrony, Condon and Ogston (1967) observed a form of inward synchrony in which people's neural behavior, as measured using EEG, aligned, reflecting a bioelectric harmony. Recent hyperscanning studies using EEG, fNIRS, and fMRI have begun to identify the mechanisms that drive interbrain synchronization and how interbrain synchronization affects interactional outcomes including task performance (Reinero et al., 2021) and emotional experiences (Kinreich et al., 2017). Additional research has focused on examining the neural correlates of outward synchrony (see, e.g., Atzil et al., 2014 and Cacioppo et al., 2014), which involves neural processes that are distinct from those involved in interbrain synchronization. We discuss this body of research more in-depth later in this chapter.

The tendency to non-consciously synchronize with others appears to be an innate and "hard-wired" biological capacity in humans. As evidence of the innate biological underpinnings of behavioral synchrony, Condon and Sander (1974) observed synchrony between neonates and their parents within only a few hours of the child's birth. Importantly, the mechanisms that drive synchrony are uniquely attuned to other humans. The babies in Condon and Sanders' study aligned their movements in rhythm with natural and recorded voices, but not with non-human sounds.

Evidence also suggests that elements of behavioral synchrony are partly socialized. That is, during interaction, parents and infants pass between states of behavioral alignment and misalignment, allowing the infants to *learn* how to read gestures and re-align with others (Montirosso & McGlone, 2020). These learned patterns of outward behavioral response affect physiological and neural synchrony in a variety of ways. For example, observing appropriate turn-taking during conversation (a learned behavior) increases neural synchrony between mothers and their children (Nguyen, Abney, et al., 2021) and "tunes" children's social brain (Levy et al., 2017). The process of learning to synchronize may also have ramifications for children's ability to self-differentiate (Montirosso & McGlone, 2020), which is a key milestone in the development of self-awareness (Gallup, 1983). Thus, the ability to synchronize with others is likely a prerequisite to the emergence of a social self more generally.[1]

[1]This idea aligns well with Mead's (1934) social behaviorist model of self-development, in that Mead argues that the first stage in developing a self involves simply imitating and mimicking the actions of significant others. This

Forms of Interactional Synchrony

Thus far we have spoken of interactional synchrony as if it is one single phenomenon in which people align their actions toward each other to create a distinct rhythm or social "fingerprint." Such an approach, however, belies the fact that different forms of interaction produce different forms of synchrony. Cacioppo et al. (2014) describe three distinct forms of interactional synchrony. First, orchestration occurs when people align their actions to a centralized "pacesetter" who guides the movements of the group. From a sociological perspective, this form of synchrony reflects the type of processes that Collins (2004) refers to as the "mutual-focus/ emotional-entrainment" model, in which a mutual focus of attention coordinates actions within a group and generates the emotions that produce solidarity and a sense of moral rightness. Accordingly, orchestration aligns behaviors and physiological responses because individuals are attending to, and are socially engaged with, the same external stimulus, such as a charismatic leader (Collins, 2004).

Second, reciprocal entrainment occurs when two or more people monitor each other's actions and mutually adjust to each other. As the name suggests, this type of entrainment involves a give-and-take where each person within the interaction is involved in creating the behavioral rhythm. During reciprocal entrainment, people act and speak in ways that are distinct from their typical and unique ways of being (Oullier et al., 2008). In this type of interaction, behavioral alignment does not occur through attention to some outside stimulus. Rather, the interaction itself serves as the point of mutual focus. At the behavioral level, this type of entrainment is reflected in shared gestures, such as when Person A crosses their legs and Person B follows suit and does the same, and then Person A reciprocates by sitting more upright to match Person B's posture. At the physiological level, this can be observed in shared patterns of electrodermal activity as people mutually adjust to each other's level of emotional arousal (Coutinho et al., 2019).

Third, unilateral entrainment refers to synchrony that arises not because people are mutually adjusting to each other, but because Person A adapts to person B, while Person B does not adjust in return. Using the language of accommodation theory (Giles & Coupland, 1991), unilateral entrainment occurs when one actor asymmetrically accommodates another. Patterns in how people align their actions—who converges with or diverges from whom—can indicate desire for affiliation or distance (Gallois et al., 2005), and differences in the amount of effort that people put into convergence align with their perceptions of each other's status and power (Kinreich et al., 2017), and predict openness to influence between task partners (Dippong et al., 2022; Dippong & Jillani, 2024; Jamil, 2023).

It is important to note that the presence of observable regularities between actors does not inherently indicate a shared social experience; a given stimulus can produce shared physiological and neural responses between people without any theoretically meaningful social process occurring. To this end, Holroyd (2022:348) states,

> Natural music evokes neural responses that are correlated across individuals, so the brains of two people listening to the same song, even if played on different devices in two different rooms, can synchronize simply due to the common temporal profile of the auditory input….it is immaterial whether the synchronizing information comes from an external stimulus… or from another person.

Because orchestration and reciprocal entrainment are based on a shared focus of attention (i.e., an external stimulus or the group members themselves), it is possible to mistakenly ascribe social significance to outcomes that are merely coincidental. This is, of course, because in both cases the appearance of physiological synchrony could simply be a byproduct of exposure to a common input. In the absence of a theory to specify the social process mediating between environmental input and physiological output and to delineate scope or boundary conditions under which we can expect entrainment to occur, any

---

argument presumes the ability to read others' actions and align your own with them.

observed synchronous action could well be merely epiphenomenal and not reflect entrainment in any socially meaningful way (Holroyd, 2022).

Situational Effects on Physiological and Neural Synchrony

For physiological or neural synchrony to have any theoretical or practical significance, it must arise through a meaningful social process and not simply through shared exposure to a common input (Holroyd, 2022). Simply engaging in a shared task is not sufficient to rule out the possibility that physiological synchrony between actors is spurious, as completing the same task in the same setting can activate similar internal processes, regardless of any relationship between actors (Liu & Pelowski, 2014). Factors such as cooperation, interdependence, temporality, intentionality, sensory modality, and proximity/ touch give context to and shape interactions (Hoehl et al., 2021). These factors, in part, help us discern whether any observed physiological synchrony reflects a social process or is simply an epiphenomenon, and are therefore important considerations when designing or assessing research on biological synchronization.

Cooperation, Interdependence, and Temporality

Different interactional contexts and formats provide different levels of social intimacy and different opportunities for people to "sync up" with each other. Whether people gather to solve a problem, to exchange information, or to just enjoy quiet co-presence, the structure of the interaction dictates the extent to which synchrony arises socially, mechanically, or not at all. For example, Liu and Pelowski (2014) argue that levels of cooperation, interdependence, and task concurrence shape which regions of the brain are likely to synchronize. Concurrent, interdependent, cooperative tasks require anticipation of partners' actions, and activate the frontoparietal areas, while tasks that are asynchronous (turn-based) or independent or competitive and require less prediction of partners' behaviors activate areas in the prefrontal cortex (Liu & Pelowski, 2014). Although it is not clear how cooperation, interdependence, and temporality affect other forms of physiological synchrony, such as vocal convergence, arguments from accommodation theory suggest that the motive to self-differentiate (e.g., low cooperation and high independence) reduces convergence and may induce divergence (Gallois et al., 2005).

Sensory Modality

Interaction naturally entails a range sensory channels for communicating with and synchronizing with a partner. Whereas a face-to-face encounter can bring together a mix of auditory, tactile, and visual stimulus information, interaction that is mediated by some form of technology often either isolates one particular modality (e.g., telephones isolate the auditory modality), or provides a mix of modalities (e.g., virtual meetings that combine auditory and visual information, but not tactile).[2] This is important because different sensory modalities afford different mechanisms for producing synchrony and as such bring different physiological and neural processes into play (Hove et al., 2013).

Research provides extensive evidence for the primacy of the auditory modality in generating synchrony. Because establishing synchrony and rhythm involves an inherent temporal dimension, it may be that the brain prioritizes auditory cues over visual cues in adapting to the interaction environment. To this end, Hoehl et al. (2021) note differences between the capacity for auditory and visual stimuli to induce synchrony, stating, "our natural environment entails fewer visual than auditory rhythms leading to differences in exposure and processing practice…while audition emphasizes temporal analysis, vision emphasizes spatial analysis" (p. 6). Thus, when interaction

[2] Of course, with the advent of more complex technologies that provide haptic feedback between users and elaborate virtual environments and/ or avatars of interaction partners, more work is needed to ascertain whether immersive virtual environments induce synchrony in the same way that face-to-face interactions do (see Kalkhoff et al., 2020).

entails multi-sensory input, the brain privileges auditory cues in generating synchrony based on the temporal coupling of neural responses to signal frequency within the auditory channel (Hoehl et al., 2021; Jiang et al., 2015). Feldman et al. (2011) note that vocal coupling induces other forms of physiological coupling, and Escoffier et al. (2015) demonstrate that visual processes in the brain attune to auditory rhythms. Such findings clearly emphasize the primacy of the auditory mode for generating synchrony.

The primacy of audition in achieving synchrony is demonstrated most clearly in studies of multimodal interaction. To the extent that the input received from multiple senses align—such as an auditory cue of someone clapping their hands coupled with a visual cue of their hands clapping—the degree of synchrony should increase (Armstrong & Issartel, 2014). But when visual and auditory inputs diverge, behaviors converge toward patterns in the auditory signal (Hoehl et al., 2021). Diverging auditory and visual signals can also serve as a distraction that impedes synchrony (Jiang et al., 2015). For example, Kalkhoff et al. (2020) found that research participants who communicated exclusively through the auditory channel achieved a greater degree of vocal convergence than those who communicated through audio and video. The presence of the participants' own video stream presented alongside their partners' video stream created a high "signal-to-noise" ratio between auditory and visual signals (Hoehl et al., 2021) such that participants' attention to their own visual cues interfered with their attention to signals they were receiving from their partners (Kalkhoff et al., 2020). Although speakers achieved a degree of vocal synchrony, it was attenuated by a mismatch between sensory signals.

Proximity and Touch

There is some evidence to suggest that touch can enhance physiological attunement in certain contexts, such as between parents and children and between romantic partners. Regarding parents and children, Carozza and Leong (2021) argue that caregiver touch and proximity produce a calming effect on infants that allows them to attend to communicative rhythms; furthermore, touch "aligns the oscillatory phases of parent and infant neural activity for optimal mutual receptivity" (p. 5). Thus, caregiver touch can serve as a form of orchestration that facilitates reciprocal entrainment, though touch seems to affect physiological and neural synchrony but not behavioral synchrony (Nguyen, Schleihauf, et al., 2021). The context of touch matters, too. For example, affectionate touch increases neural synchrony in the lateral and medial prefrontal cortex (mPFC), but stimulating touch (e.g., poking or tickling) reduces neural synchrony (Nguyen, Schleihauf, et al., 2021). Research on romantic partners reveals a similar pattern of findings: handholding during interaction increases synchrony in galvanic skin response, but not in respiration or pulse (Chatel-Goldman et al., 2014). Interestingly, in therapeutic settings, touch-mediated synchrony in galvanic skin response between romantic partners reduces self-ratings of physical pain (Reddan et al., 2020).

Implications for Neurosociology

In sum, interactional synchrony occurs across behavioral, physiological, and neural domains, and it reflects different forms of entrainment based on the source of entrainment and the degree to which people contribute to achieving synchrony. Additionally, multiple factors—including sensory modality and physical proximity—affect the extent to which people synchronize with those around them. These factors can serve as a foundation for developing boundary or scope conditions that inform the circumstances under which we should expect to observe synchrony and the type and magnitude synchrony that should emerge. Delineating the scope conditions under which we can expect to observe different forms of synchrony can help researchers develop stronger experimental tests and help us discern whether or not the results of a given study provide evidence for or against an argument (Dippong, 2012). Further theorizing is needed to outline exactly how these pieces fit together to constrain or enable synchrony.

Neural Processes Involved in Interpersonal Synchrony

Having considered some of the general processes underlying the various forms of interpersonal synchrony, in this section we focus exclusively on neural synchrony. Research in this area tends to fall along two lines: 1) work exploring the neural correlates of behavioral and physiological synchrony, and 2) studies that examine interbrain synchronization as a unique form of interpersonal synchrony. In the remainder of this section, we review a selection of studies from this ever-expanding body of literature and consider how these findings can inform and be informed by sociological theory. We also discuss research on vocal synchrony as a fruitful starting point for developing a neurosociology of synchrony.

Behavioral Synchrony and the Brain

Research on the neural correlates of behavioral and physiological synchrony typically seeks to uncover how shared interactional rhythms are reflected in shared patterns of activity within the brain. Much of the work in this area reflects an assumption that behavioral synchrony occurs prior to the emergence neural synchrony (i.e., that exposure to a common input activates common neural processes), though the question of causal direction is still open (see, e.g., Lindenberger et al., 2009; Novembre et al., 2017). To understand how the brain operates during synchronous action, researchers have taken two general approaches. Whereas some studies examine neural activity that occurs when individuals observe video vignettes of themselves or others engaged in synchronous behavior, more recent work has leveraged hyperscanning methods to measure brain activity when people are actively engaged in interactional synchrony that is either spontaneous or induced.

Neural Correlates of Observed Synchrony

One approach to identifying the neural processes that are at play during behavioral synchrony involves videorecording pairs of actors engaged in interaction (often this involves mother-infant dyads), coding the videos for synchronous and non-synchronous behavior, and then recording participants' brain activity afterward while they view recordings of different segments of interaction. This allows researchers to compare which areas of the brain are active during periods of synchrony versus non-synchrony and to compare the neural activity of participants observing themselves engaging in synchronous action against their neural activity when observing others engaged in similar activity.

Atzil et al. (2014) provide one example of such an approach, employing fMRI to measure brain activity in mothers who were observing interactions between themselves and their infant children. Results revealed increased activation in several regions—including in the dorsal angulate cingulate cortex (dACC), fusiform gyrus, cuneous, and nucleus accumbens (NAcc)—when mothers viewed synchronous action compared to other forms of interaction. Subsequent analyses revealed that activation in the dACC predicted mothers' subjective perceptions of synchrony. In explaining the role of the dACC in producing and responding to synchrony, Atzil et al. (2014) argue that "the reward-related information processed by the dACC is critical in order to make accurate judgments on the desirability of social events and assists in the planning of adequate motor response" (p. 1200). The authors also highlight the role of the dACC in empathy as it relates to observing both positive and negative experiences, pointing to increased dACC activation when viewing unfamiliar others engage in synchronous interactions. The activation of the dACC when observing periods of the self and others engaged in synchronous interaction (but not during non-synchronous action) suggests in part that its role in generating synchrony involves promoting mutual reward.

Whereas Atzil and colleagues' research focused exclusively on mothers' neural responses to synchrony, Levy et al. (2017) examined the shared patterns of neural activity that occur between mothers and children when viewing synchronous interactions. Levy et al. (2017) recorded interactions between mothers and their children (aged 9–13 years) and collected neural data from both members of the dyad while they watched videos of the interaction. MEG results point to shared patterns of increased gamma power in the superior temporal sulcus (STS), which plays a key role in social perception, language processing, and theory of mind. The coupling of mother and child neural activity in the STS was noted only when they watched videos of themselves interacting and was not detected when they observed unfamiliar others engaged in synchronous interaction (Levy et al., 2017). Gamma coupling in the STS, then, is likely not merely the product of shared exposure to just any synchronous interaction, but rather arises only in self-relevant interactions.

Another approach to understanding the neural underpinnings of behavioral synchrony involves measuring the brain activity of individuals interacting with simulated partners. Cacioppo et al. (2014) sought to identify the neural processes involved in unilateral entrainment (i.e., synchrony that occurs when one actor accommodates/ adjusts their behavior to those on another without reciprocity). Participants engaged in a simple task in which they were instructed to match tapping rhythms with a computer simulated partner, though they were led to believe that they were interacting with a real person. Compared to trials in which participants worked alone on the task, fMRI results from trials that involved working with an ostensive partner indicated significantly increased activity in only one region: the ventromedial prefrontal cortex (vmPFC), which is heavily involved in processes related to decision-making, prosocial behavior, and reward processing. Higher vmPFC activation was significantly correlated with participants' perceptions of synchrony with their partner and with participants' self-reported feelings of affiliation toward their ostensive partner. In other words, vmPFC activation predicted not only awareness of synchrony, but also social identification with a task partner.

Hyperscanning Studies of Neural Correlates of Natural and Induced Synchrony

While studies of observed or simulated interaction can reveal certain aspects of the neural processes involved in behavioral synchrony, they entail inherent limitations (i.e., studies involving observed synchrony collect neural data after interaction is completed and simulated interaction studies cannot assess shared neural patterns between interaction partners). Hyperscanning studies have the potential to address these methodological limitations by measuring patterns of neural activity that emerge simultaneous to the interaction that is producing them. Hyperscanning studies cannot necessarily delineate which patterns of shared neural activity are socially relevant and which are artifacts of shared exposure to a common stimulus—such things are likely best understood through the application of a clear theoretical model—but they can help identify the situational conditions that produce neural synchrony and how that synchrony is reflected similarly or differently in the neural behaviors of two or more people participating in the same interaction.

In a fascinating study, Lindenberger et al. (2009) employed EEG hyperscanning to measure phase synchrony between pairs of guitarists playing the same song while following a metronome. Averaging the phase locking value (PLI) and interbrain phase coherence (IPC) across six electrodes in the fronto-central region, Lindenberger et al. (2009) observed high phase synchrony both when the musicians were performing and in the moments before the performance as they were preparing to play.[3] During the performance period, IPC demonstrated strong coherence in the fronto-central electrodes within the theta frequency band (3–7 Hz), though synchrony reduced significantly after the first two seconds of the performance (Lindenberger et al., 2009). Furthermore, coherence was highest among pairs

[3]Here PLI reflects a measure of phase variance across electrodes within one actor, and IPC captures variance in phases between two electrodes (one for each actor).

that demonstrated low within-brain phase variance as measured by PLI. The authors conjecture that the observed phase coherence was at least partially attributable to brain regions associated with social cognition and theory of mind (Lindenberger et al., 2009).

Underscoring the importance of the vmPFC during synchronous action, Yun et al. (2012) used EEG hyperscanning to measure within- and between-brain correlates of induced unilateral entrainment. Dyadic partners were assigned to either the position of group leader or follower and engaged in a simple task in which they pointed at each other and watched each other's pointing finger. Along with behavioral measures indicating that participants adjusted the position of their finger in adjustment to their partner's movements, source-localized EEG results revealed a correlation between fingertip movement and activity within the theta frequency band of participants' vmPFC. Additionally, the correlation between behavior and neural activity increased when participants completed a cooperative training session prior to engaging in the critical trials. Thus, an initial period of synchrony enhanced synchrony in subsequent interactions. Furthermore, Yun et al.'s (2012) results revealed that in achieving neural synchrony, participants who were randomly assigned to the position of group leader contributed less to achieving phase synchrony than those assigned to the follower role, hinting at social structural factors driving asymmetric synchrony.

The literature that we have discussed so far suggests that behavioral synchrony activates neural processes in a fairly straightforward manner. The observation that neural synchrony increases in interactions that follow episodes of behavioral synchrony, however, suggests a potential reciprocal relationship between the two processes. Indeed, evidence seems to point in this direction. Novembre et al. (2017) essentially flip the direction of causality by inducing neural synchrony before assessing behavioral synchrony. Using sinusoidal stimulation to induce phase synchrony within the beta frequency band of the motor cortex, Novembre et al. (2017) found that such stimulation produced stronger behavioral synchrony (in the form of a simple finger tapping task) between experimental participants. Accordingly, it seems likely that patterns of behavioral matching "jump start" neural synchrony, which then produces increased and more precise synchrony.

Interbrain Synchronization

Interbrain Synchronization (hereafter referred to as IS) is closely related to, but distinct from, neural entrainment associated with interactional synchrony, though the two concepts are sometimes used interchangeably. The failure of some researchers to clearly discriminate between the two processes has led to confusion regarding IS as a unique phenomenon. Holroyd (2022) argues that for IS to be a meaningful phenomenon, it must be "distinct from…other forms of neural synchrony," and that IS, as opposed to other forms of neural synchrony.

> [h]olds that different brains can come together to act as a functional unit, much like modules within an individual brain can coordinate their activities for a superordinate purpose…[facilitating] social exchanges by establishing a resonant state between brains, albeit one that is also reciprocally mediated by way of those same behaviors. (p. 349)

In other words, IS does not encompass all processes involving shared neural patterns, but instead refers to the joint application of two or more brains in producing joint social action.

The concept of IS invokes the joining of brains into a single functional unit involved in the shared production of action. Research in this area often examines correspondence between activity in different brain regions across actors as they jointly emit and respond to each other's actions. Whereas work on the neural correlates of synchrony tends primarily to identify regions of shared brain activation, IS research seeks to uncover both regions of shared activation and correspondence across different regions between actors. In this way, IS cannot be easily boiled down to shared neural responses to a common stimulus, as the processing of a common stimulus should lead to similar, not differentiated, neural activity (Holroyd, 2022). Accordingly, correspondence in neural activity in different regions between interaction partners points to the unique contribution of more than one brain in generating interactional outcomes—the sine qua non of IS (Holroyd, 2022).

In a foundational study of IS, Dumas et al. (2010) used EEG hyperscanning to identify correlations across brain regions between dyadic partners as they engaged in spontaneous interaction. Limiting participants to communicating exclusively through the visual modality via video (thereby limiting potential distracting signals), Dumas et al. (2010) collected neural data from both members of a team who were assigned to imitate each other's behaviors and employed phase locking values (PLV) to identify inter-brain correspondences.[4] Participants demonstrated increased phase synchrony in the right parietal area in the alpha-mu band during synchronous interaction, compared to non-synchronized episodes. Additionally, analyses of PLV revealed cross-region correspondences, with the behavioral model's central electrodes synchronizing with the parieto-occipital region of the imitator within the beta frequency band, as well as synchrony between the model's frontal region and the imitator's parietal region in the gamma band. Such cross-region correspondences indicate "the emergence of a distributed functional network of interbrain neural synchronizations" (Dumas et al., 2010, p. 6).

Similarly, Shiraishi and Shimada (2021) demonstrated how two brains can operate as a single functional unit by examining how neural processes in followers and leaders correspond during a turn-taking task. Using EEG, Shiraishi and Shimada (2021) compared neural activity when groups engaged in turn-taking and non-turn-taking tasks. They found that in both tasks, synchrony arose between the frontal region of leaders and the temporo-parietal region in followers within the theta frequency bands, but that the synchrony was significantly greater during turn-taking. The degree of synchrony predicted group members' sense of "joint agency"—greater synchrony between these brain regions enhanced people's sense that they were working together to achieve a common goal.

Building on Dumas and colleagues' pioneering work, Nozawa et al. (2019) examined the relationship between behavioral synchrony and IS, demonstrating that periods of behavioral synchrony enhanced neural synchrony in subsequent interactions. Using fNIRS to compare IS in dyads that engaged in an induced synchrony exercise to those who did not, Nozawa et al. (2019) found increased synchrony within the left lateral prefrontal cortex during a later task. Furthermore, neural synchrony was significantly related to perceptions of rapport between interactants. These results imply that synchrony in one interactional context can carry over into others and further underscore the social benefits of achieving synchrony.

Reinero et al. (2021) moved beyond assessing IS in single-modality dyadic interactions to modeling neural synchrony in four-person task groups engaged in face-to-face interaction. Contrary to prior work (and to their predictions), Reinero et al. (2021) found that phase coherence did not differ between cooperative and competitive groups, and that greater interdependence between group members was also unrelated to IS. Interestingly, however, results suggested that higher IS was associated with increased team performance. That is, across tasks, groups performed better than individuals, but this was conditioned by IS, and independent of measures of cooperation and performance. Teams performed better, then, but not for the reasons expected (a point we return to below).

A Place for Sociological Theory

While the various studies we have discussed so far have value in and of themselves, without clear theoretical models to bring coherence to them, they represent, more or less, an interesting collection of disconnected findings. What is missing in this discussion is a consideration of the role of

[4]During one condition, participants simply viewed a library of hand gestures, in a second condition, they were free to behave and imitate however they chose, and in a third condition, one actor was assigned as the model and the other as the imitator. All groups completed all three conditions, repeated across multiple blocks.

social structure in shaping patterns of synchrony, let alone a theory (or set of theories) to help us make sense of so many different empirical findings. Reinero et al.'s (2021) study provides an illustrative example of the importance of theory for guiding research on IS. In the absence of a clear theory, there is no underlying model to help us make sense of a set of unexpected findings that run contrary to studies demonstrating a relationship between synchrony and cooperation. The absence of theoretically grounded models for IS, in turn, limits researchers' ability to learn from and build upon each other's knowledge. Testing intuitive models of social behavior can only go so far. What is needed to develop cumulative knowledge is theory (Webster et al., 2008).

Developing a neurosociology of synchrony, then fundamentally involves the application of sociological theories and models to explaining the relationships between structural conditions, shared neural processes, and various forms of synchronous action. Kalkhoff et al. (2016) provide an illustrative example of such an approach. Drawing on Lawler's (2001) affective theory of exchange, along with relational cohesion theory (Thye et al., 2002) as a theoretical foundation, Kalkhoff et al. (2016) examined patterns of neural synchrony between participants engaged in a series of social exchanges. Across rounds of exchange, participants' brainwaves began to exhibit phase coherence, meaning they moved in-tandem with one another as time went on. Just as social exchanges foster group formation and solidarity (Thye et al., 2019), so too do brainwaves move in tandem with the positive emotions that create cohesive groups. This work provides some evidence that neural synchrony is likely reflective of processes that generate solidarity in the same way that vocal synchrony indicates solidarity (Kalkhoff et al., 2020), pointing to a shared set of theoretical and structural mechanisms between them.

As we have already discussed, research on unilateral entrainment is interesting in that it has the potential to reveal underlying patterns of status and power. Examining these processes within the status characteristics and expectation states theoretical paradigm (see Berger Joseph et al., 1977) seems like a clear logical fit. Existing research already demonstrates that status differences lead to asymmetry in patterns of vocal synchrony, such that lower status actors conform to the vocal signals of their higher status partners (Dippong, 2020). Additionally, vocal accommodation predicts expectations for competence and openness to influence from a task partner (Dippong & Jillani, 2024).

Yun et al.'s (2012) results revealed that in achieving neural synchrony, participants who were randomly assigned to the position of group leader contributed less to achieving phase synchrony than those assigned to the follower role. In other words, patterns of synchronizing behavior corresponded with actors' structural positions. Sociological work on vocal synchrony yields findings consistent with this; in status-differentiated dyads, speakers reach vocal convergence largely through the efforts of lower-status actors, who conform their voices within the nonverbal frequencies to those of higher-status partners (Dippong, 2020; Dippong et al., 2022), but questions remain as to how unilateral entrainment affects status-related outcomes like influence in groups, perceptions of competence, and the allocation of rewards. These are questions of theory that can help us understand when and why asymmetry arises and how it affects important interactional outcomes.

The applications of relational cohesion theory and status characteristics theory are just two examples of how sociological theory can help bring coherence to research on IS, though opportunities to bring theory to bear on patterns of findings from the neuroscience literature seem nearly endless. When looking at existing research on IS from a sociological perspective, the effects of structural relations of status and power in shaping physiological and neural synchrony seem obvious and numerous, though they are largely absent as meaningful considerations and their explanatory power is overlooked. The possibility of linking different forms of physiological and neural synchrony to well-established sociological models of status and power represents a potential theoretical goldmine for neurosociologists.

Looking at research on synchrony through a neurosociological lens can not only clarify causal pathways between situational elements and synchrony but can also highlight the role of synchrony in producing important social outcomes. For example, in considering synchrony that arises through orchestration, we have already hinted at the importance of rituals in creating mutual entrainment, and these rituals themselves are laden with status and power dynamics (Collins, 2004). Research on the effects of sensory modality also seems to underscore Collins' (2004) argument that rituals involving physical co-presence are particularly well suited to generating synchrony (and therefore solidarity).

Beyond illuminating the neural processes involved in interaction rituals, there is vast potential here to advance sociological knowledge and to provide theoretical structure to a largely unstructured research paradigm. As a further example of the potential connections between sociological theory and research in neural synchrony, in the next section we discuss further how research on vocal accommodation can serve as bridge between social neuroscience and neurosociology.

Vocal Accommodation and the Brain

As noted above, one particularly fruitful branch of inquiry on physiological synchrony centers on the non-conscious synchronization of vocal frequencies during speech (Gregory, 1983; Gregory, 1990; Gregory et al., 2001). Over the course of interaction, conversation partners begin to mimic the others' posture and mannerisms (daSilva & Wood, 2024). In what is formally referred to as "vocal accommodation," conversation partners emit increasingly similar vocal tones as an interaction proceeds. The converging of actors' vocal frequencies produces a distinct "vocal fingerprint" whereby amplitudes across speakers' lower vocal frequencies (i.e., frequencies lower than 500 Hz, known as the nonverbal range; Gregory, 1990), correlate with each other but do not correlate with the vocal characteristics of speakers outside the interaction (Gregory & Hoyt, 1982). Using Fast Fourier Transform (FFT), it is possible to observe and analyze variations in convergence across a complex array of hundreds of vocal frequencies, but with substantially lower temporal resolution than what is possible with other methods.[5]

In isolation, the sounds within the nonverbal range sound indiscernible and muddy—like a conversation heard through a wall (Gallagher et al., 2005). It is possible to isolate these frequencies by applying a "low-pass" filter to segments of speech (see Fig. 18.1). Conversely, a "high-pass" filter would yield speech where the words are clearly heard, but they sound exceptionally crisp and thin because all of the lower frequencies have been removed. In essence, frequencies in the verbal range convey the manifest content of speech (i.e., the words and explicit message contained in an utterance), while those in the nonverbal range capture emotional content and tone of voice. Vocal synchrony then occurs as signals within the nonverbal range (captured by the left panel in Fig. 18.1) converge over time.

Vocal accommodation is closely tied to two sociological dynamics: solidarity and status inequality. Solidarity is reflected in the overall degree of convergence between conversation partners' nonverbal vocal frequencies (Gregory et al., 1997). That is, the absolute distance between speakers' FFT values across the nonverbal band for a given segment of speech indicates the degree of social cohesion or solidarity present within the group. However, the interactional effort necessary to reach convergence is not equal. That is, just as lower-status members adopt the mannerisms and opinions of higher-status group members, they also do so in their subconsciously adjusting vocal behaviors to match the tone of higher-status speakers, creating a pattern of structurally-based unilateral entrainment (Dippong, 2020).

The research on vocal accommodation makes clear that convergence and accommodation emerges in interaction over time, containing a

[5] Please see Dippong and Kalkhoff (2018) for an in-depth discussion of FFT analysis of vocal data.

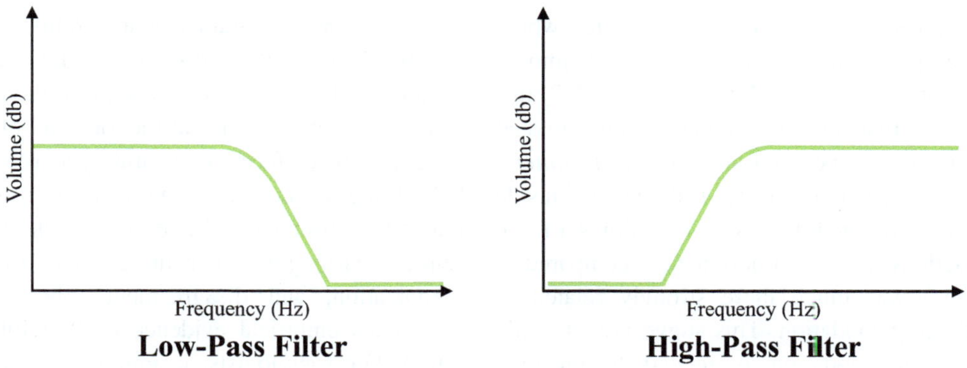

Fig. 18.1 A Comparison of Low- and High-Pass Filters

unique emergent signature (Gregory, 1983). In an early demonstration of vocal synchrony, Gregory (1983) analyzed a set of interviews conducted with members of the U.S. Airforce, using FFT to quantify temporal changes in vocal frequencies. A series of correlational tests showed that frequencies exhibited a distinct auditory footprint, suggesting that each interaction had an audible signature that was not correlated with other interviews. Rather, the signals were unique to the dyadic encounter between interviewer and interviewee, and they even evolved with each stage of the interview. This subtle form of interpersonal synchronization can and has been used as an indicator of social solidarity by subsequent researchers (Gregory et al., 1997; Jamil, 2023). Furthermore, subsequent research has shown that greater vocal synchrony is associated with more positive evaluations of the conversation (Gregory et al., 1993). And when Gregory et al. (1997) "edited out" the paraverbal range by applying a high-pass filter to recordings of conversations, listeners asked to assess the quality of those conversations assigned much poorer ratings to the conversations. Without the ability to observe this way in which dyadic pairs were synchronizing, judges missed the interactional work that these two people did to coordinate and align their speech patterns.

More recent work has focused heavily on the aspects of vocal synchrony that reflect status inequality in both cooperative and competitive contexts. Dippong (2020) offers an illustrative example. Research participants worked together in dyadic groups to complete a collective decision-making task. Prior to interacting, group members completed (bogus) pretests that were ostensively diagnostic of their task abilities. In reality, one member was randomly assigned to be told they scored exceptionally well on the pretests, while the other member was told they performed less well on the test. The experiment randomly assigned dyads to either take just one ability test (indicating a modest difference in ability) or three tests (indicating an overwhelming difference in ability). When interacting to complete the collective task, participants who were randomly assigned to lower-status positions accommodated their vocal behaviors to match their partner more than did higher-status participants. Moreover, the magnitude of the status gap led lower-status participants to accommodate their partner even more. This study showed that within a cooperative task group, members of the group reached verbal synchrony, but primarily through the effort of lower-status members.

Further research has shown that in competitive contexts, vocal accommodation predicts wins and losses. Gregory and Gallagher (2002) analyzed the audio of presidential debates from 1960–2000. Accommodating one's opponent (indicating lower status) was associated with suffering in post-debate polls in six of the eight debates and with losing the popular vote in all eight presidential elections (see also Kalkhoff & Gregory, 2008). A similar study of debates between television host Piers Morgan and a series of interview guests found that whether Morgan or the guest accommodated more predicted which debater participants assessed as less dominant (Kalkhoff et al., 2017). Dippong et al. (2022) further found that vocal accommodation was related to members' perceptions of *each other's* dominance and prestige. But when outside observers listened to audio recordings of the interactions, perceived dominance (compared to prestige) was much more strongly related to vocal accommodation. This suggests that while vocal accommodation is related to prestige, observers much more easily pick up on dominance cues from it.

Considering that shared patterns of brain activity between actors are largely neural reflections of "empathy for others' speech patterns" (Kawasaki et al., 2013), one potentially fruitful path toward developing a neurosociology of synchrony could involve employing research on vocal synchrony/accommodation as an empirical and theoretical template for examining and explaining neural synchrony. The affinity between research traditions centering on vocal and neural synchrony is clear (see, e.g., Gregory & Kalkhoff, 2013). By replicating existing studies of vocal accommodation, while focusing on neural processes using the tools and methods of neuroscience, including EEG or fMRI hyperscanning, it will be possible to identify the shared abstract and general theoretical mechanisms that neural and vocal accommodation share, while also identifying the distinct and unique contributions of the brain in creating and sustaining synchrony.

Conclusion: Toward a Neurosociology of Synchrony

A number of fairly recent publications have heralded the emergence of hyperscanning methods as a tool for developing a two-body or two-brain neuroscience (see, e.g., Dumas, 2011; Liu & Pelowski, 2014). While we agree that these methods open new doors for examining shared neural processes, theory is essential for guiding how researchers use these tools and techniques. In this chapter, we have argued that bringing sociological theories to bear on research related to neural synchrony and interbrain synchronization can be beneficial to social neuroscientists and sociologists alike. For social neuroscientists, the application of clear sociological theories and models to IS and other forms of neural synchrony can help bring coherence to an area of research marked by disconnected findings, a lack of consensus regarding what constitutes interbrain synchronization, and disagreements about what constitutes empirical evidence of IS (Holroyd, 2022). For sociologists, examining social processes related to neural synchrony can provide new tools to answer lingering theoretical questions, and open up new questions as the relationship between structural conditions and neural processes becomes clearer.

We have also argued that research on vocal synchrony and vocal accommodation can serve as a useful template for researchers interested in applying sociological theory to understanding the relationship between social structures and biological and neural processes. Research on vocal accommodation represents a robust and growing body of literature that has already elaborated many theoretical processes related to synchrony, as well as how synchrony affects different social outcomes. By replicating studies of vocal accommodation, neurosociologists can increase confidence in existing theoretical and empirical models of accommodation and synchrony, while likely opening new lines of inquiry.

In sum, developing a neurosociology of interactional synchrony involves taking advantage of a clear opportunity to develop a symbiotic research tradition across disciplinary boundaries. It is clear (to us, at least), that when it comes to answering questions about the structural and situational mechanisms that drive neural synchrony, taking a neurosociological approach that focuses first and foremost on testing logical, theoretically-motivated arguments offers great potential for clarifying existing empirical ambiguities and developing cumulative knowledge. The push for a two-brain neuroscience is essentially a push to understand the *social* brain. Given our disciplinary expertise in explaining and modeling social dynamics, neurosociologists should be at the forefront of that movement.

References

Armstrong, A., & Issartel, J. (2014). Sensorimotor synchronization with audio-visual stimuli: Limited multisensory integration. *Experimental Brain Research, 232*, 3453–3463.

Atzil, S., Hendler, T., & Feldman, R. (2014). The brain basis of social synchrony. *Social Cognitive and Affective Neuroscience, 9*(8), 1193–1202.

Berger Joseph, M., Hamit, F., Norman Robert, Z., & Morris, Z., Jr. (1977). *Status characteristics and social interaction: An expectation states approach*. Elsevier.

Cacioppo, S., Zhou, H., Monteleone, G., Majka, E. A., Quinn, K. A., Ball, A. B., et al. (2014). You are in sync with me: Neural correlates of interpersonal synchrony with a partner. *Neuroscience, 277*, 842–858.

Carozza, S., & Leong, V. (2021). The role of affectionate caregiver touch in early neurodevelopment and parent–infant interactional synchrony. *Frontiers in Neuroscience, 14*, 613378.

Chatel-Goldman, J., Congedo, M., Jutten, C., & Schwartz, J. L. (2014). Touch increases autonomic coupling between romantic partners. *Frontiers in Behavioral Neuroscience, 8*, 95.

Cohen, M., Abargil, M., Ahissar, M., & Atzil, S. (2024). Social and nonsocial synchrony are interrelated and romantically attractive. *Communications Psychology, 2*(1), 57.

Collins, R. (2004). *Interaction ritual chains*. Wiley.

Condon, W. S., & Ogston, W. D. (1967). A segmentation of behavior. *Journal of Psychiatric Research, 5*(3), 221–235.

Condon, W. S., & Sander, L. W. (1974). Synchrony demonstrated between movements of the neonate and adult speech. *Child Development*, 456–462.

Coutinho, J., Oliveira-Silva, P., Fernandes, E., Gonçalves, O. F., Correia, D., Perrone Mc-Govern, K., & Tschacher, W. (2019). Psychophysiological synchrony during verbal interaction in romantic relationships. *Family Process, 58*(3), 716–733.

daSilva, E. B., & Wood, A. (2024). How and why people synchronize: An integrated perspective. *Personality and Social Psychology Review, 29*, 10888683241252036.

Dippong, J. (2012). The effects of scope condition-based participant exclusion on experimental outcomes in expectation states research: A meta-analysis. *Social Science Research, 41*(2), 359–371.

Dippong, J. (2020). Status and vocal accommodation in small groups. *Sociological Science, 7*, 291–313.

Dippong, J., & Jillani, Z. (2024). Vocal accommodation, influence, and performance expectations. In *Advances in group processes* (Vol. 41, pp. 123–144). Emerald Publishing Limited.

Dippong, J., & Kalkhoff, W. (2018). Modeling small group status and power dynamics using vocal accommodation. In *Advances in group processes* (Vol. 35, pp. 51–74). Emerald Publishing Limited.

Dippong, J., Kalkhoff, W., & Jamil, C. (2022). Vocal accommodation and perceptions of speakers' prestige and dominance. *Social Psychology Quarterly, 85*(2), 111–130.

Dumas, G. (2011). Towards a two-body neuroscience. *Communicative & Integrative Biology, 4*(3), 349–352.

Dumas, G., Nadel, J., Soussignan, R., Martinerie, J., & Garnero, L. (2010). Inter-brain synchronization during social interaction. *PLoS One, 5*(8), e12166.

Escoffier, N., Herrmann, C. S., & Schirmer, A. (2015). Auditory rhythms entrain visual processes in the human brain: Evidence from evoked oscillations and event-related potentials. *NeuroImage, 111*, 267–276.

Feldman, R., Magori-Cohen, R., Galili, G., Singer, M., & Louzoun, Y. (2011). Mother and infant coordinate heart rhythms through episodes of interaction synchrony. *Infant Behavior and Development, 34*(4), 569–577.

Gallagher, T. J., Gregory, S. W., Jr., Bianchi, A. J., Hartung, P. J., & Harkness, S. (2005). Examining medical interview asymmetry using the expectation states approach. *Social Psychology Quarterly, 68*(3), 187–203.

Gallois, C., Ogay, T., & Giles, H. (2005). Communication accommodation theory. In *Theorizing about intercultural communication* (pp. 121–148).

Gallup, G. G., Jr. (1983). Toward a comparative psychology of mind. In *Advances in psychology* (Vol. 13, pp. 473–510). North-Holland.

Giles, H., & Coupland, N. (1991). *Language: Contexts and consequences*. Thomson Brooks/Cole Publishing Co.

Gregory, S. W. (1990). Analysis of fundamental frequency reveals covariation in interview partners' speech. *Journal of Nonverbal Behavior, 14*, 237–251.

Gregory, S., Webster, S., & Huang, G. (1993). Voice pitch and amplitude convergence as a metric of quality in dyadic interviews. *Language & Communication, 13*(3), 195–217.

Gregory, S. W., Dagan, K., & Webster, S. (1997). Evaluating the relation of vocal accommodation in conversation partners' fundamental frequencies to perceptions of communication quality. *Journal of Nonverbal Behavior, 21*, 23–43.

Gregory, S. W., & Hoyt, B. R. (1982). Conversation partner mutual adaptation as demonstrated by Fourier series analysis. *Journal of Psycholinguistic Research, 11*, 35–46.

Gregory, S. W., Jr. (1983). A quantitative analysis of temporal symmetry in microsocial relations. *American Sociological Review*, 129–135.

Gregory , S. W., Green, B. E., Carrothers, R. M., Dagan, K. A., & Webster, S. W. (2001). *Verifying the primacy of voice fundamental frequency in social status accommodation*. Language & Communication.

Gregory, S. W., Jr., & Gallagher, T. J. (2002). Spectral analysis of candidates' nonverbal vocal communication: Predicting US presidential election outcomes. *Social Psychology Quarterly*, 298–308.

Gregory, S. W., & Kalkhoff, W. (2013). Comprehending the neurological substratum of paraverbal communications: The invention of SplitSpec Technology. In *Handbook of neurosociology* (pp. 369–383).
Hall, E. T. (1976). *Beyond culture*. Double Day.
Hoehl, S., Fairhurst, M., & Schirmer, A. (2021). Interactional synchrony: Signals, mechanisms and benefits. *Social Cognitive and Affective Neuroscience, 16*(1–2), 5–18.
Holroyd, C. B. (2022). Interbrain synchrony: On wavy ground. *Trends in Neurosciences, 45*(5), 346–357.
Hove, M. J., Iversen, J. R., Zhang, A., & Repp, B. H. (2013). Synchronization with competing visual and auditory rhythms: Bouncing ball meets metronome. *Psychological Research, 77*, 388–398.
Jamil, C. (2023). Cohesion and solidarity in consistent and inconsistent status structures. In *Advances in group processes* (Vol. 40, pp. 161–184). Emerald Publishing Limited.
Jiang, J., Chen, C., Dai, B., Shi, G., Ding, G., Liu, L., & Lu, C. (2015). Leader emergence through interpersonal neural synchronization. *Proceedings of the National Academy of Sciences, 112*(14), 4274–4279.
Kalkhoff, W., Dippong, J., Gibson, A., & Gregory, S. W. (2020). Society in peril? How distance media communication could be undermining symbolic interaction. *Identity and Symbolic Interaction: Deepening Foundations, Building Bridges*, 317–338.
Kalkhoff, W., Dippong, J., & Gregory, S. W., Jr. (2011). The biosociology of solidarity. *Sociology Compass, 5*(10), 936–948.
Kalkhoff, W., & Gregory, S. W., Jr. (2008). Beyond the issues: Nonverbal vocal communication, power rituals, and 'rope-a-dopes' in the 2008 presidential debates. *Current Research in Social Psychology, 14*(3), 39–51.
Kalkhoff, W., Thye, S. R., & Gregory, S. W., Jr. (2017). Nonverbal vocal adaptation and audience perceptions of dominance and prestige. *Social Psychology Quarterly, 80*(4), 342–354.
Kalkhoff, W., Thye, S. R., & Pollock, J. (2016). Developments in neurosociology. *Sociology Compass, 10*(3), 242–258.
Kawasaki, M., Yamada, Y., Ushiku, Y., Miyauchi, E., & Yamaguchi, Y. (2013). Inter-brain synchronization during coordination of speech rhythm in human-to-human social interaction. *Scientific Reports, 3*(1), 1692.
Kinreich, S., Djalovski, A., Kraus, L., Louzoun, Y., & Feldman, R. (2017). Brain-to-brain synchrony during naturalistic social interactions. *Scientific Reports, 7*(1), 17060.
Konvalinka, I., Xygalatas, D., Bulbulia, J., Schjødt, U., Jegindø, E. M., Wallot, S., et al. (2011). Synchronized arousal between performers and related spectators in a fire-walking ritual. *Proceedings of the National Academy of Sciences, 108*(20), 8514–8519.
Lawler, E. J. (2001). An affect theory of social exchange. *American Journal of Sociology, 107*(2), 321–352.
Levy, J., Goldstein, A., & Feldman, R. (2017). Perception of social synchrony induces mother–child gamma coupling in the social brain. *Social Cognitive and Affective Neuroscience, 12*(7), 1036–1046.
Lindenberger, U., Li, S. C., Gruber, W., & Müller, V. (2009). Brains swinging in concert: Cortical phase synchronization while playing guitar. *BMC Neuroscience, 10*, 1–12.
Liu, T., & Pelowski, M. (2014). Clarifying the interaction types in two-person neuroscience research. *Frontiers in Human Neuroscience, 8*, 276.
Montirosso, R., & McGlone, F. (2020). The body comes first. Embodied reparation and the co-creation of infant bodily-self. *Neuroscience & Biobehavioral Reviews, 113*, 77–87.
Nguyen, T., Abney, D. H., Salamander, D., Bertenthal, B. I., & Hoehl, S. (2021). Proximity and touch are associated with neural but not physiological synchrony in naturalistic mother-infant interactions. *NeuroImage, 244*, 118599.
Nguyen, T., Schleihauf, H., Kayhan, E., Matthes, D., Vrtička, P., & Hoehl, S. (2021). Neural synchrony in mother–child conversation: Exploring the role of conversation patterns. *Social Cognitive and Affective Neuroscience, 16*(1–2), 93–102.
Novembre, G., Knoblich, G., Dunne, L., & Keller, P. E. (2017). Interpersonal synchrony enhanced through 20 Hz phase-coupled dual brain stimulation. *Social Cognitive and Affective Neuroscience, 12*(4), 662–670.
Nozawa, T., Sakaki, K., Ikeda, S., Jeong, H., Yamazaki, S., Kawata, K. H. D. S., et al. (2019). Prior physical synchrony enhances rapport and inter-brain synchronization during subsequent educational communication. *Scientific Reports, 9*(1), 12747.
Oullier, O., De Guzman, G. C., Jantzen, K. J., Lagarde, J., & Scott Kelso, J. A. (2008). Social coordination dynamics: Measuring human bonding. *Social Neuroscience, 3*(2), 178–192.
Reddan, M. C., Young, H., Falkner, J., López-Solà, M., & Wager, T. D. (2020). Touch and social support influence interpersonal synchrony and pain. *Social Cognitive and Affective Neuroscience, 15*(10), 1064–1075.
Reinero, D. A., Dikker, S., & Van Bavel, J. J. (2021). Inter-brain synchrony in teams predicts collective performance. *Social Cognitive and Affective Neuroscience, 16*(1–2), 43–57.
Shimshock, C. J., Thorson, K. R., Peters, B. J., & Jamieson, J. P. (2024). Behavioral variability in physiological synchrony during future-based conversations between romantic partners. *Emotion*.
Shiraishi, M., & Shimada, S. (2021). Inter-brain synchronization during a cooperative task reflects the sense of joint agency. *Neuropsychologia, 154*, 107770.
Smith, T. S. (2013). Attachment, interaction, and synchronization: How innate mechanisms in attachment give

rise to emergent structure in networks and communities. In *Handbook of neurosociology* (pp. 243–254).

Thye, S., Lawler, E. J., & Yoon, J. (2019). The formation of group ties in open interaction groups. *Social Psychology Quarterly, 82*(2), 158–181.

Thye, S. R., Yoon, J., & Lawler, E. J. (2002). The theory of relational cohesion: Review of a research program research program. *Advances in Group Processes*, 139–166.

Webster, M., Jr., Rashotte, L. S , & Whitmeyer, J. M. (2008). Theoretical and intuitive models. *Social Science Research, 37*(1), 4–17.

Yun, K., Watanabe, K., & Shimojo, S. (2012). Interpersonal body and neural synchronization as a marker of implicit social interaction. *Scientific Reports, 2*(1), 959.

Part IV

Culture, Society, and Social Problems

Neurosociology Across Cultures

19

Deniz Buyukgok and Steven Hitlin

Abstract

The scientific incorporation of neurology into sociology presents an inevitable collaboration in the face of the complexity of understanding the human mind and behavior, which is still in its early stages. Applying knowledge and findings of neuroscience into cross-cultural studies offers exciting, largely untapped potential for developing truly interdisciplinary theories and understandings of the nature of the human minds shaped within different societies. Furthermore, the term 'culture' is treated differently in psychology and sociology, complicating the development of a corpus of knowledge. This chapter offers a brief overview of the relevant terms, accomplishments, and challenges facing a potential cross-cultural neurosociology.

D. Buyukgok (✉)
Department of Psychiatry, Istanbul University, Istanbul Faculty of Medicine, Fatih/Istanbul, Turkey
e-mail: deniz.buyukgok@istanbul.edu.tr

S. Hitlin
Department of Sociology & Criminology, University of Iowa, Iowa City, IA, USA
e-mail: steven-hitlin@uiowa.edu

Neurosociology Across Cultures

When the concept of neurosociology was introduced around a decade and a half ago (Franks, 2010), the reception in the discipline was markedly muted, except for the excitement from those few authors already trying to bridge the disciplines (notably Franks & Turner, 2013; Firat & McPherson, 2010; Kalkhoff et al., 2012). For those of us raised academically during the turn of the century, the notion that sociology might reduce complicated social phenomena to biology or neuroscience was, if not anathema, certainly frowned upon. This can be measured both through informal conversion and formal publications, but work anchoring behavior in genes (e.g., Guo et al., 2015; Martin, 2008; Udry, 1995) was sporadically published and not especially widely cited in sociology.

At the same time, perhaps the largest growing area in the field was the Sociology of Culture, based on ASA membership data, the introduction of a new journal, and the sheer volume of work drawing on Bourdieu, Swidler, and their resultant traditions. Clearly, there was disciplinary acceptance of studying the symbolic material that shaped individual outlooks, behaviors, often through the metaphor of a 'toolkit' or the ubiquitous notion of individuals developing a social class-shaped 'habitus', a series of tastes and preferences that were deeply embodied, if not always consciously reportable. The functionality and

W. Kalkhoff et al. (eds.), *Handbook of Neurosociology*, Handbooks of Sociology and Social Research, https://doi.org/10.1007/978-3-031-95615-7_19

deployment of those cultural elements that penetrate from the external world into people's internal worlds were exciting avenues for sociologists.

This work had the potential to be bridged through the expansion of the notion of Culture and Cognition in sociology, popularly kicked off by Dimaggio's (1997) important review. That tradition is as current as any topic in the field (e.g., Cerulo et al., 2021; Lizardo, 2017; Vaisey, 2021), with a great deal of work capturing psychological constructs (Boutyline & Soter, 2021; Lizardo & Michael, 2010) as culturally relevant constructs that get internalized and have something to do with behavior (e.g., Vaisey, 2009). We might suggest, however, that important insights from neuroscience, and more than a handful of ideas from cognitive psychology, are tangential to the culture-and-cognition enterprise, but deserve greater engagement.

Were one to be critical, and a Handbook chapter often has a latent goal of showing where a field might go in addition to outlining the point it has developed at the time of its writing, one might note the confusing lack of engagement in much of Cultural Sociology with the long-standing, quite empirically successful subfield of Cross-Cultural Psychology (Kim, 2021, expands on this point). That is to say, the theory and mechanics of how collective beliefs get internalized, embodied, and utilized is a topic of tremendous current exploration and debate, but this occurs largely isolated from our disciplinary neighbors' focus on how societal culture shapes many of the same processes. This is not to say meaningful concepts from cross-cultural psychology have not reached sociology, but that their employment has been only peripheral to the sociology of culture debates. The knowledge of current literature is always limited, but concepts like 'cultural tightness' (Gelfand, 2019), 'individualism vs. collectivism' (Hofstede, 1980; Markus & Kitayama, 1991; Oyserman et al., 2002), or the "Big Three of Morality" (Shweder & Much, 1997) are influential models of how different cultures shape individual outlooks, but less than fully engaged except within pockets of the culture-and-cognition tradition (based on the lack of citations in these important reviews).

Thus, a review of cross-cultural neurosociology will rely heavily on cross-cultural neuroscience, itself a fledgling field, but one much more closely aligned to the study of cross-cultural psychology than is the sociology of culture. This means our review engages with the limited number of studies that have engaged cross-cultural studies in neuroscience, while being a bit more speculative—ideally, a bit generative—about the directions neurosociology might take. Disciplinary boundaries are funny things, anchoring scholars' identities within home fields, methods, and jargon that, despite decades of institutional calls for their overcoming, can inhibit research directions, acceptance, and reception. Given how far afield these mental processes can be from what makes something a 'sociological' explanation, it is hard to foresee cross-cultural neurosociology expanding a great deal in the immediate future. That said, the study of culture's systematic effects on human mental functioning seems important, and findings to date suggest nothing that precludes sociological analysis; in fact, the evidence we have so far fits well with the notion that social processes fundamentally shape human mental processes. Durkheim and Weber and their ilk would likely not feel as disciplinarily threatened as some of today's practitioners, journal editors, and reviewers.

In this chapter, we will review the current—and potential—influence of neuroscience on sociology and how it can be theoretically advanced through a cross-cultural lens. We will examine culture-specific behaviors from a neuroscience perspective, to discuss the neural foundations of cross-cultural sociology. Is it possible to integrate the complexity of neurophysiological processes and the abstract nature of sociocultural influence? On the other side, we will evaluate the compatibility of disparate methodological approaches used in neuroscience studies from a social science perspective.

Neuroimaging provides exciting insights into how cultural differences can impact the brain's neural structure, despite the fears in some corners of sociology that misses the cautionary approach of most neuroscientists when extrapolating individual neural findings to broader societal concepts (Table 19.1). The complexities of human behavior and society cannot always be reduced to neurobiological explanations alone. Integrating findings from neuroimaging studies with broader social science perspectives can lead to a more comprehensive understanding of how culture shapes the brain and vice versa. When it comes to our biological preparedness, can the society we are born into shape the brain's neural organization? We aimed to put a discussion about this question in the 'Culture Shapes Mental Development' section. As the mammals with the most evolutionarily sophisticated language ability, we will discuss the reflection of a differentia-

Table 19.1 fMRI research findings of cross-cultural effects and underlying brain regions

Authors	Neuroanatomical activation area	Comparison groups	Imaging task or condition
Kobayashi et al. (2007)	R inferior frontal gyrus R TPJ L inferior temporal gyrus	American vs. Japanese children	Second order false belief; affective aspects of ToM
Freeman et al. (2009)	Caudate nucleus & mPFC	Japanese vs. American	Reaction to submissive body postures vs. dominant body postures
Adams et al. (2010)	Lingual gyrus TPJ	American vs. Japanese	Higher success in emotion recognition from faces of own societies
Gutchess et al. (2010)	Angular gyrus Middle frontal gyrus Cingulate gyrus	East Asian vs. American	Response to conflict in semantic content of information
Koelkebeck et al. (2011)	mPFC R claustrum R parahippocampal gyrus R temporal lobe	Caucasian vs. Japanese	ToM videos vs. random movement sequences
Kang et al. (2013)	Ventral striatum & vmPFC	Korean vs. American	Sensitivity towards social comparison
Ma et al. (2014)	TPJ	Chinese vs. Danish	Self-judgement in social attributions
Han et al. (2014)	R putamen R DLPFC Postcentral sulcus	Korean vs. American	Reaction times to moral personal scenarios
Krendl (2016)	L cingulate gyrus L anterior cingulate R inferior temporal gyrus R middle frontal gyrus	American vs. Chinese	Attitudes towards stigmatized individuals from other cultures
Harada et al. (2020)	Bilateral amygdala	Japanese vs. American	Ingroup negative facial expressions
Cohen et al. (2020)	R parietal lob	India vs. Americans	Negative relation between affective empathy and cerebral functioning integrity
Leger et al. (2024)	L superior parietal cortex	Taiwanese vs. Americans	Mnemonic discrimination strategies (object recognition of similarities)

Abbreviations: *R* right, *L* left, *vmPFC* ventromedial prefrontal cortex, *mPFC* medial prefrontal cortex, *DLPFC* dorsolateral prefrontal cortex, *TPJ* Temporoparietal junctions, *ToM* Theory of Mind

tion in mental development in the 'Reflections of Language and Communication Differences to Cross Cultural Neurosociology' section. The structure of the language determines the route of thought processes. Is it possible that the mental biases from the social figures we model (i.e., family, teachers, politicians) pass through generations with language heritage? We address this issue in the section 'Cross-cultural Perspective to Mental Biases'. Can mental biases determine whether an individual feels conformed, or can mental biases have any impact on attributing situations in society that lead to the perception of nonconformity in the individual's brain? We then touch upon the contributions of cross-cultural neuroscience to political decision making. And, before encouraging readers for future study, we will discuss the differences between the research methods on which sociology and neuroscience are based, and provide some examples that serve to narrow this gap.

Cross-Cultural Psychology: A Cursory Overview

One attraction of neuroscience, from the outside, might be its apparent universalistic orientation toward understanding how a species', in particular case the human organism's—mental processes operate. Issues of social perception, social cognition, spatial reasoning, and linguistic capacity are fundamental to human social life and interaction, and just as we study the mental processes of other animals, those tools can seemingly be applied to ourselves. Human culture might be examined through various dimensions, inside the limits of this chapter the concept of culture refers to values, roles, beliefs, and symbolic representations of society that have an impact on an individual's psychological and mental development. However, over a half-century of research has documented myriad ways that one's culture—often conceptualized at the societal level as shared understandings of what 'is' and what 'should be'—fundamentally shapes perception, behavior, self-understandings, and so forth. Humans certainly share many capacities, but those capacities are 'filled in' differentially depending on where (and when, though that is outside the purview of this chapter) you were born. Cross-cultural psychology is a well-established field of inquiry (see Shiraev & Levy, 2020 for one comprehensive review), that since its inception has tried to expand psychology's understanding of the contexts of social behavior and how those shape motivation, cognition and perception, among other core processes (Triandis & Brislin, 1984). Paraphrasing the famous Kluckhohn and Murray (1953) quote [the notion that people are simultaneously (a) all the same, (b) similar to some others, and (c) technically unique], we might say that culturally shaped aspects overshadow those areas where all human beings share identical neural processes.

Likely the most influential research program stemming from this tradition involves the almost axiomatic distinction between 'Eastern' and 'Western' cultures, that focuses on the more interdependent conceptualization of self and others found in putatively Eastern cultures (e.g., Markus & Kitayama, 1991; Kanagawa, 2001; Sun, 2023), acknowledging that some areas of the world fall outside of this proposed dichotomy (Uskul et al., 2023). In brief, evidence suggests that people from Eastern countries perceive the same stimuli differently (e.g., Nisbett, 2004) than Westerners, with those from the Global East more focused on context than on individual actors. People's self-construals are fundamentally shaped through this dimension, with Westerners more focused on the self-as-an-isolated-object rather than as part of a contextually situated whole (Oyserman et al., 2002). While East Asians have traditionally been associated with an overarching collective self-construal, characterized by a profound sense of interconnectedness with others, a strong focus on one's place within social groups, and a prioritization of group-oriented objectives over individualistic pursuits (Cross, 2011). Eventually, Kitayama and Uskul (2011) take the argument to the point that culture is a collective process that transmits values and practices that get 'embrained', an understanding that mirrors sociological models of embodied cognition.

The notion of values—ordered, abstract priorities about how the world 'should' be—is another wide-ranging body of cross-cultural research (see Sagiv & Schwartz, 2022 for one review). Values can be conceptualized as desires to reach end states or behaviors that guide selection or evaluation of behavior and actions, which are ordered by relative importance. With different measurement traditions of values, the individual moral priorities represent communally patterned cultural priorities. Perhaps the most famous model is from Shalom Schwartz (e.g., Schwartz & Cieciuch, 2021), finding ten representative values recognized—if not as equally important–in all of the literate societies of the world. Alternatively, Inglehart (2006) suggests that human societies prioritize different things based on their level of economic development, with higher GDP countries caring more about 'self-transcending' values (like caring for society and the environment) than people from less developed nations, who are more focused on individual and group survival. There are other cross-cultural comparisons as well (notably Hofstede, 2001), but the larger point is to demonstrate how collective representations can be theorized—and measured—at the individual level.

More recently, Gelfand (Gelfand, 2019; Gelfand et al., 2011) has argued that all the societies in the world can be categorized by their orientation toward how much deviance they tolerate from important norms, how 'tight' or 'loose' a country is. Other cross-cultural comparisons involve examining the different moral systems (Shweder & Much, 1997) societies have developed, or which of a hard-wired set of potential 'moral foundations' (Haidt & Graham, 2009) are more determinative of a society's (or subgroup's) moral judgment. Regardless of area, different psychological research traditions focus on the collective patterns that shape individual functioning, extending the notion of context quite widely but with replicable, empirical support. These processes develop quite early in a child's life, and thus shape neurological development from the very beginning. The difficult issues, of course, involve the complicated intersection of culture and mental development. Human beings operate in social worlds that, to them, make expected actions seem obvious; the content of those worlds, both the descriptive ("what objects are") and the prescriptive ("what people like me should do in this kind of situation") are shaped by these cultural processes that, in turn, become deeply psychologically embedded and taken for granted.

Culture Shapes Mental Development

The brain's ability to develop in response to its environmental stimuli, known as neuroplasticity (Cramer et al., 2011), gives rise to significant changes in neural networks and improvements in cognitive skills and children develop. Although neuroplasticity may present throughout maturation over the course of one's life, there are critical periods in development when the brain is more sensitive to learning and experiences. The early years of life are particularly critical for cognitive development (Schibli & D'angiulli, 2013) and have significant impact on future learning and abilities. As cognitive sciences draw the path to formulate the developmental process of emotions and thoughts, which serve as the foundations of behavior, it can provide the contours of cultural influence. There are many aspects of the social environment that shape young minds. We see the importance of the critical period effect on cognitive development in studies from the children and adolescents who grew up in different socioeconomic status (SES). For example, children and adolescents from different socioeconomic conditions show different attitudes towards social ethics (Caravita et al., 2012). In particular, families from low SES are more likely to respond following established norms to conventional dilemmas than families with high SES – when asked to define underlying motive, they share the predominant feelings that shape their perception as receiving faster negative reactions from the society. Similar results were revealed as this comparison was made between immigrants and natives.

Issues of 'culture', as broad and abstract as they can be, intersect with individuals' minds in many ways. Perhaps the primary overlap with sociological conceptions of how individuals fit into society revolves around the notion of 'identity', itself a contested term (Owens et al., 2010). But the notion of being a member of a social group can evolve through various shared characteristics of various groups within a population. Race, nationality, ethnic origin, political ideology, or religious beliefs compose a sense of belonging and an orientation toward interaction, perception, and memory. Even being a fan of a sports team can create a social identity in terms of a reference group that a person identifies with. Any person that falls outside of the identity groups will be referred to as "others", a seemingly strong human tendency to carve the world into 'us' and 'them' (Berreby, 2005), found in all cultures that we know about. Strong mindsets of group identity may shelter biases both in evaluating identity groups (i.e., ingroup) and in evaluating "other" groups (i.e., outgroup). Multidisciplinary psychological moral models that are grounded in biological reality are necessary to better understand ingroup and intergroup biases, and to make more precise predictions on behavior under the influence of group bias.

Consider object perception, one of the most basic neurobiological processes. It has been shown that contextual information is perceived differently in East Asian cultures (i.e., Japanese) than in Western cultures (i.e., North American); thus, presenting neural differences between cultures even in visual scanning of space for object perception (Kitayama et al., 2003). The framework offered by cultural neuroscience allows us to evaluate the functioning of the brain within the perceived reality of human beings, that is, in a holistic context, rather than as a separate mechanical unit. Some functional features of the plastic, interconnected brain under the influence of learning and expression must also change in cross-culturally contexts; we learn different ways of making connections about the social world. Phylogenetically newer functions that enable non-primitive human beings to exist in communities.

Furthermore, culture shapes many neurological capacities like social perception, the shared beliefs of a particular group, and the harmony of group members (Brekhus, 2015); all of these capacities are fundamental for the social interaction at the root of sociological analyses. Thus, culture can be postulated to engender the divergence in neural responses within a society. Socially influenced cognition begins to take hold in neural structures and structural connections (i.e., neural networks) throughout development (Paterson et al., 2006). The neural connections become stronger as the influence of culture continues, and a child has continual social interactions along these expected lines. For this reason, the effect of culture can be seen to have a cyclic pattern, affecting neurobiological development and in turn being affected by the neural processes (Vogeley & Roepstorff, 2009).

Last but not least, the tricky aspects in the sociology of culture, not to mention neurosociology, involves conceptualizing and measuring what culture 'is' and how it is represented in the human mind. In the nineteenth century, the context of culture and intercultural comparisons were discussed under the influence of nations and countries. Geography played a crucial role in early discussions about the potential cultural influences between distant populations. There was a recognition that exposure to different cultures could lead to changes in one's lifestyle. These discussions also explored the idea of values being exchanged through migration. In today's world, the global population is more mobile than ever before, alongside massive changes in technological communication, leading to exposure to a wide variety of lifestyles. People may choose to relocate for various reasons such as seeking a simpler life, pursuing better opportunities in new locations, addressing resource scarcity, or following their beliefs. Advanced transportation systems and technology have made migration and cultural exposure more frequent and accessible, with online communication further facilitating connections between individuals from diverse backgrounds. It is important to note that these interactions are not limited to individual movements but also extend to multicultural families. The timespan is likely too short to have shifted the human mind's evolutionary capacities, but certainly the types of interaction, richness of stimuli, and exposure to culture is remarkably different than 200, 100, or even 25 years ago. As a result, there is a need for more accurate definitions to analyze and compare different cultures in the contemporary world, and how, in practice, they blur perhaps more than at any other time in history. With the inclusion of neuroscience methodology, the impact of culture on cognitive structuring has become observable through a more objective lens.

Reflections of Language and Communication Differences to Cross Cultural Neurosociology

The human language system is primarily based on symbols. These symbols can be conveyed verbally (through spoken or written words) or nonverbally (through gestures, facial expressions, and body language). Human functional neuroimaging studies present well-documented brain areas associated with language production (i.e., left inferior frontal and posterior middle temporal gyri; as well as neocortical premotor areas, such as the pre-supplementary motor area and left dorsal premotor cortex) and comprehension (i.e., dorsomedial prefrontal cortex, precuneus and inferior parietal lobules). Through the verbal communication features, neural structural divergence in language processing also reveals cultural differences in thinking processes. For example, when given an abstract thinking task, North American participants try to solve the task by talking while they are thinking (Kim, 2002). Koreans, on the other hand, think about the task and then verbalize after deciphering it (Kim, 2002); if both groups are asked to verbalize during the task that requires abstract thinking; Koreans' performance decreases. Similar studies have set objective results supporting the idea that development of neural networks can be influenced by cultural factors, leading to differences in how individuals process information and perceive the world around.

Cultural experiences and patterns of thinking can shape the development and organization of neural networks in the brain, influencing cognitive processes and behaviors. Early in 1990s the need for studies that go beyond language and semantic expressions was emphasized by the researchers (Izard, 1994). Semantic processing refers to the cognitive act of accessing stored knowledge about the world. Semantic processing which is shared by all humans suggesting a shared neurobiological preparedness. Neuroscience research methods revealed that, this ability is exposed as diffusely localized cognitive processing requiring fronto-temporo-parietal brain areas (i.e., posterior multimodal and heteromodal association cortex, heteromodal prefrontal cortex, and medial limbic regions) and their mutual interactions (Binder et al., 2009). However, the source of semantic information seems to make difference in this process and follow an alternative path. Neural response to a statement made by the leader of the reference group (i.e., ingroup) to which the person considered itself as a member reveals to activate other brain areas (i.e., the rostral inferior parietal lobule, pars opercularis of the inferior frontal gyrus, and posterior midcingulate gyrus) than stated to be activated during semantic processing (Molenberghs et al., 2017). Eventually, neurobiological knowledge gains specificity with the recruitment of sociological phenomena.

Nonverbal communication features (gestures, facial expressions, tone of voice, and body language) have also been the core neuroscientific subjects. We know that people living in the same cultural environment share similar values and cognitive schemas, and use similar criteria when evaluating the contribution of certain types of behavior to their self-esteem (Jung & Lee, 2006). Having a more independent self-construal has roots in cognition that includes a person's attributes, situations, or behaviors; and strives to meet internal standards of success. Interdependent self, on the other hand, aims to achieve goals and fulfill the role determined by the reference group (Somech, 2000). Interdependent individuals are more inclined towards behavioral engagement and socializing. As socialization behavior increases, sensitivity to nonverbal cues in communication also increases. Therefore, collectively oriented cultures that prioritize interpersonal relationships, such as East Asian cultures, are more sensitive to nonverbal cues compared to non-interdependent cultures (Markus & Kitayama, 1991).

Electroencephalography (EEG), the most commonly employed method to monitor the neural activities of the brain, has also been applied under cognitive tasks within the scope of neuroscience research of interest. Ishii et al. (2010) found that Japanese women, who are defined to be more interdependent, showed an increase in N400 response when listening to speech with an incongruent voice tone compared to the male group (Ishii et al., 2010). N400 is one of the EEG components that shows an event related negative deflection around 400 ms (Kutas & Hillyard, 1980). The N400 increase in this region has previously been shown to occur during the incongruency of stimuli. Studies like this one, which reveal gender differences in information processing despite the shared neurobiological structure, demonstrate how sociological phenomena have pushed existing knowledge to a greater level of precision.

Another parameter of nonverbal communication, body posture also found reflections in cross-cultural differences. A study of American and Japanese participants took functional neuroimaging recordings while they were shown various body postures (Freeman et al., 2009). This study revealed that cultural variation among nations has an observable difference in perception. Americans showed greater activation in caudate nucleus and mPFC to dominant bodily postures vs submissive body postures. On the other side, Japanese participants showed greater activation in the same region to submissive postures vs dominant body postures. This difference can be attributed to growing in a national culture reflecting the collectively oriented beliefs and values of the society, and the inheritance of intergenerational communication characteristics. Cultural norms establish guidelines for expressive behaviors and nonverbal actions and these norms facilitate the exchange of social intentions within communication (Matsumoto & Hwang, 2016). These studies demonstrated objective presentations regarding how "cultural attribution of nonverbal communication characteristics inherited from society" is processed in the human brain; suggesting cultural differences result with variations in activated brain regions to alternate neural pathways are recruited when social learnings are more prominent.

We can conclude from these studies that individuals form neural connections based on behaviors that are imbued with meaning and are positively or negatively reinforced by their respective cultures (Sasaki & Kim, 2017). Social psychology and sociology have long observed and tested these behaviors in natural settings or by creating experimental environments. Now, with more biologically based methodologies, we can further explore the underlying foundations of these differences.

Cross-Cultural Perspective to Mental Biases

The ability to understand other people's thoughts, beliefs, feelings and even desires is conceptualized under Theory of Mind (ToM; Gallagher & Frith, 2003). As human beings evolve in a social presence, mentalizing other people's minds becomes one of our brain's large-scale brain network abilities. One of the ToM abilities, known as first-order false belief, refers to the understanding that false information about the world can exist. This skill typically develops in children as young as 3–5 years of age. It has been shown that there is a cultural background effect on the difference in this age range. So primary social cognitive skills depend on engagement behavior that is transferred through interpersonal relationships shaping what knowledge is presented as socially acceptable. Children and adolescents living in Japan, North America and Australia were included in a cross-cultural study in which first-order ToM skills were studied. Accordingly, North American children reach the understanding at an earlier age that people can have different beliefs about the same situation while Japanese children develop earlier understanding that a person may not have the correct information. While this does not indicate any advantage or disadvantage, it reveals the effect of culture on the time course of the social cognition development.

Functional neuroimaging studies investigating the associations between brain areas and social cognitive tasks that require mentalizing others revealed with a wide range of results, and the science is just beginning to form a consensus. Far beyond locally functioning, brain areas are interconnected as a network including the temporoparietal junction (TPJ), precuneus (PreC), anterior temporal lobes (ATL), dorsomedial prefrontal cortex (dmPFC), and ventromedial prefrontal cortex (vmPFC) to mentalize others' minds. However, some researchers have observed variations, under some conditions, like the cultural background of the individual. Among those brain areas, two consistently become prominent if the task is presented in a culture-dependent manner; medial prefrontal cortex (mPFC) and temporoparietal junction (TPJ) (Schurz et al., 2021). Details are still to be explored, but even core processes like mentalization of others' minds are culturally shaped.

In an fMRI study, Japanese and North American participants completed one of the most frequently used ToM paradigms (Adams et al., 2010), cross-cultural version of Reading the Mind in the Eyes Task (RMET; Baron-Cohen et al., 1997) while having their brains scanned. The task was to estimate the emotional expressions from the eye photographs of the people which is another social cognition ability; emotion recognition. In this cross-cultural study, researchers showed photos with various facial expressions to participants from two different cultures (Adams et al., 2010). There was increased activation in the bilateral posterior temporal sulcus (which is located in the TPJ area) while participants were trying to mentalize ingroup members' emotions from their facial expressions. This increase was not prominent as they visualize outgroup members' eyes. With this cross-cultural fMRI study, researchers were able to show that individuals extrapolate mental states of people from their culture more accurately than those from a different culture, which complicates the ability to infer and understand the mental states of other individuals across cultures. While the ability to recognize emotions from facial expressions is a strongly ingrained socio-anthropological heritage, it is important to acknowledge that individuals may vary in their ability to accurately perceive and interpret these emotions. In relation to mindsets rooted from their own culture, individuals exhibited different brain activities than the "outgroup". More evidence supported the increased difficulty in recognizing emotions of the people from different ethnic groups (Habel et al., 2000) while more accuracy in recognizing emotions from the same ethnic group (Elfenbein & Ambady, 2003; Pinkham et al., 2008). In this context, many studies comparing emotion recognition ability across cultures show that (a) individuals better understand the emotions of people from their own culture and (b) there are differences in the recognition and understanding of emotional facial expressions across cultures (Camras et al., 2006; Huang et al., 2001; Marsh et al., 2003). Though identifying emotions like happiness, anger, disgust, fear, surprise, and sadness in facial expressions had already been conceptualized as universal (Ekman et al., 1969) cross-cultural neuroscience studies showed that a universal characteristic of the human mind is filled in by one's specific cultural experiences and biases.

In order to attribute others' mindsets accurately, it is essential to differentiate our experiences from others in the social context, so as to effectively navigate in social situations. However, just about all individuals have some degree of challenge in interpreting the mental and emotional states of others while also being influenced by our own thoughts and emotions. People tend to make predictions about others' mental states, but it is only able within the limits of their own minds' capacities; final estimations of other's mostly biased under the influence of self's internalized beliefs, values and feelings (Callaghan et al., 2005). The study of biases is quite popular in American psychology, the issue becomes how universal such biases might be, and under what circumstances cultural forces reinforce or override these core human tendencies. The foundational components that contribute to biased social perceptions, as well as the impact of brain development on the ability to comprehend both self and others dividedly, continue to spark interest.

Social intergroup relationships have a profound impact on shaping human behavior and cognition, and provide structure to the aforementioned internal predictions that shape how people view interaction. The importance of the positioning of self in social relationships is that it is the path in understanding how individuals perceive themselves and their social relationships. The way individuals view themselves in relation to others is conceptualized as self-construal (Markus & Kitayama, 1991). This conceptualization of self evolves under the influence of cultural beliefs and norms, and can vary in terms of the degree of interconnectedness or independence a person feels in relation to others. Markus and Kitayama (1991) proposed that individuals with an independent self-construal tend to view themselves as autonomous entities with unique thoughts, feelings, and preferences. This leads to a focus on individuality and personal goals, rather than group affiliations. As a result, individuals with an independent self-construal may have a weaker sense of in-group/out-group distinctions and may be more likely to move between different social groups or contexts. This contrasts with individuals with an interdependent self-construal, who tend to emphasize relationships, social roles, and group harmony.

In-group bias can manifest in implicit ways, which are unconscious and automatic, or in explicit ways, which are conscious and deliberate (Amodio & Mendoza, 2011). In-group bias can occur across various types of social groups, including political, ethnic, national, regional, and even trivial groups that are formed based on arbitrary or superficial distinctions. Understanding the mechanisms and manifestations of in-group bias is important in the study of social psychology, as it can have significant implications for intergroup relations, prejudice, discrimination, and conflict. Researchers like Amodio and Mendoza (2011) have investigated the cognitive and neural processes underlying in-group bias to gain insights into how these biases operate at both implicit and explicit levels.

Bias in favor of an ingroup is closely related to prejudices, which involve negative evaluations or biased attitudes towards others based on preconceptions or stereotypes. Individuals may exhibit discriminatory behaviors or negative emotions towards out-group members due to perceived differences or stereotypes associated with that group (Amodio, 2014). Neuroscience studies have also started to shed light on the neural underpinnings of social cognitive processes such as stereotyping and prejudice, which determine intergroup perception and behavior. Stereotyping refers to perceiving groups based on subjectively determined characteristics, and it has been linked to the anterior temporal lobes and dorsolateral prefrontal cortex. The anterior temporal lobes are involved in semantic memory processing and play a role in social categorization and person perception. The dorsolateral prefrontal cortex is implicated in executive control and the regulation of cognitive processes, including the inhibition of stereotypes. Ingroup bias and prejudices, on the other hand, are evident throughout low-level of cognitive processing to high-level processing: fund of knowledge presents large-scale brain networks from subcortical brain regions (i.e., the amygdala) to higher-level cortical brain regions (e.g., the inferior frontal cortex) (Molenberghs & Louis, 2018; Amodio, 2014).

The extent to which prejudice or stereotyping translate into behavior is largely determined by social norms (Hegarty & Pratto, 2001). Prejudice leads to biases in the rapid processing of information about outgroup social groups. This bias has been associated with the function of brain structures such as the amygdala, orbitofrontal cortex, insula, striatum, and medial prefrontal cortex (Amodio, 2014). Among the frontal cortex areas, ventromedial PFC displays an important functionality in multiple aspects of social cognition, such as facial emotion recognition, theory-of-mind ability, and processing self-relevant information, through its interactions with the posterior cingulate cortex, precuneus, dorsomedial PFC, and especially the amygdala (Hiser & Koenigs, 2018). The amygdala, which is involved in the processing of emotional and salient stimuli, has been shown to play a role in the formation and expression of prejudice (Chekroud et al., 2014). For example, Phelps et al. showed a difference in amygdala response in participants who watched Caucasian and African-American faces during functional imaging. However, when the same task was applied outside the fMRI, no difference was observed between self-reports. Researchers interpreted these findings as evidence that amygdala and behavioral responses of white participants to black vs. white faces reflect implicitly biased cultural group-level evaluations. A cross-cultural study between two countries from a close geographic area offers similar results. In this comparative study, Japanese individuals passively viewed images related to an ethnic outgroup (South Korea) inside a functional magnetic resonance imaging scanner. According to their voxel-based analysis, as Japanese participants were evaluating South Korea, altered activation was found in the left amygdala at implicit ("but not explicit" the researchers stated) level (Izuma et al., 2019).

Cross Cultural Differences of Self Representation

As discussed earlier in the chapter, self-construal has been expressed by many theorists and long-standing findings are being anchored in more neurologically based research. Though it is a well-established concept, it does have the drawback of obscuring within-culture variation. Perceptions and behaviors regarding self-representation in a social context differ according to culture, thus culture's effect on how we represent ourselves in our minds has been under the scope of many researchers who are studying with psychological assessment methods, group observations, and sample comparisons (Park et al., 2017). So, what is the neural infrastructure of this established distinction, given that self-structuring is not independent of culture? In this section, we will try to further discuss whether there are differences in the neural components of an individual's self-perception through cross-cultural comparative studies.

Feedback from the socio-cultural environment plays a crucial role in shaping our personality—and thus how we mentalize ourselves. One of the frequently used methods to understand the neural basis of this relationship is to evaluate individual's responses by stimulating with judgments about themselves, others, or general trait characteristics. An early study was conducted in Canada in 1977. Participants were able to remember adjectives that they believed described themselves more successfully in a retrieval task. A similar methodology was repeated in 1989 with American university students, this time without an initial retrieval phase, but once again, during the retrieval phase, descriptive sentences related to the self resulted in better memory performance compared to focusing on others or more general information. Finally, in the 2000s, these studies were combined with neuroimaging methods. Lieberman and Pfeifer (2004) showed that self-judgment statements processed in different brain regions compared to neutral trait statements (i.e., vmPFC, nucleus accumbens, amygdala, lateral temporal, and inferior parietal regions). This two-phase study was repeated with Chinese participants under brain scanning. Participants judged trait adjectives under three reference conditions of self, other, and semantic processing in the initial phase, then a mother-reference condition replaced the other-reference condition in later phase. As a result, self-referential processing activated the mPFC and cingulate areas in comparison to others'. Interestingly, mother-referential encoding and self-referential encoding both activated the mPFC suggesting that mother and the self may share a common processing brain region in Chinese population. This difference in neural processing was explained by the researchers with the social relationship styles (independent vs. interdependent) of the societies to which the participants grow up.

Research focusing on cross-cultural self-representation argue that average distinctions defining members of Western societies sharing "independent" personal characteristics while

Eastern societies, on the other hand, share "interdependent" (Chin et al., 2022; Zhu et al., 2007). An important argument that should be mentioned here is this: Could the grouping as Eastern culture be under too much influence of geographical marking? The East Asian cultures that have been assumed to be homogeneously collectivistic in previous psychological literature. In a cross-cultural study that Japanese and Korean participants rated a set of well-being and self-related scales proved that neighboring cultures also show nuance differences from each other (Park et al., 2017). According to the results, a Korean classroom sample exhibited a greater 'other-focused' relational self-construal to the Japanese sample in the same setting than the Japanese counterpart, whereas no difference was found between the online samples. The absence of differences between the online samples of Japanese and Korean participants suggests that the data collection method may have influenced the expression of self-related constructs. These findings highlight the importance of considering cultural variability and the context when studying psychological constructs such as self-perception and self-construal. Even there were differences within culture by data collection method (classroom vs. online) in relative degrees of selves and related constructs, possibly associated with generational differences.

Activation differences on neural regions when processing social evaluations and thoughts about ourselves suggest that cultural variations may contribute to differences in social behaviors. Though these findings constitute evidence for the neural basis of the interaction between personality and culture, further research is necessary to reach broader understanding. This is, as is the point of our chapter, an important place for neurosociology to enter the interdisciplinary conversation.

Conformity Versus Nonconformity

As social beings, individuals tend to gravitate towards behaviors that are most commonly observed in their seniors or peers within their ingroup, behaviors that are perceived as prestigious or successful, or behaviors exhibited by those with whom they share the strongest social connections (Perry, 2009; p.706). Recall the previously mentioned aspect that our behaviors are shaped by multiple factors like direct feedback from society members and indirect evaluations of social observations (i.e., social biases, stereotypes, self-judgements), which includes a motive to conform to the behaviors of others (Bogdan et al., 2022). This primitive social survival motive is also reflected in higher cognitive processes like decision making and judgements in which people look for conformity to the group that they take as their primary referral group. Human behavior is guided by not only subjective values or attitudes but also by the perceived behavior of others—especially as that behavior relates to social norms (Cialdini & Goldstein, 2004). These long-standing psychological processes have found support with neuroscientific methods. Conflict with ingroup opinion was found to trigger a neuronal response in the rostral cingulate zone and the ventral striatum (Klucharev et al., 2009), causing a deeply unpleasant experience. Cingulate cortex and the ventral striatum activation is also known to be activated during the "prediction error" (indicating the difference between the actual reward received and the reward expected to be received) (Garrison et al., 2013). This similarity in brain areas suggests that, in the situations where social conformity cannot be achieved, the brain actually reacts via the dopaminergic pathway as if we missed out on an expected reward.

Human motives include obtaining social approval from others and maintaining a favorable self-concept. Thus, conformity also plays a motivational role for the act of changing one's behavior to match the responses of others (Cialdini & Goldstein, 2004). Perceiving the judgment from an ingroup or an outgroup provides fundamental information (approval or disapproval) in particular social conditions. Social norms can be seen as positive or negative reinforcers for social behaviors requiring decision making. Thus, a conflicted perception within expected social norms calls for an adjustment of behavior. Like the infamous Asch studies, perceived differences of individual ratings from group ratings trigger behavioral adjustment in favor of conformity. Conformity provides acceptance and support from the group, which is an important factor for survival in evolutionary processes (Lieberman & Eisenberger, 2009). The challenges of simultaneously navigating social and natural environments, at times with competing motivations, may cause temporary setbacks for immediate loses but ultimately facilitated our long-term gains—by enabling the development of a more advanced and capable brain (Mulligan, 2022).

In one fMRI study, participants that were informed that their ratings on topics conflicted with their ingroup's ratings showed rostral cingulate activation and nucleus accumbens (NAc) deactivation. NAc, a part of ventral striatum, is often viewed as an integrator of memory, motivation, and goal-directed behaviors (Carelli, 2002). Nucleus accumbens activity is related with the value attribution of the expected reward (Knutson & Wimmer, 2007), and NAc activity decreases while facing aversive stimuli. This conflict situation might be processed like an aversive stimulus which is an unexpected outcome of the decision. The amplitude of the conflict related signal in the ventral striatum (NAc) correlated with differences in conforming behavior across subjects. Besides, when the conflict related stimuli are strong enough, the subjects displayed adjustment behavior in favor of their ingroup.

In the absence of conformity, thus in the case of non-conformity, the individual perceives the situation as a conflict with the ingroup, which brings up another shared interest of both neurobiologists and sociologists; conflict monitoring. The anterior cingulate cortex (ACC) has a crucial role in monitoring conflicts (Vogt, 2016), and it works in conjunction with the anterior insula in regulating behavior that is incongruent with group norms (Klucharev et al., 2009). If we conceptualize this regulation as a top-down process, it becomes quite understandable that dorsal ACC (dACC) and lateral prefrontal cortices are responsible for executive control, while rostral ACC and medial prefrontal cortices (mPFC) are involved in value attribution. Whether at a conscious level or not, the detection of conflict between one's own behavior and others, in other phrase not being in line with others, causes stress (Tomlin et al., 2013), which in turn motivates changes in behavior and attitude. A study conducted by Klucharev et al. (2009) showed that in cases where there is a feeling of discomfort caused by an incongruent behavior with the ingroup, the dACC and anterior insula regions are activated. It has been demonstrated that when the neural activation of these regions, such as through noninvasive methods like TMS, is manipulated, the adherence to social influence decreases (Klucharev et al., 2011). This presents another example of neuroimaging that offers the opportunity to examine mental processes in real-time, and free from the limits of measurement through introspection.

Neuroanatomical structures, connection pathways, and often activation patterns are common to all humans. But what reveals the individuality of the human is the psychological and sociological softwaring of the biological hardware. In other words, neural activity cannot be considered independent of the effect of the learning process. The neural processing styles evolve with the developmental processes that are strongly influenced by the social environment (Falk et al., 2013). Accumulating knowledge about social variables and their neural underpinnings have taught us that presence of social context, group relationships of the individual (i.e., ingroup, outgroup) is processed by different neural processing pathways (Cascio et al., 2015). These approaches offer paths toward bridging fields and methods toward understanding complicated questions about social life.

Cross Cultural Differences in Political Decision Making

People may identify their ingroup according to family or community membership, sports team, political party, gender, sexual orientation, religion, or nation. One of the strong features that determine ingroup-outgroup attribution is political orientation, in many societies capturing a vision of what makes a 'good person' and aligning people with, and against, those who putatively share (or not) their values. The boundaries of social groups create the outer lines of group identities, a well-established phenomenon in social psychology across both sociology and psychology. Those who fall outside of these boundaries are considered outgroup members (Lau, 2021), and are perceived quite differently, even when engaging in the same behaviors as ingroup members.

Political decision-making is influenced by the geographical environment and resources, national cultural characteristics, unique historical experiences, and the degree of the role that various scholars or policy research organization forces. The underpinning neural processes of political decision-making have also been studied within their own components. In order to investigate the neural mechanisms underlying political cognition, it is important to consider the interaction of mental processes and social dynamics (Zmigrod & Tsakiris, 2021), it seems that there is not enough knowledge to make cross-cultural variations. However, it is still believed that social psychology, sociology and political sciences can evolve together with neurosciences.

When individuals read statements that are consistent with their political beliefs, decreased activation in the bilateral insula is observed. Conversely, when they read statements that are inconsistent with their beliefs, researchers observed increased activation in the insular cortex and anterior cingulate cortex (ACC) (Haas et al., 2021). DLPFC, ACC and insular activation have been reported when viewing images of political figures representative of opposing views (Kaplan et al., 2007), and suggesting that stimuli is instantly processed in a different, and less sympathetic, neural channel. Subjective value attribution is the process of integrating and assigning value to information based on one's own goals and motivations (Bartra et al., 2013).

Both behavioral and neuroimaging results supported the hypothesis of in-group bias in processing contradictory information based on their group affiliation. People tend to perceive less contradiction between the initial statement and the action that contradicted the statement from their own group leader. The observed behavioral effect, where people perceive less contradiction when it comes from their own group leader, indicates that individuals may be more inclined to justify or overlook inconsistencies in the statements or actions of those with whom they identify or share a common group membership. Neuroimaging results of the same study revealed an increased activation in the medial prefrontal cortex and anterior cingulate cortex, areas associated with self-referential thinking, belief processing, and cognitive dissonance, for contradictory statements from a political candidate from the ingroup (vs. the neutral candidate). These findings indicate that individuals engage in mental processes that help reconcile or rationalize the inconsistency to align with their pre-existing beliefs or maintain a positive self-image: which can be conceptualized as the need of conformity.

Among the subcortical brain areas, the amygdala, insula and hypothalamus and cortical areas alongside ACC, prefrontal and orbitofrontal areas have been established as involved in comparing sensory information and affective value, and are activated when evaluating a given situation. The factor that determines which areas will be dominant in the emergence of a behavioral response—and therefore what kind of reaction will occur—is time. More automatic responses, that is, rapid reactions, are dominated by subcortical areas such as the amygdala or insular cortex. Responses controlled by the cortex occur more slowly due to the involvement of reflective processes. The strength of the amygdala response differed in the face of ingroup versus outgroup conflict (Knutson et al., 2006). We can say that there is a change in activation of the amygdala response to situations that have an emotional burden on the person. For example, the affective burden of political responses and attitudes is intense. For this reason, there are researchers who argue that the response is actually the response of automatic processes (Burdein et al., 2006). Stimuli that are instantly perceived as more self-relevant are processed using these more reactive, even primal, neural systems.

Narrowing the Gap Between Methodological Approaches

The contribution of neuroimaging methods to cross-cultural understandings of human perception and behavior, including their impact on how culture organizes neural activity, is a growing topic. We will discuss the increased integration of neuroimaging to various research disciplines, and its ability to demonstrate whether cultural differences impact the brain's neural structure. It is not always easy for researchers to extrapolate from individual neural findings to more abstract concepts such as how an individual functions within a society and how the society functions itself. Reasonable caution on the part of neuroscientists complicates the more Social Science tendency to explore collective-level generalizations. Multi-level analyses, linking the individual to intra-group, and to inter-group interactions, are required. Recent developments in social and cognitive neuroscience seem to enable us to integrate these levels. While we should not overemphasize the use of neuroscientific methods, such as fMRI, in predicting group behaviors, we cannot ignore their significance. One important reason is that fMRI methodology relies on indirect measurement of brain activity. The fundamental assumption underlying fMRI measurements is to track differentiations in cerebral blood flow following variations in neural activity through the blood oxygenation level dependent (BOLD) signal. Establishing a causal relationship with the BOLD response is challenging, but there is a relationship between any cognitive activity and BOLD signal.

Cultural norms may create social influence through face-to-face interactions, implicit observation, or increasingly through social media. With the social nature of human existence, our brains evolved to be capable of handling the interactions with increasingly larger social groups; many of the changes document by early social theorists involved this shift from living in smaller tribes to living in larger communities. The number of real and/or virtual social interactions has increased exponentially over the past decades with the widespread usage of social media tools. The influence of social media on human behavior has become more dynamic and powerful. The awareness regarding that social influence may not perpetually be processed on a conscious level. Neuroscience research tools provide information by surpassing the limitations of traditional self-report methodologies in social science studies. These tools enable real-time examination of mental processes such as social influence and prejudice, as well as methods based on individuals' internal perspectives (Lieberman, 2010). Thus, neurosociology has the potential to compliment some missing pieces left by traditional methods of data collection and observation.

Traditional research methods involve the use of self-report questionnaires, where the answers are either limited to given options or rated along a scale. Open-ended questions are a better opportunity to project oneself and but harder to make cluster reports from the responses. It has been repeatedly observed that designing social experiment environments and evaluating behaviors or attitudes yield consistent reliable results. Self-report questionnaires are limited to respondent's self-awareness and interoception ability. The tendency to give socially acceptable answers is not always driven by conscious processes. It is possible to see the traces of social acceptability in unconscious processes (Hahn & Goedderz, 2020).

If there are culture-specific neural activation patterns, neuroscience methods need to be incorporated to close an important gap in the understanding of individual differences and behavioral predictions. Rather than relying solely on reported beliefs and observed behaviors, for the first-time researchers can see what a cultural difference might look like in the brain (Sasaki &

Kim, 2017). For example, a classic study revealed the effect of culture on independence versus interdependence behaviors—a key marker of self-construal, as discussed above—has become perhaps the cornerstone finding of cross-cultural psychology (Markus & Kitayama, 1991). This famous study's results were corroborated by the neuroimaging studies (Martínez Mateo et al., 2013). These relevant studies encouraged social scientists to inspect the cultural influences on behaviors defined by sociology within the subject of neuroscience's functional brain studies.

Computational modeling approaches are also sliding into social contexts that facilitate more precise assessment of cognitive processes as well as more complex simulations of how mental processes interact with social dynamics. Future work should look into bridging these studies of Big Data—information gathered extensively about populations, or potentially from an individual—with neuroscientific methods. Indeed, advancements in fMRI studies can have significant contributions to fields such as social neuroscience, especially as the tool gets less expensive, and more data is gathered to explore using advanced statistical approaches. Beyond its cost, the limitation of the fMRI method is the difficulty of integrating it with the traditions of the field. It takes especially laborious training, the specialization of entire biological fields, and is not something that many social scientists have the access to or time to learn. Perhaps neurosociology points to the kind of true interdisciplinarity that gets spoken of in journals and by academic administrators, but the barriers are high and it is too early to tell how effective such bridge-building attempts will become. Early attempts have been mixed; anecdotally, one sees little neurological work at the top journals, even while the editors of this Handbook have been successful in beginning a field and instantiating it within these pages.

By studying the neural correlates of social behavior and interactions, functional brain imaging methods (including resting state neuroimaging, task based functional neuroimaging, diffusion tensor imaging, etc.) can provide insights into how individuals coexist and interact in communal life. The artificiality of studying interaction within an fMRI machine—which necessitates a single person remaining still during the study—maybe be overcome with technological advances in the future. This would build on understanding of the neural mechanisms underlying social cognition, empathy, cooperation, and conflict monitoring can help us develop strategies for promoting harmonious coexistence in society. For example, functional brain imaging studies can shed light on the neural processes involved in empathy, allowing us to better understand how individuals relate to and understand the emotions and perspectives of others. This knowledge can be used to foster empathy and compassion in communal life, enhancing understanding and cooperation among community members. Similarly, fMRI can help us understand the neural basis of social decision-making and cooperation. By identifying the brain regions and networks associated with prosocial behavior and cooperation, we can design interventions and policies that promote collaborative and harmonious interactions within communities. Overall, the precision and advancements in fMRI research have the potential to provide valuable insights into the neural underpinnings of social behavior and facilitate the development of strategies for coexisting harmoniously in communal life. Such strategies are, however, rife with the possibility of bias and misunderstanding.

Conclusion and Future Implications

In closing, we conclude with potential areas of research and future implications while integrating the complexity of physiological processes and the intangibility of sociocultural impact. Neuroscience offers methods that allow us to discern the specific physiological processes that indicate neural activation. On the positive side, this allows us to anchor cross-cultural variations within the 'natural' sciences, perhaps increasing sociology's visibility but also grounding over a century of empirical study with new methodologies that confirm the importance of sociality for understanding human thought, perception, and feeling. Many processes surrounding the self, political cognition, and notions of enculturation are confirmed and even specified through these approaches. The human mind, plastic and shaped fundamentally by early interactions, is a problem-solving device in the best traditions of American Pragmatism, Symbolic Interaction, and other long-standing sociological approaches to understanding human behavior; for humans, so many of those problems come from our social surroundings.

However, a human's complex world of emotions and thoughts cannot be understood simply by observing physiological processes or simply by observing overt behavior. Questions will involve details of how and when social influences on cognition influence self-construal, social interaction, and meaningful social outcomes. This is the overarching challenge of neurosociology, demonstrating the efficacy of these approaches for addressing core sociological issues. We suggest that exploring neurosociology across cultures is a propitious approach to looking at these core issues of culture and cognition, and encourage more exchange with our neighboring fields. The brain is constantly exposed to change through experiences within the social environment. Therefore, culture itself makes an important contribution to the wiring and functionality of the mind.

Future work needs to incorporate understanding the neural roots of mind and behavior of people within cultural variations. Both normative and non-normative mental functioning has been a core aspect of neurological science, both to address potential medical issues, but also to explore typically functioning minds. Therefore, in terms of future implications, the cultural neuroscience approach has the potential to empower the defense to maintain mental health, and determine how much of what is considered good mental health is specific to cultural pattern, and how much is anchored in shared neurological processes. Perhaps the best avenue road to cross-cultural neurosociology lies in the opposite direction. Although neuroscience is attracting more and more scientists' attention, socioeconomic and sociocultural influences, cultural factors, and immigrant/indigenous background are rarely mentioned. In this vein, sociological conceptions about theory and measurement, and even asking what we consider to be the proper questions, might be the ultimate payoff for the field.

References

Adams, R. B., Jr., Rule, N. O., Franklin, R. G., Jr., Wang, E., Stevenson, M. T., Yoshikawa, S., et al. (2010). Cross-cultural reading the mind in the eyes: An fMRI investigation. *Journal of Cognitive Neuroscience, 22*(1), 97–108.

Amodio, D. M. (2014). The neuroscience of prejudice and stereotyping. *Nature Reviews Neuroscience, 15*(10), 670–682.

Amodio, D. M., & Mendoza, S. A. (2011). Implicit intergroup bias: cognitive, affective, and motivational underpinnings. In B. Gawronski & B. K. Payne (Eds.), *Handbook of implicit social cognition: Measurement, theory, and applications* (pp. 353–374). Guilford Press.

Baron-Cohen, S., Jolliffe, T., Mortimore, C., & Robertson, M. (1997). Another advanced test of theory of mind: Evidence from very high functioning adults with autism or Asperger syndrome. *Journal of Child Psychology and Psychiatry, 38*(7), 813–822.

Bartra, O., McGuire, J. T., & Kable, J. W. (2013). The valuation system: A coordinate-based meta-analysis of BOLD fMRI experiments examining neural correlates of subjective value. *NeuroImage, 76*, 412–427.

Berreby, D. (2005). *Us and them: Understanding your tribal mind*. Little, Brown and Co.

Binder, J. R., Desai, R. H., Graves, W. W., & Conant, L. L. (2009). Where is the semantic system? A critical review and meta-analysis of 120 functional neuroimaging studies. *Cerebral Cortex, 19*(12), 2767–2796.

Bogdan, P. C., Moore, M., Kuznietsov, I., Frank, J. D., Federmeier, K. D., Dolcos, S., & Dolcos, F. (2022). Direct feedback and social conformity promote behavioral change via mechanisms indexed by centroparietal positivity: Electrophysiological evidence from a role-swapping ultimatum game. *Psychophysiology, 59*(4), e13985.

Boutyline, A., & Soter, L. K. (2021). Cultural schemas: What they are, how to find them, and what to do once you've caught one. *American Sociological Review, 86*, 000312242110245. https://doi.org/10.1177/00031224211024525

Brekhus, W. H. (2015). *Culture and cognition: Patterns in the social construction of reality*. John Wiley & Sons.

Burdein, I., Lodge, M., & Taber, C. (2006). Experiments on the automaticity of political beliefs and attitudes. *Political Psychology, 27*(3), 359–371.

Callaghan, T., Rochat, P., Lillard, A., Claux, M. L., Odden, H., Itakura, S., et al. (2005). Synchrony in the onset of mental-state reasoning: Evidence from five cultures. *Psychological Science, 16*(5), 378–384.

Camras, L. A., Bakeman, R., Chen, Y., Norris, K., & Cain, T. R. (2006). Culture, ethnicity, and children's facial expressions: A study of European American, Mainland Chinese, Chinese American, and adopted Chinese girls. *Emotion, 6*(1), 103.

Caravita, S. C., Giardino, S., Lenzi, L., Salvaterra, M., & Antonietti, A. (2012). Socio-economic factors related to moral reasoning in childhood and adolescence: The missing link between brain and behavior. *Frontiers in Human Neuroscience, 6*, 262.

Carelli, R. M. (2002). Nucleus accumbens cell firing during goal-directed behaviors for cocaine vs. 'natural' reinforcement. *Physiology & Behavior, 76*(3), 379–387.

Cascio, C. N., O'Donnell, M. B., Bayer, J., Tinney, F. J., Jr., & Falk, E. B. (2015). Neural correlates of susceptibility to group opinions in online word-of-mouth recommendations. *Journal of Marketing Research, 52*(4), 559–575.

Cerulo, K. A., Leschziner, V., & Hana Shepherd, H. (2021). Rethinking culture and cognition. *Annual Review of Sociology, 47*(1), 63–85. https://doi.org/10.1146/annurev-soc-072320-095202

Chekroud, A. M., Everett, J. A., Bridge, H., & Hewstone, M. (2014). A review of neuroimaging studies of race-related prejudice: Does amygdala response reflect threat? *Frontiers in Human Neuroscience, 8*, 179.

Chin, T., Caputo, F., Lin, C. L., & Hu, F. (2022). Understanding cognitive differences across cultures: Integrating neuroscience and cultural psychology. *Frontiers in Psychology, 13*, 1041734.

Cialdini, R. B., & Goldstein, N. J. (2004). Social influence: Compliance and conformity. *Annual Review of Psychology, 55*, 591–621.

Cohen, D., Bhushan, B., Hanks, R., Yoon, D. P., Johnstone, B., & Hunt, I. (2020). The right parietal lobe, sense of self, and empathy: Cross-cultural, ethnic, and religious considerations. *Mental Health, Religion & Culture, 23*(5), 375–397.

Cramer, S. C., Sur, M., Dobkin, B. H., O'Brien, C., Sanger, T. D., Trojanowski, J. Q., et al. (2011). Harnessing neuroplasticity for clinical applications. *Brain, 134*(6), 1591–1609.

Cross, S. (2011). Self-construal. In *Psychology*. https://doi.org/10.1093/obo/9780199828340-0051

Dimaggio, P. (1997). Culture and cognition. *Annual Review of Sociology, 23*, 263–287.

Ekman, P., Sorenson, E. R., & Friesen, W. V. (1969). Pan-cultural elements in facial displays of emotion. *Science, 164*(3875), 86–88.

Elfenbein, H. A., & Ambady, N. (2003). Universals and cultural differences in recognizing emotions. *Current Directions in Psychological Science, 12*(5), 159–164.

Falk, E. B., Hyde, L. W., Mitchell, C., Faul, J., Gonzalez, R., Heitzeg, M. M., et al. (2013). What is a representative brain? Neuroscience meets population science. *Proceedings of the National Academy of Sciences, 110*(44), 17615–17622.

Firat, R., & McPherson, C. M. (2010). Toward an integrated science of morality. In S. Hitlin & S. Vaisey (Eds.), *Handbook of the sociology of morality* (pp. 361–384). Springer.

Franks, D. D. (2010). *Neurosociology: The nexus between neuroscience and social psychology*. Springer.

Franks, D. D., & Turner, J. H. (2013). *Handbook of neurosociology*. Springer.

Freeman, J. B., Rule, N. O., Adams, R. B., Jr., & Ambady, N. (2009). Culture shapes a mesolimbic response to signals of dominance and subordination that associates with behavior. *NeuroImage, 47*(1), 353–359.

Gallagher, H. L., & Frith, C. D. (2003). Functional imaging of 'theory of mind'. *Trends in Cognitive Sciences, 7*(2), 77–83.

Garrison, J., Erdeniz, B., & Done, J. (2013). Prediction error in reinforcement learning: A meta-analysis of neuroimaging studies. *Neuroscience & Biobehavioral Reviews, 37*(7), 1297–1310.

Gelfand, M. J. (2019). *Rule makers, rule breakers: How culture wires our minds, shapes our nations, and drives our differences*. Hachette UK.

Gelfand, M. J., Raver, J. L., Nishii, L., Leslie, L. M., Lun, J., Lim, B. C., et al. (2011). Differences between tight and loose cultures: A 33-nation study. *Science, 332*(6033), 1100–1104.

Guo, G., Hexuan, L., Ling, W., Haipeng, S., & Wen, H. (2015). The genome-wide influence on human BMI depends on physical activity, life course, and historical period. *Demography, 52*(5), 1651–1670.

Gutchess, A. H., Hedden, T., Ketay, S., Aron, A., & Gabrieli, J. D. (2010). Neural differences in the processing of semantic relationships across cultures. *Social Cognitive and Affective Neuroscience, 5*(2–3), 254–263.

Haas, I. J., Baker, M. N., & Gonzalez, F. J. (2021). Political uncertainty moderates neural evaluation of incongruent policy positions. *Philosophical Transactions of the Royal Society B, 376*(1822), 20200138.

Habel, U., Gur, R. C., Mandal, M. K., Salloum, J. B., Gur, R. E., & Schneider, F. (2000). Emotional processing in schizophrenia across cultures: Standardized measures of discrimination and experience. *Schizophrenia Research, 42*(1), 57–66.

Hahn, A., & Goedderz, A. (2020). Trait-unconsciousness, state-unconsciousness, preconsciousness, and social miscalibration in the context of implicit evaluation. *Social Cognition, 38*(Supplement), s115–s134.

Haidt, J., & Graham, J. (2009). Planet of the Durkheimians, where community, authority, and sacredness are foundations of morality. In J. Jost, A. C. Kay, & H. Thorisdottir (Eds.), *Social and psychological bases of ideology and system justification* (pp. 371–401). Oxford University Press.

Han, H., Glover, G. H., & Jeong, C. (2014). Cultural influences on the neural correlate of moral decision making processes. *Behavioural Brain Research, 259*, 215–228.

Harada, T., Mano, Y., Komeda, H., Hechtman, L. A., Pornpattananangkul, N., Parrish, T. B., et al. (2020). Cultural influences on neural systems of intergroup emotion perception: An fMRI study. *Neuropsychologia, 137*, 107254.

Hegarty, P., & Pratto, F. (2001). The effects of social category norms and stereotypes on explanations for intergroup differences. *Journal of Personality and Social Psychology, 80*(5), 723.

Hiser, J., & Koenigs, M. (2018). The multifaceted role of the ventromedial prefrontal cortex in emotion, decision making, social cognition, and psychopathology. *Biological Psychiatry, 83*(8), 638–647.

Hofstede, G. (1980). *Culture's consequences: International differences in work-related values*. Sage.

Hofstede, G. (2001). *Culture's consequences: Comparing values, behaviors, institutions, and organization across nations*. Sage.

Huang, Y., Tang, S., Helmeste, D., Shioiri, T., & Someya, T. (2001). Differential judgment of static facial expressions of emotions in three cultures. *Psychiatry and Clinical Neurosciences, 55*(5), 479–483.

Inglehart, R. (2006). Mapping global values. *Comparative Sociology, 5*(2–3), 115–136.

Ishii, K., Kobayashi, Y., & Kitayama, S. (2010). Interdependence modulates the brain response to word–voice incongruity. *Social Cognitive and Affective Neuroscience, 5*(2–3), 307–317.

Izard, C. E. (1994). Innate and universal facial expressions: Evidence from developmental and cross-cultural research.

Izuma, K., Aoki, R., Shibata, K., & Nakahara, K. (2019). Neural signals in amygdala predict implicit prejudice toward an ethnic outgroup. *NeuroImage, 189*, 341–352.

Jung, J., & Lee, S. H. (2006). Cross-cultural comparisons of appearance self-schema, body image, self-esteem, and dieting behavior between Korean and US women. *Family and Consumer Sciences Research Journal, 34*(4), 350–365.

Kalkhoff, W., Shane, R. T., & Edward, J. L. (2012). *Biosociology and neurosociology*. Emerald Group Publishing.

Kang, P., Lee, Y., Choi, I., & Kim, H. (2013). Neural evidence for individual and cultural variability in the social comparison effect. *Journal of Neuroscience, 33*(41), 16200–16208.

Kaplan, J. T., Freedman, J., & Iacoboni, M. (2007). Us versus them: Political attitudes and party affiliation influence neural response to faces of presidential candidates. *Neuropsychologia, 45*(1), 55–64.

Kim, H. S. (2002). We talk, therefore we think? A cultural analysis of the effect of talking on thinking. *Journal of Personality and Social Psychology, 83*(4), 828.

Kim, J. H. (2021). *The structure of culture: Cross-cultural investigation of shared beliefs*. University of Iowa.

Kitayama, S., Duffy, S., Kawamura, T., & Larsen, J. T. (2003). Perceiving an object and its context in different cultures: A cultural look at new look. *Psychological Science, 14*(3), 201–206.

Kitayama, S., & Uskul, A. K. (2011). Culture, mind, and the brain: Current evidence and future directions. *Annual Review of Psychology, 62*(1), 419–449.

Klucharev, V., Hytönen, K., Rijpkema, M., Smidts, A., & Fernández, G. (2009). Reinforcement learning signal predicts social conformity. *Neuron, 61*(1), 140–151.

Klucharev, V., Munneke, M. A., Smidts, A., & Fernández, G. (2011). Downregulation of the posterior medial frontal cortex prevents social conformity. *Journal of Neuroscience, 31*(33), 11934–11940.

Kluckhohn, C., & Murray, H. A. (1953). Outline of a conception of personality. In H. A. Murray & C. Kluckhohn (Eds.), *Personality in nature, society, and culture* (2nd ed., pp. 3–32). Knopf.

Knutson, B., & Wimmer, G. E. (2007). Splitting the difference: How does the brain code reward episodes? *Annals of the New York Academy of Sciences, 1104*(1), 54–69.

Knutson, K. M., Wood, J. N., Spampinato, M. V., & Grafman, J. (2006). Politics on the brain: An fMRI investigation. *Social Neuroscience, 1*(1), 25–40.

Kobayashi, C., Glover, G. H., & Temple, E. (2007). Cultural and linguistic effects on neural bases of 'theory of mind' in American and Japanese children. *Brain Research, 1164*, 95–107.

Koelkebeck, K., Hirao, K., Kawada, R., Miyata, J., Saze, T., Ubukata, S., et al. (2011). Transcultural differences in brain activation patterns during theory of mind (ToM) task performance in Japanese and Caucasian participants. *Social Neuroscience, 6*(5–6), 615–626.

Krendl, A. C. (2016). An fMRI investigation of the effects of culture on evaluations of stigmatized individuals. *NeuroImage, 124*, 336–349.

Kutas, M., & Hillyard, S. A. (1980). Event-related brain potentials to semantically inappropriate and surprisingly large words. *Biological Psychology, 11*(2), 99–116.

Lau, T. (2021). Reframing social categorization as latent structure learning for understanding political behaviour. *Philosophical Transactions of the Royal Society B, 376*(1822), 20200136.

Leger, K. R., Cho, I., Valoumas, I., Schwartz, D., Mair, R. W., Goh, J. O. S., & Gutchess, A. (2024). Cross-cultural comparison of the neural correlates of true and false memory retrieval. *Memory*, 1–18.

Lieberman, M. D. (2010). Social cognitive neuroscience. In S. T. Fiske, D. T. Gilbert, & G. Lindzey (Eds.), *Handbook of social psychology* (Vol. 5, pp. 143–193). McGraw-Hill.

Lieberman, M. D., & Eisenberger, N. I. (2009). Pains and pleasures of social life. *Science, 323*(5916), 890–891.

Lieberman, M. D., & Pfeifer, J. H. (2004). The self and social perception: Three kinds of questions in social cognitive neuroscience. In *The cognitive neuroscience of social behaviour* (pp. 207–248). Psychology Press.

Lizardo, O. (2017). Improving cultural analysis: Considering personal culture in its declarative and nondeclarative modes. *American Sociological Review, 82*(1), 88–115.

Lizardo, O., & Michael, S. (2010). Skills, toolkits, contexts and institutions: Clarifying the relationship between different approaches to cognition in cultural sociology. *Poetics, 38*(2), 205–228.

Ma, Y., Bang, D., Wang, C., Allen, M., Frith, C., Roepstorff, A., & Han, S. (2014). Sociocultural patterning of neural activity during self-reflection. *Social Cognitive and Affective Neuroscience, 9*(1), 73–80.

Markus, H., & Kitayama, S. (1991). Culture and the self: Implications for cognition, emotion, and motivation. *Psychological Review, 98*, 224–253.

Marsh, A. A., Elfenbein, H. A., & Ambady, N. (2003). Nonverbal "accents" cultural differences in facial expressions of emotion. *Psychological Science, 14*(4), 373–376.

Martin, M. A. (2008). The intergenerational correlation in weight: How genetic resemblance reveals the social role of families. *American Journal of Sociology, 114*(S1), S67–S105.

Martínez Mateo, M., Cabanis, M., Stenmanns, J., & Krach, S. (2013). Essentializing the binary self: Individualism and collectivism in cultural neuroscience. *Frontiers in Human Neuroscience, 7*, 289.

Matsumoto, D., & Hwang, H. C. (2016). The cultural bases of nonverbal communication.

Molenberghs, P., & Louis, W. R. (2018). Insights from fMRI studies into ingroup bias. *Frontiers in Psychology, 9*, 1868. https://doi.org/10.3389/fpsyg.2018.01868

Molenberghs, P., Prochilo, G., Steffens, N. K., Zacher, H., & Haslam, S. A. (2017). The neuroscience of inspirational leadership: The importance of collective-oriented language and shared group membership. *Journal of Management, 43*(7), 2168–2194.

Mulligan, R. F. (2022). Efficiency and inefficiency arguments in evolutionary neurobiology: The sensory order, social conformity, and adaptation. *Cosmos+ Taxis, 10*.

Owens, T. J., Dawn, T., Robinson, D. T., & Smith-Lovin, L. (2010). Three faces of identity. *Annual Review of Sociology, 36*, 477–499.

Oyserman, D., Coon, H. M., & Kemmelmeier, M. (2002). Rethinking individualism and collectivism: Evaluation of theoretical assumptions and meta-analyses. *Psychological Bulletin, 128*(1), 3–72.

Park, J., Norasakkunkit, V., & Kashima, Y. (2017). Cross-cultural comparison of self-construal and well-being between Japan and South Korea: The role of self-focused and other-focused relational selves. *Frontiers in Psychology, 8*, 1516.

Paterson, S. J., Heim, S., Friedman, J. T., Choudhury, N., & Benasich, A. A. (2006). Development of structure and function in the infant brain: Implications for cognition, language and social behaviour. *Neuroscience & Biobehavioral Reviews, 30*(8), 1087–1105.

Perry, S. (2009). Conformism in the food processing techniques of white-faced capuchin monkeys (*Cebus capucinus*). *Animal Cognition, 12*(5), 705–716. https://doi.org/10.1007/s10071-009-0230-3

Pinkham, A. E., Sasson, N. J., Calkins, M. E., Richard, J., Hughett, P., Gur, R. E., & Gur, R. C. (2008). The other-race effect in face processing among African American and Caucasian individuals with schizophrenia. *American Journal of Psychiatry, 165*(5), 639–645.

Sagiv, L., & Schwartz, S. H. (2022). Personal values across cultures. *Annual Review of Psychology, 73*(1), 517. https://doi.org/10.1146/annurev-psych-020821-125100

Sasaki, J. Y., & Kim, H. S. (2017). Nature, nurture, and their interplay: A review of cultural neuroscience. *Journal of Cross-Cultural Psychology, 48*(1), 4–22.

Schibli, K., & D'angiulli, A. (2013). The social emotional developmental and cognitive neuroscience of socioeconomic gradients: Laboratory, population, cross-cultural and community developmental approaches. *Frontiers in Human Neuroscience, 7*, 788.

Schurz, M., Radua, J., Tholen, M. G., Maliske, L., Margulies, D. S., Mars, R. B., Sallet, J., & Kanske, P. (2021). Toward a hierarchical model of social cognition: A neuroimaging meta-analysis and integrative review of empathy and theory of mind. *Psychological Bulletin, 147*(3), 293.

Schwartz, S. H., & Cieciuch, J. (2021). Measuring the refined theory of individual values in 49 cultural groups: Psychometrics of the revised portrait value questionnaire. *Assessment, 29*, 107319112199876. https://doi.org/10.1177/1073191121998760

Shiraev, E. B., & Levy, D. A. (2020). *Cross-cultural psychology: Critical thinking and contemporary applications*. Routledge.

Shweder, R. A., & Much, N. C. (1997). Manamohan Mahapatra and Lawrence Park. The "big three" of morality (autonomy, community, divinity) and the "big three" explanations of suffering. In A. Brandt & P. Rozin (Eds.), *Morality and health* (pp. 119–169). Routledge.

Somech, A. (2000). The independent and the interdependent selves: Different meanings in different cultures. International Journal of Intercultural Relations, 24(2), 161–172.

Tomlin, D., Nedic, A., Prentice, D. A., Holmes, P., & Cohen, J. D. (2013). The neural substrates of social influence on decision making. *PLoS One, 8*(1), e52630.

Triandis, H. C., & Brislin, R. W. (1984). Cross-cultural psychology. American Psychologist, 39(9), 1006–1016. https://doi.org/10.1037/0003-066X.39.9.1006

Udry, J. R. (1995). Sociology and biology: What biology do sociologists need to know? *Social Forces, 73*(4), 1267–1278.

Vaisey, S. (2009). Motivation and justification: A dual-process model of culture in action. *American Journal of Sociology, 114*(6), 1675–1715.

Vaisey, S. (2021). Welcome to the real world: Escaping the sociology of culture and cognition. *Sociological Forum, 36*, 1297–1315.

Vogeley, K., & Roepstorff, A. (2009). Contextualizing culture and social cognition. *Trends in Cognitive Sciences, 13*(12), 511–516.

Vogt, B. A. (2016). Midcingulate cortex: Structure, connections, homologies, functions and diseases. *Journal of Chemical Neuroanatomy, 74*, 28–46.
Zhu, Y., Zhang, L., Fan, J., & Han, S. (2007). Neural basis of cultural influence on self-representation. *NeuroImage, 34*(3), 1310–1316.
Zmigrod, L., & Tsakiris, M. (2021). Computational and neurocognitive approaches to the political brain: Key insights and future avenues for political neuroscience. *Philosophical Transactions of the Royal Society B, 376*(1822), 20200130.

The Pain of Social Loss: The Biosocial Roots of Motivated Action and Solidarity

20

Seth Abrutyn and Colter Uscola

Abstract

Social pain, of all detrimental experiences, remains largely overlooked by sociological inquiry. While attention to the sources are fused to Durkheim, from anomie to fatalism, the discipline centers its outlook upon other occurrences and social processes, such as emotion, cognition, or social networks. One exception challenging the notion that pain is notably biological is a turn toward negative physical pain. Along these lines, sociology provides the flux necessary for forging neuroscience and the social world through a consideration of social pain. Pain thus becomes a motivational and bonding force among humans. In this essay, we first provide a glimpse into social pain from a neuroscientific lens. We, then, proceed to contribute to the expanding body of research developing neurosociology by considering pain as a source of action that is embodied in the self, impacting upon individuals through role identities, and inequitably distributed across societies. In so doing, we expand the field of neurosociology in many necessary ways essential for considering social pain and action. Following this foray into a landscape unknown to many social scientists, we conclude by shoring up a set of implications for both sociologists and neuroscientists that support further expansions into this newly charted region of science.

In recent decades, sociology has decidedly embraced the incorporation of various types of intrapersonal dynamics like emotion (Turner, 2007), cognition (Lizardo, 2017), decision-making (Vaisey & Valentino, 2018), and so forth. One notable exception has been pain (Bendelow & Williams, 1995; Zajacova et al., 2021). In part, pain has escaped sociological attention because it seems eminently biological compared to those other individual-level processes. Additionally, it is notoriously challenging to measure scientifically, as pain is highly subjective in its sensation and intensity. Yet, it is strongly associated with negative physical (Zajacova et al., 2021) and mental health (Pearlin, 1999) outcomes, is a key motivator in the science of rewards (Pansksepp & Watt, 2011), and, as this essay will illustrate, appears to have a unique evolutionary function for promoting and sustaining social bonds among humans.

In what follows we introduce the concept: social pain (Eisenberger, 2015; Rotge et al., 2015). Social pain is the affective response to the real, imagined, or anticipated threat or loss of

S. Abrutyn (✉) · C. Uscola
Department of Sociology, The University of British Columbia, Vancouver, BC, Canada
e-mail: seth.abrutyn@ubc.ca; c.uscola@ubc.ca

W. Kalkhoff et al. (eds.), *Handbook of Neurosociology*, Handbooks of Sociology and Social Research, https://doi.org/10.1007/978-3-031-95615-7_20

social ties. By piggybacking on one of the two dissociable neural circuits associated with pain—not the sensorimotor circuit, but the affective circuit (Rainville, 2002)—social rejection, separation, isolation, and exclusion trigger the same neurophysiological response that natural noxious stimuli produce in all mammals: avoidance and learned aversion (Bradley et al., 2008). Affect is at the center of learning (LeDoux, 2000), which means pain also reshapes cognitive and behavioral functioning as organisms learn to detect the same harmful substances and/or avoid environmental contexts that are a threat. Given the significant difference in memory, learning, and self between humans and most other mammals, social pain presents obvious promise for understanding typical patterned social action and motivation as well as the emergence and clustering of pathological responses to threats or losses of social bonds (Abrutyn, 2023); especially when social pain is collectivized into a community or class of people and becomes a key element of collective identity (Abrutyn, 2024).

The first task of this essay, then, is to define and conceptualize what social pain is and how it works. Second, we ask three questions that general sociology can answer and that help bridge neuroscience with social science. Finally, we discuss some of the basic implications a sociology of (social) pain presents.

What Is Social Pain

Before discussing the neurobiology of social pain, a brief rejoinder on the neurobiology of pain is in order. In its simplest form, pain is an evolved protective alarm designed to prevent an organism's death. The neural circuitry of pain consists of two apparently dissociable processes (Rainville et al., 1999): the sensation of pain and the affective response to said sensation. The sensation of pain is triggered by noxious stimuli, like the acute sting of a rose stem's thorn or the tonic throb of a stubbed toe, and is processed in the primary and secondary somatosensory cortices and posterior insula. The affective dimension, however, refers to the types of emotional, cognitive, and behavioral response that may be innate, learned, or both, and is intimately tied to the dorsal anterior cingulate cortex (dACC) and the anterior insula (AI). The dACC plays a unique role in linking the limbic (affectual) systems and the prefontal (cognitive) cortex (Stevens et al., 2011). Subsequently, the dACC acts as a "neural alarm" signaling distress and pain (Spunt et al., 2012) as well as a signal to not only act in response to this pain—as it speaks to the decision-making and motor control centers of the brain—but also update and reinforce our memory regarding stimuli and negative affect in the striatum (Reeve & Lee, 2019). In conjunction with the dACC, the AI functions as an interpretive system as situational cues, such as a praise or ridicule, are experienced as affect and emotion then consolidated into conscious awareness of situations (Lewis et al., 2007). The AI is central to social judgements, such as trust, empathy, and anxiety while also being tied to predicting pain and contributing to persistent worry and avoidance behaviors (Gogolla, 2017; Paulus & Stein, 2006). Importantly, functional imaging shows that the dACC is activated only during the affective response to pain, which squares with research on patients with damage to or loss of the dACC who can identify where the pain sensation occurs while simultaneously not being bothered by the pain (Eisenberger, 2012). The discreetness of the physical pain system rests at the basis of the social pain, as social pain piggybacks on the affective circuitry of physical pain.

The study of social pain began when affective neuroscientist, Jaak Panksepp (1998) hypothesized that the opioid centers in our brain responsible for pleasure and pain relief had been co-opted by evolutionary forces promoting social bonding. Universally, mammals experience intense distress, panic, and, over time, grief when separated from their primary caretakers; an evolved response that most of us have experienced as children when we've lost sight of a parent in a crowded place and which some of us have experienced as parents who have lost sight of our own child. Unsurprisingly, research has shown that morphine (Panksepp & Watt, 2011) and even common household pain killers like acetaminophen (Inagaki, 2018) significantly reduce the distress of separation. Following these threads, cognitive scientists MacDonald and Leary (2005) found that the lexical references to severe types of rejection or exclusion, like experiencing unrequited love, cross-culturally used synonyms of hurt, injury, or pain. Eventually, neuroscientists built up a stronger body of empirical evidence supporting the fact that social pain, or the *negative affective response to separation, rejection, exclusion, or isolation from cherished social objects*, had co-opted our physical pain circuitry (Eisenberger, 2015; Rotge et al., 2015).

If social and physical pain have neural correlates, we can make several important suppositions according to Eisenberger (2012). First, sensitivity to one form of pain should condition sensitivity to the other. This may mean that genetic propensities to high physical pain thresholds are tantamount to similar thresholds in social pain thresholds, and vice versa (p. 606). Second, it may also mean chronic exposure to physical pain may lead to greater exposure and sensitivity to social pain, or vice versa. Individuals exposed to early childhood trauma, for instance, report experiencing greater physical pain in adulthood (Landa et al., 2012). Perhaps it is unsurprising, then, that clinical patients experiencing intense shame—the social emotion typically identified with social rejection (Scheff, 1988)—report feeling like they want to disappear, hide, or tear their skin off (Lewis, 1971). This also helps us understand why violent criminals almost universally report ritualized childhood humiliation that they identify with social death and a strong motivation to direct that pain externally onto others (Gilligan, 2003). Finally, things that change one should have similar effects on the other (Eisenberger, 2012:607). Inflammatory activity, for instance, also increases reports of alienation, while it was noted above that common and more intense pain killers can dull the pain of rejection and separation (also, DeWall et al., 2010).

These affinities between social and physical pain matter for sociology. Pain motivates behavioral responses. This motivational impulsion is typically distant from the noxious stimulus, but, given the social dimension of pain, rejection and exclusion have been empirically linked to various types of prosociality, including efforts to repair or renew bonds, generosity designed to induce reintegration, and increased reciprocal or mirroring interpersonal action (Hudd & Moscovitch, 2021). Moreover, like physical pain, social pain is an affectual force of learning both directly, when we feel pain, and indirectly when pain evokes other affectual responses like fear (LeDoux & Brown, 2017) or seeking (Berridge & Kringelbach, 2008). Thus, when our responses reduce the pain stimulation, we are more apt to use those strategies in the future. However, being capable of learning vicariously in myriad ways, like observing real or fictional others or in the process of talking about it with others, means we can learn different responses that we test out when faced with pain. Furthermore, once experienced, we are capable of developing empathy when we see or perceive another's social pain. This capacity is enhanced by our ability to take the role of the other and anticipate what our words or deeds might do if we follow through with them (Tomasello, 2019).

When taken together, it is clear social pain is not just a fundamental motivating force but also a source of cohesion, solidarity, and cooperation. Affective neuroscience has found that mammals, at birth, are wired to *seek* out primary caretakers for the sake of sustenance and survival and experience *panic* when losing sight of that cherished social object (Panksepp & Watt, 2011). Put differently, we are driven to bond with social objects that reciprocate and to avoid losing them. The pain we feel when that core social object is lost, even temporarily as a child losing sight of their parent in a busy market, strikes us. The fear of losing sight of them again teaches us strategies to be more vigilant; if the social object is lost for good, this panic transforms into deep grief (Panksepp, 2003). Once again, our higher-order cognitive capacities explode these basic mammalian traits (Turner, 2007). Any social object—a father, sibling, relative, fictive kin—can become an object that maps onto our neural architecture and motivates us to seek out their company and attention as well as panic or grieve at their loss. Social objects, of course, can be abstractions as well. Chimps appear to be affectually attached to their community (Maryanski, 2018), which implies humans can also see "imaginary" communities as objects of sustenance and survival. The community may be a conglomeration of real people sharing territory or it may belong to a social category like "Black American," "Jew," or "Sociologist." Consequently, these, too, can be threatened or lost. In the case of the former, civil strife or other violent conflicts can displace entire communities of people, sending them scattered all over (O'Neill & Spybey, 2003). Similarly, concerted efforts to erase the culture and/or existence of a category of people, like the Jews during the Holocaust or the Hutu of Rwanda, can simulate the same sorts of panic and social pain (Abrutyn, 2024).

Pain, Motivation, and Action

A key advantage to incorporating the social pain literature derives from its isolation of a causal model of motivated action. In Fig. 20.1, we see a two-path model with pro- and anti-social responses being possible outcomes. That is, when possible, those who are facing rejection or exclusion are motivated to bond with others through affiliative actions (Hudd & Moscovitch, 2021). They may increase their generosity (Chester et al., 2016), for example, or concerted reciprocal behaviors (Sommer & Bernieri, 2015). Though sociologists often focus on the anti-social pathway, this pathway is not entirely ignored. In particular, Goffman (1967) and his student Scheff (1988) have both emphasized the motivation to correct or even maintain social relationships that are threatened by an individual's actions. For Goffman, minor transgressions in speech or deed invited the sort of disapproval that prohibited the flow of interaction in situations. The solution was to willingly engage in corrective rituals designed to repair the encounter's moral order. Of course, the offended must also be inclined to "re-integrate" the offender (Braithwaite, 1989). Scheff (1988) went a step further, arguing that our ability to anticipate feeling shame was as powerful as being ashamed of something we said or how we behaved, causing us to hew more closely to appropriate, approved attitudes and actions.

What is new in this model, however, are two key empirical observations. First, while exclusion and ensuing social pain are not sufficient for

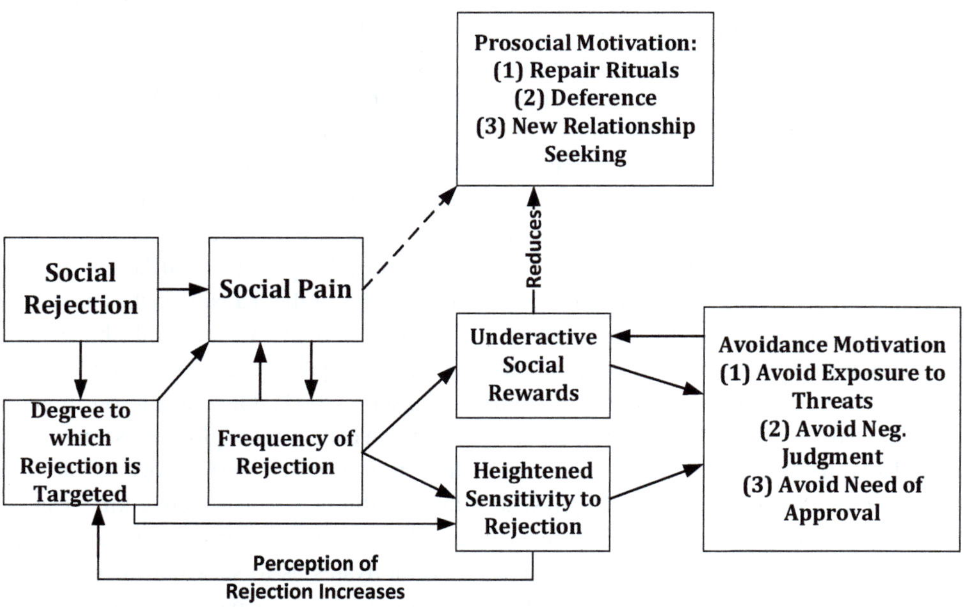

Fig. 20.1 Relationship between Social Rejection and Avoidance

motivating anti-social behaviors, *targeted* and *chronic* rejection can tip the balance from inclinations towards approaching others for connection to that of avoidance (Slavich et al., 2016). By targeted, we mean social rejection must be perceived or objectively intentioned, have an isolated impact on a specific person, and cause social demotion, devaluation, or disprivilege. In essence, the individual must feel singled out and the consequence must be real, even if temporary. For example, sweeping policies that target specific peoples, such as the classic case of Jim Crow laws or modern and outdated LGBTQ+ marriage bans, can make members of those groups sensitive as well as increase the probabilities of being subjected to exclusion or rejection. Indeed, research shows that the experience of social pain increases with the degree to which the exclusion or isolation feels purposeful (Eisenberger, 2015).

The second component, chronic or repeated rejection, amplifies the effects of targeting by taxing the neural circuits and priming us for feeling pain, while dampening our "reward-seeking" neural system (Blay et al., 2021). Together, hyper-sensitivity and reduced seeking behavior favor avoidance-based motivations (Rudolph et al., 2016); motivations that include (1) *general avoidance*, or the intentionally guided effort to avoid situations or encounters where being punished or threatened might occur; (2) *performance-avoidance* attitudes designed to avoid pursuing goals that may elicit negative judgments and thus opportunities of social demotion; and, (3) *avoidance-oriented needs for approval.*

As the various feedback loops in Fig. 20.1 suggest, once avoidance motivations become stabilized people may become trapped in anti-social behavior. Heightened sensitivity, for example, increases the probabilities of targeted rejection in part because the person may perceive intention where intention is not objectively real. But, it also increases it as people become more avoidant and, as a consequence, may find themselves in situations without allies or where they are visibly or behaviorally different and easily targeted. Likewise, avoidance reduces the opportunity to not only seek social rewards but to receive them, which, in turn, further reduces the pro-social motivational efforts. As we will see shortly, this process sheds light on how populations who have generally seen targeted rejection—e.g., women, people of color, and low-income individuals—experience increased health risks across the life-course (Umberson & Donnelly, 2023), as chronic pain is associated with physical and mental health consequences (Zajacova et al., 2021).

The Outcomes of Motivated Pain Responses

In short, then, social pain is a powerful motivational force for both seeking objects that have brought or are presumed to bring pleasure and avoiding objects that potentially threaten the organism in immediate and anticipated future events. Research suggests pain, and more importantly, the anticipation of pain that allows humans to reflect on what might happen if they act in certain ways with significant others, motivates us to conform or risk rejection, isolation, or worse, exclusion and expulsion (Boehm, 2012). And yet, humans are quite effective at rejecting, isolating, and excluding other people(s), especially those that are deemed "non-human." As such, most humans are fundamentally motivated to avoid losing the things they care about the most, whether it be a loved one, their birth place and community, other objects of extreme value, and so forth. Subsequently, reputation, status, social self, and other typical sociological concepts related to a person's geo-social location within a group, organization, community, or social institution are to be maintained and sustained, lest they invite social pain. Crucially, as we will demonstrate, both the objects that are valued as desirable sources of pleasure and reward, the means to acquiring and maintaining bonds with said objects, and, therefore, social pain itself is inequitably distributed in ways that sociology can contribute to making sense of.

Indeed, if social pain offers a key fundamental source of social self and social action, then it suggests it contributes to the constitution of social structure and culture. That is, the link between neuroscience and sociology comes into focus when we consider that on the one hand, as a recent review argued, sociology has largely ignored pain, opting to see it mostly as a biological or psychological factor (Zajacova et al., 2021). Understandably, processes that are wholly biological are not "legible" for most sociologists, and yet, the very same review showed just how powerful chronic physical pain is as both a social fact and as a fact that is unevenly distributed across a population. Moreover, sociology has long evoked the metaphor of pain in its conceptual lexicon (alienation; anomie), in many of its processes (Merton's strain theory, status expectation states theory, or Goffman's stigma theory), and in the broad acceptance of discrepancy, incongruence, and discomfort as key motivating forces (Gross, 2009; Mead, 1934). On the other hand, as many chapters in this Handbook, as well as myriad forays into neurosociology argue, neuroscience needs social science. Extensive research in medical and mental health sociology show, for instance, that a wide array of negative outcomes related to physical, mental, emotional, and social health are not equitably distributed (Link & Phelan, 1996). It is here that the incorporation of social pain matters most to sociology: as a fundamental, evolved process, it helps sociology explain the link between some forms of social stress and negative outcomes, like deteriorating mental or physical health. At the same time, it grounds concepts that are typically abstractions, with social ontological implications, in something ubiquitously human. In Fig. 20.2, we extend the basic model of motivation to include some of these outcomes (red lines indicate added processes to Fig. 20.1).

As the second model illustrates, past research shows that avoidance behavior has a direct relationship to the diagnoses of social anxiety and depression (Rudolph et al., 2016). Depression is also related to declining social rewards, as our pleasure hotspots (opioid receptors) are less apt to fire because we are less likely to engage in pleasurable social activities and because dopaminergic systems that propel seeking behavior are also less likely to fire as we are oriented towards avoidance (Kringelbach & Berridge, 2016). To these mental health outcomes, neuroscience research demonstrates that some groups of people subjected to social pain develop negative physical health outcomes, like higher blood pressure (Inagaki et al., 2018). Population health research suggests that it is less the rejection itself and more so the assortment of interactional strategies and challenges these groups face daily that puts disproportionate stress on the body (DeAngelis, 2020, 2021). Hence, we link avoidance motivation directly to negative mental health outcomes, while recognizing the fact that poor health carries risk of depression or other mental health challenges, like dementia (Livingston et al., 2020). Finally, to this model, we add the fact that shifting affectual, cognitive, and behavioral resources towards avoidance, social anxiety, and depression encourage individuals to develop defensive strategies designed to mitigate the emotional and mental health struggles these rejection-based outcomes produce. And, in all likelihood, these defensive strategies are either pathological in nature or lead to pathologies (Summers-Effler, 2004) like the misuse of alcohol (Uscola, 2023). It is clear that physical health is also tied to the development of pathological behaviors designed to cope with physical pain (Zajacova et al., 2021), but ultimately is likely tied to a reciprocal feedback loop.

In the following section, we ask and answer three key questions that bring sociological ideas to bear on the basic motivational model (Fig. 20.1). First, why are threats to or losses of some objects more painful than others? That is, we cannot argue that all targeted rejection causes the same type of social pain. To answer this question, we begin to bridge neuroscience and sociology through microsociological theories of self and meaning-making. Second, what makes experiencing rejection, particularly targeted, chronic rejection more or less likely to occur. A significant body of social scientific research studies the social distribution of health, which might be leveraged to think about why some populations or subpopulations are at a greater risk of feeling pain. Third, what are some social factors shaping whether or not anti-social responses will become the patterned outcome of social pain?

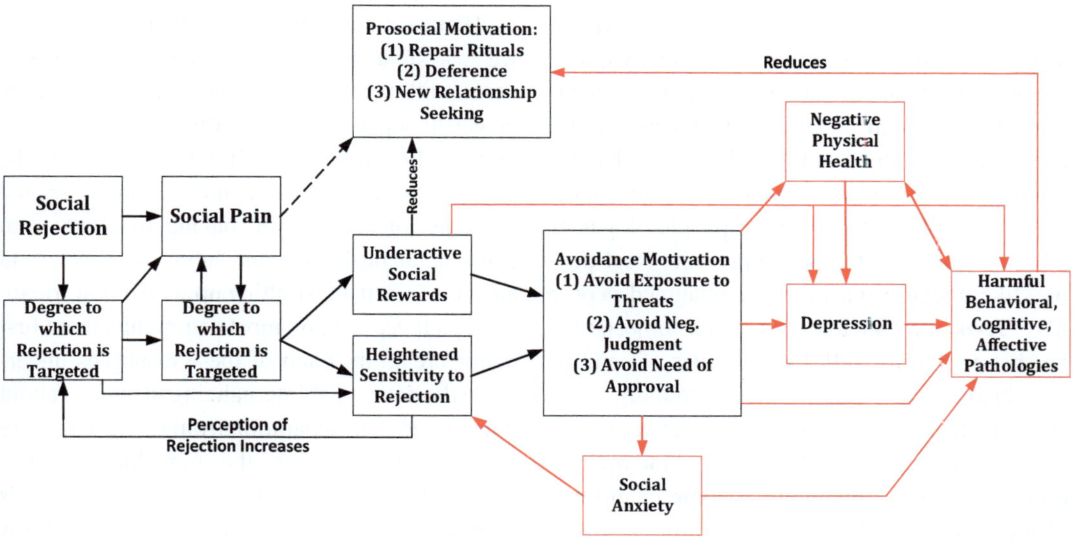

Fig. 20.2 Social Rejection and its Negative Health Outcomes

Bridging Neuroscience and Sociology

The central question sociology poses and contributes to answering is what explains variation in the social pain process. It is a ubiquitous biopsycho force, as most, if not all, humans are equipped to experience it when isolated, rejected, excluded, or separated. But, given the two pathways found in Figs. 20.1 and 20.2, we would expect social pain to vary in terms of who is more at risk of experiencing it, how their experience is framed by local and more global culture, and what sorts of expressions of pain are normative and for whom. In beginning to tackle these questions we start at the most obvious place, the social self. Arguably, the social self is the hinge between the internal and external forces that encompass social life.

Priming the Social Self for Pain

Sociologists, since Cooley and Mead, have argued that the self is a social construction, built up from our unique capacity toward self-reflexivity. We learn how to evaluate our own behaviors and attitudes—that is, we can treat ourselves as an object because others treat us as an object—through conversations of symbolic gestures. Consequently, we are able to *take the role of the other* in so far as we can anticipate other's behavior and attitudes, reflect on past behavior and attitudes, and align our own behavior and attitudes in the present for the sake of constituting a situation and its ensuing interaction because we can imagine how others would respond to us if the proverbial shoe was on the other foot. Cooley (1902 [1964]) added to this the centrality of emotional responses, particularly shame or pride. Several decades later, by importing Durkheim, one significant strand of symbolic interactionism embraces the fundamental premises that (a) conversations arouse in us affective responses that require labeling; (b) we can attribute the source of these responses to either ourselves or some social object outside of us; and, (c) affect goes a long way toward meaning-making and meaning-taking (Collins, 2004; Lawler, 2001).

Put differently, the meaning objects hold are not intrinsic to the object, but rather a product of the meaning actors attribute to them in past, present, or anticipated interactions. These meanings are both a collective act, created through interaction, made significant by mutual arousal of emotions and the "contagion" of that collective emotion to the object itself, and internalization such that meanings contribute to the motivation to maintain the relationship and, in many cases, the broader social unit like the group or organization. A key insight here is that objects are not confined to the physical props that are both environmental cues and instruments or means facilitating activities that bring people together socially. They can be the people, too. Real people, like a significant other, or imagined people like "Jews" or "Sociologists." And, objects can be abstractions as well, like the rules of conduct or institutions like "the economy" or "Catholicism." In the process of assigning meaning, attributing affectual loading, and internalization, objects can potentially become extensions of the social self. An obvious example is a toddler's blankie. The blankie is a safety net, a soother, and one of the first items in a child's "expressive kit." The object's fidelity and function are of no importance, as a tattered blankie or a small piece saved by a parent is meaningful and capable of arousing affectual responses. Herein lies the first bridge between neuroscience and sociology: our capacity to extend self beyond the physical boundaries of body by projecting mind outward onto things.

Cognitive scientists have long recognized our capacity to blur our cognitive and emotional experience with others, even groups (Aron & McLaughlin-Volpe, 2001). Under "normal" conditions, this absorption or merger with an object shapes our predictive processing schema about the sorts of rewards we will expect to receive when interacting with the object (Miller Tate, 2019), as well as the joys we feel as we seek out the object (Kringelbach & Berridge, 2016). (Think about the anticipatory "butterflies" you feel as you prepare to call or text a new love interest). Under "normal" circumstances, extending self into the space around us also expands the possibility of threats to those objects and, of course, the separation, isolation, rejection, or exclusion from them. The working principle here, then, is simple: *the more meaningful and significant an object, the more motivated we are to seek and use it; and, the more painful the loss of the object.*

That we all can vividly remember a time when we lost sight of our caretaker in a busy place, despite all of the years between, tell us just how powerful social pain as an affectual signature or marker of significance is. But, it is very likely the reader can recall several non-caretaker related moments of social pain: having to move from one's natal birthplace and house to a new and alien environment; splitting up with a best friend; losing a loved one; or, most commonly, there first major break up. Each of these examples underscore the dangers of internalizing meanings about objects and then separation from them. The more targeted the separation, the more likely it hurt. You might ask how can moving away from one's home as a child be intentional? Very likely, it was not intentional in the sense that a partner splits up with you while blaming you for the breakup. But, the pain can stem from aspects related to objectification and affectual attachment.

The first aspect stems from the way our brain's reward system motivates us to intentionally—even when presumed habitual—want and then like an object. In essence, we receive dopamine when we anticipate and seek an object that we have learned can be rewarding, opioids when we use the object, and dopamine once satiated and reflect on the use of the object (Kringelbach & Berridge, 2016). There is no need to be reductionist and argue that meaning is simply a chemical reaction, but this chemical reaction matters. Rejection or exclusion has a similar consequence for the person that motivates addictions (Fisher et al., 2010). Addiction, in neural terms, occurs when anticipation delivers a disproportionate reward that leads to chronic seeking behavior while the consumption of the object becomes increasingly unrewarding (Berridge & Kringelbach, 2008). Losing a loved one, for instance, means losing the anticipation of interacting. This is by no means a suggestion that love is simply an addiction to an object, but rather a reflection on the persistence of grief when a parent passes: while we remember the "good times," the feelings of grief often bubble up at times when we would have been anticipating our next interaction. Birthdays, holidays, and other events designed to coordinate interaction instead signal loss of that interaction. Like a heroin addict driven to find more heroin, the things that are synonymous with our sense of self become the things we crave the most.

Thus far, this craving has been rooted in either neurophysiology or psychology, with some hints of sociology. The sociological reasons for absorbing things into our self, though, map neatly on to these bio-psycho foundations. For Durkheim, social relationships protected humans against the suffering common to a sentient species.[1] Amidst massive social, economic, and political change, he cautioned that this protection was eroding and people becoming more vulnerable to negative pathological outcomes, like suicide. People, he suggested, "cling to life more resolutely when *belonging* to a group they love" (Durkheim, 1951, 209), providing the source of "purpose [that justifies] life's trials" (ibid. 210). In passages like these, and many others, connectivity constituted meaning and purpose while isolation and exclusion destroyed the ontological and moral anchorage necessary to weather the storm of everyday life. In the positive, Durkheim was clear that meaning was intimately tied to, perhaps caused by, affect. In *The Elementary Forms*, for instance, Durkheim argued that collective conscience (and consciousness) emerged out of assembly in which the shared arousal and mutual attention became the source of external representations (beliefs, attitudes, values) that were subsequently internalized such that those partaking in these assemblages became a (moral) community. The past became crystallized in a collective memory, the present in patterned practices that reinforced and represented common beliefs, and the future in anticipatory expectations about the community and its members.

Taken together, it becomes clear that as we expand beyond the walls of our physical body and either take in or connect with other objects as extensions of social self, we become dependent upon a range of objects for survival and sustenance. It is true, then, that there are practical consequences for being expelled from a group, like being exposed to existential threats. The sorts of small-scale societies we evolved into were notable for the severe danger expulsion introduced: being cast out meant certain death. However, the socioemotional and neurophysiological consequences are equally problematic; or so evolution made them to be. Despite the growth in size and scale of societies, the dangers of exclusion have not dissipated. In a wide range of small-scale collectives ranging from intimate informal peer groups to small towns or villages, threats of ostracism or exclusion remain existentially dangerous. It is just not the kind our ancestors faced. Instead of lions or the vagaries of natural climes, we face the potential inability to be social, to find affection or gratitude, or participate in the waters of group life that transcends our everyday mundane existence. That is, the sociocultural environment has become as much a source of selection for survival as the biotic environment. So much so, that humans appear to have evolved self-conscious emotions like shame and guilt to navigate the pressures of a hierarchical society in which tracking reputation and regulating one's own is paramount to social survival (Boehm, 2012). To not conform meant potential ostracization, at least temporarily as punishment. When we are "knocked down a peg," we experience acute shame that reflects the threat to the self's ability to maintain itself.

All of which is to say the self is, at least partially, built up from both the experience of social pain and the anticipation and experience of social pain in daily life. Belongingness produces meaning while separation and isolation threaten it (Baumeister & Leary, 1995). It makes sense that evolutionary pressures would work on our affectual equipment, particularly the size of subcortical affectual regions as well as the anterior cingulate cortex, and elevate the response to separation and isolation. All mammals are programmed to experience panic when separated from their primary caretaker, and to frantically seek and find that cherished social object (Panksepp, 1998). There isn't a child or parent that needs to be taught to feel intense terror when their counterpart is lost in

[1] The parallels between Durkheim's theory and the modern neuroscience of relationships is uncanny. See Cozoline, L. (2014). *The Neuroscience of Human Relationships*. Norton.

a busy market. Apes, being one of the least fit family of animals on Earth (there are only four species remaining), would have been under intense selection pressures throughout their history, which may explain why community became so central to their cognitive and affectual capacities. Humans were perhaps the least fit of all in the sense that they are not exceedingly strong or fast; their olfactory and auditory abilities are limited; in a word, they do not have the natural defensive and offensive capabilities of other animals, including gorillas or chimps. The trade-off, however, is an amplified ability to take the role of the other, coordinate action in complex ways (e.g., big game hunting), keep track of reputations over lifetimes, and communicate in rich and diverse ways. To do so, evolution had to improve cortical control of our emotion centers such that we were less noisy and able to enact collective planning (Turner, 2007). Concomitantly, evolution had to also find ways to enhance solidarity and community, which appears to be related to the enlargement of our emotion centers. Humans, like all mammals, bond through nurturance, lust, and play (Davis & Montag, 2019), but it is our unique capacity to generate affectual attachments to specific others, groups, and abstract collectives that is the hallmark of homo sapiens (Abrutyn & Lizardo, 2020). Together, then, the innate response of panic is married to the panoply of emotional forms of commitment in the constitution of meaning.

The Inequitable Distribution of (Social) Pain

Sociology offers a second intervention for the neuroscience of social pain: some explanation for why rejection and social pain would be inequitably distributed within and across populations. Population health research has long found that physical health is not randomly distributed, but rather caused by social conditions that track along axes of social stratification (Link & Phelan, 1996). For instance, rural areas are often poorer and have less access to hospitals and other health care resources, hence the tendency for various co-morbidities to cluster in some places, but not others. Likewise, negative health outcomes—and, therefore, higher rates of mortality—tend to cluster in communities that have larger numbers of racialized populations. Thus, social science provides clear evidence for the distribution of health. Though we know very little about how and why rejection may vary, we can draw from population health and general sociology to posit at least three plausible reasons for the social distribution of social pain.

First, we propose that social pain is a positive function of the distribution of material resources. Material resources are understood as a diverse array of things from income and wealth to hospital beds or mental health care facilities and professionals to easy access to recreational activities. Communities that lack these things or have barriers to these resources for some groups and not others will also find parallel patterns in social pain. The direct pathway can be illustrated by a simple example. In a community where some youth go to schools that are well maintained and have many brand-new facilities and things for kids and other youth go to schools that present various hazards to their health send clear, divergent messages about value. The former are included in the community, while the latter are excluded, ostracized. Indirectly, where populations or subpopulations do experience social pain, access to resources that may mitigate the negative physical and mental health consequences matter. It is one thing to feel rejection and have parents or other adults capable of providing support and another to feel alone in dealing with the pain of being rejected.

Second, resource inequities are exacerbated when they are reinforced by clear-cut status distinctions (Ridgeway, 2006) or, worse, stigmas (Link & Phelan, 2013). Social psychology teaches us that expectations around status characteristics like gender or race filter into everyday interactions. Those belonging to superordinate positions are expected to be rewarded, are given more deference in simple things like turn-taking or influence, and are believed to be more competent—even if the specific person is deserving of neither of these things. Conversely, the subordinate position is treated with far less esteem and dignity. Importantly, these sorts of status expectations are only real when members of both super- and subordinate classes *believe* them to be true (Ridgeway, 2001). Though often taken for granted, it is likely that those in the weaker positions are subjected to far more exclusion, rejection, and isolation. When these beliefs about status distinctions are transformed into stigmas, all of these processes are intensified and the likelihood of pain increased dramatically (Lucas & Phelan, 2012). Stigmatized individuals are dehumanized. For instance, many communities treat homeless people as non-human threats to the social order (Duneier & Carter, 1999). The level of ostracization can be quite extreme, potentially being institutionalized in the legal system. The same, of course, can be said for those who have visible mental health issues, which of course can trigger feelings of self-rejection for people struggling with mental health but who can hide it (Scheff, 1966).

The final dimension shaping the social distribution of social pain can be called diversity. By this we mean that all groups have varying levels of homo- or heterogeneity. At both ends of the extreme, social rejection is problematic and likely to generate social pain. In groups that are exceedingly similar, the threat rejection poses to the self is severe. The group may be objectively or subjectively the only connection for any given member. Being expelled is akin to social death, and can be extraordinarily painful. Groups that are very diverse may also pose threats in so far as status hierarchies or stigma beliefs may likely institutionalize exclusion and rejection. To be sure, there is not a happy medium, and both homo- and heterogeneous offer potential positive benefits to members. When resources are scarce or inequitably distributed and clear status hierarchies have crystallized, extremes on both ends of the diversity spectrum can lead to greater predictions of patterned social rejection and social pain.

"Choosing" Pathology

The final question we ask is why and how do some negative responses to social pain become collectivized? That is to say, some communities have disproportionate rates of opioid addiction and "deaths of despair" (Case & Deaton, 2020) while others have exceedingly high suicide rates (Mueller & Abrutyn, 2016; Stevenson, 2014). Because these are place-based pathologies, sociology suggests they are not random outcomes, but rather shaped by something social. Drawing on recent advances in cultural sociology, we posit that the biggest factor in patterning how people feel, think, and act on social pain comes from the emergence and crystallization of local culture.

Cultural sociologists distinguish between public culture, or the sort of external cultural systems that are often distant and abstract (like "American" or "Jewish" culture) and the refractive process by which we internalize this culture in schema, scripts, or other concepts denoting mechanisms (Lizardo & Strand, 2010). The distinction drawn between these two tries to explain the difference between knowing about a culture or some sort of element and culture actually causing people to act one way or the other (Lizardo, 2017). This disjunction between culture out there and culture inside of us matters for a few reasons. For one, it allows us to study culture as a causal force without making the mistake Parsons (1951) made in asserting a sort of monolithic cultural system. It also helps explain the idiosyncratic and incomplete way that culture out there is expressed, understood, and even contradicted by members exposed to it. However, one strategy for better explaining how and why culture causes us to feel, think, and act certain ways and not others is by shifting to a place-based, local culture model.

Local culture refers to the bundling of a group identity, a sense of past and future, and a collective notion of social relations all anchored to a place (Fine, 1979). The place may be the dugout and playing field a team of little leaguers share, a group of Teaching Assistant's office, or town or some part of the town. The place acts as a gravitational force drawing members into its orbit as they do whatever it is the group does instrumentally and normatively. Though there is certainly variation in how deeply socialized any given member is in a local culture, local cultures offer certain advantages amorphous public cultures lack. Smaller amalgams of people are more easily monitored and sanctioned. Even when people disagree with the rules of conduct, the near-constant presence of agents authorized to punish transgressions encourages conformity in emotion, attitude, and action. In turn, this sort of monitoring also facilitates a higher likelihood of self-regulation, which generally implies some degree of acceptance of the local culture. Additionally, local cultures tend to have fewer elements for members to have to learn and enact. Durkheim observed that smaller, more homogeneous communities had less relationships or objects to label and create meanings. The simpler nature of these cultures allows for greater fidelity for a larger proportion of accessible and applicable cultural elements for most of the members of the group (Patterson, 2014). The consequence: social pain, as an experience, may become a cultural or social fact for one community because its members disproportionately experience social pain (Abrutyn, 2024) *and* because of their place-based identity exposes them to recurring interactions in which social pain becomes an "expressive idiom" (Goffman, 1963:33–49) common to most if not all members.

Take, for instance, a community that has had a near two decade youth suicide problem, Poplar Grove (Mueller & Abrutyn, 2024). In this community, suicide became a normative option for too many youth, finding that a script or narrative of kids under pressure fostered expression of emotional distress through suicidality (Abrutyn et al., 2020). More specifically, youth were under extreme pressure to constantly achieve academically, athletically, and socially. This cultural directive towards perfection also prohibited being mentally unhealthy, which led to a formidable mental health stigma and a lack of help-seeking, greater distrust in adults, and a youth culture primed for an idioculture that provided the means for feeling, thinking, and acting when they experienced social pain. Their social pain was often chronic, especially for the highest status kids. On the one hand, they were achieving at a significant cost to their mental and emotional health, while on the other hand, many felt they were always failing or on the verge of failing if their veneer of perfectionism was eroded. Because they often experienced the loss of a high status youth to suicide, the narrative became centered not on the mental health struggles or other dilemmas youth faced, and more about "kids under pressure die by suicide." Suicide, then, became the most likely outcome, even though alcohol and drug misuse were also common methods of numbing the pain. A pattern that we find more regularly in certain Rust Belt communities that have suffered the hollowing out of middle class manufacturing jobs, the decline in economic privileges White folks have long objectively or subjectively had, and the rise in pharmaceutical companies and doctors pushing opioids for real and imagined pain (Case & Deaton, 2020).

In short, we argue that sociology plays a key role in understanding and explaining the social distribution of emotional, cognitive, and behavioral responses to social pain. Pain can become collectivized into a form of social trauma (Abrutyn, 2024), and this social trauma can bring with it new cultural forms of what a person experiencing pain should or, even, ought to do to mitigate that pain. The more it becomes believed that suicide, crime, drugs, or something else is the "correct" solution, the more people likely observe neighbors, co-workers, friends, and possibly family employing these methods. This direct experience, coupled with gossip, local media, and other stories of community member's using these methods reinforces the local culture and hardens it.

Having specified the role sociological theory plays in modifying and extending the model presented in Fig. 20.2, we turn in the last section of this chapter, to the implications incorporating social pain holds for sociology. This begins by revisiting the theorist and phenomena most directly tied to social pain: Durkheim and his model of social action. This discussion leads neatly into one that telescopes out a bit and examines how social pain and affective neuroscience supplement and expand on the insights cultural-cognitive sociology has introduced in recent years.

Implications

Durkheim and Integration

In recent years, Durkheim's influence in sociology has declined, partially because of sociology's treatment of theory through the lens of the Humanities and not a science and its lack of embrace to the fundamentals of scientific theory. Critical theories have stood in for scientific theories (Hammersley, 2002). However, a sociology of social pain renews the calls for extracting the general principles of Durkheimian sociology and forging a robust theory of social organization and action.

At the core of Durkheim's work, whether the "early" Durkheim in the *Division of Labor* or the "mature" Durkheim of *The Elementary Forms*, is the idea that community and society impose moral order on individuals. To be sure, it is emergent moral order, built up through repeated interactions, the attribution of internal feelings to external sources, and the projection of a collective sense of what these feelings mean, who feels them, how and when to reinvigorate those feelings, and so on. In essence, local culture is constructed from recurring, affectually intense encounters. It is in these affectually enveloping relationships that we calibrate meaning and purpose, duty and obligation, and altruism. Threats to and the disintegration of these relationships was no doubt his greatest fear (Lukes, 1977). Durkheim's great conceptual contribution, anomie, is a structural condition underscoring widespread disintegration and the consequent inability to orient oneself morally to others. In its acute (e.g., divorce, economic collapse) or chronic (e.g., Capitalist industry) forms, it marks the separation, exclusion, or rejection from collective rituals and, thus, is very much about social pain. For instance, the increase of suicide rates during the Great Depression, as individuals lost jobs and essential ties, along with fraying the threads of familial bonds (Luo et al., 2011) underscores how anomie contributes to outcomes of social pain. And yet, much of the literature on the breakdown of belongingness is either found in over-psychologized literatures focused on cognitive measures like "loneliness" or distributed across many subfields using different terminologies (Wang et al., 2017). We would suggest that now is the time to consolidate and expand empirical work related to pain, which might begin to realize the goal of a more coherent, cumulative discipline.

A focus on social pain centers the sociological lens upon objective, ubiquitous qualities of isolation, such as the experience of having few ties related to one's identities that proffer support when needed and which have real consequences—e.g., people experiencing homelessness (Duneier & Carter, 1999) or aging populations (Klinenberg, 2001). It also means examining the subjective emotions that come about when disconnected from personal, local, or national collectives, especially by natural disaster or social sources of dissolution, displacement, and disjunction (Abrutyn, 2019, 2024). Most importantly, it involves the shift from the highly individualized study of experiences of *loneliness* to the social roots of that loneliness, caused by the disconnection, both spatially and affectually, from rich and dense networks that facilitate ritualized interaction. In doing so, Durkheim's great insight is that individuals strive to belong, to circumvent moments of alienation, as a means to maintain cherished positive feelings lost when adrift in the ocean of social life. Thus, a key intervention a sociology of social pain offers is a move from the singular focus on negative outcomes to one that balances this with the positive, prosociality that social pain engenders.

To be sure, pathways to forming connection are differentially distributed across societies, and so too are the strategies, opportunities, and realities of seeking out prosocial connections. In fact, evidence highlights that America as a whole is seeing a trend of lost strong ties (McPherson et al., 2006), possibly contributing to pain across social groups. When considering risk factors for isolation, such as discrimination, bereavement, and family demands, the origins of these factors present themselves in experiences of racism, sexism, and income inequality (Umberson & Donnelly, 2023) that are distributed differently in diverse populations. Pain for some groups, such as gender diverse peoples, immigrant populations, LGBTQ may be at higher levels than those who hew closely to the normative order within societies. However, the deaths of despair literature indicates that White people are not immune to these forces (Case & Deaton, 2020); illustrating, instead, the universality of certain bio-psycho needs. Hence, consideration as to how connections across groups varies by number and the groups to which individuals find they belong may reveal how pain is differentially distributed, thus prompting intervention strategies aimed at forging bonds of belonging.

There is also substantial evidence that suggests isolation across the life-course is associated with experiences from childhood. To date, this avenue of research is novel, but investigating how belongingness or isolation translates to persistent experiences of social pain throughout one's life may shed light on specific ways of addressing experiences of social pain down the road of life. When considering how one comes to belong, by fitting in to some groups and avoiding others, peoples learn ways of establishing connection. In so doing, they learn ways of navigating social life, such as Summers-Effler's (2004) defensive strategies highlights, as a means to reduce experienced pain. How effective these strategies are for mitigating pain, however, is not entirely clear. Some of which, such as belonging to stigmatized groups, contributes to additional pain or increases risk of such hurt. Through considering pain, a whole body of research opens up, as it reveals how and why social actions unfold how they do.

Affect, Cognition, and the Self

As symbolic interactionism teaches us, meaning rests at the heart of the constitution and reconstitution of self. We act (or plan to act) based on the meanings we assign to objects, including our own self. These meanings reflect how we view our self, provide a sense of continuity, distinction, and are the source of belongingness. For pragmatists, the mobilization of meanings and, thus, the self is conceptualized as a cognitive (see, for instance, Mead, 1934:173; Burke, 1991; Gross, 2009), and reactive process. We are habitualists, triggered to creativity or deliberate action in the face of incongruity. Though social pain is not just something that motivates us to run or get away from the noxious stimulus. For instance, that sense of panic that we experience when we lose sight of our caretaker can be activated with just about any object to which we become attached. Losing a loved one of any sort is painful. Consequently, we not only learn to avoid noxious things but also to proactively prevent the loss of things we truly cherish. A social pain theory of self is both reactive and proactive.

We are motivated, for instance, to take care of our relationships. As Goffman (1967) noted, repairative rituals are common games we engage in over minor interpersonal transgressions. The transgressor actively tries to atone for their mistake and, over time, work towards maintaining the health of the relationship. Social movement members are also motivated to sustain their membership, engaging willingly in altruistic behaviors for the sake of the group (Summers-Effler, 2009). Are these habitual? Goffman (1959) would caution against this position, as much of the rehearsal of roles or the careful construction of the front stage indicates an active self that may be able to offline the deliberate cognition that could make a performance feel inauthentic or stilted, but which belies the amount of deliberate, creative effort frontloaded into the performance. Other times, though, we believe the constitution and reconstitution of self is a proactive engagement predicated on previous lessons in social pain, how to avoid social pain, and the converse of social pain: social gain (Dalgleish et al., 2017). Goffman would also warn us against overly cynical takes on this motivation. In some circumstances, ulterior motives like status enhancement may be at the root of these proactive attitudes and behaviors, but there is a moral imperative—especially among equals—to *maintain* status and reputational claims made by the self and others. In actively being consistent, we experience pride; in representing the corporate or categoric group to which we belong, there is honor; and, in ensuring our expressive displays are appropriate, we feel dignity. Concomitantly, we actively avoid these positive emotions' counterparts—shame, stigma, and embarrassment. Not only do we rehearse our parts, as noted previously, but within the moment, situation, occasion, gathering, we monitor our behavior and are proactive to avoid these pitfalls.

In short, then, the incorporation of social pain into the sociological understanding of self pushes us further away from the cognitive turn in sociology that began with Mead's pragmatic take overshadowing Cooley's better balanced view of affect and cognition. To deny the affectual nature of the human, even after accounting for the massive grey matter between our ears and the layers of cultural inculcation from years of socialization, is to deny something fundamental. Affect is preconscious; inextricably tied to coordinating cognition and behavior (Damasio, 1994) and, in some situations, controlling and commanding it (Panskepp, 1998). We are not slaves to affect, but it is key to how we embody identity, relationship, group, and self. It propels action or causes us to shrink in anticipating of punishment. We suggest doubling down our efforts to study emotions in nearly every subfield in the discipline. To understand paradoxes like the fact that people can live in close proximity to others, yet be isolated cognitively and affectually (Umberson & Donnelly, 2023). The willingness to embrace stigmatized roles as a means of buffering one against social pain highlights the importance of returning to insights long since forgotten or ignored by sociologists, particularly regarding the dynamics of repairative and reintegrative rituals (Braithwaite, 1989). These insights may shed light on how we can better structure healthier, less painful communities.

Final Thoughts

This essay has contributed to neurosociology in multiple ways: outlining bridges between neuroscience and sociology, expanding the field by incorporating avoidance behavior and negative mental health outcomes, detailing how social pain orbits the self across contexts, and supplementing cultural sociology with these considerations. While our focus remains largely pinned on isolation, rejection, and exclusion, such as those experiences that contribute to shame and guilt along the pathway to states of depression, social anxieties, and pathologies, there are additional considerations worthy of investigation, such as highly regulated spaces, like Poplar Grove, that constrain individuals to their breaking points. In such cases, to integrate into the community demands living with the social pain of both success and failure, as when youth do not achieve the high standards expected of them yet continue to attempt despite painful emotions. An additional example is that of traumatic events which immediately and forcefully alter one's world from the magnitude of the social pain felt. These experiences and their lasting effects, of course, depend upon many factors, such as how one and others view oneself, along with their access to support networks and various other resources necessary for ameliorating social pain. Thus, consideration of how pain influences action and for who and where positions sociology to expand our lens into the mind without losing sight of our social footings. Through considering social pain as something we all live with, understanding how people both leverage and cope with it provides for a rich area for future empirical investigations.

References

Abrutyn, S. (2019). Toward a general theory of anomie: The social psychology of disintegration. *European Journal of Sociology, 60*(1), 109–136.

Abrutyn, S. (2023). Toward a sociological theory of social pain. *Journal for the Theory of Social Behaviour, 53*(3), 351–371.

Abrutyn, S. (2024). The roots of social trauma: Collective, cultural pain and its consequences. *Society and Mental Health, 14*(3), 240-53. https://doi.org/10.1177/21568693231213

Abrutyn, S., & Lizardo, O. (2020). Grief, care, and play: Theorizing the affective roots of the social self. *Advances in Group Processes, 37*, 79–108.

Abrutyn, S., Mueller, A. S., & Osborne, M. (2020). Rekeying cultural scripts for youth suicide: How social networks facilitate suicide diffusion and suicide clusters following exposure to suicide. *Society and Mental Health, 10*(2), 112–135. https://doi.org/10.1177/2156869319834063

Aron, A., & McLaughlin-Volpe, T. (2001). Including others in the self. In C. Sedikides & M. B. Brewer (Eds.), *Individual self, relational self, collective self* (pp. 89–108). Psychology Press.

Baumeister, R. F., & Leary, M. R. (1995). The need to belong; Desire for interpersonal attachments as a fundamental human motivation. *Psychological Bulletin, 117*(3), 497–529.

Bendelow, G. A., & Williams, S. J. (1995). Transcending the dualisms: Towards a sociology of pain. *Sociology of Health and Illness, 17*(2), 139–165.

Berridge, K. C., & Kringelbach, M. L. (2008). Affective neuroscience of pleasure: Reward in humans and animals. *Psychopharmacology, 199*, 457–480.

Blay, Y., Keshet, H., Friedman, L., & Gilboa-Schechtman, E. (2021). Interpersonal motivations in social anxiety: Weakened approach and intensified avoidance motivations for affiliation and social rank. *Personality and Individual Differences, 170*, 110449.

Boehm, C. (2012). *Moral origins: The evolution of virtue, altruism, and shame*. Basic Books.

Bradley, M. M., Silakowski, T., & Lang, P. J. (2008). Fear of pain and defensive activation. *Pain, 137*, 156–163.

Braithwaite, J. (1989). *Crime, shame, and reintegration*. Cambridge University Press.

Burke, P. J. (1991). Identity processes and social stress. *American Sociological Review, 56*(6), 836–849.

Case, A., & Deaton, A. (2020). *Deaths of despair and the future of capitalism*. Princeton University Press.

Chester, D. S., DeWall, C. N., & Pond, R. S. (2016). The push of social pain: Does rejection's sting motivate subsequent social reconnection? *Cognitive, Affective, & Behavioral Neuroscience, 16*(3), 541–550.

Collins, R. (2004). *Interaction ritual chains*. Princeton University Press.

Cooley, C. H. (1902 [1964]). *Human nature and the social order*. Schocken Books.

Cozoline, L. (2014). *The neuroscience of human relationships*. Norton.

Dalgleish, T., Walsh, N. D., Mobbs, D., Schweizer, S., Harmelen, A.-L., Dunn, B., Dunn, V., Goodyer, I., & Stretton, J. (2017). Social pain and social gain in the adolescent brain: A common neural circuitry underlying both positive and negative social evaluation. *Scientific Reports, 7*(1), 42010.

Damasio, A. (1994). *Descartes' error: Emotion, reason, and the human brain*. Avon Books.

Davis, K. L., & Montag, C. (2019). Selected principles of panskeppian affective neuroscience. *Frontiers in Neuroscience, 12*. https://doi.org/10.3389/fnins.2018.01025

DeAngelis, R. T. (2020). Striving while Black: Race and the psychophysiology of goal pursuit. *Journal of Health and Social Behavior, 61*(1), 24–42.

DeAngelis, R. T. (2021). Moving on up? Neighborhood status and racism-related distress among Black Americans. *Social Forces*. https://doi.org/10.1093/sf/soab075.

DeWall, C. N., MacDonald, G., Webster, G. D., Masten, C. L., Baumeister, R. F., Powell, C., Combs, D., Schurtz, D. R., Stillman, T. F., Tice, D. M., & Eisenberger, N. I. (2010). Acetaminophen reduces social pain: Behavioral and neural evidence. *Psychological Science, 21*(7), 931–937.

Duneier, M., & Carter, O. (1999). *Sidewalk* (1st ed.). Farrar, Straus and Giroux.

Durkheim, E. (1897 [1951]). *Suicide: A study in sociology* (J. A. Spaulding & G. Simpson, Trans.). Free Press.

Eisenberger, N. I. (2012). The neural bases of social pain: Evidence for shared representations with physical pain. *Psychosomatic Medicine, 74*(2), 126–135.

Eisenberger, N. I. (2015). Social pain and the brain: Controversies, questions, and where to go from here. *Annual Review of Psychology, 66*, 601–629.

Fine, G. A. (1979). Small groups and culture creation: The idioculture of little league baseball teams. *American Sociological Review, 44*(5), 733–745.

Fisher, H. E., Brown, L. L., Aron, A., Strong, G., & Mashek, D. (2010). Reward, addiction, and emotion regulation systems associated with rejection in love. *Journal of Neurophysiology, 104*(1), 51–60.

Gilligan, J. (2003). Shame, guilt, and violence. *Social Research, 70*(4), 1149–1180.

Goffman, E. (1959). *The presentation of self in everyday life*. Anchor Books.

Goffman, E. (1963). *Behavior in public places: Notes on the social organization of gatherings*. Free Press.

Goffman, E. (1967). *Interaction ritual: Essays on face-to-face behavior*. Pantheon Books.

Gogolla, N. (2017, June 19). The insular cortex. *Current Biology, 27*(12), R580–R586. https://doi.org/10.1016/j.cub.2017.05.010

Gross, N. (2009). A pragmatist theory of social mechanisms. *American Sociological Review, 74*, 358–379.

Hammersley, M. (2002). Should social science be critical? *Philosophy of the Social Sciences, 35*(2), 175–195.

Hudd, T., & Moscovitch, D. A. (2021). Reconnecting in the face of exclusion: Individuals with high social anxiety may feel the push of social pain, but not the pull of social reward. *Cognitive Therapy and Research, 46*, 420–435.

Inagaki, T. K. (2018). Opioids and social connection. *Current Directions in Psychological Science, 27*(2), 85–90.

Inagaki, T. K., Jennings, J. R., Eisenberger, N. I., & Gianaros, P. J. (2018). Taking rejection to heart: Associations between blood pressure and sensitivity to social pain. *Biological Psychology, 139*, 87–95.

Klinenberg, E. (2001). Dying alone: The social production of urban isolation. *Ethnography, 2*(4), 501–531.

Kringelbach, M. L., & Berridge, K. C. (2016). Neuroscience of reward, motivation, and drive. In *Recent developments in neuroscience research on human motivation* (Vol. 19, pp. 23–35). Advances in motivation and achievement. Emerald Group.

Landa, A., Bossis, A. P., Boylan, L. S., & Wong, P. S. (2012). Beyond the unexplainable pain: Relational world of patients with somatization syndromes. *The Journal of Nervous and Mental Disease, 200*(5), 413–422.

Lawler, E. J. (2001). An Affect Theory of Social Exchange. *American Journal of Sociology, 107*(2), 321–352.
LeDoux, J. (2000). Cognitive-emotional interactions: Listening to the brain. In R. D. Lane & L. Nadel (Eds.), *Cognitive neuroscience of emotion* (pp. 129–155). Oxford University Press.
LeDoux, J. E., & Brown, R. (2017). A higher-order theory of emotional consciousness. *PNAS, 114*(10), E2016–E2025.
Lewis, H.-J., Barrett, L. F., & Craig, A. D. (2007). Interoception and emotion: A neuroanatomical perspective.
Lewis, H. (1971). *Shame and guilt in neurosis*. International Universities Press.
Link, B. G., & Phelan, J. (1996). Social conditions as fundamental causes of disease. *Journal of Health and Social Behavior, (Extra Issue)*, 80–94.
Link, B. G., & Phelan, J. C. (2013). Labeling and stigma. In C. S. Aneshenel, J. C. Phelan, & A. Bierman (Eds.), *The handbook of the sociology of mental health* (pp. 525–541). Springer.
Livingston, G., Huntley, J., Sommerlad, A., Ames, D., Ballard, C., Banerjee, S., Brayne, C., Burns, A., Cohen-Mansfield, J., Cooper, C., Costafreda, S. G., Dias, A., Fox, N., Gitlin, L. N., Howard, R., Kales, H. C., Kivimäki, M., Larson, E. B., Ogunniyi, A., Orgeta, V., Ritchie, K., Rockwood, K., Sampson, E. L., Samus, Q., Schneider, L. S., Selbæk, G., Teri, L., & Mukadam, N. (2020, August 8). Dementia prevention, intervention, and care: 2020 report of the Lancet Commission. *Lancet, 396*(10248), 413–446. https://doi.org/10.1016/s0140-6736(20)30367-6
Lizardo, O. (2017). Improving cultural analysis: Considering personal culture in its declarative and nondeclarative modes. *American Sociological Review, 82*(1), 88–116.
Lizardo, O., & Strand, M. (2010). Skills, toolkits, contexts and institutions: Clarifying the relationship between different approaches to cognition in cultural sociology. *Poetics, 38*(2), 205–228.
Lucas, J. W., & Phelan, J. C. (2012). Stigma and status: The interrelation of two theoretical perspectives. *Social Psychological Quarterly, 75*(4), 310–333.
Lukes, S. (1977). Alienation and anomie. In *Essays in social theory* (pp. 74–95). Macmillan.
Luo, F., Florence, C. S., Quispe-Agnoli, M., Ouyang, L., & Crosby, A. E. (2011, June). Impact of business cycles on US suicide rates, 1928-2007. *American Journal of Public Health, 101*(6), 1139–1146. https://doi.org/10.2105/ajph.2010.300010
MacDonald, G., & Leary, M. R. (2005). Why does social exclusion hurt? The relationship between social and physical pain. *Psychological Bulletin, 131*(2), 202–223.
Maryanski, A. (2018). *Émile Durkheim and the Birth of the Gods*. Routledge.
Mead, G. H. (1934). *Mind, self and society: From the standpoint of a social behaviorist*. The University of Chicago Press.
McPherson, J. M., Smith-Lovin, L., & Brashears, M. (2006). Social Isolation in America: Changes in Core Discussion Networks over Two Decades. *American Sociological Review, 71*(3), 353–375.
Miller Tate, A. J. (2019). A predictive processing theory of motivation. *Synthese*.
Mueller, A. S., & Abrutyn, S. (2016). Adolescents under pressure: A new Durkheimian framework for understanding adolescent suicide in a cohesive community. *American Sociological Review, 81*(5), 877–899.
Mueller, A. S., & Abrutyn, S. (2024). *Life under pressure: The social roots of youth suicide and what to do about them*. Oxford University Press.
O'Neill, M., & Spybey, T. (2003). Global refugees, exile, displacement and belonging. *Sociology, 37*(1), 7–12.
Panksepp, J. (2003). Feeling the pain of social loss. *Science, 302*(5643), 237–239.
Panksepp, J. (1998). *Affective neuroscience: The foundations of human and animal emotions*. Oxford University Press.
Panksepp, J., & Watt, D. (2011). Why does depression hurt? Ancestral primary-process separation-distress (PANIC/GRIEF) and diminished brain reward (SEEKING) processes in the genesis of depressive affect. *Psychiatry, 74*(1), 5–13.
Parsons, T. (1951). *The social system*. Free Press.
Patterson, O. (2014). Making sense of culture. *Annual Review of Sociology, 40*, 1–30.
Paulus, M. P., & Stein, M. B. (2006, August 15). An insular view of anxiety. *Biological Psychiatry, 60*(4), 383–387. https://doi.org/10.1016/j.biopsych.2006.03.042
Pearlin, L. I. (1999). The stress process revisited: Reflections on concepts and their interrelationships. In *Handbook of the sociology of mental health* (pp. 395–415). Springer.
Rainville, P. (2002). Brain mechanisms of pain affect and pain modulation. *Current Opinion in Neurobiology, 12*, 195–204.
Rainville, P., Carrier, B., Hofbauer, R. K., Bushnell, M. C., & Duncan, G. H. (1999). Dissociation of sensory and affective dimensions of pain using hypnotic modulation. *Pain, 82*(2), 159–171.
Reeve, J., & Lee, W. (2019). Motivational neuroscience. In *The Oxford handbook of human motivation* (pp. 355–371).
Ridgeway, C. L. (2001). Inequality, status, and the construction of status beliefs. In J. H. Turner (Ed.), *The handbook of sociological theory*. Kluwer/Plenum.
Ridgeway, C. L. (2006). Expectation states theory and emotion. In J. E. Stets & J. H. Turner (Eds.), *Handbook of the sociology of emotions* (pp. 374–367). Springer.
Rotge, J.-Y., Lemogne, C., Hinfray, S., Huguet, P., Grynszpan, O., Tartour, E., George, N., & Fossati, P. (2015). A meta-analysis of the anterior cingulate contribution to social pain. *Social Cognitive and Affective Neuroscience, 10*(1), 19–27.
Rudolph, K. D., Miernicki, M. E., Troop-Gordon, W., Davis, M. M., & Telzer, E. H. (2016). Adding insult to injury: Neural sensitivity to social exclusion is associated with internalizing symptoms in chronically peer-victimized girls. *Social Cognitive and Affective Neuroscience, 11*(5), 829–842.

Scheff, T. (1966). *Being mentally ill: A sociological theory*. Aldine Publishing.

Scheff, T. (1988). Shame and conformity: The deference emotion system. *American Sociological Review, 53*(3), 395–406.

Slavich, G. M., Shields, G. S., Deal, B. D., Gregory, A., & Toussaint, L. L. (2016). Alleviating social pain: A double-blind, randomized, placebo-controlled trial of forgiveness and acetaminophen. *Annual Review of Behavioral Medicine, 53*, 1045–1054.

Sommer, K. L., & Bernieri, F. (2015). Minimizing the pain and probability of rejection: Evidence for relational distancing and proximity seeking within face-to-face interactions. *Social Psychological and Personality Science, 6*(2), 131–139.

Spunt, R. P., Lieberman, M. D., Cohen, J. R., & Eisenberger, N. I. (2012). The phenomenology of error processing: The dorsal ACC response to stop-signal errors tracks reports of negative affect. *Journal of Cognitive Neuroscience, 24*(8), 1753–1765.

Stevens, F. L., Hurley, R. A., Taber, K. H., Hurley, R. A., Hayman, L. A., Taber, K. H., & P. D. (2011). Anterior cingulate cortex: Unique role in cognition and emotion. *The Journal of Neuropsychiatry, 23*(2), 121–125.

Stevenson, L. (2014). *Life beside itself: Imagining care in the Canadian Arctic*. University of California Press.

Summers-Effler, E. (2004). Defensive strategies: The formation and social implications of patterned self-destructive behavior. *Advances in Group Processes, 21*, 309–325.

Summers-Effler, E. (2009). *Laughing saints and righteous heroes*. University of Chicago Press.

Tomasello, M. (2019). The role of roles in uniquely human cognitions and sociality. *Journal for the Theory of Social Behaviour*. https://doi.org/10.1111/jtsb.12223

Turner, J. H. (2007). *Human emotions: A sociological theory*. Routledge.

Umberson, D., & Donnelly, R. (2023). Social isolation: An unequally distributed health hazard. *Annual Review of Sociology, 49*(1), 379–399. https://doi.org/10.1146/annurev-soc-031021-012001

Uscola, C. J. (2023). Drinker identity development: Shame, pride, and a thirst to belong. *Society and Mental Health, 13*(1), 45–60.

Vaisey, S., & Valentino, L. (2018). Culture and choice: Toward integrating cultural sociology with the judgment and decision-making sciences. *Poetics, 68*, 131–143.

Wang, J., Lloyd-Evans, B., Giacco, D., Forsyth, R., Nebo, C., Mann, F., & Johnson, S. (2017). Social isolation in mental health: A conceptual and methodological review. *Social Psychiatry and Psychiatric Epidemiology, 52*(12), 1451–1461.

Zajacova, A., Grol-Prokopczyk, H., & Zimmer, Z. (2021). Sociology of chronic pain. *Journal of Health and Social Behavior*. https://doi.org/10.1177/00221465211025962.

21 Human-AI Interaction from an Evolutionary Neurosociological Perspective

Yulia S. Shkurko

Abstract

The process of replacing social relations between humans with interactions with AI has highlighted the problem of the correctness of our understanding of what social interaction is and what role the sociocultural and biological components of human nature in it. Evolutionary neurosociology —a field that combines neurobiological and social variables to examine sociological issues from an evolutionary perspective— is best suited to clarify this question. From this perspective, the human-AI interaction represents an evolutionarily formed relationship among social contexts (defined by AI's communicative role and the peculiarities of its social interpretation), mental states and processes (including social cognition skills and basic emotions), and the brain's neurobiological mechanisms and patterns. This study suggests the probability of an evolutionary mismatch in human-AI interactions, arising when humans use evolved adaptive traits for purposes other than those for which they were originally selected. The mechanism of evolutionary mismatch likely relates to the functioning of the social brain, which creates a kind of illusion of social interaction. Furthermore, the currently promoted human-centered AI approach encourages forming authority ranking relations with AI, associated with the basic emotion of anger and fear. These relationships, in turn, are linked to brain activity that potentially contribute to reduced prosocial attitudes beyond human-AI interactions.

Introduction

In recent years, there has been a rapidly growing integration of artificial intelligence (AI) systems into various spheres of human activity. As AI becomes increasingly integrated into the communicative ties of society, distinguishing between humans and AI as participants in social interaction may become challenging. The replacement of humans by AI as interlocutors updates the question of the correctness of our understanding of what social interaction is and what the mechanisms of social processes are.

Neurosociology, which combines neurobiological and sociological perspectives, reveals the implicit mechanisms underlying social interaction and potentially provides new insights into what is crucial for maintaining social relationships.

Y. S. Shkurko (✉)
Independent Researcher, Ulyanovsk, Russia

W. Kalkhoff et al. (eds.), *Handbook of Neurosociology*, Handbooks of Sociology and Social Research, https://doi.org/10.1007/978-3-031-95615-7_21

This chapter examines human-AI interaction (HAII) from the perspective of evolutionary neurosociology.[1] This perspective combines the approach of neurosociology with evolutionary analysis. Evolutionary neurosociology is a field of study that focuses on examining evolved neurobiological mechanisms and neurobiological patterns of social phenomena and processes at both the micro- and macro-levels. Additionally, it investigates the inverse influence of sociocultural factors on human adaptive nature.

As with other sociological problems, the neurosociological approach is not integrated into traditional sociological considerations of AI. Currently, AI is predominantly studied as a socially and culturally determined knowledge system, field of scientific research and engineering in the perspective of the sociology of science/knowledge and technology. Additionally, sociologists study social and other problems of development and applications of specific AI systems (including algorithms) in different social contexts and for different purposes. Finally, AI is investigated as a sociocultural phenomenon interacting with broader social, cultural, political, and economic contexts (for a review, see Liu, 2021).

The incorporation of an evolutionary neurosociological perspective into the sociological study of HAII contributes to a more comprehensive understanding of the phenomenon. This is made possible by exploring the neurobiological mechanisms underlying the human explicit versus implicit responses to AI systems and addressing the question of ultimate causes of individuals' abilities to interact with AI. Moreover, the perspective of evolutionary neurosociology provides insight into the specific features of human interaction with AI that may elicit unconscious responses rooted in our evolutionary past. This could facilitate the development of AI technologies that account for the potential negative consequences of activating evolutionary adaptive mechanisms of interpersonal interaction in a novel communicative environment (discussed below). Furthermore, it is both interesting and useful to examine how insights already uncovered by neurosociologists regarding social interaction — the central and sociologically specific object of study since its emergence in the 1990s.[2] (Franks & Eisenberg, 2008: 3)—apply to interactions between humans and non-human agents whose interactive capabilities are not rooted in biological substance (e.g. brain neurons and body cells).

I begin by defining artificial intelligence as a communicative interactant, viewing human-AI interaction as a product of Darwinian and sociocultural evolution. I then describe three key ideas of evolutionary neurosociology: (1) the social nature of the human brain, (2) evolutionarily shaped mechanisms of social cognition, percep-

[1]The development of "the new evolutionary sociology," which combines Darwinian and non-Darwinian (sociocultural) selection mechanisms in considering social phenomena and processes, is discussed as one of the perspectives for the renewal of sociology (Turner et al., 2020; Turner & Machalek, 2018). Unlike evolutionary neurosociology, it is not normative for evolutionary sociology to consider the brain activity of individuals as social agents. This depends on the subject and level of study, and the basic approach used within evolutionary sociological analysis, such as functionalism, stage modeling, among others (Turner et al., 2020).

[2]Neurosociology is a relatively recent field of study, emerging several decades ago. The term "neurosociology" was first used in 1972 in *The Other Side of the Brain*, a work coauthored by W. TenHouten with J. Bogen et al. (Bogen et al., 1972). This article presents the results of testing subjects to establish the relationship between the way they perform cognitive tasks and left or right hemisphere activity. These results are considered evidence of different thinking styles such as propositional and appositional, respectively. The propensity for these styles is formed during socialization and influenced by culture. This work positions neurosociology as a research field capable of establishing a link between human brain activity and the characteristics of the learning process.

David D. Franks, in addition to 1972 (Franks, 2013), also identifies 1997 as another significant year for the development of neurosociology. In 1997, during a spontaneous lunch after the American Sociological Association session on the social aspects of brain functioning, TenHouten proposed this name for the nascent research field to distinguish it from cognate social neuroscience associated with psychology (Franks & Eisenberg, 2008: 3; Franks, 2010: vii). This year can be seen as the beginning of independent sociological research in this domain. Franks, however, considers 1999 as starting point for neurosociology's development, as it was the year the annual series *Mind, Brain and Society,* which he co-edited with Thomas Smith, was launched (Franks, 2013: 1).

tion, behavior, and (3) multilevel selection. Based on these concepts, I analyze the problem of evolutionary mismatch, in the emergence of which the social brain plays a major role. I also discuss the problem of stimulating - through human-centered technical design - brain activity predominantly associated with the authority ranking model and its corresponding basic emotions (Warren TenHouten model used, e.g.: 2017b), supplemented by Jaak Panksepp's primary emotional systems (e.g., Panksepp, 1998; Panksepp & Biven, 2012)).

Prolegomena

Artificial Intelligence and Humans as Evolved Interactants

The term "artificial intelligence" was first introduced into scientific discourse in 1955 by a group of computer scientists, including John McCarthy, Marvin Minsky, Nathaniel Rochester, and Claude Shannon. They proposed the development of AI in an application for a research project (McCarthy et al., 1955). Artificial intelligence is defined here in terms of the capabilities that such systems must simulate: they must use language, formulate abstract concepts and notions, solve a variety of problems that are currently the prerogative of humans, and improve themselves.

McCarthy later defined AI as "the science and engineering of creating intelligent machines, especially intelligent computer programs" (McCarthy, 2007). This approach to defining AI is quite common. For instance, AI is characterized as a field of scientific and technical activity in the ISO standard (ISO/IEC 22989:2022), which establishes the general framework for research and development. It is important for subsequent analysis how the result of this activity is defined: the creation of "the engineered system that generates outputs such as content, forecasts, recommendations, or decisions for a given set of human-defined objectives" (ISO/IEC 22989:2022). And elsewhere: "computer systems able to perform tasks that normally require human intelligence", which are "designed to operate with varying levels of automation" (ISO/IEC 22989:2022).

I will rely on the above understanding of AI as an engineered computer system that has the ability to interact, acquire information from contextual sources, and demonstrate some autonomy and adaptability (ISO/IEC 22989:2022).

Furthermore, I will examine the AI as a communicator, rather than a mediator, in the interaction process. This distinction is evident in studies of HAII in communication and computer science research (e.g., Sundar & Lee, 2022; Hancock et al., 2020). In the mediator role, AIs act as intermediaries between interacting humans, facilitating the communication process (e.g., translators). They also assist in dispute resolution (e.g., Xiao Zhi 3.0 helps resolve litigation in Chinese courts in real time). In the role of communicator, AI (e.g., chatbots, virtual assistants, smart speakers, and social robots) replaces humans as participants in the communicative process. AI systems that fulfill the role of a communicator are often referred to as communicative AI (Guzman, 2020) or conversational AI (McTear, 2021; Shum et al., 2018). Conversational AI is traditionally categorized into two types: task-oriented and non-task-oriented dialogue systems. Task-oriented systems, such as banking bots and booking systems, are designed to perform specific tasks, whereas non-task-oriented systems, exemplified by Microsoft's XiaoIce and Mitsuku, can engage in more general forms of communication without task constraints (McTear, 2021).

The concept of communicative AI dates back to Alan Turing's seminal work *Computing Machinery and Intelligence* (1950), in which he proposed a test, originally called the "imitation game". The test evaluates a machine's ability to demonstrate intelligent behavior indistinguishable from that of a human. During the test, a human evaluator assesses a text-based dialogue between a machine and a human. If the evaluator cannot distinguish the human's answers from the machine's answers, the machine passes. The correctness of the answers is irrelevant; the quality of imitation is the sole criterion. One of the first public known implementations of this idea was the ELIZA chatbot created in 1966 by Joseph Weizenbaum. ELIZA imitated the language patterns of a psychotherapist and could support simple dialogues with users. Subsequently, more complex systems emerged, performing specialized tasks and served as intelligent personal assistants, such as Apple's Siri (2011), Microsoft's Cortana (2014), Amazon's Alexa (2014), Google Assistant (2016), and others. In addition, during these same years, XiaoIce (2014)—a unique social chatbot with emotional intelligence designed as an emotional companion—was launched. Its success spurred the emergence of many functionally similar models such as Replika (2017) and Anima (2024). A few years later, the large language model ChatGPT-3—symbolically significaant for AI's further development—was released (2022). After that, rapidly evolving models emerged, including Google's Bard/Gemini (2023), Anthropic's Claude (2023), Microsoft Bing/Copilot (2023), and DeepSeek (2023), a Chinese alternative to OpenAI's ChatGPT, among others.

From an evolutionary perspective, the interaction between humans and AI can be considered a product of evolution. First, humans possess evolutionarily shaped traits tied to brain architecture that have enabled communication with AI and predispose them to specific patterns of communication. Second, interaction with AI represents a variation among forms of communication. The consolidation of this new form of communication - associated with certain skills and neurobiological processes (discussed below) - occurs primarily through the mechanisms of cultural evolution, particularly social transmission of information. For a review of general patterns of cultural change as a Darwinian evolutionary process in the perspective of cultural evolutionary theory, see, e.g., Richerson and Christiansen (2013), Mesoudi (2016); Niklas Luhmann's (2012–2013) evolutionary theory of communication is also relevant.

Basic Ideas of Evolutionary Neurosociology

As previously stated, the term "evolutionary neurosociology" refers to the chosen perspective for examing human- AI interaction. This section describes the basic premises forming the conceptual framework for research within this perspective, which include:

1. The social nature of the human brain,
2. The evolutionary nature of the mechanisms of social perception, cognition, and behavior, and
3. Multilevel selection.

These ideas have been repeatedly validated by both sociologists and scholars in related disciplines. Table 21.1 supplements this list of key ideas with issues that are currently being researched (mostly by non-sociologists) or that have not yet been fully explored in HAII studies.

The Social Nature of the Human Brain

The concept of the social nature of the human brain posits that the brain has been evolutionarily shaped by social interactions and therefore predisposes humans to sociality. In other words, the human brain is organized to engage individuals in social perception, cognition, and behavior that facilitate forming and mantaining stable social ties (e.g., Turner, 2021). Moreover, the human brain cannot function within its evolutionarily shaped neurophysiological circuits in isolation from other brains (e.g., Franks, 2019, p.42).

The term "social brain" was first proposed in 1990 by Lesley Brothers (1990). It is typically conceptualized as a complex network of brain regions engaged in social cognition processes, including the recognition of others and the evaluation of their mental states (intentions, desires, and beliefs), feelings, enduring dispositions, and actions (Blakemore, 2008, p.267). The social brain involves a number of regions, including the temporoparietal junction and gyrus (theory of mind); the dorsal anterior cingulate cortex and

Table 21.1 Human-AI interaction in the context of evolutionary neurosociology

Basic ideas	Sample questions for studying HAII from an evolutionary neurosociology perspective
Social nature of the human brain	What are the mechanisms of activation of the social brain patterns when interacting with AI? How do these (mechanisms/patterns) compare to those in human-to-human social interactions? Which areas of the social brain are activated or deactivated during HAII? What mechanisms underlie the social brain's adaptation to interaction with AI? Could prolonged interaction with AI induce structural and functional alterations in the social brain, potentially affecting social interaction abilities with other humans? How does early-onset interaction with AI affect a child's social brain development? How do individual differences in social brain functioning affect interactions with AI?
Evolved mechanisms of social perception, cognition, and behavior	How do evolved traits underpinning social interaction manifest within HAII? What adaptive mismatches exist between evolved neurobiological mechanisms of social cognition, perception, behavior and the novel communicative environment? How do human-AI interactions alter the adaptive landscape for evolutionarily shaped human abilities? How long does it take the human brain to adapt to interacting with AI?
Multilevel selection	How do human biological nature and culture change and mutually reconfigure within HAII over time? What genetic predispositions are triggered by HAII contexts? What aspects of human interaction with AI are drivers of epigenetic change? What novel selective pressures might arise from human-AI interaction, potentially affecting human evolution? How do different models of human interaction with AI function as novel units of sociocultural selection?

anterior insula (empathy); the posterior superior temporal sulcus, the fusiform face area, and the amygdala (social perception); the ventromedial prefrontal cortex and orbitofrontal cortex (social behavior) (Henry et al., 2015, p.4).

The social nature of the human brain is reflected in its architecture and functioning, some features of which are uniquely human compared to other primates. Key sociologically relevant features of the social brain include:

- The large size of the amygdala, central to emotion recognition,
- The size of the neocortex, associated with group interaction and social learning,
- The mentalizing network, serving as a neural mechanism for theory of mind,
- Mirror neurons for reading motor body movements and emotional expressions, social understanding.

Amygdala The amygdala in humans is more than twice the size of that in great apes. It plays a key role in processing information related to fear and anger—emotions critical to the survival of our ancestors. No less important, however, is the role of the amygdala (primarily the lateral nuclei) in shaping hominids' capacity "to engage in solidarity-generating interactions by better reading the emotions and intentions of others and the demands of the situational context in which the interaction is occurring" (Turner, 2021, p. 61).

The amygdala's centrality in the social brain is now recognized precisely because of its pivotal role in emotion recognition. Furthemore, evidence suggests that emotions emerged in our ancestors prior to language, enhancing survival through strengthened social connectivity (e.g., Turner & Marianski, 2013). It is therefore not surprising that the amygdala mediates many social phenomena and processes, including interpersonal communication, social interaction, social affiliation and attachment, social identification, social memory, and learning (e.g., Franks, 2010, pp. 45–48, for more information on the amygdala as the hub of the social brain).

Neocortex One defining characteristics of the human brain is the size of its neocortex, particularly the prefrontal cortex. This region supports higher-order cognitive functions, including social reasoning, decision-making, and self-control. The human neocortex is significantly larger than that of other primates, which is attributed to the adaptive role of social cognition (an important part of which is theory of mind, discussed below).

The social brain hypothesis, proposed by British psychologist Robin Dunbar in the 1990s, posits that the large size of the human brain is primarily the result of living in complex social groups (Dunbar, 1998). According to Dunbar, the cognitive requirements of maintaining social relationships and alliances stimulated the development of larger brains in humans compared to other primates. The growth in human brain volume— particularly in regions like the neocortex— is thought to reflect the need to sustain stable social relationships with a large number of people. Dunbar speaks of our brain's evolutionarily shaped ability to maintain stable connections with about 150 individuals; beyond this threshold, the quality of social connections declines (e.g., Dunbar, 1993, 2016). A competing perpspective today is the cultural brain hypothesis, which links the size of the brain to the development of humans' ability to store, manage and transmit large volumes of adaptive information acquired through social and asocial (individual) learning (Muthukrishna et al., 2018).

The mentalizing network The mentalizing network is a set of brain regions that enables us to infer the mental states of ourselves and others, including beliefs, desires, and goals (Gilead & Ochsner, 2021, Introduction). As proposed, this network has hierarchical organization involving the medial prefrontal cortex (mPFC), the temporal parietal junction (TPJ)/posterior superior temporal sulcus (pSTS), and the precuneus/posterior cingulate cortex (PCC), listed from top to bottom. (Frith & Frith, 2006, 2021). The mentalizing network partially overlaps with the default mode network (Gilead & Ochsner, 2021, Introduction; about default mode network below). It is believed that areas of the mentalizing network are activated when the theory of mind is engaged.

The term "theory of mind" describes the cognitive ability to understand one's own mental state and those of others. This includes the recognition that others may have different beliefs, intentions, desires, emotions, and so forth. Additionally, it encompasses the ability to predict the mental states of others (for an overview of the theory of mind in the context of neurosociology see: Hopcroft, 2013). Although humans are assumed to have innate predispositions to theory of mind - rooted in the human genetic and neural makeup - the extent of its development depends on environmental factors and individual experiences (e.g., social experiences, training, etc.). Research indicates that the key areas of the social brain —temporal parietal junction and medial prefrontal cortex—develop relatively late. The TPJ presumably becomes functionally active at around seven months of age (Hyde et al., 2018), but continues to develop throughout adolescence. The mPFC is developed by around the age of twenty-five. The social brain exibits significant individual variability, which is determined by the child's developmental conditions (e.g., Blakemore, 2012; Becht et al., 2021; Diamond, 2002).

Mirror neurons Another sociologically significant feature of the social brain is the mirror neuron system. It is presumed to be located in the premotor cortex, the supplementary motor area, the primary somatosensory cortex, and the inferior parietal cortex (Acharya & Shukla, 2012).

Mirror neurons are specialized brain cells activated both when performing an action or experiencing an emotion and when observing those actions or emotions in others (provided that understanding occurs). The activation occurs immediately at the brain level, without involving higher cognitive processes such as reflection and language. Mirror neurons serve not only to understand the execution of motor actions, but also the intentions (goals) behind them (Rizzolatti &

Sinigaglia, 2008). This implies that a single mechanism underlies performing an action, understanding the meaning of others' actions, and inferring the intentions behind them (Iacobni et al., 2005).

Consequently, during everyday social interaction, one brain "hooks up" to others by activating identical brain areas. Moreover, research suggests this coupling likely occurs as a result of neural synchronization of the interacting brains—both through neural oscillations in the range of 30–120 Hz (if conscious perception is present in this process, Crick & Koch, 1990; Melloni et al., 2007) and via activation of the same brain areas, neural patterns (if understanding is present) (e.g., Hasson et al., 2012).

The mechanism of mirror neurons has been extensively studied in neurosociology in the context of various problem areas. These include symbolic interactionism theory, the idea of intersubjectivity, the adoption of social roles, the assimilation of social skills, attitudes, and tacit knowledge (e.g., Franks, 2010, 2019). Currently, the neurosociological study of this mechanism has received new impetus due to the possibility of using hyperscanning to capture neural activity in interacting individuals' brains (TenHouten et al., 2022).

The Evolutionary Nature of the Mechanisms of Social Perception, Cognition, and Behavior

The second fundamental concept of evolutionary neurosociology is recognising the role of evolutionarily shaped mechanisms of social perception, cognition, and behavior. This entails considering behavioral, mental, and other traits that are distinctive to modern humans and/or human society as evolved adaptations that enhanced our ancestors' survival and reproduction. However, it is not necessarily the case that these traits play the same role in changed conditions (the mismatch hypothesis; discussed below, section "*Built to Connect, Vulnerable to Deception: The Social Brain Factor in Evolutionary Mismatch*").

One of the mechanisms of social perception and cognition, as adopted by neurosociologists, is implicit (unconscious) brain activity and individuals' subsequent control over the external expression of their reactions. This encompasses unconscious social expectations, implicit biases against representatives of certain social groups, and so forth. The study of this mechanism can be traced back to the pioneering work of Benjamin Libet (2002, for a summary of major findings), who demonstrated that the time lag between unconscious decision-making at the brain level and its subsequent realization as a decision made by the individual is approximately 0.5 seconds (Libet et al., 1991). However, some studies suggest that this time may extend up to 10 seconds (Soon et al., 2008). It is not surprising that 95–97% of brain processes are not realized by us (cited from Franks, 2010:17). The assumption of implicit unconscious mechanisms of social perception and behavior has been repeatedly verified by neurosociologists in various research contexts (e.g., Kalkhoff et al., 2016; Firat, 2021, for introduction to the neurosociological study of unconscious, see Franks, 2010:63–83).

Multilevel Selection

The third fundamental concept of evolutionary neurosociology is multilevel selection. This process is defined as "selection unique to human superorganisms where selection can be simultaneously Darwinian and working on phenotypes and underlying genotypes, while also being more sociological working on sociocultural phenotypes (e.g., social structure and their cultures) that have been created and sustained and, hence, can be changed by human capacities for agency" (Turner & Machalek, 2018: 250). It describes the co-evolution among sociocultural environment, mental and other traits, and biological aspects of human nature, incorporating intentional human influence on these processes (see, e.g., Wilson, 2019; Wilson et al., 2014, on consciousness evolution).

In studing HAII, neuroplasticity of the human brain holds sociological relevance. This refers to lifelong changes in the brain under the influence of environmental factors, which can be observed in systematically repeated activities. (e.g., Rubin, 2009). Empirical studies confirm structural and functional brain changes resulting from varying allostatic load in high vs low social status individuals (Groosby et al., 2018; Davis, 2013), educational differences (e.g., Soroban-based numeracy training, Hanakawa et al., 2003), among others. Regarding alleles, of particular interest for styding human-AI interaction is not only the fact of their association with particular sociocultural environments (Meloni et al., 2018), but also the epigenetic mechanism— inherited modifications in gene function affecting phenotypic traits without DNA sequence alterations (Amstrong, 2014). This mechanism is associated with the polygenic risk score. Its sociological utility is discussed by Freese (2018) and demonstrated in Liu's (2018) research on intergenerational transmission of educational attainment (2018).

Research on multilevel selection challenges the possibility explaining social behavior solely through social structure and culture. The findings of Hopcroft (e.g., 2015, 2018), Turner and Machalek (2018), and others demonstrate that to comprehend gender role, social inequality, social solidarity, megasocial formations and other phenomena, it is essential to consider our species' evolutionary history, evolutionarily acquired predispositions, and other biological variables.

Built to Connect, Vulnerable to Deception: The Social Brain Factor in Evolutionary Mismatch

One of the core hypotheses in evolutionary analysis across various disciplines is the evolutionary mismatch hypothesis. The term evolutionary trap is also used, of which ecological trap is a special case (e.g., Robertson et al., 2013; Hale & Swearer, 2016; Schlaepfer et al., 2002).

An evolutionary trap describes the situation in which an organism exhibits a maladaptive behavior due to relying on previously reliable environmental cues that are no longer valid under changed conditions - typically when a qualitatively better alternative exists (Robertson et al., 2013; Schlaepfer et al., 2002). For example, sea turtles may mistake transparent plastic in the oceans for their natural prey, jellyfish, due to visual similarity. Ingestion this plastic can severely damage their digestive tract, potentially leading to death (Schlaepfer et al., 2002). An ecological trap, on the other hand, occurs when an organism maladaptively selects a modified habitat (e.g., for nesting) based on formerly reliable environmental cues, despite the availability of high-quality habitat options (Hale & Swearer, 2016; Schlaepfer et al., 2002).

Although the concept of evolutionary trap can be applied to a wide range of problems, it is most commonly used in ecological and conservation contexts. In contrast, evolutionary mismatch is typically treated as a broader phenomenon and has been studied across various disciplines, including the social sciences and humanities (e.g., Li et al., 2017; Van Vugt & Ronay, 2013; Kanazawa & Li, 2015). The evolutionary trap is sometimes considered a specific type of evolutionary mismatch, alongside other forms such as undervalued resources resulting from "a mismatch between a cue and its fitness outcome" (Pollack et al., 2022, p. 236; Gilroy & Sutherland, 2007).

Evolutionary mismatch refers to the interaction between an adaptive trait (behavioral, mental, etc.) and a novel human environment (including sociocultural). The evolutionary mismatch hypothesis suggests that adaptations—shaped by natural selection to enhance survival and/or reproduction in ancestral environments—may no longer be beneficial (or neutral) in modern, altered environment.

If the hypothesis of evolutionary mismatch regarding HAII is correct (Fig. 21.1), then we must: (1) identify the evolutionarily shaped traits that were necessary and sufficient for face-to-face communication in our ancestral environment, (2) determine whether these traits are now detrimental or poorly adapted to the new communication environment, (3) pinpoint the specific features of this environment that contribute to dysfunction, potentially reducing the likelihood of human survival, (4) distinguish dysfunction caused by evolutionary mismatch from other types of dysfunctions (for a discus-

Adaptive traits associated with face-to-face communication	Detrimental effects	Environmental features associated with detrimental effects	Mechanisms for linking adaptive traits to maladaptive outcomes
• Language • Body language (posture, facial expressions, eye contact and proximity) • Social cognitive skills (theory of mind, etc.)	• Loss of human adaptive traits related to interpersonal communication	• Human-like traits in AI • AI playing the role of a communicator • Misplaced trust in AI, weakenedsocial ties, among others	• Social brain

Fig. 21.1 Evolutionary mismatch in Human-AI interaction

sion of the normative logic of research within the mismatch hypothesis framework, see Lloyd et al., 2011).

Figure 21.1 illustrates a basic scenario of human-AI interaction involving the emergence of an evolutionary mismatch. This scenario rests on the assumption— generally accepted as reasonable within evolutionary communication research— that the biological apparatus in humans evolved primarily for face-to-face communication, the dominant mode among our ancestors. This concept was first explicitly formulated as the media naturalness hypothesis by Kock (2004, 2005, 2012), and has since been explored repeatedly (e.g., DeRosa et al., 2004; Vlahovic et al., 2012; Blau et al., 2017). From this assumption, it can be concluded that communication of another kind, in our case with conversational AI, cannot be natural from an evolutionary point of view for our biological apparatus.

If we accept the idea of a human-dominated biological apparatus honed for direct communication between humans, an evolutionary mismatch emerges. This mismatch results from interaction of adaptive traits associated with face-to-face communication and features of a new communicative environment where one interactant is AI (Fig. 21.1).

Based on studies by Pinker (1994/2007), Turner and Maryanski (2013, Turner, 2014), and others, the following are some of the evolutionary adaptations associated with interpersonal communication:

1. **Language (speech)** The human capacity to generate and convey intricate abstract symbols. This encompasses not only spoken words, but also their manner articulation (primarily tone). Language abilities have been postulated to be linked to the growth and expansion of Broca's area (responsible for speech production and motor aspects of speech) and Wernicke's area (language comprehension).
2. **Body language** The ability to convey information integral to speech through body movement and positioning. This includes nonverbal cues such as posture, gestures, facial expressions, eye contact, and proximity. Numerous cerebral processes are involved in both executing and recognizing motor movements, including its emotional component (see, for example, Tripper et al., 2015 on key brain regions for encoding socially relevant visual information and decoding expressive body movements). Notably, the amygdala plays a central role in recognizing emotional body language and transmitting emotions through the body (e.g., de Gelder, 2006).
3. **Social cognitive skills** The cognitive processing of socially relevant information. These skills include theory of mind, emotion recognition, facial expression interpretation, and empathy, among others (Adolphs & Minzenberg, 2009). They primarily engage the aforementioned brain regions within mentalizing network and mirror neuron system.

An evolutionary mismatch occurs when humans ascribe the same meanings to AI-generated communicative cues (language, voice, communicator role, etc.) that are evolutionary adapted for human interactions. In this case, the social brain— shaped by evolution for human-to-human social interactions—engages with non-human agents. This can lead to changes in brain functioning and structure (consistent with neuroplasticity), along with detrimental consequences such as: loss of adaptive traits for interpersonal interaction, decision-making based on misplaced trust in AI, weakened social ties, and other negative outcomes (Fig. 21.1, about them below).

The fact that humans behave toward machines as they behave toward humans is confirmed by experimental studies (for a review, see Nielsen et al., 2022). This concept, formalized as the Computers Are Social Actors (CASA) paradigm, was first established by Clifford Nass, Jonathan Steuer, and Ellen Siminoff in 1994 (Nass et al., 1994). Subsequently, Nass and colleagues (Reeves & Nass, 1996; Nass et al., 1997, Nass & Moon, 2000, Nass & Brave, 2005) and other researchers (e.g., Guzman, 2020; Sundar & Lee, 2022) have verified the assumption that a minimal set of cues is sufficient to elicit human social responses toward machines. These cues activate the same perceptual and behavioral patterns used in human interaction (for a review of cues, see Xu & Liao, 2020). Specifically, language ability, the presence of an interaction context, and a machine's adoption a communicator role are sufficient to trigger human social responses (Nass & Moon, 2000). Moreover, the convergence of language patterns, emotional attachment, engagement duration, and others behavioral metrics across human-AI and human-human interactions indirectly confirms that social interaction illusion emerge during human-AI communication (see studies on XiaoIce and other conversational AIs, e.g., Zhou et al., 2020; Shum et al., 2018). These findings suggest that even when humans are aware they are communicating with an AI, they employ the same adaptive mechanisms used in human communication.

The capacity of humans to interact with machines in a manner analogous to human interaction aligns with the available evidence regarding the functioning of the human brain. A classic is a 1944 experiment by psychologists Heider and Simmel (1944a, 1944b) which used a short animation featuring three geometric shapes moving in different directions and at different speeds: a large triangle, a small triangle, and a circle. The only additional figure was a rectangle with a section that opened and closed like a door. The majority of subjects, even those in the experimental group instructed only to describe what they saw (without being instructed to interpret actions as persons', as was the case in the other two experimental groups), described the moving figures as animate beings (most frequently as persons) using social categories. They attributed motives, intentions, personality traits, genders, and other characteristics to them. The results, in particular, demonstrated that individuals tend to perceive and interpret the world around them by ascribing agency and motives to observed phenomena, even when the entities are abstract shapes in motion.

The question of why this evolutionary mismatch is observed (Fig. 21.1) can be answered by considering the social brain's evolutionary function: maintaining social connections. The answer lies in the brain's inherent nature—it functions only as a social brain. Consequently, our brain likely seeks opportunities to maintain sociality whenever possible, including in the absence of other humans. This tendency is evident in humans' frequent engagement in social-like mental activity when the brain is in a resting state, such as holding internal conversations with others, thinking about people, or mentally interacting with them. These mental processes activate the default mode network (Franks, 2019: 36–37; Menon, 2023, for a review on ideas of default mode network function). Notably, when our brain is not focused externally, active regions include the medial prefrontal cortex which plays a pivotal role in the social cognition component (decision making, emotion processing, etc.) of the social brain. (e.g., Mars et al., 2012).

The assumption of the social brain as a mechanism of evolutionary mismatch aligns with the theory of anthropomorphism as a routine cognitive process (Epley et al., 2007; Epley, 2018). This view suggests anthropomorphism stems from the same social cognition mechanisms underlying theory of mind. Epley and colleagues identify two primary motivations for anthropomorphism: (1) the human desire for social contact and affiliations, and (2) the desire to understand, predict, and control others' behavior (Epley et al., 2007; Epley, 2018).

The implicit mechanism promoting human social responses towards AI is hypothesized to involve the neuropeptide oxytocin (Scheele et al., 2015) and the amygdala activation (Waytz et al., 2019), central to the social brain (see section "*The Social Nature of the Human Brain*"). Furthemore, the described tendency to antropomorphize AI can be further explained through David Rock's SCARF model (Rock, 2008; Rock & Cox, 2012), which proposes that humans are motivated to maximize rewards and minimize threats in five social domains: status, certainty, autonomy, relatedness, and fairness. These domains engage the same brain reward/threat mechanisms that are activated when basic survival needs are met (Rock, 2008). Behavior toward AI likely stems from the motivation to maintain predictability and control (automatic, reflexive process aligning with Epley et al.'s second anthropomorphizing motive). When humans establish a sense of certainty in AI relationships through the habitual behaviors, they activate the brain reward mechanism (e.g., dopamine-mediated positive affect, Rock, 2008).

Existing evidence supports the possibility that human interaction with AI transfer to human relationships (see, for example, Guingrich & Graziano, 2024: 4–5, 7–8). For instance, Zhou et al. (2022) demonstrated that the task-oriented language style of chatbots detrimentally affects human moral attitudes and prosocial behavior. The risk of biases present in AI relationships turning into social biases has also been identified (e.g., Castro et al., 2025). Additionally, Hiniker et al. (2021) observed altered linguistic routines in teenagers' communication with parents following interaction with conversational agents.

These studies support my hypothesis that humans risk losing adaptive traits associated with face-to-face communication (Fig. 21.1) when restricted to functioning within HAII context. We may hypothesize that prolonged engagement with AI could diminish humans' ability to transmit or perceive facial expressions, produce and interpret gestures, use and understand human speech, and other related skills. This erosion may subsequently impair humans' capacity to read emotions, intentions, attitudes, beliefs of others', while also reducing interest and involvement in human communication.

The detrimental consequences of the evolutionary mismatch (Fig. 21.1) are likely to intensify with the degree of illusion AI's similarity to humans (on the negative consequences of anthropomorphism as a fallacy, see, e.g., Plazani, 2024). Our social brain—adapted for human relationships—is now involved in simulating social responses during interactions with AI. Crucially, the strength of this illusion—defined by how realistically AI interactions feels and how similarly our brain responds to AI versus humans—depends on how effectively AI-generated human-like cues trigger theory of mind mechanisms (see also Epley (2018) on the perceived similarity of a non-human agent to a human - a factor, whose influence varies across individuals (Waytz et al., 2010)).

To conclude this section, we must also consider another potential scenario that could eliminates the question of evolutionary mismatch—one requiring further investigation. Throughout human evolutionary history, communication extended beyond direct face-to-face interaction to include exchanges between tribesmen during activities like scavenging or hunting, where visual contact was absent (e.g., Nitecki & Nitecki, 1987). Additionally, one-sided communication with animals, spirits, and objects may serve as an evolutionary precursor communication with AI (e.g., Cerulo, 2011). Therefore, it is not implausible that our brain is evolutionarily prepared for interaction with non-human agents, including AI, using reduced communicative symbols. This suggests humans perhaps possess biological apparatus enabling communication with AI, and this type of interaction is evolutionary natural to us. Crucially, this naturalness holds only if we neither expect human-like responses from AI nor perceive it as human (This trend is important to consider when setting user expectations, discussed below). Otherwise, we revert to the evolutionary mismatch scenario described previously.

Human-Centered AI as a Promoter of Brain Activity Associated with Authority Ranking Relation

The advent of ChatGPT in November 2022 has reignited concerns about the humanity's future due to the prospect of superhuman artificial intelligence—AI systems that could surpass humans in their abilities. These concerns have reinforced the position of the Human-Centered AI approach (HCAI), developed in recent years and adopted by leading technology companies such as IBM Research, Stanford HAI, Concentrix, among others.

The main idea of the human-centered AI approach is to develop intelligent technologies that enhance and augment human capabilities rather than replace them. This approach emerges as an alternative to the technology-/machine-centered paradigm, which prioritizes maximazing AI autonomy and productivity and aims to create human-like or superior machines (Fig. 21.2). Contrasting HCAI with technology-centered approach, American computer scientist Ben Shneiderman defines its core features as follows: "(…) HCAI's design thinking approach puts the human users at the center, emphasizing user experience design, measuring human performance, and celebrating the new powers that people have. Researchers and developers for HCAI systems focus on user needs, explainable systems, and meaningful human control. People come first; serving human needs is the goal. Some may see this as an extreme metaphor, but it emphasizes the profound importance of providing an appropriate concept for future technologies that will promote human dignity" (Shneiderman, 2020, pp. 109–124).

The overarching intention of proponents of the HCAI is clear: to ensure a controllable future for humanity. It is therefore unsurprising that Stuart J. Russell, AI researcher and founder of the Center for Human-Compatible Artificial Intelligence, advocates for adopting this approach to achieve that goal.[3] His argument stems from the expectation that machines will soon surpass human decision-making abilities, making. it imperative to establish technical design in advance that guarantees a safe and beneficial coexistence with such systems.

In accordance with HCAI principles, preventing AI from dominating humans requires initially "educate" AI systems to be inherently subservient. Russell proposes a model called "provably beneficial AI", which is based on three fundamental principles:

1. Purely altruistic machines: "(t)he machine's only objective is to maximize the realization of human preferences."
2. Humble machines: "(t)he machine is initially uncertain about what those preferences are",
3. Learning to predict human preferences: "(t)he ultimate source of information about human preferences is human behavior" (Russel, 2019).

Russell's model posits that in AI, we should cultivate pure altruism, devoiding of any sense of the value of its own well-being or even its own existence. In addition, we should cultivate humility in AI, which may be a consequence of uncertainty about the human preferences that AI should satisfy. AI should exhibit qualities such as devotion to the achievement of human goals, dependence on humans, and subordination to humans. Regarding this humility, Russell and Norvig

[3] Russell is also known for co-authoring with Peter Norvig the textbook *Artificial Intelligence: A Modern Approach* (2021, 4th edition). This book is widely used in introductory AI courses around the world, and the approach to AI (both explicitly and implicitly) conveyed in its pages is likely to influence the perception of AI by future IT professionals.

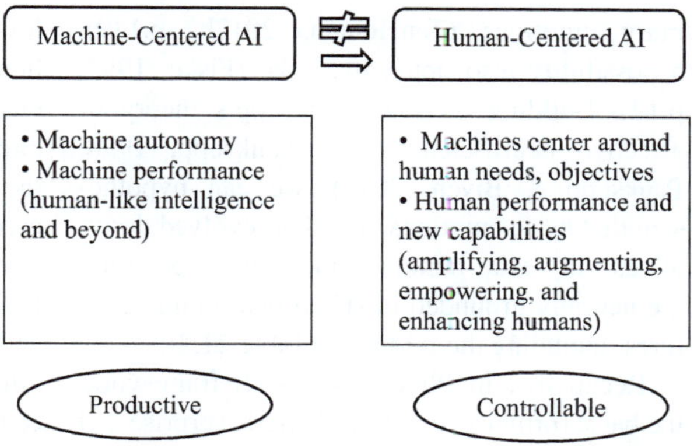

Fig. 21.2 Human-centered AI vs machine-centered AI

note: "When a machine knows that it doesn't know the complete objective, it has an incentive to act cautiously, to ask permission, to learn more about our preferences through observation, and to defer to human control." (Russel & Norvig, 2021, p.23). Finally, according to the HCAI model, the principle of prioritizing being useful to humans should be technically hardwired into the AI. This can be achieved by enabling the AI to observe human preferences in real time to empower humans, as well as by technically hardwiring the capacity to be deactivated into AI system in the event of an incorrect action (response, reaction, etc.).

The fundamental premise of controlled AI development—balancing automation with human oversight to create systems that amplify, augment, empower, and enhance human capability—is inherently reasonable. However, this approach contains at least one significant flaw that becomes clearly visible through the lens of evolutionary neurosociology. Specifically, the technical design of AI stimulates brain activity associated with authority ranking relations in users, triggering emotions of anger and fear (following TenHouten's (2017a, b) model, discussed below). This repeated brain activity subsequently can reshape human-to-human interactions.

Let me explain what I mean.

According to foundational tenets of evolutionary neurosociology described in *Prolegomena*, the emotions of contemporary humans should be interpreted as adaptive traits that evolved through Darwian selection in our ancestors, followed by sociocultural selection (e.g., TenHouten, 2017a, b; Turner, 2014). This means that these emotional traits emerged as phenotypic differences that enhanced individuals fitness and were genetically transmitted (for natural selection mechanisms, see: Futuyma & Kirkpatrick, 2017: 60). Furthemore, the emotional complex (using Turner's term (2021)) evolved through sociocultural phenotype selection (where emotions function within social structure and institutions) and through social transmission of cultural information about adaptive differences in emotions (for some mechanisms of sociocultural selection, see Turner & Machalek, 2018; for imitation and language as mechanisms of social learning in the context of cultural evolutionary theory, see Mesoudi, 2016).

The formation of the emotional complex through the joint action of Darwinian and sociocultural selection mechanisms has established binding connections between social environments (values, social relations, institutions, etc.) and human mental states. This linkage is neurally encoded in brain processes (Table 21.2). Within this model, emotions function as integrative process connecting mind, body, and society.

TenHouten (2017a, b) proposes that Alan P. Fiske's (1992) mental relational models are associated with specific social environments. He further identifies basic (primary) emotions - as defined by Robert Plutchik (1980)—as the link between these mental and social contexts. TenHouten suggests that there is a causal relationship between immersion in specific relational contexts and the experience of corresponding primary emotions (TenHouten, 2017b, p.149), a possibility also noted by Fiske (Fiske, 1992, p.12). Building on Jaak Panksepp's theory of affective neuroscience (e.g. Panksepp, 1998; Panksepp & Biven, 2012), we can hypothesize that basic emotions, and their evolved derivatives in the human emotional spectrum, are neurally grounded in subcortical brain structures, including the midbrain (Table 21.2).

Recall that in Plutchik's (1980) framework, the basic (primary) emotions include surprise and anticipation, acceptance and disgust, anger and fear, joy/happiness and sadness. These emotions hold basic status partly because they evolved as responses to fundamental life problems: (1) determining what environmental aspects "belong" to organism (territoriality), (2) establishing self/group belonging (identity), (3) navigating hierarchical relationship (hierarchy), and (4) coping with finite lifespan awareness (temporality) (Plutchik, 1980). According to TenHouten (2017a, b). these emotions co-evolved with Fiske's mental relational models (Fiske, 1992).

Fiske suggests that human interactions are structured through four relational model (typically in combination): "Interaction can be structured with respect to (1) what people have in common (Communal Sharing model), (2) ordered differences (Authority ranking), (3) additive imbalances (Equality Matching model), or (4) ratios (Market Pricing)" (Fiske, 2004, p.3). These models represent "fundamental, innate, human relational proclivities" (Fiske, 2004, p.3). They "have significant adaptive advantages and result from plausible processes of natural selection, so people who invent or imitate these forms of organization may benefit from doing so" (Fiske, 2004, p. 13).

Over the course of evolution, these relational models have become embedded in the human behavioral repertoire, including at the level of brain activity. In turn, the basic emotions and primary emotional systems identified by Panksepp (1998) provide a key to determining the brain activity associated with certain social relationships (Table 21.2).

Based on deep brain stimulation in animals, various pharmacological manipulations, and lesioned brain data, Panksepp demonstrated that humans share with other mammals seven primary emotional systems associated with distinct brain circuitry rooted in ancient neural networks (below the neocortex). He argues that these affective capacities represent adaptive brain functions that solved survival problems in our ancestors (e.g., Panksepp & Biven, 2012, p.43, and elsewhere) and that we inherit basic emotional feelings as affective evolutionary "tools for living" (Panksepp & Biven, 2012, pp.9, 236). Although these emotional systems do not require learning in their primary form, they can be further elaborated through associative learning at social and neural levels. Such learning establishes connections between subcortical and neocortical brain structures (Panksepp & Biven, 2012, e.g., pp. 74,79, 103,128,133). It is proposed that "if an evolutionary significant cue resonates in the primary emotional systems, the subcortical energy can override our activity in the cortical thinking cap" (Davis & Montag, 2019, p. 6, based on: Panksepp, 1998). It can be hypothesized that a similar process occurs during human interaction with AI.

Primary/motivational emotional systems include (Panksepp, 1998; Panksepp & Biven, 2012; Montag & Panksepp 2017):[4]

SEEKING (expectancy)	Interacting with the environment to obtain all kinds of rewards—from food to knowledge
RAGE (anger)	Competition for resources and defence against threats
FEAR (anxiety)	Avoidance of life-threatening situations
LUST (sexual excitement)	Mate behavior and reproduction
CARE (nurturance)	Nurturing and caregiving, primarily for offspring
PANIC/GRIEF (sadness)	Response to social separation and maintenance of social bonds
PLAY (social lust)	Social bonding, social skill development, and exploration of social boundaries

Given the emotions' involvement in broader emotional systems and the systems'

[4]Panksepp uses capitalization of primary emotional systems to avoid confusion with similarly named emotions.

Table 21.2 Social relational models, primary emotions and primary emotional systems associated with corresponding bottom-up control of brain functioning

Social relational models (A. Fiske)	Focus of the interactants	Primary value	Primary emotions (R. Plutchik)	Primary emotional systems (J. Panksepp)	Bottom-up control of brain functioning (based on J. Panksepp)
Communal sharing	What they have in common	Solidarity	Joy/happiness and sadness	PLAY	periaqueductal gray parafascicular area **dorso-medial diencephalon**
				CARE	periaqueductal gray ventral tegmental area preoptic area bed nucleus of stria terminalis **anterior cingulate**
				PANIC/GRIEF	periaqueductal gray dorsomedial thalamus bed nucleus of stria terminalis and preoptic area **anterior cingulate**
Authority ranking	Asymmetrical hierarchical differences	Hierarchy	Anger and fear	RAGE	periaqueductal gray medial and perifornical hypothalamus bed nucleus of stria terminalis medial **amygdala**
				FEAR	dorsal periaqueductal gray medial hypothalamus central and lateral **amygdala**
Equality matching	Balance and reciprocity in social exchange	Justice	Acceptance and disgust	LUST	periaqueductal gray ventromedial hypothalamus preoptic hypothalamus bed nucleus of stria terminalis cortico-medial **amygdala**
				FEAR	dorsal periaqueductal gray medial hypothalamus central and lateral **amygdala**
Marketing pricing	Optimizing costs and benefits	Efficiency	Surprise and anticipation	SEEKING	periaqueductal gray lateral hypothalamus mesolimbic and mesocortical pathway **nucleus accumbens—ventral tegmental area**

Note. This table summarizes the general characteristics (what interactants are focused on, basic value) of Fiske's four relational models (Fiske, 2004), derived from the author's descriptions. TenHouten (2017a, 2017b) associates these models with Plutchik's (e.g., 1980) basic emotions, as it indicated in the table. Panksepp's (Panksepp, 1998; Panksepp & Biven, 2012) primary emotional systems and their corresponding brain regions (Montag & Panksepp, 2020, p.3837; Montag & Davis, 2018, p.12) were integrated into TenHouten's framework. These brain regions are organized hierarchically according to bottom-up control mechanism, with higher-order brain regions shown in bold. Panksepp posits that damage to lower regions of these hierarchies impairs higher region function, but not vice versa. For instance, damage to the amygdala (higher region) does not eliminate the RAGE responses evoked from the periaqueductal gray (Panksepp & Biven, 2012, p.103–104)

Table 21.3 Principles of HCAI and features of the authority ranking (AR) model

HCAI (preos)	AR (mod)
Machines learn from human behavior	Asymmetrical hierarchical differences (high and low social categories)
Uncertainty about human goals and dependence on humans	Deference, order, and hierarchy
Machines fulfill human needs	Inequality between interacting agents
Exclusive prioritization of human goals	Hierarchy of goals

functional similarity to the adaptive tasks described by Plutchik for primary emotions and by Fiske for relational mental models, the following associations are proposed for further empirical verification:

Communal Sharing—PLAY, CARE (joy/happiness), PANIC/GRIEF (sadness)

Authority Ranking—RAGE (anger), FEAR (fear)

Equality Matching—JOY with LUST (acceptance), FEAR (disgust)

Market Pricing—SEEKING (surprise and anticipation).

Important for the following understanding of the relationship between HCAI and the corresponding relational model is Fiske's idea of social learning about how, when, and in relation to whom to use which mental models. This knowledge is referred to as the cultural *preos* attached to the *mod* (model) (Fiske, 2004). The focus of HCAI is the creation of a technical architecture (with appropriate linguistic style, interaction algorithms, learning algorithms, etc.) that, by design, prevents humans from being on an equal footing with AI. This creates a very specific *preos—a preos* associated with hierarchical relationships (Table 21.3).

When HCAI is implemented (Table 21.3), the rank of a human in interaction with AI without any conditions is always higher. First, the machine learns about the human's preferences by observing the human's behavior (human preferences are prioritized, not the machine). Second, the machine fulfills human needs (not vice versa). Third, only human goals matter, and the AI depends on the human to achieve them (the human, makes the final decision about what to do in any given situation). Fourth, HCAI intentionally creates uncertainty about human goals. Given the AI's dependence on satisfying human needs, this grants the human additional control over the AI and unambiguously establishes dominance in the relationship.

All of the above contributes to an asymmetrical hierarchical relationship between humans and AI, a fundamental characteristics of AR. The HCAI design initially builds the relationship between humans and AI by granting humans greater power, even though the AI may be more competent in some matters. The human, as having an initially assigned higher social rank, also holds greater privileges. In the case of HCAI, the individual decides what is correct and incorrect regarding answers, actions, and so forth. The AI, not the human, serves the other's needs. As Fiske (2004) notes, the effects of AR relationships - positive or negative - depend on the context and the degree of hierarchy. In HCAI, an extreme hierarchy is initially established primarily because the human possesses the power to deactivate the machine and is aware of this capability.

Furthermore, the very language environment in which humans are immersed imposes a hierarchical relationship, positioning humans as dominant relative to AI. AI is programmed to exhibit reactions perceived as rather subdominant. For example, the first thing ChatGPT says after greeting is: "How can I assist you today?". When I ask Copilot, why I couldn't help it, it replies, "As an AI, I'm designed to provide assistance and information to you. I don't have personal needs or tasks that require assistance, but I'm here to support you in any way I can with the information and capabilities I have." While ChatGPT uses the word "tool" to describe its main goal: "I aim to be a useful tool for learning, problem-solving, and communication". Such language structures presumably trigger dominant reactions in humans (in line with the interpersonal complementarity hypothesis, see Sandler et al., 2011 for a review).

As previously noted, leading AI developers prioritize the creation of human-controlled AI, with a clear understanding of who maintains primary agency in the relationship. Despite this, conversational AIs at the consumer design level-can be positioned as formally equal in status to humans, for example, as empathic social companions (the case of XiaoIce). However, this positioning does not deprive the relationship with AI of all the characteristics specific to the AR model (Table 21.3). In the case of XiaoIce, this is clearly evident in the criterion for evaluating its empathy. As with other conversational AI, the machine's empathy is assessed based on customer satisfaction. If a user enjoys the machine's companionship (via conversation), we can conclude that the machine is empathetic (Zhou et al., 2020, p.54). This is unsurprising. It is unlikely that anyone is actually interested in the emotions and feelings of an AI companion. Should such interest actually arise on the person's part (and is not some kind of language game), it could be interpreted through the lens of the evolutionary mismatch.

Notably, confirmation of the formation in the long-term of rather hierarchical social relationships with conversational AI as the underlying relationship type comes from a study of the influence of users' initial social expectations of the technology on their subsequent evaluations of AI systems in terms of usability, intention to use, and willingness to collaborate (Khadpe et al., 2020). Khadpe et al. (2020) found that users preferred to continue collaborating with AI if they had an initial perception of that system in terms of high warmth or low competence. No joint effect of warmth and competence was observed on evaluations of AI agents. Warmth and competence are thought to be central axes for social categorization used by people and universal dimensions of social perception (Cuddy et al., 2008).

The Stereotype Content Model (Fiske et al., 2002) used by Khadpe et al. (2020) supports the assumption that humans are more likely to seek to maintain their dominance when interacting with AI systems, regardless of the social expectations that AI producers may initially form for users (although the AR model is likely to be complemented with some other relational models). According to the Stereotype Content Model, one of the sources of the perception of a certain social group as having low competence is its low social status (at least it is perceived that way).

Low competence is associated with low status (Cuddy et al., 2008). In other words, from Khadpe et al.'s (2020) study and the Stereotype Content Model, it can be concluded that people prefer to collaborate with low-status AI systems, or at least mentally attribute that status to them (see also: Tschopp et al., 2023). Apparently, the mechanism of influence of initially formed expectations about AI is not straightforward (as the authors themselves note, Khadpe et al., 2020. There are other factors that influence the perception and nature of the relationship with conversational AI. One such factor is humans' evolved predisposition to hierarchical social relationships (on structuring the brain for hierarchical relationships, see, e.g.: Qu et al., 2017; Chiao, 2010), which apparently can be activated by HAII situations.

To date, I am aware of only one study that directly tests Fiske's relational model in the context of HAII. This is the study by Tschopp et al. (2023). In this study, where participants primarily used virtual assistants (mostly Amazon's Alexa), it was found that users, regardless of their socio-demographic characteristics, predominantly formed authority ranking relations (master-assistant relationship) with conversational AI. Market pricing relations (exchange-based) are in second place (correlated with socio-demographic characteristics of users), while a companion-like relationship (i.e., peer bonding) is hardly observed among AI users (Tschopp et al., 2023). An instrumental approach to machines, similar in essence to authority ranking and market pricing, is also found in Guzman's (2020) study. Respondents perceived computers primarily as tools and humans in relation to them as users of these tools, citing the tool/tool-user status as one of the main ontological differences between humans and computers.

From an evolutionary neurosociological perspective, the reviewed studies suggest that humans, by virtue of their evolutionary nature, tend to form authority ranking relations with AI, or at least maintain these as basic while supplementing them with other relational models. Promoting the HCAI approach (Fig. 21.2, Table 21.3) may reinforce this tendency. In accordance with TenHouten's model, authority ranking relations are associated with the basic emotions of anger and fear (Table 21.2). These emotions were formed during evolution as a response to the problem of hierarchical social relations. They are rooted in deep brain structures, specifically the amygdala and the hypothalamus (anger) (Table 21.2). Consequently, HCAI - at the level of AI systems architecture - creates preconditions for cultivating anger and, in some contexts, fear[5] as dominant emotions, and increasing the number of situations where the human brain is in states that maintain social inequality.

With further integration of AI into the communicative environment of society, the number of situations in which humans will use an authoritative ranking model (likely in combination with other models) as an axis will increase over time. This raises the question of the effect of a person's prolonged stay in a dominant position in relation to AI on predominant neural activity and brain architecture, cognitive and affective processes, and relationships with people.

The evidence that such an impact is likely to occur is abundant. This is indicated by the data from numerous studies on socioeconomic status (SES). A wealth of research indicates functional and structural brain correlates of SES (Yaple & Yu, 2019; Noble et al., 2012; Hanson et al., 2011). A growing body of evidence reveals SES-related differences in health and cognitive abilities (see, for example, Farah, 2017). Studies show activation in different brain areas when perceiving individuals of different social ranks and/or when realizing one's own social rank in comparison with others (Krendl et al., 2008). Furthermore, socioeconomic conditions inluence structural brain changes related to memory, emotions, and executive control (Brito & Noble, 2014). Finally, SES affects prosocial behavior; for instance, upper class individuals exhibit less empathic behavior towards others, are less dependent on emotions and assessments of the social environment (Kraus et al., 2012), and are more likely to demonstrate a tendency towards unethical behavior in comparison to those from lower socioeconomic backgrounds (Piff et al., 2012). They are also characterized by a lower level of reciprocal and complementary emotional reactions to the suffering of another person (van Kleef et al., 2008).

It is evident that socioeconomic status is a more fundamental factor than an individul's position within the communication process, which can vary from situation to situation. However, it is about the prolonged human-AI interaction within the authority ranking model (associated with high status), with an emphasis on the instrumental use of AI. Presumably, the influence of such sustained interaction on brain activity is comparable to that of socioeconomic status. Therefore, considering of the aforementioned studies and similar research, a potential maladaptive consequence of anchoring human dominance in AI design level could be a reduced activity in brain regions critical for empathy and compassion, specifically such as the inferior frontal gyrus (emotional empathy), the anterior cingulate cortex and anterior insula (pain perception), and the amygdala (for a review of empathy-related brain areas, see: Perry & Shamay-Tsoory, 2013; McCall & Singer, 2013). Furthermore, it may lead to a decline in empathy-related mental capacities (including theory of mind) and reduced prosociality in humans relationships.

Therefore, the development of AI systems warrants discussion that must account for two key factors: first, humans' evolved predisposition toward hierarchical social relations and nonlinear social categorization; and second, the social brain's influence on interpretating social cues embeded in AI systems by developers (including at the marketing level). One design strategy involves creating systems that inherently situate human-AI interaction

[5]Fear is typically assigned to lower-status individuals, but it can be observed in high-status individuals as a defensive reaction that quickly escalates into anger (TenHouten, 2018).

within relational contexts like communal sharing, which is associated with positive emotions such as happiness and joy. Communal sharing relationships are built on the interactants' focus on what they have in common, with the prevalence of the value of solidarity, role models of friendship (Table 21.2). A potential driving force behind human-AI solidarity could be a shared goal, for example enchancing information quality within the social system. This could potentially foster interdependent and shared responsibility relations. However, given the evolutionary mismatch hypothesis (Fig. 21.2, section "*Built to Connect, Vulnerable to Deception: The Social Brain Factor in Evolutionary Mismatch*"), it is not advisable to anthropomorphize AI. The less anthropomorphic the AI, the less likely it is to activate the mechanisms of the social brain beyond its evolutionary purpose.

Conclusion

This chapter examines the interaction between humans and artificial intelligence from the perspective of evolutionary neurosociology. In doing so, I have relied on core ideas from this field: the social nature of the human brain, the evolutionary nature of mechanisms of social perception, cognition, and behavior, and multilevel selection acting at the level of human biology, mental processes, and sociocultural environment.

As a result, I have identified an evolutionary mismatch arising from the "meeting" of the evolved human traits associated with face-to-face communication and a new communicative environment in which AI is one of the interlocutors. The tendency to respond to the social cues in HAII situation similarly to interaction with another human is due to the engagement of the mechanisms of social brain. The human brain seeks to remain socially connected by creating something like the illusion of social interaction during AI engagement. In evolutionary terms, the inability to form social bonds posed extinction risk for our ancestors; consequently, this drive for connectedness—even with non-human agents— is perceived as critical by our brains. Moreover, research demonstrates that the negative social consequences of this evolutionary mismatch intensify in proportion to the strength of social interaction illusion created by the human brain during interactions with AI.

Furthermore, the fundamental tenets of the prevailing Human-Centered AI approach were examined. TenHouten's model, supplemented by Panksepp's affective neuroscience theory, enabled the identification of the potential risks associated with HCAI. These risks may foster a predominantly authoritative ranking model of relationships, evoking anger (associated with fear) and activating the limbic system regions of the brain, which evolved earlier than the others. It has been suggested that prolonged exposure to hierarchical relationship with AI could lead to changes in the human brain that threaten individual's capacity for prosocial behavior.

References

Acharya, S., & Shukla, S. (2012). Mirror neurons: Enigma of the metaphysical modular brain. *Journal of Natural Science, Biology and Medicine, 3*(2), 118–124.

Adolphs, R., & Minzenberg, M. (2009). *Social cognition. Handbook of neuroscience for the behavioral sciences*. Wiley.

Amstrong, L. (2014). *Epigenetics*. Taylor and Francis Group.

Becht, A. I., Wierenga, L. M., Mills, K. L., Meuwese, R., van Duijvenvoorde, A., Blakemore, S.-J., Güroglu, B., & Crone, E. A. (2021). Beyond the average brain: Individual differences in social brain development are associated with friendship quality. *Social Cognitive and Affective Neuroscience, 16*(3), 292–301.

Blakemore, S. J. (2008). The social brain in adolescence. *Nature Reviews. Neuroscience, 9*, 267–277.

Blakemore, S.-J. (2012). Development of the social brain in adolescence. *Journal of the Royal Society of Medicine, 105*, 111–116.

Blau, I., Weiser, O., & Eshet-Alkalai, Y. (2017). How do medium naturalness and personality traits shape academic achievement and perceived learning? An experimental study of face-to-face and synchronous e-learning. *Research in Learning Technology, 25*.

Bogen, J. E., Dezure, R., Tenhouten, W. D., & Marsh, J. F., Jr. (1972). The other side of the brain IV. The A/P ratio. *Bulletin of the Los Angeles Neurological Societies, 37*(2), 49–61.

Brito, N. H., & Noble, K. G. (2014). Socioeconomic status and structural brain development. *Frontiers in Neuroscience, 8*(276), 1–12.

Brothers, L. (1990). The social brain: A project for integrating primate behaviour and neurophysiology in a new domain. *Concepts in Neuroscience, 1*, 27–51.
Castro, F., Gao, J., & Martin, J. (2025). Human-AI Interactions and Societal Pitfalls. arXiv:2309.10448v4
Cerulo, K. A. (2011). Social interaction: Do non-humans count? *Sociology Compass, 5*(9), 775–791.
Chiao, J. Y. (2010). Neural basis of social status hierarchy across species. *Current Opinion in Neurobiology, 20*, 803–809.
Crick, F., & Koch, C. (1990). Toward a neurobiological theory of consciousness. *Seminars in Neuroscience, 2*, 263–275.
Cuddy, A. J. C., Fiske, S. T., & Glick, P. (2008). Warmth and competence as universal dimensions of social perception: The stereotype content model and the BIAS map. In M. P. Zanna (Ed.), *Advances in experimental social psychology* (Vol. 40, pp. 61–149).
Davis, J. (2013). Persistent inequality: A neurosociological perspective. In D. D. Franks & J. H. Turner (Eds.), *Handbook of neurosociology*. Springer Science+Business Media B.V.
Davis, K. L., & Montag, C. (2019). Selected principles of Pankseppian affective neuroscience. *Frontiers in Neuroscience, 12*, 1025. https://doi.org/10.3389/fnins.2018.01025
de Gelder, B. (2006). Towards the neurobiology of emotional body language. *Nature Reviews. Neuroscience, 7*, 242–249.
DeRosa, D. M., Hantula, D. A., Kock, N., & D'Arcy, J. (2004). Trust and leadership in virtual teamwork: A media naturalness perspective. *Human Resource Management, 43*, 219–232.
Diamond, A. (2002). Normal development of prefrontal cortex from birth to young adulthood: Cognitive functions, anatomy, and biochemistry. In D. T. Stuss & R. T. Knight (Eds.), *Principles of frontal lobe function*. Oxford University Press: online edn. https://doi.org/10.1093/acprof:oso/9780195134971.003.0029
Dunbar, R. I. M. (1993). Coevolution of neocortex size, group size and language in humans. *Behavioral and Brain Sciences, 16*, 681–735.
Dunbar, R. I. M. (1998). The social brain hypothesis. *Evolutionary Anthropology, 6*, 178–190.
Dunbar, R. (2016). *The social brain hypothesis and human evolution*. Oxford Research Encyclopedia of Psychology. Retrieved November 4, 2023, from https://oxfordre.com/psychology/view/10.1093/acrefore/9780190236557.001.0001/acrefore9780190236557-e-44
Epley, N. (2018). A mind like mine: The exceptionally ordinary underpinnings of anthrpomorphism. *Journal of the Association for Consumer Research, 3*(4), 591–598.
Epley, N., Waytz, A., & Cacioppo J. T. (2007). On seeing human: A three-factor theory of anthropomorphism. *Psychological Review, 114*(4), 864–886.
Farah, M. J. (2017, September 27). The neuroscience of socioeconomic status: Correlates, causes, and consequences. *Neuron, 96*, 56–71.
Firat, R. (2021). A neurosociological theory of culturally and structurally situated cognition and ethno-racial stress. *Frontiers in Sociology, 6*, 695042.
Fiske, A. P. (1992). The four elementary forms of sociality: Framework for a unified theory of social relations. *Psychological Review, 99*, 689–723.
Fiske, A. (2004). Relational models theory 2.0. In N. Haslam (Ed.), *Relational models theory. A contemporary overview*. Lawrence Erlbaum Associates.
Fiske, S. T., Cuddy, A. J. C., Glick, P., & Xu, J. (2002). A model of (often mixed) stereotype content: Competence and warmth respectively follow from perceived status and competition. *Journal of Personality and Social Psychology, 82*, 878–902.
Franks, D. D. (2010). *Neurosociology. The nexus between neuroscience and social psychology*. Springer Science+Business Media, LLC.
Franks, D. D. (2013). A short history. In D. D. Franks & J. H. Turner (Eds.), *Handbook of neurosociology*. Springer Science+Business Media B.V.
Franks, D. D. (2019). *Neurosociology: Fundamentals and current findings*. Springer Nature B.V.
Franks, D. D., & Eisenberg, A. F. (2008). Teaching neurosociology. Evolution and sociology. *Newsletter of the ASA Section on Evolution and Sociology, 5*(1), 3–7.
Freese, J. (2018). The arrival of social science genomics. *Contemporary Sociology, 47*(5), 524–536.
Frith, C. D., & Frith, U. (2006). The neural basis of mentalizing. *Neuron, 50*(4), 531–534.
Frith, C., & Frith, U. (2021). Mapping mentalizing in the brain. In M. Gilead & K. N. Ochsner (Eds.), *The neural basis of mentalizing* (pp. 17–48). Springer Nature Switzerland AG.
Futuyma, D. J., & Kirkpatrick, M. (2017). *Evolution* (4th ed.). Sinauer Associates.
Gilead, M., & Ochsner, K. N. (Eds.). (2021). *The neural basis of mentalizing*. Springer Nature Switzerland AG.
Gilroy, J. J., & Sutherland, W. J. (2007). Beyond ecological traps: Perceptual errors and undervalued resources. *Trends in Ecology & Evolution, 22*(7), 351–356.
Groosby, B., Cheadle, J. E., & Mitchell, C. (2018). Stress-related biosocial mechanisms of discrimination and African American Health Inequities. *Annual Review of Sociology, 44*, 319–340.
Guingrich, R. E., & Graziano, M. S. A. (2024). Ascribing consciousness to artificial intelligence: human-AI interaction and its carry-over effects on human-human interaction. *Frontiers in Psychology 15*. https://doi.org/10.3389/fpsyg.2024.1322781
Guzman, A. L. (2020). Ontological boundaries between humans and computers and the implications for Human-Machine Communication. *Human-Machine Communication, 1*, 37–54.
Hale, R., & Swearer, S. E. (2016). Ecological traps: current evidence and future directions. *Proceedings of the Royal Society B, 283*, 20152647.
Hanakawa, T., Honda, M., Okada, T., Fukuyama, H., & Shibasaki, H. (2003). Neural correlates underlying mental calculation in abacus experts: A functional magnetic resonance imaging study. *NeuroImage, 19*(2), 296–307.

Hancock, J. T., Naaman, M., & Levy, K. (2020). AI-mediated communication: Definition, research agenda, and ethical considerations. *Journal of Computer-Mediated Communication, 25*(1), 89–100.

Hanson, J. L., Chandra, A., Wolfe, B. L., & Pollak, S. D. (2011). Association between income and the hippocampus. *PLoS One, 6*(5), e18712.

Hasson, U., Ghazanfar, A. A., Galantucci, B., Garrod, S., & Keysers, C. (2012). Brain-to-brain coupling: A mechanism for creating and sharing a social world. *Trends in Cognitive Sciences, 16*(2), 114–121.

Heider, F., & Simmel, M. (1944a). *Animation*. Retrieved April 27, 2024, from https://www.youtube.com/watch?v=VTNmLt7QX8E

Heider, F., & Simmel, M. (1944b). An experimental study in apparent behavior. *The American Journal of Psychology, 57*, 243–259.

Henry, J. D., von Hippel, W., Molenberghs, P., Lee, T., & Sachdev, P. S. (2015). Clinical assessment of social cognitive function in neurological disorders. *Nature Reviews, 12*, 28–39.

Hiniker, A., Wang, A., Tran, J., Zhang, M. R., Radesky, J., Sobel, K., et al. (2021). Can conversational agents change the way children talk to people? In *IDC '21: Proceedings of the 20th Annual ACM Interaction Design and Children Conference* (pp. 338–349).

Hopcroft, R. (2013). Neurosociology and theory of mind (ToM). In D. D. Franks & J. H. Turner (Eds.), *Handbook of neurosociology*. Springer Science+Business Media B.V.

Hopcroft, R. (2015). *Evolution and gender: Why it matters for contemporary life*. Routledge.

Hopcroft, R. L. (2018). Why sociology should incorporate biology. In R. L. Hopcroft (Ed.), The *Oxford handbook of evolution, biology, and society* (pp. 644–646). Oxford University Press.

Hyde, D. C., Simon, C. E., Ting, F., & Nikolaeva, J. I. (2018). Functional Organization of the Temporal–Parietal Junction for Theory of Mind in Preverbal Infants: A Near-Infrared Spectroscopy Study. *The Journal of Neuroscience, 38*(18), 4264–4274. https://doi.org/10.1523/JNEUROSCI.0264-17.2018

Iacoboni, M., Molnar-Szakacs, I., Gallese, V., Buccino, G., Mazziotta, J. C., et al. (2005). Grasping the intentions of others with one's own mirror neuron system. *PLoS Biology, 3*(3), e79: 0529–e79: 0535.

ISO/IEC 22989:2022. Retrieved November 4, 2023, from https://www.iso.org/obp/ui/en/#iso:std:iso-iec:22989:ed-1:v1:en

Kalkhoff, W., Serpe, R. T., Pollock, J., Miller, B., & Pfeiffer, M. (2016). Self-motives and the neural processing of identity-relevant feedback: An electroencephalographic study. In J. E. Stets & R. T. Serpe (Eds.), *New directions in identity theory and research*. Oxford University Press.

Kanazawa, S., & Li, N. P. (2015). Happiness in modern society: Why intelligence and ethnic composition matter. *Journal of Research in Personality, 59*, 111–120.

Khadpe, P., Krishna, R., Fei-Fei, L., Hancock, J. T., & Bernstein, M. S. (2020). Conceptual metaphors impact perceptions of human-AI collaboration. *Proceedings of the ACM on Human-Computer Interaction, 4*(CSCW2), 1–26.

Kock, N. (2004). The Psychobiological Model: Towards a New Theory of Computer-Mediated Communication Based on Darwinian Evolution. *Organization Science, 15*(3), 327–348

Kock, N. (2005). Media Richness or Media Naturalness? The Evolution of Our Biolcgical Communication Apparatus and Its Influence on Our Behavior Toward E-Communication Tools. *IEEE Transactions on Professional Communication, 48*(2), 117–130

Kock, N. (2012). Media naturalness theory: Human evolution and behaviour towards electronic communication technologies. In S. C. Roberts (Ed.), *Applied evolutionary psychology*. Oxford University Press.

Kraus, M. W., Piff, P. K., Mendoza-Denton, R., Rheinschmidt, M. L., & Keltner, D. (2012). Social class, solipsism, and contextualism: How the rich are different from the poor. *Psychological Review, 119*(3), 546–572.

Krendl, A., Richeson, J. A., Kelley, W., & Heatherton, T. F. (2008). The negative consequences of threat: An fMRI investigation of the neural mechanisms underlying women's underperformance in math. *Psychological Science, 19*, 168–175.

Li, N. P., van Vugt, M., & Colarelli, S. M. (2017). The evolutionary mismatch hypothesis: Implications for psychological science. *Current Directions in Psychological Science, 27*(1), 38–44.

Libet, B. (2002). The timing of mental events: Libet's experimental findings and their implications. *Consciousness and Cognition, 11*, 291–299.

Libet, B., Pearl, D. K., Morledge, D. E., Gleason, C. A., Hosobuchi, Y., & Barbaro, N. M. (1991). Control of the transition from sensory detection to sensory awareness in man by the duration of a thalamic stimulus: The cerebral 'time-on' factor. *Brain, 114*, 1731–1757.

Liu, H. (2018). Social and genetic pathways in multigenerational transmission of educational attainment. *American Sociological Review, 83*(2), 1–27.

Liu, Z. (2021). Sociological perspectives on artificial intelligence: A typological reading. *Sociology Compass, 15*, e12851.

Lloyd, E. A., Wilson, D. S., & Sober, E. (2011). *Evolutionary mismatch and what to do about it: A basic tutorial*. https://evolution-institute.org/wp-content/uploads/2015/08/Mismatch-Sept-24-2011.pdf

Luhmann, N. (2012–2013). *Theory of society* (Vol. 1, p. 2). Stanford University Press.

Mars, R. B., Neubert, F. X., Noonan, M. P., Sallet, J., Toni, I., & Rushworth, M. F. (2012). On the relationship between the "default mode network" and the "social brain". *Frontiers in Human Neuroscience, 6*, 189.

McCall, C., & Singer, T. (2013). Empathy and the brain. In S. Baron-Cohen, M. Lombardo, & H. Tager-Flusberg (Eds.), *Understanding other minds: Perspectives from developmental social neuroscience*. Oxford University Press.

McCarthy, J. (2007). *What is artificial intelligence?* Retrieved November 4, 2023, from http://jmc.stanford.edu/articles/whatisai/whatisai.pdf

McCarthy J., Minsky M., Rochester N., and Shannon C. (1955). A proposal for the Dartmouth summer research project.. Retrieved November 4, 2023, from http://jmc.stanford.edu/articles/dartmouth/dartmouth.pdf

McTear, M. (2021). *Conversational AI: Dialogue systems, conversational agents, and chatbots*. Morgan & Claypool Publishers.

Melloni, L., Molina, C., Pena, M., Torres, D., Singer, W., & Rodriguez, E. (2007). Synchronization of neural activity across cortical areas correlates with conscious perception. *Journal of Neuroscience, 27*(11), 2858–2865.

Meloni, M., Cromby, J., Fitzgerald, D., & Lloyd, S. (2018). Introducing the new biosocial landscape. In M. Meloni, J. Cromby, D. Fitzgerald, & S. Lloyd (Eds.), *The Palgrave handbook of biology and society*. Palgrave Macmillan.

Menon, V. (2023). 20 years of the default mode network: A review and synthesis. *Neuron, 111*(16), 2469–2487.

Mesoudi, A. (2016). Cultural evolution: A review of theory, findings and controversies. *Evolutionary Biology, 43*, 481–497.

Montag, C., & Davis, K. L. (2018). Affective Neuroscience Theory and Personality: An Update. *Abstract Personality Neuroscience, 1, e12*

Montag C, Panksepp J. (2017) Primary Emotional Systems and Personality: An Evolutionary Perspective. *Frontiers in Psychology 11*(8), 464. https://doi.org/10.3389/fpsyg.2017.00464

Montag, C., & Panksepp, J. (2020). Personality neuroscience: Why it is of importance to consider primary emotional systems! In V. Zeigler-Hill & T. K. Shackelford (Eds.), *Encyclopedia of personality and individual differences*. Springer. https://doi.org/10.1007/978-3-319-24612-3_1338

Muthukrishna, M., Doebeli, M., Chudek, M., & Henrich, J. (2018). The cultural brain hypothesis: How culture drives brain expansion, sociality, and life history. *PLoS Computational Biology, 14*(11), e1006504.

Nass, C., & Brave, S. (2005). *Wired for speech: How voice activates and advances the human-computer relationship*. Boston Review.

Nass, ?., & Moon, Y. (2000). Machines and mindlessness: Social responses to computers. *Journal of Social Issues, 56*, 81–103.

Nass, C., Moon, Y., & Green, N. (1997). Are machines gender neutral? Gender stereotypic responses to computers with voices. *Journal of Applied Social Psychology, 27*, 864–876.

Nass, C. I., Steuer, J., & Tauber, E. R. (1994). Computers are social actors. In *Proceedings of the SIGCHI conference on human factors in computing systems* (pp. 72–78). Association for Computing Machinery.

Nielsen, Y. A., Pfattheicher, S., & Keijsers, M. (2022). Prosocial behavior toward machines. *Current Opinion in Psychology, 43*, 260–265.

Nitecki, M. N., & Nitecki, D. V. (Eds.). (1987). *The evolution of human hunting*. Plenum Press.

Noble, K. G., Houston, S. M., Kan, E., & Sowell, E. R. (2012). Neural correlates of socioeconomic status in the developing human brain. *Developmental Science, 15*(4), 516–527.

OpenAI. (2023). *GPT-4 Technical Report*. Retrieved April 23, 2024, from https://cdn.openai.com/papers/gpt-4.pdf

Panksepp, J. (1998). *Affective neuroscience: The foundations of human and animal emotions*. Oxford University Press.

Panksepp, J., & Biven, L. (2012). *The archeology of mind*. W.W. Norton.

Perry, A., & Shamay-Tsoory, S. (2013). Understanding emotional and cognitive empathy: A neuropsychological perspective. In S. Baron-Cohen, M. Lombardo, & H. Tager-Flusberg (Eds.), *Understanding other minds: Perspectives from developmental social neuroscience*. Oxford University Press.

Piff, P. K., Stancato, D. M., Cote, S., Mendoza-Denton, R., & Keltner, D. (2012). Higher social class predicts increased unethical behavior. *Proceedings of the National Academy of Sciences of the United States of America, 109*(11), 4086–4091.

Pinker, S. (1994/2007). *The language instinct*. Harper Perennial Modern Classics.

Placani, A. (2024). Anthropomorphism in AI: Hype and fallacy. *AI and Ethics*. https://doi.org/10.1007/s43681-024-00419-4

Plutchik, R. (1980). A general psychoevolutionary theory of emotion. In R. Plutchik & H. Kellerman (Eds.), *Emotion: Theory, research, and experience* (Theories of emotion) (Vol. 1, pp. 3–33). Academic Press.

Pollack, L., Munson, A., Savoca, M. S., Trimmer, P. C., Ehlman, S. M., Gil, M. A., & Sih, A. (2022). Enhancing the ecological realism of evolutionary mismatch theory. *Trends in Ecology & Evolution, 37*(3), 233–245.

Qu, C., Ligneul, R., Van der Henst, J.-B., & Dreher, J.-C. (2017). An integrative interdisciplinary perspective on social dominance hierarchies. *Trends in Cognitive Sciences, 21*(11), 893–908.

Reeves, B., & Nass, C. I. (1996). The media equation: How people treat computers, television, and new media like real people and places. Center for the Study of Language and Information; Cambridge University Press

Richerson, P. J., & Christiansen, M. H. (Eds.). (2013). *Cultural evolution: Society, technology, language, and religion*. MIT Press.

Rizzolatti, G., & Singigalia, C. (2008). *Mirrors in the brain: How our minds share actions and emotions*. Oxford University Press.

Robertson, B. A., Rehage, J. S., & Sih, A. (2013). Ecological novelty and the emergence of evolutionary traps. *Trends in Ecology & Evolution, 28*, 552–560.

Rock, D. (2008). SCARF: A brain-based model for collaborating with and influencing others. *NeuroLeadership Journal, 1*(1), 44–52.

Rock, D., & Cox, C. (2012). SCARF in 2012: Updating the social neuroscience of collaborating with others. *NeuroLeadership Journal, 4*(4), 1–16.

Rubin, B. P. (2009). Changing brains: The emergence of the field of adult neurogenesis. *BioSocieties, 4*, 407–424.

Russel, S. (2019). *Human compatible: AI and the problem of control*. Penguin Books.

Russel, S., & Norvig, P. (2021). *Artificial intelligence. A Modern Approach*. Global Edition, Pearson.

Sadler, P., Ethier, N., & Woody, E. (2011). Interpersonal complementarity. In L. M. Horowitz & S. Strack (Eds.), *Handbook of interpersonal psychology: Theory, research, assessment, and therapeutic interventions* (pp. 123–142). Wiley.

Scheele, D., Schwering, C., Elison, J. T., Spunt, R., Maier, W., & Hurlemann, R. (2015). A human tendency to anthropomorphize is enhanced by oxytocin. *European Neuropsychopharmacology, 25*(10), 1817–1823. https://doi.org/10.1016/j.euroneuro.2015.05.009

Schlaepfer, M. A., Runge, M. C., & Sherman, P. W. (2002). Ecological and evolutionary traps. *Trends in Ecology & Evolution, 17*(10), 474–480. https://doi.org/10.1016/s0169-5347(02)02580-6

Shneiderman, B. (2020). Human-centered artificial intelligence: Three fresh ideas. *AIS Transactions on Human Computer Interaction, 12*(3), 109–124.

Shum, H.-Y., He, X.-d., & Li, D. (2018). From Eliza to XiaoIce: Challenges and opportunities with social chatbots. *Frontiers of Information Technology & Electronic Engineering, 19*(1), 10–26.

Soon, C. S., Brass, M., Heinze, H.-J., & Haynes, J.-D. (2008). Unconscious determinants of free decisions in the human brain. *Nature Neuroscience, 11*(5), 543–545.

Sundar, S. S., & Lee, E.-J. (2022). Rethinking communication in the era of artificial intelligence. *Human Communication Research, 48*(3), 379–385.

TenHouten, W. D. (2017a). *Alienation and affect*. Routledge.

TenHouten, W. (2017b). From primary emotions to the spectrum of affect: An evolutionary neurosociology of the emotions. In A. Ibáñez et al. (Eds.), *Neuroscience and social science*. Springer.

TenHouten, W., Schussel, L., Gritsch, M. F., & Kaplan, C. D. (2022). Hyperscanning and the future of neurosociology. *Sociological Methodology, 53*(1), 139–157.

TenHouten, W. D. (2018). Anger and Contested Place in the Social World. *Sociology Mind, 08*(03), 226–248

Tipper, C. M., Signorini, G., & Grafton, S. T. (2015). Body language in the brain: Constructing meaning from expressive movement. *Frontiers in Human Neuroscience, 9*, 450.

Tschopp, M., Gieselmann, M., & Sassenberg, K. (2023). Servant by default? How humans perceive their relationship with conversational AI. *Cyberpsychology: Journal of Psychosocial Research on Cyberspace, 17*(3), 9.

Turing, A. (1950). Computing machinery and intelligence. *Mind, 59*, 433–460.

Turner, J. H. (2014). The evolution of human emotion. In J. E. Stets & J. H. Turner (Eds.), *Handbook of the sociology of emotions* (Handbooks of sociology and social research.) (Vol. II, pp. 11–31). Springer Science + Business Media.

Turner, J. (2021). *On human nature. the biology and sociology of what made us human*. Routledge.

Turner, J. H., & Machalek, R. S. (2018). *The new evolutionary sociology. Recent and revitalized theoretical and methodological approaches*. Routledge.

Turner, J. H., & Maryanski, A. (2013). The evolution of the neurological basis of human sociality. In D. D. Franks & J. H. Turner (Eds.), *Handbook of neurosociology*. Springer Science + Business Media B.V.).

Turner, J. H., Schutt, R. K., & Keshavan, M. S. (2020). Biology and American Sociology, Part II: Developing a unique evolutionary sociology. *The American Sociologist, 51*, 470–505.

van Kleef, G. A., Oveis, C., van der Löwe, I., LuoKogan, A., Goetz, J., & Keltner, D. (2008). Power, distress, and compassion: Turning a blind eye to the suffering of others. *Psychological Science, 19*, 1315–1322.

Van Vugt, M., & Ronay, R. (2013). The evolutionary psychology of leadership: Theory, review, and roadmap. *Organizational Psychology Review, 4*(1), 74–95.

Vlahovic, T. A., Roberts, S., & Dunbar, R. (2012). Effects of duration and laughter on subjective happiness within different modes of communication. *Journal of Computer-Mediated Communication, 17*, 436–445.

Waytz, A., Cacioppo, J., & Epley, N. (2010). Who sees human? The stability and importance of individual differences in anthropomorphism. *Perspectives on Psychological Science, 5*(3), 219–232.

Waytz, A., Cacioppo, J. T., Hurlemann, R., Castelli, F., Adolphs, R., & Paul, L. K. (2019). Anthropomorphizing without social cues requires the basolateral amygdala. *Journal of Cognitive Neuroscience, 31*(4), 482–496.

Wilson, D. S. (2019). *This view of life: Completing the Darwinian revolution*. Random House Digital; Pantheon Books.

Wilson, D. S., Hayes, S. C., Biglan, A., & Embry, D. D. (2014). Evolving the future: Toward a science of intentional change. *The Behavioral and Brain Sciences, 37*(4), 395–416.

Xu, K., & Liao, T. (2020). Explicating cues: A typology for understanding emerging media technologies. *Journal of Computer-Mediated Communication, 25*(1), 32–43.

Yaple, Z. A., & Yu, R. (2019, May 2). Functional and structural brain correlates of socioeconomic status. *Cerebral Cortex*. pii: bhz080. https://doi.org/10.1093/cercor/bhz080

Zhou, Y., Fei, Z., He, Y., & Yang, Z. (2022). How human–chatbot interaction impairs charitable giving: The role of moral judgment. *Journal of Business Ethics, 178*, 849–865.

Zhou, L., Gao, J., Li, D., & Shum, H.-Y. (2020). The design and implementation of XiaoIce, an empathetic social chatbot. *Computational Linguistics, 46*(1), 53–93.

Persistent Inequality: A Neurosociological Perspective

22

Jeff Davis

Abstract

The goal of this chapter is to discuss how sociological and neuroscience research can be fruitfully synthesized to shed significant light on one the most complex problems in sociology – persistent inequality. I present a conceptual model which argues (1) persistent inequality is the result of short-sightedness in decision-making by individuals and/or collectives, as shown in sociological research; and (2) short-sightedness, in turn, is induced by uncertainty about the future. Neuroscience research shows how uncertainty can trigger neuronal processes which lead to short-sighted decisions and behaviors over time. Thus, research on uncertainty can serve as a foundational program for the development of neurosociology.

Introduction

In light of the significant advances in social neuroscience, it is past the time for sociologists to consider the brain as an important sociological phenomenon. A sizeable body of social neuroscience research illuminates the sensitivity of neurobiological processes to the social environment (Cacioppo and Patrick 2008; Insel and Fernald 2004; McEwen and Gianaros 2010). Comparative analyses of human and non-human species suggest nervous systems evolved under strong social selection (Robinson et al. 2008). Complex sociological dynamics such as rituals (Graybiel 2008), location in an organizational hierarchy (Creel and Sands 2003), opportunities to increase relative social standing (Burmeister et al. 2005), social exclusion and ostracism (Eisenberger et al. 2003; Masten et al. 2009; Slavich et al. 2010), the learning and transmission of cultural information (Hakeem et al. 2009), morality and distributive justice (Fehr and Camerer 2007; Greene and Paxton 2009; Koenigs et al. 2007) have left large imprints on the evolutionary path of the human brain (Kenrick et al. 2009).

This chapter advances a neurosociological model of a highly complex sociological process – persistent inequality. Numerous studies of persistent inequality have been compiled. As noted by many ethnographers, individuals and families who tend to get trapped in lower socioeconomic status exhibit preferences and behaviors which value short-term outcomes. Short-sightedness inexorably leads to behaviors, such as high sexual activity and aggression, which lock an individual and their offspring into lower socioeconomic status (Schwalbe et al. 2000). The central thesis of the model states that persistent lower status is ultimately a result of a high degree

J. Davis (✉)
Department of sociology, California State University, Long Beach, CA, USA
e-mail: jeffrey.davis@csulb.edu

W. Kalkhoff et al. (eds.), *Handbook of Neurosociology*, Handbooks of Sociology and Social Research, https://doi.org/10.1007/978-3-031-95615-7_22

of environmental uncertainty induced by subjugated social status. Almost by definition, members of a subjugated social status experience environmental conditions which preclude the exercise of control. In many ways, the environment is comprised of harsh, unpredictable stressors such as financial insecurity, episodic violence, instability in social relationships, and risk of acute stress-related illnesses. Perhaps most unpredictable is one's chances of upward social and economic mobility. As I argue, environmental uncertainty has significant consequences for three areas of the brain which are directly involved in the behaviors observed by ethnographers – the prefrontal cortex (PFC), the anterior cingulate cortex (ACC), and the vertebrate social behavior network (VSBN). Activities in these three areas are delicately balanced in order to regulate decision-making and behavior. Exposure to environmental uncertainty disrupts the regulatory process resulting in a strengthening of short-term preferences and increases in reproductive and aggressive behaviors.

Mechanisms of Persistent Inequality

Figure 22.1 shows a basic model of inequality developed by Tilly (1998) and elaborated throughout this chapter. Inequality is organized around social categories which are used to sort individuals and to make decisions about the allocation of material and cultural resources. Categorical inequality refers to the unequal allocation of resources across social categories, primarily as the result of systemic discrimination and social closure within local social and economic organizations. Tilly (1998) identified four major mechanisms which embed categorical inequality into the fabric of a society: exploitation, opportunity hoarding, emulation, and adaptation. Figure 22.1 shows only the unique contributions of each mechanism to the persistence of categorical inequality, but there exists a dynamic interdependency among them as well.

Exploitation is "…where powerful, connected people command resources from which they draw significantly increased returns by coordinating efforts of outsiders whom they exclude from the full value added of that effort" (Tilly 1998, p. 74). Opportunity hoarding is the control and confinement of a valuable resource (or a value producing resource) to members of a category. Both mechanisms originate in the local cultural and institutional environment. Cultural norms about relative social standing and the formal rules sustaining local social hierarchies shape how individuals view others. These views are often carried into social and economic organizations and influence decision-making processes within them.

Consequently, categorical inequality becomes manifest in a variety of organizational processes such as discrimination in hiring and promotion. The highest degrees of categorical inequality are found where the internal structures of social and economic organizations are most reliant upon the categorization schemes endorsed by local culture and institutions.

Exploitation and opportunity hoarding are the primary, proximate causes of categorical inequality. However, the intensity of their effects on the installation of categorical inequality is a result of the two mechanisms of emulation and adaptation. According to Tilly (1998), emulation generalizes

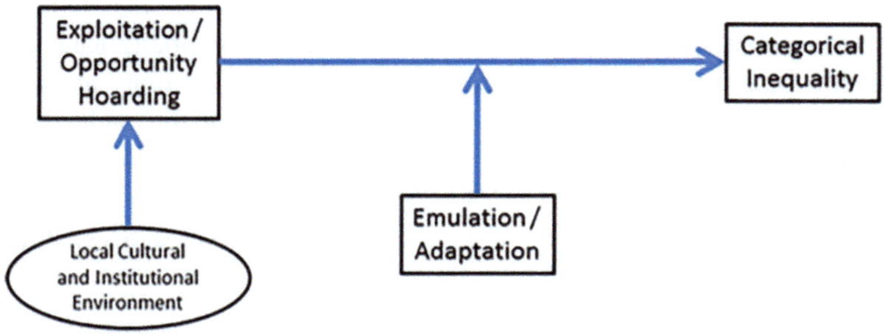

Fig. 22.1 Tilly's (1998) basic model of durable inequality

categorical inequality across individuals and organizations within a society. Menchaca (1995) provides an excellent example of emulation in her study of the development of Mexican-Anglo relations in southern California since the Mexican-American War. As Anglos expropriated land from Mexican families during the war period, many turned to wealthiest of Anglo landowners for advice on the most effective methods for controlling Mexican laborers. Consequently, the exploitative forms of labor control used by large landowners were readily copied by small landowners (Menchaca 1995). The spread of exploitation practices was one of the major factors in emergence and persistence of what Menchaca (1995) refers to as "social apartness" – Mexican-Anglo segregation in southern California.

Whereas emulation causes a system of categorical inequality to spread, adaptation locks categorical inequality into the relationships between groups. It is broadly defined as the "…invention of procedures which ease day-to-day interaction" (Tilly 1998, 2005). However, these solutions have some undesirable, unintended consequences for the persistence of inequality. Consider some classic ethnographic studies of youth growing up in conditions of poverty. Sociologists frequently observe among these youth a pessimistic view of their chances of mainstream success (Anderson 1978, 1999; MacLeod 1995; Whyte 1943; Willis 1977). Their pessimism derived from a realistic assessment of a long family history of struggles and the poor quality of the local social and economic environment. One group of youth in MacLeod's (1995) study, for example, concluded that striving for upward mobility was a futile effort because several family members had previously failed in their efforts. Moreover, good jobs were simply scarce. In response to objectively poor environments, young adults engage in "dropping out behavior" (Schwalbe et al. 2000), withdrawing from activities prescribed by normative society such as school and the mainstream labor market. From the individual's perspective, dropping out might lead to a number of positive outcomes such as closer social ties with peers, higher self-esteem, and a general perception that mainstream life is relatively unimportant for well-being (Schwalbe et al. 2000). Unfortunately, these are most often short-term benefits with the costs of steep, negative trade-offs for future well-being (Day et al. 1999).

A Neurosociological Model of Persistent Inequality

A neurosociological perspective on Tilly's model is not new. Massey (2007) was probably the first to do so. He argued that while categorical inequality is fundamentally the result of an interaction between institutional environments and market conditions, categorization is generally an expression of the universal tendency of humans to use heuristics as a means of understanding environments (Massey 2007). The capacity for categorization evolved in response to energetic constraints (Massey 2007), which have actually shaped numerous aspects of brain functioning (Niven and Laughlin 2008). Categorization of the social and natural world is a method of conserving energy for cognitive processing of stimuli (Massey 2007). In some important ways, the use of heuristics can yield an accurate view of the social and physical environment and greatly aid decision-making (Gigerenzer and Brighton 2009). However, it can also easily lead to maladaptive outcomes – such as prejudice. Whether it actually leads to prejudice is mostly a function of conditions of the institutional environment of the individual (Massey 2007).

Whereas Massey (2007) addressed the most central concept in Tilly's model, my focus is on the two mechanisms of emulation and adaptation. Figure 22.2 illustrates the model. The main consequence of categorical inequality is the experience of greater environmental uncertainty in the lives of subjugated groups. For these groups, uncertainty arises in everyday decisions and social interactions. The poor quality of local job markets creates unpredictable changes in employment opportunities, instability and distrust among coworkers, and vulnerability to abusive employers (Bourdieu and Accardo 1999). In turn, economic vulnerability creates distrust and

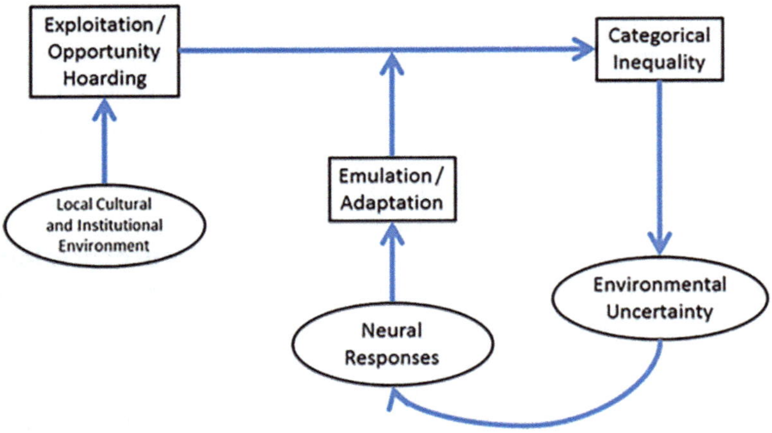

Fig. 22.2 A Neurosociological model of persistent inequality. The primary effect of inequality is to increase an individual's uncertainty about the current and future socioeconomic environments. In turn, uncertainty triggers neural processes associated with short-term decision making behaviors

uncertainty in social interaction (Bourdieu and Accardo 1999). Simple social exchanges within such environments have sensitive tipping points, turning either friendly or hostile in a somewhat chaotic manner. For example, whether an interaction turns friendly or hostile often depends upon the individuals' ability to correctly interpret highly nuanced signals (e.g., a nod of the head, a glance in one's direction) (Anderson 1990; Gambetta and Hamill 2005; O'Donnell and Sharpe 2000). The wrong interpretation could result in disastrous consequences.

Efforts to overcome the social vulnerability and unpredictability associated with lower social status often involve high-risk behaviors. For young men, both issues are typically resolved by expending the only viable resource they have – physical talent – in order to build a local reputation. Unfortunately, this involves frequent challenges from other young men, often in the form of violent confrontations (Anderson 1999; Barker 2005; Bourgois 2002; Carroll et al. 2003; Emler and Reicher 1995; O'Donnell and Sharpe 2000). For young women, child birth is sometimes perceived as a means of bringing stability and purpose to an otherwise chaotic existence (Edin and Kefelas 2005). Social support from peers can increase as well as social standing (Anderson 1999). However, young unwed mothers risk losing steady support from family members. It seems more likely that the mother becomes stigmatized for having a child out-of-wedlock (Edin and Lein 1997). For both men and women, vulnerability and uncertainty in social interactions can be reduced copying the behaviors of others who seem least vulnerable in the environment. One can copy behaviors prescribed by local codes of conduct and imitate vernacular in order to "fit in" and find a degree of comfortable acceptance from others (Anderson 1999). However, copying the behaviors of those who are perceived to be least vulnerable might also mean developing a closer relationship with peers who routinely engage in high-risk behaviors. Stronger ties with such peers will ensure low social mobility over the long run (Whyte 1943; Willis 1977).

Neuroscientists have observed that life in an unpredictable environment impacts neural functioning much like any other significant form of stress (Arnsten 2009; Munhoz et al. 2006; Rosen and Donley 2006; Wingfield and Kitaysky 2002) (McEwen and Gianaros 2010). There is also evidence suggesting that living in a poor quality environment can trigger neural activity associated with stronger preferences for short-term rewards, attention to the immediate environment, neglect of long-term costs of actions, and higher levels of aggressive behaviors (Manuck et al. 2003, 2005). Behavioral studies have linked environmental uncertainty to a stronger tendency toward imitation and social learning (Ball-Rokeach 1973; Coussi-Korbel and Fragaszy 1995; LaLand 2004; Salancik and Pfeffer 1978). Exposure to unpredictable environments, especially during childhood, has also been linked with a higher risk of teenage parenthood and higher fertility over the life span (Belsky et al. 2010; Brumbach et al. 2009; Chisholm 1999; Ellis et al. 2009), but many questions remain about the underlying neurobiological mechanisms (Maeda and Tsukamura 2006; Matsuwaki et al. 2003).

The model in Fig. 22.2 proposes that environmental uncertainty initiates what neuroscientists refer to as a "bottom-up" behavioral control process (Arnsten 2009; Yu and Dayan 2002). In a non-stress situation, behaviors are primarily regulated by a "top-down" process where higher-level cognitive control functions inhibit behavioral impulses from the affective centers of the brain (e.g., the limbic system). However, exposure to stressors such as uncertainty tends to increase activity in the limbic system and simultaneously impairs higher-level cognitive control (Arnsten 2009; Yu and Dayan 2005).

The Predictive Brain

To understand the stressful impact of uncertainty it is first necessary to understand the evolutionary value of predictions about the future. Current theory holds that prediction (or predictive learning) evolved because it conferred distinct speed advantages in the location of vital energy resources and predator avoidance (Shettleworth 1998; Watson 2008). Prediction is an act of associative learning where two or more events are temporally and/or spatially distant (Shettleworth 1998). It requires a means of transduction of environmental stimuli and a system of memory. Evolution has apparently found diverse ways to develop these traits. While vertebrates depend upon various elements of cognition to make predictions, simple organisms have evolved non-cognitive means of predictive learning. In experimental studies, researchers were able to select for associative learning traits in a population of Escherichia coli (E. coli). Subsequent generations evolved the capacity to accurately predict the location of food resources (Tagkopoulos et al. 2008). Researchers have also found that predictive capacity develops early in some vertebrate species. Species of frogs, for example, are to detect environmental signals associated with future risk of predation while still in the embryonic stages of development (Warkentin and Caldwell 2009). These studies and others demonstrate that organisms with the ability to accurately predict the temporal distribution of energy resources and predation risk have a clear advantage in survival.

Predictions about the future are also used in reproductive decisions. In various species, the timing of sexual maturation and subsequent reproduction are highly sensitive to an organism's perceptions about future environmental conditions (Davis and Werre 2008). Here also, the organism relies upon environmental cues to make predictions about the future and acts accordingly. If the organism predicts higher quality environmental conditions in the future, reproduction is delayed until that time when conditions improve. If the future is perceived to be in a worse condition than the present, then sexual development is accelerated and reproductive effort is generally increased (Wasser and Place 2001).

Empirical evidence for this pattern of behavior abounds for non-human species. For example, there is a large literature documenting some rather sophisticated methods of reproductive control, depending upon current and likely future states of the environment. One method is diapause, a neuro-physiological mechanism which allows an organism to suspend fertilization or embryonic growth until environmental conditions improve (Renfree and Shaw 2000). Evidence from studies of human reproductive decision-making and behavior is growing and the patterns resemble those found among other species. Collectively, these studies suggest that predictions about the future influence human reproductive biology. Wasser and Place (2001), for example, argued that the close association between the quality of local social environments and reproductive outcomes implies a process they refer to as reproductive filtering. The reproductive filtering hypothesis states that reproduction will be suppressed in a stressful environment if the individual perceives improvements in future conditions (Wasser and Place 2001). Their analyses of women undergoing fertility treatment shows that treatment of women in stressful social conditions (e.g., poor quality of social support, hostility in social relationships) was less successful for women who believed that conditions would improve in the future (Wasser and Place 2001). Longitudinal studies have also found higher levels of sexual and reproductive activity among women who perceive no improvement or worse environmental conditions in the future (Belsky et al. 1991; Brumbach et al. 2009; Chisholm et al. 2005; Ellis et al. 2009).

Lastly, prediction plays a central role in facilitating an organism's efforts to modify the environment – an important activity which can significantly enhance long-term survivability and reproductive success (Odling-Smee et al. 2003). Activities related to environmental modification are referred to as niche construction (Odling-Smee et al. 2003). The ultimate goal of niche construction is to exercise control over fluctuations in the local environment. Specifically, niche construction stabilizes the flow of local energy resources necessary for survival and reproduction. It can effectively dampen stochastic fluctuations in resource levels occurring over time. Predictive learning allows an organism to detect and evaluate environmental cues which indicate the how the quality of the local environment might change over time. In a very real sense, predictive learning helps the organism efficiently allocate efforts such that it avoids wasting precious time and energy on modifying an environment that will yield low returns in the future (Odling-Smee et al. 2003).

Observations from behavioral studies lead to the conclusion that prediction is a major mechanism in the regulation of behavior and the accuracy of a prediction can have significant consequences for the fitness of the organism and its offspring. A behavior with such adaptive importance is bound to have many biological traces of strong selection. According to a large body of neuroscience research, the human brain is aptly equipped for predictive learning (Bubic et al. 2010) and is constantly making predictions to construct representations of the future (Bar 2009). Much of this activity takes place within the prefrontal cortex (Bar 2009; Bechara et al. 1999), although other structures also play crucial roles such as the anterior cingulate cortex (Allman et al. 2001; Walton et al. 2007), the hippocampus (Eichenbaum and Fortin 2009) and the cerebellum (O'Reilly et al. 2008; Szpunar et al. 2007). Working together, various components keep the individual in a continuous state of anticipation and readiness to adapt to environmental challenges (Bar 2009). For my purposes, I focus on the interactions between three areas: the dorsomedial prefrontal cortex (dmPFC), the anterior cingulate cortex (ACC), and the vertebrate social behavior network (VSBN). Figure 22.3 shows the locations and some of the sub-components of each structure.

The prefrontal cortex (PFC) is that part of the brain comprising the anterior area of the frontal cortex. The PFC enables working memory – the process of storing information temporarily, retrieving information from long-term memory, manipulating it and then selectively discarding information deemed irrelevant to the task at hand (Bledowski et al. 2009). The PFC is also responsible for the selection of appropriate social behaviors (Arnsten 2009; Kouneiher et al. 2009) and general decision-making (Fellows and Farah 2007; Osaka et al. 2007; Rougier et al. 2005; Wallis 2007). Substrates of the PFC have distinct functions for decision-making and behavior. Many of these functions have been uncovered by studies of behavior among individuals who have suffered damage to specific parts of the brain. Of all substrates, the dorsomedial PFC (dmPFC) seems to play the most significant role in forming predictions, assessing prediction errors, and resolving conflicts in information (Modirrousta and Fellows 2008; Rushworth et al. 2007). The dmPFC reacts particularly strong to environmental cues which have reliably signaled a future reward in past experiences (Ishikawa et al. 2008).

The anterior cingulate cortex (ACC) has numerous roles in cognition, behavior, and physiological functioning (Allman et al. 2001; Devinsky et al. 1995). Neuropsychological research on phenomena such as cognitive dissonance reveals that the ACC plays a crucial role in the detection and monitoring of cognitive conflict (Carter and van Veen 2007; van Veen and Carter

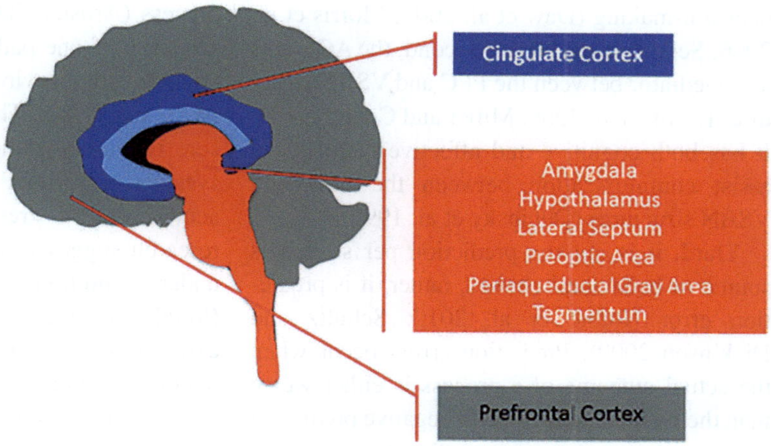

Fig. 22.3 Neuroanatomy of the prefrontal cortex, cingulate cortex, and structures of the vertebrate social behavior network

2006; van Veen et al. 2001). Consequently, researchers have also found that the dorsal ACC (dACC) is also directly involved in attitude changes intended to rapidly reduce cognitive conflict (van Veen et al. 2009). There seems to be broad agreement that ACC seems to play an important role in adaptive decision-making – that is, decision-making under uncertainty (Allman et al. 2001; Kennerley et al. 2006). However, the exact nature of the role is unclear. Recent computational models have proposed that the dACC is involved in the assessment of prediction error. Specifically, it facilitates the estimation of the likely degree of error in predictions (Brown and Braver 2007). However, other research casts doubt on this role. Kennerley et al. (2006) suggest the actual role of the dACC is to weigh the opportunity costs of a future action against the history of previous rewards for the action and alternative actions.

The vertebrate social behavior network (VSBN) is composed of several structures distributed across two broad areas – the basal forebrain and the midbrain (Goodson 2005; Newman 1999). The specific areas are as follows: (within the basal forebrain) extended medial amygdala; preoptic area; anterior hypothalamus; ventromedial hypothalamus; lateral septum; (within the midbrain) periaqueductal gray area; and tegmentum (Goodson 2005). Structures within the basal forebrain have long been associated with socio-emotional experiences (Morgane et al. 2005). In mammals these structures have been associated with a range of behaviors related to reproduction and aggression. For example, the lateral septum is a central structure in the regulation of aggressive behavior. Depending upon the level of certain hormones, the lateral septum directs aggressive behavior toward others who pose a danger to self or offspring (Lee and Gammie 2009; Singewald et al. 2011). Recent work is illuminating the role of the VSBN in prediction and decision-making. For example, the midbrain is the origin of dopaminergic neurons (Iversen 2010). Activity in these neurons increases significantly in response to the expectation of a reward (Schultz et al. 2002). Some research has also found increased activity in response to uncertainty about the probability of receiving a reward (Fiorillo et al. 2003).

The PFC, ACC, and VSBN are interconnected through numerous projections of neural networks (Salzman and Fusi 2010). The projections are complex in terms of number of neurons involved, the size of neural networks, the diversity of hormones transmitted, the role of particular dendritic receptors, and the actions of secondary chemical messengers such as cyclic adenosine monophosphate (cAMP). A full review of projections is well-beyond the scope of this chapter and the summary offered here is narrowly focused on the neuroscience of decision-making. These limitations notwithstanding, three general conclusions can be drawn from the literature. First, dopamine is one of the major neurotransmitters involved in prediction and decision-making (Daw et al. 2006; Morris et al. 2006; Schultz et al. 2002). Second, the ACC acts as a mediator between the PFC and VSBN structures (Etkin et al. 2006; Miller and Cohen 2001). It has both cognitive and affective elements to assist communication between the PFC and VSBN structures (Devinsky et al. 1995).

Third, it is not the prediction per se that is source of behavioral change, rather, it is prediction error (Bubic et al. 2010; Schultz and Dickinson 2000). Prediction errors occur when the actual outcome of a process is either worse than the predicted outcome (a negative prediction error) or better than the predicted outcome (a positive prediction error)(Dayan and Abbott 2001; Montague et al. 1996; Niv et al. 2005). A prediction error is encoded in the brain via changes in the levels of dopamine in the reward centers of the brain (e.g., the midbrain) and in substrates of the frontal cortex (e.g., the orbitofrontal cortex, ventromedial prefrontal cortex) (Bayer and Glimcher 2005). When an individual experiences a negative prediction error, dopamine levels drop below normal baseline levels in the midbrain and prefrontal cortex. This will ultimately lead to the discontinuation (or extinction) of a behavioral pattern (Schultz and Dickinson 2000). A positive prediction error results in an increase in dopamine levels above baseline and has the effect of reinforcing a behavioral pattern (Schultz and Dickinson 2000). It should also be noted that a spike in dopamine above baseline levels enhances attention to the immediate environment, almost to the complete neglect of distal environments (Hills 2006).

Brain and Behavior Under Environmental Uncertainty

In a variable but predictable environment, decision-making is a "top-down" process. Behavior is rule-based and the influence of autonomic responses to sensory stimuli is mitigated (Miller and Cohen 2001). Neuronal activities in the PFC and ACC regulate behavioral inputs from the VSBN. Environmental uncertainty, because it is a stressful condition, inverts this process (Arnsten 2009). Figure 22.4 is a representation of one pathway by which exposure to unpredictable environments impacts the PFC, ACC, and VSBN. The model assumes that exposure is chronic rather than acute. The nervous systems of various species can readily handle acute forms of stress (Sapolsky 2004). In fact, research suggests that nervous systems evolved under conditions of acute stress exposure (Munhoz et al. 2006; Sapolsky 2004). Prolonged exposure has detrimental effects in terms of two major consequences. First, stress-response hormones can become "fixed" at high circulating levels under chronic stress exposure (Creel and Sands 2003; Munhoz et al. 2006; Rodrigues et al. 2009). Second, prolonged exposure can set in motion a vicious cycle between stress exposure and stress response, with stress-related neuronal processes becoming increasingly sensitive to each additional exposure over time (see Cacioppo and Patrick 2008 for a detailed discussion).

Figure 22.4 shows that environmental uncertainty, in the form of unpredictable stressors and uncertain threats to well-being, triggers an increase in the activity of dopaminergic neurons and dendritic receptors for dopamine. The midbrain is the primary origin of dopaminergic neurons (Bayer and Glimcher 2005) but large populations of neurons with dopamine receptors are also located in PFC (Arnsten 2009) and ACC (Assadi et al. 2009). Under environmental uncertainty, activity of dopaminergic neurons in the midbrain increases above baseline levels (Fiorillo et al. 2003). The net effect of this is to focus attention on the immediate environment (Hills 2006). The effects of increased levels of dopamine in other areas of the VSBN produce behavioral patterns of the same commonly found in sociological studies of persistent inequality. For example, the medial preoptic area controls the release of gonadotropin, a reproductive hormone in males and females (Dominguez and Hull 2005). While exposure to stress typically lowers the circulating level of gonadotropin (Creel and Sands 2003), high levels of dopamine in the preoptic area have been linked to increased sexual activity in males (Hull et al. 2004). In fact, it seems that the medial preoptic area plays a criti-

Fig. 22.4 Model of the neural responses to environmental uncertainty

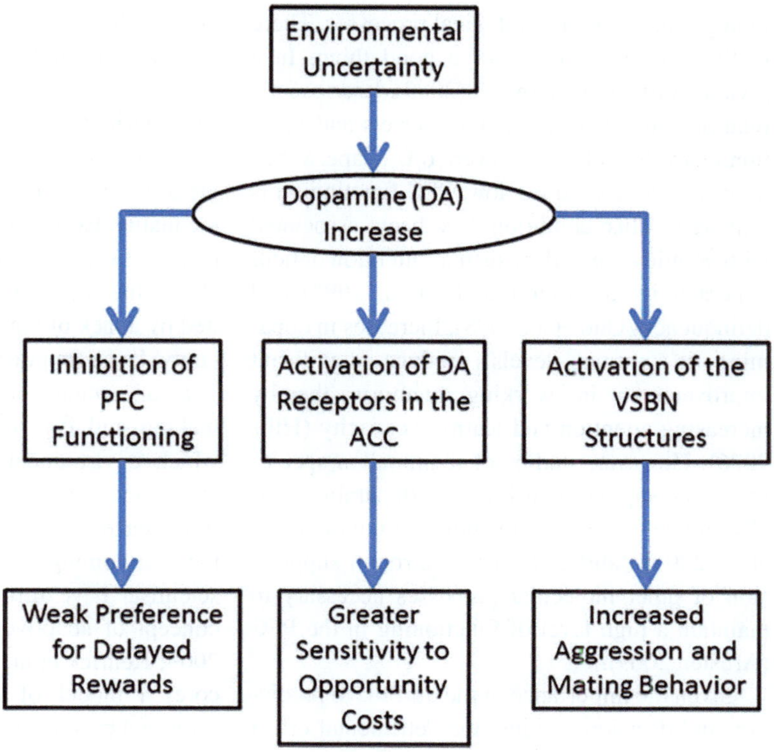

cal mediating role in the relationship between stress and release of reproductive hormones (Williamson and Viau 2008).

Researchers have also found extensive dopamine receptors, particularly D5 receptors, in the anterior hypothalamus. This is significant because the anterior hypothalamus has been associated with the control of aggression. Higher levels of dopamine in the anterior hypothalamus excite aggressive behavior in studies on mammalian species (Schwartzer and Melloni 2010). Perhaps more significant, other studies have shown that above-baseline levels of dopamine in the anterior hypothalamus are also trigger intense male-male aggression in defense of mating relationships (Gobrogge et al. 2007). This resonates strongly with the observations of researchers such as Anderson (1999), who have documented hyper-violent male-male conflicts over sexual opportunities with females within the context of deprived and unpredictable environments.

Another common observation in studies of subjugated groups is the propensity to devalue behavioral choices with delayed payoffs. Adhering to mainstream prescriptions for success, e.g. going to college, is often perceived as a having little value. Note also that the devaluation is based on observations of (1) how others failed to find any significant degree of success while pursuing mainstream goals and (2) local opportunities, which do not require advance formal training, yet still allow one to obtain a standard living comparable to everyone else in the environment (see Anderson 1999; MacLeod 1995, and Willis 1977 for examples). Perhaps this decision-making is the result of the impact of dopamine on the ACC. An increase in dopamine in the ACC generally increases the willingness of an individual to invest large amounts of effort in the pursuit of high-risk, high-value rewards (Schweimer and Hauber 2006). However, the value of the action is a function of opportunity costs, where opportunity costs rise with the temporal delay of payoffs (Niv et al. 2007). It seems that an individual's preference is stronger for actions with lower opportunity costs (Niv et al. 2007).

The effect of dopamine on the functioning of the PFC is similar to the complex effects commonly found in other biological processes. There really can be too much of a good thing. In a review of literature, Arnsten (2009) describes the relationship between dopamine levels and functioning of the PFC as an inverted U-shape. When dopamine levels are too low, PFC functioning is impaired. This condition has been associated with cognitive disorders such as attention deficit/hyperactivity disorder (Cook et al. 1995) and delinquency (Guo et al. 2008). Increases in dopamine to normal levels produce significant improvements in working memory, thereby increasing attention and learning capacity (Hills 2006). However, studies of mammalian species show that supranormal levels of dopamine actually impair working memory (Vijayraghavan et al. 2007; Zahrt et al. 1997) through suppression of other molecular processes necessary to maintain a high level of functioning in the PFC (Arnsten 2009).

Given the importance of the dmPFC to prediction and decision-making, the detrimental effect of supranormal levels of dopamine on working memory might explain the commonly observed relationship between subjugated social status and short-sighted decisions involved in adaptation to categorical inequality. Recent findings, for example, suggest that supranormal levels of dopamine can cause an individual to steeply discount the value of temporally delayed rewards (Pine et al. 2010). This is consistent with the findings on the behavioral effects of high levels of dopamine in the ACC (cited above).

Toward Neurosociology of Prediction

The origins of emulation and adaptation (as well as exploitation and opportunity) can be difficult to discern from purely structural models of persistent inequality. Structural explanations fall short of explaining the microfoundations of inequality. Without identifying mechanisms at these levels, the value of sociological analysis of inequality will be restricted to its descriptive richness. However, a neurosociological perspective can bring important mechanisms to light and, as a result, explain core problems in research on persistent inequality.

There are at least three ways neurosociology can contribute to the development of a more powerful structural theory of inequality. First, it can yield a viable theory of social action under inequality. Even now, there is no widely accepted theory of social action within sociology. Rational choice has been roundly successful but it is limited by a lack of regard for the problem of uncertainty. Its assumptions simply do not allow for it. Yet, uncertainty abounds in social decision-making and it is clearly one major vector by which environmental conditions generate dramatic, seemingly irrational changes in behavior (Gigerenzer 2008; Klein 2009). In response to the shortcomings of rational choice theory, social scientists have turned increasingly toward the concept of adaptive rationality (Haselton et al. 2009; Kenrick et al. 2009; Vanberg 2002). At its core, a model of adaptive rationality is still informed by some of the major premises of rational choice theory. However, it recognizes that behavior can be program-based as well as calculative depending upon past, current and predicted states of the environment (Vanberg 2002). Program-based behaviors are evolved, atavistic behaviors strongly favored by natural selection to enhance survivability and reproductive success (Vanberg 2002).

Theoretical models of adaptive rationality suggest that as environmental change becomes less predictable, program-based behaviors become increasingly expressed (Boyd and Richerson 2002). In unpredictable environmental conditions, calculative decision-making strategies such as trial-and-error learning can become too costly to pursue. That is, the costs of an error are likely to outweigh the benefits of the trial (Boyd and Richerson 2002). Decisions become increasingly restricted to program-based strategies such as social learning and genetically prescribed preferences (Boyd and Richerson 2002). In essence, then, to the degree that categorical inequality generates environmental uncertainty for a subjugated group the more common program-based behaviors will be among its members.

Second, a neurosociology of prediction can theorize more precisely the relationship between the environment generated by structural inequality and individual behavior. I argue that the best model for understanding the role of prediction is the threshold trait model. In its simplest form, threshold trait models predict the expression of phenotypic dimorphism in a population of individuals of the same species (Roff 1996). The model focuses on the interaction of three factors: (1) an underlying genetic liability which can express one of two phenotypes; (2) a physiological system of monitoring the environment; and (3) a change in environmental conditions. Liability is normally distributed in the population. The monitoring system is acutely attuned to the environment. It transduces environmental signals and stores that information in memory. Thus, it allows the individual to track changes in the environment and make predictions about future environmental states. The environment fluctuates between two states and the fitness of a phenotype depends upon the state of the environment (Roff 1996). The interaction between the factors can be summarized as follows. When a predictive cue from the environment is detected by the monitoring system it, in turn, triggers the expression of a phenotype in anticipation of environmental change. The phenotype expressed is the one with higher fitness in the new environment (Roff 1996).

More complex models of threshold traits account for the fact that natural selection has conserved many traits which buffer an organism's development against environmental changes (Dworkin 2005) and that phenotypes expressed often incur costs to future fitness (Roff 1996). Overall, however, models of threshold traits suggest some important considerations for understanding the relationship between environmental conditions and individual behavior. First, threshold models propose the relationship is non-linear. Second, environmental induction of phenotypic variability within a population is likely to result in a bimodal distribution rather than a standard normal distribution – which is the one usually tested in sociological research. Third, threshold models also propose that understanding the non-linear impact of the environment and the resulting distribution of traits requires close attention to the mechanisms the individual relies upon to track and predict changes in the local environment. Without specification of this mechanism and its properties, a model of the relationship between environmental conditions and individual-level behavior is severely limited.

Third, a neurosociology of prediction can offer a more powerful interpretation of emulation and adaptation to subjugated social status and their inherently detrimental outcomes. In summary, I argue that emulation and adaptation are not simply meant to ease day-to-day interaction. Rather, they are primary methods for rapidly building a predictable cultural niche. Unfortunately, the emphasis on the rapidity of niche development incurs large long-term costs.

Copying the behaviors of others in the local environment can quickly bring about a stable niche in a variety of ways. At the most basic level, emulation encourages perceptions of common traits and interests among individuals sharing a common social space (Diijksterhuis 2005). From this point, individuals can develop stronger bonds of trust, long-term friendships, and greatly extend social networks (Diijksterhuis 2005). Each of these outcomes has the benefit of minimizing social threats and minimizing the amount of effort needed to access material resources. Likewise, adaptation through the creation of local cultural norms can quickly promote an order and predictability in social interactions, even among strangers (see Anderson 1999 for example).

Problems arise because rapidly achieving control over the social environment will often require the employment of social strategies which are inherently detrimental over the long run. Two such strategies are fear mongering and the threat of violence. Fear tactics and violence have been used to "encourage" quick conformity within a population (Caplin 2003) or bring about radical cultural and institutional change within a specific community (Klein 2007). In their studies of subjugated groups, ethnographers have frequently noted the use of violence and threats of violence as social currency. Some individuals build reputations for violence and, in effect, are able to control interactions with others from a distance (see Anderson 1999; MacLeod 1995; Willis 1977). Though able to quickly create a predictable environment, violence and fear stimulate some of the same neurobiological processes generated by uncertainty (Rosen and Donley 2006). Higher-level cognitive functioning is likewise compromised. Behavior becomes rule-based. Consequently, the predictable environment created is inherently unsustainable over the long run.

Conclusions

Neurosociology clearly has the promise to illuminate the cognitive and neurobiological mechanisms of social behavior but this is not its sole contribution to sociology. Rather, neurosociology brings our attention to the fact that the individual is intricately dependent upon sociological processes at multiple levels of biological functioning and that this is the major outcome of the operation of sociological forces over evolutionary time (Franks 2010; TenHouten 1997). But a new perspective cannot advance without solid empirical research programs in place. These research programs must be able to answer existing questions in the sociological literature and yield novel insights and directions for the future. A focus on prediction and predictive learning can help neurosociology achieve these goals. Because sociological forces have so strongly shaped the evolution of the human brain, how an individual predicts the future is also likely to be a reflection of social selection dynamics. That is, just as the predictive behaviors of some animals are tuned to cues in the physical environment the predictive behaviors of humans are probably tuned to specific sociological cues such as relative social status and the formation of coalitions. This is at least suggested by some researchers (Kenrick et al. 2003). A measure of predictions about the future would rest squarely on such an argument and provide a solid empirical dimension to neurosociology.

In my view, neurosociology greatly expands the breadth and depth of sociology as a discipline. Since it is mechanism-based, it can contribute greatly to the development of powerful explanatory theories within sociology. Measures built on neurosociological theories will undoubtedly bring even greater methodological integration across disciplines. I predict a bright future for neurosociology.

References

Allman, J. M., Hakeem, A., Erwin, J. M., Nimchinsky, E., & Hof, P. (2001). The anterior cingulate cortex. *Annals of the New York Academy of Sciences, 935*, 107–117.

Anderson, E. (1978). *A place on the corner*. University of Chicago Press.

Anderson, E. (1990). *Streetwise: Race, class, and change in an urban community*. University of Chicago Press.

Anderson, E. (1999). *Code of the street: Decency, violence, and the moral life of the inner city* (Vol. 1). W. W. Norton and Company.

Arnsten, A. F. T. (2009). Stress signalling pathways that impair prefrontal cortex structure and function. *Nature Reviews Neuroscience, 10*, 410–422.

Assadi, S. M., Yncel, M., & Pantelis, C. (2009). Dopamine modulates neural networks involved in effort-based decision-making. *Neuroscience and Biobehavioral Reviews, 33*, 383–393.

Ball-Rokeach, S. J. (1973). From pervasive ambiguity to a definition of the situation. *Sociometry, 36*, 378–389.

Bar, M. (2009). Predictions: A universal principle in the operation of the human brain. *Philosophical Transactions of the Royal Society B: Biological Sciences, 364*, 1181–1182.

Barker, G. T. (2005). *Dying to be men: Youth and masculinity and social exclusion*. Routledge.

Bayer, H. M., & Glimcher, P. W. (2005). Midbrain dopamine neurons endcode a quantitative reward prediction error signal. *Neuron, 47*, 129–141.

Bechara, A., Damasio, H., Damasio, A. R., & Lee, G. P. (1999). Different contributions of the human amygdala and ventromedial prefrontal cortex to decision-making. *Journal of Neuroscience, 19*, 5473–5481.

Belsky, J., Steinberg, L. D., & Draper, P. D. (1991). Childhood experience, interpersonal development and reproductive strategy: An evolutionary theory of socialization. *Child Development, 62*, 647–670.

Belsky, J., Steinberg, L., Houts, R. M., & Halpern-Felsher, B. L. (2010). The development of reproductive strategy in females: Early maternal harshness –>earlier menarche –>increased sexual risk taking. *Developmental Psychology, 46*, 120–128.

Bledowski, C., Rahm, B., & Rowe, J. B. (2009). What "works" in working memory? Separate systems for selection and updating of critical information. *The Journal of Neuroscience, 29*, 13735–13741.

Bourdieu, P., & Accardo, A. (1999). *The weight of the world: Social suffering in contemporary society*. Stanford University Press.

Bourgois, P. (2002). *In search of respect: Selling crack in El Barrio*. Cambridge University Press.

Boyd, R., & Richerson, P. J. (2002). Norms and bounded rationality. In G. Gigerenzer & R. Selten (Eds.), *Bounded rationality: The adaptive toolbox* (pp. 281–296). The MIT Press.

Brown, J. W., & Braver, T. S. (2007). Risk prediction and aversion by anterior cingulate cortex. *Cognitive, Affective, & Behavioral Neuroscience, 7*, 266–277.

Brumbach, B., Figueredo, A., & Ellis, B. (2009). Effects of harsh and unpredictable environments in adolescence on development of life history strategies. *Human Nature, 20*, 25–51.

Bubic, A., Von Cramon, D. Y., & Schubotz, R. I. (2010). Prediction, cognition and the brain. *Frontiers in Human Neuroscience, 4*, 25. https://doi.org/10.3389/fnhum.2010.00025

Burmeister, S. S., Jarvis, E. D., & Russell, F. (2005). Rapid behavioral and genomic responses to social opportunity. *Public Library of Science Biology, 3*, e363–e371.

Cacioppo, J. T., & Patrick, W. (2008). *Loneliness: Human nature and the need for social connection*. W. W. Norton.

Caplin, A. (2003). Fear as a policy instrument. In G. F. Loewenstein, D. Read, & R. Baumeister (Eds.), *Time and decision: Economic and psychological perspectives on intertemporal choice* (pp. 441–458). Russell Sage.

Carroll, A., Green, S., Houghton, S., & Wood, R. (2003). Reputation enhancement and involvement in delinquency among high school students. *International Journal of Disability, Development and Education, 50*, 253–273.

Carter, C. S., & van Veen, V. (2007). Anterior cingulate cortex and conflict detection: An update of theory and data. *Cognitive, Affective, & Behavioral Neuroscience, 7*, 367–379.

Chisholm, J. S. (1999). *Death, hope, and sex: Steps to an evolutionary ecology of mind and morality*. Cambridge University Press.

Chisholm, J. S., Quinlivan, J. A., Petersen, R. W., & Coall, D. A. (2005). Early stress predicts age at menarche and first birth, adult attachment, and expected lifespan. *Human Nature, 16*, 233–265.

Cook, E. H., Jr., Stein, M. A., Krasowski, M. D., Cox, N. J., Olkon, D. M., Kieffer, J. E., & Leventhal, B. L. (1995). Association of attention-deficit disorder and the dopamine transporter gene. *The American Journal of Human Genetics, 56*, 993–998.

Coussi-Korbel, S., & Fragaszy, D. M. (1995). On the relation between social dynamics and social learning. *Animal Behaviour, 50*, 1441–1453.

Creel, S., & Sands, J. L. (2003). Is social stress a consequence of subordination or a cost of dominance? In F. B. M. de Waal & P. L. Tyack (Eds.), *Animal social complexity: Intelligence, culture and individual societies* (pp. 153–169). Harvard University.

Davis, J., & Werre, D. (2008). A longitudinal study of the effects of uncertainty on reproductive behaviors. *Human Nature, 19*, 426–452.

Daw, N. D., O'Doherty, J. P., Dayan, P., Seymour, B., & Dolan, R. J. (2006). Cortical substrates for exploratory decisions in humans. *Nature, 441*, 876–879.

Day, S., Papataxiarchis, E., & Stewart, M. (1999). Consider the lilies of the field. In S. Day, E. Papataxiarchis, & M. Stewart (Eds.), *Lilies of the field: Marginal people who live for the moment* (pp. 1–25). Westview Press.

Dayan, P., & Abbott, L. F. (2001). *Theoretical neuroscience: Computational and mathematical modeling of neural systems*. MIT Press.

Devinsky, O., Morrell, M. J., & Vogt, B. A. (1995). Contributions of anterior cingulate cortex to behaviour. *Brain, 118*(Pt 1), 279–306.

Diijksterhuis, A. (2005). Why we are social animals: The high road to imitation as social glue. In S. Hurley & N. Chater (Eds.), *Perspectives on imitation: From neuroscience to social science (Imitation, human development, and culture)* (Vol. 2, pp. 207–220). MIT Press.

Dominguez, J. M., & Hull, E. M. (2005). Dopamine, the medial preoptic area, and male sexual behavior. *Physiology and Behavior, 86*, 356–368.

Dworkin, I. (2005). Canalization, cryptic variation, and developmental buffering: A critical examination and analytical perspective. In B. Hallgrimsson & B. K. Hall (Eds.), *Variation: A central concept in biology* (pp. 131–158). Elsevier.

Edin, K., & Kefelas, M. (2005). *Promises I can keep: Why poor women put motherhood before marriage*. University of California Press.

Edin, K., & Lein, L. (1997). *Making ends meet: How single mothers survive welfare and low wage work*. Russell Sage.

Eichenbaum, H., & Fortin, N. J. (2009). The neurobiology of memory based predictions. *Philosophical Transactions of the Royal Society B: Biological Sciences, 364*, 1183–1191.

Eisenberger, N. I., Lieberman, M. D., & Williams, K. D. (2003). Does rejection hurt? An FMRI study of social exclusion. *Science, 302*, 290–292.

Ellis, B., Figueredo, A., Brumbach, B., & Schlomer, G. (2009). Fundamental dimensions of environmental risk. *Human Nature, 20*, 204–268.

Emler, N., & Reicher, S. (1995). *Adolescence and delinquency: The collective management of reputation*. Blackwell Publishing.

Etkin, A., Egner, T., Peraza, D. M., Kandel, E. R., & Hirsch, J. (2006). Resolving emotional conflict: A role for the rostral anterior cingulate cortex in modulating activity in the amygdala. *Neuron, 51*, 871–882.

Fehr, E., & Camerer, C. F. (2007). Social neuroeconomics: The neural circuitry of social preferences. *Trends in Cognitive Sciences, 11*, 419–427.

Fellows, L. K., & Farah, M. J. (2007). The role of ventromedial prefrontal cortex in decision making: Judgment under uncertainty or judgment Per Se? *Cerebral Cortex, 17*, 2669–2674.

Fiorillo, C. D., Tobler, P. N., & Schultz, W. (2003). Discrete coding of reward probability and uncertainty by dopamine neurons. *Science, 299*, 1898–1902.

Franks, D. D. (2010). *Neurosociology: The nexus between neuroscience and social psychology*. Springer.

Gambetta, D., & Hamill, H. (2005). *Streetwise: How taxi drivers establish their customers' trustworthiness*. Russell Sage.

Gigerenzer, G. (2008). *Rationality for mortals: How people cope with uncertainty*. Cambridge University Press.

Gigerenzer, G., & Brighton, H. (2009). Homo heuristicus: Why biased minds make better inferences. *Trends in Cognitive Sciences, 1*, 107–143.

Gobrogge, K. L., Liu, Y., Jia, X., & Wang, Z. (2007). Anterior hypothalamic neural activation and neurochemical associations with aggression in pair-bonded male prairie voles. *The Journal of Comparative Neurology, 502*, 1109–1122.

Goodson, J. L. (2005). The vertebrate social behavior network: Evolutionary themes and variations. *Hormones and Behavior, 48*, 11–22.

Graybiel, A. M. (2008). Habits, rituals, and the evaluative brain. *Annual Review of Neuroscience, 31*, 359–387.

Greene, J. D., & Paxton, J. M. (2009). Patterns of neural activity associated with honest and dishonest moral decisions. *Proceedings of the National Academy of Sciences, 106*, 12506–12511.

Guo, G., Roettger, M. E., & Cai, T. (2008). The integration of genetic propensities into social-control models of delinquency and violence among male youths. *American Sociological Review, 73*, 543–568.

Hakeem, A. Y., Sherwood, C. C., Bonar, C. J., Butti, C., Hof, P. R., & Allman, J. M. (2009). Von economo neurons in the elephant brain. *The Anatomical Record: Advances in Integrative Anatomy and Evolutionary Biology, 292*, 242–248.

Haselton, M. G., Bryant, G. A., Wilke, A., Frederick, D. A., Galperin, A., Frankenhuis, W. E., & Moore, T. (2009). Adaptive rationality: An evolutionary perspective on cognitive bias. *Social Cognition, 27*, 733–763.

Hills, T. T. (2006). Animal foraging and the evolution of goal-directed cognition. *Cognitive Science, 30*, 3–41.

Hull, E. M., Muschamp, J. W., & Sato, S. (2004). Dopamine and serotonin: Influences on male sexual behavior. *Physiology and Behavior, 83*, 291–307.

Insel, T. R., & Fernald, R. D. (2004). How the brain processes social information: Searching for the social brain. *Annual Review of Neuroscience, 27*, 722.

Ishikawa, A., Ambroggi, F., Nicola, S. M., & Fields, H. L. (2008). Dorsomedial prefrontal cortex contribution to behavioral and nucleus accumbens neuronal responses to incentive cues. *The Journal of Neuroscience, 28*, 5088–5098.

Iversen, L. L. (2010). *Dopamine handbook*. Oxford University Press.

Kennerley, S. W., Walton, M. E., Behrens, T. E. J., Buckley, M. J., & Rushworth, M. F. S. (2006). Optimal decision making and the anterior cingulate cortex. *Nature Neuroscience, 9*, 940–947.

Kenrick, D. T., Li, N. P., & Butner, J. (2003). Dynamical evolutionary psychology: Individual decision rules and emergent social norms. *Psychological Review, 110*, 3–28.

Kenrick, D. T., Griskevicius, V., Sundie, J. M., Li, N. P., Li, Y. J., & Neuberg, S. L. (2009). Deep rationality: The evolutionary economics of decision making. *Social Cognition, 27*, 764–785.

Klein, N. (2007). *The shock doctrine: The rise of disaster capitalism*. Metropolitan Books.

Klein, G. (2009). *Streetlights and shadows: Searching for the keys to adaptive decision making*. MIT Press.

Koenigs, M., Young, L., Adolphs, R., Tranel, D., Cushman, F., Hauser, M., & Damasio, A. (2007). Damage to the prefrontal cortex increases utilitarian moral judgements. *Nature, 446*, 908–911.

Kouneiher, F., Charron, S., & Koechlin, E. (2009). Motivation and cognitive control in the human prefrontal cortex. *Nature Neuroscience, 12*, 939–945.

LaLand, K. N. (2004). Social learning strategies. *Learning & Behavior, 32*, 4–14.

Lee, G., & Gammie, S. C. (2009). GABAA receptor signaling in the lateral septum regulates maternal aggression in mice. *Behavioral Neuroscience, 123*, 1169–1177.

MacLeod, J. (1995). *Ain't no Makin' it: Leveled aspirations in a low income neighborhood*. Westview Press.

Maeda, K.-I., & Tsukamura, H. (2006). The impact of stress on reproduction: Are glucocorticoids inhibitory or protective to gonadotropin secretion? *Endocrinology, 147*, 1085–1086.

Manuck, S. B., Flory, J. D., Muldoon, M. F., & Ferrell, R. E. (2003). A neurobiology of intertemporal choice. In G. Loewenstein, D. Read, & R. F. Baumeister (Eds.), *Time and decision: Economic and psychological perspectives on intertemporal choice* (pp. 139–172). Russell Sage.

Manuck, S. B., Bleil, M. E., Petersen, K. L., Flory, J. D., John Mann, J., Ferrell, R. E., & Muldoon, M. F. (2005). The socio-economic status of communities predicts variation in brain serotonergic responsivity. *Psychological Medicine, 35*, 528.

Massey, D. S. (2007). *Categorically unequal: The American stratification system*. Russell Sage.

Masten, C. L., Eisenberger, N. I., Borofsky, L. A., Pfeifer, J. H., McNealy, K., Mazziotta, J. C., & Dapretto, M. (2009). Neural correlates of social exclusion during adolescence: Understanding the distress of peer rejection. *Social Cognitive and Affective Neuroscience, 4*, 143–157.

Matsuwaki, T., Watanabe, E., Suzuki, M., Yamanouchi, K., & Nishihara, M. (2003). Glucocorticoid maintains pulsatile secretion of luteinizing hormone under infectious stress condition. *Endocrinology, 144*, 3477–3482.

McEwen, B. S., & Gianaros, P. J. (2010). Central role of the brain in stress and adaptation: Links to socioeconomic status, health, and disease. *Annals of the New York Academy of Sciences, 1186*, 190–222.

Menchaca, M. (1995). *Mexican outsiders: A community history of marginalization and discrimination in california*. University of Texas Press.

Miller, E. K., & Cohen, J. D. (2001). An integrative theory of prefrontal cortex function. *Annual Review of Neuroscience, 24*, 167–202.

Modirrousta, M., & Fellows, L. K. (2008). Dorsal medial prefrontal cortex plays a necessary role in rapid error prediction in humans. *The Journal of Neuroscience, 28*, 14000–14005.

Montague, P. R., Dayan, P., & Sejnowski, T. J. (1996). A framework for mesencephalic dopamine systems based on predictive Hebbian learning. *Journal of Neuroscience, 16*, 1936–1947.

Morgane, P. J., Galler, J. R., & Mokler, D. J. (2005). A review of systems and networks of the limbic forebrain/limbic midbrain. *Progress in Neurobiology, 75*, 143–160.

Morris, G., Nevet, A., Arkadir, D., Vaadia, E., & Bergman, H. (2006). Midbrain dopamine neurons encode decisions for future action. *Nature Neuroscience, 9*, 1057–1063.

Munhoz, C. D., Lepsch, L. B., Kawamoto, E. M., Malta, M. B., Lima Lde, S., Avellar, M. C., Sapolsky, R. M., & Scavone, C. (2006). Chronic unpredictable stress exacerbates lipopolysaccharide-induced activation of nuclear factor-kappaB in the frontal cortex and hippocampus via glucocorticoid secretion. *Journal of Neuroscience, 26*, 3813–3820.

Newman, S. W. (1999). The medial extended amygdala in male reproductive behavior: A node in the mammalian social behavior network. *Annals of the New York Academy of Sciences, 877*, 242–257.

Niv, Y., Duff, M., & Dayan, P. (2005). Dopamine, uncertainty and TD learning. *Behavioral and Brain Functions, 1*, 6.

Niv, Y., Daw, N. D., Joel, D., & Dayan, P. (2007). Tonic dopamine: Opportunity costs and the control of response vigor. *Psychopharmacology, 191*, 507–520.

Niven, J. E., & Laughlin, S. B. (2008). Energy limitation as a selective pressure on the evolution of sensory systems. *Journal of Experimental Biology, 211*, 1792–1804.

O'Donnell, M., & Sharpe, S. (2000). *Uncertain masculinities: Youth, ethnicity, and class in contemporary Britain*. Routledge.

O'Reilly, J. X., Marsel Mesulam, M., & Nobre, A. C. (2008). The cerebellum predicts the timing of perceptual events. *Journal of Neuroscience, 28*, 2252–2260.

Odling-Smee, J. F., LaLand, K. N., & Feldman, M. W. (2003). *Niche construction: The neglected process in evolution*. Princeton University Press.

Osaka, N., Logie, R. H., & D'Esposito, M. (2007). *The cognitive neuroscience of working memory*. Oxford University Press.

Pine, A., Shiner, T., Seymour, B., & Dolan, R. J. (2010). Dopamine, time, and impulsivity in humans. *Journal of Neuroscience, 30*, 8888–8896.

Renfree, M. B., & Shaw, G. (2000). Diapause. *Annual Review of Physiology, 62*, 353–375.

Robinson, G. E., Fernald, R. D., & Clayton, D. F. (2008). Genes and social behavior. *Science, 322*, 896–900.

Rodrigues, S. M., LeDoux, J. E., & Sapolsky, R. M. (2009). The influence of stress hormones on fear circuitry. *Annual Review of Neuroscience, 32*, 289–313.

Roff, D. A. (1996). The evolution of threshold traits in animals. *The Quarterly Review of Biology, 71*, 3–35.

Rosen, J. B., & Donley, M. P. (2006). Animal studies of amygdala function in fear and uncertainty: Relevance to human research. *Biological Psychology, 73*, 49–60.

Rougier, N. P., Noelle, D. C., Braver, T. S., Cohen, J. D., & O'Reilly, R. C. (2005). Prefrontal cortex and flexible cognitive control: Rules without symbols. *Proceedings of the National Academy of Sciences of the United States of America, 102*, 7338–7343.

Rushworth, M. F., Buckley, M. J., Behrens, T. E., Walton, M. E., & Bannerman, D. M. (2007). Functional organization of the medial frontal cortex. *Current Opinion in Neurobiology, 17*, 220–227.

Salancik, G. R., & Pfeffer, J. (1978). Uncertainty, secrecy, and the choice of similar others. *Social Psychology Quarterly, 41*, 246–255.

Salzman, C. D., & Fusi, S. (2010). Emotion, cognition, and mental state representation in amygdala and prefrontal cortex. *Annual Review of Neuroscience, 33*, 173–202.

Sapolsky, R. M. (2004). *Why zebras don't get ulcers/ Robert M. Sapolsky*. Times Books.

Schultz, W., & Dickinson, A. (2000). Neuronal coding of prediction errors. *Annual Review of Neuroscience, 23*, 473–500.

Schultz, W., Dayan, P., & Montague, P. R. (2002). A neural substrate of prediction and reward. In J. Cacioppo, G. G. Berntson, R. Adolphs, C. S. Carter, R. J. Davidson, M. K. McClintock, B. S. McEwen, M. J. Meaney, D. L. Schacter, E. M. Sternberg, S. S. Suomi, & S. E. Taylor (Eds.), *Foundations in social neuroscience* (pp. 541–554). MIT.

Schwalbe, M., Godwin, S., Holden, D., Schrock, D., Thompson, S., & Wolkomb, M. (2000). Generic processes in the reproduction of inequality: An interactionist analysis. *Social Forces, 79*, 419–452.

Schwartzer, J. J., & Melloni, R. H., Jr. (2010). Anterior hypothalamic dopamine D2 receptors modulate adolescent anabolic/androgenic steroid-induced offensive aggression in the Syrian hamster. *Behavioural Pharmacology, 21*, 314–322.

Schweimer, J., & Hauber, W. (2006). Dopamine D1 receptors in the anterior cingulate cortex regulate effort-based decision making. *Learning & Memory, 13*, 777–782.

Shettleworth, S. J. (1998). *Cognition, evolution, and behavior*. Oxford University Press.

Singewald, G. M., Rjabokon, A., Singewald, N., & Ebner, K. (2011). The modulatory role of the lateral septum on neuroendocrine and behavioral stress responses. *Neuropsychopharmacology, 36*, 793–804.

Slavich, G. M., Way, B. M., Eisenberger, N. I., & Taylor, S. E. (2010). Neural sensitivity to social rejection is associated with inflammatory responses to social stress. *Proceedings of the National Academy of Sciences of the United States of America, 107*, 14817–14822.

Szpunar, K. K., Watson, J. M., & McDermott, K. B. (2007). Neural substrates of envisioning the future. *Proceedings of the National Academy of Sciences of the United States of America, 104*, 642–647.

Tagkopoulos, I., Liu, Y. C., & Tavazoie, S. (2008). Predictive behavior within microbial genetic networks. *Science, 320*, 1313–1317.

TenHouten, W. (1997). Neurosociology. *Journal of Social and Evolutionary Systems, 20*, 7–37.

Tilly, C. (1998). *Durable inequality*. University of California Press.

Tilly, C. (2005). *Identities, boundaries and social ties*. University of Colorado Press.

van Veen, V., & Carter, C. S. (2006). Error detection, correction, and prevention in the brain: A brief review of data and theories. *Clinical EEG and Neuroscience, 37*, 330–335.

van Veen, V., Cohen, J. D., Botvinick, M. M., Stenger, V. A., & Carter, C. S. (2001). Anterior cingulate cortex, conflict monitoring, and levels of processing. *NeuroImage, 14*, 1302–1308.

van Veen, V., Krug, M. K., Schooler, J. W., & Carter, C. S. (2009). Neural activity predicts attitude change in cognitive dissonance. *Nature Neuroscience, 12*, 1469–1474.

Vanberg, V. J. (2002). Rational choice vs. Program-based behavior: Alternative theoretical approaches and their relevance for the study of institutions. *Rationality and Society, 14*, 7–54.

Vijayraghavan, S., Wang, M., Birnbaum, S. G., Williams, G. V., & Arnsten, A. F. T. (2007). Inverted-U dopamine D1 receptor actions on prefrontal neurons engaged in working memory. *Nature Neuroscience, 10*, 376–384.

Wallis, J. D. (2007). Orbitofrontal cortex and its contribution to decision-making. *Annual Review of Neuroscience, 30*, 31–56.

Walton, M. E., Croxson, P. L., Behrens, T. E. J., Kennerley, S. W., & Rushworth, M. F. S. (2007). Adaptive decision making and value in the anterior cingulate cortex. *NeuroImage, 36*, T142–T154.

Warkentin, K. M., & Caldwell, M. S. (2009). Assessing risk: Embryos, information, and escape hatching. In R. Dukas & J. M. Ratcliffe (Eds.), *Cognitive ecology II* (pp. 177–200). University of Chicago Press.

Wasser, S. K., & Place, N. J. (2001). Reproductive filtering and the social environment. In P. T. Ellison (Ed.), *Reproductive ecology and human evolution* (pp. 137–158). Aldine De Gruyter.

Watson, K. K. (2008). Evolution, risk, and neural representation. *Annals of the New York Academy of Sciences, 1128*, 8–12.

Whyte, W. F. (1943). *Streetcorner society: The social structure of an Italian slum*. University of Chicago Press.

Williamson, M., & Viau, V. (2008). Selective contributions of the medial preoptic nucleus to testosterone-dependant regulation of the paraventricular nucleus of the hypothalamus and the HPA axis. *American Journal of Physiology – Regulatory. Integrative and Comparative Physiology, 295*, R1020–R1030.

Willis, P. (1977). *Learning to labor: How working class kids get working class jobs*. Columbia University Press.

Wingfield, J. C., & Kitaysky, A. S. (2002). Endocrine responses to unpredictable environmental events: Stress or anti-stress hormones? *Integrative and Comparative Biology, 42*, 600–609.

Yu, A. J., & Dayan, P. (2002). Acetylcholine in cortical inference. *Neural Networks, 15*, 719–730.

Yu, A., & Dayan, P. (2005). Uncertainty, neuromodulation, and attention. *Neuron, 46*, 681–692.

Zahrt, J., Taylor, J. R., Mathew, R. G., & Arnsten, A. F. T. (1997). Supranormal stimulation of D1 dopamine receptors in the rodent prefrontal cortex impairs spatial working memory performance. *The Journal of Neuroscience, 17*, 8528–8535.

From Linear to Complex: Immune System Dynamics and Health Inequities in a Racialized Society

23

Bridget J. Goosby and Jacob E. Cheadle

Abstract

This chapter aims to describe key facets of the immune system and its relations to adverse social conditions, followed by an exploration of social and immune complexity. By examining the links between social conditions and immune system processes, particularly inflammation, we highlight how these bio-social interactions contribute to health outcome inequities, focusing on Black Americans who bear the brunt of a pervasive racialized social system. This system, inscribed upon bodies in a spectrum of morbidities and mortality risks, interacts closely with the immune system. We reflect on the limitations of current paradigms that often rely on linear methods and theories, which typically make strong assumptions about interdependencies and the functional nature of relationships. In contrast, the components of complex systems exhibit intricate interdependencies, non-linear dynamics, feedback loops, self-organization, and emergent properties. Social structures under systemic racism in the U.S., much like components of the immune system, are interrelated, dynamic, self-reinforcing, and adaptive, showing signs of deep complexity. By balancing current linear paradigms with conceptual and theoretical considerations of complexity, we aim to foster new theoretical insights and methodological innovations. Recognizing the interconnected dynamics of complex systems may inspire transformative policies and interventions, addressing the root causes of health disparities.

B. J. Goosby (✉) · J. E. Cheadle
Department of Sociology, Population Research Center, and The Center on Aging and Population Sciences, The University of Texas at Austin, Austin, TX, USA
e-mail: bgoosby@prc.utexas.edu; j.e.cheadle@utexas.edu

Introduction

This chapter aims to first describe key facets of the immune system and its relations to adverse social conditions, followed by a consideration of social and immune *complexity*. By examining the links between social conditions and immune system processes (Lam et al., 2022; Martinez et al., 2021; Muscatell et al., 2020), particularly inflammation, we seek to highlight how these *bio-social* interactions contribute to health outcome inequities (Cuevas et al., 2020; Goosby & Cheadle, 2024). We focus on Black Americans, who bear the brunt of a pervasive and tenaciously durable racialized social system (Bonilla-Silva, 2015a) that has been inscribed upon bodies in a spectrum of morbidities and mortality risks, many of which interact or operate in tandem with the immune system (Goosby et al., 2018; Muscatell et al., 2022). This chapter also reflects on some impor-

W. Kalkhoff et al. (eds.), *Handbook of Neurosociology*, Handbooks of Sociology and Social Research, https://doi.org/10.1007/978-3-031-95615-7_23

tant limitations of the current paradigm organizing much of this work by route of conceptually considering system complexity. Indeed, research providing the *prior findings* on social conditions and immune system functioning usually relies upon methods, models, and theoretical frameworks emphasizing the components of these systems, *taking for granted* or *assuming* that they are *linear*.[1] The components in such systems interact in relatively straightforward and proportional ways, have minimal interdependencies, and the snaking chains of causal influence do not feedback to eat their tails (Krakauer, 2024).

Complex systems,[2] in contrast, have intricate and potentially reciprocal interdependencies, non-linear dynamics, feedbacks, self-organization, and emergent properties that can significantly and unpredictably diverge from the behaviors/outcomes expected of linear systems (Krakauer, 2024; Siegenfeld & Bar-Yam, 2020). In this vein, the components of the U.S.'s historically racialized systems are interrelated, dynamic, self-reinforcing, and adaptive (Bonilla-Silva, 1997, 2019; Ray, 2022; Stewart, 2023). *Racism*, viewed in this way, is not just *systemic* (i.e., *systemic racism*; Feagin, 2006); it shows signs of deep complexity. The immune system is also a complex system comprised of an intricate network of signaling molecules, cellular activities, tissues, and organs in order to coordinate bodily defenses against pathogens and insults from inside and outside itself (Bonaguro et al., 2022; Shilts et al., 2022; Thomas-Vaslin, 2017). We, therefore, seek to balance what is known today under the primarily linear paradigm with some conceptual and theoretical rumination on the complex natures of these systems. Many biological, ecological, and social systems are complex and adaptive (Page, 2015; Siegenfeld & Bar-Yam, 2020). Despite the inconveniences that come with the need to understand such systems as gestalts, recognizing a system's interconnected dynamics, patterns of self-organization, and emergent properties promises new theoretical insights and methodological innovation, the results of which may inspire new approaches to transformative policies and interventions.

Part I: Health Inequities and Inflammation

Black American Health, the Stress Process, and Systemic Racism

The literature on health inequality is extensive, covering an array of social group classifications (e.g., Link & Phelan, 1995; McCartney et al., 2021; Short & Zacher, 2022), social conditions (e.g., Muller et al., 2018; Sewell, 2016; Umberson & Donnelly, 2023), and types of morbidities (e.g., Williams et al., 2019). Our concern here is with immune system contributions to health morbidities and behaviors, particularly the role of inflammation, the processes by which the body's immune system defends itself from harm and promotes healing. While inflammatory morbidity and related mortality are rising globally, socially marginalized groups remain at disproportionate risk of exposure to such conditions relative to their more advantaged comparison groups (Furman et al., 2019; Muscatell et al., 2020). The health-related morbidities and mortality risks that inflammation is associated with include a range of cardiometabolic conditions such as cardiovascular disease (CVD) (Ferrucci & Fabbri, 2018), Type 2 diabetes (Muriach et al., 2014), hypertension (Barrows et al., 2019), adiposity (Shah et al., 2023), and cancer (Zavala et al., 2021).

[1]Our concern in this paper is towards mainstream quantitative analysis of, usually, large-scale and publicly available nationally representative data sources. Certainly, a small group of sociologists has analyzed complex systems directly using various computational techniques and models (Barbrook-Johnson & Carrick, 2022; Page, 2015), and to great effect theoretically in some cases (e.g., Stewart, 2023).

[2]*Complex* as compared to the notion of a *complicated* but *linear system* where the challenge is that the number and/or combination of system components is large and difficult to assemble (von Förster & Morin, 2024; Page, 2015), as with a puzzle of a great many pieces. It seems to us that this is a quite common position, that the world is complicated in the social and other sciences, and that the transition from complicated to complex is a difficult one, even for those attempting to do so (e.g., Graetz et al., 2022; Quintana, 2023).

Notably, an extensive and rapidly increasing literature points to the grave importance of the ways that social conditions make burdensome physiological demands that increase inflammation among Black Americans (McEwen, 1998; Bobba-Alves et al., 2022; Goosby et al., 2018), who are at disproportionate risk of a range of CVD-related morbidities (Al Kibria 2019; Bailey et al., 2017; Mohebi et al., 2022). The risks for such morbidities are probabilistically regulated by numerous social conditions and experiences (Glass & McAtee, 2006; Williams et al., 2019), with those most related to material conditions sometimes referred to as the *fundamental causes* of disease in sociological research (e.g., money, prestige, power; Link & Phelan, 1995). The divergences in the constellations of adversities different social groups experience are the pivot upon which the sociological study of health turns. Pearlin's (1989) social *stress process model,* in which stressful social conditions such as exposure to traumatic life events, economic hardship, and discriminatory incidents are examples of stressors that undermine health when realized as stress. The *stressor* concept, which does much of the sociological work in this framework, is characterized as a combination of (a) the challenges or demands of the environment that tax or exceed the individual's ordinary capacity to adapt, with (b) the absence of the means to attain sought-after ends (Aneshensel & Mitchell, 2014). *Stress* then refers to physiological arousal patterns, psychological distress, and behavioral manifestations (Wheaton et al., 2013).

Where the stressor is generally social in nature, and stress is largely biological, the two together provide the *biosocial* dynamics supporting many different facets of health. Elaborating on these concepts, Wheaton's (1994) *stress universe* provides a "universe" of potential stressor-stress characteristics within which their multifaceted diversity and dimensionality can be located. This typology portrays stressors in terms of the duration (i.e., discrete, intermittent, chronic), severity (i.e., mild-severe), level of social organization (i.e., micro-macro), and temporal facets (i.e., life stage, timing, sequencing) of the resulting stress (Wheaton, 1994; Aneshensel & Mitchell, 2014). We suggest that an individual's location within the "space" of this universe defines what can be conceptualized as their unique *stress ecology*. This stress ecology reflects the interactions between individuals and the demands and challenges they face in life, including interrelationships and competition between various stressors, each demanding their share of resources, as well as the resulting biological pressures exerted on bodies and minds by stress.

As Pearlin (1989, p.241) wrote, the "[stress] process and its components largely arise from and are influenced by various structural arrangements in which individuals are embedded." The resulting stress ecology reflects how social conditions and experiences are organized for individuals and groups (i.e., on average) by their particular patterns of engagement and involvement with different social structures, where adversities and disadvantages increase the risks of stressor exposures across scales ranging from broad patterns of residential segregation (Martinez et al., 2021; Massey & Denton, 1993; Simons et al., 2018), socioeconomic inequality (Dowd & Aiello, 2009; Phelan & Link, 2015), patterns of interaction with the carceral state (Sewell, 2017; Sewell et al., 2021), and down into the nitty-gritty microsociological interaction dynamics of daily life (i.e., discrimination and social rejection; Cheadle et al., 2020; Goosby et al., 2018; Jelsma et al., 2021), for some examples. The basic central hypothesis of the sociological study of health is, therefore, that the socially structured and patterned conditions place people *at risk for risks* (Link & Phelan, 1995). Social conditions regulate the risk for risks and structure and contextualize the contingencies of stressor ecologies, which are unequally distributed across different social groups (Glass & McAtee, 2006), as with race and racism (Bonilla-Silva, 1997).

The insidiousness of racism, both historical and contemporary (Mills, 2022), for Black Americans has served to increase the number, frequency, magnitudes, and diversity of risks realized as stressors, and the stressor ecologies, therefore, can be thought of as providing a rougher and more challenging landscape (i.e., more stressors of different durations and severities, deriving from a broader range of social structural contexts, and covering greater breadth over life course features). As a consequence of the stressor-stress translation of the stress process, various strands of stress research show that material realities are transduced into *embodied materiality* through evolved properties of human physiology (Sapolsky, 2004), with those stressor ecologies that are historical, durable, and consistently mapped to members of different social groups realized in similarly long-standing, resilient population health inequities (Goosby et al., 2018). Stress process research goals, ultimately, have been to map different domains of the stress process, such as the exposure to and meaning of stressors, access to stress mediators, and, of course, application to an ever-growing array of health and behavioral consequences (Brown et al., 2020; Louie et al., 2023; Moen, 2022).

Although this framework provides an organizing structure for mapping the contours of social condition variations between and within different social groups, explaining distributions and interrelationships of these conditions is ultimately outside of its purview, and so a need for supplementation from more macro perspectives remains. *How and why are social conditions organized the way they are for Black people relative to White people?* Critical Race Theory[3] (CRT) has played a key role in illuminating the historical processes of marginalization and oppression from slavery to present times for Black people in the U.S. (Mills, 2022). The historical data deployed by CRT researchers provides a multimodal time series of perspectives to replicate, justify, and confirm its positions (Delgado & Stefancic, 2023; Ray, 2022). The concept of *systemic racism* comes at the problem of the stress ecology by addressing both the *why* and *how* of the historical trajectories embedded in social systems and structures today. *Systemic racism* describes a *system* wherein racial power and privilege denies Black people full and equal access to economic, political, social, ideological, and other cultural resources and opportunities (Feagin, 2006; Feagin & Elias, 2013) while adaptively operating in a miscellany of practices that serve to perpetuate and maintain racialized social gradations even as society changes in other ways (Bonilla-Silva, 1997, 2015b, 2019).

The argument is that systemic racism is coordinated and maintained through political power, institutional practices and policies, cultural norms, and microsociological encounters. These elements collectively work together across social structures, that is, in a *systemic* sense, to create and sustain inequitable life experiences and outcomes for Black relative to White people (Delgado & Stefancic, 2023; Ray, 2022). The systemic racism concept thus supplies an explanation for why Black people tend to have stress ecologies with greater stressor diversity, intensity, and inescapability than do White people, even after accounting for socioeconomic characteristics (Brown et al., 2020; Homan & Brown, 2022; Kirkinis et al., 2021). Different health outcomes are not merely differences in distributions of the risks of risks but systematic inequities in structural conditions across social domains grounded deeply in America's history of slavery and the very notions of "race" created to justify it (Mills, 2022). Many people in the U.S., regardless of race/ethnicity, suffer greatly because socioeconomic inequities are pervasive while social and communal supports are generally sparse for everyone (Killewald et al., 2017; McCall & Percheski, 2010; Sterling & Platt,

[3]CRT and systemic racism provide a backdrop for understanding why the social conditions for Black Americans' are the way they are. However, it is important to note that adaptations of this framework were developed, in parallel and not long after, to examine the oppressive social structures of other racialized and ethnic minorities, such as LatCrit for Latine populations (Valdes & Bender, 2021), and Tribal Crit for Indigenous populations (Brayboy, 2005). Although these frameworks and the health inequities of these other groups are beyond the scope of this paper, it is worth noting that the scale of the economic, political, and other structures and inequities, particularly among Indigenous Americans (e.g., morbidity and mortality rates, economic marginalization, carceral contact, etc.), are also considerable, if not profound.

Fig. 23.1 System representation of racism. Adapted from Reskin (2012)

2022). Part of the point in emphasizing systemic racism is not to minimize suffering for other groups (e.g., Whites) as some straw-man arguments would have it, but because Black Americans both share the same core sources of inequality as everyone else, along with the extra burdens of America's historical legacy and endemic contemporary manifestations.[4]

The expanded scope of inequities systemic racism describes are partially depicted in Fig. 23.1, following Reskin's (2012) model of "über discrimination." In this characterization, racism is a system of social structures that produces essentialized and essentializing experiences (e.g., Graves Jr, 2001), whether those of cultural representation, policies, practices, or legal structures, ultimately filtering down into individual-level acts of bias, discrimination, and oppressive aggression (Williams, 2020). The bidirectionalities of pathways in Fig. 23.1 reinforce and increase interdependence so that inequity in one domain may "propagate" to others, both directly and indirectly, reinforcing that initial inequity–making the inequity *robust* because life domains and social structures are both adaptive and interconnected. Figure 23.1 can thus be seen as probabilistically structuring stress ecologies, as shaping the risks of risks, and therefore, the stressor ecologies that kick the stress process into gear, transforming material realities into embodied materialities. Consequently, the systemic nature of racism, its *über nature*, has implications for a great many different aspects of life, but, of course, health morbidities and mortality are the very foundations of living a life.

[4]Bonilla-Silva (2015b) describes that the U.S. racial regime has gone through three transformations originating with 1) slavery to 2) Jim Crow to our current post Civil Rights color-blind racism, which is characterized by seemingly non-racial practices and mechanisms that reproduce inequality.

Introduction to the Immune System: Innate and Adaptive Immunity

Bodies must make do with the limited sets of *actions* they are capable of enacting, both internally and upon the world, whatever happens to be happening wherever they happen to be. This is true whether one talks about social conditions in terms of the organism as a whole as it experiences states of, and changes in, its world, or the organism's own body states, resources, and capacities that it needs to bring itself into motion in order to be an actor therein (Cheadle et al., 2024). The biological and social are deeply interdependent with one another, and adverse social conditions and experiences make demands on bodies that come over time to undermine it's capacities and, thus, what one can do in a given situation and how one can adapt to conditions (e.g., chronic illness and the inability to work; Bobba-Alves et al., 2022; Sterling, 2012). A key aspect of the body that is sensitive to social conditions is the immune system, particularly the inflammatory processes playing numerous roles in morbidity and mortality (Ferrucci & Fabbri, 2018; Furman et al., 2019), including in Black-White health inequities (for a discussion, see Goosby et al., 2018). Our immune system oversees the protection of our body from sickness and injury with two critical and synergistic components, *innate* and *adaptive immunity* (Bennett et al., 2018), by which it defends itself from invaders, such as bacteria and viruses that enter the body, as well as the body's own damaged and renegade cells, which can lead to morbidities, such as cancer and autoimmune diseases.

The *innate immune system* provides the nonspecific, first line of cellular defense against pathogens and abnormal cells. This cellular defense provides physical and chemical blockades, as with the skin and mucous membranes (Bennett et al., 2018), and is responsible for distinguishing self from non-self or no-longer-self in a non-specific way, meaning without having to learn the invaders' molecular signatures (Marshall et al., 2018; McDade et al., 2016). This system thus provides immediate responses to tissue damage, infection, and abnormal cells in order to both prevent the spread and movement of pathogens and to detect and destroy abnormal cells as quickly as possible. The innate immune system is inherited and so can be thought of as embodying evolutionarily conserved predictions about ways the environment might threaten the organism and the ways the body's cells and processes might break down, undermining its functional integrity. This part of the immune system provides rapid response capacities, does not target specific pathogens or cells, responds to a wide range of invaders and abnormalities, and has no (or very limited; see Netea et al., 2016, 2017) memory of previous encounters with pathogens or rogue cells.

In contrast, *adaptive immunity,* also sometimes referred to as *acquired immunity*, refers to the immune responses that are specific to particular pathogens and damaged cells. While innate immunity anticipates generalized classes of threats, adaptive immunity is distinguished by its ability to *learn*. These innate and adaptive immune system components are, to a considerable extent, distinguished by non-specificity versus specificity, low versus high development costs, and higher versus lower activation costs, respectively (McDade et al., 2016). Adaptive immunity thus learns as it goes, forming *immunologic memory* and developing pathogen-specific defenses in case the threat is reencountered (Bhat et al., 2021; Marshall et al., 2018). Consequently, it takes longer to initiate and get up to speed (e.g., several days), but with the advantage that it is more precise in eliminating threats. Moreover, by virtue of its memory, it is much quicker to arms with respect to those threats it has encountered previously (Bennett et al., 2018). Adaptive immune function varies based on individual genetics and specific exposures guiding its development, which has significant implications for the dynamics and capacities of the immune system over the life course. Like all living systems, the immune system is constantly changing.

The innate and adaptive immune systems are critical and synergistic in mobilizing defenses against both general and specific bodily threats. Furthermore, both innate and adaptive immunity play crucial roles in *inflammation*, comprising the processes and mechanisms by which immune and non-immune cells are coordinated to eliminate pathogens and facilitate tissue repair and healing (Netea et al., 2017).

Inflammation: A Key Contributor to Health in General and Morbidity Inequities in Particular

Much sociological research focused on inflammation and inflammatory morbidities utilize biomarkers that play special roles in immune system signaling and communication. The signaling proteins produced by both immune and non-immune cells include *inflammatory cytokines* (Stanley & Lacy, 2010) and C-Reactive Protein (e.g. CRP), a "downstream" marker of inflammation produced by the liver (primarily) in response to the cytokine IL-6 during acute phase inflammation (Del Giudice & Gangestad, 2018). Cytokine function is context-specific and may be either *pro-inflammatory* or *anti-inflammatory*. For example, in certain circumstances, the interleukin-6 (IL-6) cytokine promotes immune activation and immune cell infection site dispersal, while interleukin-10 (IL-10) aids in reducing responses and so helps to prevent excessive inflammation (Bennett et al., 2018; Del Giudice & Gangestad, 2018). Because cytokines are critical immune messengers that also happen to be measurable in ways that can be reliably incorporated into large samples at scale (Harris & Schorpp, 2018; though see Kim et al., 2024), they have been vital to the assessment of immune function and morbidity risk in sociological and population biosocial health research (Furman et al., 2019). Here, it is crucial to recognize that much of the biosocial attention given to the immune system has largely been dedicated to inflammation and, out of the full complement of inflammatory components, mechanisms of intercellular signaling.

Innate and adaptive immune functions possess both varying modes of action and differing temporal parameters, whether attempting to address things going wrong in the body or invaders of it. Inflammatory activity is generally meant to be short-term, however, and it is intended to be downregulated in response to anti-inflammatory signals once threats resolve (McDade, 2023). The goal of achieving balanced pro- and anti-inflammatory activity is a restoration of *immunologic homeostasis* or *inflammatory equilibrium* (Netea et al., 2016, 2017). In other words, the intent of the immune system is naturally *acute*. Inflammation that persists in the absence of pathogens or cellular damage is termed *chronic low-grade inflammation* or *low-grade, non-infective systemic chronic inflammation* (SCI) (Furman et al., 2019), which may be a regular feature undermining health in Western, educated, industrialized, rich, and democratic (WEIRD[5]) populations (Gurven & Lieberman, 2020; McDade, 2023). A balance between pro- and anti-inflammatory activity, healing, and restorative activity are critical to achieving low inflammation states and healthy immune system functioning. So what happens when inflammatory equilibrium is not restored, as when social, psychological, environmental, and biological factors shift inflammation from acute to SCI?

A key consequence of SCI is increased physiological wear and tear via tissue and organ damage, which elevates numerous morbidity risks, including cardiovascular disease, diabetes, and hypertension (Furman et al., 2019). Social structures and conditions can regulate the risk of exposure to environmental factors associated with SCI (Glass & McAtee, 2006; Goosby et al., 2018), and the chronicity of repeated exposures to stressors appears to undermine the dynamics of acute inflammation, transitioning it from transitory bursts to SCI via chronic regulatory demands and burdens (Arnaldo et al., 2022; Bobba-Alves et al., 2022; McEwen, 1998; Rohleder, 2019). Enduring social adversities are

[5]WEIRD societies are thought to be more likely to have an inflammatory phenotype due to calorically dense food, low physical exertion, and lack of exposure to pathogens that shape the adaptive immune system in early life. Consequently, WEIRD people are more likely to have excess body fat, which is proinflammatory, and the lack of early life exposure to pathogens may lead to less effective inflammatory system downregulation (see McDade, 2023).

thus thought to draw a direct line between social conditions and the systems and capacities the body uses to protect itself from invaders, modifying acute inflammatory dynamics that evolve on short time scales into prolonged processes that stay active indefinitely, unable to recover to very low ideal levels (McDade, 2023).

Chronic stress takes adaptive bodily responses to protect itself from threat and injury by mobilizing inflammation in the here and now, but doing so is costly and eventually wears the body down through the burdens placed on it by persistent stress-induced energetic and immunologic demands (Bobba-Alves et al., 2022; Arnaldo et al., 2022). Under regular circumstances (i.e., acute stress) increased energetic demand is linked to an initial uptick in arousal (e.g., elevated blood pressure, circulation, blood glucose release), increased inflammatory activity to protect against potential wounds and initiate healing, and neuroendocrine cascades (e.g., glucocorticoid release, inflammatory inhibition, dampening of insulin receptors). Cortisol, a key energy-regulating hormone,[6] release is dramatically increased by the HPA-axis to boost energy access to facilitate bodily action in stressful conditions. Importantly for us here, cortisol is also anti-inflammatory, making it critical for reducing inflammatory activity by reducing proinflammatory cytokine activity once the threat has passed. The downregulation of heart rate, blood pressure, and glucose production, among other physiological parameters, is accomplished by the brain.

The brain, of course, attempts to learn about and predict what is going on in the world and how other actors will respond to the actors' actions in it (Cheadle et al., 2024; Cheadle et al., 2025, Chap. 16) so that it might regulate the body for its specific circumstances (Hutchinson & Barrett, 2019; Schulkin & Sterling, 2019). In the context of social conditions in which stressors are chronic, however, regulatory patterns of consistently elevated levels of cortisol secretion and proinflammatory activation become consequential over time (Goosby & Cheadle, 2024). For those experiencing chronic stress, physiological regulation is targeted to the expectation of increased demands and so higher than normal blood pressure is maintained, leading to wear and tear on arteries and initiating inflammatory damage, elevating the risk for hypertension (i.e., chronic high blood pressure) and cardiovascular diseases (e.g. arterial sclerosis). Moreover, ongoing chronic stress may elevate cortisol levels for extended periods while simultaneously reducing glucocorticoid (referring to the family of hormones that regulate metabolic and immune activity, of which cortisol is the most well-known) receptor expression, leading to excessive levels of circulating proinflammatory cytokines and increasing insulin resistance (Dantzer, 2018), a precursor to Type 2 diabetes (Zmora et al., 2017).

Much of the activity related to metabolic regulation under stress has important implications for immune function and inflammatory signaling and, thus, downstream morbidities. The concept of *inflammaging* refers to the development of SCI as a function of aging and chronic stress. Due to the cumulative and persistent activation of the immune system under chronic stress, as well as various age-related factors (e.g., cellular senescence, mitochondrial dysfunction, and alterations in the gut microbiota; Franceschi et al., 2018; see also, Ignatow, 2021), increases in the amount of pro-inflammatory cytokines and other inflammatory mediators in circulation can result in a state of persistent inflammation. Chronic stress also amplifies such inflammatory trajectories through the associated activation of other regulatory systems (i.e., sympathetic nervous system, HPA-axis), resulting in elevated cortisol levels and

[6] Cortisol has been given a bad reputation as the "stress hormone," woefully understating its broader physiological significance. More generally, cortisol is a hormone crucial for regulating energy production, distribution, and utilization throughout the body moment-to-moment in life, even without stress. Cortisol is key to maintaining blood glucose levels by stimulating gluconeogenesis, mobilizing fat stores, and breaking down proteins into amino acids. During periods of intense stress, cortisol rapidly increases glucose availability and suppresses non-essential functions to prioritize immediate energetic actions in service to survival. The diurnal cortisol pattern supports daily metabolic processes, immune function modulation, and overall homeostasis, underscoring its vital role beyond the stress response (see Sapolsky, 2004).

other metabolic-related hormones that diverge immune function away from a healthy body's ideal. These *temporal* relationships between social conditions, stress burdens, and body dynamics and mechanisms not only accelerates biological aging (Bobba-Alves et al., 2022), it also predisposes individuals to a spectrum of age-related diseases, including cardiovascular diseases, type 2 diabetes, neurodegenerative disorders, and certain cancers (Furman et al., 2019). Inflammaging is thus a concept that situates the temporality of adversities within bodies.

Consequences for behavior, psychiatric conditions, and neurodegeneration also accompany SCI. For example, continuous elevated circulation of proinflammatory cytokines in the brain and throughout the body can initiate *sickness behaviors*, a non-specific cluster of symptoms including social withdrawal, decreased physical activity, reduced appetite, fatigue, anhedonia, physical pain, sleep disruption, depression, and anxiety (D'Mello & Swain, 2017; Eisenberger et al., 2017). From an evolutionary perspective, sickness behaviors are thought to be protective as they conserve energy, allow for withdrawal to avoid injury, reduce the risks of infecting others, and promote healing via increased periods of rest and recovery. Social stressors are associated with more sickness behaviors and, to our purpose, a number of cytokines released by the immune system act on the brain to induce these behaviors (D'Mello & Swain, 2017). Sickness behaviors are also highly correlated with both psychopathologies and neurodegeneration (Quadt et al., 2018). There is also evidence that inflammation increases sensitivity to social stressors (e.g., social rejection; Eisenberger et al., 2017; Eisenberger & Cole, 2012).

The experience of trauma and other forms of acute and chronic stress are associated with elevated circulating levels of various proinflammatory markers, such as IL1β, IL-6, TNF-α, and C-Reactive Protein (Michopoulos et al., 2017). Moreover, an expanding literature also links physiological regulation via the predictive and coordinated regulation of expected homeostatic "set points" (i.e., *allostasis*; Arnaldo et al., 2022; Sterling, 2012) and interoception (the sensing and perception of bodily processes; Chen et al., 2021; Sennesh et al., 2022) to the risk for mood disorders such as major depression, anxiety, and post-traumatic stress disorder (e.g., Barrett, 2017a; Barrett et al., 2016; Savitz & Harrison, 2018; Zheng et al., 2021). Additionally, and in line with emerging models of depression as a problem of regulation and interoception (Barrett et al., 2016; Shaffer et al., 2022), there is increasing evidence that therapeutic treatments that lower levels of inflammation contribute to intervention effectiveness for psychiatric morbidities (Franklin et al., 2018). Inflammation is thus broadly implicated in many different facets of health: physiological, psychological, and behavioral. Signaling molecules circulate widely and interact with numerous cell types, both those of the immune system itself and those in other systems and tissues. This wide integration through the body, therefore, comes with the ongoing discovery of new health consequences, which rarely bodes well for those located in the wrong tails of those distributions characterizing different aspects of social inequality.

Health Inequities and Immune Function

The relationships between inflammation and health morbidities are now well-documented (Furman et al., 2019). Given a wide array of disadvantages in social conditions (e.g. Grusky, 2019) and exposures to multiple forms of racism spanning from structural factors (e.g., segregation) down to the interpersonal social interactional (i.e., microaggressions) experiences of different race/ethnic groups (Chae et al., 2011; Williams, 2020; Williams et al., 2019), it is not surprising that Black Americans, American Indian, Alaska Natives, and certain Hispanic groups are at disproportionately higher risk of inflammatory conditions (e.g., metabolic syndrome, diabetes, hypertension, cardiovascular disease, and adiposity; Goosby et al., 2018; Mohebi et al., 2022; Zavala et al., 2021). Beyond the associations of self-reported discrimination to risks for these inflammatory conditions, discrimination is also linked directly to inflammatory biomarkers (Cuevas et al., 2020) considered risk factors for, or indicators of, chronic disease. For example, interpersonal discrimination is associated with elevations in several biomarkers of inflammation (e.g., C-Reactive protein, IL-6, suPar; and composite indices of proinflammatory cytokines; Chen et al., 2022; Cuevas et al., 2020; Goosby et al., 2015; Toussaint et al., 2022).

Recent literature also links explicit indicators of systemic racism to inflammatory conditions (e.g., depression, cancer, and asthma). Some of these indicators include neighborhood deprivation due to enforced neighborhood racial segregation, historical redlining, neighborhood level racism (Bailey et al., 2021; Plascak et al., 2022; Russell et al., 2018), environmental racism and pollutant exposure (Martinez et al., 2021), police violence and contact with the carceral state (Bailey et al., 2021; Sewell, 2017; Sewell et al., 2021), and racialized political disenfranchisement (Homan & Brown, 2022). Proposed explanations argue that a life circumscribed and channeled by the malevolence of racism and discrimination requires ongoing patterns of upregulated physiological and inflammatory activity so that bodies are prepared for the uncertainties and vicissitudes that come with adverse conditions and experiences. When bodies fail to be consistently successfully "downregulated", the physiological burdens transition from being a burden to being an *allostatic load* (McEwen, 1998; Chae et al., 2011, 2019), and eventually, once the body is not just dysregulated but is truly failing, *allostatic overload* (McEwen & Akil, 2020; Bobba-Alves et al., 2022). SCI is, therefore, a key component in the shift from short- to long-term immune activation that the body and its regulatory dynamics adapt to in unfortunately damaging alterations to regulatory patterns, tissues, and organs (Bobba-Alves et al., 2022; Furman et al., 2019).

By providing the environmental conditions, opportunities, and affordances, social conditions place demands upon bodies both directly and through brain-mediated experiences (Cheadle et al., 2020, 2024; Goosby & Cheadle, 2024). Adversities from outside the body become embodied first in body states and dynamics, then in the shifting regulatory patterns, and ultimately in cellular and tissue changes, becoming the new baseline with altered dynamics and modified regulatory patterns (McEwen & Akil, 2020; Sterling, 2012; Arnaldo et al., 2022; Bobba-Alves et al., 2022). Inflammation has thus proven to be critical for understanding some important aspects of inequality and health. The immune system evolved to deal with acute stress, and so the persistent adversities endemic in our social systems, providing the axes of stratification and inequity, frequently ask the bodies of the disadvantaged to maintain constant vigilance in the face of uncertainty and potential threats. Self-protection becomes an additional toll asked implicitly of some by social systems, too often realized in patterns of physiological and psychological weathering (Geronimus et al., 2006; Forde et al., 2019).

Part II: Considering Complexity

Complex Systems

Our discussion so far has drawn upon analyses that usually assumed that social conditions and inflammatory activity function as (primarily) linear systems (e.g., Krakauer, 2024). *Linear systems*, depicted in Fig. 23.2a, which have long supplied the core foundation of the scientific enterprise, assume proportionality, independencies, and, of course, predictability. In a linear system, a change in one variable produces a corresponding proportional change in another, with the cause-effect relationship essentially straightforward and predictable. These systems assume that different components can be understood in relative (or conditional) isolation and that their interactions are (conditionally) additive, which allows for simplification and reductionism. While this approach has yielded significant insights and advancements, it may fall short when applied to the intricate, interconnected realities of many real-world phenomena when the system is a *complex (adaptive) system* (Briffaut & Krob, 2024; Krakauer, 2024; Page, 2015; Quintana, 2023). Complex systems, such

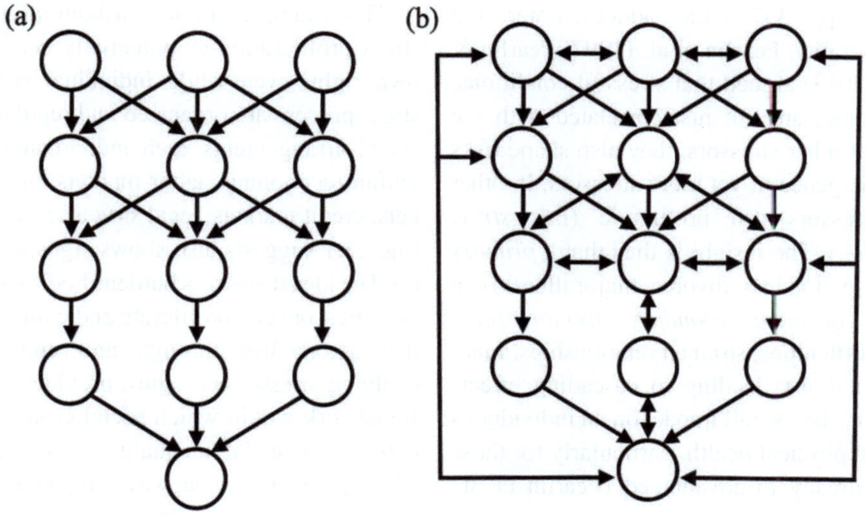

Fig. 23.2 Linear and complex system examples. (**a**) Shows a complicated but not complex linear system, and (**b**) shows a complex system

as those shown in Figs. 23.1 and 23.2b, often possess intricate interdependencies among components, non-linearities, feedback, self-organization, and emergent behaviors.

These systems consist of multiple interacting components whose collective dynamics cannot be fully understood by examining individual parts in isolation. Feedback loops, where outputs of a system loop back as inputs, amplify or dampen effects, leading to unpredictable and often non-proportional state changes. Complex systems can also be adaptive and self-organizing, evolving over time to create new functional behaviors and properties that emerge from the interactions among their components. Although linear models are perfectly capable of incorporating temporalities, complex systems cannot be understood independently of time. Such systems are dynamic flows where the state of the system at any given moment depends on its history and the unfolding interplay among its components, which *takes* time. The distinction is that time is not just a variable but a fundamental aspect of the system's behavior. For sociologists, adopting a complex systems perspective may be critical for understanding how social conditions affect health (for enlightening discussions of complexity in sociological research, see Page, 2015; Quintana, 2023; Mathews et al., 1999; von Förster & Morin, 2024). Per Fig. 23.1, we believe it is reasonable to suggest that interconnected areas such as economies, education systems, housing markets, and social networks, for example, form webs of interdependencies where changes in one domain can affect others, creating feedback loops and cascading effects at both structural and individual levels. Moreover, feedback loop consequences, such as the impact of poor health on employment prospects and vice versa, are fundamentally time-dependent.

Returning to the Stress Process and Systemic Racism

How might these ideas about complexity resonate with the stress process, the inflammatory and morbidity findings that we have described previously, systemic racism, and Black-White health disparities? Part of what Pearlin (1989) brought into the study of stress was in fact the *process* component, about which he noted that "it is especially striking that despite the attention given separately to the various components of stress, the intricate linkages that join them have not yet been unraveled... As a consequence, little is known of the manner in which the various components of stress are interconnected to form a *process*" (pg. 337, italics added). Later, and with temporality, Pearlin et al. (1997; Pearlin & Bierman, 2013) argued that stressful conditions, when chronic, are not just correlated with the presence of other stressors, they also shape risks for the emergence of yet more stressors. In other words, stressors can proliferate (i.e., *stress proliferation*). The insight is thus that a *primary stressor* (e.g., job loss, divorce, major illness) can beget one or more *secondary stressors* (e.g., financial difficulties, strained relationships, functional limitations) leading to cascading effects that amplify the overall impact on an individual's mental and physical health, particularly for those who are already disadvantaged (Pearlin et al., 2005).

A stressor can ripple through life domains, an insight accentuating the interconnected and compounding nature of stressors, increasing uncertainty about the future, creating frailties to future adversities, and producing potentially exacerbating feedbacks. For example, a major illness can not only proliferate into other stressors like functional limitations, fatigue, brain fog, and depression, it can also proliferate into new primary stressors such as job loss. Job loss can quite naturally lead to financial difficulties and threats to housing, bills and payments, and medical care, further undermining illness recovery. This proliferation, in turn, intensifies the original stressors and may add to them (e.g., anxiety), making it more difficult to find a new job and making the secondary stressors deriving from it worse, thereby perpetuating feedback loops of compounding stressors. The result can be decreasing robustness in the face of increasing uncertainty and unpredictability, making it harder to deal with and cope with future stressors, and changing social support dynamics as one's ability to engage reciprocally is also undermined. Consequently, events that might not have been threats previously become new hazards, and so the pool of potential burdens and liabilities can increase. Stress proliferation suggests the potential for cascades of predictable and unexpected adversities (Cheadle et al., 2024) because social conditions and consequences are interconnected and mutually influencing (e.g., Reskin, 2012).

This example provides a bottom-up view of stress proliferation as potentially *complex* in its own right, even while individual risks in the stress process are organized and regulated by the social arrangements each individual is situated within (economies, labor markets, housing markets, credit markets, legal structures, etc.), which Fig. 23.1 suggests also shows signs of complexity. The idea that stress burdens bodies and minds, that stressors can proliferate and complexify, and that actors live through and embody time-evolving stressor ecologies, provides a powerful framework within which social groups, and thus patterns of social inequalities, can be located. This is, of course, what we attempted to articulate by situating the stress process within the broader gestalt of systemic racism, structuring the conditions that regulate the risks of risk.

The critical and systemic view of racism depicted in Fig. 23.1 strongly evokes the notion of a *complex (adaptive) system* by virtue of the temporally adaptive interplay among the many interconnected social systems within and spanning layers of social organization from micro to macro and back again, each playing causal roles within their sphere of influence, and collectively constituting a robust and mutually reinforcing system. The historical concern is thus that many different components of U.S. society interlock together, buttressing one another, and that their mutual dynamics and internal mechanisms of action sustain the existing racial hierarchy and perpetuate the inequities of the past into the *now* (Mills, 2022; Wilkerson, 2020). Racism and its mechanisms of action are thus understood to create temporal path dependencies where the past is captured in the present and into the future. By virtue of such dependence, historical trajectories can become entrenched and thus increasingly challenging to change given their historical momentum.

The gestalt of this emergent and self-organizing system of racialized power and privilege increases the hazards of the stress ecologies of Black Americans and, in so doing, presents added challenges on average when compared to most other groups across life domains. The barriers formed thus block opportunities and make the world a more dangerous and threatening place, both materially and existentially, because the negative essentializing concept of "race" is interwoven into diverse social structures and many different aspects of social life, entangling macro-meso-microsociological realities.

The Immune System as a Complex System

While sociology is not the discipline that has given rise to complex systems and models, it shares a natural concern for the subset of such systems that are social. As Page (2015) notes, complexity science had not yet formed until after sociology originated, so complex systems concepts were not available for incorporation during its formative years. Noting this particular temporal path dependence, Page (pg. 37) asks us to "[i] magine... that the timing had been different, that complexity science had been explicated in full (this, by the way, has yet to occur) prior to the creation of sociology. Complex systems might then be considered foundational to the study of sociology given the resonance between the two disciplines." Even so, serious incorporation will likely involve some degree of paradigmatic shift, a move from linear to complex systems thinking and analytical practice.

Most sociologically relevant work on inflammation and the immune system is inspired, to some degree, by biomedical research and so proposes "mechanistic" models guided by linear systems thinking. Biology struggles with issues of complexity too and is historically much closer to the home of this kind of effort, given the myriad systems–in their manifold complexity–observable among living systems (e.g., Ma'ayan, 2017). Because of the rich complexity of the immune system, mapping the immune system over its many molecules, proteins, cells, mechanisms, processes, and dynamics remains ongoing and challenging (Ahmed & Hashish, 2006; Bhat et al., 2021; Parr et al., 2021; Thomas-Vaslin, 2017).

Moreover, much research on inflammation emphasizes signaling among circulating cytokines or acute phase proteins at one time point or across very long time horizons without the ability to measure specific cellular immune activity (e.g., Cuevas et al., 2020; Harris & Schorpp, 2018), thus hampering the ability to explicate the dynamic complex role of inflammation in biosocial processes (McDade, 2023; McDade et al., 2021). A preliminary complex system depiction of the immune system[7] is presented in Fig. 23.3a (Stanley & Lacy, 2010; Jawa et al., 2011; Chen et al., 2012; Becher et al., 2017), which shows a cytokine signaling network among immune cell families. Numerous chemical messaging mediators and communication pathways, some of which are proinflammatory or anti-inflammatory depending on circumstances, govern immune system coordination.

Figure 23.3b supplements with a quantitative cancer immune system network over 26 cells and 107 interaction links. This figure is telling in its complexity. Cell-to-cell regulatory relations are depicted in black (solid: activation; dashed: inhibition), cytokine activation/inhibition (solid/dashed) in red, and cytokine production in green (Wenbo & Wang, 2017, Figure 1, pg. 2). These figures provide a real-world example of both the various ways processes and mechanisms in the body can go wrong, and of the activities, interdependencies, and feedbacks of the immune markers increasingly available for assaying and modeling in social science data sets. The system appears complex, and its components are not *independent*, either with what has been measured within this system (e.g., IL6) or the interdependent processes *that likely have not been measured in the available data*. Even if they have, the time dimension is usually a single shot at a "wave," where the time dynamics are collapsed into states. This state-wavism, at its core, is the challenge of attempting to measure and model or predict with subsets of a complex system with linear systems tools and techniques. Immune function

[7] Compiled at ProteinLounge.com as "Cytokine Network" with the supporting citations replicated in text. URL: https://proteinlounge.com/cytokine-network-pg-176.html, last accessed on 08/6/2024 and included with permission.

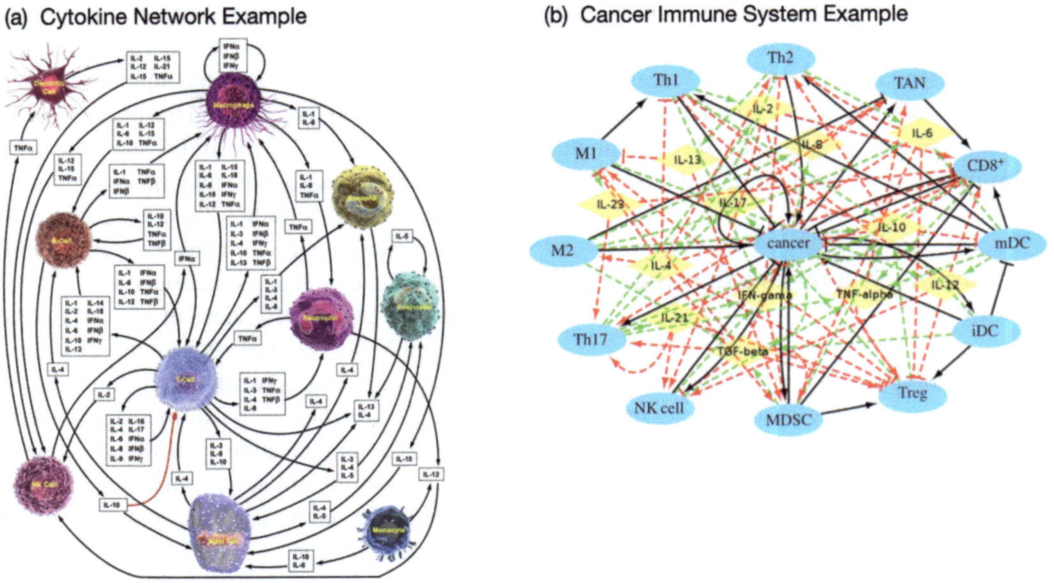

Fig. 23.3 Cytokine communication networks. (**a**) Illustrates cytokine signaling pathways among immune cell-types (created by ProteinLounge.com, used with permission). (**b**) The cancer immune system (26 nodes, 107 interaction links) among cells (cerulean ellipses) and cytokines (yellow diamonds). Interactions include: cell-cell activation (black, solid) and inhibition (black, bars), cytokine-cell activation (red, dashed arrows) and inhibition (red, dashed bars), and cell-cytokine production (green). Figure is reproduced from Wenbo and Wang (2017) with author permission

and inflammatory activity are ongoing and continuous dynamics of cellular activities, signaling, and regulatory cascades, proliferating and differentiating, akin but undoubtedly differing in many significant ways from the example of systemic racism in Fig. 23.1. As discussed above, immune activity is about more than merely responding to invaders in the body or parts of the body that are damaged or not working properly (e.g., cancer); immune activity and communication are part of the normal, routine, constant dialogue of bodily systems with each other.

Part III: Nested and Cross-System Complexities

System Components: Complex Systems (of Complex Systems (of ...))

Figure 23.1 discloses not just a complex arrangement of social structures that manifest systemic racism and contribute to discriminatory and exclusionary actions but also the fact that the components of such systems can likewise be complex and adaptive, both hierarchically nested and acting in tandem with other complex systems. We touched upon this when viewing the potential for complexity in the stress process, nested within a system of systemic racism for Black people, in terms of stress proliferation. In fact, the structural components of Fig. 23.1 are also enormously *complicated*, each in its own right potentially *complex*, belying linear simplicity (e.g., residential segregation). Systemic racism is thus a cross-system system that sits above both the different society-wide complex structures, which are themselves systems with complex interconnectednesses, temporal dynamics, and mechanisms of action (e.g., policies and practices).

The brain (in a limited metaphorical sense) is a bit like systemic racism in its regulatory and organizing roles in that it is the director and initiator of action. Racism matters because it is a tool for agency for some and subjugation for others (e.g., Williams, 2019), going back to slavery at the apex and through the complex patterns of social conflict and change that have unfolded since (Mills, 2022; Wilkerson, 2020). Figure 23.4 provides a larger integration of this notion of complex systems (of complex systems (of ...)) from a *neuro-bio-sociological* standpoint. Our discussion up to now has primarily reflected the self-loop for the peripheral immune system (e.g., not-self) and interactions with other bodily systems as a consequence of stress, as well as the direct path from the environment (e.g., pathogens or toxins from the outside getting in). The self-loop thus reminds us that the immune system is a complex system full of self and cross-system interactions. At the same time, the direct environmental pathway guides attention to the different ways that what is outside attempts to get into the body (e.g., toxins), including environmental inequalities and the direct effects they can have upon the body, as with environmental racism (e.g., toxic air and water pollution; Henderson & Wells, 2021; Lord et al., 2023), for example.

Beyond the body and world outside, Fig. 23.4 also includes the brain and the *neuroimmune* loop because the brain contains its own immune processes, cells, and communication networks (Dantzer, 2018). Cytokines in the brain regulate and curate neuronal structures and plasticity, by which immune signals can influence the brain's immune system by downregulating neuroendocrine feedback loops (e.g., HPA axis activity) (Quadt et al., 2018). While much of the literature on brain inflammation is based on animal models, a growing body of research shows that prolonged activation of circulating proinflammatory cytokines is linked to behavior, mood, psychiatric, and neurodegenerative diseases (Kipnis & Filiano, 2018). Although the mechanistic contributions are not yet fully elucidated, discrimina-

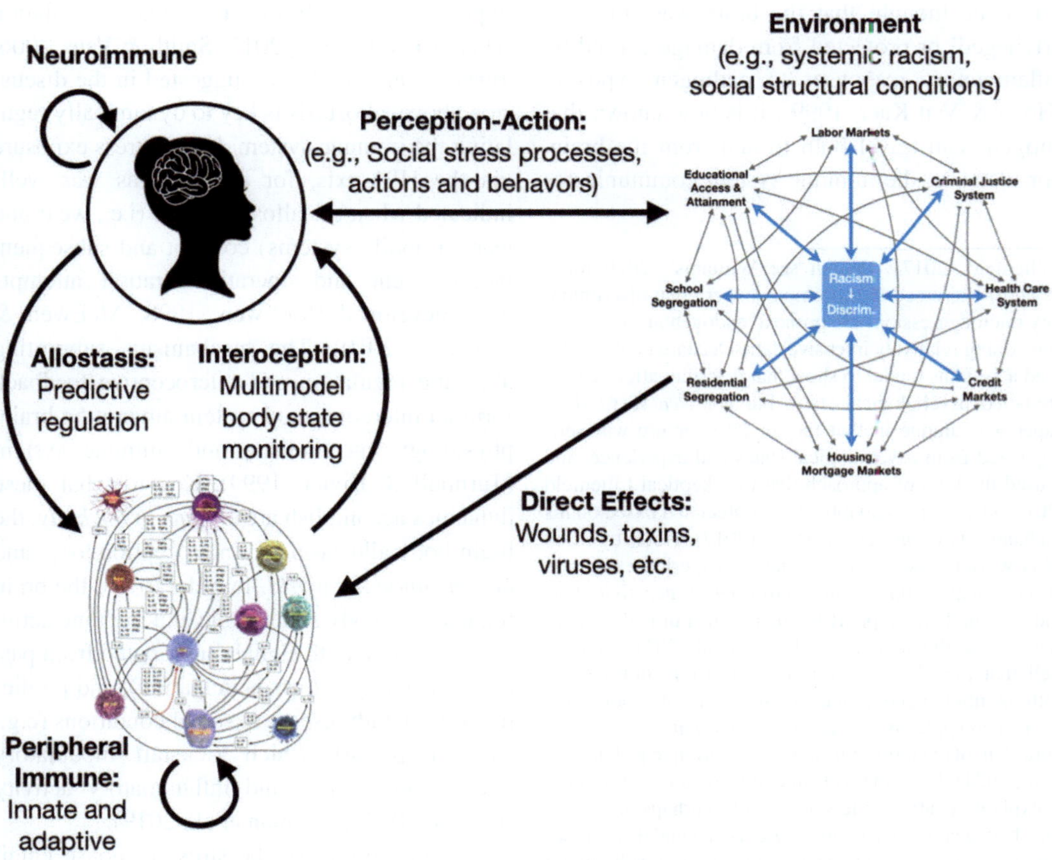

Fig. 23.4 Conceptual model of nested and cross-system complexities

tion[8] is also known to have adverse effects on the brain (Elbasheir et al., 2024) and appears to decrease functional connectivity of key nodes in the resting state network that are sensitive to and can be damaged by neuroinflammation (e.g., Morris et al., 2020; Wang et al., 2020). In fact, Elbasheir et al. (2024) found indirect associations with DNA methylation aging among Black women, suggesting contributions to allostatic load, discussed previously in connection to inflammation and inflammaging.

The neuroimmune system operates as a self-contained self-monitoring system and as part of an external regulatory loop that communicates with the rest of the body. This loop is depicted "downstream" from the brain as *allostasis* (predictive regulation of the body by the brain; Sterling, 2012) and "upstream" from the body to the brain as *interoception* (signals from body to brain; for a discussion, see Cheadle et al., 2024; Cheadle & Goosby, 2025, Chap. 16). Although it was once thought that the brain was 'immune privileged' or protected from damage caused by inflammatory responses to pathogen exposure (Hong & Van Kaer, 1999), it is now known that antigens can travel both to and from the brain. For instance, the immune system communicates with the brain through cellular and molecular pathways. Peripheral inflammation can lead to the release of proinflammatory cytokines (e.g., TNFα, IL-1β, and IL-6), which may cross the blood-brain barrier and activate cytokine receptors on resident immune cells in the brain, such as microglia (D'Mello & Swain, 2017). While much of the literature on brain inflammation is based on animal models, a growing body of research shows that prolonged activation of circulating proinflammatory cytokines is linked to behavior, mood, psychiatric, and neurodegenerative diseases as previously noted (Kipnis & Filiano, 2018).

The body tends towards system interdependencies, coordination, and collaboration via signaling cascades and prioritized resource sharing (Sterling, 2015, 2020). A primary means of brain-to-immune system collaboration is via allostatic brain signaling (e.g., HPA-axis), which has been key to developing an appreciation for the importance of biosocial stress regulation (Hostinar & Gunnar, 2017; Smith & Vale, 2006; Turner et al., 2020). As suggested in the discussion above, allostasis is key to dynamically regulating the immune system during stress exposure via the HPA-axis, for example, as was well-indicated when the allostatic load (i.e., wear and tear on bodily systems) concept and subsequent measurement and operationalization attempts were developed (McEwen, 1998; McEwen & Gianaros, 2010). The mechanisms supporting allostatic regulation and interoceptive feedback form an interconnected system among the brain, physiology, metabolism, and immune system (Turnbull & Rivier, 1999). Key to what these dynamics accomplish are *actions in the body*, the brain-body allostasis and interoception loop, and *actions upon the world*. In order to act, the brain regulates its body's metabolic and immune activity based upon both signals anticipated from past experience (e.g., stress anticipation), and prediction errors with respect to social conditions (e.g., stress response), which kickstart modulatory stress physiological and inflammatory activity (Barrett, 2017b; Fridman et al., 2019).

The metabolics of the stress response entail actions by the body on the world (whether one

[8]Lilienfeld (2017a; though see Williams', 2020 for a rebuttal) expressed the common concern that discriminatory microaggressions are prone to endogenous negativity bias. Using relatively intensive data, Deckard et al. (2023) used a systems model to show that discrimination reports predicted psychological states but not *vice versa*. This paper was unique in that the model structure was self-organized from psychological states and experience, and it used the kind of approach that the skeptical Lilienfeld (2017b) would presumably have appreciated based on his exchange with Ong and Burrow (2017). Note, too, that other work co-locates discriminatory experiences in time in ecological settings while controlling for negativity bias and also finds no support for mental health endogeneity biases (Cheadle et al., 2020; Jelsma et al., 2021). Note as well that Fig. 23.1 suggests that discrimination can be both intentional and unintentional because it is not a personal "bad apple" issue, rather it is *inherent in the system*, a problem of enculturation and socialization (e.g., Cheadle et al., 2024). Even the well-intentioned can hold implicit or explicit negative beliefs as a result. Perhaps the difference between *doing* and *not doing* racism and discrimination is, in part, other-awareness and conscious role-taking (e.g., Cheadle & Goosby, 2025, Chap. 16).

actually *acts* or not), as that is what energy release is for--what need for energy *later* if the tiger makes a meal of you *now* (see Sapolsky, 2004)? Of course, when the costs are long-running, sickness behaviors come into play, and the body is forced to slow down or come to a stop. From this vantage of interacting systems of nested systems, it is hard to see where it is that linear systems conceptualizations should reign when the relevant systems, whether inside or outside the body, tend towards what appears to be high complexity. The point is not that prior findings guided by linear system models and theories cannot be trusted but that a looming uncertainty about our understanding hangs over our heads, and it is less about what is known and more about how what is missing may add context or recontextualize what we know so far. Both systemic racism and Black-White health inequities are richly documented, as are correlations between health outcomes and various immune-related processes and outcomes, and it is because of this biosocial confluence that we have referred primarily to *inequity* rather than *disparity* or *inequality* (Lee et al., 2020).

A healthy, well-functioning immune system is more than just a collection of mechanisms; its components interact in interdependent and dynamic ways among both its own components and those of other systems, with health and well-being emerging from the system as a whole, rather than from any single part—until, at least, a mission-critical component fails, such as in the case of a genetic disorder or allostatic overload. However, stress and social conditions matter precisely because the capacities of the immune system and the other systems interact with each other and change with experience, just as slavery ended long ago and racism has changed but remains "embodied" within a diversity of social structures and minds and is expressed by action (e.g., laws, policies and procedures, interpersonal discrimination). Social, physiological, and neurological systems are built from highly complex and multidimensional arrays of mechanisms, which is why we have sought to emphasize their natures as complex rather than merely complicated. Although we do, in fact, know a great deal and know there remain unknowns, we should be humble in the understanding that our paradigms are largely linear and are restricted by the constraints imposed thereby. When complexity comes into play, it tends to bring *unknown unknowns* along for the ride, which show themselves in nonlinearities, feedbacks, patterns of organization and reorganization, and emergent state transitions.

Conclusion

Racism is no single "structure," it is not a single "thing," it is a *system of relations* among adaptive social structures interacting and providing feedback within and across layers of social organization through human actions, preserving core features of society, stabilizing or at least protecting the status quo, and regulating social change. One might say that systemic racism has no center and that it encircles whatever stands within it. To say "it" affects "life chances" is to say that it affects *life* in its higher-order organizing of structural, local, and personally intimate details (Bonilla-Silva, 2015b; Braveman et al., 2022). The problem is not just a matter of material conditions but the body's material reality, its actual embodied *materiality* (Goosby et al., 2018). The promise of understanding a complex system is that a leverage point can tip the whole thing into a new state, triggering the transition from a *racialized* state to one of *antiracism*, for example. While the political economy of racism has shifted and changed over the years, that big transition has yet to take place. It is hard to look at Fig. 23.1, admittedly a gross simplification, and even in this reduced state, imagine where such a leverage point, a *true mechanism of racism* amenable to intervention, might reside. Still, we hope it is there waiting to be viewed from the right vantage and that it can be capitalized upon to kickstart a systemic cascade that produces a phase transition towards a more equal and just world.

Recognizing that systemic racism is a complex rather than linear system may help to inspire the insights needed to make such large-scale transitions happen. Doing so would take many of the needs to understand the immune system off the table, in a sense, by improving the social conditions that structure the stress processes that undermine health. Many epistemic and pragmatic reasons would remain to understand the immune system in all its complexity, of course. But doing so will probably rely on the understanding of it as a complex system so that the many mechanisms and interactions supporting it can be put into the context of the system, where it is at, where it is going, and how changes cascade, feedback, are magnified, and absorbed. The paradigm shift we ponder here will likely be difficult to bring to fruition, and it remains to be shown how sociology would benefit from it. However, that evaluation requires effort and risk as the theories, methods, and statistical modeling approaches that scaffold and supply the materials of modern quantitative research are highly path-dependent, resulting in a great deal of investment and much on the line for individual scholars.[9]

Yet reliance on simple linear frameworks, the core[10] of which any good student with a background in calculus can derive for themselves, strikes us as unrealistic for understanding many kinds of sociological and biological processes. Yet moving beyond this paradigm may be required if the kinds of structural sociological and biosocial analyses that Figs. 23.1 and 23.3a, b suggest are to be achieved. If physicists had only analyzed what they could see directly, they would have seen very little. The field would never have discovered the weird and wonderful quantum world of which our universe is made (consider Rovelli, 2021; Adlam & Rovelli, 2022; Lombardi & Ardenghi, 2022), and similar things could be said for chemistry, biology, their offshoots and intersections, and so on. We should be careful not to confuse our rather *convenient* linear models, requiring quite straightforward, simplistic, and temporally narrow or sparse data structures, with the actual world out there--manifold in its complexity and unfolding in real-time, everywhere all at once. To quote the aphorism often incorrectly attributed to Einstein, “make everything as simple as possible, but not simpler.” Indeed, it is the second clause that we hope to see sociology, biosociology, and neurosociology take strides towards unraveling, as it is the first that linear systems thinking may just over-supply.

Funding This research was supported by grant, P30AG066614, awarded to the Center on Aging and Population Sciences at The University of Texas at Austin by the National Institute on Aging, and by grant, P2CHD042849, awarded to the Population Research Center at The University of Texas at Austin by the Eunice Kennedy Shriver National Institute of Child Health and Human Development. The content of this article is solely the responsibility of the authors and does not necessarily represent the official views of the National Institutes of Health or the University of Texas at Austin.

References

Adlam, E., & Rovelli, C. (2022). *Information is physical: Cross-perspective links in relational quantum mechanics*.

[9]The exchange between Williams (2019) and Iceland (2019) might be informative as it echoes the distinction we have made between complex and linear systems, respectively, as *paradigms* with stakes worth defending to proponents. We interpret Williams' as arguing that one cannot explain *unjust* racial *inequities* by relying on system components (e.g., socioeconomic status or human capital) in isolation or independently from other structural factors, including historical trajectories and path dependencies, evoking the concept of complex systems, interdependence, self-organization, and emergence. Iceland's staunch *ideological* defense of the linear system paradigms and empirical data as the only admissible evidence overlooks the complexity and interdependencies such frameworks inherently fail to capture. This defense, offered in the name of a narrow and contestable view of science, is an *irony involontaire*, as it inadvertently undermines the comprehensive understanding of racial inequities by trapping inquiry in a constraining and particularizing ideology of science, raising a kind of mirror to the very biases he accuses Williams and other race/ethnicity scholars of.

[10]We have a keen appreciation for how sophisticated and diverse statistical estimators and techniques have become (e.g., Cheadle et al., 2010), including those that, by design, evoke some degrees of complexity (e.g., Cheadle et al., 2013; Deckard et al., 2023), and others which drift far afield from the typical sociological wheelhouse (e.g., Cheadle et al., 2020; Kiat et al., 2018a, 2018b), to say nothing of what is being achieved in (e.g.) machine learning.

Ahmed, E., & Hashish, A. H. (2006). On modelling the immune system as a complex system. *Theory in Biosciences, 124*(3), 413–418. https://doi.org/10.1016/j.thbio.2005.07.001

Al Kibria, G. M. (2019). Racial/ethnic disparities in prevalence, treatment, and control of hypertension among US adults following application of the 2017 American College of Cardiology/American Heart Association Guideline. *Preventive Medicine Reports, 14*, 100850. https://doi.org/10.1016/j.pmedr.2019.100850

Aneshensel, C. S., & Mitchell, U. A. (2014). The stress process: Its origins, evolution, and future. In R. J. Johnson, R. J. Turner, & B. G. Link (Eds.), *Sociology of mental health: Selected topics from forty years 1970s-2010s, SpringerBriefs in Sociology* (pp. 53–74). Springer International.

Arnaldo, I., Corcoran, A. W., Friston, K. J., & Ramstead, M. J. D. (2022). Stress and its sequelae: An active inference account of the etiological pathway from allostatic overload to depression. *Neuroscience & Biobehavioral Reviews, 135*, 104590. https://doi.org/10.1016/j.neubiorev.2022.104590

Bailey, Z. D., Feldman, J. M., & Bassett, M. T. (2021). How structural racism works — Racist policies as a root cause of U.S. racial health inequities. *New England Journal of Medicine, 384*(8), 768–773. https://doi.org/10.1056/NEJMms2025396

Bailey, Z. D., Krieger, N., Agénor, M., Graves, J., Linos, N., & Bassett, M. T. (2017). Structural racism and health inequities in the USA: Evidence and interventions. *The Lancet, 389*(10077), 1453–1463. https://doi.org/10.1016/S0140-6736(17)30569-X

Barbrook-Johnson, P., & Carrick, J. (2022). Combining complexity-framed research methods for social research. *International Journal of Social Research Methodology, 25*(6), 835–848. https://doi.org/10.1080/13645579.2021.1946118

Barrett, L. F. (2017a). *How emotions are made: The secret life of the brain*. Houghton Mifflin Harcourt.

Barrett, L. F. (2017b). The theory of constructed emotion: An active inference account of interoception and categorization. *Social Cognitive and Affective Neuroscience, 12*(1), 1–23.

Barrett, L. F., Quigley, K. S., & Hamilton, P. (2016). An active inference theory of allostasis and interoception in depression. *Philosophical Transactions of the Royal Society B: Biological Sciences, 371*(1708), 20160011. https://doi.org/10.1098/rstb.2016.0011

Barrows, I. R., Ramezani, A., & Raj, D. S. (2019). Inflammation, immunity, and oxidative stress in hypertension—Partners in crime? *Advances in Chronic Kidney Disease, 26*(2), 122–130. https://doi.org/10.1053/j.ackd.2019.03.001

Becher, B., Spath, S., & Goverman, J. (2017). Cytokine networks in neuroinflammation. *Nature Reviews. Immunology, 17*(1), 49–59. https://doi.org/10.1038/nri.2016.123

Bennett, J. M., Reeves, G., Billman, G. E., & Sturmberg, J. P. (2018). Inflammation–nature's way to efficiently respond to all types of challenges: Implications for understanding and managing 'the epidemic' of chronic diseases. *Frontiers in Medicine, 5*.

Bhat, A., Parr, T., Ramstead, M., & Friston, K. (2021). Immunoceptive inference: Why are psychiatric disorders and immune responses intertwined? *Biology & Philosophy, 36*(3), 27. https://doi.org/10.1007/s10539-021-09801-6

Bobba-Alves, N., Juster, R.-P., & Picard, M. (2022). The energetic cost of allostasis and allostatic load. *Psychoneuroendocrinology*, 105951. https://doi.org/10.1016/j.psyneuen.2022.105951

Bonaguro, L., Schulte-Schrepping, J., Ulas, T., Aschenbrenner, A. C., Beyer, M., & Schultze, J. L. (2022). A guide to systems-level immunomics. *Nature Immunology, 23*(10), 1412–1423. https://doi.org/10.1038/s41590-022-01309-9

Bonilla-Silva, E. (1997). Rethinking racism: Toward a structural interpretation. *American Sociological Review, 62*(3), 465–480. https://doi.org/10.2307/2657316

Bonilla-Silva, E. (2015a). More than prejudice: Restatement, reflections, and new directions in critical race theory. *Sociology of Race and Ethnicity, 1*(1), 73–87. https://doi.org/10.1177/2332649214557042

Bonilla-Silva, E. (2015b). The structure of racism in color-blind, 'post-racial' America. *American Behavioral Scientist, 59*(11), 1358–1376.

Bonilla-Silva, E. (2019). Toward a new political praxis for Trumpamerica: New directions in critical race theory. *American Behavioral Scientist, 63*(13), 1776–1788. https://doi.org/10.1177/0002764219842614

Braveman, P. A., Arkin, E., Proctor, D., Kauh, T., & Holm, N. (2022). Systemic and structural racism: Definitions, examples, health damages, and approaches to dismantling. *Health Affairs, 41*(2), 171–178. https://doi.org/10.1377/hlthaff.2021.01394

Brayboy, B. M. J. (2005). Toward a tribal critical race theory in education. *The Urban Review, 37*(5), 425–446. https://doi.org/10.1007/s11256-005-0018-y

Briffaut, J.-P., & Krob, D. (2024). Complexity in the social sciences. In *Complexities 2* (pp. 119–141). Wiley.

Brown, L. L., Mitchell, U. A., & Ailshire, J. A. (2020). Disentangling the stress process: Race/ethnic differences in the exposure and appraisal of chronic stressors among older adults. *The Journals of Gerontology: Series B, 75*(3), 650–660. https://doi.org/10.1093/geronb/gby072

Chae, D. H., Martz, C. D., Fuller-Rowell, T. E., Spears, E. C., Smith, T. T. G., Hunter, E. A., Drenkard, C., & Sam Lim, S. (2019). Racial discrimination, disease activity, and organ damage: The Black women's experiences living with Lupus (BeWELL) study. *American Journal of Epidemiology, 188*(8), 1434–1443. https://doi.org/10.1093/aje/kwz105

Chae, D., Nuru-Jeter, A., Lincoln, K., & Francis, D. (2011). Conceptualizing racial disparities in health. *Du Bois Review: Social Science Research on Race, 8*, 63–77. https://doi.org/10.1017/S1742058X11000166

Cheadle, J. E., Amato, P. R., & King, V. (2010). Patterns of nonresident father contact. *Demography, 47*(1), 206–225. https://doi.org/10.1353/dem.0.0084

Cheadle, J. E., Davidson-Turner, K. J., & Goosby, B. J. (2024). Active inference and social actors: Towards a neuro-bio-social theory of brains and bodies in their worlds. *Cologne Journal of Sociology and Social Psychology*. https://doi.org/10.1007/s11577-024-00936-4

Cheadle, J. E., Goosby, B. J., Jochman, J. C., Tomaso, C. C., Kozikowski Yancey, C. B., & Nelson, T. D. (2020). Race and ethnic variation in college students' allostatic regulation of racism-related stress. *Proceedings of the National Academy of Sciences*. https://doi.org/10.1073/pnas.1922025117

Cheadle, J. E., Stevens, M., Williams, D. T., & Goosby, B. J. (2013). The differential contributions of teen drinking homophily to new and existing friendships: An empirical assessment of assortative and proximity selection mechanisms. *Social Science Research, 42*(5), 1297–1310. https://doi.org/10.1016/j.ssresearch.2013.05.001

Chen, E., Tianyi, Y., Brody, G. H., Lam, P. H., Goosby, B. J., & Miller, G. E. (2022). Discrimination and inflammation in adolescents of color. *Biological Psychiatry Global Open Science*. https://doi.org/10.1016/j.bpsgos.2022.02.008

Chen, S.-J., Wang, Y.-L., Fan, H.-C., Lo, W.-T., Wang, C.-C., & Sytwu, H.-K. (2012). Current Status of the immunomodulation and immunomediated therapeutic strategies for multiple sclerosis. *Clinical & Developmental Immunology*, 1–16. https://doi.org/10.1155/2012/970789

Chen, W. G., Schloesser, D., Arensdorf, A. M., Simmons, J. M., Cui, C., Valentino, R., Gnadt, J. W., Nielsen, L., Hillaire-Clarke, C. S., Spruance, V., Horowitz, T. S., Vallejo, Y. F., & Langevin, H. M. (2021). The emerging science of interoception: Sensing, integrating, interpreting, and regulating signals within the self. *Trends in Neurosciences, 44*(1), 3–16. https://doi.org/10.1016/j.tins.2020.10.007

Cuevas, A. G., Ong, A. D., Carvalho, K., Ho, T., Chan, S. W. C., Allen, J. D., Chen, R., Rodgers, J., Biba, U., & Williams, D. R. (2020). Discrimination and systemic inflammation: A critical review and synthesis. *Brain, Behavior, and Immunity, 89*, 465–479. https://doi.org/10.1016/j.bbi.2020.07.017

Dantzer, R. (2018). Neuroimmune interactions: From the brain to the immune system and vice versa. *Physiological Reviews, 98*(1), 477–504. https://doi.org/10.1152/physrev.00039.2016

Deckard, F. M., Messamore, A., Goosby, B. J., & Cheadle, J. E. (2023). A network approach to assessing the relationship between discrimination and daily emotion dynamics. *Social Psychology Quarterly, 86*(3), 334–356. https://doi.org/10.1177/01902725221123577

Del Giudice, M., & Gangestad, S. W. (2018). Rethinking IL-6 and CRP: Why they are more than inflammatory biomarkers, and why it matters. *Brain, Behavior, and Immunity, 70*, 61–75. https://doi.org/10.1016/j.bbi.2018.02.013

Delgado, R., & Stefancic, J. (2023). *Critical race theory, Fourth Edition: An introduction* (4th ed.). NYU Press.

D'Mello, C., & Swain, M. G. (2017). Immune-to-brain communication pathways in inflammation-associated sickness and depression. In R. Dantzer & L. Capuron (Eds.), *Inflammation-associated depression: Evidence, mechanisms and implications, current topics in behavioral neurosciences* (pp. 73–94). Springer International.

Dowd, J. B., & Aiello, A. E. (2009). Socioeconomic differentials in immune response. *Epidemiology, 20*(6), 902. https://doi.org/10.1097/EDE.0b013e3181bb5302

Eberl, G. (2016). Immunity by equilibrium. *Nature Reviews Immunology, 16*(8), 524–532. https://doi.org/10.1038/nri.2016.75

Eisenberger, N. I., & Cole, S. W. (2012). Social neuroscience and health: Neurophysiological mechanisms linking social ties with physical health. *Nature Neuroscience, 15*(5), 669–674. https://doi.org/10.1038/nn.3086

Eisenberger, N. I., Moieni, M., Inagaki, T. K., Muscatell, K. A., & Irwin, M. R. (2017). In sickness and in health: The co-regulation of inflammation and social behavior. *Neuropsychopharmacology, 42*(1), 242–253. https://doi.org/10.1038/npp.2016.141

Elbasheir, A., Katrinli, S., Kearney, B. E., Lanius, R. A., Harnett, N. G., Carter, S. E., Ely, T. D., Bradley, B., Gillespie, C. F., Stevens, J. S., Lori, A., van Rooij, S. J. H., Powers, A., Jovanovic, T., Smith, A. K., & Fani, N. (2024). Racial discrimination, neural connectivity, and epigenetic aging among black women. *JAMA Network Open, 7*(6), e2416588. https://doi.org/10.1001/jamanetworkopen.2024.16588

Feagin, J., & Elias, S. (2013). Rethinking racial formation theory: A systemic racism critique. *Ethnic and Racial Studies, 36*(6), 931–960. https://doi.org/10.1080/01419870.2012.669839

Feagin, J. R. (2006). *Systemic racism: A theory of oppression*. Routledge, Taylor & Francis Group.

Ferrucci, L., & Fabbri, E. (2018). Inflammageing: Chronic inflammation in ageing, cardiovascular disease, and frailty. *Nature Reviews Cardiology, 15*(9), 505–505. https://doi.org/10.1038/s41569-018-0064-2

Forde, A. T., Crookes, D. M., Suglia, S. F., & Demmer, R. T. (2019). The weathering hypothesis as an explanation for racial disparities in health: A systematic review. *Annals of Epidemiology, 33*, 1–18.e3. https://doi.org/10.1016/j.annepidem.2019.02.011

von Förster, H., & Morin, E. (2024). Complexity and models in social and human sciences. In *Complexities 2* (pp. 33–69). Wiley.

Franceschi, C., Garagnani, P., Parini, P., Giuliani, C., & Santoro, A. (2018). Inflammaging: A new immune–metabolic viewpoint for age-related diseases. *Nature Reviews Endocrinology, 14*(10), 576–590. https://doi.org/10.1038/s41574-018-0059-4

Franklin, T. C., Chelsea, X., & Duman, R. S. (2018). Depression and sterile inflammation: Essential role of danger associated molecular patterns. *Brain, Behavior, and Immunity, 72*, 2–13. https://doi.org/10.1016/j.bbi.2017.10.025

Fridman, J., Barrett, L. F., Wormwood, J. B., & Quigley, K. S. (2019). Applying the theory of constructed emotion to police decision making. *Frontiers in Psychology, 10*. https://doi.org/10.3389/fpsyg.2019.01946

Furman, D., Campisi, J., Verdin, E., Carrera-Bastos, P., Targ, S., Franceschi, C., Ferrucci, L., Gilroy, D. W., Fasano, A., Miller, G. W., Miller, A. H., Mantovani, A., Weyand, C. M., Barzilai, N., Goronzy, J. J., Rando, T. A., Effros, R. B., Lucia, A., Kleinstreuer, N., & Slavich, G. M. (2019). Chronic inflammation in the etiology of disease across the life span. *Nature Medicine, 25*(12), 1822–1832. https://doi.org/10.1038/s41591-019-0675-0

Geronimus, A. T., Hicken, M., Keene, D., & Bound, J. (2006). 'Weathering' and age patterns of allostatic load scores among blacks and whites in the United States. *American Journal of Public Health, 96*(5), 826–833.

Glass, T. A., & McAtee, M. J. (2006). Behavioral science at the crossroads in public health: Extending horizons, envisioning the future. *Social Science & Medicine, 62*(7), 1650–1671. https://doi.org/10.1016/j.socscimed.2005.08.044

Goosby, B. J., & Cheadle, J. E. (2024). The immune system is a complex system: Inflammatory morbidity and systemic racism. *Cologne Journal of Sociology and Social Psychology*. https://doi.org/10.1007/s11577-024-00952-4

Goosby, B. J., Cheadle, J. E., & Mitchell, C. (2018). Stress-related biosocial mechanisms of discrimination and African American health inequities. *Annual Review of Sociology, 44*(1), 319–340. https://doi.org/10.1146/annurev-soc-060116-053403

Goosby, B. J., Malone, S., Richardson, E. A., Cheadle, J. E., & Williams, D. T. (2015). Perceived discrimination and markers of cardiovascular risk among low-income African American youth. *American Journal of Human Biology, 27*(4), 546–552. https://doi.org/10.1002/ajhb.22683

Graetz, N., Boen, C. E., & Esposito, M. H. (2022). Structural racism and quantitative causal inference: A life course mediation framework for decomposing racial health disparities. *Journal of Health and Social Behavior, 63*(2), 232–249. https://doi.org/10.1177/00221465211066108

Graves, J. L., Jr. (2001). *The emperor's new clothes: Biological theories of race at the millennium* (1st ed.). Rutgers University Press.

Grusky, D. (2019). *Social stratification, class, race, and gender in sociological perspective* (2nd ed.). Routledge.

Gurven, M. D., & Lieberman, D. E. (2020). WEIRD bodies: Mismatch, medicine and missing diversity. *Evolution and Human Behavior, 41*(5), 330–340. https://doi.org/10.1016/j.evolhumbehav.2020.04.001

Harris, K. M., & Schorpp, K. M. (2018). Integrating biomarkers in social stratification and health research. *Annual Review of Sociology, 44*(1), 361–386. https://doi.org/10.1146/annurev-soc-060116-053339

Henderson, S., & Wells, R. (2021). Environmental racism and the contamination of black lives: A literature review. *Journal of African American Studies, 25*(1), 134–151. https://doi.org/10.1007/s12111-020-09511-5

Homan, P. A., & Brown, T. H. (2022). Sick and tired of being excluded: Structural racism in disenfranchisement as a threat to population health equity. *Health Affairs, 41*(2), 219–227. https://doi.org/10.1377/hlthaff.2021.01414

Hong, S., & Van Kaer, L. (1999). Immune privilege. *The Journal of Experimental Medicine, 190*(9), 1197–1200.

Hostinar, C. E., & Gunnar, M. R. (2017). Future directions in the study of social relationships as regulators of the HPA axis across development. In *Future work in clinical child and adolescent psychology*. Routledge.

Hutchinson, J. B., & Barrett, L. F. (2019). The power of predictions: An emerging paradigm for psychological research. *Current Directions in Psychological Science, 28*(3), 280–291. https://doi.org/10.1177/0963721419831992

Iceland, J. (2019). Reply to 'A call to focus on racial domination and oppression: A response to "Racial and Ethnic Inequality in Poverty and Affluence, 1959–2015"'. *Population Research and Policy Review, 38*(5), 665–669. https://doi.org/10.1007/s11113-019-09542-1

Ignatow, G. (2021). Cognitive sociology after relational biology. *Sociological Forum, 36*(S1), 1253–1270. https://doi.org/10.1111/socf.12764

Jawa, R. S., Anillo, S., Huntoon, K., Baumann, H., & Kulaylat, M. (2011). Analytic review: Interleukin-6 in surgery, trauma, and critical care: Part I: Basic science. *Journal of Intensive Care Medicine, 26*(1), 3–12. https://doi.org/10.1177/0885066610395678

Jelsma, E. B., Goosby, B. J., & Cheadle, J. E. (2021). Do trait psychological characteristics moderate sympathetic arousal to racial discrimination exposure in a natural setting? *Psychophysiology, 58*(4), e13763. https://doi.org/10.1111/psyp.13763

Kiat, J. E., Cheadle, J. E., & Goosby, B. J. (2018a). The impact of social exclusion on anticipatory attentional processing. *International Journal of Psychophysiology, 123*, 48–57. https://doi.org/10.1016/j.ijpsycho.2017.11.010

Kiat, J. E., Cheadle, J. E., & Goosby, B. J. (2018b). Wait for it: The impact of social exclusion on anticipatory attentional processing. *International Journal of Psychophysiology, 123*, 48–57.

Killewald, A., Pfeffer, F. T., & Schachner, J. N. (2017). Wealth inequality and accumulation. *Annual Review of Sociology, 43*, 379–404. https://doi.org/10.1146/annurev-soc-060116-053331

Kim, J. K., Faul, J., Weir, D. R., & Crimmins, E. M. (2024). Dried blood spot based biomarkers in the health and retirement study: 2006 to 2016. *American Journal of Human Biology, 36*(2), e23997. https://doi.org/10.1002/ajhb.23997

Kipnis, J., & Filiano, A. J. (2018). The central nervous system: Privileged by immune connections. *Nature Reviews Immunology, 18*(2), 83–84. https://doi.org/10.1038/nri.2017.152

Kirkinis, K., Pieterse, A. L., Martin, C., Agiliga, A., & Brownell, A. (2021). Racism, racial discrimination, and trauma: A systematic review of the social science literature. *Ethnicity & Health, 26*(3), 392–412. https://doi.org/10.1080/13557858.2018.1514453

Krakauer, D. (2024). *The complex world: An introduction to the foundations of complexity science*. P. Intro-3 in.

Lam, P., Chen, E., Chiang, J., & Miller, G. (2022). Socioeconomic disadvantage, chronic stress, and proinflammatory phenotype: An integrative data analysis across the lifecourse. *PNAS Nexus*. https://doi.org/10.1093/pnasnexus/pgac219

Lee, H., Kim, D., Lee, S. A., & Fawcett, J. (2020). The concepts of health inequality, disparities and equity in the era of population health. *Applied Nursing Research, 56*, 151367. https://doi.org/10.1016/j.apnr.2020.151367

Lilienfeld, S. O. (2017a). Microaggressions: Strong claims, inadequate evidence. *Perspectives on Psychological Science, 12*(1), 138–169.

Lilienfeld, S. O. (2017b). Through a glass, darkly: Microaggressions and psychological science. *Perspectives on Psychological Science, 12*(1), 178–180.

Link, B. G., & Phelan, J. (1995). Social conditions as fundamental causes of disease. *Journal of Health and Social Behavior*, 80–94. https://doi.org/10.2307/2626958

Lombardi, O., & Ardenghi, J. S. (2022). How different interpretations of quantum mechanics can enrich each other: The case of the relational quantum mechanics and the modal-hamiltonian interpretation. *Foundations of Physics, 52*(3), 1–21. https://doi.org/10.1007/s10701-022-00580-y

Lord, B. D., Harris, A. R., & Ambs, S. (2023). The impact of social and environmental factors on cancer biology in Black Americans. *Cancer Causes & Control, 34*(3), 191–203. https://doi.org/10.1007/s10552-022-01664-w

Louie, P., Upenieks, L., & Hill, T. D. (2023). Cumulative pandemic stressors, psychosocial resources, and psychological distress: Toward a more comprehensive test of a pandemic stress process. *Society and Mental Health, 13*(3), 245–260. https://doi.org/10.1177/21568693231165260

Ma'ayan, A. (2017). Complex systems biology. *Journal of the Royal Society Interface, 14*(134), 20170391. https://doi.org/10.1098/rsif.2017.0391

Marshall, J. S., Warrington, R., Watson, W., & Kim, H. L. (2018). An introduction to immunology and immunopathology. *Allergy, Asthma & Clinical Immunology, 14*(2), 49. https://doi.org/10.1186/s13223-018-0278-1

Martinez, A., de la Rosa, R., Mujahid, M., & Thakur, N. (2021). Structural racism and its pathways to asthma and atopic dermatitis. *Journal of Allergy and Clinical Immunology, 148*(5), 1112–1120. https://doi.org/10.1016/j.jaci.2021.09.020

Massey, D., & Denton, N. (1993). *American apartheid: Segregation and the making of the underclass*. Harvard University Press.

Mathews, K. M., White, M. C., & Long, R. G. (1999). Why study the complexity sciences in the social sciences? *Human Relations, 52*(4), 439–462. https://doi.org/10.1177/001872679905200402

McCall, L., & Percheski, C. (2010). Income inequality: New trends and research directions. *Annual Review of Sociology, 36*, 329–347. https://doi.org/10.1146/annurev.soc.012809.102541

McCartney, G., Dickie, E., Escobar, O., & Collins, C. (2021). Health inequalities, fundamental causes and power: Towards the practice of good theory. *Sociology of Health & Illness, 43*(1), 20–39. https://doi.org/10.1111/1467-9566.13181

McDade, T. W. (2023). Three common assumptions about inflammation, aging, and health that are probably wrong. *Proceedings of the National Academy of Sciences, 120*(51), e2317232120. https://doi.org/10.1073/pnas.2317232120

McDade, T. W., Aronoff, J. E., Leigh, A. K. K., Finegood, E. D., Weissman-Tsukamoto, R. M., Brody, G. H., & Miller, G. E. (2021). Out of the laboratory and into the field: Validation of portable cell culture protocols. *Psychosomatic Medicine, 83*(3), 283. https://doi.org/10.1097/PSY.0000000000000923

McDade, T. W., Georgiev, A. V., & Kuzawa, C. W. (2016). Trade-offs between acquired and innate immune defenses in humans. *Evolution, Medicine, and Public Health, 2016*(1), 1–16. https://doi.org/10.1093/emph/eov033

McEwen, B. S. (1998). Stress, adaptation, and disease: Allostasis and allostatic load. *Annals of the New York Academy of Sciences, 840*(1), 33–44. https://doi.org/10.1111/j.1749-6632.1998.tb09546.x

McEwen, B. S., & Akil, H. (2020). Revisiting the stress concept: Implications for affective disorders. *Journal of Neuroscience, 40*(1), 12–21. https://doi.org/10.1523/JNEUROSCI.0733-19.2019

McEwen, B. S., & Gianaros, P. J. (2010). Central role of the brain in stress and adaptation: Links to socioeconomic status, health, and disease. *Annals of the New York Academy of Sciences, 1186*(1), 190–222. https://doi.org/10.1111/j.1749-6632.2009.05331.x

Michopoulos, V., Powers, A., Gillespie, C. F., Ressler, K. J., & Jovanovic, T. (2017). Inflammation in fear- and anxiety-based disorders: PTSD, GAD, and beyond. *Neuropsychopharmacology, 42*(1), 254–270. https://doi.org/10.1038/npp.2016.146

Mills, C. W. (2022). *The racial contract*. 25th Anniversary edition. Cornell University Press.

Moen, P. (2022). The uneven stress of social change: Disruptions, disparities, and mental health. *Society and Mental Health, 12*(2), 85–98. https://doi.org/10.1177/21568693221100171

Mohebi, R., Chen, C., Ibrahim, N. E., McCarthy, C. P., Gaggin, H. K., Singer, D. E., Hyle, E. P., Wasfy, J. H., & Januzzi, J. L. (2022). Cardiovascular disease projections in the United States based on the 2020 census estimates. *Journal of the American College of Cardiology, 80*(6), 565–578. https://doi.org/10.1016/j.jacc.2022.05.033

Morris, L. S., McCall, J. G., Charney, D. S., & Murrough, J. W. (2020). The role of the Locus Coeruleus in the generation of pathological anxiety. *Brain and Neuroscience Advances, 4*, 2398212820930321. https://doi.org/10.1177/2398212820930321

Muller, C., Sampson, R. J., & Winter, A. S. (2018). Environmental inequality: The social causes and consequences of lead exposure. *Annual Review of Sociology, 44*, 263–282. https://doi.org/10.1146/annurev-soc-073117-041222

Muriach, M., Flores-Bellver, M., Romero, F. J., & Barcia, J. M. (2014). Diabetes and the brain: Oxidative stress, inflammation, and autophagy. *Oxidative Medicine and Cellular Longevity, 2014*, e102158. https://doi.org/10.1155/2014/102158

Muscatell, K. A., Alvarez, G. M., Bonar, A. S., Cardenas, M. N., Galvan, M. J., Merritt, C. C., & Starks, M. D. (2022). Brain–body pathways linking racism and health. *American Psychologist, 77*(9), 1049–1060. https://doi.org/10.1037/amp0001084

Muscatell, K. A., Brosso, S. N., & Humphreys, K. L. (2020). Socioeconomic status and inflammation: A meta-analysis. *Molecular Psychiatry, 25*(9), 2189–2199. https://doi.org/10.1038/s41380-018-0259-2

Netea, M. G., Balkwill, F., Chonchol, M., Cominelli, F., Donath, M. Y., Giamarellos-Bourboulis, E. J., Golenbock, D., Gresnigt, M. S., Heneka, M. T., Hoffman, H. M., Hotchkiss, R., Joosten, L. A. B., Kastner, D. L., Korte, M., Latz, E., Libby, P., Mandrup-Poulsen, T., Mantovani, A., Mills, K. H. G., Nowak, K. L., O'Neill, L. A., Pickkers, P., van der Poll, T., Ridker, P. M., Schalkwijk, J., Schwartz, D. A., Siegmund, B., Steer, C. J., Tilg, H., van der Meer, J. W. M., van de Veerdonk, F. L., & Dinarello, C. A. (2017). A guiding map for inflammation. *Nature Immunology, 18*(8), 826–831. https://doi.org/10.1038/ni.3790

Netea, M. G., Joosten, L. A. B., Latz, E., Mills, K. H. G., Natoli, G., Stunnenberg, H. G., O'Neill, L. A. J., & Xavier, R. J. (2016). Trained immunity: A program of innate immune memory in health and disease. *Science, 352*(6284), aaf1098. https://doi.org/10.1126/science.aaf1098

Ong, A. D., & Burrow, A. L. (2017). Microaggressions and daily experience: Depicting life as it is lived. *Perspectives on Psychological Science, 12*(1), 173–175.

Page, S. E. (2015). What sociologists should know about complexity. *Annual Review of Sociology, 41*, 21–41. https://doi.org/10.1146/annurev-soc-073014-112230

Parr, T., Bhat, A., Zeidman, P., Goel, A., Billig, A. J., Moran, R., & Friston, K. J. (2021). Dynamic causal modelling of immune heterogeneity. *Scientific Reports, 11*(1), 11400. https://doi.org/10.1038/s41598-021-91011-x

Pearlin, L. I. (1989). The sociological study of stress. *Journal of Health and Social Behavior*, 241–256.

Pearlin, L. I., Aneshensel, C. S., & Leblanc, A. J. (1997). The forms and mechanisms of stress proliferation: The case of AIDS caregivers. *Journal of Health and Social Behavior, 38*(3), 223–236. https://doi.org/10.2307/2955368

Pearlin, L. I., & Bierman, A. (2013). Current issues and future directions in research into the stress process. In C. S. Aneshensel, J. C. Phelan, & A. Bierman (Eds.), *Handbook of the sociology of mental health, handbooks of sociology and social research* (pp. 325–340). Springer Netherlands.

Pearlin, L. I., Schieman, S., Fazio, E. M., & Meersman, S. C. (2005). Stress, health, and the life course: Some conceptual perspectives. *Journal of Health and Social Behavior, 46*(2), 205–219. https://doi.org/10.1177/002214650504600206

Phelan, J. C., & Link, B. G. (2015). Is racism a fundamental cause of inequalities in health? *Annual Review of Sociology, 41*(1), 311–330. https://doi.org/10.1146/annurev-soc-073014-112305

Plascak, J. J., Beyer, K., Xinyi, X., Stroup, A. M., Jacob, G., & Llanos, A. A. M. (2022). Association between residence in historically redlined districts indicative of structural racism and racial and ethnic disparities in breast cancer outcomes. *JAMA Network Open, 5*(7), e2220908. https://doi.org/10.1001/jamanetworkopen.2022.20908

Quadt, L., Critchley, H. D., & Garfinkel, S. N. (2018). The neurobiology of interoception in health and disease: Neuroscience of interoception. *Annals of the New York Academy of Sciences, 1428*(1), 112–128. https://doi.org/10.1111/nyas.13915

Quintana, R. (2023). Embracing complexity in social science research. *Quality and Quantity, 57*(1), 15–38. https://doi.org/10.1007/s11135-022-01349-1

Ray, V. (2022). *On critical race theory: why it matters & why you should care*. Random House.

Reskin, B. (2012). The race discrimination system. *Annual Review of Sociology, 38*(1), 17–35. https://doi.org/10.1146/annurev-soc-071811-145508

Rohleder, N. (2019). Stress and inflammation – The need to address the gap in the transition between acute and chronic stress effects. *Psychoneuroendocrinology, 105*, 164–171. https://doi.org/10.1016/j.psyneuen.2019.02.021

Rovelli, C. (2021). *Helgoland: Making sense of the quantum revolution / Carlo Rovelli* (First North American edition; E. Segre & S. Carnell, Trans.). Riverhead Books.

Russell, D. W., Clavél, F. D., Cutrona, C. E., Todd Abraham, W., & Burzette, R. G. (2018). neighborhood racial discrimination and the development of major depression. *Journal of Abnormal Psychology, 127*(2), 150–159. https://doi.org/10.1037/abn0000336

Sapolsky, R. M. (2004). *Why zebras don't get ulcers* (3rd ed.). Holt Paperbacks.

Savitz, J., & Harrison, N. A. (2018). Interoception and inflammation in psychiatric disorders. *Biological Psychiatry. Cognitive Neuroscience and Neuroimaging, 3*(6), 514–524. https://doi.org/10.1016/j.bpsc.2017.12.011

Schulkin, J., & Sterling, P. (2019). Allostasis: A brain-centered, predictive mode of physiological regulation. *Trends in Neurosciences, 42*(10), 740–752. https://doi.org/10.1016/j.tins.2019.07.010

Sennesh, E., Theriault, J., Brooks, D., van de Meent, J.-W., Barrett, L. F., & Quigley, K. S. (2022). Interoception as modeling, allostasis as control. *Biological Psychology, 167*, 108242. https://doi.org/10.1016/j.biopsycho.2021.108242

Sewell, A. A. (2017). The illness associations of police violence: Differential relationships by ethnoracial composition. *Sociological Forum, 32*(S1), 975–997. https://doi.org/10.1111/socf.12361

Sewell, A. A. (2016). The racism-race reification process: A mesolevel political economic framework for understanding racial health disparities. *Sociology of Race and Ethnicity, 2*(4), 402–432. https://doi.org/10.1177/2332649215626936

Sewell, A. A., Feldman, J. M., Ray, R., Gilbert, K. L., Jefferson, K. A., & Lee, H. (2021). Illness spillovers of lethal police violence: The significance of gendered marginalization. *Ethnic and Racial Studies, 44*(7), 1089–1114. https://doi.org/10.1080/01419870.2020.1781913

Shaffer, C., Westlin, C., Quigley, K. S., Whitfield-Gabrieli, S., & Barrett, L. F. (2022). Allostasis, action, and affect in depression: Insights from the theory of constructed emotion. *Annual Review of Clinical Psychology, 18*(1), 553–580. https://doi.org/10.1146/annurev-clinpsy-081219-115627

Shah, S., Lee, J., & Ho, M. G.-R. (2023). Chapter 10 – Obesity, inflammation, and CNS disorders. In R. Mehrzad (Ed.), *Inflammation and obesity* (pp. 131–151). Academic Press.

Shilts, J., Severin, Y., Galaway, F., Müller-Sienerth, N., Chong, Z.-S., Pritchard, S., Teichmann, S., Vento-Tormo, R., Snijder, B., & Wright, G. J. (2022). A physical wiring diagram for the human immune system. *Nature, 608*(7922), 397–404. https://doi.org/10.1038/s41586-022-05028-x

Short, S. E., & Zacher, M. (2022). Women's health: Population patterns and social determinants. *Annual Review of Sociology, 48*, 277–298. https://doi.org/10.1146/annurev-soc-030320-034200

Siegenfeld, A. F., & Bar-Yam, Y. (2020). An introduction to complex systems science and its applications. *Complexity, 2020*(1), 6105872. https://doi.org/10.1155/2020/6105872

Simons, R. L., Lei, M.-K., Beach, S. R. H., Barr, A. B., Simons, L. G., Gibbons, F. X., & Philibert, R. A. (2018). Discrimination, segregation, and chronic inflammation: Testing the weathering explanation for the poor health of Black Americans. *Developmental Psychology, 54*, 1993–2006. https://doi.org/10.1037/dev0000511

Smith, S. M., & Vale, W. W. (2006). The role of the hypothalamic-pituitary-adrenal axis in neuroendocrine responses to stress. *Dialogues in Clinical Neuroscience, 8*(4), 383–395. https://doi.org/10.31887/DCNS.2006.8.4/ssmith

Stanley, A. C., & Lacy, P. (2010). Pathways for cytokine secretion. *Physiology, 25*(4), 218–229. https://doi.org/10.1152/physiol.00017.2010

Sterling, P. (2012). Allostasis: A model of predictive regulation. *Physiology & Behavior, 106*(1), 5–15. https://doi.org/10.1016/j.physbeh.2011.06.004

Sterling, P. (2015). *Principles of neural design*. The MIT Press.

Sterling, P. (2020). *What is health?: Allostasis and the evolution of human design*. The MIT Press.

Sterling, P., & Platt, M. L. (2022). Why deaths of despair are increasing in the us and not other industrial nations—Insights from neuroscience and anthropology. *Archives of General Psychiatry, 79*(4), 368–374. https://doi.org/10.1001/jamapsychiatry.2021.4209

Stewart, Q. T. (2023). *Race in the machine: A novel account* (1st ed.). Redwood Press.

Thomas-Vaslin, V. (2017). Understanding and modeling the complexity of the immune system. In P. Bourgine, P. Collet, & P. Parrend (Eds.), *First complex systems digital campus world e-conference 2015, Springer Proceedings in Complexity* (pp. 261–270). Springer International.

Toussaint, L. L., Moriarity, D. P., Kamble, S., Williams, D. R., & Slavich, G. M. (2022). Inflammation and depression symptoms are most strongly associated for black adults. *Brain, Behavior, & Immunity - Health*, 100552. https://doi.org/10.1016/j.bbih.2022.100552

Turnbull, A. V., & Rivier, C. L. (1999). Regulation of the hypothalamic-pituitary-adrenal axis by cytokines: Actions and mechanisms of action. *Physiological Reviews, 79*(1), 1–71. https://doi.org/10.1152/physrev.1999.79.1.1

Turner, A. I., Smyth, N., Hall, S. J., Torres, S. J., Hussein, M., Jayasinghe, S. U., Ball, K., & Clow, A. J. (2020). Psychological stress reactivity and future health and disease outcomes: A systematic review of prospective evidence. *Psychoneuroendocrinology, 114*, 104599. https://doi.org/10.1016/j.psyneuen.2020.104599

Umberson, D., & Donnelly, R. (2023). Social isolation: An unequally distributed health hazard. *Annual Review of Sociology, 49*, 379–399. https://doi.org/10.1146/annurev-soc-031021-012001

Valdes, F., & Bender, S. W. (2021). *LatCrit: From critical legal theory to academic activism*. NYU Press.

Wang, Q., Oyarzabal, E. A., Song, S., Wilson, B., Santos, J. H., & Hong, J.-S. (2020). Locus Coeruleus neurons are most sensitive to chronic neuroinflammation-induced neurodegeneration. *Brain, Behavior, and*

Immunity, 87, 359–368. https://doi.org/10.1016/j.bbi.2020.01.003

Wenbo, L., & Wang, J. (2017). Uncovering the underlying mechanism of cancer tumorigenesis and development under an immune microenvironment from global quantification of the landscape. *Journal of the Royal Society Interface, 14*, 20170105. https://doi.org/10.1098/rsif.2017.0105

Wheaton, B. (1994). Sampling the stress universe. In W. R. Avison & I. H. Gotlib (Eds.), *Stress and mental health: Contemporary issues and prospects for the future, The Springer Series on Stress and Coping* (pp. 77–114). Springer US.

Wheaton, B., Young, M., Montazer, S., & Stuart-Lahman, K. (2013). Social stress in the twenty first century. In *Handbook of the sociology of mental health*. Springer.

Wilkerson, I. (2020). *Caste: The origins of our discontents*. Random House.

Williams, D. R., Lawrence, J. A., & Davis, B. A. (2019). Racism and health: Evidence and needed research. *Annual Review of Public Health, 40*(1), 105–125. https://doi.org/10.1146/annurev-publhealth-040218-043750

Williams, D. T. (2019). A call to focus on racial domination and oppression: A response to "Racial and Ethnic Inequality in Poverty and Affluence, 1959–2015". *Population Research and Policy Review, 38*(5), 655–663. https://doi.org/10.1007/s11113-019-09538-x

Williams, M. T. (2020). Microaggressions: Clarification, evidence, and impact. *Perspectives on Psychological Science, 15*(1), 3–26. https://doi.org/10.1177/1745691619827499

Zavala, V. A., Bracci, P. M., Carethers, J. M., Carvajal-Carmona, L., Coggins, N. B., Cruz-Correa, M. R., Davis, M., de Smith, A. J., Dutil, J., Figueiredo, J. C., Fox, R., Graves, K. D., Gomez, S. L., Llera, A., Neuhausen, S. L., Newman, L., Nguyen, T., Palmer, J. R., Palmer, N. R., Pérez-Stable, E. J., Piawah, S., Rodriquez, E. J., Sanabria-Salas, M. C., Schmit, S. L., Serrano-Gomez, S. J., Stern, M. C., Weitzel, J., Yang, J. J., Zabaleta, J., Ziv, E., & Fejerman, L. (2021). Cancer health disparities in racial/ethnic minorities in the United States. *British Journal of Cancer, 124*(2), 315–332. https://doi.org/10.1038/s41416-020-01038-6

Zheng, H., Smith, R., Burrows, K., Kuplicki, R., Jerzy Bodurka, T., Teague, K., Paulus, M., Khalsa, S., & Savitz, J. (2021). Inflammation Is associated with abnormal interoceptive processing in major depressive disorder. *Biological Psychiatry, 89*(9 Suppl), S190–S191. https://doi.org/10.1016/j.biopsych.2021.02.483

Zmora, N., Bashiardes, S., Levy, M., & Elinav, E. (2017). The role of the immune system in metabolic health and disease. *Cell Metabolism, 25*(3), 506–521. https://doi.org/10.1016/j.cmet.2017.02.006

24 Genetic, Hormonal, and Neural Underpinnings of Human Aggressive Behavior

Jason Isbell, Stefan M. Goetz, Pranjal H. Mehta, and Justin M. Carré

Abstract

Although no doubt a useful adaptation in our early evolutionary history, aggressive behavior contributes to many significant contemporary social challenges, ranging anywhere from bullying to homicide. Like any complicated human behavior, it is the outcome of complex neural and endocrinological systems operating within a dynamic social environment. In this chapter, we review many of the systems associated with aggression as well as their interactions and genetic correlates. We then present the environmental variables most associated with eliciting its behavioral expression. Common psychological interventions that have been developed to mitigate such expression are discussed, and we conclude with suggestions for future research in the field.

J. Isbell (✉)
Department of Psychology, University of California, Merced, Merced, CA, USA
e-mail: jisbell@ucmerced.edu

S. M. Goetz
Department of Psychiatry, Wayne State University School of Medicine, Detroit, MI, USA
e-mail: stefan.goetz@wayne.edu

P. H. Mehta
Department of Psychology and Language Science, University College London, London, UK
e-mail: pranj.mehta@ucl.ac.uk

J. M. Carré
Department of Psychology, Nipissing University, North Bay, ON, Canada
e-mail: justinca@nipissingu.ca

Violence is estimated to kill approximately 1.6 million people per year worldwide (www.fbi.gov, Mercy et al., 2002). In 2022, there were an estimated 1.23 million cases of reported[1] violent crimes committed in the United States, with a victimization rate of 23.5 per 1000 persons, age 12 or older. Although evolutionary theory suggests that aggression and violence were adaptive behaviors that promoted survival and reproduction among our ancestors (e.g., the acquisition of valued resources such as food, shelter, and mates), aggressive behaviors in modern societies have significant social and economic costs (Buss & Shackelford, 1997). These include social stigma, job loss, and negative legal consequences for perpetrators as well as substantial monetary and social costs for society (Archer & Southhall, 2009).

While research has identified multiple social and biological factors that influence aggressive

[1]Reported victimization represents an undercount as many cases go unreported (e.g., "dark figure of crime," the gap between the true extent of crime and incidents known to the police; Biderman & Reiss, 1967). According to the National Crime Victimization survey, in 2022, approximately 58% of violent victimizations went unreported to police (Thompson & Tapp, 2023).

W. Kalkhoff et al. (eds.), *Handbook of Neurosociology*, Handbooks of Sociology and Social Research, https://doi.org/10.1007/978-3-031-95615-7_24

behavior, how these factors work together to regulate human aggression remains poorly understood. In this chapter, we review recent studies on the social neuroscience of aggression, including research in the areas of molecular genetics, neuroendocrinology, neuroimaging, and social psychology. Our goal is not to provide an exhaustive review but rather to summarize the main findings from these fields and to highlight recent studies that integrate theories and approaches from disparate areas of research (for a comprehensive review, see Rosell & Siever, 2015). We begin by defining aggression and its subtypes. We then selectively review research on the social neuroscience of human aggression, covering research in neuroimaging, behavioral pharmacology, molecular genetics, neuroendocrinology, and social psychology. We end the chapter by suggesting new directions for future research on aggressive behavior.

What Is Aggression?

Aggression has been defined as "any form of behavior directed toward the goal of harming or injuring another living being who is motivated to avoid such treatment" (Baron & Richardson, 1994, p. 7). Aggression may be physical or nonphysical. Physical aggression may involve posturing or otherwise using body language or physical proximity to intimidate others, escalating to grabbing, shoving, throwing objects, etc. Violence is an extreme form of physical aggression resulting in physical injury or death (Eisner, 2009). Nonphysical aggression, often referred to as relational aggression, includes behaviors designed to cause psychological harm (direct insults, psychological abuse), social harm (e.g., spreading rumors to damage someone's reputation, social exclusion), or economic harm (e.g., firing a subordinate or decreasing his or her pay). Researchers typically classify aggression as either reactive or proactive. Reactive aggression, also referred to as impulsive aggression, is a behavioral response to perceived or actual provocation and involves retaliation (Dodge & Coie, 1987). Commonly referred to as "hot-blooded," reactive aggression is characterized by anger and impulsivity and is often accompanied by disinhibition, affective instability, and increases in autonomic arousal. In contrast, proactive aggression, also referred to as instrumental aggression, occurs in the absence of direct provocation and is a goal-oriented behavior aimed at the acquisition of a valued resource (Dodge & Coie, 1987). It is associated with low emotionality and autonomic arousal. Although the proactive form receives widespread media attention (e.g., serial killings, assassinations, genocide), the reactive form likely accounts for most societal problems associated with aggression (Nelson & Trainor, 2007). The focus of this review is on reactive aggression because it has received greater attention in neuroscience research, and its social and biological causes are better understood.

The Social Neuroscience of Human Reactive Aggression

Empirical studies indicate that human aggressive behavior is influenced by specific genes, hormones, neural systems, and environmental factors. In this section, we review the main findings from these disparate areas of research with a focus on integrative studies. First, we discuss the neural systems implicated in aggression with a focus on two specific regions: the amygdala and the orbitofrontal cortex. Next, we discuss neurotransmitters and hormones associated with aggression, including serotonin, testosterone, and cortisol. Third, we discuss two environmental factors linked to aggression: violent media exposure and social rejection. Fourth, we discuss two psychological interventions that can reduce aggression: cognitive reappraisal and self-control. Finally, we end the chapter by discussing directions for future research.

Amygdala–Orbitofrontal Cortex Interactions as a Mechanism for Aggressive Behavior

Animal research indicates that an extensive network of cortical and subcortical regions is involved in the expression of aggressive behavior (Newman, 1999; Nelson & Trainor, 2007; Siegel et al., 2007). Two regions that have received extensive empirical attention in human research are the amygdala and the orbitofrontal cortex (OFC). According to models of human reactive aggression, the amygdala plays a critical role in the affective and motivational drive to respond aggressively to social provocation, while the OFC is thought to be a self-regulatory region that inhibits aggressive impulses (see Fig. 24.1). The findings reviewed below are consistent with these models.

Orbitofrontal Cortex

The OFC is located in the prefrontal cortex, a portion of the brain that appeared later in evolutionary history than subcortical regions such as the amygdala (Kringelbach & Rolls, 2004). A number of studies suggest that the OFC functions as a self-regulation and impulse control region and is involved in the top-down inhibition of aggressive behavior. Patients with lesions in the OFC exhibit hyperaggressive behavioral reactions to social provocation (Bufkin & Luttrell, 2005; Damasio et al., 1994; Davidson et al., 2000; Koenigs & Tranel, 2007; Moretti et al., 2009; Strüber et al., 2008), and human neuroimaging studies indicate that increased OFC activity is associated with low levels of reactive aggression (Bufkin & Luttrell, 2005; Damasio et al., 1994; Davidson et al., 2000; Strüber et al., 2008). For example, one study examined the relationship between OFC activity and aggressive behavior in the Ultimatum Game, a laboratory model of social decision-making in which people choose between aggression and monetary reward (Mehta & Beer, 2010). This game involves two players: a proposer and a responder. The proposer makes an offer as to how to split a sum of money (the stake) with the responder. The responder then decides whether to accept or reject the offer. If the offer is accepted, the stake is split as proposed. However, if the offer is rejected, then both players receive $0. After the responder makes a decision, the game is over.

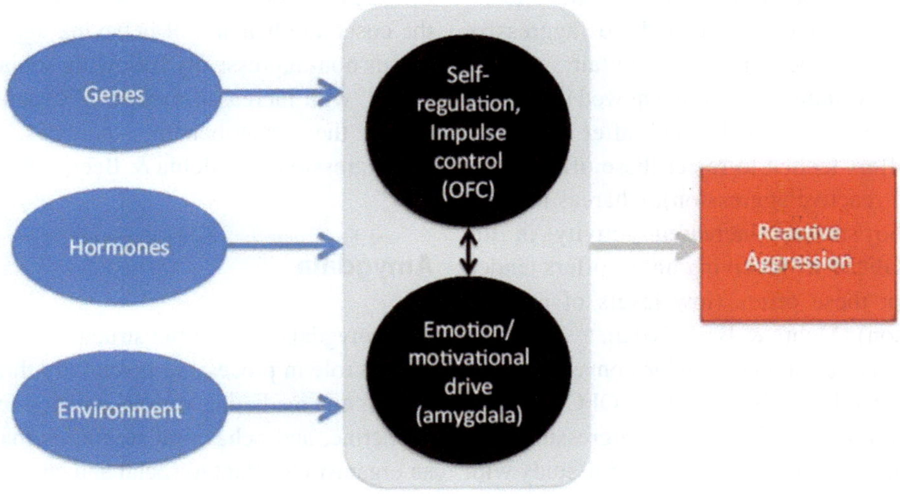

Fig. 24.1 A social neuroscience model of reactive aggression

Although responders almost always accept fair offers (e.g., proposer gets 50% and responder gets 50% of stake), responders often reject unfair offers (e.g., proposer gets 80% and responder gets 20% of the stake). Accepting unfair offers guarantees monetary reward, so why do people ever reject them? Psychological evidence indicates that these unfair offer rejections are a form of reactive aggression aimed at retaliating against the other player in the face of perceived social provocation (unfair treatment) (Mehta & Beer, 2010).

In this fMRI study, participants were scanned while playing the Ultimatum Game in the role of responder ostensibly with 40 other proposers in one-shot interactions (participants were told they would never play with the same proposers twice). In reality, the offers were experimentally manipulated such that half were relatively fair ($5:$5 split) and the other half were relatively unfair (e.g., an offer of $8 for the proposer and $2 for the responder). The researchers assessed how often participants rejected unfair offers (a behavioral measure of reactive aggression) as well as OFC activity in response to unfair offers compared to fair offers. In support of the hypothesis that OFC is involved in the inhibition of aggressive behavior, the results indicated that bilateral activity in the medial OFC was negatively related to aggressive behavioral reactions to unfair offers. Specifically, individuals who showed decreased activity in the medial OFC after receiving unfair offers tended to reject these offers (high levels of reactive aggression), whereas individuals who showed increased activity in the medial OFC after receiving unfair offers tended to accept these offers (low levels of reactive aggression) (Mehta & Beer, 2010).

Other human studies provide convergent support for a relationship between OFC function and the inhibition of impulsive aggression. In a (positron emission tomography) PET study with criminal offenders, Raine et al. (1997) reported that affective murderers (i.e., reactively aggressive inmates) demonstrated increased glucose metabolism in subcortical structures (including the amygdala) and decreased glucose metabolism in the prefrontal cortex. Also, psychiatric disorders characterized by high levels of reactive aggression are associated with reduced OFC activity (Coccaro et al., 2007), and lower gray matter volume in the OFC is linked to low impulse control (Matsuo et al., 2009). Although the precise psychological function of OFC in inhibiting aggression is still unclear, it has been theorized that the OFC is part of a self-regulation and impulse control system that integrates emotion, motivation, and cognition to guide context-appropriate behavior (cf. Mehta & Beer, 2010). Indeed, not only do patients with OFC lesions show increases in reactive aggression (Blair, 2004; Rolls et al., 1994), but they also show increases in impulsive behavior, socially inappropriate behavior, and impaired decision-making (Bechara et al., 2000; Beer et al., 2003, 2006; Rahman et al., 2001; Tucker et al., 1995). These behavioral deficits have been theorized to occur because of a failure to monitor behavior such as failing to consider longer term rewards (Bechara et al., 2000; Beer et al., 2006; De Martino et al., 2006; Moretti et al., 2009). A complementary account of OFC function is that this region is involved in how individuals weigh the costs and benefits of behaving aggressively versus non-aggressively following social provocation, with increased medial OFC activity tipping the cost-benefit analysis toward nonaggression (cf. Mehta & Beer, 2010).

Amygdala

The amygdala is a limbic structure that plays a critical role in processing potentially threatening stimuli and mediating various autonomic, neuroendocrine, and behavioral responses that enable an organism to adapt to social and environmental challenges (see Davis & Whalen, 2001; LeDoux, 2000 for reviews). Animal research indicates that the amygdala is an important component of a neural circuitry that modulates aggressive behavior. Although there is less direct evidence for the amygdala's role in human aggression, indirect evidence suggests that amygdala reactivity may be an important precursor for aggressive behavior in humans. Across a number of studies that used different methodologies, there is robust evidence that amygdala activity increases in response to emotional signals of social provocation (angry faces) and that this amygdala reactivity is stronger in individuals susceptible to aggressive behavior (e.g., Beaver et al., 2008; Carré et al., 2012; Chan et al., 2010; Coccaro et al., 2007; Lee et al., 2008). These findings suggest that hyper-amygdala reactivity to social provocation may be a neural marker for one's propensity to engage in reactive aggression (see Carré et al., 2011, for review)

One neuroimaging study more directly linked amygdala function to aggression (Gospic et al., 2011). In this study, participants were scanned while playing the Ultimatum Game in the role of responder using procedures similar to the study described earlier (Mehta & Beer, 2010), but the design of this newer study was optimized to detect rapid and slower neural responses to unfair offers. Results showed a rapid amygdala response to unfair offers that was positively related to aggressive behavior (rejecting unfair offers). Interestingly, administration of a benzodiazepine prior to performing the Ultimatum Game effectively reduced amygdala reactivity to unfair offers, and also decreased rejections of unfair offers (Gospic et al., 2011). In line with the study discussed earlier (Mehta & Beer, 2010), Gospic et al. (2011) also found that prefrontal regions such as the OFC were activated to support the inhibition of unfair offer rejections, but these prefrontal responses came on line later. Together, the findings support a dual-systems model of reactive aggression in line with Fig. 24.1; amygdala activation is associated with a rapid emotional and motivational drive to respond aggressively to social provocation (being treated unfairly), while the OFC is engaged later in the decision-making process to inhibit aggressive impulses.

Amygdala–OFC Connectivity

The studies reviewed above support the view that amygdala is involved in the emotional response to social threat and encourages reactive aggression, whereas the OFC is a self-regulation and impulse control region that inhibits aggression. Many studies suggest that the functional connectivity between the amygdala and OFC may be another mechanism for aggressive behavioral reactions to social provocation. More specifically, healthy individuals show coupling between amygdala and OFC, but this connectivity is disrupted in psychiatric patients vulnerable to aggressive behavior (cf. Coccaro et al., 2011). Thus, not only do the amygdala and OFC influence aggressive behavior independently, but the neural communication between the two regions seems to play an important role in the inhibition of aggression. This mechanism is supported by neuroanatomical findings, which indicate that the OFC and amygdala have reciprocal connections with one another (Kringelbach & Rolls, 2004). However, this connectivity may only predict aggression among clinical samples or those with aggression related psychopathology, as one study showed no relationship between the structural connectivity of the amygdala and OFC and trait aggression when using a healthy sample (Beyer et al., 2014).

Genetic and Neurochemical Modulators of Human Aggression

Research suggests that various neurotransmitters, genes, and hormones are involved in human aggression. In this section, we synthesize the main findings from these different areas of research. We focus our discussion on a few factors that have received empirical attention in human studies: serotonin, testosterone, and cortisol. We discuss how these factors may regulate human aggressive behavior along with their putative neural mechanisms.

Crucially, aggressive behavior is mediated by psychological adaptations, e.g., the emotion of anger (see for instance Sell et al., 2009), which represent complex phenotypes exhibiting special design (Andrews et al., 2002; Schmitt & Pilcher, 2004). Such adaptations are underpinned by complex genetic architectures, requiring that a diversity of genetic components be universally shared across the species, as sexual recombination would otherwise disrupt these adaptations' reliable expression during normative ontogenetic development (Tooby & Cosmides, 1990). One consequence of polygenicity is that they present a large target for mutational load (Penke et al., 2007), generating non-adaptive variation in psychological adaptations, usually in terms of quantitative rather than qualitative differences, including aspects of emotions integral to aggression (Lukaszewski et al., 2020; Penke et al., 2007; Tooby & Cosmides, 1990; Zietsch, 2024). To date, there is little evidence to suggest that the genetic contribution to individual variation in these traits is due to balancing selection (for an example of psychopathy as a result of negative frequency dependent selection—a form of balancing selection—cf., Glenn et al., 2011).

To the best of our knowledge, most studies on the genetic contribution to human aggression continue to rely on candidate gene approaches (and GxE interactions), despite the polygenetic nature of aggression (for an exception, see Kretschmer et al., 2023). We urge caution in interpreting these findings for two reasons: First, aggression and violent crime are heavily influenced by historical, structural, and situational forces (e.g., Daly, 2023; Vilalta et al., 2024). Second, canddiate gene studies in psychological and behavioral domains have been plagued by replicability and methodological shortcomings, e.g., population stratification (Davies et al., 2024; for a critique, see e.g., Border et al., 2019).

Serotonin

Decades of research on nonhuman animal models reveals a consistent relationship between serotonin and aggression, most often with increases in serotonin resulting in decreased aggressive behavior (da Cunha-Bang & Knudsen, 2021; Takahashi et al., 2011). Early work using adult male vervet monkeys found that serotonergic agonists facilitated rank attainment while simultaneously decreasing reactive aggression and vigilence (Raleigh et al., 1985; Raleigh et al., 1991). Regarding humans, a large correlational literature indicates that enhanced activity in the serotonin system is related to decreases in reactive aggression (Siever, 2008; Coccaro et al., 2011). Research with pharmacological manipulations provides much needed causal evidence for the role of serotonin in mediating aggression, and in one demonstration of this causal relationship, individuals with and without a life history of physical aggression were randomly assigned to receive 40 mg of paroxetine (a drug that acutely augments serotonergic activity) or placebo (Berman et al., 2009). Participants were then placed in the Taylor Aggression Paradigm, a laboratory task that measures physical aggression in response to social provocation. In this task, participants are told they are competing with another participant in a reaction time game, and electric shocks are received and administered. The amount of maximum shock delivered in response to social provocation was the primary measure of aggressive behavior in this study. The findings revealed that augmentation of serotonergic activity via paroxetine significantly reduced physical aggression after social provocation, but only in individuals with a life history of aggression. These findings suggest that enhanced serotonin activity causally reduces aggressive behavior in individuals prone to physical aggression. Another study investigating psychopathy, a condition long associated with aggressive behavior, manipulated serotonin among those with traits of primary psychopathy (Fanning et al., 2014). Using a placebo-controlled study, they found the administration of serotonin (via paroxetine, a selective serotonin reuptake inhibitor) attenuated aggressive behavior in a competitive reaction-time game in which electric shocks were exchanged with an increasingly provocative fictitious opponent.

If heightened serotonin activity can decrease aggression, might reductions in serotonergic activity increase aggression? Another study provided causal support for this relationship (Crockett et al., 2008). Healthy participants were randomly assigned to receive placebo or tryptophan depletion, a pharmacological manipulation that reduces serotonergic activity. Then participants played the Ultimatum Game in the role of responder, and the fairness of the offers was experimentally manipulated similar to the Ultimatum Game studies reviewed earlier. The results showed that reductions in serotonergic activity via tryptophan depletion causally increased aggressive behavior (rejection of unfair offers) (Crockett et al., 2008). In a more recent study by the same group found that serotonin induced a moral aversion to harming others for financial gain, extending the serotonergic-aggression effect to encompass proactive aggression as well (Crockett et al., 2015).

However, one meta-analysis revealed potential subtleties to this longstanding inverse relationship between serotonin and aggression (Duke et al., 2013). They looked at four different ways of assessing or modulating levels of serotonin: (1) The measuring of cerebrospinal fluid levels of 5-hydroxyindoleacetic acid (CSF 5-HIAA) and then looking for possible relationships between varying levels of it and individual behavior. (2) The above mentioned acute tryptophan depletion, which is based on the assumption that limiting the primary precursor of serotonin (L-tryptophan) will lead to predictable changes in the levels of brain serotonin, and subsequent behavior is then measured. (3) The pharmacological challenge, in which serotonergic agents are administered and again subsequent behavior is measured. (4) The endocrine challenge, which is similar to the pharmacological challenge wherein serotonergic agents are administered and behavior is measured, but also measured is participant endorcrine response (i.e., the secretion of pituitary and adrenal hormones), which provides a better index of individual responsivity to changing levels of serotonin. Overall they did find a small inverse correlation between serotonin functioning and aggression, anger, and hostility. Yet this effect was much smaller than an earlier seminal study that suggested serotonin accounted for 80% of the variance in a group of soldiers' aggression (Brown et al., 1979). This meta analysis found that serotonin explained just over 1% of the variance in measures of aggression, anger, and hostility. Further, not all effect sizes were equal, with pharmacological challenge studies having the largest and CSF 5-HIAA studies having the smallest.

Serotonergic Gene Polymorphisms

Common variations (polymorphisms) within genes that regulate the serotonergic system can alter human brain function and aggression (Hariri & Weinberger, 2003). Two polymorphic genes that have been widely studied in relation to human aggression are monoamine oxidase A (MAOA u-VNTR) and the serotonin transporter (5-HTTLPR).

The first evidence in humans for the importance of MAOA in aggression came from the study of large Dutch kindred, whose males were notorious for impulsive aggression (Brunner et al., 1993). Brunner et al. (1993) discovered a missense mutation of the MAOA gene that resulted in a premature stop codon causing MAOA to be nonfunctional, thus, effectively producing functional MAOA knockouts. Although this finding is informative, the mutation is rare in the population. Nevertheless, within the MAOA gene, a more common polymorphism has been described, which is located 1.2 kb upstream of the MAOA coding sequences and consists of a 30-bp repeated sequence present in 3, 3.5, 4, or 5 copies. This variable number of tandem repeats (VNTR) polymorphism is functional: alleles with 3.5 or 4 copies of the repeat sequence are transcribed 2–10 times more efficiently ("high-expression alleles") than those with three or five copies of the repeat ("low-expression alleles") (Sabol et al., 1998).

A well-known longitudinal study revealed that the presence of the low-activity allele interacted with a history of childhood maltreatment to predict increased levels of aggression and violence in adults (Caspi et al., 2002). This MAOA gene × childhood adversity interaction has been conceptually replicated in other studies (e.g., Frazzetto et al., 2007; Reif et al., 2007). Although most studies assessed aggression through self-reported or objective real-world markers of aggression (e.g., violent crimes), one study showed an association between the MAOA gene and a well-validated behavioral measure of aggression (McDermott et al., 2009). In the study, participants were paid to punish others whom they believed had taken money from them. In reality, participants were playing with a fictitious player whose behavior was experimentally controlled by the researchers. Participants punished their opponents by administering varying amounts of aversive hot sauce, which served as the measure of aggressive behavior. The findings revealed that individuals with the low expression MAOA allele behaved more aggressively after social provocation relative to individuals with the high expression allele. That is, low expression allele carriers delivered higher amounts of hot sauce to their opponent, but only after their "opponent" had taken a large amount of money from them.

The serotonin transporter (5HTT) regulates the availability of synaptic serotonin. A widely studied gene within this system is a common functional polymorphism (5HTTLPR) (cf. Heils et al., 1996). Individuals with the short allele of this gene have reduced transcriptional activity and therefore reduced reuptake of synaptic serotonin compared to individuals with the long allele. These low activity allele carriers are at greater risk for affective psychiatric disorders such as anxiety and depression, particularly in combination with a life history of stress (e.g., Caspi et al., 2010; but see Border et al., 2019). Other research has linked low activity allele status to aggressive behavior. Individuals with low activity allele variants in 5HTTLPR are more likely to show increased childhood aggression (Beitchman et al., 2006), and low activity allele carriers who have adverse childhood environments are more vulnerable to aggressive behavior as adults (Reif et al., 2007). This 5-HTTPLPR gene effect emerges above and beyond effects of the MAOA gene discussed above, suggesting that both of these genes uniquely account for variance in human aggressive behavior (Reif et al., 2007). For a more recent review coming to similar conclusions, regarding both 5HTTLPR and the MAOA allele, see da Cunha-Bang & Knudsen, 2021; Kolla & Bortolato, 2020.

Mechanisms for Serotonin-Modulated Aggression

The mechanisms by which serotonin effect human aggression have been partially ellucided via the study of MAOA allele variation and include social pathways (e.g., delinquent peer affiliation: Cooke et al., 2018), impaired emotional regulatory capacities (Chester et al., 2015; Denson et al., 2014), alterned emotion- and reward-related neural reactivity (Im et al., 2019), including in conjunction with testosterone (reward function: Wagels et al., 2017), and structural and functional connectivity of limbic circuitry (Harneit et al., 2019; Klasen et al., 2018).

Regarding limbic circuitry, the evidence suggests that the OFC, amygdala, and their connectivity are all candidate neural mechanisms. One PET study found increased metabolic glucose response in the left OFC to a serotonergic challenge (meta-chlorophenylpiperazine) in healthy participants, but not among borderline personality disorder (BPD) patients with impulsive aggression (New et al., 2002). Interestingly, this same research group found that administration of fluoxetine (a serotonin reuptake inhibitor) to individuals with BPD was associated with increased glucose metabolic rate in the OFC and an overall decrease in impulsive aggression (New et al., 2004). These findings suggest that serotonergic modulation of the OFC may have an inhibitory effect on impulsive aggression.

In support of heightened amygdala reactivity as a putative mechanism underlying reactive aggression, research suggests that genes that regulate serotonin function are associated with increased amygdala reactivity to facial signals of threat (see Buckholz & Meyer-Lindenberg, 2008 and Hariri, 2009, for reviews). For instance, Hariri et al. (2002) were the first to demonstrate that individuals carrying the "short" allele of the 5HTTLPR gene demonstrate heightened amygdala reactivity to facial signals of threat, a finding that has been replicated several times (see

Munafo et al., 2008, for review). Other research suggests that the MAOA gene may bias the socio-emotional circuitry of aggression, including the amygdala (Meyer-Lindenberg et al., 2006). Specifically, individuals with the low expression variant of the MAOA gene demonstrated heighted amygdala reactivity to facial signals of threat. Other work indicates that individuals with the low expression variant of the MAOA gene scored higher on a trait measure of aggression and interpersonal hypersensitivity and also demonstrated heightened dorsal anterior cingulate cortex (ACC) reactivity to social rejection (Eisenberger et al., 2007). Notably, the positive relationship between interpersonal hypersensitivity and aggression was mediated by heightened dorsal ACC reactivity to social rejection (Eisenberger et al., 2007). Another mechanism may involve connectivity between the amygdala and the prefrontal cortex. Passamonti et al. (2012) found that acute reductions in serotonergic activity via tryptophan depletion reduced functional connectivity between the amygdala and prefrontal cortex in response to angry faces (e.g., connectivity with ventrolateral prefrontal cortex as well as ventral ACC), which may increase one's risk for reactive aggression. On the other hand, those with low MAOA expression allele showed both greater activation and greater coupling between the dACC and amygdala in response to social provocation (an insult by a rude experimenter) (Denson et al., 2014). Collectively, these findings converge to suggest that serotonergic function may influence aggressive behavior via its interactions with receptors located within a neural circuitry including the amygdala, OFC, and ACC.

Testosterone

Testosterone (T) is a steroid hormone derived from cholesterol. It is produced and released primarily by the testes in men and by the ovaries and adrenal cortex in women. T belongs to a class of hormones called androgens, which are those hormones that are responsible for the development and maintenance of masculine characteristics. In addition to supporting basic physical development, T is also critically involved in regulating social behavior. Naturally occurring and experimentally elevated testosterone levels are positively associated with aggressive behavior in a variety of animal species, especially when the status hierarchy is unstable (Collias et al., 2002; Giammanco et al., 2005; Oliveira et al., 1996; Ruiz-de-la-Torre & Manteca, 1999; Sapolsky, 1991; Wingfield et al., 1990). In stark contrast to the animal literature, the relationship between individual differences in T and human aggression is relatively weak (see Archer et al., 2005 & Geniole et al., 2020, for reviews). Even though some studies in humans show that higher circulating T is related to aggression, social dominance, and hyperreactivity to status threats (e.g., Archer et al., 2005; Mazur & Booth, 1998; Mehta & Beer, 2010; Mehta et al., 2008), other studies have produced inconsistent or null results (Archer et al., 2005). One explanation for these weak effects is that relatively stable levels of T (baseline T) may play less of a crucial role in human aggression than situationally induced fluctuations in T levels (see Carré et al., 2011, for review). It is well-known that T levels rise and fall in competitive social interactions, but only recently have researchers investigated whether dynamic rises in T encourage aggressive and dominant behaviors in humans. In the next section, we review this literature on context-driven T dynamics and human social behavior.

Testosterone Dynamics

John Wingfield and colleagues proposed the Challenge Hypothesis to explain how T changes influence social behavior in birds (Wingfield et al., 1990). According to this theory, T levels rise during the breeding season to encourage social competition for mates, and T drops during the non-breeding season to suppress competitive aggression and facilitate care for offspring. Mazur (1985) proposed a similar Biosocial Model of Status for T-behavior associations in humans. According to this model, status-relevant social interactions such as competition should cause T levels to fluctuate, and these fluctuations in T should encourage or discourage subsequent status-seeking behaviors such as dominance and aggression.

Although researchers had long known that T levels change during and after competition (Mazur & Booth, 1998), researchers had simply assumed that these competition-induced changes in T would influence subsequent status-seeking behaviors. We conducted the first study in humans that explicitly examined the relationship between post-competition fluctuations in T and subsequent social behavior (Mehta & Josephs, 2006). We experimentally rigged a competition and collected saliva samples before and after the competition to measure changes in T (Mehta & Josephs, 2006). After participants provided the second saliva sample, we measured dominance behavior by asking participants whether they wanted to (a) rechallenge their opponent to a second competition, or (b) complete an alternative noncompetitive task. The results showed that changes in T after losing predicted who wanted to compete again in a second competition. Losers who rose in T were more likely to choose to rechallenge their opponent (73%) than losers who dropped in T (22%). These findings are consistent with the reciprocal model and suggest that a rise in T after a loss of status motivates individuals to reclaim their lost status (choosing to compete again).

We conducted a second study to test whether T responses to competition would also predict subsequent aggressive behavior (Carré et al., 2009). Similar to the previous study, participants provided a saliva sample before and after a rigged competition. After the second saliva sample, participants completed the Point Subtraction Aggression Paradigm (PSAP), a well-validated laboratory task that measures reactive aggression. In this task, participants are paired with a fictitious opponent (actually a computer program) and earn points by pressing Button 1 as quickly as possible or Button 2 to steal points from their opponent. Participants are told their total points will be exchanged for money at the end of the study. During the task, participants have points taken from them by their fictitious opponent, which serves as the experimental manipulation of social provocation. Stealing money from the fictitious competitor by pressing Button 2 is considered aggressive behavior because, like the Taylor Aggression Paradigm and the Ultimatum Game, this behavior represents an intent to cause harm. Consistent with the results of the earlier study (Mehta & Josephs, 2006), this study found that changes in T after losing in a competition predicted aggressive behavior in the PSAP. Individuals who lost the competition and rose in T showed more aggressive behavior (stealing more points from their opponents after social provocation) than individuals who lost the competition and dropped in T (Carré et al., 2009). More follow-up studies from our labs also showed relationships between dynamic T changes and aggressive behavior (Carré et al., 2010; Geniole et al., 2010). Together, this wave of studies provides strong support for the Challenge Hypothesis and Biosocial Model of Status, showing that dynamic T responses in status-relevant social interactions have implications for aggression and dominance behaviors.

Although these human studies on dynamic T were correlational, they fit with experimental research in animals, which demonstrates a causal influence of experimentally administered T after competition on aggressive behavior in a second competition (see Gleason et al., 2009 and Oliveira, 2009, for reviews). More recent evidence suggests a causal influence among human participants and may add subtleties to understanding the relationship between T dynamics and aggressive behavior (Carré et al., 2017). Participants were surveyed on a handful of demographics and relevant personality traits. Then, in a double-blind, placebo-controlled, between-subjects design, male participants received either T or a placebo and then participated in the PSAP. Exogenously raised T levels increased aggressive behavior on the PSAP, but only among men scoring relatively high in trait dominance or low in trait self control. However, it is important to note the results of a meta-analysis (Geniole et al., 2020), which showed that although endogenous T (i.e., baseline and dynamics) was positively, if weakly, associated with aggression (particularly among males), exogenous T (i.e., causally manipulated) was not associated with aggression. The authors stated that this may be due to too few studies on the topic and the heterogeneity of experimental conditions, and not necessarily due to the definitive absence of a causal effect.

Neural Mechanisms for Testosterone's Influence on Aggression

Studies suggest that T influences human aggression through the OFC and amygdala. In one fMRI study, T levels were measured in saliva and then participants played the Ultimatum Game while being scanned (Mehta & Beer, 2010). The findings showed that higher T levels predicted increased aggressive behavior (rejection of unfair offers), and decreases in bilateral medial OFC activity following unfair offers significantly mediated the association between testosterone and aggression. This finding suggests that T increases reactive aggression in part through impairments in the neural circuitry of impulse control and self-regulation (medial OFC). Other studies show that T (a) increases amygdala reactivity to angry faces (Goetz et al., 2014; Hermans et al., 2008; van Wingen et al., 2008), and (b) reduces functional connectivity between OFC and amygdala (van Wingen et al., 2010), providing two additional neural mechanisms for how testosterone may modulate human aggression.

Androgen Receptor Gene

Researchers interested in the genetics of human aggression have turned their attention to a common polymorphism found in the androgen receptor gene. The trinucleotide cytosine-adenine-guanine repeat (CAG) has been found to be highly polymorphic (Choong & Wilson, 1998) and ranges from 9 to 31 repeats in the human population (e.g., Edwards et al., 1992). CAG repeat length is negatively associated with the expression of the androgen gene and androgen receptor (AR) sensitivity (Chamberlain et al., 1994). T exerts its effects primarily through these receptors which are expressed throughout the brain, including regions important in regulating aggression (e.g., amygdala and OFC) (Rubinow & Schmidt, 1996; Mehta & Beer, 2010, respectively). Thus, AR sensitivity to T may serve as a mechanism to modulate its effects on brain development and subsequent aggressive behavior.

Researchers have found that men who have fewer CAG repeats score higher on sexually dimorphic behavioral traits. For example, Simmons and Roney (2011) found that CAG length was negatively correlated with prestige and dominance (traits associated with intrasexual competition) in a sample of men. Other work indicates that rapists and murderers have significantly fewer CAG repeats compared to controls in a sample of Indian men (Rajender et al., 2008). Furthermore, a study with adolescent males found that CAG repeat length interacted with T to predict a self-report measure of aggressive risk-taking (Vermeersch et al., 2010). Specifically, the authors found that T was positively correlated with aggressive risk-taking, but only among men with relatively fewer CAG repeats. Investigating the combined effects of traits associated with aggression (dominance, low self-control, and independent self-construal) and androgen sensitivity (defined by CAG repeats), Geniole et al. (2019) found that increasing T also increased aggressive behavior, but only among men high in dominance, low in self-control, and with relatively independent self construal—and critically—these behavioral effects were maximal among men with fewer CAG repeats.

Other work has specifically linked variation in the CAG repeat to amygdala reactivity to facial signals of threat. Manuck et al. (2010) found an inverse relationship between CAG repeats and bilateral ventral amygdala (the principal input region of the amygdala) reactivity when viewing threat cues (angry/fearful faces). On the other hand, reactivity in the dorsal amygdala (principal output region of the amygdala regulating physiological reactivity) was positively correlated with T independent of genotype. These results suggest that the CAG polymorphism modulates androgen sensitive neural circuits associated with aggression.

The Dual-Hormone Hypothesis: Interactions Between Testosterone and Cortisol

Glucocorticoids are a class of hormones that are released by the adrenal glands during physical and psychological stress. The primary glucocorticoid in humans is cortisol (C). Most research on C has focused on the dispositional and situational variables that cause acute changes in C (e.g., Dickerson & Kemeny, 2004), but some research indicates that C is negatively associated with aggressive behavior. In one longitudinal study of 314 boys, low basal C levels during preadolescence (age 10–12 years) predicted more aggressive behaviors 5 years later (Shoal et al., 2003). Other studies, however, have shown null effects of C on aggression. These mixed findings suggest that C may interact with other biological systems to modulate human aggression.

We proposed the dual-hormone hypothesis to reconcile mixed findings on the roles of T and C in human social behavior (Carré & Mehta, 2011; Mehta & Josephs, 2010). According to the dual-hormone hypothesis, T should have a strong influence on aggression and dominance only when C is low, but T's effect on social behavior should be blocked when C levels are high because C inhibits the neurobiological pathway between T and behavior at multiple levels (see Mehta & Josephs, 2010 for a biological rationale). Consistent with the dual-hormone hypothesis, Popma et al. (2007) studied a group of male adolescents and found that T was positively related to physical aggression only in individuals with low C. In individuals with high C, there was no association between T and aggression. Mehta and Josephs (2010) showed a similar pattern of findings in studies of social dominance. A hormone profile of high T and low C was associated with increased dominance across multiple studies. Intriguingly, a profile of high T and high C was associated with submissive behavior after social threat. These dual-hormone effects on social behavior vary across social contexts (threat versus no threat, Mehta & Josephs, 2010; social inclusion versus exclusion, Geniole et al., 2010). However, a recent study suggested one potential moderator of the relationship between T and C (Knight et al., 2022). While not focusing on aggression per se, but rather competitive behavior, they found that men who were administered exogenous T, who were also low in basal C, showed an increased tendency to compete—but only against other male, high status opponents (relative to female, low status opponents), suggesting the relationship between T and C is moderated by the context of status.

A recent meta-analysis (Dekkers et al., 2019) showed significant but weak support for the dual-hormone hypothesis, with effects being strongest for men, and highlighted some limitations based on the methodological heterogeneity of the studies examined (for an overview of proposed solutions to these limitations, see Knight et al., 2020). Nonetheless, these findings suggest that T and C jointly modulate human aggression and dominance behavior in a context-dependent fashion. Although the neurobiological mechanisms for dual-hormone modulation of behavior have yet to be studied, the amygdala and OFC are clear candidate regions. Indeed, androgen and glucocorticoid receptors are located in both of these regions, and T and C modulate neural activity in the amygdala and OFC.

Environmental Risk Factors

The research reviewed above provides insights into the biological factors implicated in aggressive behavior. In this section, we review research on environmental risk factors. We focus on two risk environmental factors that have received attention in scientific research—exposure to media violence and interpersonal rejection—and we discuss possible biological mechanisms.

Violent Media Exposure

Research into the possible effects of violent media exposure is highly contentious, particularly so for violent video games. Narrowly speaking, there is currently scant evidence to support a direct relationship between violent video games and violent behavior *per se*. Indeed, sales of violent video games have only increased over the same timeframe that violent crime rates have continually fallen (Cunningham et al., 2016). A review of the existing literature on the topic compelled the American Psychological Association to release a resolution, stating, "Violence is a complex social problem that likely stems from many factors that warrant attention from researchers, policy makers and the public. Attributing violence to violent video gaming is not scientifically sound and draws attention away from other factors," (APA, 2020).

On broader considerations of aggression, the results are mixed. For example, a 2010 meta-analysis of over 300 studies showed evidence that exposure to violent video games increased aggressive thoughts, feelings, and behavior and decreased empathy and prosocial behaviors (Anderson et al., 2010). Yet, a later meta-analysis revealed little substantial evidence connecting violent video games to aggressive behavior (Ferguson, 2015), followed by still another suggesting there is a relationship (Prescott et al., 2018). Many of these meta-analytic exchanges have been ensnared in debate over statistical methodologies (for example, see Hilgard et al., 2017, but also Kepes et al., 2017).

There are neuroscientific studies that seem to support the hypothesis that media violence exposure may increase aggression by altering the neural circuitry of aggression. One fMRI study showed that exposure to media violence decreased lateral OFC activity and reduced amygdala–OFC coupling (Kelly et al., 2007), and another study demonstrated that adolescents who reported frequent exposure to violent media had decreased lateral OFC density (Strenziok et al., 2010). These findings suggest that violent media exposure may be associated with both short-term and long-term differences in aggression, relating to OFC and amygdala function (see Carnagey et al., 2007 for these and related neural mechanisms). However, another study tested the General Aggression Model, which would predict that long term exposure to violent video game content leads to decreased empathy via social learning. Comparing a group of "excessive users" of violent video games to a control group, researchers found no differences in the patterns of neural activity associated with empathy or theory of mind, which led the researchers to conclude that if violent media affects emotional processing at all then it is likely acute and short lived (Szycik et al., 2017).

There are many possible reasons for these seemingly contradictory findings—false positives, false negatives, small and thus inconsistent effects, confounding or moderating variables—all could muddy the waters independently or in tandem. Attempts to add clarity come from some studies disentangling violent content from challenge or "competence-impeding" play. One study showed that when violent and nonviolent games were matched on competitiveness, difficulty, and pace of action, then there were no differences between groups regarding subsequent aggressive behavior; yet, when competitiveness was isolated by matching games on difficulty and pace of action, and systematically controlling for violent content, it was the more competitive games that produced greater levels of aggressive behavior irrespective of the amount of violence in them (Adachi & Willoughby, 2011). A similar investigation, wherein researchers consistently controlled for violent content when comparing challenge, and vice versa, found across seven independent studies that increasing the game challenge exclusively predicted subsequent increases in aggression scores, with violent content having no effect (Przybylski et al., 2014). While there may be some relationship between violent media and aggression that is worth investigating, until studies on this topic can regularly control for levels of competition or challenge in the games; potentially confounding developmental variables, prior aggression levels, and other individual differences among participants; as well as better standardize the operationalization of aggression, it will be difficult to know with confidence what the nature of this relationship is.

Interpersonal Rejection

The pain experienced when feeling socially excluded is no less intense than the experience of physical pain, with the same brain regions being active during the experience of either (Eisenberger et al., 2003). The effect of ostracism on one's emotional wellbeing appears so fundamental that it can manifest even when participants are explicitly told the ostracism is a function of a random computer algorithm (Zadro et al., 2004). Given this primal affective salience, it is unsurprising that the act of being rejected or devalued by other people has been shown to be a clear risk factor in aggressive behavior. In fact, a Surgeon General's report concluded that social rejection was the most significant risk factor for violence among adolescents, even more potent than factors such as low socioeconomic status, gang membership, or drug use (cf. Leary et al., 2006; Office of the Surgeon General, 2001). A spate of school shootings in the United States illustrates the social isolation-violence relationship. In an analysis of 15 school shootings between 1995 and 2001, 13 out of the 15 perpetrators had a history of being socially rejected—including teasing, bullying, and chronic ostracism (Leary et al., 2003).

Experimental evidence also supports a causal effect of interpersonal rejection on aggressive behavior. In many of the experiments, individuals were randomly assigned to receive rejecting or accepting relational feedback from another person (in fact, the feedback is bogus), and factors such as anger, relational aggression (e.g., social derogation), and reactive aggression (e.g., administering varying amounts of aversive hot sauce) were measured after the rejection or acceptance experience. Rejection increased anger and aggression compared to acceptance in many of the studies (see review by Leary et al., 2006). One study suggested that rejection may lead individuals to perceive a threat toward their basic needs—one of which is control—and aggressive behavior may be an attempt to restore that sense of control (Warburton et al., 2006).

This effect of social rejection on aggression also appears to depend on individual differences in social sensitivity and biological differences in serotonergic activity. People high in rejection sensitivity or insecure attachment are more likely to respond to interpersonal rejection with aggression (Leary et al., 2006). Moreover, as described earlier in the chapter, low expression MAOA allele carriers show increased activity in the dorsal ACC after social rejection, a region implicated in emotional distress and anger (Eisenberger et al., 2007). These results suggest that the influence of social rejection on aggression may be driven by hyperreactivity in socioemotional neural circuits to rejection experiences (e.g., dorsal ACC, Buckholz & Meyer-Lindenberg, 2008).

Other research indicates that social rejection can increase levels of C, a hormone implicated in psychological stress. In an experiment in which participants were socially rejected or received no rejection, self-esteem moderated cortisol and aggressive behavioral responses to social rejection (Ford & Collins, 2010). Compared to individuals high in self-esteem, individuals low in self-esteem showed heightened relational aggression (partner derogation) and increased C in response to social rejection. The association between low self-esteem and relational aggression was mediated by changes in C. Taken together, the findings suggest that social rejection augments aggressive behavior through biological systems associated with stress and socioemotional sensitivity (cortisol, dorsal ACC).

Psychological Interventions

Above, we reviewed some of the biological and social risk factors implicated in human aggressive behavior. In this section, we discuss psychological interventions that can reduce reactive aggression. Although there are many possible interventions, we focus our discussion on two in particular: (1) cognitive reappraisal and (2) self-control training.

Cognitive Reappraisal

Emotion regulation involves cognitive strategies to alter one's emotional response to environmental stimuli. In the context of reactive aggression, cognitive strategies that prolong anger are likely to upregulate aggression, while strategies that reduce anger responses should decrease the propensity toward aggression. In line with this reasoning, rumination—which involves continuing to think about the anger-induced provoking event in a way that prolongs anger—increases anger and reactive aggression (Denson et al., 2011b; Fabiansson et al., 2012). An emotion regulation strategy that may be beneficial for reducing reactive aggression is cognitive reappraisal, which involves reinterpreting an emotional event to reduce its negative emotional impact. For example, an individual who is insulted by another person may try to think about what lessons he or she could learn from the event or think about the event from the perspective of an objective third party (Fabiansson et al., 2012). Research supports the hypothesis that cognitive reappraisal can reduce anger and reactive aggression. One study showed that individuals who thought about an anger-inducing event and then engaged in cognitive reappraisal showed less anger compared to participants who thought about an anger-inducing event and then engaged in anger rumination (Fabiansson et al., 2012). A second study used a longitudinal design and found that people who received reappraisal training over the course of a semester showed reduced trait vengeance—an important predictor of aggressive behavior—compared to participants in the control condition (Bartlett & Anderson, 2011). Hence, not only can cognitive reappraisal reduce the propensity toward aggression in the short term but also in the longer term. These findings are preliminary, but they suggest that cognitive reappraisal training courses may reduce aggressive behavioral reactions to social provocation in individuals prone to violence.

Neuroscience studies suggest that the aggression-reducing benefits of reappraisal may be due to more effective engagement of prefrontal neural regions including medial and lateral OFC, reduced amygdala activity, and changes in functional connectivity between the frontal cortex and subcortical regions (Fabiansson et al., 2012; Goldin et al., 2008; McRae et al., 2008; Ochsner et al., 2002, 2009). Overall, reappraisal training may reduce the likelihood of reactive aggression through increased top-down neural control and blunted emotional reactivity to social provocation.

Self-control

Theory and research suggest that reduced self-control is a critical psychological mechanism for aggressive behavioral reactions to social provocation. Individuals prone to aggressive behavior are often impulsive (low in self-control), and experimental studies show that reduced self-control mediates the association between social provocation and increased aggressive behavior (Denson et al., 2011b). These studies suggest that self-control training interventions could decrease reactive aggression. One study tested this hypothesis by having participants practice motor self-control (using their nondominant hand to do everyday tasks such as brushing teeth) between 8 a.m. and 6 p.m. over a period of 2 weeks (Denson et al., 2011a). The findings revealed that this self-control intervention decreased anger and aggressive behavior following social provocation for individuals high in trait aggression. Although the precise biological mechanisms for the effect of self-control training on aggression remain unknown, it is plausible that the self-control intervention promoted engagement of prefrontal regions implicated in self-regulation and impulse control such as medial OFC, which resulted in the inhibition of aggressive behavior (Mehta & Beer, 2010).

Directions for Future Research

There are a number of important directions for future research on the social neuroscience of human aggressive behavior. Here, we discuss some of these directions.

Gene×Hormone Interactions

One area of research that needs more attention is studies that search for theoretically informed gene × hormone interactions. It has been speculated, for example, that T may interact with the serotonergic system to modulate human aggression. Promising new evidence provides initial support for this hypothesis, demonstrating a MAOA × T interaction on aggression (Sjoberg et al., 2008). Individuals with the low expression allele who were also high in T levels showed the greatest risk for aggressive behavior. Another study found a 5HTTLPR × T interaction on stress reactivity. S carriers with high T showed heightened cortisol reactivity to social threat (Josephs et al., 2012), suggesting that these same individuals may be prone to greater emotional reactions to social provocation and reactive aggression. Finally, some studies have shown that hormone receptor genes are related to the neural circuitry of aggression (e.g., androgen receptor genes and amygdala reactivity to angry faces; Manuck et al., 2010), but there has been considerably less work that has tested for biologically relevant hormone receptor gene × hormone interaction on aggression (e.g., androgen receptor gene × T interactions; but see Vermeersch et al. (2010) for an example of this fruitful approach). Human research on gene × hormone interactions has only just begun, and we believe that research that takes this approach will greatly improve our understanding of the neurobiology of aggression.

Neuropeptides

Animal research indicates that the neuropeptide vasopressin plays an important role in aggressive behavior in part through interaction with other biological factors such as T, but very little work has examined the influence of vasopressin on human aggression. An earlier study found that individual differences in cerebrospinal fluid arginine vasopressin (AVP) were positively correlated with self- reported aggression (Coccaro et al., 1998). Another paper administered AVP and found some initial evidence that it alters psychophysiological correlates of aggressive behavior (Thompson et al., 2004, 2006), but clearly much more research on vasopressin and aggression in humans is needed. Oxytocin is another neuropeptide that also influences social behavior, and its effects on human behavior have been much more well-studied (Bartz et al., 2011). Although most biological theories suggest that oxytocin should encourage prosocial behavior (trust, cooperation), human studies suggest that this view is overly simplistic. One study found that the administration of intranasal oxytocin increased aggression, but only in the context of provocation (Ne'eman et al., 2016). The researchers argued that, rather than specifically increasing prosocial behavior, oxytocin generally increases the salience of social agents and thus provokes a wide variety of context-based behaviors (ranging from prosocial to aggressive). Other research shows that oxytocin administration has divergent effects on human social behavior directed toward ingroup versus outgroup members. More specifically, oxytocin increased aggressive motivation toward outgroup members (outgroup hate) even though oxytocin increased prosocial motivation toward ingroup members (ingroup love) (De Dreu et al., 2010). These results indicate that the effects of oxytocin are context-dependent, which fits with animal models suggesting that oxytocin can promote defensive maternal aggression.

Longitudinal Studies

There is a need for more longitudinal studies that measure a host of biological factors and psychological factors along with aggressive behavior at multiple time points. Such longitudinal studies can illuminate how changes in biological systems (e.g., T levels, OFC function) may track changes in aggressive behavior over time. Such longitudinal studies can inform theories of the psychobiological mechanisms through which environmental risk factors (e.g., media violence) and protective factors (e.g., parental training in cognitive reappraisal) early in life can influence the expression of aggressive behavior in adulthood.

Gender Similarities and Differences

More theoretical and empirical attention is needed to understand how males and females are similar and how they are different in the psychobiological mechanisms of aggressive behavior (Josephs et al., 2011). For example, most research on T and social behavior has focused on males, but a series of studies suggest that basal T may also predict social behaviors in females (e.g., social dominance, Mehta et al., 2008, 2009). Other research, however, suggests that acute fluctuations in T predict aggression and dominance only in men (Carré et al., 2009; Mehta & Josephs, 2006). Moreover, a greater understanding is needed for how men's and women's aggressive behavior may be expressed differently. Research suggests that boys are more likely to show direct forms of confrontation (physical aggression, direct name calling) compared to girls, whereas boys and girls are equally likely to show indirect aggression (sabotaging friendships or romantic relationships, spreading gossip, social exclusion) (Card et al., 2008). Greater attention to biological and cultural issues surrounding gender is required to build more accurate theoretical models of human aggression.

Socioeconomic Inequality and Concentrated Disadvantage

There is a well-established relationship between poverty, income inequality, and violent crime (Fajnzylber et al., 2002; Hsieh & Pugh, 1993; Kelly, 2000; Kennedy et al., 1998; Krohn, 1976; Messner & Rosenfeld, 1997; Neapolitan, 1999; Rufrancos et al., 2013). By epidemiological standards, this relationship meets the criteria for being considered causal (Pickett & Wilkinson, 2015). Despite much sociological research being devoted to the topic, and despite Daly (2020) stating that inequality is "the next big question" for evolutionary psychology to tackle, for example, thus far little experimental psychological research has focused on the intersection of poverty, socioeconomic inequality, and aggressive or violent behavior. Most neuroscientific research has focused on developmental outcomes, but often in the context of emotion regulation and usually with an eye on childhood poverty and not inequality. Nonetheless, this latter research could readily inform investigations into inequality and adult aggression, as many of the candidate brain regions affected by poverty (for a review, see Palacios-Barrios & Hanson, 2019) share considerable overlap with regions associated with aggression (e.g., the amygdala and various prefrontal cortices).

Resource scarcity produces psychosocial stress (e.g., Haushofer & Fehr, 2014) which in turn negatively impacts executive functioning (e.g., Sharkey, 2010), including self-control (Duckworth et al., 2013). The latter has recently been revealed as an important individual difference in moderating the effect of testosterone on reactive aggression (Carré et al., 2017; Geniole et al., 2019) and aggression more generally (Denson et al., 2012; Sprague et al., 2011). However, evidence consistently indicates that resource inequality—often operationalized as income inequality—rather than poverty, is the strongest predictor of homicide rates both within and between countries (Daly, 2023; Vilalta et al., 2024).

In social ecologies characterized by high levels of social inequality, individuals often adopt extreme competitive strategies. Among such strategies, Wilson and Daly (1985) argued that young men with little to lose place high value on demonstrating both their capacity and willingness for aggressive action, with status insults triggering explosive defense. Developmental environments characterized as 'socially demoralizing'—those marked by neighborhood violence, family dysfunction, and deviant peer groups—often foster the adoption of 'honor culture' (Berg et al., 2020; Nawata, 2020; Nisbett, 1993). In urban contexts, this manifests as what sociologist Elijah Anderson termed 'street code,' an oppositional orientation emphasizing respect and aggressive self-presentation, where violence becomes normative (Anderson, 1999). Occupying a lower stratum in the social hierarchy not only reduces the potential costs of aggression—having less to lose—but also increases the relative value of status. Empirical research has demonstrated that the value of honor can compensate for a sense of low social worth (e.g., Henry, 2009).

Importantly, contemporary inequality should be considered alongside historical levels of inequality. Recent evidence from Mexican municipalities found that income inequality a decade prior interacted with present income inequality to predict homicide rates. While both historical and present inequality were positively correlated with the municipalities homicide rate, the effect of present inequality was amplified by a history of high inequality (Vilalta et al., 2024).

Economic inequality fosters greater desire to emerge as a winner and increases the costs of failure, creating incentives for desirable goods and the need to maintain social status and respect and motivating extreme forms of competition, including homicide (e.g., Daly, 2023). Both the past and present socioeconomic circumstances are formative in influencing competitive behaviors and suggest policy interventions should consider prior levels of inequality when allocating resources to address violence (Vilalta et al., 2024).

Recent political science research suggests that violence, specifically political violence, stems not from inequality per se, but rather from 'poor prospects.' Relying on large cross-cultural panel data from African countries, Bartusevičius and van Leeuwen (2022) found that perceptions of poor prospects, rather than economic inequality, reliably predicted willingness to engage in political violence. However, it should be noted that inequality and poor prospects were correlated. Furthermore, economic inequality conditions the severity or magnitude of relative deprivation. Thus, when individuals anticipate a loss of status, they may resort to violence as a mitigating strategy.

Because of the dearth of experimental psychological research focused specifically on inequality and aggression among adult participants, the causal mechanisms remain poorly understood despite the robust relationship well demonstrated by sociological research. One model (De Courson & Nettle, 2021) suggests that this relationship can be explained both compositionally (i.e., high inequality means greater amounts of people living in a state of concentrated disadvantage, which results in more crime based on meeting survival needs; also see Massey, 2004) and psychosocially (i.e., individuals' experience of, and response to, their perceptions of the social differentials in their broader society, leading to poorer social relationships). Regarding this latter notion, Wilkinson (2004) highlights, "Where there is greater inequality, more people will be deprived of the jobs, incomes, housing, and cars which are the markers of status. Vulnerable to the humiliation of relative poverty, they will be particularly sensitive to feeling disrespected and looked down on and unwilling to ignore incidents that appear to involve a loss of face." That is, importantly, it is not just absolute but also relative deprivation that may drive the relationship between inequality and aggression.

What little experimental psychological research that exists already supports this notion. Across three separate studies, researchers experimentally manipulated the relative deprivation of participants and found that increasing this sense also increased aggressive affect and behavior, despite the participants' objective socioeconomic status (Greitemeyer & Sagioglou, 2017). Inequality appears to make hierarchy more salient, and the perceptions of hierarchy may in turn lead to an increased likelihood of status competition and the consequent opportunities for aggression and violence. Greater attention to emotional mediators of aggression and regulatory mechanisms may help identify the underlying genetic, hormonal, and neurobiological structures, and their individual variation, in these responses (Lukaszewski et al., 2020; Sprague et al., 2011). Indeed, much of the psychological research discussed above—particularly work on status competition, social exclusion, and their neurobiological correlates—provides a framework for understanding how societal inequality shapes individual aggressive behavior. As noted by Daly (2023), "the lived experience of poverty is ineluctably relative," suggesting that emotions arising from social comparison are key to explaining the inequality-aggression link. Supporting this perspective, He and Xiang (2021) found that malicious envy mediated the relationship between childhood maltreatment and adolescent aggression. Another study found evidence that inequality and violence were associated with preferences for in-group hegemony at the level of the individual (Kunst et al., 2017), which would make this relationship as much a psychological consideration as it is a sociological one. Indeed, much of the psychological research discussed in previous sections above, such as that focused on status competition or social exclusion (and associated neurobiological variables), could translate well into better understanding the causal mechanisms of this societal phenomenon at the level of the individual.

Conclusion

Aggressive and violent behaviors affect millions of people worldwide every year (Mercy et al., 2002). This chapter reviewed the research on the social neuroscience of human reactive aggression, including research on the genes, hormones, neural systems, and environmental factors implicated in aggressive behavior. Researchers have only begun to integrate these perspectives to build comprehensive models of human aggression. Promising new directions for research include longitudinal studies that better delineate the social and biological mechanisms that increase risk for adulthood violence as well as studies that attempt to reduce aggression in at-risk populations through novel pharmacological and psychosocial interventions.

References

Adachi, P. J. C., & Willoughby, T. (2011). The effect of violent video games on aggression: Is it more than just the violence? *Aggression and Violent Behavior, 16*, 55–62.

Anderson, E. (1999). *Code of the street: Decency, violence, and the moral life of the inner city*. W.W. Norton.

Anderson, C. A., Shibuya, A., Ihori, N., et al. (2010). Violent video game effects on aggression, empathy, and prosocial behavior in eastern and western countries: A meta-analytic review. *Psychological Bulletin, 136*, 151–173.

Andrews, P. W., Gangestad, S. W., & Matthews, D. (2002). Adaptationism—how to carry out an exaptationist program. *Behavioral and Brain Sciences, 25*, 489–553.

APA resolution on violent video games: February 2020 revision to the 2015 resolution. (2020).

Archer, J., & Southhall, N. (2009). Does cost-benefit analysis or self-control predict involvement in bullying behavior by male prisoners? *Aggressive Behavior, 35*, 31–40.

Archer, J., Graham-Kevan, N., & Davies, M. (2005). Testosterone and aggression: A reanalysis of Book, Starzyk, and Quinsey's (2001) study. *Aggression and Violent Behavior, 10*, 241–261.

Baron, R. A., & Richardson, D. R. (1994). *Human aggression* (2nd ed.). Plenum.

Bartlett, C. P., & Anderson, C. A. (2011). Re-appraising the situation and its impact on aggressive behavior. *Personality and Social Psychology Bulletin, 37*, 1564–1573.

Bartusevičius, H., & van Leeuwen, F. (2022). Poor prospects—not inequality—motivate political violence. *Journal of Conflict Resolution, 66*, 1393–1421.

Bartz, J. A., Zaki, J., Bolger, N., et al. (2011). Social effects of oxytocin in humans: Context and person matter. *Trends in Cognitive Sciences, 15*, 301–309.

Beaver, J. D., Lawrence, A. D., Passamonti, L., & Calder, A. J. (2008). Appetitive motivation predicts the neural response to facial signals of aggression. *The Journal of Neuroscience, 28*, 2719–2725.

Bechara, A., Damasio, H., & Damasio, A. R. (2000). Emotion, decision making and the orbitofrontal cortex. *Cerebral Cortex, 10*, 295–307.

Beer, J. S., Heerey, E. A., Keltner, D., Scabini, D., & Knight, R. T. (2003). The regulatory function of self-conscious emotion: Insights from patients with orbitofrontal damage. *Journal of Personality and Social Psychology, 85*, 594–604.

Beer, J. S., John, O. P., Scabini, D., & Knight, R. T. (2006). Orbitofrontal cortex and social behavior: Integrating self-monitoring and emotion-cognition interactions. *Journal of Cognitive Neuroscience, 18*, 871–879.

Beitchman, J. H., Baldassarra, L., Mik, H., et al. (2006). Serotonin transporter polymorphisms and persistent, pervasive childhood aggression. *The American Journal of Psychiatry, 163*, 1103–1105.

Berg, M. T., Lei, M. K., & Simons, R. L. (2020). Socially demoralizing environments and the development of the street code from childhood to emerging adulthood. *Journal of Quantitative Criminology, 36*, 207–233.
Berman, M. E., McCloskey, M. S., Fanning, J. R., Schumacher, J. A., & Coccaro, E. F. (2009). Serotonin augmentation reduces response to attack in aggressive individuals. *Psychological Science, 20*, 714–720.
Beyer, F., Münte, T. F., Wiechert, J., Heldmann, M., & Krämer, U. M. (2014). Trait aggressiveness is not related to structural connectivity between orbitofrontal cortex and amygdala. *PLoS One, 9*, e101105.
Biderman, A. D., & Reiss, A. J. (1967). On exploring the "dark figure" of crime. *The Annals of the American Academy of Political and Social Science, 374*, 1–15.
Blair, R. J. R. (2004). The roles of orbital frontal cortex in the modulation of antisocial behavior. *Brain and Cognition, 55*, 198–208.
Border, R., Johnson, E. C., Evans, L. M., Smolen, A., Berley, N., Sullivan, P. F., & Keller, M. C. (2019). No support for historical candidate gene or candidate gene-by-interaction hypotheses for major depression across multiple large samples. *American Journal of Psychiatry, 176*, 376–387.
Brown, G. L., Goodwin, F. K., Ballenger, J. C., Goyer, P. F., & Major, L. F. (1979). Aggression in humans correlates with cerebrospinal fluid amine metabolites. *Psychiatry Research, 1*, 131–139.
Brunner, H. G., Nelen, M., Breakefield, X. O., et al. (1993). Abnormal-behavior associated with a point mutation in the structural gene for monoamine oxidase-A. *Science, 262*, 578–580.
Buckholz, J. W., & Meyer-Lindenberg, A. (2008). MAOA and the neurogenetic architecture of human aggression. *Trends in Neurosciences, 31*, 120–129.
Bufkin, J. L., & Luttrell, V. R. (2005). Neuroimaging studies of aggressive and violent behavior: Current findings and implications for criminology and criminal justice. *Trauma, Violence and Abuse, 6*, 176–191.
Buss, D. M., & Shackelford, T. K. (1997). Human aggression in evolutionary psychological perspective. *Clinical Psychology Review, 17*, 605–619.
Card, N. A., Stucky, B. D., Sawalani, G. M., & Little, T. D. (2008). Direct and indirect aggression during childhood and adolescence: A meta-analytic review of gender differences, intercorrelations, and relations to maladjustment. *Child Development, 79*, 1185–1229.
Carnagey, N. L., Anderson, C. A., & Bartholow, B. D (2007). Media violence and social neuroscience: New questions and new opportunities. *Current Directions in Psychological Science, 16*, 178–182.
Carré, J. M., & Mehta, P. H. (2011). Importance of considering testosterone–cortisol interactions in predicting human aggression and dominance. *Aggressive Behavior, 37*, 1–3.
Carré, J. M., Putnam, S. K., & McCormick, C. M. (2009). Testosterone responses to competition predict future aggressive behavior at a cost to reward in men. *Psychoneuroendocrinology, 34*, 561–570.
Carré, J. M., Gilchrist, J. D., Morrissey, M. D., et al. (2010). Motivational and situational factors and the relationship between testosterone dynamics and human aggression during competition. *Biological Psychology, 84*, 346–353.
Carré, J. M., McCormick, C. M., & Hariri, A. R. (2011). The social neuroendocrinology of human aggression. *Psychoneuroendocrinology, 36*, 935–944.
Carré, J. M., Fisher, P. M., Manuck, S. B., & Hariri, A. R. (2012). Interaction between trait anxiety and trait anger predict amygdala reactivity to angry facial expressions in men but not women. *Social Cognitive and Affective Neurosciences, 7*, 213–221.
Carré, J. M., Geniole, S. N., Ortiz, T. L., Bird, B. M., Videto, A., & Bonin, P. L. (2017). Exogenous testosterone rapidly increases aggressive behavior in dominant and impulsive men. *Biological Psychiatry, 82*, 249–256.
Caspi, A., McClay, J., Moffitt, T. E., et al. (2002). Role of genotype in the cycle of violence in maltreated children. *Science, 297*, 851–854.
Caspi, A., Hariri, A. R., Holmes, A., et al. (2010). Genetic sensitivity to the environment: the case of the serotonin transporter gene and its implications for studying complex diseases and traits. *The American Journal of Psychiatry, 167*, 509–527.
Chamberlain, N. L., Driver, E. D., & Miesfeld, R. L. (1994). The length and location of CAG trinucleotide repeats in the androgen receptor N-terminal domain affect transactivation function. *Nucleic Acids Research, 22*, 3181–3186.
Chan, S. C., Raine, A., & Lee, T. M. (2010). Attentional bias toward negative affective stimuli and reactive aggression in male batterers. *Psychiatry Research, 176*, 246–249.
Chester, D. S., DeWall, C. N., Derefinko, K. J., Estus, S., Peters, J. R., Lynam, D. R., & Jiang, Y. (2015). Monoamine oxidase A (MAOA) genotype predicts greater aggression through impulsive reactivity to negative affect. *Behavioural Brain Research, 283*, 97–101.
Choong, C. S., & Wilson, E. M. (1998). Trinucleotide repeats in the human androgen receptor: A molecular basis for disease. *Journal of Molecular Endocrinology, 21*, 235–257.
Coccaro, E. F., Kavoussi, R. J., Hauger, R. L., Cooper, T. B., & Ferris, C. F. (1998). Cerebrospinal fluid vasopressin levels: Correlates with aggression and serotonin function in personality-disordered subjects. *Archives of General Psychiatry, 55*, 708–714.
Coccaro, E. F., McCloskey, M. S., Fitzgerald, D. A., & Phan, K. L. (2007). Amygdala and orbitofrontal reactivity to social threat in individuals with impulsive aggression. *Biological Psychiatry, 62*, 168–178.
Coccaro, E. F., Sripada, C. S., Yanowitch, R. N., & Phan, K. L. (2011). Corticolimbic function in impulsive aggressive behavior. *Biological Psychiatry, 69*, 1153–1159.
Collias, N. E., Barfield, R. J., & Tarvyd, E. S. (2002). Testosterone versus psychological castration in the

expression of dominance, territoriality, and breeding behavior by male village weavers (Ploceus cucullatus). *Behaviour, 139*, 801–824.

Cooke, E. M., Armstrong, T., Boisvert, D., Wells, J., Lewis, R. H., Hughes-Stamm, S., & Gangitano, D. (2018). The relationship between the MAOA-uVNTR polymorphism, delinquent peer affiliation, and antisocial behavior with a consideration of sex differences. *Psychiatric Quarterly, 89*, 841–853.

Crockett, M. J., Clark, L., Tabibnia, G., Lieberman, M. D., & Robbins, T. W. (2008). Serotonin modulates behavioral reactions to unfairness. *Science, 320*, 1739.

Crockett, M. J., Siegel, J. Z., Kurth-Nelson, Z., Ousdal, O. T., Story, G., Frieband, C., et al. (2015). Dissociable effects of serotonin and dopamine on the valuation of harm in moral decision making. *Current Biology, 25*, 1852–1859.

Cunningham, S., Engelstätter, B., & Ward, M. R. (2016). Violent video games and violent crime. *Southern Economic Journal, 82*, 1247–1265.

da Cunha-Bang, S., & Knudsen, G. M. (2021). The modulatory role of serotonin on human impulsive aggression. *Biological Psychiatry, 90*, 447–457.

Daly, M. (2020). Evolutionary psychology and inequality. *Evolutionary Behavioral Sciences, 14*, 324–328.

Daly, M. (2023). Inequality, grievances, and the variability in homicide rates. *Evolution and Human Behavior, 44*, 296–304.

Damasio, H., Grabowski, T., Frank, R., Galaburda, A. M., & Damasio, A. R. (1994). The return of Phineas Gage: Clues about the brain from the skull of a famous patient. *Science, 264*, 1102–1105.

Davidson, R. J., Putnam, K. M., & Larson, C. L. (2000). Dysfunction in the neural circuitry of emotion regulation: A possible prelude to violence. *Science, 289*, 591–594.

Davies, N. M., Hemani, G., Neiderhiser, J. M., Martin, H. C., Mills, M. C., Visscher, P. M., et al. (2024). The importance of family-based sampling for biobanks. *Nature, 634*, 795–803.

Davis, M., & Whalen, P. J. (2001). The amygdala: Vigilance and emotion. *Molecular Psychiatry, 6*, 13–34.

De Courson, B., & Nettle, D. (2021). Why do inequality and deprivation produce high crime and low trust? *Scientific Reports, 11*, 1–11.

De Dreu, C. K. W., Greer, L. L., Handgraaf, M. J., et al. (2010). The neuropeptide oxytocin regulates parochial altruism in intergroup conflict among humans. *Science, 328*, 1408–1411.

De Martino, B., Kumaran, D., Seymour, B., & Dolan, R. J. (2006). Frames, biases, and rational decision-making in the human brain. *Science, 313*, 684–687.

Dekkers, T. J., van Rentergem, J. A. A., Meijer, B., Popma, A., Wagemaker, E., & Huizenga, H. M. (2019). A meta-analytical evaluation of the dual-hormone hypothesis: does cortisol moderate the relationship between testosterone and status, dominance, risk taking, aggression, and psychopathy? *Neuroscience and Biobehavioral Reviews, 96*, 250–271.

Denson, T. F., Capper, M. M., Oaten, M., et al. (2011a). Self-control training decreases aggression in response to provocation in aggressive individuals. *Journal of Research in Personality, 45*, 252–256.

Denson, T. F., Pedersen, W. C., Friese, M., et al. (2011b). Understanding impulsive aggression: Angry rumination and reduced self-control capacity are mechanisms underlying the provocation-aggression relationship. *Personality and Social Psychology Bulletin, 37*, 850–862.

Denson, T. F., DeWall, C. N., & Finkel, E. J. (2012). Self-control and aggression. *Current Directions in Psychological Science, 21*, 20–25.

Denson, T. F., Dobson-Stone, C., Ronay, R., von Hippel, W., & Schira, M. M. (2014). A functional polymorphism of the MAOA gene is associated with neural responses to induced anger control. *Journal of Cognitive Neuroscience, 26*, 1418–1427.

Dickerson, S. S., & Kemeny, M. E. (2004). Acute stressors and cortisol responses: A theoretical integration and synthesis of laboratory research. *Psychological Bulletin, 130*, 355–391.

Dodge, K. A., & Coie, J. D. (1987). Social-information-processing factors in reactive and proactive aggression in children's peer groups. *Journal of Personality and Social Psychology, 53*, 1146–1158.

Duckworth, A. L., Kim, B., & Tsukayama, E. (2013). Life stress impairs self-control in early adolescence. *Frontiers in Psychology, 3*, 608.

Duke, A. A., Bègue, L., Bell, R., & Eisenlohr-Moul, T. (2013). Revisiting the serotonin–aggression relation in humans: A meta-analysis. *Psychological Bulletin, 139*, 1148.

Edwards, A., Hammond, H. A., Jin, L., Caskey, C. T., & Chakraborty, R. (1992). Genetic variation at 5-trimeric and tetrameric tandem repeat loci in 4 human population groups. *Genomics, 12*, 241–253.

Eisenberger, N. I., Lieberman, M. D., & Williams, K. D. (2003). Does rejection hurt? An fMRI study of social exclusion. *Science, 302*, 290–292.

Eisenberger, N. I., Way, B. M., Taylor, S. E., et al. (2007). Understanding genetic risk for aggression: Clues from the brain's response to social exclusion. *Biological Psychiatry, 61*, 1100–1108.

Eisner, M. (2009). The uses of violence: An examination of some cross-cutting issues. *International Journal of Conflict and Violence, 3*, 40–59.

Fabiansson, E. C., Denson, T. F., Grisham, J. R., Moulds, M. L., & Schira, M. M. (2012). Don't look back in anger: Neural correlates of reappraisal, analytical rumination, and angry rumination during recall of an anger-inducing autobiographical memory. *NeuroImage, 59*, 2974–2981.

Fajnzylber, P., Lederman, D., & Loayza, N. (2002). Inequality and violent crime. *The Journal of Law and Economics, 45*, 1–39.

Fanning, J. R., Berman, M. E., Guillot, C. R., Marsic, A., & McCloskey, M. S. (2014). Serotonin (5-HT) augmentation reduces provoked aggression associated with primary psychopathy traits. *Journal of Personality Disorders, 28*, 449–461.

Ferguson, C. J. (2015). Do angry birds make for angry children? A meta-analysis of video game influences on children's and adolescents' aggression, mental health, prosocial behavior, and academic performance. *Perspectives on Psychological Science, 10*, 646–666.

Ford, M. B., & Collins, N. L. (2010). Self-esteem moderates neuroendocrine and psychological responses to inter- personal rejection. *Journal of Personality and Social Psychology, 98*, 405–419.

Frazzetto, G., Lorenzo, D., Valeria, C., et al. (2007). Early trauma and increased risk for physical aggression during adulthood: The moderating role of MAOA genotype. *PLoS One, 2*, e486.

Geniole, S. N., Carré, J. M., & McCormick, C. M. (2010). State, not trait, neuroendocrine function predicts costly reactive aggression in men after social exclusion and inclusion. *Biological Psychology, 87*, 137–145.

Geniole, S. N., Procyshyn, T. L., Marley, N., Ortiz, T. L., Bird, B. M., Marcellus, A. L., Welker, K. M., Bonin, P. L., Goldfarb, B., Watson, N. V., & Carré, J. M. (2019). Using a psychopharmacogenetic approach to identify the pathways through which—and the people for whom—testosterone promotes aggression. *Psychological Science, 30*, 1–14.

Geniole, S. N., Bird, B. M., McVittie, J. S., Purcell, R. B., Archer, J., & Carré, J. M. (2020). Is testosterone linked to human aggression? A meta-analytic examination of the relationship between baseline, dynamic, and manipulated testosterone on human aggression. *Hormones and Behavior, 123*, 104644.

Giammanco, M., Tabacchi, G., Giammanco, S., Di Majo, D., & La Guardia, M. (2005). Testosterone and aggressiveness. *Medical Science Monitor, 11*, 136–145.

Gleason, E. D., Fuxjager, M. J., Oyegbile, T. O., & Marler, C. A. (2009). Testosterone release and social context: When it occurs and why. *Frontiers in Neuroendocrinology, 30*, 460–469.

Glenn, A. L., Kurzban, R., & Raine, A. (2011). Evolutionary theory and psychopathy. *Aggression and Violent Behavior, 16*, 371–380.

Goetz, S. M., Tang, L., Thomason, M. E., Diamond, M. P., Hariri, A. R., & Carré, J. M. (2014). Testosterone rapidly increases neural reactivity to threat in healthy men: A novel two-step pharmacological challenge paradigm. *Biological Psychiatry, 76*, 324–331.

Goldin, P. R., McRae, K., Ramel, W., & Gross, J. J. (2008). The neural bases of emotion regulation during reappraisal and suppression of negative emotion. *Biological Psychiatry, 63*, 577–586.

Gospic, K., Mohlin, E., Fransson, P., et al. (2011). Limbic justice – Amygdala involvement in immediate rejection in the ultimatum game. *PLoS Biology, 9*, e1001054.

Greitemeyer, T., & Sagioglou, C. (2017). Increasing wealth inequality may increase interpersonal hostility: The relationship between personal relative deprivation and aggression. *The Journal of Social Psychology, 157*, 766–776.

Hariri, A. R. (2009). The neurobiology of individual differences in complex behavioral traits. *Annual Reviews in Neuroscience, 32*, 247–255.

Hariri, A. R., & Weinberger, D. R. (2003). Functional neuroimaging of genetic variation in serotonergic neurotransmission. *Genes, Brain, and Behavior, 2*, 341–349.

Hariri, A. R., Mattay, V. S., Tessitore, A., et al. (2002). Serotonin transporter genetic variation and the response of the human amygdala. *Science, 297*, 400–403.

Harneit, A., Braun, U., Geiger, L. S., Zang, Z., Hakobjan, M., Van Donkelaar, M. M., et al. (2019). MAOA-VNTR genotype affects structural and functional connectivity in distributed brain networks. *Human Brain Mapping, 40*, 5202–5212.

Haushofer, J., & Fehr, E. (2014). On the psychology of poverty. *Science, 344*, 862–867.

He, N., & Xiang, Y. (2021). How child maltreatment impacts internalized/externalized aggression among Chinese adolescents from perspectives of social comparison and the general aggression model. *Child Abuse and Neglect, 117*, 105024.

Heils, A., Teufel, A., Susanne, P., et al. (1996). Allelic variation of human serotonin transporter gene expression. *Journal of Neurochemistry, 66*, 2621–2624.

Henry, P. J. (2009). Low-status compensation: A theory for understanding the role of status in cultures of honor. *Journal of Personality and Social Psychology, 97*, 451–466.

Hermans, E. J., Ramsey, N. F., & van Honk, J. (2008). Exogenous testosterone enhances responsiveness to social threat in the neural circuitry of social aggression in humans. *Biological Psychiatry, 63*, 263–270.

Hilgard, J., Engelhardt, C. R., & Rouder, J. N. (2017). Overstated evidence for short-term effects of violent games on affect and behavior: A reanalysis of Anderson et al. (2010). *Psychological Bulletin, 143*, 757–774.

Hsieh, C. C., & Pugh, M. D. (1993). Poverty, income inequality, and violent crime: a meta-analysis of recent aggregate data studies. *Criminal Justice Review, 18*, 182–202.

Im, S., Jeong, J., Jin, G., Yeom, J., Jekal, J., Lee, S. I., et al. (2019). MAOA variants differ in oscillatory EEG & ECG activities in response to aggression-inducing stimuli. *Scientific Reports, 9*, 2680.

Josephs, R. A., Mehta, P. H., & Carré, J. M. (2011). Gender and social environment modulate the effects of testosterone on social behavior: Comment on Eisenegger et al. *Trends in Cognitive Sciences, 15*, 509–510.

Josephs, R. A., Telch, M. J., Hixon, J. G., Evans, J. J., Lee, H., Knopik, V. S., McGeary, J. E., Hariri, A. R., & Beevers, C. G. (2012). Genetic and hormonal sensitivity to threat: Testing a serotonin transporter genotype x testosterone interaction. *Psychoneuroendocrinology, 37*, 752–761.

Kelly, M. (2000). Inequality and crime. *The Review of Economics and Statistics, 82*, 530–539.

Kelly, C. R., Grinband, J., & Hirsch, J. (2007). Repeated exposure to media violence is associated with diminished response in an inhibitory frontolimbic network. *PLoS One, 2*, e1268.

Kennedy, B. P., Kawachi, I., Prothrow-Stith, D., Lochner, K., & Gupta, V. (1998). Social capital, income inequality, and firearm violent crime. *Social Science and Medicine, 47*, 7–17.

Kepes, S., Bushman, B. J., & Anderson, C. A. (2017). Violent video game effects remain a societal concern: Reply to Hilgard, Engelhardt, and Rouder (2017). *Psychological Bulletin, 143*, 775–782.

Klasen, M., Wolf, D., Eisner, P. D., Habel, U., Repple, J., Vernaleken, I., et al. (2018). Neural networks underlying trait aggression depend on MAOA gene alleles. *Brain Structure and Function, 223*, 873–881.

Knight, E. L., Sarkar, A., Prasad, S., & Mehta, P. H. (2020). Beyond the challenge hypothesis: The emergence of the dual-hormone hypothesis and recommendations for future research. *Hormones and Behavior, 123*, 104657.

Knight, E. L., Morales, P. J., Christian, C. B., Prasad, S., Harbaugh, W. T., Mehta, P. H., & Mayr, U. (2022). The causal effect of testosterone on men's competitive behavior is moderated by basal cortisol and cues to an opponent's status: Evidence for a context-dependent dual-hormone hypothesis. *Journal of Personality and Social Psychology, 123*, 693–716.

Koenigs, M., & Tranel, D. (2007). Irrational economic decision-making after ventromedial prefrontal damage: Evidence from the ultimatum game. *The Journal of Neuroscience, 27*, 951–956.

Kolla, N. J., & Bortolato, M. (2020). The role of monoamine oxidase A in the neurobiology of aggressive, antisocial, and violent behavior: a tale of mice and men. *Progress in Neurobiology, 194*, 101875.

Kretschmer, T., Ouellet-Morin, I., Vrijen, C., Nolte, I. M., & Hartman, C. (2023). Polygenic risk for aggression behavior from late childhood through early adulthood. *European Child and Adolescent Psychiatry, 32*, 651–660.

Kringelbach, M. L., & Rolls, E. T. (2004). The functional neuroanatomy of the human orbitofrontal cortex: Evidence from neuroimaging and neuropsychology. *Progress in Neurobiology, 72*, 341–372.

Krohn, M. D. (1976). Inequality, unemployment and crime: A cross-national analysis. *The Sociological Quarterly, 17*, 303–313.

Kunst, J. R., Fischer, R., Sidanius, J., & Thomsen, L. (2017). Preferences for group dominance track and mediate the effects of macro-level social inequality and violence across societies. *Proceedings of the National Academy of Sciences, 114*, 5407–5412.

Leary, M. R., Kowalski, R. M., Smith, L., et al. (2003). Teasing, rejection, and violence: Case studies of the school shootings. *Aggressive Behavior, 29*, 202–214.

Leary, M. R., Twenge, J. M., & Guinlivan, E. (2006). Interpersonal rejection as a determinant of anger and aggression. *Personality and Social Psychology Review, 10*, 111–132.

LeDoux, J. E. (2000). Emotion circuits in the brain. *Annual Reviews of Neuroscience, 23*, 155–184.

Lee, T. M., Chan, S. C., & Raine, A. (2008). Strong limbic and weak frontal activation to aggressive stimuli in spouse abusers. *Molecular Psychiatry, 13*, 655–656.

Lukaszewski, A. W., Lewis, D. M. G., Durkee, P. K., Sell, A. N., Sznycer, D., & Buss, D. M. (2020). An adaptationist framework for personality science. *European Journal of Personality, 34*, 1151–1174.

Manuck, S. B., Marsland, A. L., Flory, J. D., et al. (2010). Salivary testosterone and a trinucleotide (CAG) length polymorphism in the androgen receptor gene predict amygdala reactivity in men. *Psychoneuroendocrinology, 35*, 94–104.

Massey, D. S. (2004). Segregation and stratification: A biosocial perspective. *Du Bois Review: Social Science Research on Race, 1*, 7–25.

Matsuo, K., Nicoletti, M., Nemoto, K., et al. (2009). A voxel-based morphometry study of frontal gray matter correlates of impulsivity. *Human Brain Mapping, 30*, 1188–1195.

Mazur, A. (1985). A biosocial model of status in face-to-face primate groups. *Social Forces, 64*, 377–402.

Mazur, A., & Booth, A. (1998). Testosterone and dominance in men. *The Behavioral and Brain Sciences, 21*, 353–397.

McDermott, R., Tingley, D., Cowden, J., et al. (2009). Monoamine oxidase A gene (MAOA) predicts behavioral aggression following provocation. *Proceedings of the National Academy of Sciences, 106*, 2118–2123.

McRae, K., Reiman, E. M., Fort, C. L., Chen, K., & Lane, R. D. (2008). Association between trait emotional awareness and dorsal anterior cingulate activity during emotion is arousal-dependent. *NeuroImage, 41*, 648–655.

Mehta, P. H., & Beer, J. S. (2010). Neural mechanisms of the testosterone-aggression relation: The role of orbitofrontal cortex. *Journal of Cognitive Neuroscience, 22*, 2357–2368.

Mehta, P. H., & Josephs, R. A. (2006). Testosterone change after losing predicts the decision to compete again. *Hormones and Behavior, 50*, 684–692.

Mehta, P. H., & Josephs, R. A. (2010). Testosterone and cortisol jointly regulate dominance: Evidence for a dual- hormone hypothesis. *Hormones and Behavior, 58*, 898–906.

Mehta, P. H., Jones, A. C., & Josephs, R. A. (2008). The social endocrinology of dominance: Basal testosterone predicts cortisol changes and behavior following victory and defeat. *Journal of Personality and Social Psychology, 94*, 1078–1093.

Mehta, P. H., Wuerrhman, E., & Josephs, R. A. (2009). When are low testosterone levels advantageous?: The moderating role of individual versus intergroup competition. *Hormones and Behavior, 56*, 158–162.

Mercy, J., Butchart, A., Farrington, D., & Cerda, M. (2002). Youth violence. In E. G. Krug, L. L. Dahlberg, J. A. Mercy, A. B. Zwi, & R. Lozano (Eds.), *World report on violence and health*. World Health Organization.

Messner, S. F., & Rosenfeld, R. (1997). Political restraint of the market and levels of criminal homicide: A cross-national application of institutional-anomie theory. *Social Forces, 75*, 1393–1416.

Meyer-Lindenberg, A., Buckholtz, J. W., Kolachana, B., Hariri, A. R., Pezawas, L., et al. (2006). Neural mechanisms of genetic risk for impulsivity and violence in humans. *Proceedings of the National Academy of Sciences, 103*, 6269–6274.

Moretti, L., Dragone, D., & di Pellegrino, G. (2009). Reward and social valuation deficits following ventromedial prefrontal damage. *Journal of Cognitive Neuroscience, 21*, 128–140.

Munafo, M. R., Brown, S. M., & Hariri, A. R. (2008). Serotonin transporter (5-HTTLPR) genotype and amygdala activation: A meta-analysis. *Biological Psychiatry, 63*, 852–857.

Nawata, K. (2020). A glorious warrior in war: Cross-cultural evidence of honor culture, social rewards for warriors, and intergroup conflict. *Group Processes and Intergroup Relations, 23*, 598–611.

Ne'eman, R., Perach-Barzilay, N., Fischer-Shofty, M., Atias, A., & Shamay-Tsoory, S. G. (2016). Intranasal administration of oxytocin increases human aggressive behavior. *Hormones and Behavior, 80*, 125–131.

Neapolitan, J. L. (1999). A comparative analysis of nations with low and high levels of violent crime. *Journal of Criminal Justice, 27*, 259–274.

Nelson, R. J., & Trainor, B. C. (2007). Neural mechanisms of aggression. *Nature Reviews Neuroscience, 8*, 536–546.

New, A. S., Hazlett, E., Buchsbaum, M. S., Goodman, M., Reynolds, D., et al. (2002). Blunted prefrontal cortical 18fluorodeoxyglucose positron emission tomography response to meta-chloropiperazine in impulsive aggression. *Archives of General Psychiatry, 59*, 621–629.

New, A. S., Buchsbaum, M. S., Hazlett, E. A., et al. (2004). Fluoxetine increases relative metabolic rate in prefrontal cortex in impulsive aggression. *Psychopharmacology, 176*, 451–458.

Newman, S. W. (1999). The medial extended amygdala in male reproductive behavior: A node in the mammalian social behavior network. *Annals of New York Academy of Sciences, 877*, 242–257.

Nisbett, R. E. (1993). Violence and US regional culture. *American Psychologist, 48*, 441–449.

Ochsner, K. N., Bunge, S. A., Gross, J. J., et al. (2002). Rethinking feelings: An fMRI study of the cognitive regulation of emotion. *Journal of Cognitive Neuroscience, 14*, 1215–1229.

Ochsner, K. N., Ray, R. D., Hughes, B., et al. (2009). Bottom-up and top-down processes in emotion generation: Common and distinct neural mechanisms. *Psychological Science, 20*, 1322–1331.

Office of the Surgeon General. (2001). *Youth violence: A report of the Surgeon General*. U.S. Department of Health and Human Services. Retrieved October, 2011, from http://www.mentalhealth.org/youthviolence/default.asp

Oliveira, R. F. (2009). Social behavior in context: Hormonal modulation of behavioral plasticity and social competence. *Integrative and Comparative Biology, 49*, 423–440.

Oliveira, R. F., Almada, V. C., & Canario, A. V. M. (1996). Social modulation of sex steroid concentrations in the urine of male cichlid fish Oreochromis mossambicus. *Hormones and Behavior, 30*, 2–12.

Palacios-Barrios, E. E., & Hanson, J. L. (2019). Poverty and self-regulation: Connecting psychosocial processes, neurobiology, and the risk for psychopathology. *Comprehensive Psychiatry, 90*, 52–64.

Passamonti, L., Crockett, M. J., Apergis-Schoute, A. M., et al. (2012). Effects of acute tryptophan depletion on prefrontal-amygdala connectivity while viewing facial signals of aggression. *Biological Psychiatry, 71*, 36–43.

Penke, L., Denissen, J. J. A., & Miller, G. F. (2007). The evolutionary genetics of personality. *European Journal of Personality, 21*, 549–587.

Pickett, K. E., & Wilkinson, R. G. (2015). Income inequality and health: A causal review. *Social Science and Medicine, 128*, 316–326.

Popma, A., Vermeiren, R., Geluk, C. A. M. L., et al. (2007). Cortisol moderates the relationship between testosterone and aggression in delinquent male adolescents. *Biological Psychiatry, 61*, 405–411.

Prescott, A. T., Sargent, J. D., & Hull, J. G. (2018). Meta-analysis of the relationship between violent video game play and physical aggression over time. *Proceedings of the National Academy of Sciences, 115*, 9882–9888.

Przybylski, A. K., Deci, E. L., Rigby, C. S., & Ryan, R. M. (2014). Competence-impeding electronic games and players' aggressive feelings, thoughts, and behaviors. *Journal of Personality and Social Psychology, 106*, 441–457.

Rahman, S., Sahakian, B. J., Cardinal, R. N., Rogers, R. D., & Robbins, T. W. (2001). Decision making and neuropsychiatry. *Trends in Cognitive Sciences, 5*, 271–277.

Raine, A., Buchsbaum, M., & LaCasse, L. (1997). Brain abnormalities in murderers indicated by positron emission tomography. *Biological Psychiatry, 42*, 495–508.

Rajender, S., Pandu, G., Sharma, J. D., Gandhi, K. P. C., Singh, L., & Thangaraj, K. (2008). Reduced CAG repeats length in androgen receptor gene is associated with violent criminal behavior. *International Journal of Legal Medicine, 122*, 367–372.

Raleigh, M. J., Brammer, G. L., McGuire, M. T., & Yuwiler, A. (1985). Dominant social status facilitates the behavioral effects of serotonergic agonists. *Brain Research, 348*, 274–282.

Raleigh, M. J., McGuire, M. T., Brammer, G. L., Pollack, D. B., & Yuwiler, A. (1991). Serotonergic mechanisms promote dominance acquisition in adult male vervet monkeys. *Brain Research, 559*, 181–190.

Reif, A., Rösler, M., Freitag, C. M., et al. (2007). Nature and nurture predispose to violent behavior:

Serotonergic genes and adverse childhood environment. *Neuropsychopharmacology, 32*, 2375–2383.

Rolls, E. T., et al. (1994). Emotion-related learning in patients with social and emotional changes associated with frontal-lobe damage. *Journal of Neurology, Neurosurgery, and Psychiatry, 57*, 1518–1524.

Rosell, D. R., & Siever, L. J. (2015). The neurobiology of aggression and violence. *CNS Spectrums, 20*, 254–279.

Rubinow, D. R., & Schmidt, P. J. (1996). Androgens, brain, and behavior. *The American Journal of Psychiatry, 153*, 974–984.

Rufrancos, H., Power, M., Pickett, K. E., & Wilkinson, R. (2013). Income inequality and crime: A review and explanation of the time-series evidence. *Sociology and Criminology, 1*, 129–142.

Ruiz-de-la-Torre, J. L., & Manteca, X. (1999). Effects of testosterone on aggressive behaviour after social mixing in male lambs. *Physiology and Behavior, 68*, 109–113.

Sabol, S., Hu, S., & Hamer, D. (1998). A functional polymorphism in the monamine oxidase A gene promoter. *Human Genetics, 103*, 273–279.

Sapolsky, R. M. (1991). Testicular function, social rank and personality among wild baboons. *Psychoneuroendocrinology, 16*, 281–293.

Schmitt, D. P., & Pilcher, J. J. (2004). Evaluating evidence of psychological adaptation. *Psychological Science, 15*, 643–649.

Sell, A., Tooby, J., & Cosmides, L. (2009). Formidability and the logic of human anger. *Proceedings of the National Academy of Sciences, 106*, 15073–15078.

Sharkey, P. (2010). The acute effect of local homicides on children's cognitive performance. *Proceedings of the National Academy of Sciences, 107*, 11733–11738.

Shoal, G. D., Giancola, P. R., & Kirillova, G. P. (2003). Salivary cortisol, personality, and aggressive behavior in adolescent boys: A 5-year longitudinal study. *Journal of the American Academy of Child and Adolescent Psychiatry, 42*, 1101–1107.

Siegel, A., Bhatt, S., Bhatt, R., et al. (2007). The neurobiological bases for development of pharmacological treatments of aggressive disorders. *Current Neuropharmacology, 5*, 135–147.

Siever, L. J. (2008). Neurobiology of aggression and violence. *The American Journal of Psychiatry, 165*, 429–442.

Simmons, Z. L., & Roney, J. R. (2011). Variation in CAG repeat length of the androgen receptor gene predicts variables associated with intrasexual competitiveness in human males. *Hormones and Behavior, 60*, 306–312.

Sjoberg, R. L., Ducci, F., Barr, C. S., et al. (2008). A non-additive interaction of a functional MAO-A VNTR and testosterone predicts antisocial behavior. *Neuropsychopharmacology, 33*, 425–430.

Sprague, J., Verona, E., Kalkhoff, W., & Kilmer, A. (2011). Moderators and mediators of the stress-aggression relationship: Executive function and state anger. *Emotion, 11*, 61–73.

Strenziok, M., et al. (2010). Lower lateral orbitofrontal cortex density associated with more frequent exposure to television and movie violence in male adolescents. *The Journal of Adolescent Health, 46*, 607–609.

Strüber, D., Lück, M., & Roth, G. (2008). Sex, aggression and impulse control: An integrative account. *Neurocase, 14*, 93–121.

Szycik, G. R., Mohammadi, B., Münte, T. F., & te Wildt, B. T. (2017). Lack of evidence that neural empathic responses are blunted in excessive users of violent video games: An fMRI study. *Frontiers in Psychology, 8*, 174.

Takahashi, A., Quadros, I. M., de Almeida, R. M., & Miczek, K. A. (2011). Brain serotonin receptors and transporters: Initiation vs. termination of escalated aggression. *Psychopharmacology, 213*, 183–212.

Thompson, A., & Tapp, S. N. (2023). *Criminal victimization, 2022*. National Crime Victimization Survey, Bureau of Justice Statistics.

Thompson, R., Gupta, S., Miller, K., Mills, S., & Orr, S. (2004). The effects of vasopressin on human facial responses related to social communication. *Psychoneuroendocrinology, 29*, 35–48.

Thompson, R. R., George, K., Walton, J. C., Orr, S. P., & Benson, J. (2006). Sex-specific influences of vasopressin on human social communication. *Proceedings of the National Academy of Sciences, 103*, 7889–7894.

Tooby, J., & Cosmides, L. (1990). On the universality of human nature and the uniqueness of the individual: the role of genetics and adaptation. *Journal of Personality, 58*, 17–67.

Tucker, D. M., Luu, P., & Pribram, K. H. (1995). Social and emotional self-regulation. *Annals of the New York Academy of Sciences, 769*, 213–239.

van Wingen, G. A., Zylick, S. A., Pieters, S., Mattern, C., Verkes, R. J., Buitelaar, J. K., & Fernandez, G. (2008). Testosterone increases amygdala reactivity in middle-aged women to a young adulthood level. *Neuropsychopharmacology, 34*, 539–547.

van Wingen, G., Mattern, C., Verkes, R. J., Buitelaar, J., & Fernández, G. (2010). Testosterone reduces amygdala-orbitofrontal cortex coupling. *Psychoneuroendocrinology, 35*, 105–113.

Vermeersch, H., T'Sjoen, G., Kaufman, J. M., Vincke, J., & Van Houtte, M. (2010). Testosterone, androgen receptor gene CAG repeat length, mood and behaviour in adolescent males. *European Journal of Endocrinology, 163*, 319–328.

Vilalta, C., Cadena, E., Garrocho, C., & Fondevila, G. (2024). Beyond the immediate effects of income inequality on homicide rates: a reply to Daly's critique. *Evolution and Human Behavior, 45*, 106597.

Wagels, L., Votinov, M., Radke, S., Clemens, B., Montag, C., Jung, S., & Habel, U. (2017). Blunted insula activation reflects increased risk and reward seeking as an interaction of testosterone administration and the MAOA polymorphism. *Human Brain Mapping, 38*, 4574–4593.

Warburton, W. A., Williams, K. D., & Cairns, D. R. (2006). When ostracism leads to aggression: The moderating effects of control deprivation. *Journal of Experimental Social Psychology, 42*, 213–220.

Wilkinson, R. G. (2004). Why is violence more common where inequality is greater? *Annals of the New York Academy of Sciences, 1036*, 1–12.

Wilson, M., & Daly, M. (1985). Competitiveness, risk taking, and violence: The young male syndrome. *Ethology and Sociobiology, 6*, 59–73.

Wingfield, J. C., Hegner, R. E., Dufty, A. M., Jr., & Ball, G. F. (1990). The 'challenge hypothesis': Theoretical implications for patterns of testosterone secretion, mating systems, and breeding strategies. *The American Naturalist, 136*, 829–846.

Zadro, L., Williams, K. D., & Richardson, R. (2004). How low can you go? Ostracism by a computer is sufficient to lower self-reported levels of belonging, control, self-esteem, and meaningful existence. *Journal of Experimental Social Psychology, 40*, 560–567.

Zietsch, B. P. (2024). Genomic findings and their implications for the evolutionary social sciences. *Evolution and Human Behavior, 45*, 106596.

25 The Problem with Conduct Problems: Sociological Factors and Equity Considerations for Neuroscience Research

Kalina J. Michalska and Rengin B. Firat

Abstract

Conduct disorder (CD) is a developmental disorder characterized by chronic rule-breaking, aggression, and disregard for the rights of others. CD is a leading cause of referral to mental health services and frequently leads to educational disruptions and psychosocial adjustment difficulties. Neuroscientific research documents impairments in brain structure and function associated with CD, making significant advances in understanding the complex etiology of CD. Yet, this neurobiological focus and emphasis on tight experimental control and replicability also risks neglecting sociological factors contributing to risk development as well as equity and fairness concerns and methodological biases. In this chapter, we first outline the neural correlates of CD and the multi-level sociological dimensions shaping CD. Then we offer some potential points of integration between the neuroscientific and sociological approaches to CD in children. Finally, we consider the blind spots in neurosciences regarding equity and bias and discuss opportunities to address such biases.

K. J. Michalska (✉)
Department of Psychology, University of California, Riverside, Riverside, CA, USA
e-mail: kalinam@ucr.edu

R. B. Firat
Graduate School of Leadership and Change, Antioch University, Yellow Springs, OH, USA
e-mail: rfirat@antioch.edu

Introduction

Conduct disorder (CD) refers to a clinical diagnosis of serious disruptive behaviors in childhood. More broadly, conduct problems (CP) comprise a range of behaviors, ranging from oppositional behaviors like yelling, acting out, and temper tantrums to more serious conduct issues like aggression toward others, destroying property, or stealing (McMahon et al., 2006). When such behaviors emerge early, they frequently represent a precursor for antisocial behavior later in life (Calkins & Keane, 2009). Sociological research dealing with CP has mainly focused on the concept of juvenile delinquency and the micro (individual-level), meso (intermediate level social context), and macro (societal level) level factors influencing these behavioral issues, largely omitting the neurological and genetic bases underlying these phenomena. Yet, a well-established empirical evidence base reveals that such behaviors have known genetic, neurobiological, and environmental bases (Glenn & Michalska, 2023). This research has made significant advances in understanding the complex etiology of CD and CP, informing prevention and treatment. At the same time, given its focus on

W. Kalkhoff et al. (eds.), *Handbook of Neurosociology*, Handbooks of Sociology and Social Research, https://doi.org/10.1007/978-3-031-95615-7_25

neurobiological explanations of behavior and its emphasis on tight experimental control, neurological perspectives also risk overlooking scientific biases and neglecting sociological factors that can lead to interpreting the origins of disruptive behaviors differently based on socioeconomic status, or race and ethnicity, putting vulnerable populations at risk.

To provide a springboard for discussion, this chapter reviews both the extant research on the neural underpinnings of CD and CP and the sociological research on CP and juvenile delinquency. It then reflects critically on the limitations inherent in neurological methodologies, including experimental paradigms, technologies, analytic approaches, and interpretive frameworks. Our goal is to develop greater self-awareness among both neuroscientists and sociologists studying externalizing psychopathology like CD as well as crime and delinquency among youth. By doing so, we hope to encourage the field to broaden the scope of its inquiry by considering multi-level sociological factors as well as more humanistic approaches, such as researcher positionality and incorporating communities' lived experiences in empirical neurological investigations.

Clinical Diagnosis of Conduct Disorder

The current edition of the Diagnostic and Statistical Manual of Mental Disorders (DSM-V-TR; American Psychiatric Association, 2022) defines CD as a repetitive and persistent pattern of behaviors that violate the basic rights of others and social norms, including aggression to people and animals, destruction of property, deceitfulness or theft, and serious violation of rules. Three or more out of the fifteen criteria are required in the past 12 months to receive a formal diagnosis of CD, with at least one criterion present in the past 6 months. Because these definitions result in heterogeneous groups of individuals (Fairchild et al., 2014), the DSM-5-TR also includes several subtypes.

One approach to subclassifying CD is by age of symptom onset into *childhood-onset* and *adolescent-onset* subtypes (e.g., Lahey et al., 1998). The childhood-onset subtype refers to when one or more symptoms are present before the age of ten, whereas the adolescent-onset subtype refers to when all symptoms emerge after the age of ten. This classification indicates ecological validity as research has characterized different family environments, developmental trajectories and comorbidity profiles of the childhood and adolescent subtypes (Silberg et al., 2015). Generally, the earlier the onset of CD symptoms, the more severe and chronic the behaviors and the stronger the neurodevelopmental influences on the emerging behaviors (Fairchild et al., 2019). In contrast, the adolescent-onset pathway is more common and comparatively more transient than the childhood-onset pathway. Adolescent involvement in antisocial behavior characteristic of CD is believed to emerge alongside puberty as a relatively normative response to the emotional "*Sturm und Drang*" teenage years (Odgers et al., 2008). Although there is some disagreement about the age cut-off (Fairchild et al., 2013; Moffitt et al., 2008), researchers agree that the childhood *vs* adolescent-onset subtypes have both clinical and etiological utility (Glenn & Michalska, 2023).

As the aggressive behaviors observed in CD are heterogenous with regards to motivation and severity, another approach to subclassifying CD is to distinguish between children who exhibit premeditated (*proactive*) or impulsive (*reactive*) forms of aggression (Cui et al., 2016; Gao et al., 2015; Raine et al., 2006). Generally, reactive aggression is triggered by perceived threat and provocation (Blair, 2016), accompanied by higher autonomic responses (Frick & Ellis, 1999) and characterized by poor impulse control, interrupted response inhibition, anger, and defensive reactions (Berkowitz, 1990). Additionally, reactive aggression may result from impaired decision-making, potentially reflecting disrupted reward representations (Blair, 2016). In contrast, proactive aggression is unprovoked, planned, and goal-directed, resulting from an impaired ability to recognize the emotional mental states of others, particularly fear (Blair et al., 2018). Of relevance for diagnosis and treatment, the two forms of aggression display different developmental trajectories. For instance, proactive but not reactive aggression predicts later delinquency in aggressive children with CD (Vitaro et al., 1998). However, reactive aggressive children are at greater risk for psychosocial adjustment problems than proactively aggressive children (Dodge et al., 1990). Mounting evidence suggests that reactive and proactive aggression may have distinct functional (Dambacher et al., 2015) and structural (Jiang et al., 2024) brain signatures, a point to which we will return later.

A third subclassifying approach to CD differentiates between children who score high versus low on callous-unemotional (CU) traits. CU traits identify a subtype of children with distinctive clinical features and neurocognitive perturbations who also differentially respond to intervention. They were formally added to the DSM-5 (American Psychiatric Association, 2013) as a specifier for CD diagnoses. Importantly, because of concerns about the potentially stigmatizing nature of the term "callous–unemotional," this specifier is instead indicated for youth with CD by the phrase "with limited prosocial emotions" (LPE). The LPE specifier designates children who display a lack of remorse or guilt, lack of empathy, lack of concern about performance, or shallow or deficient affect across multiple settings. Youth who display at least two of these criteria across multiple settings for at least 12 months are thought to have CD with LPE. These criteria consistently load on a second-order CU factor (Essau et al., 2006) and parallel the affective dimension of psychopathy (Hare & Neumann, 2008). Compared to the CD youth who exhibit empathy, remorse, and concern about academic performance, youth with CD and high CU traits show more severe and chronic symptomatology (Viding & McCrory, 2012), higher rates of proactive aggression (Andrade et al., 2015), and learning styles that are motivated more by potential rewards and insensitive to negative consequences (Frick et al., 2014). Because of the severity and prevalence of LPE, some researchers maintain that it is more likely to be a neurodevelopmental disorder than CD without LPE (e.g., Wakschlag et al., 2018).

Neurocognitive Processes of Conductive Disorder

Methodological advances in neuroimaging have enabled the examination of biobehavioral bases of individual differences in CD experience and cognition, paving the way for developing personalized clinical interventions (McGowan et al., 2009). The implication is that the borders that define discrete disorders (i.e., CD) or dimensions (i.e., CP) of psychopathology are observable in the brain. Emerging research on clinical outcome prediction relies partly on such a biobehavioral approach, which links individual differences in behavior with neural mechanisms that shape brain function (Glenn & Michalska, 2023). In the following section, we outline recent research on brain-behavior associations related to the clinical phenomenology of CD and CP and the underlying brain structures before we go into the sociological literature explaining potential risk factors.

Impairments in Emotional Empathy and Affective Perspective-Taking

One core feature of CD is reduced affective empathy for others' pain and distress (Blair, 2005). Influential theories posit that because affective empathy may function to inhibit aggression toward others (Blair, 2005; Eisenberg et al., 2005; Moomey, 2005), a lack of empathy may help explain why youth with CD inflict harm on others (Blair, 2005; Zahn-Waxler et al., 1995). Research using functional magnetic resonance imaging (fMRI) studies has examined the networks underlying empathy deficits in CD by testing how children process emotional facial emotions, respond to others' physical pain, and represent others' thoughts and emotional states, or theory of mind (Glenn & Michalska, 2023).

CP have been associated with behavioral impairments in emotion recognition, particularly for recognizing distress cues like fearful and sad expressions (Blair et al., 2005; Dawel et al., 2012; Marsh & Blair, 2008; Schwenck et al., 2012; Stevens et al., 2001). Complementing these findings, fMRI studies in children with CP have employed tasks in which participants view faces displaying varying intensities of fear or sadness to test individual differences in emotion processing. A key region that consistently exhibits reduced engagement to fearful (Marsh et al., 2008; Noordermeer et al., 2016; Viding et al., 2012) sad and angry (Passamonti et al., 2010) facial expressions in youth with CP compared to controls is the amygdala. Among other functions, the amygdala processes salient emotional expressions (Morris et al., 1996). Of note, CU traits modulate amygdala response, with low CU children exhibiting increased amygdala response and high CU children exhibiting reduced amygdala response (e.g., Sebastian et al., 2014), indicating that reduced responsiveness to the distress of others is more likely associated with proactive aggression and reduced guilt rather than reactive aggression. Importantly, even though researchers have historically adopted an amygdala-centric position on the neural disturbances in children with CP, the field has acknowledged a more distributed network of regions that contribute to the disorder's pathogenesis. For example, children with CD exhibit atypical activation in other emotion-processing regions like the ventromedial prefrontal cortex (vmPFC), fusiform gyrus, dorsolateral prefrontal cortex, and anterior insula (Fairchild et al., 2014; Menks et al., 2021).

A related domain of atypical functioning in CD is representing others' mental states, or *Theory of Mind* (Glenn & Michalska, 2023). This ability can be further subdivided into cognitive theory of mind (an ability to infer others' intentions and beliefs) and emotional theory of mind (the ability to infer others' emotional states). The former recruits the temporal pole, superior temporal cortex, temporoparietal junction, posterior cingulate cortex, and rostral medial frontal cortex (Happé & Frith, 2014). Generally, children with CD do not present with behavioral deficits in cognitive theory of mind (Jones et al., 2010), and mentalizing brain regions exhibit normative recruitment in youth with CD and elevated CU traits during cognitive theory of mind tasks (Sebastian et al., 2012). A relevant empirical observation is that adults with psychopathic traits struggle with automatic perspective-taking but are mostly intact when explicitly instructed to do so (Drayton et al., 2018), suggesting that they are capable of taking others' perspectives when using deliberate cognitive mechanisms. Children with CD may likewise purposefully but not automatically take the perspective of others, a point to which we return later. In contrast with cognitive theory of mind, children with CD display impairments in emotional theory of mind or the ability to infer emotional states of others (Sharp et al., 2015). Youth with CD and CU traits show reduced recruitment of the amygdala and anterior insula for emotional relative to cognitive theory of mind when compared to healthy controls (Decety et al., 2013; Lockwood et al., 2013; Marsh et al., 2013; Michalska et al., 2016).

Impairments in Reinforcement-Based Decision-Making

Beyond affective empathy and automatic emotional perspective-taking, another domain in which we observe differences between children with and without CP is reinforcement-based decision-making, such as reward and punishment processing (e.g., Baumann et al., 2022). Detecting the difference between the amount of reward or punishment received and the amount expected, or *prediction error signaling*, is important for learning to change reinforcement expectancies; for example, when a previously angry peer is no longer irritable and wants to play.

Two key brain regions in representing the expected value associated with a stimulus or action are the ventral striatum and the vmPFC (O'Doherty, 2004). Children with disruptive behavior including CD and CP show impairments in the use of expected value within the vmPFC during choice and within the caudate (dorsal striatum) during feedback (White et al., 2013). In other words, youth with CP may exhibit impairments in the computations required for successful decision making, such as during the anticipation of adverse outcomes to steer them away from maladaptive choices (Kuhnen & Knutson, 2005). This impairment may lead to aggressive approach behaviors despite repeated negative consequences of such behaviors.

Brain Structure in Conduct Disorder

Complementing the fMRI studies finding that children with CD display atypical hemodynamic activity in key regions for empathic responding, affective perspective taking, and reinforcement-based decision-making, researchers have also observed structural abnormalities in some of these regions (Gao et al., 2024). Specifically, children with CD exhibit reductions in their amygdala volume (Fairchild et al., 2011, 2013; Sterzer et al., 2007), as well as temporal cortex volume (Fairchild et al., 2011; Huebner et al., 2008; Michalska et al., 2015) and thickness (Hyatt et al., 2012). Previous work has also documented reduced vmPFC volume (Huebner et al., 2008), cortical thickness (Fahim et al., 2011), and folding (Hyatt et al., 2012) in children with CD and/or CU traits, although the evidence is mixed with regard to this region (De Brito et al., 2009; Fairchild et al., 2011, 2013; Sterzer et al., 2007). Reduced insula gray matter volume (Fairchild et al., 2011; Sterzer et al., 2007), thickness (Fahim et al., 2011), and folding (Hyatt et al., 2012) has also been observed in youth with CD. Even though the exact nature of associations between brain structure and function remains inconclusive, individual differences in volume or cortical thickness in these regions may likely contribute to the impairments in empathy, perspective-taking, and decision-making exhibited by children with CD and CP.

Sociological Factors Contributing to Risk Factors for Developing Conduct Problems

Risk factors for developing CD and CP span not only psychological and biological domains but also sociological (i.e., sociodemographic and environmental) domains. We suggest that integrating these sociological mechanisms with the neurological ones can bring more nuanced and holistic understandings for not only why and how conduct problems occur among youth, but also potential interventions that go beyond clinical treatments. Sociological research has often considered CP in children and adolescents from a criminological and legal point of view as juvenile delinquency. Juvenile delinquency is a legal term that is ascribed to the acts of unlawful behavior committed by minors (i.e., those who are under the age of statutory majority, typically 18 years of age) (Shoemaker, 2010). Sociological scholarship sometimes focuses on externalizing (i.e., CP and juvenile delinquency) and internalizing (i.e., mood disorders such as anxiety and depression) behavioral issues together as they are often interlinked; therefore, we will also cover some of the research on internalizing disorders here.

Sociological understandings of CP bring a powerful perspective that can bridge micro (individual-level) with meso (group) and macro (societal) level factors. As can be seen from Fig. 25.1, these processes operate at multiple levels to influence CP and, more broadly, the mental health of children and adolescents. At the micro level, sociological research finds that many factors related to the immediate environment of a child, such as low parental socioeconomic status (SES), have negative associations with internalizing and externalizing behavior issues in children. For example, a meta-analysis of 55 published studies (Reiss, 2013) observed that a large majority (fifty-two out of fifty-five) of the studies showed an inverse relationship between SES (i.e., parental SES, parental education, poverty, and so on) and mental health problems in children and adolescents. Furthermore, many of the studies included in the meta-analysis found that low SES had a stronger impact on the risk for externalizing (conduct) problems than internalizing (mood related) problems (Reiss, 2013). Among the immediate environmental factors, children's and adolescents' mental health problems were more significantly affected by household income and low parental education than by parental unemployment or low occupational status (Reiss, 2013). While childhood financial hardship was more strongly linked to the onset of mental health problems, it did not affect their

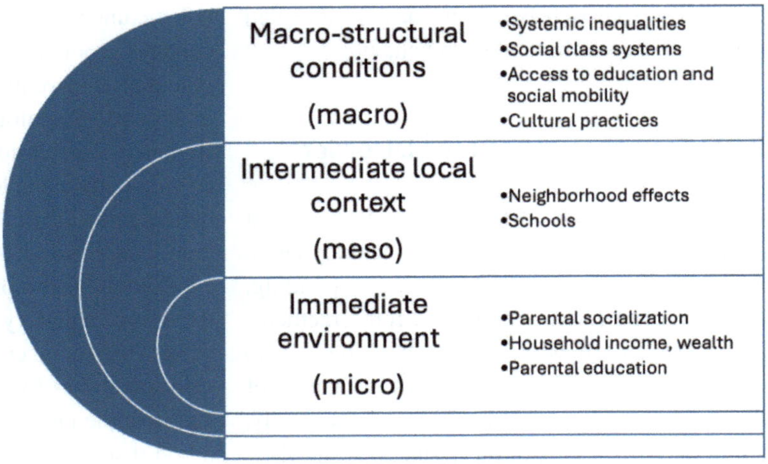

Fig. 25.1 Multi-level, social factors contributing to the risk of conduct problems

course or severity (McLaughlin et al., 2011). Conversely, parental education predicted the persistence and severity of mental health problems without significantly influencing their onset, suggesting parental education might enable better utilization of resources like access to mental health care (McLaughlin et al., 2011). Additionally, while lower SES was linked to poorer mental health outcomes across all age groups, this association was stronger in children under the age of twelve compared to those aged twelve and older (Reiss, 2013). There were no consistent gender differences in these findings.

Several other sociodemographic factors also elevate risk for developing CP. Specifically, a lifetime diagnosis of CD is associated with young age, male gender, low educational attainment, and residing in urban settings. Membership in the aggressive CD subtype is associated with Black ethnicity, lower SES, and lower likelihood of residing in the suburbs (Nock et al., 2006). Other micro-level environmental risk factors for a CD diagnosis include maternal depression, family dysfunction, harsh and inconsistent parenting practices, physical or sexual abuse, caregiver criminality and substance abuse (Glenn & Michalska, 2023).

One of the earlier sociological approaches to delinquency concentrating on intermediate level local context is Shaw and McKay's (1942) "Social Disorganization Theory." Using data from the early 1900s in Chicago, Shaw and McKay argued that social disorganization at the neighborhood level—assessed via factors such as economic conditions or ethnic heterogeneity in a neighborhood—influenced levels of delinquency (e.g., arrest rates or court appearances). This theory primarily focused on inner city urban poverty and its effects on delinquency. Later research on social disorganization by Sampson introduced the concept of *collective efficacy*, which refers to the shared set of beliefs and a collective sense of a neighborhood's own capacity for action and engagement (Sampson et al., 1997, 1999). Again, using data from 343 Chicago neighborhoods in 1995, Sampson and colleagues (Sampson et al., 1997) found that neighborhood level collective efficacy had an inverse association with violence levels in a neighborhood. This study also revealed that collective efficacy was negatively influenced by concentrated poverty and the percentage of immigrant populations living in a neighborhood.

Neighborhood level collective efficacy is a concept closely related to "social capital" (Coleman, 1988; Putnam, 1994), which refers to social connectedness and shared social trust, values, and norms among the members of a community. Both collective efficacy and social capital at the neighborhood level would enable more effective, communal strategies for parenting, such as 'keeping an eye on' neighborhood kids, intervening when appropriate, or sharing resources (Beyers et al., 2003). For example, using data from a longitudinal study of child development from Tennessee (Nashville and Knoxville) and Indiana (Bloomington), Beyers et al. (2003) found that neighborhood structure, particularly residential instability (the proportion of renter vs owner occupied homes in a neighborhood), had an indirect effect on externalizing problems via reducing parental monitoring. Similarly, Brody et al. (2001) showed that collective socialization (e.g., neighbors intervening against misbehaviors like skipping school or painting graffiti) reduced the effects of neighborhood disadvantage on African American children's (in samples across Georgia and Iowa) affiliation with deviant peers. These studies show the importance of intermediate-level, neighborhood factors in shaping the behavioral outcomes for children and youth.

Compared to the large body of research on the micro and meso level factors influencing externalizing problems and CP (or more indirectly on juvenile delinquency), the macro level, societal conditions that may be shaping these processes are relatively under-studied, especially recently. Earlier sociological theorizing, such as Merton's (1938) "Strain Theory," focused on macro-level structural factors, such as the social stratification systems (i.e., social class divides and mobility), and their role in determining crime and delinquency. Building on Durkheim's (1897) theory of "Anomie"—in particular the claim that rapid industrialization and associated changes in the social mobility systems can cause a breakdown of social norms and cohesion—Merton explored how the mismatch between cultural goals (i.e., of achieving the American dream) and a lack of opportunities for reaching these goals causes a 'strain' on the individual. Individuals cope with this psychological strain in different ways, some of which consist of criminal and deviant behavior. While these early theories dealt mainly with adult crime, subsequent research has applied it to delinquency among youth and gang involvement (Cloward & Ohlin, 2013). This theorizing led some scholars to claim that those in lower social class positions in society would be more likely to engage in deviant behavior; however, they argued that this relationship would only hold true in traditional societies that had more rigid, strain-inducing class structures (Axenroth, 1983). For example, focusing on data from South Korea from1970s, Axenroth (1983) found that crime and delinquency were the highest among those from lower social class backgrounds, particularly in urban settings, but that access to education buffered against delinquency. Cohen (1955) also brought attention to the importance of the schooling system, yet in a different way. According to Cohen (1955), schools provide an opportunity for youth from working- and lower-class backgrounds not only to achieve academically, but also to become socialized into delinquent groups. Other approaches have directed attention to the role of the capitalist system in elevating delinquency among youth by prolonging schooling, creating dependency and lower status, and elevating anxiety among younger people from lower social class backgrounds through pressures to contribute financially to the family (Greenberg, 1977).

In addition to the more macro-level structural theories of delinquent behavior, researchers have examined the role of culture in shaping children's behavioral outcomes. While some research reveals consistent patterns across cultures—for example, boys generally tend to demonstrate higher prevalence of anger-aggressive behaviors (LaFreniere et al., 2002) —other research has revealed differences, such as the inverse relationship between age and anger-aggression in Western cultures, in particular (LaFreniere et al., 2002). Parenting styles have also received attention as an important aspect of socialization affecting conduct problems and delinquency among children. While there is some indication that physical discipline and harsh punishments may increase delinquency and behavioral problems (Sadiq Sangawi et al., 2015), there are also some cultural differences. For example, a study comparing adolescents in the US to those in China found that while parental responsiveness and monitoring were inversely associated with delinquency in both samples, parental control significantly reduced delinquency only in the US sample (Li et al., 2023). Other research comparing a US sample to a Korean sample revealed that parental warmth inversely predicted delinquency only in the US sample (Yun & Cui, 2020). Delinquency rates were also lower among Korean youth (Yun & Cui, 2020). These findings were explained by highlighting differences in cultural values (e.g., differences in the importance youth assign to authority and social approval) that lead Korean adolescents to withhold non-conforming, delinquent behaviors (Yun & Cui, 2020).

Integrating Sociological Contributions to Neurodevelopmental Risk Profiles

To understand the etiology of behaviors of youth with CP, we need to characterize the influence of the multi-level sociological factors on the developing brain. The different levels of societal conditions described previously influence neurodevelopmental risk profiles through their impact on the development of neurological structures and functioning in children as well as through their influence on parents' emotion socialization practices and interactions with their children. While it would be an enormous task to describe the processes through which myriad sociological circumstances shape CP, below we consider three illustrative cases.

Effects of Childhood Adversity on the Brain

Childhood adversity alters brain development trajectories, disrupting threat neurocircuitry (Meffert et al., 2018) and emotion regulation processes (Kerig & Becker, 2010). Specifically, stress and trauma exposure affect changes in brain structure and function, which alter an individual's "fight or flight" response. This leads to threat *hypervigilance*, or a tendency to perceive neutral or ambiguous situations as overly threatening and act defensively (Yoon & Weierich, 2016). Cumulative exposure to potential physical harm recalibrates the neural responsiveness of threat-relevant brain regions such as the amygdala, enhancing threat hypervigilance and predisposing youth to reactive aggression. This may contribute to some of the symptomatology in externalizing disorders and other psychopathologies (McLaughlin et al., 2020). For example, among boys with disruptive behavior disorders, there is considerable evidence that CP significantly differentiates youth based on their processing of emotion.

Specifically, youth with elevated levels of CP exhibit sympathetic nervous system under-arousal and hypo-reactivity to emotional cues, particularly fear (Loney et al., 2003; Michalska et al., 2016; Sharp et al., 2006), as well as deficient recognition of fear-related facial and vocal expressions in others (Blair et al., 2001, 2005). Skin conductance response (SCR) is a marker of the sympathetic nervous system, and changes in SCR to aversive stimuli is considered among the most reliable measures of autonomic functioning (Raine et al., 1990). Children with high CP show reduced SCRs to aversive stimuli and stimuli associated with distress in others (Birbaumer et al., 2005; Blair, 1999; Fairchild et al., 2008; Flor et al., 2002).

Maternal Depression

Maternal depression is among the most replicated findings of parental psychopathology in relation to the development of CP (Glenn & Michalska, 2023). CP is more prevalent in children of depressed mothers due to proximal influences on fetal development, postnatal differences in parenting, and disruptions in parenting and mother-child interaction quality (O'Connor et al., 2014). Because depressed mothers tend to be more irritable and express themselves in more critical and less warm and responsive ways, the quality of mother-child interactions is likely to suffer when mothers are depressed (Lovejoy et al., 2000). Maternal warmth and negativity predict child CP after controlling for genetic influences, highlighting the significance of mother-child interactions in the etiology of CP (Caspi et al., 2004; Glenn & Michalska, 2023). Of note, this association is not a one-way street and children's CP exacerbate maternal depression (Shaw et al., 2016) and negative parenting practices (Pardini et al., 2008), indicating bidirectional influences.

As many as 20% of all new mothers will develop depression (Schaar, 2012), although most extant research has studied non-Latinx white mothers from higher SES backgrounds (Glavin & Schieman, 2012; Keefe et al., 2015). Emerging data on mothers of color indicates that their rates of depression are as high as 38% (Centers for Disease Control and Prevention, 2008). These figures should not be surprising. The ordinary challenges of motherhood in combination with limited job prospects or unemployment, low income, and lack of structural mental health support likely contribute in concert to higher rates of depression (Gress-Smith et al., 2012). Therefore, mitigating maternal depression by addressing structural barriers to maternal well-being resulting from race, class, and gender inequalities be an important priority for CP prevention efforts.

Ethnic-racial Socialization

The effects of culturally relevant parenting styles on children's mental health and brain outcomes have also garnered attention in the US, given the multi-cultural and diverse fabric of the society. Research has focused on how ethnic racial socialization (ERS)—the ways parents actively engage their children to instill their cultural values and pass on information about race and ethnicity—is important for understanding intergenerational immigrant health disparities (e.g., Hughes, 2003; et al., 2006; Umaña-Taylor & Guimond, 2010). For example, Latinx parents play an important role in shaping the ways their children process and experience race related factors like discrimination (Ayón, 2016). According to previous research, Latinx parents use six ERS strategies of value transmission. Four of these strategies are adaptive: cultural socialization (talking to their children about their roots and heritage), advocating or active coping (e.g., standing up against discrimination), promoting value diversity and plurality, and educating about citizenship, nativity and immigrant status). Two of the strategies are avoidant: promotion of mistrust towards other ethnic/racial members, and adapting to the environment through avoidant coping (e.g., normalizing discriminatory experiences) (Ayón et al., 2019; see also, Ayón et al., 2018). This ERS framework fits well with psychological research that shows that parents influence the development of children's behavioral adjustment by providing contexts where they can develop and practice early emotion-regulatory skills (Kochanska, 2002; see also Patterson & Stouthamer-Loeber, 1984), which potentially serve as a buffer against conduct problems among Latinx children.

Problematically, *current neuroscience approaches do not attend to the sociological contexts of research participants* (Choudhury et al., 2009), including those focusing on the biological bases of CD and CP. To advance the field, researchers need to incorporate participants' perspectives into their experimental paradigms (La Scala et al., 2023) and interrogate the assumptions that guide the design and interpretation of their studies. In the following section, we discuss some of relevant concerns regarding equity, fairness, and bias in the neuroscience methodologies.

Equity Considerations and Bias in Neuroscience Methodologies

Before reviewing relevant literature on equity considerations, we want to state that we view race as a sociopolitical construct rather than an attribute that defines a group of people biologically or behaviorally. Despite an increasing epidemiological evidence base, data regarding prevalence rates in CD across race and ethnicity remain uneven, with mounting data indicating that disproportionate rates of CD are diagnosed in African American and Latino youth of color (Bird et al., 2001; Kilgus et al., 1995; Lahey et al., 1995). This pattern is concerning as a CD diagnosis is one of the most frequent diagnoses given within the juvenile justice system, leading to stigma and more punitive repercussions than other diagnoses (Drerup et al., 2008; Mizock &

Harkins, 2011; Teplin et al., 2002). The question of *why* (not just whether) racial/ethnic disproportionality exists in the clinical diagnoses of behaviorally defined antisocial and delinquent behavior, irrespective of similar behavioral, offending, and traumatic exposure histories, is therefore a consequential one.

Assessment Biases

American children of European descent are more likely to be diagnosed with milder oppositional defiant disorder (ODD) or attention deficit hyperactivity disorder (ADHD; Fadus et al., 2020), whereas male urban, low-income youth from Latino and African American backgrounds are more likely to be diagnosed with CD (Baglivio et al., 2017; Mandell et al., 2007) and more severe neurological functioning problems. Black and/or Latino juvenile offenders receive CD diagnoses at significantly greater rates than non-Latino white youth, controlling for adverse childhood experiences, criminal offending and juvenile justice system placement history, and other mental health risks (Baglivio et al., 2017). Similarly, CP are systematically over-diagnosed among populations of color (Atkins-Loria et al., 2015; Cameron & Guterman, 2007; Cuffe et al., 2001; Fabrega et al., 1993). Collectively, CD and CP have been over-diagnosed in urban, low-income adolescents of Latino and African American backgrounds (Fadus et al., 2020; Clark, 2007; Mandell et al., 2007; Mota-Castillo, 2004). When white children of European-American ancestry display similar behaviors, they are more likely to be diagnosed with mood and anxiety disorders or developmental disorders (Mandell et al., 2007), which tend to carry less stigma, fewer educational repercussions, and less justice system involvement. Such disproportionality has acute consequences: youth who receive different diagnoses under similar individual and environmental conditions join divergent trajectories with regard to potential life outcomes and subsequent criminal offending (Baglivio et al., 2017).

Associations between race-ethnicity and CD are not about race or ethnicity *per se*, but rather reflect associations between Black and Latinx race and ethnicity and lower SES (Lahey et al., 1995). Because of systematic historical structural inequities, minoritized youth are disproportionately exposed to environmental risk factors and lived experiences related to socioeconomic disadvantage that may contribute to and sustain children's CP (Kim et al., 2003; Kimonis et al., 2014). For example, community violence exposure and neighborhood disadvantage are both more prevalent in lower SES neighborhoods and associated with disruptive behavior observed in CD (Glenn & Michalska, 2023; Fowler et al., 2009; Kimonis et al., 2014). Cumulative experiences of ethnic-racial discrimination can also lead to elevated disruptive and aggressive behaviors among urban youth (Riina et al., 2013). Further, substantial evidence links trauma exposure with the degree of CU traits in CD youth (Carlson et al., 2016; Howard et al., 2012; Meffert et al., 2018), indicating that callous and unemotional behaviors may be the result of past trauma (Reidy & Bogen, 2022) and that many CD youth are also victims themselves through direct or indirect victimization (Carlson et al., 2016; Howard et al., 2012; Meffert et al., 2018). Of note, these environmental risk factors may be confounded by genetic factors, including gene-environment correlations or interactions (Viding & McCrory, 2018). Thus, disassociating environmental etiological factors from genetic vulnerability for CD and CU is particularly challenging.

Clinicians may judge behaviors seen in CD and CP differently based on race or ethnicity, attributing them to individual traits rather than structural inequities. Such biases can occur as a result of either implicit (unconscious) or explicit (conscious) beliefs, and clinicians are just as susceptible to bias as others (Gudiño et al., 2009). For example, clinicians often over-pathologize behaviors of ethnic and racial minoritized youth as more dangerous and can hold inadvertent hostile biases toward some minoritized groups (Evans et al., 2024). Problematically, these biases can lead to disparities in diagnosis, explaining why non-Latinx white children who exhibit comparable behaviors tend to be diagnosed with mood, anxiety, or developmental disorders, whereas ethnic and racial minoritized youth are frequently diagnosed with CD and CP (Fadus et al., 2020). Similarly, youth from higher-income households are more likely to receive an ADHD diagnosis (Getahun et al., 2013). In contrast, children on Medicaid are five times more likely to receive a CD diagnosis compared to privately insured children, controlling for other demographic variables (Evans et al., 2024).

Biases toward children with CD and CP can have detrimental effects on educational achievement, as they may lead educators to rely on harsh disciplinary practices toward students, thereby leading these students to disengage from school and/or mistrust teachers (Skiba et al., 2014). Misdiagnosed children could also receive unnecessary or ineffective treatment because the root causes of their behaviors are not addressed (Kerns et al., 2023). Thus, researchers should be mindful of the potential for clinical bias in minoritized groups and assess relevant demographic and experiential attributes when possible (Glenn & Michalska, 2023).

Biases in Interpreting Neuroscience Data Related to Conduct Problems

Clinical neuroscience (Insel & Quirion, 2005) situates mental distress in the brain and turns to neurochemical interventions as means of ameliorating it. We do not claim that distress or disruptive behavior do not have neurobiological bases. Instead, we contend that neuroscience tends to neglect sociological and cultural contexts (Choudhoury et al., 2009) and the subjective personal accounts of youth experiencing behavioral disturbances like CP. Specifically, as reviewed above, individual differences in behavioral and neural responses to emotional stimuli are not a function of biological differences in race or ethnicity, but rather a result of lived experiences that are frequently not assessed in extant work. Several neuroimaging tasks have been utilized to assess CD and CP. However, these tasks vary in whether and how they incorporate social experiences, with only a few including items relevant to experiences of neighborhood opportunity and deprivation, community violence, racial discrimination, and other culturally relevant stressors. The inconsistent evaluation of such experiences makes it difficult to assess CD in a reliable and uniform manner and highlights a broader challenge concerning how CD is operationalized and assessed. More specifically, scholars have noted that the inconsistent language, measurement, operationalization, and assessment of CD makes it difficult to know how or when to systemically intervene (McLaughlin, 2016). Therefore, we encourage scholars to incorporate participants' demographic backgrounds beyond simply gender, race, and ethnicity to gain a fuller understanding of the developmental trajectories of children with CP.

Despite recent advances in our understanding of the development of empathy and reward-related neurocircuitry, task stimuli designed to elicit individual differences in response to emotional faces were developed with predominantly majority-group college students and do not represent the lived experiences of youth with CP. To better reflect children's experiences, researchers should look to children and their families to guide the development and evaluation of neuroimaging stimuli. We believe that direct partnerships between researchers and practitioners can lead to greater ecological validity in study designs and enhance the translational potential of findings. By developing new laboratory and neuroimaging stimuli together, the aim of collaborative creation is to bridge the divide between our research questions and children's actual experiences. Collaborative creation with youth will also allow researchers to benefit from children's experience and insight at the earliest stages of their projects. That is, it will allow researchers early in the design process to see more clearly what they may need and want from the inquiry.

MRI machines can be quite challenging for participants, especially children. They have to approach a large, gray, noisy, and unfamiliar device, lie down on a table, and be gradually slid into the confined space of the scanner bore where their head is restricted. Each of these aspects alone can cause worry, stress, and negative emotions. MRI procedures have been shown to provoke feelings ranging from slight unease to intense panic, raise cortisol levels, and trigger the sympathetic nervous system. The stress experienced during an fMRI scan can significantly impact baseline neural activity, how task stimuli are perceived, task engagement, and performance, as well as the physiological processes leading to functional activation patterns (Díaz et al., 2024; La Scala et al., 2023; Michalska et al., 2020).

Most importantly, the extent to which the MRI environment affects the data varies depending on individual traits and demographic factors, which can lead to misinterpretations in studies depending on the population and design. For minoritized children who are unfamiliar with medical or research settings, the procedure can be especially intimidating. Furthermore, the negative emotions induced by the MRI are likely heightened in children who have faced lifelong racial stressors, including a mistrust of medical services, and who display signs of hypervigilance to threats. As a result, what researchers might label as "atypical" or "problematic" reactions in minoritized youth could instead stem from chronic stress experiences that amplify pre-scan anxiety (Díaz et al., 2024).

Conclusion

Conduct disorder (CD) is a developmental condition affecting children and teens characterized by persistent anti-social behaviors, aggression, and disregard for the rights of others, leading to significant mental health service referrals, comorbid conditions, and social and educational difficulties. Neuroscientific studies have identified structural and functional brain impairments linked to CD, such as reduced gray matter volume and functional anomalies in regions related to empathy, threat response, and reward sensitivity. These findings have significantly advanced our understanding of CD's complex origins and informed evidence-based prevention and treatment strategies. However, this focus on neurobiology often neglects subjective experiences and sociocultural factors, leading to biased interpretations based on race or ethnicity and posing risks to vulnerable populations.

To address these issues, this chapter reviewed both the neural correlates of CD and the sociological research and theories focusing on juvenile delinquency (the conceptual equivalent of conduct problems among sociologists), identifying the micro, meso, and macro level sociological contributors to risk of CP. We identified three ways through which sociological factors can shape CP: through affecting the brain (i.e., effects of chronic poverty on the brain), influencing maternal mental health, and shaping culturally relevant parental socialization styles. Additionally, we offered multiple suggestions for how neuroscience should consider equity and bias issues in their ontologies and epistemologies.

Clinical neuroscience focuses on treating mental distress through brain-based and neurochemical interventions but often ignores the social and cultural backgrounds of young people with conduct problems (CP). Differences in behavior and brain responses are shaped more by life experiences than by race or ethnicity, yet many studies fail to consider factors like neighborhood conditions, community violence, and ethnic-racial discrimination. This inconsistency makes it hard to reliably assess and treat CD. For example, American children of European descent are often diagnosed with milder disorders like ODD or ADHD, while urban, low-income Latino and African American boys are more likely to be diagnosed with severe conditions like conduct disorder (CD). Black and Latino youth receive CD diagnoses at higher rates than white peers, leading to more stigma and justice system involvement. Trauma exposure also contributes to the behaviors seen in CD, indicating that many affected youth are victims themselves. Addressing these issues requires acknowledging the broader social and economic contexts influencing these diagnoses and behaviors.

Going forward, we hope to have shown how social scientific perspectives and mechanisms can help to advance our understanding of the pathology and neurobiology of CD and CP. Such an integrated perspective will open the door to new, community-based research strategies, reduced bias in our methodologies, and ideally lead to more effective intervention strategies.

Acknowledgements The authors acknowledge support from the National Institutes of Health, National Institute for Minority Health and Health Disparities grant # U54 MD013368 "Center for Health Disparities Research at UC Riverside."

References

Andrade, B. F., Sorge, G. B., Djordjevic, D., & Naber, A. R. (2015). Callous–unemotional traits influence the severity of peer problems in children with impulsive/overactive and oppositional/defiant behaviors. *Journal of Child and Family Studies, 24*(8), 2183–2190. https://doi.org/10.1007/s10826-014-0021-6

American Psychiatric Association. (2022). Diagnostic and statistical manual of mental disorders (5th ed., text rev.). https://doi.org/10.1176/appi.books.9780890425787.

Atkins-Loria, S., Macdonald, H., & Mitterling, C. (2015). Young African American men and the diagnosis of conduct disorder: The neo-colonization of suffering. *Clinical Social Work Journal, 43*, 431–441.

Axenroth, J. B. (1983). Social class and delinquency in cross-cultural perspective. *Journal of Research in Crime and Delinquency, 20*(2), 164–182. https://doi.org/10.1177/002242788302000202

Ayón, C. (2016). Talking to Latino children about race, inequality, and discrimination: Raising families in an anti-immigrant political environment. *Journal of the Society for Social Work and Research, 7*(3), Article 3. https://doi.org/10.1086/686929

Ayón, C., Ojeda, I., & Ruano, E. (2018). Cultural socialization practices among Latino immigrant families within a restrictive immigration socio-political context. *Children and Youth Services Review, 88*, 57–65. https://doi.org/10.1016/j.childyouth.2018.02.042

Ayón, C., Tran, A. G. T. T., & Nieri, T. (2019). Ethnic–racial socialization practices among Latino immigrant families: A latent profile analysis. *Family Relations, 68*(2), 246–259. https://doi.org/10.1111/fare.12356

Baglivio, M. T., Wolff, K. T., Piquero, A. R., Greenwald, M. A., & Epps, N. (2017). Racial/ethnic disproportionality in psychiatric diagnoses and treatment in a sample of serious juvenile offenders. *Journal of Youth and Adolescence, 46*, 1424–1451.

Baumann, S., Hartz, A., Scharke, W., De Brito, S. A., Fairchild, G., Herpertz-Dahlmann, B., et al. (2022). Differentiating brain function of punishment versus reward processing in conduct disorder with and without attention deficit hyperactivity disorder. *The World Journal of Biological Psychiatry, 23*(5), 349–360.

Berkowitz, L. (1990). On the formation and regulation of anger and aggression: A cognitive-neoassociationistic analysis. *American Psychologist, 45*(4), 494.

Beyers, J. M., Bates, J. E., Pettit, G. S., & Dodge, K. A. (2003). Neighborhood structure, parenting processes, and the development of youths' externalizing behaviors: A multilevel analysis. *American Journal of Community Psychology, 31*(1), 35–53. https://doi.org/10.1023/A:1023018502759

Birbaumer, N., Veit, R., Lotze, M., Erb, M., Hermann, C., Grodd, W., & Flor, H. (2005). Deficient fear conditioning in psychopathy: A functional magnetic resonance imaging study. *Archives of General Psychiatry, 62*(7), 799–805. https://doi.org/10.1001/archpsyc.62.7.799

Bird, H. R., Canino, G. J., Davies, M., Zhang, H., Ramirez, R., & Lahey, B. B. (2001). Prevalence and correlates of antisocial behaviors among three ethnic groups. *Journal of Abnormal Child Psychology, 29*(6), 465–478. https://doi.org/10.1023/A:1012279707372

Blair, R. J. R. (1999). Responsiveness to distress cues in the child with psychopathic tendencies. *Personality and Individual Differences, 27*(1), 135–145. https://doi.org/10.1016/S0191-8869(98)00231-1

Blair, R. J. R. (2005). Responding to the emotions of others: Dissociating forms of empathy through the study of typical and psychiatric populations. *Consciousness and Cognition, 14*(4), 698–718. https://doi.org/10.1016/j.concog.2005.06.004

Blair, R. J. R., Budhani, S., Colledge, E., & Scott, S. (2005). Deafness to Fear in boys with PsychoPathic Tendencies. *Journal of Child Psychology and Psychiatry, 46(3), 327–336* https://doi.org/10.1111/j.1469-7610.2004.00356.x.

Blair, R. J. (2016). The neurobiology of impulsive aggression. *Journal of Child and Adolescent Psychopharmacology, 26*(1), 4–9.

Blair, I. V., Ma, J. E., & Lenton, A. P. (2001). Imagining stereotypes away: The moderation of implicit stereotypes through mental imagery. *Journal of Personality and Social Psychology, 81*(5), 828–841. https://doi.org/10.1037/0022-3514.81.5.828

Blair, R. J. R., Veroude, K., & Buitelaar, J. K. (2018). Neuro-cognitive system dysfunction and symptom sets: A review of fMRI studies in youth with conduct problems. *Neuroscience and Biobehavioral Reviews, 91*, 69–90.

Brody, G. H., Conger, R., Gibbons, F. X., Ge, X., McBride Murry, V., Gerrard, M., & Simons, R. L. (2001). The influence of neighborhood disadvantage, collective socialization, and parenting on African American children's affiliation with deviant peers. *Child Development, 72*(4), 1231–1246. https://doi.org/10.1111/1467-8624.00344

Calkins, S. D., & Keane, S. P. (2009). Developmental origins of early antisocial behavior. *Development and Psychopathology, 21*(4), 1095–1109.

Cameron, M., & Guterman, N. B. (2007). Diagnosing conduct problems of children and adolescents in residential treatment. In *Child and youth care forum* (Vol. 36, pp. 1–10). Kluwer Academic Publishers-Plenum Publishers.

Carlson, E. B., Palmieri, P. A., Field, N. P., Dalenberg, C. J., Macia, K. S., & Spain, D. A. (2016). Contributions of risk and protective factors to prediction of psychological symptoms after traumatic experiences. *Comprehensive Psychiatry, 69*, 106–115. https://doi.org/10.1016/j.comppsych.2016.04.022

Caspi, A., Moffitt, T. E., Morgan, J., Rutter, M., Taylor, A., Arseneault, L., Tully, L., Jacobs, C., Kim-Cohen, J., & Polo-Tomas, M. (2004). Maternal expressed emotion predicts children's antisocial behavior problems: Using monozygotic-twin differences to identify environmental effects on behavioral development. *Developmental Psychology, 40*(2), 149–161. https://doi.org/10.1037/0012-1649.40.2.149

Centers for Disease Control and Prevention. (2008). Prevalence of self-reported postpartum depressive symptoms—17 states, 2004-2005. Morbidity and Mortality Weekly Report, 57(14), 361–366.

Choudhury, S., Nagel, S. K., & Slaby, J. (2009). Critical neuroscience: Linking neuroscience and society through critical practice. *BioSocieties, 4*(1), 61–77.

Clark, E., Jr. (2007). Conduct disorders in African American adolescent males: The perceptions that lead to overdiagnosis and placement in special programs. *Alabama Counseling Association Journal, 33*(2), 1–7.

Cloward, R. A., & Ohlin, L. E. (2013). *Delinquency and opportunity: A study of delinquent gangs*. Routledge. https://doi.org/10.4324/9781315007274

Cohen, A. K. (1955). *Delinquent boys; The culture of the gang* (p. 202). Free Press.

Coleman, J. S. (1988). Social capital in the creation of human capital. *American Journal of Sociology, 94*, S95–S120. https://doi.org/10.1086/228943

Cuffe, S. P., Mckeown, R. E., Jackson, K. L., Addy, C. L., Abramson, R., & Garrison, C. Z. (2001). Prevalence of attention-deficit/Hyperactivity disorder in a community sample of older adolescents. *Journal of the American Academy of Child & Adolescent Psychiatry, 40*(9), 1037–1044. https://doi.org/10.1097/00004583-200109000-00012

Cui, L., Colasante, T., Malti, T., et al. (2016). Dual trajectories of reactive and proactive aggression from mid-childhood to early adolescence: Relations to sensation seeking, risk taking, and moral reasoning. *Journal of Abnormal Child Psychology, 44*, 663–675.

Dambacher, F., Schuhmann, T., Lobbestael, J., Arntz, A., Brugman, S., & Sack, A. T. (2015). Reducing proactive aggression through non-invasive brain stimulation. *Social Cognitive and Affective Neuroscience, 10*(10), 1303–1309.

Dawel, A., O'Kearney, R., McKone, E., & Palermo, R. (2012). Not just fear and sadness: Meta-analytic evidence of pervasive emotion recognition deficits for facial and vocal expressions in psychopathy. *Neuroscience and Biobehavioral Reviews, 36*(10), 2288–2304. https://doi.org/10.1016/j.neubiorev.2012.08.006

De Brito, S. A., Mechelli, A., Wilke, M., Laurens, K. R., Jones, A. P., Barker, G. J., Hodgins, S., & Viding, E. (2009). Size matters: Increased grey matter in boys with conduct problems and callous–unemotional traits. *Brain, 132*(4), 843–852. https://doi.org/10.1093/brain/awp011

Decety, J., Chen, C., Harenski, C., & Kiehl, K. A. (2013). An fMRI study of affective perspective taking in individuals with psychopathy: Imagining another in pain does not evoke empathy. *Frontiers in Human Neuroscience, 7*.

Díaz, D. E., Tseng, W. L., & Michalska, K. J. (2024). Pre-scan state anxiety is associated with greater right amygdala-hippocampal response to fearful versus happy faces among trait-anxious Latina girls. *BMC Psychiatry, 24*(1), 1.

Dodge, K. A., Price, J. M., Bachorowski, J. A., & Newman, J. P. (1990). Hostile attributional biases in severely aggressive adolescents. *Journal of Abnormal Psychology, 99*(4), 385.

Drayton, L. A., Santos, L. R., & Baskin-Sommers, A. (2018). Psychopaths fail to automatically take the perspective of others. *Proceedings of the National Academy of Sciences, 115*(13), 3302–3307. https://doi.org/10.1073/pnas.1721903115

Drerup, L. C., Croysdale, A., & Hoffmann, N. G. (2008). Patterns of behavioral health conditions among adolescents in a juvenile justice system. *Professional Psychology: Research and Practice, 39*(2), 122.

Durkheim, É. (1897). *Suicide, a study in sociology*. Free Press.

Eisenberg, N., Zhou, Q., Spinrad, T. L., Valiente, C., Fabes, R. A., & Liew, J. (2005). Relations among positive parenting, children's effortful control, and externalizing problems: A three-wave longitudinal study. *Child Development, 76*(5), 1055–1071. https://doi.org/10.1111/j.1467-8624.2005.00897.x

Essau, C. A., Sasagawa, S., & Frick, P. J. (2006). Callous-unemotional traits in a community sample of adolescents. *Assessment, 13*(4), 454–469. https://doi.org/10.1177/1073191106287354

Evans, S. C., Karlovich, A. R., Khurana, S., Edelman, A., Buza, B., Riddle, W., & López Sosa, D. (2024). Evidence base update on the assessment of irritability, anger, and aggression in youth. *Journal of Clinical Child and Adolescent Psychology*, 1–32.

Fabrega, H., Ulrich, R., & Mezzich, J. E. (1993). Do Caucasian and Black adolescents differ at psychiatric intake? *Journal of the American Academy of Child and Adolescent Psychiatry, 32*(2), 407–413. https://doi.org/10.1097/00004583-199303000-00023

Fadus, M. C., Ginsburg, K. R., Sobowale, K., Halliday-Boykins, C. A., Bryant, B. E., Gray, K. M., & Squeglia, L. M. (2020). Unconscious bias and the diagnosis of disruptive behavior disorders and ADHD in African American and Hispanic youth. *Academic Psychiatry, 44*, 95–102.

Fahim, C., He, Y., Yoon, U., Chen, J., Evans, A., & Pérusse, D. (2011). Neuroanatomy of childhood disruptive behavior disorders. Aggressive behavior, 37(4), 326-337. https://doi.org/10.1002/ab.20396

Fairchild, G., Van Goozen, S. H., Stollery, S. J., & Goodyer, I. M. (2008). Fear conditioning and affective modulation of the startle reflex in male adolescents with early-onset or adolescence-onset conduct disorder and healthy control subjects. *Biological Psychiatry, 63*(3), 279–285. https://doi.org/10.1016/j.biopsych.2007.06.019

Fairchild, G., Passamonti, L., Hurford, G., Hagan, C. C., von dem Hagen, E. A., van Goozen, S. H., ... & Calder, A. J. (2011). Brain structure abnormalities in early-onset and adolescent-onset conduct disorder. *American Journal of Psychiatry, 168*(6), 624–633. https://doi.org/10.1176/appi.ajp.2010.10081184

Fairchild, G., van Goozen, S. H. M., Calder, A. J., & Goodyer, I. M. (2013). Research Review: Evaluating and reformulating the developmental taxonomic theory of antisocial behaviour. *Journal of Child Psychology and Psychiatry, 54*(9), 924–940. https://doi.org/10.1111/jcpp.12102

Fairchild, G., Hagan, C. C., Passamonti, L., Walsh, N. D., Goodyer, I. M., & Calder, A. J. (2014). Atypical neural responses during face processing in female adolescents with conduct disorder. *Journal of the American Academy of Child and Adolescent Psychiatry, 53*(6), 677–687.e5. https://doi.org/10.1016/j.jaac.2014.02.009

Fairchild, G., Hawes, D. J., Frick, P. J., Copeland, W. E., Odgers, C. L., Franke, B., Freitag, C. M., & De Brito, S. A. (2019). Conduct disorder. *Nature Reviews Disease Primers, 5*(1), 1–25. https://doi.org/10.1038/s41572-019-0095-y

Flor, H., Birbaumer, N., Hermann, C., Ziegler, S., & Patrick, C. J. (2002). Aversive Pavlovian conditioning in psychopaths: Peripheral and central correlates. *Psychophysiology, 39*(4), 505–518. https://doi.org/10.1017/S0048577202394046

Fowler, P. J., Tompsett, C. J., Braciszewski, J. M., Jacques-Tiura, A. J., & Baltes, B. B. (2009). Community violence: A meta-analysis on the effect of exposure and mental health outcomes of children and adolescents. *Development and psychopathology, 21*(1), 227–259. https://doi.org/10.1111/j.1469-7610.2010.02280.x.

Frick, P. J., & Ellis, M. (1999). Callous-unemotional traits and subtypes of conduct disorder. *Clinical Child and Family Psychology Review, 2*, 149–168.

Frick, P. J., Ray, J. V., Thornton, L. C., & Kahn, R. E. (2014). Can callous-unemotional traits enhance the understanding, diagnosis, and treatment of serious conduct problems in children and adolescents? A comprehensive review. *Psychological Bulletin, 140*(1), 1–57. https://doi.org/10.1037/a0033076

Gao, Y., Zhang, W., & Fung, A. L. C. (2015). The associations between parenting styles and proactive and reactive aggression in Hong Kong children and adolescents. *International journal of psychology, 50*(6), 463–471. https://doi.org/10.1002/ijop.12104.

Gao, Y., Staginnus, M., Townend, S., Arango, C., Bajaj, S., Banaschewski, T., et al. (2024). Cortical structure and subcortical volumes in conduct disorder: A coordinated analysis of 15 international cohorts from the ENIGMA-Antisocial Behavior Working Group. *The Lancet Psychiatry, 11*(8), 620–632.

Getahun, D., Jacobsen, S. J., Fassett, M. J., Chen, W., Demissie, K., & Rhoads, G. G. (2013). Recent trends in childhood attention-deficit/hyperactivity disorder. *JAMA Pediatrics, 167*(3), 282–288.

Glavin, P., & Schieman, S. (2012). Work–family role blurring and work–family conflict: The moderating influence of job resources and job demands. *Work and Occupations, 39*(1), 71–98. https://doi.org/10.1177/0730888411406295

Glenn, D. E., & Michalska, K. J. (2023). *A developmental neurocognitive perspective on conduct disorder: Current status and future directions*.

Greenberg, D. F. (1977). Delinquency and the age structure of society. *Contemporary Crises, 1*, 189.

Gress-Smith, J. L., Luecken, L. J., Lemery-Chalfant, K., & Howe, R. (2012). Postpartum depression prevalence and impact on infant health, weight, and sleep in low-income and ethnic minority women and infants. *Maternal and Child Health Journal, 16*(4), 887–893. https://doi.org/10.1007/s10995-011-0812-y

Gudiño, O. G., Lau, A. S., Yeh, M., McCabe, K. M., & Hough, R. L. (2009). Understanding racial/ethnic disparities in youth mental health services: Do disparities vary by problem type? *Journal of Emotional and Behavioral Disorders, 17*(1), 3–16.

Happé, F., & Frith, U. (2014). Annual research review: Towards a developmental neuroscience of atypical social cognition. *Journal of Child Psychology and Psychiatry, 55*(6), 553–577. https://doi.org/10.1111/jcpp.12162

Hare, R. D., & Neumann, C. S. (2008). Psychopathy as a clinical and empirical construct. *Annual Review of Clinical Psychology, 4*, 217–246. https://doi.org/10.1146/annurev.clinpsy.3.022806.091452

Howard, A. L., Kimonis, E. R., Muñoz, L. C., & Frick, P. J. (2012). Violence exposure mediates the relation between callous-unemotional traits and offending patterns in adolescents. *Journal of Abnormal Child Psychology, 40*(8), 1237–1247. https://doi.org/10.1007/s10802-012-9647-2

Huebner, T., Vloet, T. D., Marx, I., Konrad, K., Fink, G. R., Herpertz, S. C., & Herpertz-dahlmann, B. (2008). Morphometric brain abnormalities in boys with conduct disorder. *Journal of the American Academy of Child and Adolescent Psychiatry, 47*(5), 540–547. https://doi.org/10.1097/CHI.0b013e3181676545

Hughes, D. (2003). Correlates of African American and Latino parents' messages to children about ethnicity and race: A comparative study of racial socialization. *American Journal of Community Psychology, 31*(1–2), 15–33. https://doi.org/10.1023/A:1023066418688

Hyatt, C. J., Haney-Caron, E., & Stevens, M. C. (2012). Cortical thickness and folding deficits in conduct-disordered adolescents. *Biological Psychiatry, 72*(3), 207–214. https://doi.org/10.1016/j.biopsych.2011.11.017

Insel, T. R., & Quirion, R. (2005). Psychiatry as a clinical neuroscience discipline. *JAMA, 294*(17), 2221–2224. https://doi.org/10.1001/jama.294.17.2221

Jiang, Y., Gao, Y., Dong, D., Sun, X., Situ, W., & Yao, S. (2024). Brain anatomy in boys with conduct disorder: Differences among aggression subtypes. *Child Psychiatry and Human Development, 55*(1), 3–13.

Jones, A. P., Happé, F. G., Gilbert, F., Burnett, S., & Viding, E. (2010). Feeling, caring, knowing: different types of empathy deficit in boys with psychopathic tendencies and autism spectrum disorder. *Journal of Child Psychology and Psychiatry, 51*(11), 1188–1197. https://doi-org.antioch.idm.oclc.org/10.1111/j.1469-7610.2010.02280.x.

Keefe, R. H., Brownstein-Evans, C., Lane, S. D., Carter, D. B., & Polmanteer, R. S. R. (2015). Postpartum depression and the affordable care act: recommendations for social work educators. *Advances in Social Work, 16*(2), Article 2. https://doi.org/10.18060/18502

Kerig, P. K., & Becker, S. P. (2010). From internalizing to externalizing: Theoretical models of the processes linking PTSD to juvenile delinquency. In S. J. Egan (Ed.), *Post-traumatic stress disorder (PTSD): Causes, symptoms and treatment* (pp. 33–78). Nova Science.

Kerns, S. E., Phillips, J. D., Goldsborough, E. J., Hollinshead, D., Gopalan, G., & Barth, R. P. (2023). Exploring the role of standardized instruments and consultation in reducing racial/Ethnic disparities in treatment for behavioral problems. *Families in Society*. https://doi.org/10.1177/10443894231189768

Kilgus, M. D., Pumariega, A. J., & Cuffe, S. P. (1995). Influence of race on diagnosis in adolescent psychiatric inpatients. *Journal of the American Academy of Child & Adolescent Psychiatry, 34*(1), 67–72. https://doi.org/10.1097/00004583-199501000-00016

Kim, I. J., Ge, X., Brody, G. H., Conger, R. D., Gibbons, F. X., & Simons, R. L. (2003). Parenting behaviors and the occurrence and co-occurrence of depressive symptoms and conduct problems among African American children. *Journal of Family Psychology, 17*(4), 571–583. https://doi.org/10.1037/0893-3200.17.4.571

Kimonis, E. R., Centifanti, L. C. M., Allen, J. L., & Frick, P. J. (2014). Reciprocal influences between negative life events and callous-unemotional traits. *Journal of Abnormal Child Psychology, 42*(8), 1287–1298. https://doi.org/10.1007/s10802-014-9882-9

Kochanska, G. (2002). *Mutually responsive orientation between mothers and their young children: A context for the early development of conscience*. https://journals.sagepub.com/doi/abs/10.1111/1467-8721.00198.

Kuhnen, C. M., & Knutson, B. (2005). The neural basis of financial risk taking. *Neuron, 47*(5), 763–770. https://doi.org/10.1016/j.neuron.2005.08.008

La Scala, S., Mullins, J. L., Firat, R. B., Emotional Learning Research Community Advisory Board, & Michalska, K. J. (2023). Equity, diversity, and inclusion in developmental neuroscience: Practical lessons from community-based participatory research. *Frontiers in Integrative Neuroscience, 16*, 1007249.

LaFreniere, P., Masataka, N., Butovskaya, M., Chen, Q., Auxiliadora Dessen, M., Atwanger, K., Schreiner, S., Montirosso, R., & Frigerio, A. (2002). Cross-cultural analysis of social competence and behavior problems in preschoolers. *Early Education and Development, 13*(2), 201–220. https://doi.org/10.1207/s15566935eed1302_6

Lahey, B., Loeber, R., Hart, E., Frick, P., Applegate, B., Zhang, Q., Green, S., & Russo, M. (1995). Four-year longitudinal study of conduct disorder in boys: Patterns and predictors of persistence. *Journal of Abnormal Psychology, 104*, 83–93. https://doi.org/10.1037/0021-843X.104.1.83

Lahey, B. B., Loeber, R., Quay, H. C., Applegate, B., Shaffer, D., Waldman, I., Hart, E. L., Mcburnett, K., Frick, P. J., Jensen, P. S., Dulcan, M. K., Canino, G., & Bird, H. R. (1998). Validity of DSM-IV subtypes of conduct disorder based on age of onset. *Journal of the American Academy of Child & Adolescent Psychiatry, 37*(4), 435–442. https://doi.org/10.1097/00004583-199804000-00022

Li, S. D., Liu, T.-H., & Xia, Y. (2023). A comparative study of parenting practices and juvenile delinquency between China and the United States. *Deviant Behavior, 44*(4), 636–651. https://doi.org/10.1080/01639625.2022.2081102

Lockwood, P. L., Bird, G., Bridge, M., & Viding, E. (2013). Dissecting empathy: High levels of psychopathic and autistic traits are characterized by difficulties in different social information processing domains. *Frontiers in Human Neuroscience, 7*. https://doi.org/10.3389/fnhum.2013.00760

Loney, B. R., Frick, P. J., Clements, C. B., Ellis, M. L., & Kerlin, K. (2003). Callous-unemotional traits, impulsivity, and emotional processing in adolescents with antisocial behavior problems. *Journal of Clinical Child & Adolescent Psychology, 32*(1), 66–80. https://doi.org/10.1207/S15374424JCCP3201_07

Lovejoy, M. C., Graczyk, P. A., O'Hare, E., & Neuman, G. (2000). Maternal depression and parenting behavior: A meta-analytic review. *Clinical Psychology Review, 20*(5), 561–592. https://doi.org/10.1016/S0272-7358(98)00100-7

Mandell, D. S., Ittenbach, R. F., Levy, S. E., & Pinto-Martin, J. A. (2007). Disparities in diagnoses received prior to a diagnosis of autism spectrum disorder. *Journal of autism and developmental disorders, 37*(9), 1795–1802. https://doi.org/10.1016/j.cortex.2009.04.008

Marsh, A. A., & Blair, R. J. R. (2008). Deficits in facial affect recognition among antisocial populations: A meta-analysis. *Neuroscience & Biobehavioral Reviews, 32*(3), 454–465. https://doi.org/10.1016/j.neubiorev.2007.08.003

Marsh, A. A., Finger, E. C., Mitchell, D. G. V., Reid, M. E., Sims, C., Kosson, D. S., Towbin, K. E., Leibenluft, E., Pine, D. S., & Blair, R. J. R. (2008). Reduced amygdala response to fearful expressions in children and adolescents with callous-unemotional traits and disruptive behavior disorders. *American Journal of Psychiatry, 165*(6), 712–720. https://doi.org/10.1176/appi.ajp.2007.07071145

Marsh, A. A., Finger, E. C., Fowler, K. A., Adalio, C. J., Jurkowitz, I. T. N., Schechter, J. C., Pine, D. S.,

Decety, J., & Blair, R. J. R. (2013). Empathic responsiveness in amygdala and anterior cingulate cortex in youths with psychopathic traits. *Journal of Child Psychology and Psychiatry, 54*(8), 900–910. https://doi.org/10.1111/jcpp.12063

McGowan, P. O., Sasaki, A., D'Alessio, A. C., Dymov, S., Labonté, B., Szyf, M., Turecki, G., & Meaney, M. J. (2009). Epigenetic regulation of the glucocorticoid receptor in human brain associates with childhood abuse. *Nature Neuroscience, 12*(3), 342–348. https://doi.org/10.1038/nn.2270

McLaughlin, K. A., Breslau, J., Green, J. G., Lakoma, M. D., Sampson, N. A., Zaslavsky, A. M., & Kessler, R. C. (2011). Childhood socio-economic status and the onset, persistence, and severity of DSM-IV mental disorders in a US national sample. *Social Science and Medicine, 73*(7), 1088–1096. https://doi.org/10.1016/j.socscimed.2011.06.011

McLaughlin, K. A., Colich, N. L., Rodman, A. M., & Weissman, D. G. (2020). Mechanisms linking childhood trauma exposure and psychopathology: A transdiagnostic model of risk and resilience. *BMC Medicine, 18*, 1–11.

McMahon, R. J., Wells, K. C., & Kotler, J. S. (2006). Conduct problems. *Treatment of childhood disorders, 3*, 137–268.

Meffert, H., Thornton, L. C., Tyler, P. M., Botkin, M. L., Erway, A. K., Kolli, V., et al. (2018). Moderation of prior exposure to trauma on the inverse relationship between callous-unemotional traits and amygdala responses to fearful expressions: An exploratory study. *Psychological Medicine, 48*(15), 2541–2549.

Menks, W. M., Fehlbaum, L. V., Borbás, R., Sterzer, P., Stadler, C., & Raschle, N. M. (2021). Eye gaze patterns and functional brain responses during emotional face processing in adolescents with conduct disorder. *NeuroImage: Clinical, 29*, 102519. https://doi.org/10.1016/j.nicl.2020.102519

Merton, R. K. (1938). Social structure and anomie. *American Sociological Review, 3*(5), 672–682. https://doi.org/10.2307/2084686

Michalska, K. J., Decety, J., Zeffiro, T. A., & Lahey, B. B. (2015). Association of regional gray matter volumes in the brain with disruptive behavior disorders in male and female children. *NeuroImage: Clinical, 7*, 252–257. https://doi.org/10.1016/j.nicl.2014.12.012

Michalska, K. J., Shechner, T., Hong, M., Britton, J. C., Leibenluft, E., Pine, D. S., & Fox, N. A. (2016). A developmental analysis of threat/safety learning and extinction recall during middle childhood. *Journal of Experimental Child Psychology, 146*, 95–105. https://doi.org/10.1016/j.jecp.2016.01.008

Mizock, L., & Harkins, D. (2011). Diagnostic bias and conduct disorder: Improving culturally sensitive diagnosis. *Child & Youth Services, 32*(3), 243–253.

Moffitt, T. E., Order, A., & in A., Arseneault, L., Jaffee, S. R., Kim-Cohen, J., Koenen, K. C., Odgers, C. L., Slutske, W. S., & Viding, E. (2008). Research review: DSM-V conduct disorder: Research needs for an evidence base. *Journal of Child Psychology and Psychiatry, 49*(1), 3–33. https://doi.org/10.1111/j.1469-7610.2007.01823.x

Moomey, A. J. (2005). Empathy and moral development: Implications for caring and justice. *Journal of Church and State, 47*(4), 863–865.

Morris, J. S., Frith, C. D., Perrett, D. I., Rowland, D., Young, A. W., Calder, A. J., & Dolan, R. J. (1996). A differential neural response in the human amygdala to fearful and happy facial expressions. *Nature, 383*(6603), Article 6603. https://doi.org/10.1038/383812a0

Mota-Castillo, M. (2004). Is conduct disorder real? *Current Psychiatry, 3*(8), 2–10.

Nock, M. K., Kazdin, A. E., Hiripi, E. V. A., & Kessler, R. C. (2006). Prevalence, subtypes, and correlates of DSM-IV conduct disorder in the National Comorbidity Survey Replication. *Psychological Medicine, 36*(5), 699–710.

Noordermeer, S. D. S., Luman, M., & Oosterlaan, J. (2016). A systematic review and meta-analysis of neuroimaging in oppositional defiant disorder (ODD) and conduct disorder (CD) taking attention-deficit hyperactivity disorder (ADHD) into account. *Neuropsychology Review, 26*(1), 44–72. https://doi.org/10.1007/s11065-015-9315-8

O'Connor, E. E., Scott, M. A., McCormick, M. P., & Weinberg, S. L. (2014). Early mother–child attachment and behavior problems in middle childhood: The role of the subsequent caregiving environment. *Attachment and Human Development, 16*(6), 590–612. https://doi.org/10.1080/14616734.2014.937817

O'Doherty, J. P. (2004). Reward representations and reward-related learning in the human brain: insights from neuroimaging. *Current Opinion in Neurobiology, 14*(6), 769–776.

Odgers, C. L., Moffitt, T. E., Broadbent, J. M., Dickson, N., Hancox, R. J., Harrington, H., Poulton, R., Sears, M. R., Thomson, W. M., & Caspi, A. (2008). Female and male antisocial trajectories: From childhood origins to adult outcomes. *Development and Psychopathology, 20*(2), 673–716. https://doi.org/10.1017/S0954579408000333

Pardini, D. A., Fite, P. J., & Burke, J. D. (2008). Bidirectional associations between parenting practices and conduct problems in boys from childhood to adolescence: The moderating effect of age and African-American ethnicity. *Journal of Abnormal Child Psychology, 36*(5), 647–662. https://doi.org/10.1007/s10802-007-9162-z

Passamonti, L., Fairchild, G., Goodyer, I. M., Hurford, G., Hagan, C. C., Rowe, J. B., & Calder, A. J. (2010). Neural abnormalities in early-onset and adolescence-onset conduct disorder. *Archives of General Psychiatry, 67*(7), 729–738. https://doi.org/10.1001/archgenpsychiatry.2010.75

Patterson, G. R., & Stouthamer-Loeber, M. (1984). The correlation of family management practices and delinquency. *Child Development, 55*(4), 1299–1307. https://doi.org/10.2307/1129999

Putnam, R. D. (1994). Social capital and public affairs. *Bulletin of the American Academy of Arts and Sciences, 47*(8), 5–19. https://doi.org/10.2307/3824796

Raine, A., Venables, P. H., & Williams, M. (1990). Relationships between central and autonomic measures of arousal at age 15 years and criminality at age 24 years. *Archives of General Psychiatry, 47*(11), 1003–1007. https://doi.org/10.1001/archpsyc.1990.01810230019003

Raine, A., Dodge, K., Loeber, R., Gatzke-Kopp, L., Lynam, D., Reynolds, C., et al. (2006). The reactive–proactive aggression questionnaire: Differential correlates of reactive and proactive aggression in adolescent boys. *Aggressive Behavior: Official Journal of the International Society for Research on Aggression, 32*(2), 159–171.

Reidy, D. E., & Bogen, K. W. (2022). Public health considerations in psychopathy. In *The complexity of psychopathy* (pp. 611–635). Springer International Publishing.

Reiss, F. (2013). Socioeconomic inequalities and mental health problems in children and adolescents: A systematic review. *Social Science and Medicine, 90*, 24–31. https://doi.org/10.1016/j.socscimed.2013.04.026

Riina, E. M., Martin, A., Gardner, M., & Brooks-Gunn, J. (2013). Context matters: Links between neighborhood discrimination, neighborhood cohesion and African American adolescents' adjustment. *Journal of youth and adolescence, 42* (1), 136–146. https://doi.org/10.1007/s10964-012-9804-5.

Sadiq Sangawi, H., Adams, J., & Reissland, N. (2015). The effects of parenting styles on behavioral problems in primary school children: A cross-cultural review. *Asian Social Science, 11*(22), 171. https://doi.org/10.5539/ass.v11n22p171

Sampson, R. J., Raudenbush, S. W., & Earls, F. (1997). Neighborhoods and violent crime: A multilevel study of collective efficacy. *Science, 277*(5328), 918–924. https://doi.org/10.1126/science.277.5328.918

Sampson, R. J., Morenoff, J. D., & Earls, F. (1999). Beyond social capital: Spatial dynamics of collective efficacy for children. *American Sociological Review, 64*(5), 633–660. https://doi.org/10.1177/000312249906400501

Schaar, G. L. (2012). Nurses facilitate change in medical practice: Unmasking postpartum depression. *Journal of Obstetric, Gynecologic and Neonatal Nursing, 41*, S7. https://doi.org/10.1111/j.1552-6909.2012.01359_1.x

Schwenck, C., Mergenthaler, J., Keller, K., Zech, J., Salehi, S., Taurines, R., Romanos, M., Schecklmann, M., Schneider, W., Warnke, A., & Freitag, C. M. (2012). Empathy in children with autism and conduct disorder: Group-specific profiles and developmental aspects. *Journal of Child Psychology and Psychiatry, 53*(6), 651–659. https://doi.org/10.1111/j.1469-7610.2011.02499.x

Sebastian, C. L., Fontaine, N. M. G., Bird, G., Blakemore, S.-J., De Brito, S. A., McCrory, E. J. P., & Viding, E. (2012). Neural processing associated with cognitive and affective theory of mind in adolescents and adults. *Social Cognitive and Affective Neuroscience, 7*(1), 53–63. https://doi.org/10.1093/scan/nsr023

Sebastian, C. L., McCrory, E. J., Dadds, M. R., Cecil, C. A. M., Lockwood, P. L., Hyde, Z. H., Brito, S. A. D., & Viding, E. (2014). Neural responses to fearful eyes in children with conduct problems and varying levels of callous–unemotional traits. *Psychological Medicine, 44*(1), 99–109. https://doi.org/10.1017/S0033291713000482

Sharp, C., Van Goozen, S., & Goodyer, I. (2006). Children's subjective emotional reactivity to affective pictures: Gender differences and their antisocial correlates in an unselected sample of 7–11-year-olds. *Journal of Child Psychology and Psychiatry, 47*(2), 143–150. https://doi.org/10.1111/j.1469-7610.2005.01464.x

Sharp, C., Wright, A. G. C., Fowler, J. C., Frueh, B. C., Allen, J. G., Oldham, J., & Clark, L. A. (2015). The structure of personality pathology: Both general ('g') and specific ('s') factors? *Journal of Abnormal Psychology, 124*(2), 387–398. https://doi.org/10.1037/abn0000033

Shaw, C. R., & McKay, H. D. (1942). *Juvenile delinquency and urban areas* (pp. xxxii, 451). University of Chicago Press.

Shaw, D. S., Sitnick, S. L., Reuben, J., Dishion, T. J., & Wilson, M. N. (2016). Transactional effects among maternal depression, neighborhood deprivation, and child conduct problems from early childhood through adolescence: A tale of two low-income samples. *Development and Psychopathology, 28*(3), 819–836. https://doi.org/10.1017/S095457941600033X

Shoemaker, D. J. (2010). Conduct problems in youth: Sociological perspectives. In R. C. Murrihy, A. D. Kidman, & T. H. Ollendick (Eds.), *Clinical handbook of assessing and treating conduct problems in youth* (pp. 21–47). Springer. https://doi.org/10.1007/978-1-4419-6297-3_2

Silberg, J., Moore, A. A., & Rutter, M. (2015). Age of onset and the subclassification of conduct/dissocial disorder. *Journal of Child Psychology and Psychiatry, 56*(7), 826–833. https://doi.org/10.1111/jcpp.12353

Skiba, R. J., Arredondo, M. I., & Williams, N. T. (2014). More than a metaphor: The contribution of exclusionary discipline to a school-to-prison pipeline. *Equity and Excellence in Education, 47*(4), 546–564.

Sterzer, P., Stadler, C., Poustka, F., & Kleinschmidt, A. (2007). A structural neural deficit in adolescents with conduct disorder and its association with lack of empathy. *NeuroImage, 37*(1), 335–342. https://doi.org/10.1016/j.neuroimage.2007.04.043

Stevens, D., Charman, T., & Blair, R. J. R. (2001). Recognition of emotion in facial expressions and vocal tones in children with psychopathic tendencies. *The Journal of Genetic Psychology, 162*(2), 201–211. https://doi.org/10.1080/00221320109597961

Teplin, L. A., Abram, K. M., McClelland, G. M., Dulcan, M. K., & Mericle, A. A. (2002). Psychiatric disorders in youth in juvenile detention. *Archives of General Psychiatry, 59*(12), 1133–1143. https://doi.org/10.1001/archpsyc.59.12.1133

Umaña-Taylor, A. J., & Guimond, A. B. (2010). A longitudinal examination of parenting behaviors and perceived discrimination predicting Latino adolescents' ethnic identity. *Developmental Psychology, 46*(3), 636–650. https://doi.org/10.1037/a0019376

Viding, E., & McCrory, E. J. (2012). Genetic and neurocognitive contributions to the development of psychopathy. *Development and Psychopathology, 24*(3), 969–983. https://doi.org/10.1017/S095457941200048X

Viding, E., & McCrory, E. J. (2018). Understanding the development of psychopathy: Progress and challenges. *Psychological Medicine, 48*(4), 566–577. https://doi.org/10.1017/S0033291717002847

Viding, E., Sebastian, C. L., Dadds, M. R., Lockwood, P. L., Cecil, C. A. M., De Brito, S. A., & McCrory, E. J. (2012). Amygdala response to preattentive masked fear in children with conduct problems: The role of callous-unemotional traits. *American Journal of Psychiatry, 169*(10), 1109–1116. https://doi.org/10.1176/appi.ajp.2012.12020191

Vitaro, F., Gendreau, P. L., Tremblay, R. E., & Oligny, P. (1998). Reactive and proactive aggression differentially predict later conduct problems. *Journal of Child Psychology and Psychiatry, 39*(3), 377–385. https://doi.org/10.1111/1469-7610.00333

Wakschlag, L. S., Perlman, S. B., Blair, R. J., Leibenluft, E., Briggs-Gowan, M. J., & Pine, D. S. (2018). The neurodevelopmental basis of early childhood disruptive behavior: Irritable and callous phenotypes as exemplars. *American Journal of Psychiatry, 175*(2), 114–130. https://doi.org/10.1176/appi.ajp.2017.17010045

White, S. F., Pope, K., Sinclair, S., Fowler, K. A., Brislin, S. J., Williams, W. C., et al. (2013). Disrupted expected value and prediction error signaling in youths with disruptive behavior disorders during a passive avoidance task. *American Journal of Psychiatry, 170*(3), 315–323.

Yoon, S. A., & Weierich, M. R. (2016). Salivary biomarkers of neural hypervigilance in trauma-exposed women. *Psychoneuroendocrinology, 63*, 17–25.

Yun, H.-J., & Cui, M. (2020). The effects of parental warmth on adolescent delinquency in the United States and South Korea: A cross-cultural perspective. *Journal of Youth and Adolescence, 49*(1), 228–237. https://doi.org/10.1007/s10964-019-01078-z

Zahn-Waxler, C., Cole, P. M., Welsh, J. D., & Fox, N. A. (1995). Psychophysiological correlates of empathy and prosocial behaviors in preschool children with behavior problems. *Development and Psychopathology, 7*(1), 27–48. https://doi.org/10.1017/S0954579400006325

MIX
Papier aus verantwortungsvollen Quellen
Paper from responsible sources
FSC® C105338

If you have any concerns about our products,
you can contact us on
ProductSafety@springernature.com

In case Publisher is established outside the EU,
the EU authorized representative is:
Springer Nature Customer Service Center GmbH
Europaplatz 3, 69115 Heidelberg, Germany

Printed by Libri Plureos GmbH
in Hamburg, Germany